Occupational Health & Safety

OCCUPATIONAL SAFETY AND HEALTH SERIES

The National Safety Council Press' occupational safety and health series is composed of six volumes and two study guides written to help readers establish, maintain, and improve safety, health, and environmental programs. These books when used together contain the most up-to-date and reliable information on establishing priorities, collecting and analyzing data to help identify problems, and developing methods and procedures to reduce or eliminate illness and incidents, thus mitigating injury and minimizing economic loss resulting from these events.

- *Accident Prevention Manual for Business & Industry—4 volume set*
    1. *Administration & Programs*
    2. *Engineering & Technology*
    3. *Security Management*
    4. *Environmental Management*
- *Study Guide: Accident Prevention Manual for Business & Industry: Administration & Programs and Engineering & Technology*
- *Occupational Health & Safety*
- *Fundamentals of Industrial Hygiene*
- *Study Guide: Fundamentals of Industrial Hygiene*

Some recent NSC Press additions include:
- *Injury Facts*™ (formerly *Accident Facts*®) published annually
- *Safety Culture and Effective Safety Management*
- *Safety Through Design*
- *On-Site Emergency Response Planning Guide*
- *Safety and Health Classics*
- *Lockout/Tagout: The Process of Controlling Hazardous Energy*
- *Supervisors' Safety Manual*
- *Out in Front: Effective Supervision in the Workplace*
- *Product Safety: Management Guidelines*
- *OSHA Bloodborne Pathogens Exposure Control Plan* (National Safety Council/CRC/Lewis Publication)
- *Complete Confined Spaces Handbook* (National Safety Council/CRC-LEWIS Publication)

Occupational Health & Safety

Third Edition

Edited by
Marci Z. Balge, RN, MSN, COHN-S
Gary R. Krieger, MD, MPH, DABT

Chief Editors: Marci Z. Balge, RN, MSN, COHN-S
Gary R. Krieger, MD, MPH, DABT
Technical Adviser: Leela Murthy, MSc, PhD
Project Editor: Jodey B. Schonfeld

NATIONAL SAFETY COUNCIL MISSION STATEMENT

The mission of the National Safety Council is to educate and influence society to adopt safety, health, and environmental policies, practices, and procedures that prevent and mitigate human suffering and economic losses arising from preventable causes.

COPYRIGHT, WAIVER OF FIRST SALE DOCTRINE

The National Safety Council's materials are fully protected by the United States copyright laws and are solely for the noncommercial, internal use of the purchaser. Without the prior written consent of the National Safety Council, purchaser agrees that such materials shall not be rented, leased, loaned, sold, transferred, assigned, broadcast in any media form, publicly exhibited or used outside the organization of the purchaser, or reproduced, stored in a retrieval system or transmitted in any form or by any means, electronic, mechanical, photocopying, recording or otherwise. Use of these materials for training for which compensation is received is prohibited, unless authorized by the National Safety Council in writing.

DISCLAIMER

Although the information and recommendations contained in this publication have been compiled from sources believed to be reliable, the National Safety Council makes no guarantee as to, and assumes no responsibility for, the correctness, sufficiency, or completeness of such information or recommendations. Other or additional safety measures may be required under particular circumstances.

Copyright © 1986, 1994, 2000 by the National Safety Council
All Rights Reserved
Printed in the United States of America
05 04 03 02 01   5 4 3 2 1

Library of Congress Cataloging-in-Publication Data
Occupational health & safety / edited by Gary R. Krieger, Marci Z. Balge. – 3rd ed.
    p. cm. – (Occupational safety & health series)
  Includes bibliographical references (p.) and index.
  ISBN 0-87912-203-X (alk. paper)
  1. Industrial hygiene.   2. Industrial safety.   I. Krieger, Gary R.   II. Balge, Marci Z., 1954–   .   III: Title: Occupational health and safety.   IV. Series: Occupational safety and health series (Chicago, Ill.)
RC967.026 1999
616.9'803—dc21
                                      99-30897
                                          CIP

5C600   NSC Press Product Number: 12220-0000

Contents

Foreword vii
Preface ix

Part 1 Background
1. Growth of Occupational Health 3
2. Early History and Post-World War II Trends..........................25

Part 2 Occupational Health and Safety Professions
3. Occupational Medicine Profession 47
4. Occupational Health Nursing Profession 61
5. Safety Profession 77
6. Industrial Hygiene Profession 87

Part 3 Occupational Health and Safety Programs
7. Occupational Health and Safety Program Design 109
8. Occupational Medicine Programs 117
9. Occupational Health Nursing Programs 127
10. Safety Programs 153
11. Industrial Hygiene Programs 167
12. Environmental Regulations 207
13. Radiation Safety Programs 221
14. Workers' Compensation Management Programs................235
15. Travel Health and Remote Work Programs 249
16. Ergonomics Programs 277
17. Employee Safety and Security Programs............................315
18. Emergency Response Programs..325
19. Community Involvement Programs 345
20. Program Assessment and Evaluation 369

Part 4 Human Resources Issues
21. Preplacement Testing 381
22. Stress Management 391
23. Substance Abuse 409
24. Scheduling Shiftwork 431
25. Employee Education 443
26. Gender Issues in the Workplace..451
27. Workers with Disabilities and the ADA..............................471
28. Outsourcing Occupational Health and Safety Services 485

Part 5 Future Issues
29. Infectious Diseases and Occupational Health in Developing Countries....499
30. Information Management 513
31. Sources of Help...517

Appendix—Contributors..561
Index 571

Foreword

We all learn certain axioms that we use to make judgements. For me, one of those is, "Quality stands the test of time." This is the third edition of the *Occupational Health & Safety* book. The National Safety Council first published it 14 years ago.

Since then more than 15,000 copies are in circulation all across world. We feel strongly that this reaction by Safety & Health professionals is the strongest sort of consumer endorsement for which we could ask. It is the most productive way for the Council to meet its mission:

> *The mission of the National Safety Council is to educate and influence society to adopt safety, health and environmental policies, practices and procedures that prevent and mitigate human suffering and economic losses arising from preventable causes.*

I believe that Marci Z. Balge and Gary R. Krieger and their many contributors have written material that can form the basis of an occupational health and safety program. As the fourth element of the National Safety Council's OCCUPATIONAL SAFETY AND HEALTH SERIES, this book, *Occupational Health & Safety,* 3rd edition, offers Occupational Safety and Health professionals, Human Resources professionals, managers, and students the last component in a completely updated series. This series provides a solid library of Occupational Safety and Health basics, and serves as the foundation for exceptional safety and health programs in facilities of all sizes.

I would like to acknowledge the work and dedication of Marci Z. Balge, RN, MSN, COHN-S; Gary R. Krieger, MD, MPH, DABT; and Leela Murthy, MSc, PhD, and their contributors in revising this book. A special thanks to 3M, Mine Safety Appliances, and Argonne National Laboratory for supplying many of the internal photos.

JERRY F. SCANNELL
PRESIDENT AND CEO, NATIONAL SAFETY COUNCIL

Preface to the Third Edition

Occupational Health and Safety is a continually evolving profession. Those who enter the field of study in the 21st century are confronted with workplace hazards and exposures that require multidisciplinary areas of expertise to manage and control. Regulatory requirements affecting occupational safety and health continually expand and can significantly vary across the United States and even more so from country to country. Multinational companies establish work forces in ever-challenging international remote working environments, creating the need for expertise in new areas of emerging diseases, public health planning, and environmental protection.

Addressing the span of expertise necessary to develop health and safety programs for any size company can be daunting and often requires a multi-specialty team of professionals. This edition of the National Safety Council's publication, *Occupational Health & Safety,* is designed for those who are responsible for implementing health and safety programs but who have not been formally trained in the field. It is also intended to serve as a general guidance for corporate managers, human resource professionals, and others who interface with health and safety managers. The overall goal was to describe the professionals needed, the programs they can provide, methods of evaluating programs, and to discuss new and emerging issues in the field of Occupational Health and Safety.

MARCI Z. BALGE, RN, MSN, COHN-S
GARY R. KRIEGER, MD, MPH, DABT
EDITORS

Part 1

Background

chapter 1

Growth of Occupational Health

by Leela I. Murthy, MSc, PhD

4	**1970 Williams-Steiger Occupational Safety and Health Act** Safety and health standards ▪ Recent findings
10	**Occupational Illnesses, Injuries, and Deaths** Occupational deaths ▪ Occupational injuries ▪ Occupational illnesses ▪ Incidence rates of disease or injury ▪ Workers' compensation
17	**Occupational Health and Safety Programs** Recognizing occupational disease ▪ Importance of occupational health and safety programs ▪ Medical surveillance programs ▪ Occupational health and safety professionals ▪ Occupational health and safety team ▪ Health and safety committees
20	**Starting an Occupational Health Program** Medical and industrial hygiene services
20	**NIOSH Education and Training Programs**
21	**Small Businesses**
23	**Summary**
23	**References**

After passage of the Occupational Safety and Health Act (OSHAct) in 1970, the field of occupational health and safety continues to grow. In the United States, more experts are involved in the process of maintaining a safe and healthy workplace. Among them are physicians, nurses, safety professionals, and industrial hygienists. These professionals are collectively involved and work as a team. Today these professionals receive more training in occupational health than ever before.

Today, workers are more aware of the types of exposures they encounter in the workplace, and management better recognizes the need for employing trained personnel to maintain a safe and healthy work environment. However more needs to be done, particularly, in small businesses, which comprise more than 99% of the total U.S. work force (CBP, 1999). Many of these businesses cannot afford to hire trained occupational health and safety personnel on a full-time basis. Their only recourse may be to use these professional services on a contractual or shared basis with other small businesses.

According to the 1999 County Business Patterns of the U.S. Bureau of the Census, a total of 6,894,869 establishments employed 105,299,123 employees. Of these, approximately 6,877,784 (99.75%) employed fewer than 500 employees; another 10,903 (0.16%) establishments employed 500–999 employees, and 6,182 (0.09%) employed more than 1,000 employees (CBP, 1999). The percent distribution of establishments and employment by four employment-size classes are presented in Table 1-A.

Table 1-A. Percent Distribution of Employment in Four Employment-Sized Classes—1997, County Business Patterns, October 1999.

A. BY ESTABLISHMENTS

Employment Size Class	No. of Establishments	Percent Total
<100	6,734,738	97.68
100–499	143,046	2.07
500–999	10,903	0.16
1,000 or more	6,182	0.09
TOTAL	**6,894,869**	**100.00**

B. BY EMPLOYMENT

Employment Size Class	No. of Employees	Percent Total
<100	57,513,655	54.62
100–499	26,993,124	25.63
500–999	7,422,258	7.05
1,000 or more	13,370,086	12.70
TOTAL	**105,299,123**	**100.00**

Source: County Business Patterns, September 1999

Even with the progress made in using skilled professionals to identify and control occupational risks, occupational diseases and injuries continue to occur in the workplace. The latest detailed statistics on injuries and illnesses in the workplace are available from the Bureau of Labor Statistics (BLS), U.S. Department of Labor. The number of nonfatal occupational illnesses and injuries reported in the private sector in 1997 was 6.15 million. This was down from 6.2 million reported in 1996. Of the 6.15 million cases, about 1.92 million cases (31.2%) were from the Manufacturing industry. Wholesale & Retail Trade, and Services industries each accounted for about 23%–25% of total cases (BLS, 1998a). According to a recent report by the National Safety Council, the estimated number of occupational injury deaths (including homicide and suicide) for 1997 for all industries was 6,218, slightly higher than the 6,202 noted in the previous year (NSC, 1999).

For 1998, the National Safety Council estimated that 5,100 unintentional injury deaths (not including homicides and suicides) were attributed to injuries on the job versus 37,600 deaths off the job. The latter include motor-vehicle, public nonmotor-vehicle, and home deaths (NSC, 1999). Of the total 9.5 million disabling injuries in 1998, 3.8 million were occupationally related. Production time lost due to off-the-job injuries were approximately 140 million days in 1998 compared with 80 million days lost by workers injured on-the-job (NSC, 1999).

A part of the employer's burden of occupational deaths, illnesses, and injuries is the cost for workers' compensation claims. The National Safety Council estimated that the total costs of occupational deaths and injuries in 1998 were $125.1 billion, down from $127.7 billion for 1997. Total costs for 1998 included wage and productivity losses of $62.9 billion, medical costs of $19.9 billion, and administrative expenses of $25.6 billion. There were other losses ($12.0 billion) that included monetary value of time lost by workers other than those with disabling injuries, damages to motor vehicles, and fire losses (NSC, 1999).

The costs for workers' compensation claims are even more critical for small businesses. For them to remain financially competitive, they often have to reduce costs, particularly for health care and workers' compensation claims.

1970 WILLIAMS-STEIGER OCCUPATIONAL SAFETY AND HEALTH ACT

The Williams-Steiger Occupational Safety and Health Act (OSHAct) passed in 1970 (Figure 1-1). Its passage followed the recognition of work-related illnesses and injuries by the unions and by industries. The Occupational Safety and Health Administration (OSHA) in the Department of Labor was created to administer this act. According to this act,

> ...the OSHA Administrator shall, among other purposes, promulgate and revise occupational safety and health standards; encourage employers and employees in their effort to reduce the number of occupational safety and health hazards; provide for research in the field of occupational health; provide medical criteria which will ensure that no employee will suffer diminished health; and provide for training programs to increase the number and competence of personnel engaged in the field of occupational safety and health.

The National Institute for Occupational Safety and Health (NIOSH) also was created by this act "to assure so far as possible every man and woman in the nation safe and healthful working conditions." NIOSH was placed in the Department of Health, Education, and Welfare, now renamed the Department of Health and Human Services. NIOSH conducts research for OSHA and provides

Public Law 91-596
91st Congress, S. 2193
December 29, 1970

An Act

84 STAT. 1590

To assure safe and healthful working conditions for working men and women; by authorizing enforcement of the standards developed under the Act; by assisting and encouraging the States in their efforts to assure safe and healthful working conditions; by providing for research, information, education, and training in the field of occupational safety and health; and for other purposes.

Be it enacted by the Senate and House of Representatives of the United States of America in Congress assembled, That this Act may be cited as the "Occupational Safety and Health Act of 1970".

Occupational Safety and Health Act of 1970.

CONGRESSIONAL FINDINGS AND PURPOSE

SEC. (2) The Congress finds that personal injuries and illnesses arising out of work situations impose a substantial burden upon, and are a hindrance to, interstate commerce in terms of lost production, wage

An Act

To assure safe and healthful working conditions for working men and women; by authorizing enforcement of the standards developed under the Act; by assisting and encouraging the States in their efforts to assure safe and healthful working conditions; by providing for research, information, education, and training in the field of occupational safety and health; and for other purposes.

Be it enacted by the Senate and House of Representatives of the United States of America in Congress assembled, That this Act may be cited as the "Occupational Safety and Health Act of 1970".

(3) by authorizing the Secretary of Labor to set mandatory occupational safety and health standards applicable to businesses affecting interstate commerce, and by creating an Occupational Safety and Health Review Commission for carrying out adjudicatory functions under the Act;

(4) by building upon advances already made through employer and employee initiative for providing safe and healthful working conditions;

(5) by providing for research in the field of occupational safety and health, including the psychological factors involved, and by developing innovative methods, techniques, and approaches for dealing with occupational safety and health problems;

(6) by exploring ways to discover latent diseases, establishing causal connections between diseases and work in environmental conditions, and conducting other research relating to health problems, in recognition of the fact that occupational health standards present problems often different from those involved in occupational safety;

(7) by providing medical criteria which will assure insofar as practicable that no employee will suffer diminished health, functional capacity, or life expectancy as a result of his work experience;

(8) by providing for training programs to increase the number and competence of personnel engaged in the field of occupational safety and health;

Figure 1-1. The OSHAct passed in 1970 after work-related illnesses and injuries were widely recognized by industries and unions.

"criteria for recommended standards for chemicals and mixtures of chemicals." NIOSH also provides information to the public on occupational topics via an 800-telephone number (800/35-NIOSH or 800/356-4674). They have a fax-on-demand service at 888/232-3299 and their web site address is http://www.cdc.gov/niosh. According to Part 85a of the act, NIOSH conducts health hazard evaluations in industries upon request by an employer, employee representatives, or by a union representative. NIOSH experts suggest innovative ways to control the particular substance (chemical or physical) from causing illness or injury to employees in the workplace.

Safety and Health Standards

Passage of the 1970 OSHAct resulted in promulgation of various safety and health standards (Figure 1-2). Throughout the years some regulations have been contested by specific industries; and the Circuit Court of Appeals decides the outcome of the appeal. In some instances, only a section of the regulation has been vacated. One example was the updating by OSHA of Subpart Z—Toxic and Hazardous Substances, specifically, Table Z-1, which provides permissible exposure limits (PELs) for various chemicals designated in the regulations as "Limits for Air Contaminants."

OSHA published a revised Table Z-1 in the July 1, 1989, issue of Code of Federal Regulations (CFR, 1989). The PELs that were originally promulgated by OSHA for the substances known to be hazardous were updated; the original limits were designated "transitional" and the updated limits were termed "final rule limits." Upon contest by the industries in the Court of Appeals, the "final rule limits" were vacated by the Court and the "transitional limits" were maintained as the current Permissible Exposure Limits (PELs).

In addition to promulgating standards, OSHA is authorized by the act to conduct inspections at worksites and to issue citations and civil penalties to enforce the standard. The act also allows states to have even more stringent standards than those of the national standard and to administer their own safety and health standards. States are not required to submit to the federal OSHA program, provided they establish and administer their own occupational safety and health standards, which are at least in as stringent as those of the federal program.

Section 8c of the OSHAct requires that employers maintain a record of work-related illnesses and injuries in the OSHA 200 Log. A copy of the OSHA 200 log (Figure 1-3) should be made available to employees, former employees, or their representatives, upon request. This record is to be posted in the workplace in February for a month. The employees are also responsible under Section 2(b) of the act by being encouraged (1) to reduce the number of occupational safety and health hazards in their place of employment and (2) to achieve safe and healthful working conditions (OSHAct, 1970 and 1991 [revised]).

In 1981, the U.S. Congress mandated that OSHA would not inspect businesses with less than 10 employees unless (1) a fatality occurred or (2) there was an employee complaint. Businesses with less than 10 employees are also exempt from OSHA record-keeping requirements.

In many facilities, employers and staff are not aware of the chemical or physical hazards that may be present; or what occupational diseases can occur as a result of overexposure to these hazards. Some reasons for this include the following:

- Today there are so many trade-name products used in the workplace.
- Manufacturers' product information may not provide toxicity data on all the chemicals used in a facility. This makes it difficult to know what chemicals to sample there.

OSHA addressed this problem via the Hazard Communication Standard (29 *CFR* 1910.1200) for all manufacturing industries in the private sector, effective Nov. 25, 1985 (OSHA, 1985, 1987). This standard requires all manufacturers to evaluate the hazards of chemicals they use and produce.

The standard requires all facilities that use and produce chemicals to maintain a Material Safety Data Sheet (MSDS) for each chemical they use and produce. This information enables those who either use chemicals or formulate other chemicals from these chemicals to be aware of the toxicity of these chemicals. All facility employees must be informed about the toxicity of the chemicals being handled in their day-to-day work.

The Hazard Communication Standard requires the industries that use or produce chemicals to accomplish the following:

- Have a complete knowledge of the toxicity of these chemicals.
- Label the chemicals appropriately.
- Prepare a MSDS for each of the chemicals used in the workplace and place each in specific locations accessible to the employees.

JOB SAFETY & HEALTH PROTECTION

The Occupational Safety and Health Act of 1970 provides job safety and health protection for workers by promoting safe and healthful working conditions throughout the Nation. Provisions of the Act include the following:

Employers

All employers must furnish to employees employment and a place of employment free from recognized hazards that are causing or are likely to cause death or serious harm to employees. Employers must comply with occupational safety and health standards issued under the Act.

Employees

Employees must comply with all occupational safety and health standards, rules, regulations and orders issued under the Act that apply to their own actions and conduct on the job.

The Occupational Safety and Health Administration (OSHA) of the U.S. Department of Labor has the primary responsibility for administering the Act. OSHA issues occupational safety and health standards, and its Compliance Safety and Health Officers conduct jobsite inspections to help ensure compliance with the Act.

Inspection

The Act requires that a representative of the employer and a representative authorized by the employees be given an opportunity to accompany the OSHA inspector for the purpose of aiding the inspection.

Proposed Penalty

The Act provides for mandatory civil penalties against employers of up to $7,000 for each serious violation and for optional penalties of up to $7,000 for each nonserious violation. Penalties of up to $7,000 per day may be proposed for failure to correct violations within the proposed time period and for each day the violation continues beyond the prescribed abatement date. Also, any employer who willfully or repeatedly violates the Act may be assessed penalties of up to $70,000 for each such violation. A minimum penalty of $5,000 may be imposed for each willful violation. A violation of posting requirements can bring a penalty of up to $7,000.

There are also provisions for criminal penalties. Any willful violation resulting in the death of any employee, upon conviction, is punishable by a fine of up to $250,000 (or $500,000 if the employer is a corporation), or by imprisonment for up to six months, or both. A second conviction of an employer doubles the possible term of imprisonment. Falsifying records, reports, or applications is punishable by a fine of $10,000 or up to six months in jail or both.

Voluntary Activity

While providing penalties for violations, the Act also encourages efforts by labor and management, before an OSHA inspection, to reduce workplace hazards voluntarily and to develop and improve safety and health programs in workplaces and industries. OSHA's Voluntary Protection Pro...

Figure 1-2. The Job Safety & Health Protection poster is shown. This is posted in the workplace.

- Maintain a Hazard Communication Program (HCP) that includes training of industrial hygienists and safety officers about specific hazards in the workplace.

In 1987, the Hazard Communication Standard was amended to include all nonmanufacturing industries in the private sector (OSHA, 1985, 1987, 1994), see Figure 1-4.

Recent Findings

According to a report published in April 1998, the most cited OSHA General Industry Standard during fiscal year 1996/97 was the Hazard Communication Standard (29 *CFR* Part 1910.1200). Violations for three of the top five (by number of violation) were for hazard communication lapses totaling close to $4.4 million in initial penalty. There were 1,094 violations for the OSHA 200 log and summary with an initial penalty of approximately $857,000, which was later adjusted to $55,146. There were 892 violations for the OSHAct general duty clause—employers must provide a safe workplace for employees—which resulted in an initial penalty of $4,537,000 (Keller, 1998).

At its 103rd session, the U.S. Congress found the following:

1. Despite the progress made in the reduction of occupational illnesses, injuries, and deaths during the past two decades, work-related injuries, illnesses, and deaths continue to occur at rates that are unacceptable and impose a substantial burden upon employees, employers, and the

Bureau of Labor Statistics
Log and Summary of Occupational Injuries and Illnesses

NOTE: This form is required by Public Law 91-596 and must be kept in the establishment for 5 years. Failure to maintain and post can result in the issuance of citations and assessment of penalties. *(See posting requirements on the other side of form.)*

RECORDABLE CASES: You are required to record information about every occupational **death**; every nonfatal occupational **illness**; and those nonfatal occupational **injuries** which involve one or more of the following: loss of consciousness, restriction of work or motion, transfer to another job, or medical treatment (other than first aid). *(See definitions on the other side of form.)*

Case or File Number	Date of Injury or Onset of Illness	Employee's Name	Occupation	Department	Description of Injury or Illness
Enter a nonduplicating number which will facilitate comparisons with supplementary records.	Enter Mo./day.	Enter first name or initial, middle initial, last name.	Enter regular job title, not activity employee was performing when injured or at onset of illness. In the absence of a formal title, enter a brief description of the employee's duties.	Enter department in which the employee is regularly employed or a description of normal workplace to which employee is assigned, even though temporarily working in another department at the time of injury or illness.	Enter a brief description of the injury or illness and indicate the part or parts of body affected. Typical entries for this column might be: Amputation of 1st joint right forefinger; Strain of lower back; Contact dermatitis on both hands; Electrocution--body.
(A)	(B)	(C)	(D)	(E)	(F)

PREVIOUS PAGE TOTALS →

TOTALS (Instructions on other side of form.) →

OSHA No. 200   ☆U.S.GPO:1990-262-256/15419

Figure 1-3 (continued)

Figure 1-3. The OSHA 200 Log contains a record of work-related illnesses and injuries.

nation in terms of lost production, wage loss, medical expenses, compensation payments, and employee disability.

2. Employers and employees are not sufficiently involved in working together to identify and correct occupational safety and health hazards and thus require better training.

3. Enforcement of occupational safety and health hazards has not been adequate to bring about timely abatement of hazardous conditions or to

deter violations of occupational safety and health hazards.
4. Millions of employees exposed to serious occupational safety and health hazards were excluded from the full coverage under the Occupational Safety and Health Act of 1970.
5. The lack of accurate data and information has impeded efforts to prevent work-related deaths, injuries, and illnesses.
6. Injury and illness statistics from the Construction industry showed that current regulations under the 1970 OSHAct did not adequately address the safety and health hazards of construction work (OSHRA, 1994).

Most of the titles discussed in the Reform Act of 1994 deal with OSHA; however, Title IX highlights NIOSH research in occupational safety and health. Some of the main requirements of this title for NIOSH are health hazard evaluations, safety research, contractor rights, a national surveillance program, and training, all of which have been conducted since its inception. Title XII dis-

Figure 1-4. Facilities should have a Hazard Communication Program in the workplace. This includes training of industrial hygienists and safety practitioners about the specific hazards in the workplace.

cusses construction safety in great detail (OSHRA, 1994).

Some of these congressional findings are already being addressed. The Bureau of Labor Statistics (BLS) implemented a new plan in 1992 to conduct a systematic verifiable count of all fatal occupational injuries and to obtain descriptive data on the circumstances surrounding these events (NSC, 1994). This would yield better data on actual deaths than the estimated deaths that were used previously. The formation of occupational safety and health teams to identify occupational injuries and illnesses is also under way; data are being collected and reported. In addition to the requirement in Title IX, under the Small Business Initiative NIOSH is involved in the identification of problems in small businesses. A report on Identifying Hazardous Small-Business Industries — The Basis for Preventing Occupational Injury, Illness, and Fatality was published in May 1999 (NIOSH, 1999).

OSHA was criticized for not adequately addressing the safety and health hazards of construction work. To remedy this, OSHA and the Associated General Contractors of America (AGC) pledged to work together under a newly developed partnership charter to reduce workplace injuries and illnesses in the construction industry (BNA, 1998). Such partnerships help reduce illnesses and injuries since both industry and government are involved. Complete findings of the 103rd U.S. Congress can be obtained from the Comprehensive Occupational Safety and Health Act (COSHA) (OSHRA, 1994).

OCCUPATIONAL ILLNESSES, INJURIES, AND DEATHS

Occupational health and safety professionals use specific terms in their reporting systems. These terms are defined by the Bureau of Labor Statistics (BLS), U.S. Department of Labor, in the *Record Keeping Guidelines for Occupational Injuries and Illnesses* (BLS, 1986). The following definitions are used:

Nonfatal recordable injuries and illnesses are either

1. nonfatal occupational illnesses, or
2. nonfatal occupational injuries that involve one or more of the following:
   - loss of consciousness
   - restriction of work or motion
   - transfer to another job
   - medical treatment (other than first aid).

Unless otherwise mentioned, the terms *occupational illness* and *occupational injury* mentioned in this chapter mean they are of the nonfatal type.

- *Occupational injury* is an injury (such as a cut, fracture, sprain, strain, amputation, etc.) that results from a work event or from a single instantaneous exposure in the work environment.
- *Occupational illness* is any abnormal condition or disorder, other than one resulting from an occupational injury, caused by exposure to factors associated with employment. It includes acute and chronic illnesses or diseases that can be caused by inhalation, absorption, ingestion, or direct contact. Repetitive motion diagnoses are considered occupational illnesses since there is not an instantaneous exposure event.
- *Lost workday* cases are cases that involve days away from work, or days of restricted work activity, or both.
- *Lost workday cases involving days away from work* are those resulting in days away from work, or a combination of days away from work and days of restricted work activity.

- *Incidence rates* represent the number of injuries and /or illnesses per 100 full-time workers and are calculated as:

    (N/EH) × 200,000
    where:
    N = number of injuries and/or illnesses
    EH = total hours worked by all employees during the calendar year
    200,000 = base for 100 full-time equivalent workers (working 40 hours per week, 50 weeks per year).

The National Safety Council has its own definition of *injury*:

A disabling injury is defined as one that results in death, some degree of impairment, or renders the injured person unable to perform his or her regular duties or activities for a full day beyond the day of injury. This definition applies to all categories of incidents—motor vehicles (moving), work, home, other, and not classified.

Occupational Deaths

The number of fatal occupational injuries in 1996 for all industries was 6,112, down from 6,275 the previous year, according to the BLS (Toscano & Windau, 1998). However, the 1999 BLS report shows an increase in fatal occupational injuries to 6,238 in 1997, but these decrease to 6,026 in 1998. Details of these data are still pending (BLS, 1999). The 1998 report showed that highway traffic incidents and homicides continued to lead all other events in the number of fatal work injuries in 1996, contributing to more than one-third of the total that occurred during the year. The 1998 report also noted that work-related highway deaths accounted for one-fifth of fatal work injuries though the number of fatal work injuries fell to the lowest level in five years in 1996. Table 1-B presents the trend of occupational injury deaths (total deaths, assaults & violent acts, homicide, and suicide) for all industries for the years 1994 through 1998. Table 1-B shows that there was a steady decrease in homicide deaths—from 1,080 in 1994 to 709 homicides in 1998. Suicide deaths, on the other hand, increased from 214 in 1994 to 221 in 1995, decreased to 199 in 1996; but again increased to 216 and 223 in 1997 and 1998, respectively. (Toscano & Windau, 1998; News Release, BLS, 1998b; News Release, BLS, 1999.)

The National Safety Council adopted the BLS Census of Fatal Occupational Injuries (CFOI) figures as the authoritative count of work-related deaths. The CFOI counts intentional (including homicide and suicide) as well as unintentional work injuries. Using the CFOI data, the Council published details of occupational injury deaths and death rates for the years 1993–1997, including preliminary data for 1998, which are presented in Table 1-C (NSC, 1999). An increase in injury deaths in 1997 was noted in all industries except Trade and Services. The increased trend in injury deaths for the Construction, Services, and Government industries continued in 1998. The total number of deaths for 1998 is not yet known since the number of homicide and suicide deaths are not yet available (NSC, 1999). A slight increase in unintentional deaths (excluding homicide & suicide) for all industries was noted, from 5,069 in 1996 to 5,148 in 1997, but decreased to 5,100 in 1998. When death rates were compared for 1996 and 1997, a slight increase in death rates were noted for Agriculture, Manufacturing, and for Government but decreased in all other industries as noted in Table 1-D (NSC, 1999).

Preliminary BLS data are available for unintentional injuries at work for 1998. Table 1-E presents the National Safety Council's estimated number of occupational unintentional injury deaths, disabling injuries, and death rates by industry division for 1998 (NSC, 1999). In 1998 there was a total of 5,100 unintentional deaths for all industries, with a death rate of 3.8 per 100,000 workers, and 3.8 million disabling injuries. The Mining industry employed only 618,000 workers but showed a death rate of 24.3 per 100,000 workers, the highest death rate among all industries. The next highest death rate (22.1/100,000

Table 1-B. Fatal Occupational Injury Deaths Due to Assaults and Violent Acts for All Industries—1994–1998, Bureau of Labor Statistics U.S. Department of Labor, 1999.*

Deaths	1994	1995	1996	1997	1998
Total	6,632	6,275	6,112	6,238	6,026
Assaults and Violent Acts	1,321	1,280	1,144	1,111	960
Homicide	1,080	1,036	912	860	709
Self-Inflicted Injury	214	221	199	216	223

* *Sources:* (1) Toscano & Windau, 1998; News Release, BLS, Aug 4, 1999.

Table 1-C. Occupational Injury Deaths by Industry—1993–1998, National Safety Council, 1999.

Deaths	1993	1994	1995	1996	1997	1998
Total	6,331	6,632	6,275	6,202	6,218	—
Homicide and Suicide	1,296	1,294	1,257	1,133	1,070	—
Unintentional						
All Industries	5,035	5,338	5,018	5,069	5,148	5,100
Agriculture	842	814	770	768	799	780
Mining and Quarrying	169	177	155	152	156	150
Construction	895	1,000	1,021	1,073	1,105	1,120
Manufacturing	698	734	640	663	679	660
Transportation & Public Utilities	753	819	784	923	927	920
Trade	450	492	462	454	451	450
Services	632	676	608	671	670	680
Government	528	534	529	321	339	340

Source: *Injury Facts*,™ formerly *Accident Facts*,® National Safety Council, 1999.

Table 1-D. Occupational Injury Death Rates by Industry—1993–1998, National Safety Council, 1999.

Deaths/100,000 Workers	1993	1994	1995	1996	1997	1998
Total	5.2	5.3	4.9	4.8	4.7	—
Homicide, Suicide	1.0	1.0	0.9	0.9	0.8	—
Unintentional						
All Industries	4.2	4.3	4.0	4.0	3.9	3.8
Agriculture	26.0	22.8	21.5	21.3	22.5	22.1
Mining and Quarrying	25.3	26.5	24.8	26.8	24.7	24.3
Construction	13.3	14.4	14.3	14.4	14.1	13.9
Manufacturing	3.6	3.7	3.1	3.2	3.3	3.2
Transportation & Public Utilities	11.0	11.6	11.0	12.7	12.2	11.9
Trade	1.8	1.9	1.8	1.7	1.7	1.7
Services	1.6	1.7	1.5	1.6	1.5	1.5
Government	2.6	2.7	2.7	1.6	1.8	1.7

Source: *Injury Facts*,™ formerly *Accident Facts*,® National Safety Council, 1999.

workers) was seen in the Agriculture industry, which had 3,450,000 workers on the payroll. Of the 132,772,000 workers in all industries, only 8,045,00 (58.4%) were in the Construction industry, with 1,120 deaths noted, yielding a death rate of 13.9 per 100,000 workers. These data again show that using deaths alone may be erroneous since the number of workers in some industries are fewer compared to those in Manufacturing, Wholesale & Retail Trade, and Services industries. Table 1-E also shows that the Services industry had the highest number of disabling injuries (900,000) compared to the Mining industry, which had 30,000 disabling injuries. These data show that the number of disabling injuries is greater among industries employing larger numbers of workers.

NIOSH monitors occupational injury deaths using the National Traumatic Occupational Fatalities (NTOF) surveillance system (Jenkins et al, 1993). In the Apr 24, 1998, *Morbidity and Mortality Weekly Report* (*MMWR*), NIOSH reported on the magnitude of work-injury deaths for the United States from 1980 through 1994. This NIOSH report showed that the annual total number of deaths and crude death rates decreased from 7,405 (7.5 per 100,000 workers) in 1980 to 5,406 (4.4 per 100,000 workers) in 1994. The report also identified high-risk industries and occupations both at national and state levels. National death rates were calculated using denominators from employment data from the BLS' Current Population Survey (BLS, 1980–1985). Only deaths of civilian workers were used for this survey (Jenkins et al, 1993). The Construction industry had the highest number of deaths in the period 1980–1994, with 16,091 deaths (18.2%). However, when the death rates were calculated, the Mining industry topped the list with 30.5 deaths per 100,000 workers, because there were fewer workers in the mining industry than in the construction industry. (One has to be very careful in discussing death rates!) The largest number of deaths occurred among the occupation category Precision production/crafts/repairers with 17,392 deaths (19.6%), but the highest death rates were seen among Transportation/material movers with 23.0 deaths per 100,000 workers (Jenkins et al, 1993).

Table 1-E. Unintentional Injuries at Work by Industry—1998, National Safety Council, 1999.

Industry Division	Workers[a] (000)	Deaths[a]	Change from 1997	Deaths per 100,000 Workers[a]	Disabling Injuries
All Industries	132,772	5,100	–1%	3.8	3,800,000
Agriculture[b]	3,450	780	–2%	22.1	140,000
Mining, quarrying[b]	618	150	–4%	24.3	30,000
Construction	8,045	1,120	+1%	13.9	410,000
Manufacturing	20,666	660	–3%	3.2	650,000
Transportation and public utilities	7,713	920	–1%	11.9	380,000
Trade[b]	27,087	450	–([c])%	1.7	730,000
Services[b]	45,575	680	+1%	1.5	900,000
Government	19,618	340	–([c])%	1.7	560,000

Source: National Safety Council estimates based on data from the Bureau of Labor Statistics, National Center for Health Statistics, state vital statistics departments, and state industrial commissions.
[a] Deaths include persons of all ages. Workers and death rates include persons 16 and older.
[b] Agriculture includes forestry, fishing, and agricultural services. Mining includes oil and gas extraction. Trade includes wholesale and retail trade. Services includes finance, insurance, and real estate.
[c] Less than 0.5%.

Occupational Injuries

The BLS also noted that wide variations existed in the frequency of nonfatal workplace incidents (illnesses and injuries) by industry, even for industries that produced similar types of goods and services. This variation may be due to the fact that different processes, and thereby different agents, are used to produce the same product; hence the nonfatal incidents may be different in two industries producing the same product. For example, using the Standard Industrial Classification, SIC 243, for the industry group Millwork, Plywood, and Structural members, the incidence rate for illnesses and injuries was 13.6 per 100 full-time workers in 1996. At the detailed industry level (4-digit level), the incidence rates ranged from 5.3 in Softwood veneer and plywood (SIC 2436) to 13.8 in Hardwood veneer and plywood (SIC 2435) to 14.5 in Millwork (BLS, 1998b).

Table 1-F presents the number of nonfatal occupational injuries reported by the Bureau of Labor Statistics for 1996 and 1997. It can be seen from Table 1-F that 78.1% of the total nonfatal injuries (5,799,900) occurred in three industrial sectors in 1996: Manufacturing (1,668,000–28.8%), Wholesale & Retail Trade (1,491,100–25.5%), and Services (1,380,800–23.8%) industries. The same three industries also accounted for 78% of the injuries in 1997 and for total cases with and without lost workdays, in both 1996 and 1997.

For the surveillance of nonfatal occupational injuries, NIOSH uses the National Electronic Injury Surveillance System (NEISS) developed by the Consumer Product Safety Commission (CPSC) (McDonald, 1994). National estimates of the magnitude and risk for nonfatal occupational injuries treated in hospital emergency departments for 1996 showed that workers at highest risk were young and male (*MMWR*, 1998). For the NEISS, data were collected at 91 hospitals selected from a stratified probability sample of all hospitals in the United States and its territories; NIOSH used data from 65 hospitals for its survey. The *MMWR* report showed that of the 3.3 million workers treated for occupational injuries in emergency departments in 1996, approximately 765,762 (23.2%) workers were in the age-group, 16–24 years; of these, 544,374 (71.1%) were men. Hands and fingers were the anatomic sites sustaining the most injuries, accounting for 30% of total injuries treated in emergency hospitals (*MMWR*, 1998).

In addition to the NEISS data, other data systems provide occupational morbidity surveillance systems, e.g., the 1988 National Health Interview Occupational Health Supplement, the 1996 National Hospital Ambulatory Medical Care Survey (NHAMCS), and the Annual Survey of Occupational Injuries and Illnesses maintained by the Bureau of Labor Statistics. Each data system has its drawbacks and one must be careful while discussing or comparing morbidity data using the various sources. For example, the NHAMCS system lacks industrial and occupational data though it provides comparisons between work-related and other injuries treated in the hospitals. The BLS system excludes self-employed people and farms with fewer than 11 employees; thus age-specific injury rates cannot be calculated from this survey. The NEISS system, on the other hand, is a continuous, ongoing surveillance system that includes industrial and occupational data and thus readily provides a mechanism for follow-up telephone interviews with injured workers (McDonald, 1994).

Table 1-F. Number of Nonfatal Occupational Injuries* by Industry and Lost Workday Cases—1996 and 1997, Bureau of Labor Statistics, U.S. Department of Labor, 1998.

Industry	Total Cases 1996	Total Cases 1997	Lost Workday Cases Total 1996	Lost Workday Cases Total 1997	With Days Away from Work 1996	With Days Away from Work 1997	Cases Without Lost Workdays 1996	Cases Without Lost Workdays 1997
Private Industry	5,799.9	5,715.8	2,646.3	2,682.6	1,785.8	1,746.5	3,153.6	3,033.2
Agricultural, Forestry & Fishing	108.4	106.9	49.0	53.8	37.0	39.0	59.4	53.1
Mining	32.4	35.1	19.5	22.6	14.7	17.7	12.9	12.5
Construction	476.1	485.6	216.8	227.4	179.1	187.1	259.3	258.3
Transportation & Public Utilities	494.6	477.1	293.0	281.3	217.3	213.3	201.6	195.8
Manufacturing	1,668.0	1,662.1	782.9	785.4	419.5	405.4	905.1	876.1
Wholesale & Retail Trade	1,491.1	1,480.1	637.8	657.6	452.3	439.4	853.3	822.5
Finance, Insurance, & Real Estate	128.5	124.6	49.5	47.6	38.0	37.8	79.0	77.0
Services	1,380.8	1,344.2	597.8	606.9	428.1	406.8	783.0	737.2

Source: News Release, BLS, Aug 12, 1998.
* = in thousands

Occupational Illnesses

Information on occupational illnesses is far less quantitative than that for occupational injuries, for the following reasons:

1. Illnesses related to work exposures may not have been diagnosed as being work-related.
2. In chronic exposure of workers, the recording of such exposures may not have been included in the BLS Annual Survey estimates. Then these may have underestimated the magnitude of the occupational disease problem.

Similar to definitions provided by BLS for the recording of occupational injuries, specific definitions for some categories of occupational disease and disorder are used to classify recordable illnesses along with a few examples (BLS, 1986):

- *Dust diseases of the lungs (pneumoconioses).* Examples: silicosis, asbestosis, and other asbestos-related diseases; coal worker's pneumoconiosis; byssinosis; siderosis; and other pneumoconioses.
- *Occupational skin diseases or disorders.* Examples: contact dermatitis, eczema or rash caused by primary irritants and sensitizers or poisonous plants; oil acne; chrome ulcers; chemical burns or inflammations.
- *Respiratory conditions due to toxic agents.* Examples: pneumonitis; pharyngitis; rhinitis or acute congestion due to chemicals, dusts, gases, or fumes; farmer's lung.
- *Poisoning (systemic effects).* Examples: poisoning by lead, mercury, cadmium, arsenic, or other metals; poisoning by carbon monoxide, hydrogen sulfide, or other gases; poisoning by benzene, carbon tetrachloride, or other organic solvents; poisoning by insecticide sprays such as parathion, lead arsenate; poisoning by other chemicals such as formaldehyde, plastics, and resins.
- *Disorders due to physical agents.* Examples: heatstroke, sunstroke, heat exhaustion, and other effects of environmental heat; freezing, frostbite, and effects of ionizing radiation (isotopes, x-rays, radium); effects of nonionizing radiation (welding flush, ultraviolet rays, microwaves, sunburn).
- *Disorders associated with repeated trauma.* Examples: noise-induced hearing loss; synovitis, tenosynovitis, and bursitis; Raynaud's phenomenon; and other conditions due to repeated motion, vibration, or pressure, such as carpal tunnel syndrome.
- *All other occupational illnesses.* Examples: anthrax, brucellosis; infectious hepatitis; malignant and benign tumors; histoplasmosis; coccidiomycosis.

According to the National Safety Council, in 1997 the number of occupational illnesses recognized or diagnosed in the private sector was 429,800, down from 493,000 noted in 1996. Of the 429,800 cases, about 259,300 cases (60.3%) were noted in the Manufacturing industries. Wholesale & Retail Trade accounted for 43,800 cases (10.2%); and the Services industries for 75,200 cases (17.5%) of total cases (NSC, 1999).

In 1996, the number of occupational illnesses recognized or diagnosed in the private sector was 439,900; of these, 264,900 (60.3%) were in the Manufacturing industries (NSC, 1998). About 281,100 (64%) illnesses reported in 1996 belonged to the repeated trauma disorders category and 203,000 (72.2%) of them were noted in the Manufacturing industries (NSC, 1998).

Table 1-G presents the number of occupational illnesses published by BLS for 1997. A total of 429,800 cases were reported for all industries combined. As mentioned earlier, the Manufacturing industry reported 259,300 (60.3%) cases; the Services industry accounted for 75,200 (17.5%) cases; and the Wholesale & Retail Trade industry showed 43,800 (10.2%) cases. Total lost workday cases in the Manufacturing industry accounted for 63.5% of all cases (116,600 of 183,600); the total lost workday cases for Services industry was 24,100 (13.1%); and 21,300 (11.6%) for Wholesale & Retail Trade industry. Again, as noted in the occupational injury statistics, 400 of the 600 nonfatal occupational illnesses in the Mining industry were due to lost workdays, and another 600 cases without lost workdays. Table 1-G also shows that the Manufacturing industry topped the list with 198,600 (71.9%) of the 276,000 cases in private industry for disorders associated with repeated trauma. Detailed statistics within each industry are available from BLS *Survey of the Occupational Injuries and Illnesses for 1996* (BLS, 1998a).

Incidence Rates of Disease or Injury

Incidence rates of disease or injury are sometimes used rather than actual numbers; for example, the total number of illnesses may not put the industry in the top 10 high-risk industries, but if we look at the rate of illness or the rate of injury, it may be high. A good example is that of the Shipping industry where the rate of injuries may be higher than that found in the Manufacturing industry though the number of workers in the former may be one-tenth that in the latter industry. Similarly, the incidence rate in the Mining industry may be quite high compared to the Manufacturing industry though the Mining industry has only 3% of workers compared to those in the Manufacturing industry. Therefore, when summarizing such data it may be better to use both actual numbers and rates when making comparisons.

The incidence rate can be determined either in terms of *number of injuries and illnesses* or in terms of *lost workdays*. When discussed as the *number of injuries and illnesses*, the incidence rate is defined as the number of injuries and illnesses *times* 200,000 *divided* by the total hours worked by all employees during the period covered. The 200,000 used is the base for 100 full-time equivalent workers, i.e., for work conducted 40 hours per week for 50 weeks in a year. When discussed as the number of lost workdays, the incidence rate is defined as the *number of lost workdays* multiplied by 200,000 divided by the total hours worked by all employees during the period covered. The 200,000 is again the base for 100 full-time equivalent workers, i.e., for work conducted 40 hours per week for 50 weeks in a year.

Table 1-H presents the incidence rate of nonfatal occupational illnesses and injuries by industry and case type for 1997 (News Release, BLS, 1998c). The data in Table 1-H show that the Manufacturing industry topped the list with an incidence rate of 10.3 per 100 full-time workers followed closely by the Construction industry, Agriculture industry, and Transportation & Public Utilities industry with incidence rates of 9.5, 8.4, and 8.2 per 100 full-time workers, respectively. Table 1-H also shows that the incidence rate for cases without loss of work-

Table 1-G. No. of Nonfatal Occupational Illnesses* by Industry and Lost Workday Cases—1997, Bureau of Labor Statistics, U.S. Department of Labor, 1999.

| Industry | Total Cases | Lost Workday Cases | | Cases Without Lost Workdays | Total Cases of Disorders Associated with Repeated Trauma |
		Total	With Days Away from Work		
Private Industry	429.8	183.6	86.9	246.2	276.6
Agricultural, Forestry & Fishing	—	2.0	1.5	—	1.4
Mining	1.2	0.5	0.4	0.6	0.5
Construction	6.9	3.4	2.7	3.5	2.0
Manufacturing	259.3	116.6	40.7	142.8	198.6
Transportation & Public Utilities	20.4	9.2	7.4	11.2	10.6
Wholesale & Retail Trade	43.8	21.3	12.7	22.5	23.1
Finance, Insurance, & Real Estate	17.4	6.6	4.9	10.8	13.1
Services	75.2	24.1	16.7	51.1	27.1

Source: News Release, BLS, Aug 4, 1999.
* = in thousands

days followed the trend of total cases of injuries and illnesses (News Release, BLS, 1998c).

Table 1-I presents the incidence rates of nonfatal occupational injuries by Industry Division for 1997 (News Release, BLS, 1998c). Table 1-I shows that though the Construction industry had 5.6 million workers compared to 101.7 million workers in all industries, the injury rate in the Construction industry was the highest among all industries (9.3 per 100 full-time workers). This was followed by the Manufacturing industry (8.9 per 100 full-time workers), and the Transportation & Public Utilities and Agriculture industries, each with an incidence rate of 7.9 injuries per 100 full-time workers. (News Release, BLS, 1998c). This is a cause for concern for it shows that safety conditions must be improved in such industries. The same situation was noted in these three industries with regard to total lost workday cases; the total workday cases was also high in the Mining industry with an incidence rate of 3.7 injuries per 100 full-time workers. Intensive surveys should be conducted to determine the cause of such injuries in these industries and intervention procedures instituted to prevent further injury among the workers.

Workers' Compensation

In addition to time lost to the employer and to the worker (due to productivity and wage loss) because of injury or illness nationwide, the true cost of work-related deaths and injuries is much greater than the cost of workers' compensation insurance alone (NSC, 1999). The total cost of workers' compensation claims in 1998 was $125.1 billion, according to the National Safety Council estimate. This included wage and productivity losses of $62.9 billion, medical costs of $19.9 billion, and administrative expenses of $25.6 billion. Other employer costs amounting to $12 billion were also included, such as time lost by workers other than the injured or disabled worker, the time taken for investigation of the injury, write-up of reports, etc. In 1998, the cost per worker was $940, the cost per death was $910,000, and cost per disabling injury was $28,000 (NSC, 1999).

More small businesses are becoming aware of the problem. This is evident from a study showing the most effective solutions for intensive smoking cessation programs came from partnerships among the medical provider, employer, and employee (patient). Results of the study showed that though the partnerships possessed a weak knowledge base of health care cost containment methods, they had a strong level of confidence (85%) that lifestyle modification programs such as smoking cessation could help control health care costs (Lesmes, 1992).

Data published by the National Council on Compensation Insurance (NCCI) showed that the most costly lost-time workers' compensation claims for 1996 and 1997 by *Part of Body* were for those involving multiple body parts. The claims filed in 1996 and 1997 averaged nearly $21,500 per workers' compensation claim. Other high claims were those that involved the head or central nervous system ($20,614), neck ($14,756), leg ($13,214), knee ($12,252), arm or shoulder ($10,726) and lower back ($10,833). The average cost for all claims by *Part of Body* was $10,488 (NSC, 1999).

Regarding the *Nature of Injury*, the NCCI reported that the most costly lost-time workers'

Table 1-H. Incidence Rate Per 100 Full-time Workers of Nonfatal Occupational Injuries and Illnesses by Industry and Case Type—1997, Bureau of Labor Statistics, U.S. Department of Labor, 1999.

Industry	1997 Employment (000's)	Total Cases	Injuries & Illnesses Lost Workday Cases Total	With Days Away From Work	Cases Without Loss of Workdays
Private Industry	101,666.5	7.1	3.3	2.1	3.8
Agriculture, Forestry, & Fishing	1,765.4	8.4	4.1	3.0	4.2
Mining	595.9	5.9	3.7	2.9	—
Construction	5,637.1	9.5	4.4	3.6	5.0
Manufacturing	18,656.9	10.3	4.8	2.4	5.4
Transportation & Public Utilities	6,170.8	8.2	4.8	3.7	3.4
Wholesale & Retail Trade	28,583.6	6.7	3.0	2.0	3.7
Finance, Insurance, & Real Estate	6,952.2	2.2	0.9	0.7	1.4
Services	33,304.8	5.6	2.5	1.7	3.1

Source: News Release, Bureau of Labor Statistics, Department of Labor, Dec 17, 1998.

Table 1-I. Incidence Rate Per 100 Full-time Workers of Nonfatal Occupational Injuries by Industry Division—1997, Bureau of Labor Statistics, U.S. Department of Labor, 1999.

Industry	1997 Annual Employment (000's)	Total Cases	Lost Workday Cases		Cases Without Loss of Workdays
			Total	With Days Away From Work	
Private Industry	101,666.5	6.6	3.1	2.0	3.5
Agricultural, Forestry & Fishing	1,765.4	7.9	4.0	2.9	3.9
Mining	595.9	5.7	3.7	2.9	—
Construction	5,637.1	9.3	4.4	3.6	5.0
Manufacturing	18,656.9	8.9	4.2	2.2	4.7
Transportation & Public Utilities	6,170.8	7.9	4.7	3.5	3.2
Wholesale & Retail Trade	28,583.6	6.5	2.9	1.9	3.6
Finance, Insurance, & Real Estate	6,952.2	2.0	0.8	0.6	1.2
Services	33,304.8	5.3	2.4	1.6	2.9

Source: News Release, Bureau of Labor Statistics, Department of Labor, Dec 17, 1998.

compensation claims were for those resulting from amputation, averaging $23,600 per workers' compensation claim in 1996 and 1997. Other high claims were for injuries resulting in "other trauma" ($15,995), fracture ($13,848), and carpal tunnel syndrome ($12,611). The average cost for all claims by *Nature of Injury* was $10,488 (NSC, 1999).

Trends in the workers' compensation claims by state for 1996–1998, for the United States, and the amount of compensation paid in 1996 are presented in *Injury Facts*™, formerly *Accident Facts*®, published by the National Safety Council (NSC, 1999).

OCCUPATIONAL HEALTH AND SAFETY PROGRAMS

An effective occupational health and safety program relies on the accessibility of occupational health and safety practitioners, who are skilled in the following:

- recognizing occupational disease
- evaluating the workplace for hazardous conditions
- instituting appropriate medical surveillance programs.

Recognizing Occupational Disease

Recognizing occupational disease is crucial to establishing the correct diagnosis and further treatment of the worker, thus preventing aggravation or recurrence of the disease in the worker. This does not stop with the recognition of disease in one worker. If occupational disease is suspected, the occupational physician should do the following:

- Determine whether the disease noted is occupational.
- Determine if other workers doing the same job and using the same materials are at risk for the same disease.
- Treat the affected worker(s).
- Adopt preventive measures to prevent recurrence of the disease among the remainder of the work force.

Importance of Occupational Health and Safety Programs

An integrated occupational health and safety program is vital in all workplaces. Although industrial hygiene, safety, and environmental professionals often are asked to evaluate a workplace, their evaluations will be incomplete without access to an occupational medical professional familiar with the workplace. For example, industrial hygienists and safety practitioners may not be aware that an occupational disease or condition has occurred as a result of an occupational exposure, without access to the medical information obtained during the evaluation conducted by the occupational physician and nurse.

Medical Surveillance Programs

Preventive occupational medical surveillance programs are essential to maintaining a safe and healthy workplace. They also reduce the occurrence of occupational diseases, occupational disease conditions, and occupational injuries. Within the context of the Americans with Disabilities Act (ADA), the attending physician and occupational health nurse can perform medical evaluations on all new workers and on those who have been reassigned to new jobs. For special jobs, specific medical tests may be necessary, e.g., baseline tests.

To prevent aggravation of an occupational disease, perform periodic physical examinations on all workers. These physical examinations should include routine urinalysis, blood tests, and specific urine tests for chemicals known to be toxic. For example, a worker who is exposed to trichloroethylene (TCE) should be monitored for TCE along with its metabolites, trichloroethanol and trichloroacetic acid in urine tests, since both the chemical and the two metabolites are known to be toxic. (Note that this test differs from the routine urinalysis done in a physician's office.) Depending upon the half-life of the chemical in question, monitor the affected worker periodically, not annually as is normally done during physical examinations. In the case of trichloroethylene, the half-life of the chemical is longer than five hours; hence the worker should be monitored at the end of the workweek using the following media: exhaled air, urine, and blood. Then the consequences of the worker's exposure throughout the week are taken into consideration. Thus, an annual examination of the worker is not appropriate, but monitoring the worker at the appropriate time, i.e., at the end of the workweek, is extremely important.

Occupational Health and Safety Professionals

Since the passage of the OSHAct the number of industrial hygienists and safety professionals trained in the United States has gradually increased. Rapid changes in technology also have increased the demand for such professionals, yet in the 1980s many industries understaffed in these areas. NIOSH conducted two nationwide surveys, a decade apart. Some results of the second survey, the National Occupational Exposure Survey (NOES) (1981–1983) are described herein. A stratified sample of industries was statistically chosen excluding the following industry types: mining; certain agricultural sectors; federal, state, and local governments; financial institutions, wholesale and retail trade, and certain professions (NIOSH, 1988).

The NOES data was collected from a sample survey of 4,490 businesses employing nearly 1.8 million workers. The facilities were divided into small (8–99 employees), medium (100–499 employees), and large (500 or more employees). From NIOSH's survey, 446,700 of 505,700 (88.3%) were small facilities with an estimated 11.1M of 33.2M (33%) employees, of whom 74.2% were men. Approximately 73% of the total work force were in nonadministrative jobs (73.3% in small, 75.8% in medium, and 70.5% in large facilities).The data from NIOSH's National Occupational Exposure Survey (NOES 1981–1983) showed that only 30%–35% of the industries had hired such occupational health and safety professionals or used consultants in the 1980s (NIOSH, 1988).

Occupational Health and Safety Team

To achieve the safest and healthiest workplace, it is important to use skilled professionals from various disciplines, including physicians, nurses, safety engineers, industrial hygienists, chemists, physicists, medical technicians, ergonomists, statisticians, etc. To maintain a safe and healthy workplace, all these members should work together, perhaps in a group designated the Occupational Health and Safety Team. All members of the team should be

- technically competent
- aware of the type of hazards present in the workplace and their health consequences
- aware of the health and safety programs conducted in the workplace.

Unions and management alike recognize the need for such professionals. Increased awareness of the benefits of such a team approach to occupational safety and health resulted in the hiring of more occupational health professionals in industries. Unfortunately, in a small business, this may not be possible. In this case, consultants can be used, or one of the workers can be trained to become aware of the health and safety factors in the workplace. He or she, in turn, can train other workers, who can help to maintain a safe and healthful workplace. If needed expert consultants can support workplace efforts. In this way, all workers will be protected from the hazards that occur in that worksite.

Why is a team necessary? Can hiring only a physician be sufficient? Not really—an occupational physician can diagnose the occupational disease and describe what may have caused it if he or she is aware of the exposure of the worker. But more support is needed to maintain a safe and healthful workplace. An industrial hygienist monitors the worker (personal monitoring) and the workplace (area monitoring). A safety professional places the appropriate controls (engineering controls such as machine guards, appropriate ventilation, etc.). A chemist analyzes the samples, etc., an occupational health nurse monitors the ongoing health status of the worker. The entire team is necessary even for one occupational disease recognized in the workplace.

Consider adequate ventilation. Often, a change in the flow of air in the workplace can make a difference for the workers, particularly when large fans are used in a big room and the fan blows air right on their faces or bodies. If an exhaust fan is used in such a case, the workers will not be exposed to the air containing the fumes or gases thus preventing inhalation exposures. Noise exposure is another situation that may require enclosure to protect workers from excessive noise exposures. When large machinery is used which causes a lot of noise, an enclosure may be necessary for the worker who watches over the machinery. The worker on such machinery should wear earplugs and see that appropriate safety shields are placed on the machine. If a substance is known to cause lung effects, the industrial hygienist should evaluate the exposure, alleviate the situation, and conduct both personal and area monitoring of that worksite, etc. These examples show the importance of an integrated occupational health and safety team.

Once the disease has been found to be occupationally related, the entire team of health and safety professionals should be involved in preventing further disease from occurring among the work force. The team approach includes the following members:

- industrial hygienist(s) who can determine the causative agent by using data from the MSDS and confirm its presence by performing industrial hygiene (personal and area) monitoring of the workplace. If necessary, the industrial hygienist would recommend appropriate personal protective equipment (PPE)
- chemist(s) to perform the sampling and analytic tests
- safety practitioner(s) to determine if safety factors are involved in the causation of the disease and if so, attempt to abate them
- physician(s) to be involved in the diagnosis, treatment, and follow-up of the worker
- occupational health nurse(s) to implement the medical surveillance program.

Once the findings are available, the team can discuss the situation, make the necessary changes to prevent further disease from occurring among the remainder of the work force, and conduct continued medical surveillance of the work force.

Working as a team, the occupational physician, with the help of the occupational health nurse, industrial hygienist, and safety practitioner, can prevent occupational diseases by maintaining a safe and healthful work environment.

Recently, new technologies, employment shifts from manufacturing to service industries, and more competition among similar producers are contributing to the introduction of new substances into the workplace. That means that the current crop of occupational diseases can vary from those seen in the 1970s and 1980s. Some examples of the newer problems include the following:

- video display terminals being used in the workplace associated with both ergonomic and visual problems
- continued use of new chemicals to replace the more toxic ones used earlier (some of the newer ones without being adequately tested)
- toxic chemicals being "reintroduced" in the form of trade-name compounds, the composition of which is rarely known to users
- exposure of workers to hazardous physical and biological agents as well as chemical substances.

These examples suggest that the techniques used for recognizing and preventing occupational diseases have to be modified with every change made in the process.

Distinguishing an occupational disease from other medical problems can be difficult, and depends most importantly on a comprehensive occupational history taken by the attending physician who knows that occupational diseases are almost always preventable. Two NIOSH surveys (NIOSH: 1972–1974 and 1981–1983) may indicate a trend away from full-time, on-staff occupational health professionals. Frequently the occupational health activities are being conducted by contract physicians. Many may not be familiar with occupational medicine. For those who are taking an occupational history, the following points should be considered:

1. current job, previous jobs, longest job held
2. type of exposures (gas, liquid, mist), name of chemical(s) and level of exposures, if known
3. length of exposure
4. protective devices used, if any; if used appropriately and in working condition
5. types of control technology methods used to minimize exposure of the workers to the inciting chemical(s)
6. symptoms occurring during exposure, continued symptoms during nonworking hours, or

symptoms experienced upon reexposure after the weekend break
7. severity of symptoms
8. potential for nonoccupational exposures, e.g., hobbies, household chores, etc.

A good record-keeping system is equally important to a comprehensive occupational history. Such record keeping can be useful for determining or preventing further occupational diseases, for future litigation, or for general research.

Health and Safety Committees

In many medium-sized companies, health and safety committees provide health and safety programs. Depending upon the size and type of workplace, functions of the health and safety committee range from ascertaining health and safety hazards to complete responsibility for the health and safety of the workplace. The health and safety committee should also maintain good incident prevention programs so that workers will be trained and experienced to use the safety equipment when necessary. In the case of a small business, the owner or facility manager may be the only one in the health and safety committee (Hoover et al, 1989; Rovins, 1989).

STARTING AN OCCUPATIONAL HEALTH PROGRAM

To begin an adequate occupational health program in your facility, obtain the services of an occupational physician or occupational health nurse, an industrial hygienist and safety practitioner. If your facility has limited resources, cross training can help fill in any gaps in the team. For example, an industrial hygienist could be cross-trained in safety procedures. If the company uses more machinery than chemicals, a safety practitioner could be hired and trained in industrial hygiene techniques. Training courses in all branches of occupational health and safety including ergonomics are given by the National Safety Council and its chapters and in 15 NIOSH-sponsored Educational Resource Centers across the country. Employers can benefit from this kind of training, and with the knowledge gained in the courses will better be able to institute appropriate occupational safety and health programs.

Before beginning an occupational health program, the occupational health and safety team should discuss the current conditions of the workplace and inventory all potential chemical, physical, and biological hazards that exist in the workplace, including raw products, intermediate products, catalysts, by-products, final products, slag, etc. Including workers at an open meeting to survey the kinds of products they may be using is also a good idea. They may be using a product, chemical, or compound that may not be actually used during the production process, e.g., a special type of soap to clean the oil off their hands, or a lotion to soothe their dry hands or a brace that they may have brought from home for daily use, which can cause problems. Many of these compounds used are trade-name products; hence, input from workers regarding symptoms, exposure situations, problems of any sort, e.g., drying of hands, itching due to chemicals dripping on hands, etc., is useful to an occupational health professional.

Medical and Industrial Hygiene Services

Whatever the size of the facility, establishing and maintaining a safe and healthy workplace by providing medical and industrial hygiene services is key, since the work force is the main asset to the employer (Hoover, 1989; Rovins, 1989).

The occupational health and safety team should assess the exposure conditions in the workplace and take appropriate measures to improve working conditions, e.g., exposures to hazardous chemicals and to physical hazards; most often, exposures to noise, excessive heat, intense cold, ergonomics, etc. In two surveys conducted a decade apart, NIOSH assessed the provision of medical services, industrial hygiene, and safety programs in the United States (NIOSH, 1988). Some of the medical services, industrial hygiene, and safety programs noted during the survey are presented in Table 1-J.

Table 1-J shows that 1 in 1,000 facilities had a physician in charge, 2 in 1,000 had a nurse on site; however, 25% of the facilities had emergency health care provision, and 44% of facilities had off-site health care. More than 47% of the facilities recorded health information on workers, 28% of workers had formal safety training but only 3% monitored fumes, gases, etc. About 46% of facilities required personal protective devices but another 22% used no personal protective devices, and 45% of facilities maintained the OSHA 200 log.

NIOSH EDUCATION AND TRAINING PROGRAMS

According to the 1970 OSHAct, NIOSH was mandated to provide professional education either directly, or indirectly through contracts. To accomplish this, NIOSH established a nationwide system of Educational and Research Centers (ERC) and other Training Programs Grants (TPG). These feature both full-degree and short-course professional education programs for occupational safety and health professions. They are currently provided by a network of universities across the country. NIOSH funds the ERC and TPG at each university, and their programs are reviewed annually by NIOSH.

In addition, internal NIOSH training and educational resources are now directed to provide ancillary services not easily conducted by the ERC and TPG facilities. One service that NIOSH provides is the development and evaluation of prototype training and educational curricula for new targeted populations, e.g., the formation of preservice occupational safety and health (OSH) curricula for vocational and technical education programs and for small business operators and workers.

The vocational and technical occupational safety and health (OSH) curricula are tailored for secondary school teachers; and address trade-specific OSH training (e.g., electrical safety) and cross-cutting issues (e.g., lockout/tagout). Each curriculum includes the following:

- topic-specific background information and instructor lecture notes
- student learning activities
- an accompanying video
- pre- and post- measures, overheads, a glossary of key terms, and a list of appropriate reference materials.

A second training and education service being provided by NIOSH includes the development and evaluation of multimedia CD-ROMs and Internet-delivered occupational safety and health courses. Planned efforts include the following:

- translation of the secondary school occupational safety and health curricula discussed previously
- safety training for miners
- development of research simulations for professional education
- a consortium of ERC/TPGs, an Internet degree program to train industrial hygienists.

The third training and educational service provided by NIOSH includes the development of models and partnerships for improved occupational safety and health training. One model is the development of means to measure social outcomes and impacts that result from occupational safety and

Table 1-J. Availability of Medical Services, Industrial Hygiene, and Safety Programs in Small Business Establishments: National Occupational Exposure Survey (NOES), 1981–1983 (NIOSH, 1988).

Health and Safety Programs	%
Facilities with physician in charge	0.1
Full-time (FT) physician	0.3
Part-time (PT) physician	0.2
Facilities with nurse on site	0.2
Registered nurse on site	9.4
Licensed practical nurse on site	6.9
Facilities with emergency health care FT	24.6
Facilities with emergency health care PT	8.7
Contract physician on site or on call	1.6
Facilities with health care on off-site location	44.4
Facilities with preplacement examination	21.1
Facilities that require post-illness examination	35.5
Facilities that require exit examination	0.5
Facilities that record health information on workers	47.3
Facilities with regular safety inspection	48.7
Facilities with formal safety training	22.1
Workers with formal safety training	28.3
Facilities monitoring fumes, gases, etc.	3.7
Facilities with personal protective devices required	45.9
Facilities with personal protective devices recommended	16.0
Facilities with no personal protective devices	22.1
Facilities with scheduled preventive maintenance programs	62.3
Facilities that maintain OSHA Form 200 Log	45.2
Facilities that provide one or more screening tests	6.2
Facilities that provide blood tests	8.9
Facilities that provide urine tests	9.2

health training in the workplace. In cooperation with the Educational Foundation for the National Restaurant Association (EFNRA), NIOSH is assisting in a nationwide study to assess injury-reduction that occurs after occupational safety and health training developed by EFNRA.

NIOSH is also sponsoring community-based intervention models to increase awareness of child labor laws, and reduce injuries to young workers. In two California communities, in partnership with UCLA and UC-Berkeley, and in a third partnership with the Massachusetts Department of Health, NIOSH is exploring intervention effectiveness to increase awareness through different community-based communication channels, e.g., the business community, churches, schools, social and professional organizations, and parent and student groups.

SMALL BUSINESSES

Statistics show that more than one million small businesses were started in the United States in the

1980s (*U.S. News and World Report,* 1989). In addition, small businesses, defined as those with fewer than 100 workers, now employ nearly 60 percent of the work force and are expected to generate half of all new jobs by the year 2000 (*U.S. News and World Report,* 1989; Bureau of the Census, 1996).

According to NIOSH, of the approximate 6.5 million private industry establishments in the United States, more than 6.3 million (98%) have between 1 and 99 employees, and more than 5.6 million (87%) have between 1 and 19 employees. Since many of these small business do not have resources for occupational health and safety programs, and most of them never will be inspected by OSHA, greater focus should be made on these small businesses with regard to prevention activities (NIOSH, 1998). If we use the NIOSH category of a small business as one with fewer than 100 employees, the 1994 County Business Patterns of the U.S. Census Bureau would show that 253 establishments are identified as small business industries. These small business industries included 4 million establishments and more than 30 million workers, roughly 30% of all employees in private industry. Again, NIOSH points out that these small business industries accounted for more than 1.6 million nonfatal occupational injury cases and 58,000 cases of illnesses in 1995, and 2,287 work-related fatalities in 1994. Occupational fatality, injury, and illness data were available for only 105 of the 253 small business industries at the 4-digit industry level, i.e., Standard Industrial Classification code. One major reason for nonavailability of data in these small business industries was due to small numbers of events occurring in these industries.

Currently, a small business has to deal with regulations mandated by the various government agencies, e.g., OSHA, EPA, etc., in the same way as the medium- and large-sized industries including those regarding the health and safety of its employees. The U.S. Securities and Exchange Commission (the Commission) has therefore focused its proposals on facilitating access to the public market for start-up and developing companies, and on lowering the costs for small businesses that undertake to trade their securities in the public market. The Commission undertook a host of proposals designed to facilitate the raising of seed capital by small businesses and to reduce the compliance burdens placed on these companies (*Federal Securities Law Reports,* 1992).

If such proposals are adopted, small businesses would incur significant cost savings for issuers without compromising investor protection. The small businesses can then concentrate on implementing appropriate occupational safety and health programs, either by themselves or in groups with other companies with the same four-digit Standard Industrial Classification (SIC) industry code. In the past few years, innovations used by some small businesses have been discussed in professional meetings, which give impetus to other small businesses to do the same and thus improve the safety and health of their work force.

Though small businesses (except those with fewer than 10 employees) are subject to the overall coverage of the OSHAct of 1970, most small employers are not required to keep injury and illness records because of the exemption in 29 *CFR* 1904.15; however, a few states still require all small employers to maintain OSHA 200 logs. Small business employers must comply with the requirements of 29 *CFR* 1904.15 only under (1) obligation to report on fatalities or on multiple hospitalization incidents; and (2) obligation to maintain OSHA 200 log when selected to participate in a statistical survey of occupational injuries and illnesses (BLS, 1986).

Small businesses still must comply with OSHA regulations, display the OSHA poster, and report to OSHA within eight hours (instead of the current 48 hours) any work-related incident that results in a fatality or the hospitalization of multiple employees (OSHA, 1993).

Another law that affects small business employers is the Americans with Disabilities Act (ADA) that went into effect on July 26, 1992. According to the ADA, any business with 25 or more employees was covered under the ADA and employers with 15 or more employees became covered in 1994. The ADA requirement applies not only to hiring practices but also to all employment-related activities, including layoffs, promotions, training, and pay. It also covers benefits, mandating, for example, that companies provide the same insurance coverages to all employees, including the disabled. For thousands of employers, the ADA has raised new questions, new uncertainties, and the fear of expensive, time-consuming lawsuits. For smaller businesses, the maximum on damages is $50,000. As a result of employers trying to understand the ADA law, many changes are expected to take place in the near future.

SUMMARY

The field of occupational health and safety has grown steadily since passage of the OSHAct. However, more needs to be done as occupational illnesses and injuries continue to occur. Occupational health, safety, and environmental professionals should work as a team to reduce the number of occupational injuries and illnesses. Newer chemicals and technologies are continuously introduced to the workplace, some of them without appropriate toxicity testing. The Hazard Communication Standard helps since it requires maintaining a Material Safety Data Sheet (MSDS) on each chemical used in the workplace. If any type of hazardous exposure occurs, the industrial hygienist and safety practitioner are intricate parts of the occupational health and safety team and should be consulted. Recognizing that a disease is occupational is sometimes a problem since the symptoms can be similar to those of nonoccupational illness. This accentuates the need for trained occupational health and safety staff to evaluate each situation. This can start with a comprehensive occupational history—including appropriate questions about chemicals being used, length of exposure, symptoms noted, etc. Then, with the help of the occupational safety and health team who will confirm the use of that chemical, the occupational components of the disease can be ascertained and appropriate treatment provided. Control methods should be instituted to prevent the spread of any occupational disease among other workers.

Small businesses can be at a disadvantage since they may not have in-house occupational health and safety professionals. However, nonprofessionals can benefit from the occupational safety, health, and environmental training opportunities available, including those offered by the National Safety Council and NIOSH Educational Resource Centers (ERCs). Preventing an occupational disease is far better than treating a worker after the disease has occurred, to avoid consequences including lost work days, workers' compensation problems, litigation, etc. With a growing network of trained occupational health and safety professionals and support services, increasing numbers of workplaces will benefit from reduced injuries and illnesses and safer work conditions.

REFERENCES

Bureau of the Census. *County Business Patterns, 1994*. Washington DC: U.S. Dept. of Commerce, Bureau of the Census, CBP-94-1, 1996.

Bureau of the Census. *County Business Patterns, 1997*. Washington DC: U.S. Dept of Commerce. Economics and Statistics Administration, Bureau of the Census. CBP/97-1, 1999.

Bureau of National Affairs (BNA). *Workers' Compensation Report* 9(7), Mar 30, 1998.

Bureau of Labor Statistics (BLS). *Employment and Earnings*. Washington DC: BLS, 1980–1985, (issue no.1 of each year).

BLS. *Record Keeping Guidelines for Occupational Injuries and Illnesses*. Washington DC: BLS, September 1986.

BLS. *Survey of Occupational Injuries and Illnesses, 1996*. Washington DC: BLS, Summary 98-1, January 1998a.

BLS. *National Census of Fatal Occupational Injuries, 1997*, news release. Washington DC: BLS, Aug 12, 1998b.

BLS. *Workplace Injuries and Illnesses in 1997*, news release. Washington DC: BLS, Dec 17, 1998c.

BLS. *National Census of Fatal Occupational Injuries, 1998*, news release. Washington DC: BLS, Aug 4, 1999.

Code of Federal Regulations. 29 CFR Part 1910.1000 to End. Revised as of July 1, 1989. Washington DC: U.S. Department of Labor. 1989.

Fatal occupational injuries—United States, 1990–1994. *Morbidity and Mortality Weekly Report* 47(15):297–302, 1998.

Hoover RL, Hancock RL, Hylton KL, et al (eds). *Health, Safety, and Environmental Control*. New York: Van Nostrand Reinhold, 1989.

Jenkins EL, Kisner SM, Fosbroke DE, et al. *Fatal Injuries to Workers in the United States, 1980–1989: A Decade of Surveillance, National and State Profiles*. Atlanta: U.S. Department of Health and Human Services/Public Health Service/Centers for Disease Prevention/National Institute for Occupational Safety and Health, DHHS Publication No. (NIOSH) 93-108S, 1993.

Lesmes GR. Corporate healthcare costs and smoke-free environments. *Am J Med* 93(1A) 48S–54S, 1992.

McDonald AK. *NEISS—The National Electronic Injury Surveillance Systems: A Tool for Researchers.* Washington DC: U.S. Consumer Product Safety Commission, Division of Hazard and Injury Data Systems, 1994.

National Safety Council (NSC). *Accident Facts,*® 1994 edition. Itasca IL: NSC, 1994.

National Safety Council (NSC). *Accident Facts,*® 1997 edition. Itasca IL: NSC, 1997.

National Safety Council (NSC). *Accident Facts,*®, 1998 edition. Itasca IL: NSC, 1998.

National Safety Council. *Injury Facts*™ formerly *Accident Facts,*® 1999 edition. Itasca IL: National Safety Council, 1999.

NIOSH. NOHS—National Occupational Hazard Survey, 1972–1983, and NOES—National Occupational Exposure Survey, 1981–1983. Cincinnati: U.S. HHS, PHS, CDC, NIOSH Division of Surveillance, Hazard Evaluations, and Field Studies. Vol 1, Pub No. 74-127; Vol 2, Pub No. 77-213; Vol 3, Pub No. 78-114; Vol 1, Pub No. 88-106; Vol 2, Pub No. 89-102; Vol 3, Pub No. 89-103, March 1988.

NIOSH. *Identifying Hazardous Small-Business Industries: The Basis for Preventing Occupational Injury, Illness, and Fatality.* Cincinnati: U.S. HHS, PHS, Centers for Disease Prevention, NIOSH, Education and Information Division, DHHS (NIOSH) Pub No. 99-107, May 1999.

NIOSH. *National Occupational Exposure Survey: Analysis of Management Interview and Responses.* U.S. HHS, PHS, Centers for Disease Prevention, NIOSH Pub No. 89-103, March 1988.

OSHAct. The Williams-Steiger Occupational Safety and Health Act of 1970 (The OSHAct). Public Law 91-596. 91st U.S. Congress, S. 2193. Dec 29, 1970 and 1991 (Revised).

OSHA. U.S. Department of Labor. Occupational Safety and Health Administration. Chemical Hazard Communication Standard, 1985 and 1987 (Amendment). Washington DC: U.S. Government Printing Office, 29 *CFR* 1910.1200.

OSHA. *Small Business Handbook: Laws, Regulations and Technical Assistance Services.* U.S. Department of Labor. Occupational Safety and Health Administration. Office of the Assistant Secretary for Policy. Washington DC: U.S. Government Printing Office, 1993.

OSHRA. Comprehensive Occupational Safety and Health Reform Act. Washington DC: U.S. Government Printing Office, U.S. House of Representatives, 103rd Congress. House Report 103–825, Part 1. Oct 3, 1994.

Rovins C. *Health and Safety in Small Industry—A Practical Guide for Managers.* Chelsea MI: Lewis, 1989.

Small business initiatives: Securities and Exchange Commission. In. *Federal Securities Law Reports.* Riverwoods IL: CCH, 1992, ¶84,928, pp 82, 469–482, 473.

Special report: The 1990 guide to small business. *U.S. News and World Report.* Oct 23, 1989. pp 72–73.

Top 25 general industry violations for 1996/97. *Keller's Industrial Safety Report* 8(4), April 1998.

Toscano GA, Windau JA. *Profile of Fatal Work Injuries in 1996: Compensation and Working Conditions.* Washington DC: Office of Safety, Health, and Working Conditions, Bureau of Labor Statistics. Spring 1998.

chapter 2

Early History and Post-World War II Trends

by Jean Spencer Felton, MD

25 **Introduction**
25 **What is Past is Prologue**
Prehistory ■ The Egyptians ■ The Babylonians ■ The Greeks ■ The Romans ■ Middle ages ■ The 16th century ■ The 18th century ■ Appearance of mass manufacturing ■ Industrial revolution
31 **Growth in the United States**
32 **Occupational Disease—Early Observations**
English reformers ■ Lead poisoning ■ Other illnesses
33 **Dust Diseases of the Lung**
Silicosis—Gauley Bridge ■ Black lung disease
34 **Professional Awareness of Occupational Health Hazards**
Trade union activity ■ Other early 20th-century movements ■ Women and children at work ■ Workers' compensation ■ Early organizational growth
39 **Genesis of Occupational Health in Welfare**
Immigrant work force ■ Physical examinations
40 **20th-Century Developments**
Occupational health nursing ■ Industrial hygiene ■ Millennium's end
42 **Summary**
43 **References**

INTRODUCTION

Occupational health and safety practitioners, especially if they are new to the field, may think that this area of human endeavor began with the Occupational Safety and Health Act of 1970 (OSHAct) and that little in the way of historical development preceded it. Although strict laws are of more recent origin, the ill effects of the environment on the worker have been known for thousands of years.

But just because hazards are known does not mean that they have been corrected; correction results from a deep social concern coupled with effective legal action. Most employers often hesitate to make changes, because almost every move to create a safer and more healthful worksite can cost money and time, which contribute to higher overhead costs. Since higher costs must be avoided to remain competitive, legal action often becomes the remedy.

WHAT IS PAST IS PROLOGUE

In an era of rapid change, it is important to consider the development of occupational health and safety to get a perspective on current theory and practice. A view of the past in any profession or culture is needed to become familiar with the early errors in judgment and the mode of the succession of events leading to improvements in practice and thought. Any undertaking is part of a developmental continuum and without knowledge of a discipline's growth pattern, one is likely to recommit the wrongs of history and delay all progress in one's specialty.

Prehistory

Although there are no extant writings to offer descriptive discourses on mankind's early efforts to stay alive, we do know from archaeological findings that there was early cooperation in performing tasks, a way humans ensured continuance of the race. Industry was necessary for survival. Uncovered shell mounds and flint heaps testify to humans' early manufacture of offensive and defensive weaponry.

Preparing flint by knapping—the shaping by a series of sharp blows—led to effective, usable points for spears and arrows. These primitive knappers probably were the first humans to acquire silicosis from the fragmentation of the hard quartz—researchers have observed evidence of carbonization and silicosis with pleural adhesions.

The search for flint—by both Paleolithic man and his Neolithic successor—became an important industry, for freshly extracted flint was more easily worked than flint that had long been weathered. Remains of the mine sources of this material were discovered throughout Europe, and the presence of skeletons indicated death by falling earth. In these mines were shafts, rather primitive tunnels, and illumination was provided by crude lamps.

Artisans during the first Sumerian Dynasty—in ancient Babylonia, currently Iraq—produced socketed axes and adzes. After the discovery of copper and the use of bronze, more sophisticated tools were developed, around 3000–2500 B.C. The use of sun-dried bricks for the construction of buildings began about 3400 B.C.

In the Assyro-Babylonian civilization, excellent sanitation facilities were created—drains, sewers, and stone privies. A further sense of cleanliness was seen in the insistence upon trench burial of warriors.

The Egyptians

The monuments—temples, pyramids, tombs, and obelisks—are evidence that the early Egyptians were productive. Although the workers were slaves, there were discrete tasks, supervised groups of workers, architecture, and methods of material preparation, transport, and erection, indicating productive industry. The systems of sanitation and irrigation also required great amounts of labor with the attendant injuries depicted in various excavated tombs.

Early documentation of surgical knowledge and medical therapy is seen in the Ebers Papyrus, found in 1862, described as a miscellany of extracts, recipes, and jottings collected from at least forty different sources . . . exactly analogous to the collections of household and medical recipes of Europe in later times. There are treatments given for various traumatic events such as crocodile bite, burns, and foreign bodies (splinters). The Edwin Smith Papyrus, acquired in the same year, dating to about 3000 B.C., was more of a textbook of surgery, and it discussed injuries and treatments involving splints, dressings, and ointments.

Diodorus, an historian in about 44 B.C., described the working of the mines in his history of Egypt and neighboring countries. The miners, usually slaves, condemned criminals, prisoners of war, or even exiled innocent families, worked in chains, were practically naked, and were subjected to the cruelty and blows of overseers. The sick, the disabled, the children, the women, all worked until worn out or until they died. The heat, smoke from the oil lamps, perspiration, and the lack of any hygienic measures must have created an unbearable atmosphere. There is no remaining evidence of occupational disease, for it is believed that no one in those conditions lived long enough to develop a chronically disabling lung disorder.

However, silica and carbon particles were found in the lungs of mummies, indicating pneumoconiosis, probably caused by the inhalation of sand during desert dust storms. It is not possible to ascertain if there were symptoms. Measured mercury levels in the bodies of ancients are equal to contemporary assays. The development of early "occupational health services" is of particular interest though. Rameses II, possibly about 1500 B.C., not only revived a project of constructing a canal from the Mediterranean to the Red Sea, but also ordered a colossal mortuary temple to be built, the Rameuseum. To construct them, many workers were acquired from nearby villages; and the Pharaoh strengthened his industrial medical service by adding physicians, as civil servants, to care for the mine and quarry workers, and other working teams and settlers engaged in the projects. It was not love of his subjects that motivated such action, but the desire to retain a healthy work force.

Each worker was examined regularly and forced to bathe in the Nile several times daily. During epidemics, the sick laborers were isolated in special camps, and every year the workers had to burn their huts and erect new ones. They could not relieve themselves at the worksite.

By means of "site newspapers"—notations on fragments of limestone—it was learned that one worker "has been examined by beating with a stick," an action interpreted by the reviewer as an energetic, and possibly an efficient, procedure for reducing absenteeism. The physician was placed in the work hierarchy between a foreman and a guard, suggesting a salary about equal to that of foreman. Again, according to Diodorus, the physician at the time of the Middle and New Empires was still a civil servant of the Egyptian state, and his services were provided free, as they were during military expeditions and voyages.

The Babylonians

Hammurabi, the sixth and most eminent ruler of the first dynasty of Babylon, about 2000 B.C., systematized the old laws of the land, revised them, and produced a complete codification of all statutes in a body of law of some 280 paragraphs. The Code of Hammurabi was unusual for it covered bodily injuries and fees for physicians, and probably was the first document that bore any resemblance to today's workers' compensation insurance laws.

Two of the pertinent clauses are:
- 196. If a man has caused the loss of a gentleman's eye, his one eye shall cause to be lost.
- 199. If he has caused the loss of the eye of a gentleman's servant or has shattered the limb of a gentleman's servant, he shall pay half his price.

Hammurabi's laws, although not a code in today's sense, must be considered as representative of his concern in being a just ruler, an ideal which was followed by the Mesopotamian kings at all times. The code is the oldest in existence, and antedates the first books of the Old Testament by about 1,000 years.

The Greeks

Nicander, a Greek grammarian, poet, and physician, who lived in the second century B.C., is cited by historians as providing in his poem, "Alexipharmaca," the earliest description of lead poisoning, with its colic, paralysis, and ocular disturbances.

> *The harmful cerussa, [white lead] that most noxious thing*
> *Which foams like the milk in the earliest spring*
> *With rough force it falls and the pail beneath fills*
> *This fluid astringes and causes grave ills.*
> *The mouth it inflames and makes cold from within*
> *The gums dry and wrinkled, are parch'd like the skin*
> *The rough tongue feels harsher, the neck muscles grip*
> *He soon cannot swallow, foam runs from his lip*
> *A feeble cough tries in vain to expel*
> *He belches so much, and his belly does swell*
> *His sluggish eyes sway when he totters to bed*
> *Complains that so dizzy and heavy his head*
> *Phantastic forms flit now in front of his eyes*
> *While deep from his breast there soon issue sad cries*
> *Meanwhile there comes a stuporous chill*
> *His feeble limbs droop and all motion is still*
> *His strength is now spent and unless one soon aids*
> *The sick man descends to the Stygian shades.*
> —*Euricius Cordus, 1532 A.D.*

Although lead water systems created a considerable risk to the health of the early Greeks, a greater hazard resulted from the coating of copper vessels with lead, or lead alloys, to prevent leaching of the copper and spoiling of the taste of the food. As lead inhibited enzyme activity, the Greeks found that it would improve a poor wine and lengthen the time that any wine could be kept.

Tetanus was described by Hippocrates, born on Cos around 460 B.C., and usually termed the father of medicine. He told of a shipmaster whose second digit of his right hand was crushed by an anchor.

Seven days later a somewhat foul discharge appeared; then trouble with his tongue—he complained he could not speak properly. The presence of tetanus was diagnosed, his jaws became pressed together, his teeth were locked, then symptoms appeared in his neck: on the third day opisthotonos (spasm of the back muscles) appeared with sweating. Six days after the diagnosis was made, he died.

Galen of the second century A.D., considered the most distinguished Greek physician after Hippocrates, practiced in Pergamum, his birthplace in Asia Minor. He served as physician to the gladiators, a position possibly comparable to the physicians who attend today's professional athletes. In that assignment, Galen gained experience not only in medicine, but was able to increase his knowledge of anatomy by having to treat the extensive wounds of his patients. Further, he visited the Phoenician copper mines on the island of Cyprus, and wrote that he smelled the suffocating fumes emitted from the mines and saw the workers naked, because the acid mist and vapor destroyed their clothing.

The Romans

Lucretius, a Roman poet and philosopher, 98–55 B.C., produced his *De Rerum Naturum* (On the Nature of Things), a poem in six books. Book VI considered various natural phenomena and the plague of Athens. In this writing, he referred to mining and the ill effects of such environments on the workers.

Lucretius wrote:

*Do you not see that sulphur is generated
In the earth, like bitumen with its deadly smell?
And where there is mining for veins of gold and silver
Which men will dig for deep down in the earth
What stenches arise, a Scaptensula!
How deadly are the exhalations of gold mines!
You can see the ill effects in the miner's complexions.
Have you not heard and seen how short is the life
Of a miner compelled to remain at this terrible task?
All these exhalations come from the earth
And are breathed forth into the open light of day.
The presence of metals was acknowledged:
All things of every king: Many they serve for food friendly to life: Many with power and to reflect diseases and so hasten death.*

Lucretius also spoke of entry of toxic substances into the body:

Many harmful elements pass through the ears and through the nostrils, and many make their way that are noxious and rough; and not a few should be avoided by the touch.

Martial (Marcus Valerius Martialis), a well-known epigrammatist who wrote of the Roman manners and customs of his day with a sharp wit, moved readily from formal orations to ridicule, even in relation to his benefactors. He discussed the hazard of handling sulfur:

*. . . and Bellona's raving throng
does not rest . . .*

*. . . nor the blear-eyed
huckster of sulphur wares.*

And for those physicians in academe who, with their entourages comprising house officers and medical students, visit patients, Martial made this pertinent comment to a practitioner who visited him:

I was indisposed; and straightway you came to see me, Symmachus, accompanied by a hundred of your pupils. A hundred hands frozen by the northern blast, felt my pulse. I had not then an ague, Symmachus, but I have now.

Rome's contribution to health was the construction of aqueducts, sewerage systems, drains, public baths, and well-ventilated houses. Large blocks of cut stone were used in the sewers and the manholes that provided access to them; the latrines were made of slabs of white limestone and faced with polished marble; and the aqueducts required endless blocks of cut and fitted stone. The preparation of these building materials unquestionably created respirable dust. Workers must have suffered disabling pneumoconiosis, but there is no record of such diseases, secondary to the work with stone.

However, Vitruvius, the eminent Roman architect and engineer, known only by his writings, publicly condemned the use of lead for piping water to be used domestically. Pipes of beaten lead were introduced to convey water from the masonry aqueducts to the homes, resulting in several outbreaks of lead poisoning. As water conditions can change, so can its degree of lead solvency. Thus the source of this illness was finally recognized.

Before leaving the Romans, we must mention Pliny the Elder. The only writings extant are his *Natural History*, in 37 books. Apart from discussing copper, the purification of precious metals, and the discovery of glass, he also described use of a primitive respirator to obviate dust inhalation. He wrote of ox bladders used by workers engaged in vermillion production; the hazard was mercury. Alexander the Great was the first general to devise a type of medical service for the army and to use surgeons on the field of battle.

Middle Ages

One significant document on the diseases of workers appeared in the seventh century in ancient Lombardy—the Code of King Rothari, promulgated on November 22, 643. At age 38, this ruler codified the laws in 388 chapters; previously all existed only in the memories of unlettered judges.

The origin of the basic principles of compensation for injury was in this body of law. The edict applied to personal injuries received in brawls, fights, and feuds; and there was a sliding scale in that injured slaves and servants received less than free men. There were payments for disability and death. There was a dual purpose in establishing such a system: (1) to fine the offender, creating revenue for the court—only partial payment went to the victim; and (2) to denounce the barbaric system of resorting to combat to settle disputes.

King Canute, a Danish king of England from 1016–1035, was also King of Denmark and Norway. Not only did he establish names for the digits of the hand, but he also stated the principles of compensation for particular injuries, especially in cases of trauma involving the hand. The importance of the loss of a thumb was recognized even then, for the compensation was twice that for the loss of the second digit, and 2.5 times that given for the loss of the third digit.

In 1524, a small booklet appeared, written in 1473 by Ulrich Ellenbog, an Austrian physician. The tract was directed toward goldsmiths and other handlers of metal, offering counsel regarding the hazards of lead, antimony, silver, and mercury. Ellenbog warned against burning coal in confined spaces and inhaling the vapor—more dangerous than the parent substance—arising from the heating of the metals. As he wrote,

Workers should not hold their heads above this vapor, but should wear a mask. The vapors of quicksilver, silver, and lead are a cooling poison, for they make a heaviness and tightness in the chest, burden the limbs and often lame them, as one can often see in the foundries where men work with large quantities of these metals.

The 16th Century

As the Renaissance began, work indicating a sensitivity to diseases resulting from contact with toxic work substances and/or environments appeared. Philippus Aureolus, Theophrastus Bombastus von Hohenheim, later calling himself Paracelsus (greater than Celsus), was born in 1493, grew up in Switzerland, and later studied medicine in Italy. He learned chemistry by working in the mines and smelters. Because of a rebellious spirit, he sought out other teachers by traveling throughout Europe. While so doing, he practiced medicine as an itinerant, and visited many mines and workshops.

Although his theories of disease were unacceptable to the faculties, he wrote furiously, but eventually quarreled with everyone. However, he produced a treatise, "On the Miners' Sickness and Other Miners' Diseases." The treatise was first published in 1567 and dealt with the pulmonary diseases of miners, the disorders of smelter workers and metallurgists, and diseases caused by mercury.

Paracelsus was able to distinguish between acute and chronic poisoning. His direct observations of work in the mines and smelters were outstanding. But Paracelsus did not refer to protective devices, and disregarded dust as an etiologic factor in the miners' diseases.

About the same time, Saxon physician Georgius Agricola (1494–1555) settled among the miners, where his interest and observations led to his writing a work of prominence, which has endured for centuries. In 1561, his *De Re Metallica* was published. In it, he emphasized the need for ventilation of mines including many illustrations showing various devices that would force air below ground.

But Agricola wrote mostly of miners' afflictions, pointing out that some disorders affect the joints, lungs, and the eyes; some were fatal. He underscored the hazards of the dust produced in dry mines, ". . . for the dust which is stirred and beaten up by digging penetrates into the windpipe and lungs, and produces difficulty in breathing." He relates that, in the Carpathian mines, women were found who had married as many as seven husbands, "all of whom this terrible consumption has carried off to a premature death." More recently, we believe that these deaths resulted from lung cancer, sustained by working with rich radium-containing ores. Agricola was aware of the need for personal protective devices, and illustrated gloves, leggings, and masks in his work.

The 18th Century

Bernardino Ramazzini (1664–1714), identified with both the universities of Modena and Padua, visited innumerable shops to study the working conditions and the clinical afflictions of artisans in various trades. His advice in the classic "Discourse on the Diseases of Workers" still applies to physicians today. Ramazzini pointed out that, in addition to the questions asked of a patient, as suggested by Hippocrates, one more should be added, simply, "What is your occupation?" He wrote,

> *In medical practice, however, I find that attention is hardly ever paid to this matter, or if the doctor in attendance knows it without asking, he gives little heed to it, though for effective treatment evidence of this sort has the utmost weight.*

Ramazzini viewed medicine in a social light. In his discussion of brickmakers' disease, he derides fellow physicians for treating all patients with the same remedies of purging and bloodletting, irrespective of the occupational origin of the illness. He writes, ". . . for the doctors know nothing of the mode of life of these workers who are exhausted and prostrated by unceasing toil."

Ramazzini believed that there were two causes of occupational disease:

> *The first and most potent is the harmful character of the materials that they handle, for these emit noxious vapors and very fine particles inimical to human beings and induce particular diseases; the second cause I ascribe to certain violent and irregular motions and unnatural postures of the body, by reason of which the natural structure of the vital machine is so impaired that serious diseases gradually develop therefrom.*

Physicians employed by industry were rarities until the 1900s. However, in about 1682 a Mr. Ambrose Crowley established the sales outlet of his iron works in London. He sold irons, screws, chains, files, hammers, hoes, and locks. Crowley's employees were eligible for medical care, sick benefits, and provision for their dependents, if the employees became disabled or deceased.

The work was done under a set of extremely stringent laws governing not only the site of manufacturing but also what was the equivalent of a company town. But more particularly, the last of 113 "Orders and Instructions for Carrying on the Works at Winlaton—1692–1815" called for the appointment of a "Doctor and Chirurgeon."

John Crowley, considering the deplorable state of his "honest and laborious workmen and their families," resolved to appoint "one well skilled in Physick . . . one that's of a sober life and conversation and not addicted so much to pleasures as to be withdrawn from a due attendance on his business." His patients, in keeping with accepted company practice, were encouraged to inform against him if he were suspected of "making use of my [the owner's] Medicines or Druggs in his own private practice."

Appearance of Mass Manufacturing

The economic changes of the 1700s were significant. Europe primarily was engaged in agriculture, and most workers were serfs on manorial estates or, rarely, free men working small holdings. In England, many of those peasants, freed from serfdom, were driven into the towns because landlords preferred to enclose their estates and raise sheep for wool production. English towns were filled with unemployed vagrants, who had little experience in farming and could find no employment.

In England, there was also a growing demand for goods to export. The production methods were labor-intensive; cotton and wool still were carded, spun, and woven by hand in the scattered cottages of the rural North and West. Further, transport was slow. Then, James Hargreaves developed the spinning jenny in 1764, which soon was to be improved by Richard Arkwright in 1769, and Samuel Crompton in 1779. From 1784 on, the Rev. Edmund Cartwright perfected the power loom, for which Parliament voted him a £10,000 gift, and in 1786 he was knighted for his service to national industry. Next in 1792, the American Eli Whitney invented the cotton gin (Figure 2-1).

Other production methods were needed to make the developing machinery available. Additionally, numbers of laborers who would work regular hours, under the control of their employers, and in structures where the machinery and power were provided were also needed. This change was the beginning of what first was termed the "factory system," but later was designated by A. Toynbee as the industrial revolution.

The new system first involved the manufacture of cotton goods, but soon was extended to the production of woolens, the fabrication of metal, wooden, and leather items, and, ultimately, into all forms of production. These changes in production methods and need for masses of workers affected the history of occupational health and safety, because it brought with it hazards never before encountered (Figure 2-1).

Industrial Revolution

But what, exactly, are the characteristics of the system now affecting the lives of millions of workers, and currently being established in the developing nations? The industrial revolution implies those changes in the processes and organization of production that mark the passage from an agrarian, handicraft economy to one dominated by industry and

Figure 2-1. Inventions such as the cotton gin were the beginning of the industrial revolution. (Reprinted with permission from the Chicago Historical Society, from The Growth of Industrial Art. Washington, DC: U.S. Government Printing Office, 1988.)

machine manufacture, from what is sometimes called a premodern or traditional economy to a modern one.

Specifically, the innovations encountered in the processes and organization of production included the following:
- Substituting inanimate for animal sources of power (particularly steam power through the combustion of coal).
- Substituting machines for human skills and strength.
- Inventing new methods for transforming raw materials (particularly in the making of iron, steel, and industrial chemicals).
- Organizing work in large units, such as factories, forges, or mills, making possible the direct supervision of the process and an efficient division of labor.

Paralleling these production changes were the altered technologies used in agriculture and transportation. These same changes in infrastructure are now taking place in developing countries, where the hazards and stresses to the workers' health and safety are identical.

GROWTH IN THE UNITED STATES

In the 16th century, there were a number of chartered British companies that provided an extension of overseas trade. The objective of such companies was a monopoly, collective trading, or regulation of trade, with the Crown holding the right to control the economy.

Hoping to apply the principle of merchant trading organizations to the New World, England was motivated by the desire, through colonization, to break hated Spain's monopoly of newly discovered land.

Sir Walter Raleigh, granted a perpetual patent to discover, occupy, fortify, and enjoy in fee simple all lands "not actually possessed of any Christian prince or people," sent an expedition to America. The expedition arrived at the island of Roanoke in 1584, and christened the existing American Indian kingdom Virginia.

By 1625, the Virginia experience had encouraged additional attempts at colonization of the New World. The Plymouth Company underwent great difficulties in its new colony, as nearly half of the 102 adventurers arriving in the Mayflower died during the first winter, 1620–1621. But, by 1630, new powers of legislation were granted the colonists, along with the monopoly of the Indian trade in that area.

Efforts to establish an iron works in Virginia using iron ore from North Carolina were unsuccessful. In part of Lynn, Mass., called Saugus, the first successful works were developed in 1645. Production lasted until about 1675. There was a demand for iron in both England and the Massachusetts Bay Colony, and ore, timber, and streams were abundant. Funded by English capital, the Hammersmith works was the first large-scale corporate industrial enterprise in America. At its peak, a ton of iron was produced per day, but costs were too high, even though a larger work force of indentured servants was brought over. Although the enterprise failed, it was considered the foundation of the American iron and steel industry. Eventually, skilled workers scattered to other parts of New England.

The New World was soon engaged in typefounding; printing; shipbuilding; quarrying; cabinetmaking; bookbinding; the production of paper, chocolate, and cottonseed oil; clockmaking; and even the building of spinets. But the growth of the textile industry was the true beginning of the factory system in America.

At Lowell, Mass., where there was sufficient water power to support a large industrial center, many young women were brought from adjacent farms to work in the mills. The owners depicted advantages and benefits of such employment, and Charles Dickens wrote glowingly of the working conditions, yet these comments appeared in the *Boston Quarterly Review* (July 1840):

> . . . *They are said to be contented, healthy and happy. This is the fair side of the picture; the side exhibited to distinguished visitors. There is a dark side, moral as well as physical. Of common operatives, few, if any, by their wages, acquire a competence. A few are well paid, and now and then an agent or an overseer rides in his coach. But the great mass wear out their health, spirits, and morals, without becoming one whit better off than when they commenced labor* . . .

Pierre Didier, MD, a native of France, settled in Wilmington, Del., in 1795, and through a personal and professional relationship with the first Eleuthere Irené du Pont, the company developed an early form of in-plant occupational medicine. A retainer system was developed, families were paid $50.10 per year, with additional fees for certain treatments or medications. Thus, Didier received immediate payment, workers received costly medical care, and they could gradually repay the company.

OCCUPATIONAL DISEASE— EARLY OBSERVATIONS
English Reformers

The growth of the industrial town pointed to the rising power and wealth of the industrial and the middle class. As towns grew, members of special literary and philosophical societies discussed work-related problems and opinions regarding resolution. Physicians were prominent in these groups. They knew the harmful effects ceaseless factory operations had on working-class families; the high numbers of working women and children; and the undesirable living conditions of working-class families.

Early on, there was an expressed need to use statistics to highlight social problems. From such demands came Edwin Chadwick's Report on the Sanitary Condition of the Laboring Population of 1842.

It was a time when the conditions of the workmen mattered nothing compared with the mad rush for production. New factories were springing up everywhere. To serve the new machines, men, women and children were toiling day and night under inhuman conditions in these factories and in the mills. Accommodations had to be provided for the new workers. New houses, new streets, new towns sprang into existence to house the thousands of workmen pouring up from the South of England to seek employment in the North. Houses were erected with such speed and lack of foresight that they were for the most part scarcely fit to be lived in. Faulty construction and bad drainage were the characteristic, and no thought of hygiene or sanitation entered the heads of the builders.

Chadwick's goal was to clean up his country's slums, espousing many social reforms, particularly in public health. He understood the relationship between poverty and chronic disease, and strove to convert poorhouses into schools and hospitals. He was described as "a stubborn cantankerous man who served Great Britain for more than half a century as a public scold." His efforts led to Parliament's passing the Ten Hours Act and establishing the half-time system of education. Working children were now required to attend school three hours daily.

Keeping with the reform movement in England was Dr. Charles Turner Thackrah, Britain's pioneer in occupational medicine. His 1831 volume, entitled *The Effects of the Principal Arts, Trades, and Professions and of Civic States and Habits of Living on Health and Longevity,* stressed prevention: "Surely humanity forbids that the health of workmen, and that of the poor at large, should be sacrificed to the saving of half-pence in the price of pots." The early death of Thackrah was lamented because it slowed the progress of industrial medicine, surgery, and better working conditions for 50 years.

Lead Poisoning

By the end of the American Revolution, medicine was still primitive; purging, bleeding, emetics, and blisters were considered standard therapy. Diseases like smallpox, dysentery, jaundice, cholera, and typhoid caused more American deaths than British bullets.

Earlier in 1745, Philadelphian Thomas Cadwalader produced over the imprint of Benjamin Franklin, *An Essay on the West India Dry Gripes,* in which he described the symptoms of lead poisoning. Franklin was concerned with the effects of lead on the human body and responded to a letter from Dr. Cadwalader Evans, a Philadelphia physician, agreeing that the lead used in the coils of stills employed in making rum was the cause of the "dry bellyache."

Other Illnesses

Poisoning from mercury was mentioned by Ellenbog, Paracelsus, and Fallopius. Mattioli (mid-16th century) observed chronic mercury poisoning among miners at Idria, Slovenia, formerly part of Yugoslavia, where mercury is still being mined today.

While other writers brought to light the cause-and-effect relationship between the work environment and occupational disease, efforts were being made to improve air quality in the workplace and prevent occupational illness and death. One innovation was the invention of the coal miners' safety lamp by Sir Humphrey Davy. The invention was based on the principle that a naked flame could be exposed in an explosive mixture of gas and air, provided the flame was surrounded by wire gauze of a particular mesh. The new wire-mesh lamp was adopted throughout the world, and saved the lives of thousands of miners.

DUST DISEASES OF THE LUNG

Diseases of miners and others coming in contact with respirable particulates were known clinically for centuries. The mechanisms of how they got there, however, were yet to be clarified. Initially, from the 10th century on, it was believed that the small demons or gnomes presumably inhabiting the mines were responsible for a generic disease entity called Bergsucht (miners' phthisis), which included pathogenic change—from a neoplasm to metal disposition—in the lung.

When steam power became possible during the industrialization of England, coal was needed in large amounts. With increased coal production, the problem of "the black spit" appeared early in the 1800s. Working conditions were dangerous, hours were long, and people were kept in debt to the company store. Even women and children worked underground, pulling loaded carts or carrying 100–300 lb of coal in creels on their backs. In keeping with the ethos of Victorian England as the evils of this child and female labor racket were gradually exposed, the subsequent popular outcry arose not so much because of revulsion from the abhorrent material conditions, but because the system prevented the children's receiving religious instruction, and because the fact that nearly naked men and women worked together underground was thought to be conducive to immorality.

For many years, the important question was whether coal miners' "black spit" and pigmentation of the lungs came from coal dust that was inhaled or if the pigmentation was produced by the pulmonary tissue itself. Pathologists were concerned solely with tissue change, while local practitioners saw the problem as one with social, community, economic, and engineering aspects. In 1866 Zenker coined the term *pneumonokoniosis*, from the Greek dust. It is used today, and can be modified with anthracosis, siderosis, etc.

Ventilation in the mines improved, work hours were shortened, and by 1875, as indicated by Meiklejohn, anthracosis, the black spit, "ceased to exist as a medical problem." Attention subsequently was directed to silicosis, and its complication, tuberculosis. By the 1930s, lung disease among coal miners reappeared in Great Britain, with many claims for compensation. Radiologists reported reticulation or a network, and, by microscopic examination, pathologists saw a dust-laden tissue network, unaccompanied by tuberculosis. Gough, through his studies of large tissue sections, described the early lesions of the coal nodule—collections of coal dust—around which was a peculiar type of focal emphysema, clearly different from silicosis. A second disorder, progressive massive fibrosis (PMF) with localized opacities of larger size, was later described. A worker with simple pneumoconiosis, when removed from dust contact, would not likely develop PMF. If enough simple pneumoconiosis existed, PMF might develop, irrespective of the cessation of dust exposure. It was concluded that PMF was a progressive disease. The term *coal workers' pneumoconiosis (CWP)* was first used by British investigators in 1942. The following year, the disease became compensable, distinct from silicosis.

Silicosis—Gauley Bridge

In the depths of the Great Depression in the United States, a project originated to help bring hydroelectric power to West Virginia. The digging of a tunnel offered job opportunities to unemployed workers, and hundreds of men came across the country with their families. The tunnel was cut, however, through pure silica, and for the first time, "acute silicosis" was seen. In addition to the many deaths that resulted from the Gauley Bridge exposure, the workers were forced to endure poor working and living conditions. Yet, there still was no workers' compensation or benefits to assist the workers and their families. This tragedy was treated fictionally by Hubert Skidmore in his *Hawk's Nest,* and in scientific detail by Dr. Martin Cherniack in *The Hawk's Nest Incident—America's Worst Industrial Disaster* (Yale University Press, 1986).

In 1936 a congressional committee heard details of the construction at Gauley Bridge, and based on the information given, the Air Hygiene Foundation, now the Industrial Health Foundation, was founded under the auspices of the Mellon Institute.

Another result of the Gauley Bridge disaster was a 1936 convention of representatives of labor, industry, the public, insurance underwriters, physicians, engineers, and attorneys in a meeting designated as the National Silicosis Conference. They concluded that "silica dust is the important causative factor in silicosis and that other dusts may modify or diminish the toxic effect of the silica."

Black Lung Disease

Despite these findings and work done on CWP in Britain, the condition of CWP was not accepted as a separate occupational disease in the United States until about 1969. It took the loss of 78 lives in a 1968 coal mine explosion in Farmington, W.Va., to direct attention to both safety in the mines and the health of the miners. A strike of nearly 40,000 West Virginia miners focused on the issue of compensability of CWP. After a three-week walkout, a bill was passed by the West Virginia legislature in keeping with the miners' demands. Later, the U.S. Congress passed the Coal Mine Health and Safety Act of 1969 (CMHSAct). The term "black lung" was given legal standing in the enactment, and the bill probably aided in the passage of the Occupational Safety and Health Act (OSHAct) in 1970.

Black lung, although rampant among U.S. coal miners, was not recognized as a disease by most U.S. physicians. Recognition of black lung was forced on the medical community by the intervention of the miners themselves. In 1977 and 1978, the CMHSAct 1969 was amended to broaden the coverage and to make the benefit program permanent. It is interesting that by far, the greatest amount of research on pneumoconiosis was conducted in South Africa and Australia, resulting in governmental controls of the dust hazard.

PROFESSIONAL AWARENESS OF OCCUPATIONAL HEALTH HAZARDS

As U.S. industry developed, professional awareness of work-generated disease slowly grew. The first document covering this aspect of medicine to be published was a $50-prize-winning dissertation by Benjamin W. McCready, MD, visiting physician at New York's Bellevue Hospital. One of his recommendations involved seamstresses. He believed they should spend Sundays outdoors instead of in church, after being confined all week at their sewing tables. The first U.S. literature on decompression illness appeared in 1843; and in 1860, Freeman, of Orange, N.J., published a brief report on mercury poisoning among hat finishers.

John B. Andrews, PhD, the long-time secretary of the American Association for Labor Legislation (AALL), was an unusual leader in the move toward control of occupational health hazards. Articles in the AALL's Review directed attention to occupational disease, and Andrews' efforts produced the First National Conference on Industrial Diseases, held in Chicago in 1910. The speakers included

Andrews, Alice Hamilton, MD, statistician and author of many studies on occupationally incurred illness, Frederick L. Hoffman, and University of Chicago sociologist Charles R. Henderson, PhD, who helped establish the Illinois Occupational Disease Commission.

The Conference generated the Memorial on Occupational Diseases for U.S. President Howard Taft. This document was a carefully prepared memorial of facts and conclusions, emphasizing the urgent necessity and practical expediency of a national expert inquiry into the whole subject of industrial or occupational diseases; their relative degree of frequency in various trades and occupations, the causes responsible for their occurrence, the methods desirable and practicable for their prevention or diminution, and all other matters having a relation thereto including methods of amelioration and relief.

While the history of subsequent action is not known, another effort did produce results through the U.S. Executive Office. The AALL Secretary John B. Andrews was struck by the horrors of phosphorus poisoning in the match industry, and began a study of match workers. Under the auspices of the Federal Bureau of Labor, he reviewed 15 plants with 3,600 employees and found more than 100 affected employees.

In 1911, after much testimony and an appeal to President Taft, Congress passed the Esch Act, effective January 1, 1915. The act placed a two-cent tax on every 100 white phosphorus matches produced. Obviously, the tax burden made production of phosphorus matches unprofitable, and phosphorus matches disappeared. Despite his work, Andrews was never recognized for removing one of the most hideously disfiguring occupational diseases from the American work scene. From then on, the AALL went on to fight for workers' compensation and other protective forms of insurance.

Alice Hamilton reported on lead poisoning in the sanitary fixture industry; the Bureau of Labor presented its findings on hookworm disease ("cotton-mill anemia") among cotton-mill operatives; and by 1911, California led the list of six states (the others were Connecticut, Illinois, Michigan, New York, and Wisconsin) mandating the reporting by physicians of all cases of certain occupational diseases—poisoning by lead, phosphorus, arsenic or mercury, or their compounds; anthrax; and compressed air illness.

Trade Union Activity

Until the mid-19th century, any effort on the part of U.S. organized labor to improve working conditions was considered illegal. An individual worker could bargain legally for higher wages and shorter hours, but any joint effort to "regulate the value of labor" was considered a criminal conspiracy. In 1825, carpenters in Boston struck for higher wages, for it was believed impossible for a journeyman housewright and house carpenter to maintain a family at that time with the wages then given to the journeymen house carpenters in Boston. The resolutions were supported by neither the Master Carpenters nor, as would be expected, by the builders.

In response to such constraints and seeing the rich rewards of their labor going to a few employers, in 1827 journeymen of several trades in Philadelphia formed a Union of Trade Associations in the hopes of "establishing a just balance of power, both mental, moral, political, and scientific, between all the various classes and individuals that constitute society at large." The precedent of handling labor union cases of this type persisted until 1842, when the Massachusetts Supreme Court overturned the doctrine of criminal conspiracy.

Miners in the United States attempted to organize the American Miners' Association in January 1861, an effort to secure better mining laws in the coal-producing states. The action was given added support by the Avondale disaster in the Pennsylvania anthracite fields where, in September of the same year, 109 mine employees were killed.

The United Mine Workers of America (UMWA) was formed in 1890. It developed a unique program in health care. An industry-wide trust fund financed by royalty on production was jointly administered by the operators and the union. The fund was based on the principle expounded by organizer John L. Lewis, that the cost of maintaining the manpower in a workable and healthful condition should be considered as much a part of production costs as any other item figured into the expense of operation.

The UMWA Welfare and Retirement Fund was a child of private enterprise within the perimeters of private industry. A spin-off was "a comprehensive survey and study of the hospital and medical facilities, medical treatment, sanitary and housing conditions in the coal mining areas." The study depicted in striking words and pictures, "the enormity of the social blight coincidental to the exploitation of native wealth without obligation or compensation to the resources of the ravaged communities."

While many trade unions were not greatly concerned with occupational disease or its prevention until the 1970 passage of the OSHAct, great concern developed in organizations whose members worked with such hazardous materials as asbestos. Today, many international unions have specialists in occupational health and safety for purposes of counseling and education.

Other Early 20th-Century Movements

In the days before workers' compensation coverage, a study was conducted on work-related injuries as part of a larger Pittsburgh Survey that covered investigations of steel workers. Crystal Eastman, a member and secretary of the New York State Employer's Liability Commission, reviewed injuries in Allegheny County, Pa., occurring within three months' time, and industrial fatalities that took place within a year. The county was essentially the Steel District, with 70,000 workers in the steel mills, 20,000 in the mines, and 50,000 on the railroads. The frontispiece for the monographic report was a "death calendar" for one year (July 1906 through June 1907) with crosses indicating fatalities in industry. The deaths ranged from 35–60 per month for a total number of 526 for the year. This came to nearly 1.5 fatal injuries per day, during a seven-day workweek.

William Hard explored occupational disease and wrote of the bends, lead poisoning, chrome holes, and chlorine. He attacked the lack of workers' compensation insurance in an article, "The Law of the Killed and Wounded." The editor of the 1908 monthly *Everybody's,* in which the piece appeared, introduced the writing with this note:

> *On broadly human lines, this article should interest every citizen. On economic lines it should interest every employer of labor. It is not intended to arraign a class, nor to array one set of our people against another. It is rather to demonstrate where a great, heartbreaking waste can be minimized to the physical and financial benefit of all. Congress and the President have failed, so far, in their efforts to safeguard the man who works, and the companies for whom men work, and above all, the women and children who are dependent. But earnest men will continue to fight, with the certainty that when the evil and remedy are both known, all classes will rally in support of the needy. Mr. Hard points that way, clearly, and without prejudice.*

George Kober, MD (1850–1931), one-time Army surgeon for the military installation on Alcatraz Island and later professor of hygiene and dean at Georgetown University School of Medicine, wrote extensively of health hazards in many different industries. In his introduction to his report, Industrial Hygiene and Social Betterment, Kober wrote that during a visit to Berlin in September 1907, he met E. J. Neisser, MD. Dr. Neisser had just completed *An International Review of Industrial Hygiene*, a 352-page volume. Excepting an inspector's reports in New Jersey, no recent data concerning "factory sanitation" were available. Kober commented, "Realizing the importance of the subject, not only to wage-earners, but to all interested in the conditions under which our fellow men and women live and work, an effort has been made in the succeeding pages to supply this information."

Alice Hamilton reported on the sanitary fixture industry in 1911, and soon was assigned under the Bureau of Labor (Department of Commerce) to do a comparable investigation of all the states. Hamilton noted that, as a federal agent, every detail was left to her—finding the plants, gaining permission for entry, and designing the survey. She did not receive a salary. When the report was ready for publication, the government agreed to buy it. Dr. Hamilton later went on to become assistant professor of industrial medicine, the first female faculty member at Harvard.

William Gilman Thompson (1856–1927) made a strong contribution in 1914, with the completion of his lengthy text, The Occupational Diseases. In 1910, he was involved in the establishment of the first occupational disease clinic in connection with Cornell University. In 1912, the eminent Richard C. Cabot, MD, of Boston's Massachusetts General Hospital, rationalized the inability or unwillingness of hospitals to tackle prevention of occupational disease, because there were too many other significant problems—poor housing, alcoholism, poisoned air, and massive immigration. He did say that when physicians realized that the problem could be solved, they would call for trained investigators to assist in stopping workplace health hazards.

Women and Children at Work

As in England, there were also women and children in the U.S. work force. Before the organization of the National Child Labor Committee in 1904, there were no standards of protection against the exploitation of children. Ten-year-old boys worked in the constant dust of coal breakers, picking out slate or sweltering all night near the furnaces of a glasshouse. In the South, girls of the same age were employed in dusty cotton mills. A child working a 12-hour day in a Pennsylvania silk mill met with either approval or indifference.

John Spargo, a former worker for the Socialist cause and active in social reform, wrote a stinging, illustrated book in 1906: *The Bitter Cry of the Children.* He cited some 1900 census data on employment of children:

- 1,750,178 working children between 10 and 15 years
- 25,000 boys under 16 years working in mines and quarries
- 12,000 children under 16 years manufacturing tobacco and cigars
- 5,000 children working in sawmills
- 5,000 children working at or near steam planers and lathes
- 7,000 girls working in laundries
- 2,000 girls working in bakeries
- 138,000 children working as servants and waiters in restaurants and hotels
- 42,000 boys working as paid messengers
- 20,000 boys and girls working in stores.

Spargo believed that all of these "official" figures were too low as evidenced in a 1904 lament that "the factories will not take you unless you are eight years old."

By 1920, many states had child-labor laws, but they could be evaded easily. Attempts at federal legislation were opposed by businessmen as destructive of individual choice and states' rights. Even with age limits, who could tell a child's age with any accuracy? Furthermore, many parents, in collusion with employers, flouted the law for the additional income their children earned.

The Fair Labor Standards Act in 1938 cleared the U.S. workplace of children. The act established 16 years as the minimum age for employment; 18 years was the lower limit in occupations deemed hazardous by the U.S. Department of Labor. Nevertheless, minors aged 14 and 15 years were permitted to work outside school hours in certain nonmanufacturing and nonmining occupations, such as clerical and sales work. Children, however, could not work before 7 a.m. or after 7 p.m.

Many states passed protective acts for women, in consideration of the hours of the work, hazardous employment, and rectification of extraordinary differentials in pay. In the decades ahead, as women were liberated they moved into nontraditional employment as some "protective" laws were dropped. Any ill effects of work in the heavier trades were determined over many years. Much of this movement began in World War I, when women took jobs in war production plants. While some left such jobs in 1918, many stayed on after World War II. Women enjoyed earning good wages. Both wars brought women into the armed services, providing a new experience and, for many, new careers in the years after 1945 (Figure 2-2).

Workers' Compensation

Even though Great Britain had passed its Workmen's Compensation Act in 1897, the United States continued to operate under various employer's liability laws. The need for protective insurance had sound historic roots. The growth of capitalism increased the insecurity of the worker, and the situation had grown acute after the onset of the industrial revolution in the 18th century. Millions of people were in the labor force, working long hours, receiving wages too low to permit saving. Exposed as they were to physical and chemical hazards, workers could suddenly be unemployed by virtue of work accidents, disease, old age, or periodic economic crisis. These unemployed workers became public charges, having no reserves upon which to draw, and having needs far beyond the ability of any charity to handle.

In 1897, the British law read,

> *If in any employment to which this Act applies personal injury by accident arising out of and in the course of the employment is caused to the workman, his employer shall ... be liable to pay compensation ...*

The wording, "arising out of and in the course of the employment," remained in future amendments and was carried over to comparable legislation in the United States. By 1906, the British act was repealed, and the principle of compulsory payment extended to nearly all workers. Protection against occupational disease was included for the first

Figure 2-2. Children wait for work permits in Chicago, 1911. Before the state passed the child labor laws urged by settlement workers, it was not unusual for children to work 12-hour days. (Reprinted with permission from the Chicago Historical Society.)

time—anthrax, lead poisoning or its sequelae, mercury poisoning or its sequelae, phosphorus poisoning, arsenic poisoning, and hookworm disease.

Finally, after a long study of the German insurance plan, a scheme was submitted to the U.S. Congress by the Commissioner of Labor. After the passage of legislation protecting federal workers, and after much deliberation by many state commissions, Wisconsin passed the first workers' compensation legislation undertaken by the states on May 4, 1911. This law, in its opening sections, abrogated the old defenses of assumptions of risk, contributory negligence, and the fellow-workman principle. The validity of the act was tested within a few months and was declared constitutional by the Wisconsin Supreme Court. California led a few states the same year by passing legislation providing coverage for occupational disease.

A bill for workers' compensation (the Wainwright Law) was passed in New York in 1910. However, it was declared unconstitutional by the New York Court of Appeals on the grounds that the law violated both the federal and New York State Constitutions, "because it took property from the employer and gave it to his employee without due process of law." By the adoption of an amendment to the state constitution by the legislature and approval in 1913 at a general election, a compulsory New York Workmen's Compensation Act finally became effective in mid-1914.

At the same time the 1910 act was being declared unconstitutional, a fire occurred in a clothing factory in New York City that took the lives of 146 employees. This Triangle Fire disaster unified the demand for factory legislation. It gave greater force to reform because the fire took place on the same day (March 25, 1911) that the Court of Appeals decision was publicized. Despite earlier recommendations regarding the Triangle Shirtwaist Company's fire escapes, barred windows, and stairways, no action had been taken. The fire and the public's agitation resulted in the passage of a new industrial code, which was a model for several states. More importantly, the disaster opened the door to progress in job safety. In 1913, New York, California, Massachusetts, Ohio, and Pennsylvania granted rule-making authority to their labor commissioners. Finally, with the passage of a Workers' Compensation Act in Mississippi in 1948—37 years after Wisconsin's legislation—coverage was introduced to all of the states.

Early Organizational Growth

The year 1910 was significant in the growing concern about the health of workers. In that year, the Bureau of Mines was established, to be headed by Joseph E. Holmes, MD. Thompson had created the first clinic for occupational disease, the first occupational disease conference was held, Andrews had published his report on phosphorus poisoning, Hamilton completed her study on lead poisoning, and the Bureau of Labor issued a list of industrial poisons. These events seemed to catalyze the growth of other movements and group efforts in the immediate years ahead (Figure 2-3).

As nearly all contemporary legislation linked occupational health to safety, it was difficult to separate illness from injury in the work setting. On October 8, 1911, the United Association of Casualty Inspectors, composed of insurance company inspectors in New York, was formed. Three years later, the name was changed to the American Society of Safety Engineers. After merging with the National Safety Council, which was founded in 1912, it withdrew in 1947 and has been independent since.

One piece of legislation that did not cross over between occupational health and industrial safety

Figure 2-3. Many states passed protective acts for women in consideration of the work hours, conditions of employment, and pay differentials. (Reprinted with permission from the Chicago Historical Society.)

was the Walsh-Healey Public Contracts Act, approved in 1936. Although not to be enforced with any strength in the years following, the act represented a second effort to put a floor under wages and a ceiling over hours worked. Most activity centered about the law concerned these issues rather than the health and safety requirements. Apart from limiting workers in plants providing materials for the federal government in any amount exceeding $10,000 to eight hours per day and 40 hours per week, and granting them the minimum wage as established by the Secretary of Labor, these conditions were laid down: No work was to be performed or materials manufactured under contract ". . . in any plants, factories, buildings, or surroundings or under working conditions which are unsanitary or hazardous or dangerous to the health and safety of employees engaged . . ." Further, no work was to be carried out by any man under 16 years of age, or any woman under 18 years.

However, by 1942, the clause regarding health and safety was given evidence of intent through the publication by the Department of Labor of a booklet, "Basic Safety and Health Requirements for Establishments Subject to Walsh-Healey Public Contracts Act." In addition to standard subjects in safety, there were provisions relating to atmospheric contamination, sanitation, personal service conveniences, illumination, and first-aid facilities.

By 1960, the initial major revision was published in the Federal Register, and for the first time in the history of Walsh-Healey, the act had force and effort of law. Reaction was mixed; public hearings were held, and management, labor, and national safety associations became vociferous. Revisions followed, and standards were included relating to occupational noise exposure, medical services, and radiation. By 1969, the bombshell created by the revisions seemed to be subsiding. However, the OSHAct of 1970 paled the original corporate professional organizational response.

GENESIS OF OCCUPATIONAL HEALTH IN WELFARE

Occupational health as a medical undertaking, as a support segment of a manufacturing, service, financial, or sales enterprise, or as a governmental regulatory service, was not born exclusively from the concern of the medical establishment. In the later decades of the 19th century, millions of people emigrated from Europe and came to the United States to settle in its large cities. Steel mills, machine tool industries, railroads, canals, farms, and municipal governments hired these people. In the 1840s, 1.5 million persons came to America. By 1860, about 6 million had arrived. The influx continued up to the early years immediately before World War I, and resumed in the 1930s with the rise of Hitler.

Immigrant Work Force
In the early years, immigrants entered the work force as unskilled laborers. With comparably low wages in the decades to come, poverty was the characteristic of these urban groups. Demographic knowledge of the population at large was incomplete. Birth registration was limited and in 1912, only eight states recorded births in accordance with Census Bureau standards. An attempt to establish a federal Department of Health was strongly opposed.

Settlement houses were an attempt to reverse these unanswered needs. Jane Addams, founder of Hull House in Chicago, gave her impression of the plight of certain city dwellers in one ward of about 50,000 inhabitants:

The streets are inexpressibly dirty, the number of schools inadequate, the street-lighting bad, the pavement miserable and altogether lacking in the alleys and smaller streets, and the stables defy all laws of sanitation. Hundreds of houses are unconnected with the street sewer.

In response to medical needs, a public dispensary was established at Hull House, and a Visiting Nurses Association nurse made his headquarters there.

As the settlement houses developed, some positive occupational health and safety activities began in industry. In 1907, a group of cotton mills in the South "at the expense of the corporation," employed trained nurses to attend persons who were sick in the mill community, and get them well as quickly as possible.

At the New York Telephone Company, there was a "retiring-room" for operators who felt indisposed. If one fell ill, she would be taken to this room, and a physician called. When the employee was well enough to be moved, she would be sent home in a closed carriage, in the charge of another female employee.

Physical Examinations
Eventually, active medical practice replaced clumsy initial preventive measures in the workplace. The first attempt was the conducting of physical examinations. The National Cash Register Company examined applicants for work as early as 1901. Frank Taylor Fulton, MD, a graduate of Johns Hopkins and a practitioner in Providence, R.I., recorded what probably were the first documented medical evaluations conducted among factory workers.

In a 1906 presentation before the Providence Medical Association, he gave credit to "certain manufacturers" who gave assistance in the examination of a group of employed persons. Dr. Fulton examined at no charge employees in one of the big saw factories for the purpose of discovering tuberculosis.

Examinations were introduced at Sears, Roebuck, and Company in Chicago in 1909. By 1914, physical examinations of employees were a fixture in many large companies. Eventually, physical examinations were relatively commonplace. One commentator pointed out the gains through these examinations when he stated that examining large groups of men at work added insight into the occupational causes of disease. Another medical director, in discussing her own examination program in a plant with 3,000 female employees, answered complaints of organized labor by stating that such examinations were not conducted to turn down anyone who was fit to work anywhere. Instead, they were for the purpose of fitting any applicants, even the physically disabled, into the work for which they were best adapted physically, where they would injure themselves least, possibly benefit themselves, and continue to work longest.

In 1920, several in-house programs began, based on the World War I industrial health experience. The influenza epidemics had come and gone, and although data were available from the military camps and from population groups at large—2% of all men in the U.S. Army had the disease and one of every 67 who had it, died—no data were derived regarding attack rates among industrial facility-employed groups, although the three waves of the 1918–1919 pandemic caused 500,000 deaths in the United States alone and cost over $100 billion. Factory operations must indeed have been disrupted.

20TH-CENTURY DEVELOPMENTS
Several breakthroughs made the 20th century significant in the development of occupational health

and safety. Certain occupational diseases were identified and controlled. During World War I, young women were hired to apply luminous paint on watch faces and the dials of instruments. The practice led to radium poisoning and many fatalities demanding diagnosis after the war, when the practice of such dial painting became a fad. As another example, chronic mercury poisoning was affecting processors of the source ore and hatmakers alike.

A study of mercury poisoning in New York City hatmaking plants identified some 80 workers with typical symptoms of toxic mercury salt absorption. In Connecticut, where the hatmaking industry was centered, manufacturers, unions, the U.S. Public Health Service, and the State Department agreed to substitute hydrogen peroxide for mercury nitrate used in the carotting process.

In the early 1950s, a strange disease of the central nervous system appeared among the citizens of Minamata, a Japanese fishing village. By final reckoning, there were 103 deaths, and some 700 persons were disabled. The disease? Organic (methyl) mercury poisoning, traced to a chemical plant discharge system that dumped the waste into the Minamata Bay. The fish became contaminated, the mercury entered the food chain, and the citizens, typically fish-eaters, were poisoned by the chemical. Subsequently, waste treatment systems were installed, but, the impairments of the victims persist in many areas of the world.

Asbestos, once considered a mild fibrogenic substance causing infrequent lung tissue disease, became a subject of extensive study and epidemiological investigation. It was found to be a carcinogen. Asbestos exposure with cigarette smoking caused death from cancer of the lung, or without smoking, mesothelioma.

Effects of chromium compounds have been studied since World War II. While chromic acid mist was known to produce septal perforations, "chrome holes" of the skin, and chronic dermatitis, occupationally caused lung malignancies from certain chromates were a new epidemiological finding.

In the early post-World War II years, beryllium became a part of the nuclear energy industry. (Later it was used as a heat shield for reentry of the Project Mercury astronauts.) It was treated with care because of the earlier findings that exposure to the beryllium-zinc-silicate of the original fluorescent lamps resulted in both acute and chronic disease. Exposure to vinyl chloride produced angiosarcoma of the liver in a few patients; and chlordecone (Kepone) production with no environmental controls resulted in overexposure of more than 100 workers. The exposure produced an occupational disease involving several bodily organ systems.

Since cancer of the scrotum in chimney sweeps was described in 1775, the incidence of work-related carcinogenesis has increased. One of the first to proclaim the relationship between certain industrial substances and cancer in workers exposed to such toxic materials was Schereschewsky. In 1925, he completed a statistical review of the steady rise in cancer mortality over a 20-year period. He also organized a 1927 conference that advised the U.S. Public Health Service on a program of cancer research. Research in occupational cancer was among four recommendations submitted by a committee of eminent investigators.

Occupational Health Nursing

The early history in industry presents a fuzzy distinction among public health nurses, visiting nurses in industry, public health nurses in industry, and industrial nursing. Although the English J. and J. Colman plant in 1878 first hired a professional nurse, it was 10 years later that a Ms. Betty Moulder, trained at Philadelphia's Blockley Hospital, was credited with being the first U.S. company-based practitioner. Most frequently cited is the hiring of Ada Mayo Stewart by the Vermont Marble Works of Proctor in 1895. A strong community-minded citizen, Fletcher D. Proctor, one-time governor and president of the company, selected the 24-year old graduate of the Waltham (Mass.) Training School to carry out district-nursing activities for his employees. After a year of home visitations, Ms. Stewart became the first superintendent of the newly built company hospital in Proctor.

In 1922, the first American Association of Industrial Nurses was formed, but in 1933, was absorbed by the New England Association. With the growth of local and regional groups, some governance was needed, and in April 1942, at the fourth joint conference of the several associations, 300 nurses from 16 states voted for the creation of a national association, designated the American Association of Industrial Nurses (AAIN). The second meeting was held with the American Association of Industrial Physicians and Surgeons, a practice that materialized into the annual

American Occupational Health Conference. In keeping with the changing concepts, the AAIN became the American Association of Occupational Health Nurses (AAOHN). After the OSHAct passed in 1970, educational opportunities were provided the occupational health nurse, and many have partaken of either full-time curricula leading to the master's degree or courses in continuing education. See chapter 4, the Occupational Health Nursing Profession, for more information.

The future of nursing at various worksites is unquestionably bright, and many nurses have been granted full responsibility for in-plant programs. With the rising clamor for greater participation in health planning and health care, and as nursing becomes more specialized, equity between medicine and nursing in the larger programs may grow.

Industrial Hygiene

The greatest growth of occupational health professionals has been in industrial hygiene. With a shortage of graduates before World War II, the postwar years saw a great number of highly trained specialists emerge from the universities that had established programs in occupational health. The American Industrial Hygiene Association (AIHA) increased membership several fold after passage of the OSHAct and many women entered this technical field. See chapter 6, The Industrial Hygiene Profession, for more information.

Millennium's End

In the 1980s, the format for provision of in-house occupational health services began to change. As many large corporations in a variety of industries merged and downsized, numbers of existing in-house medical departments began to give way to the outsourcing of the services. As more "off-shore" manufacturing began to replace the production of marketable goods in the United States, the occupational health services of an organization shrank; the examination and treatment of job applicants and employees was accomplished by consultants or specialty medical facilities in the community, particularly hospitals or other managed-care organizations, e.g., HMOs, PPOs, etc.

Matching this in-house loss has been the preference of graduates from occupational medicine residency programs for private office consultant appointments. While such a change may be the current trend, provision of preventive medicine services may diminish. Though consultants can visit an organization that is working toward prevention of a particular occupational disease and aid in the resolution of that problem, their work cannot compare with an on-site physician's familiarity with the employees. Without adequate oversight by consulting occupational medicine providers, the often supportive, knowledgeable preventive care given by on-site staff may be replaced by the solution of individual environmental quandaries. These in turn may bypass the needs of the individual worker who displays some impairment in job performance.

Academic medical centers with strength in occupational medicine are providing much consultative assistance in the workplace. Hence more and more problem employees with possible work-related disorders will be referred to multispecialty centers for resolution of their medical disorders. With the outsourcing of occupational medicine services, facilities retaining their occupational health services may increasingly use the skills of occupational health nurses and nurse practitioners as the managers of the health facilities.

Research in occupational disease occurs around the world; and international bodies like the International Labor Office and the World Health Organization continue to guide developing countries in the institution of medical assistance for workers.

Introduction of managed health care in the United States has affected profoundly its delivery; and the exact format of possible universal coverage is still being worked out. How occupational health will relate to the final configuration of health care remains to be clarified in the early years of the new millennium. The integration of occupational health within the managed care universe is one of the major issues confronting occupational health professionals in the 21st century.

SUMMARY

No sketch of history can be more than just that. Events, people, legislative changes, ideas, and natural and human-caused calamities arise in too great numbers to permit detailed documenting of all episodes. The growth of occupational health was shown by the advent of specialized educational programs, periodical literature, occupational mental health programs, employee assistance programs, scope of company and national medical conferences, extent of occupational health and safety activities in the armed services, the governmental regulation of workplace safety and health, and increased health, education, and fitness programs. Occupational health and safety touch all aspects of human endeavor—group spirit at the worksite, workers' communities, recreation, spiritual beliefs, family life, and personality development.

REFERENCES

Addams J. The objective value of a social settlement, in C. Lasch (ed). *The Social Thought of Jane Addams.* Indianapolis: Bobbs-Merrill, 1965, p 46.

Agricola G. *De Re Metallica,* Trans. From the first Latin edition of 1556 by HC Hoover and LH Hoover. New York: Dover, 1950.

Clark C. *Radium Girls—Woman and Industrial Health Reform,* 1910–1935. Chapel Hill NC: University of North Carolina Press, 1997.

Corn JK. *Response to Occupational Health Hazards—A Historical Perspective.* New York: Van Nostrand Reinhold, 1992.

Felton JS. Phosphorus necrosis—A classical occupational disease. *Am J Ind Med* 3:77–120, 1982.

Felton JS. Man, medicine, and work in America: An historical series, Part IV.

Thomas Cadwalader, MD, physician, Philadelphian, and philanthropist. *Occ Med* 11: 374–380, 1969.

Goldwater LJ. From Hippocrates to Ramazzini: Early history of industrial medicine. *Ann Med Hist* 8:27–35, 1936.

Hamilton A. *Exploring the Dangerous Trades.* Boston: Little, Brown, 1943.

Legge TR. The history of industrial medicine and occupational disease. *Ind Med* 5:300–314, 1936.

MacLaury J (ed). *Protecting People At Work: A Reader in Occupational Safety and Health.* Washington DC: U.S. Department of Labor, 1980.

Major RH. *Classic Descriptions of Disease.* Springfield IL: Thomas, 1955, p 312.

Major RH. Some landmarks in the history of lead poisoning. *Ann Med Hist* 3:218–227, 1931.

Cited by E. Mastromatteo, In. From Ramazzini to occupational health today, from an international perspective. *J Occ Med* 17:289–294, 1975.

McCready BW. *On the Influence of Trades, Professions, and Occupations In the United States, in the Production of Disease.* Baltimore: Johns Hopkins Press, 1943. (Reprinted from 1837 article.)

Meiklejohn A. *The Life, Work and Times of Charles Turner Thakrah, Surgeon and Apothecary of Leeds* (1775–1833). London: Livingstone, 1957, pp 45–46.

Pendergrass EP, et al. Historical perspectives of coal workers' pneumoconiosis in the United States. *Ann NY Acad Sci* 200:835–854, 1972.

Raffle PA, Adams PH, Baxter PJ, Lee WR, eds. *Hunter's Diseases of Occupations* (8th ed.).

Ramazzini B. *Diseases of Workers,* trans. from the Latin text, *De Morbis Artificum,* of 1713, by WC Wright. New York: Hafner, 1964.

Report on the American Workforce. Washington, DC: U.S. Department of Labor, 1994.

Sellers CC. *Hazards of the Job: From Industrial Disease to Environmental Health Science.* Chapel Hill, NC: University of North Carolina Press, 1992.

Toynbee A. *The Industrial Revolution.* (First published in 1884.) Boston: Beacon, 1956.

Whittaker AH, Sobin DJ. Historical milestones in occupational medicine and surgery. *Ind Med* 10:203–205, 1941.

Part 2

Occupational Health and Safety Professions

chapter 3

Occupational Medicine Profession

by John S. Hughes, MD

47 Overview
48 Role Definition—The Physician's Perspective
49 **Historical Trends in Occupational Medicine**
Industrial revolution ■ Early 1800s ■ Civil War to present
51 **Placement in the Organization**
52 **Physician Training and Educational Requirements**
52 **Certification and Licensure**
53 **Professional Organizations**
Interaction with the regional and national medical community
55 **Scope of Practice**
Designated care of occupational illness and injury ■ Fitness-for-duty evaluation ■ Preplacement, periodic, and exit medical evaluations ■ Medical evaluation of impairment and disability ■ Medical role in drug and alcohol testing ■ Health hazard evaluation ■ General health promotion and preventive medicine ■ Corporate medical direction
59 **Role as Member of the Health and Safety Team**
Physician input to the health and safety team ■ Patient advocacy in the realm of employment and management ■ Coordination with outside medical providers
59 **Summary**
60 **References**

OVERVIEW

Physicians have provided medical care for people who work for as long as employment has existed. There was always a clear economic motivation to maintain worker health, as this allows an optimal production of goods and services. Production techniques became more complex, and workers were exposed to an increasing array of hazards. As a result, occupational injuries and illnesses became more significant, challenging all members of the health and safety team, including physicians. The specialty of occupational medicine emerged when a segment of physicians began to focus their efforts on medical problems in the workplace.

Occupational medicine spans an array of medical practice. A notable feature of the specialty is that physicians are involved in both direct patient care and nonpatient care health and safety activities. In the area of direct patient care, an occupational medicine physician's practice includes the following:

- primary care of occupational injury and illness
- worker fitness for duty evaluations
- preplacement and periodic medical evaluations of workers in safety sensitive positions; and
- medical evaluation of impairment and disability.

Nonpatient care functions include the following:

- providing medical input into drug and alcohol surveillance programs
- evaluating health hazards
- conducting preventive medicine programs
- performing corporate medical direction activities.

The primary objective of this chapter is to define occupational medicine and provide an overview of

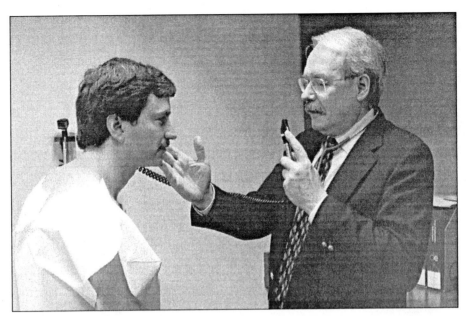

Figure 3-1. Physical examinations are common components of a physician's direct patient care obligations. When occupational illnesses and injuries occur, occupational physicians usually serve as the primary-care physicians for these patients. (Courtesy American Medical Association)

the ethical and legal framework of this type of medical practice. The chapter will also review the occupational physician's scope of practice. New occupational medicine trends constitute both threats and opportunities for the health and safety team, and these trends will be discussed.

ROLE DEFINITION—THE PHYSICIAN'S PERSPECTIVE

The first and most obvious function of the occupational medicine physician is to perform direct patient care. Physical examinations are common components of a physician's direct patient care obligations. When occupational illnesses and injuries occur, occupational physicians usually serve as the primary-care physicians for these patients. Occupational medicine specialists consult to assist other physicians with occupational issues such as employability, medical impairment, causation of condition, and disability (Figure 3-1).

Companies have much to gain from having a close affiliation with an occupational medicine physician in this direct patient care role. The physician may be given precise job task descriptions for the purpose of minimizing lost time by an injured or ill employee. The company may receive frequent updates from the physician on the status of injured and ill employees.

Employees also have much to gain by having care directed by an occupational medicine physician. There must be protection of patient rights and confidentiality, along with a physician practice pattern that follows the venerated physicians' ethic "first do no harm." Employees know they are protected when they express confidential concerns to a physician, and this provides a valuable outlet for their concerns (Figure 3-2).

There is potential for conflict when both companies and employees have so much to gain by the actions of the occupational physician. The doctor must sometimes make hard choices between loyalty to a company that provides the physician with employment, and the company's employees who are also his or her patients. There are many examples that illustrate how difficult this may be.

Senior management may develop significant concern about a potential risk of increased claims. It may seem to be an easy solution for them to suppress research findings and discourage the corporate medical director from discussing specific research findings with employees. Company management may even ask the corporate medical director for input on these matters, and then turn around and issue a "gag policy" forbidding the physician from discussing specific research findings.

This actually happened at an asbestos products company. Management asked their corporate

Figure 3-2. Employees know they are protected when they express confidential concerns to a physician, and this provides a valuable outlet for their concerns. (Courtesy American Board of Occupational Health Nurses)

physician for an opinion regarding the reliability of chest x-ray findings in predicting asbestos-related disease. The physician replied that certain chest x-ray abnormalities such as pleural plaques, while associated with asbestos exposure, were not specific markers for asbestos-related disease. The company relied on the doctor's advice and published a policy that directed the occupational medicine physician to conceal or minimize any discussion with employees regarding chest x-ray abnormalities. This policy was applied to interactions between the doctor and individual members of the labor force. Each individual meeting between the occupational physician and an employee served to convert the status of the individual employee from laborer to patient. Naturally, these patients trusted the integrity of the occupational physician. Later, many of these asbestos laborers felt betrayed that their physician and his representatives concealed specific x-ray findings from them. The courts agreed. Damages were assigned, and were significant to the company. Of special interest, however, is what the courts thought of the physician's actions (Castleman, 1984). He was found to be separately liable for medical malpractice and his losses reportedly exceeded the limits of his insurance policy.

Outside of direct patient care, the occupational medicine physician has a significant role as corporate medical advisor on medical matters. The physician may provide medical input to the health and safety team regarding scientific evidence of cause and effect relationships, as well as in the assessment of outcomes following particular interventions in the workplace. Physicians may also provide input regarding medical considerations in reasonable accommodation cases and light duty work assignments.

Occupational medicine physicians may also serve as drug testing medical review officers (MROs). This provides employees and applicants who undergo drug testing with an opportunity to have direct physician review of positive findings. Often there is an alternate medical explanation for a positive drug test. The MRO function is to ensure that positive drug test results do not reflect prescriptive medication use or other sources of false-positive test results.

HISTORICAL TRENDS IN OCCUPATIONAL MEDICINE
Industrial Revolution

Occupational medicine emerged during the industrial revolution when it became apparent that specific diseases were associated with certain occupations. In 1775, Dr. Percival Pott discovered an association between the dirty profession of chimney sweeping in England and development of cancer of the scrotum. Dr. Charles Turner Thackrah described abdominal colic in ceramic workers who used lead glazes as well as pulmonary diseases in underground coal miners (Hunter, 1955).

When industry first developed in Europe during the period from 1700 to 1800, physicians and surgeons were separate medical professions. Physicians were university educated and had strong scientific backgrounds that made them highly capable of determining cause and effect relationships through application of deductive logic. Surgeons were educated in anatomy and practical physiology, as their primary goal was the treatment of injuries. American medicine and surgery derived almost exclusively from Western Europe; England, France, and the German states had the greatest influences on American medical practice.

Early 1800s

In the early 1800s, excellent medical schools and universities had been established in the United States. Early leaders included Pennsylvania University in Philadelphia and Harvard University in Boston. However, in most regions of the country the standard of care was extremely low, as was the educational level of most physicians and surgeons. In 1856 the booming young city of Los Angeles welcomed an acclaimed physician, who arrived with a degree from Harvard University. The Harvard degree was mounted in a prominent place in his office, and he began a medical practice that flourished before a patient arrived who could read the Latin on the doctor's Harvard diploma. At this point it was discovered that the doctor's Harvard degree was an undergraduate degree and that the doctor had no formal medical training! (Packard, 1931)

Civil War to Present

In the late 1800s, the professions of medicine and surgery remained quite separate in the United States. Due to the Civil War, both industrial and academic institutions lagged in development significantly behind those in Europe. However, the war had galvanized both medical professionals and the public alike to the need for better medical care. Within a generation, medical schooling had undergone a complete transformation, and surgery was merged into mainstream medicine. By the time the First World War started in 1914, American physicians and surgeons were considered peers of the Europeans in training and knowledge (Figure 3-3). Early in the 20th century there were two ways these physicians and surgeons would be involved in occupational medicine:

- The first way followed a military model—surgeon physicians attending traumatic cases of injury sustained in the line of work.
- The second way physicians were involved with occupational issues was during their epidemiologic activities and through diagnosis and treatment of nontraumatic diseases. Medical physicians in academic settings developed essential understandings of disease processes by observing working populations, noting particular associations between occupational exposures and disease outcomes.

Occupational injury care initially was provided as a part of liability legal actions on the part of injured workers, but in Europe workers' compensation systems had been developed to provide injured workers with medical care and disability benefits. These were copied by the individual states and by certain components of the federal government of the United States as they each developed individual workers' compensation systems. Each workers' compensation statute arose separately in the 50 states, and they remain very different from each other. There is to-date no unifying federal legislation relating directly to workers' compensation systems.

As medical knowledge of occupational diseases increased, physicians became advocates for the development of occupational health and safety systems that extended beyond the simple injury-based workers' compensation systems. In the 1930s, scientific studies began to show a relationship between asbestos exposure and lung diseases that was similar to the relationship noted in coal miners. Armed with this increased understanding of occupational disease, occupational medicine physicians were poised to expand their roles beyond occupational injury into the area of occupational disease.

World War II interrupted the process of occupational disease recognition simply because worker productivity became the primary national concern. While wartime advances in surgical capability greatly improved occupational injury care during this time, there was a relative lag in the United States behind European countries in the recognition and treatment of occupational disease. The lag ended fairly abruptly during the 1960s. One stimulus to this change was a large cohort of occupationally exposed asbestos workers who brought occupational disease to the awareness of

Figure 3-3. Due to the Civil War, both industrial and academic institutions lagged in development significantly behind those in Europe. However, the war had galvanized both medical professionals and the public alike to the need for better medical care. Here a horse-drawn ambulance from Yale-New Haven Hospital is shown. (Courtesy Yale-New Haven Hospital)

Americans. The year 1964 was a watershed year, as the U.S. Surgeon General issued the first of several reports on cigarette smoking. Irving J. Selikoff noted that this was a watershed year and he later concluded that around this time, the focus in American medicine changed from infectious disease to noninfectious agents (Selikoff, 1992). Spurred by this public pressure, elected representatives to Congress passed the OSHAct; and this was passed into law in 1970. Physician involvement was integral to development of OSHA programs for preplacement, periodic, and exit examinations of certain categories of workers in safety-sensitive positions and who had potential or measured occupational exposures to particular chemicals or materials like asbestos.

In the 1970s and 1980s, the recognition of occupational disease continued to grow. While toxicologists defined occupational exposure rules for carcinogenic chemicals like benzene, recognition of musculoskeletal overuse syndromes increased. A major increase in recognition of occupational musculoskeletal disease followed the publication of a benchmark research article, Dr. Barbara Silverstein's doctoral thesis at the University of Michigan (Silverstein et al, 1986). At the National Institute of Occupational Safety and Health (NIOSH), researchers and physicians evaluated these research studies, and developed advisories regarding occupational musculoskeletal diseases (Hales et al, 1994). State workers' compensation systems were revised to include musculoskeletal occupational diseases.

At present, there is a continuing increase in medical attention being turned to the workplace. This is a critical transition time for the occupational medicine physician in the current environment of rapid change and increasing recognition of occupational disease. Today's occupational medicine physician must have a strong background in epidemiology, public health, preventive medicine, and have a clinical familiarity with both occupational trauma and occupational disease.

PLACEMENT IN THE ORGANIZATION

Occupational medicine physicians have two ways they can work with companies—either as *full time "in house" corporate medical directors,* or as *independent contractors.* Both types of placement have relative strengths and weaknesses. In general, the corporate medical director model is superior when management requires consistent physician input, and there are trade secret concerns that require all involved members of the health and safety team to be company employees. The independent contractor model is superior in settings where there is a significant requirement for direct patient care.

The corporate medical director is generally an employee of the company. Close control exists over the activities of physicians in this setting, and there can be a high degree of consistency. Occupational physicians in this position are in the best spot to perform worksite health hazard evaluations, as they are so familiar with the operations of the company they serve.

An independent contractor physician on the other hand is more familiar with the medical community of providers than the corporate medical director is. This type of occupational physician may be a member of a large physician / hospital organization and have access to services not generally available to outside physicians. These factors are why most companies, even large self-insureds, use independent contractors for direct patient care programs like workers' compensation illness and injury care.

Some independent service organizations may provide occupational physicians to companies for particular highly specialized projects like risk assessment and development of an employee health service. These periodic needs are best met by such an independent contractor, as this type of occupational physician has specialized skills that will serve to complement the capabilities of the corporate medical director.

PHYSICIAN TRAINING AND EDUCATIONAL REQUIREMENTS

Physician education begins with medical school. This is generally a four-year course of academic and practical training in medicine, culminating in the award of Medical Doctorate or Doctorate of Osteopathy, depending on the particular medical school. In United States, all medical students take a three-part medical board administered by the National Board of Medical Examiners that allows all U.S. medical students to be compared with each other and to an absolute academic standard. The third part of this test is given during the internship, now referred to as postgraduate year one. State licensure in the various states requires completion of postgraduate year one as well as passage of the three-part medical board. This is a minimum requirement, and many states have additional requirements. However, physicians at this stage generally do not work independently and most proceed with additional training.

After completing medical school, physicians undergo specific postgraduate training generally referred to as residency training. The history of this terminology is that physicians in many specialties would spend these years of training "residing" in treatment facilities such as hospitals. Highly procedural specialties such as neurosurgery or orthopedic surgery require a number of years of residency training, as there is a requirement for technical competence superimposed on the training of medical knowledge and judgment. Occupational medicine physicians do not have the technical demands of surgery, and so postgraduate training is somewhat shorter. Most occupational medicine residencies are three to four years in length. Many occupational medicine physicians have supplemented their residency training with particular training in military medicine specialties or in public health. An example of this supplemental training would be completion of a fellowship at the Centers for Disease Control in Atlanta.

CERTIFICATION AND LICENSURE

Physicians in United States are certified through member boards of the American Board of Medical Specialties (ABMS). This organization was founded in 1933; and by 1949 there were 18 specialty boards, including the American Board of Preventive Medicine. The mission of ABMS is to provide an organization of approved medical specialty boards. This is done to maintain and improve the quality of medical care by assisting the member boards in their efforts to develop and use professional and educational standards to evaluate and certify physician specialists. In 1999, the American Board of Medical Specialties had 24 member boards.

After completing a residency and with approval of the residency director, physicians become board-eligible in their particular specialties. Physicians proceed with examination, and achieve board certification in their particular specialty. It is important to note that in the United States, *both specialty and primary medical care* are considered to be specialty areas.

In occupational medicine, physicians are certified through the American Board of Preventive Medicine. This particular specialty board is unusual as it contains three specialty divisions; occupational medicine, general preventive medicine, and aerospace medicine. Most occupational medicine physicians are certified through the occupational medicine division, but many physicians enter the specialty of occupational medicine after primarily training in preventive medicine or aerospace medicine.

Significant numbers of occupational physicians also are primarily boarded in other specialties, and act as *subspecialists* in particular occupational disease subspecialties. Subspecialty occupational fields include occupational dermatology, neurology, and pulmonary medicine. The most common area where this is seen is in pulmonary medicine, a subspecialty of internal medicine. An elite few of these physicians proceed with additional training in occupational medicine during special fellowships (post residency) or they complete other equivalent training programs to make them board eligible in occupational medicine. They may sit for the occupational medicine boards and become double-boarded in occupational medicine and internal medicine with subspecialty certification in pulmonary medicine. This type of occupational medicine specialist is highly qualified to evaluate and treat occupational pulmonary disease.

PROFESSIONAL ORGANIZATIONS

Each of the medical specialties and subspecialties has a national society, college, or academy. Examples of these include the American College of Physicians for the specialty of internal medicine, the American Academy of Orthopedic Surgeons for the specialty of orthopedic surgery, and the American College of Occupational and Environmental Medicine for the specialty of occupational medicine. The general purpose of these organizations is to promote the specialties that they support. They also provide valuable information and they are instrumental in determining the specific standard of care that exists in each particular specialty. See chapter 31, Sources of Help for a list of relevant organizations.

The American College of Occupational and Environmental Medicine (ACOEM) is the national specialty organization for occupational medicine physicians. The mission of ACOEM is to provide leadership to promote optimal health and safety of workers, work places, and environments. A primary part of this mission is to educate health professionals and the public regarding health and safety issues. The ACOEM has developed a number of position statements and guidelines that define the standard of medical care in the field of occupational and environmental medicine.

An actively practicing occupational medicine physician should be licensed in the state of practice. This is important, considering the new concerns about corporate medical directors dictating medical care specifics for individuals from a state remote from where the individual lives and works. The physician should be board eligible or certified by the American Board of Preventive Medicine in Occupational Medicine. Finally, the physician should be active in the American College of Occupational and Environmental Medicine or in an equivalent subspecialty organization such as the American Thoracic Society.

A particular function of ACOEM has been to develop a code of ethical conduct pertaining to occupational medicine physicians (Figure 3-4). These were adopted in 1993. They serve to guide occupational medicine physicians in their relationships with individuals they serve; employers and worker representatives; colleagues and other health professionals; the public; and all levels of government including the judiciary. There are eight of these standards.

1. Accord the highest priority to the health and safety of individuals in both the workplace and the environment.
2. Practice on a scientific basis with integrity and strive to acquire and maintain adequate knowledge and expertise upon which to render professional service.
3. Relate honestly and ethically in all professional relationships.
4. Strive to expand and disseminate medical knowledge and participate in ethical research efforts as appropriate.
5. Keep confidential all individual medical information, releasing such information only when required by law or overriding public health considerations, or to other physicians according to accepted medical practice, or to others at the request of the individual.
6. Recognize that employers may be entitled to counsel about an individual's medical work fitness, but not to diagnoses or specific details, except in compliance with laws and regulations.

AMERICAN COLLEGE OF
OCCUPATIONAL AND
ENVIRONMENTAL MEDICINE

Code of Ethical Conduct

Code of Ethical Conduct

This code establishes standards of professional ethical conduct with which each member of the American College of Occupational and Environmental Medicine (ACOEM) is expected to comply. These standards are intended to guide occupational and environmental medicine physicians in their relationships with the individuals they serve, employers and workers' representatives, colleagues in the health professions, the public, and all levels of government including the judiciary.

Physicians should:

1. accord the highest priority to the health and safety of individuals in both the workplace and the environment;

2. practice on a scientific basis with integrity and strive to acquire and maintain adequate knowledge and expertise upon which to render professional service;

3. relate honestly and ethically in all professional relationships;

4. strive to expand and disseminate medical knowledge and participate in ethical research efforts as appropriate;

5. keep confidential all individual medical information, releasing such information only when required by law or overriding public health considerations, or to other physicians according to accepted medical practice, or to others at the request of the individual;

6. recognize that employers may be entitled to counsel about an individual's medical work fitness, but not to diagnoses or specific details, except in compliance with laws and regulations;

7. communicate to individuals and/or groups any significant observations and recommendations concerning their health or safety; and

8. recognize those medical impairments in oneself and others, including chemical dependency and abusive personal practices, which interfere with one's ability to follow the above principles, and take appropriate measures.

Adopted October 25, 1993 by the Board of Directors of the American College of Occupational and Environmental Medicine.

Figure 3-4. American College of Occupational and Environmental Medicine Physicians' Code of Ethical Conduct is shown. (Courtesy American College of Occupational and Environmental Medicine)

7. Communicate to individuals and/or groups any significant observations and recommendations concerning their health or safety.
8. Recognize those medical impairments in oneself and others, including chemical dependency and abusive personal practices, which interfere with one's ability to follow the above principles, and take appropriate measures.

Standard No. 6 is quite important. This standard states that physicians should generally keep specific diagnoses confidential, and communicate only those details of the diagnosis that pertain to the individual's medical work fitness. Physicians are often pushed to discuss medical specifics relating to a particular employee/patient by supervisors and other company personnel who are not aware of this ethical standard. The Americans with Disabilities Act (ADA) formally recognizes this ethical standard. The act requires that companies establish a "medical records coordinator" to keep medical records that contain specific diagnostic information. The act also requires that these records be kept separate from other personnel records.

Standard No. 7 is also important. This requires that physicians communicate all findings to individuals and/or groups. There is a significant history of company representatives requesting that the company physician withhold specific information. If information that is available to the physician is considered significant to the health and safety of interested individuals and/or groups, the physician has an ethical obligation under this standard to communicate the information to those who may be affected.

Interaction with the Regional and National Medical Community

An important role of the occupational medicine physician is to interact with the regional and national medical community. The occupational physician often serves as a primary-care provider for workers' compensation cases, and may refer to specific physicians for particular specialty care needs. Workplace health and safety concerns may require the occupational medicine physician to obtain position statements from other physician specialty organizations pertaining to these concerns. The perspectives of specialty physicians may vary significantly, and the occupational physician often is called on to provide physician insight into national specialty biases and philosophies.

SCOPE OF PRACTICE

The practice of occupational medicine can include the following:
- designated care of occupational illness and injury
- fitness-for-duty evaluations
- preplacement, periodic, and exit medical evaluations
- medical evaluations of impairment and disability
- medical role in drug and alcohol testing
- health hazard evaluation
- general health promotion and preventive medicine
- corporate medical direction.

These topics are explored in the following section.

Designated Care of Occupational Illness and Injury

Treatment of occupational illness and injury is the most apparent physician role for an occupational medicine physician. Many states allow employers to designate a specific physician or group for treatment of occupational illness and injury. These designated physicians generally serve as primary-care providers, and referrals are made to other physicians for specific specialty evaluation and treatment. Occupational medicine physicians often will reevaluate occupationally ill and injured employees for determination of work status and permanent impairment.

Fitness-for-Duty Evaluation

Occupational physicians often perform medical evaluations regarding an individual's fitness for duty. When this is done, the physician should be familiar with the individual's job. This is best achieved when the company provides the physician with a true essential function analysis that outlines specific tasks required in a particular job. Otherwise, the physician must depend upon a description of the job by the individual, or a general job description provided from a document such as *The Diseases of the Occupations* (Hunter, 1955).

The physician must determine if there is a direct threat posed by the presence of an individual's medical condition in the setting of particular job tasks. In the ADA, the term *direct threat* is defined as a medical condition that poses a specific current risk. For a condition to pose a direct threat, there must be a significant risk of substantial harm. This term usually refers to harm that may occur to the individual worker. In addition, the physician must determine if an individual's condition poses a direct threat to co-workers or to members of the general public. In these settings, a physician can place specific restrictions and communicate with the company regarding reasonable accommodations that may be made given these physical restrictions. The ADA defines *reasonable accommodation* as any change in the work environment or the way things are usually done that results in equal employment opportunity for an individual with a disability. Useful guidance on these matters is found in the U.S. Equal Employment Opportunity Commission's *Technical Assistance Manual on the ADA* (1992). For more information, see chapter 27, the Americans with Disabilities Act.

When the particular job in question is truly *safety sensitive,* a higher standard exists. The occupational physician must protect co-workers and members of the general public to a greater degree than the physician is obliged to protect the individual employee. Employees in safety sensitive jobs such as firefighters and airline transport pilots actually must even give up some degree of privacy for the sake of public safety. These principles are illustrated in the following examples.

A 20-year-old man applies for two jobs: (1) a beef-processing job requiring use of a sharp knife; and (2) a firefighter position requiring scaling a ladder. He also is considering going to commercial aviation school through the metropolitan college in his home city. This individual discloses an active epilepsy condition for which he is on medication. His most recent seizure was two months previously; and this event occurred suddenly and resulted in loss of consciousness causing a fall. Current guidelines would require the beef-processing company to provide accommodations for the condition, while the fire protection district would be justified in precluding employment based on the direct threat principle. If the individual applies through the FAA for a medical certificate to allow him to fly, he will have to disclose specific diagnostic information under federal aviation regulations, and he will categorically be denied a medical license to operate aircraft.

An occupational physician also can make medical recommendations that will allow the individual to become physically qualified for a particular job. An 18-year-old applicant for a construction laborer position is discovered to have an inguinal hernia. Medical recommendation is made by the occupational physician to have this surgically repaired. The individual has surgery, and returns for reevaluation—cleared for the construction labor job. In the setting of *return to work after a medical illness,* the occupational physician generally is working with the attending physician and should make medical recommendations to that particular physician. Reevaluation may follow additional medical treatment directed to restoring the individual to duty status.

Preplacement, Periodic, and Exit Medical Evaluations

Companies can realize real financial value by using occupational physicians for preplacement evaluations. These can be performed after an offer of employment has been made but before the individual actually has begun work. Under the ADA, an employer can condition a job offer on the satisfactory result of a post-offer medical examination, if this sort of evaluation is required of all entering employees in the same job category (EEOC, 1992). At this examination, the occupational physician may ask questions about previous injuries and workers' compensation claims, questions that are specifically forbidden to human resources representatives by the ADA. Answers to these questions and specific examination findings can provide a basis for apportionment of future workers' compensation claims, or even may prevent an injury completely.

For example, an individual applies for a forklift operator job and is sent for preplacement evaluation. The occupational physician discovers the applicant has had back surgery for a previous work-related injury and has ongoing stiffness and pain. While the physician finds that the individual can do the job with accommodation for lifting items in excess of 50 lb, quantification of ranges of motion provides a medical impairment baseline for future injuries. Later, this individual sustains a work-related back injury. The medical baseline exists, and it is relatively easy for the attending physician to determine when the individual has returned to this baseline level of function. Even if the individual never does return to the baseline, there is a basis for apportionment of permanent impairment in most workers' compensation jurisdictions. This saves direct expenses on partial disability awards.

Periodic medical evaluations are required by a number of OSHA standards. These include medical exposure history, physical examination, and specific laboratory and radiographic analyses. These types of periodic medical examinations are specific to the particular exposures sustained or potentially sustained by employees at risk. The purpose of these types of examinations is twofold: (1) to identify individuals who have sustained significant exposures; and (2) to identify if there might be a special risk in the event of occupational exposures or use of personal protective equipment (29 *CFR* 1910.1001).

An exit medical evaluation also may be required by particular OSHA regulations. This type of evaluation is performed to identify any exposure-related effects that may have been sustained by exposed employees.

Medical Evaluation of Impairment and Disability

Most workers' compensation systems require an evaluation for evidence of permanent impairment at a point when the individual's condition has stabilized. Occupational physicians generally recognize that *impairment* means an alteration of an individual's health status that is assessed by medical means, while *disability,* which is assessed by nonmedical means, is an alteration of an individual's capacity to meet personal, social, or occupational demands. This principle is outlined in the *AMA Guides to the Evaluation of Permanent Impairment* (AMA, 1993). The disability arises out of the interaction between impairment and external requirements, particularly those of a person's occupation. Occupational physicians perform a medical evaluation to determine if there is a deviation from normal in a body part or organ system and its functioning. This is then translated into a determination of permanent impairment.

Disability evaluations are more focused on the evaluation regarding the presence of a gap between what a person can do and what the person needs or wants to do. Occupational physicians may perform these types of evaluations in an individual's application for disability retirement or other similar benefits.

Medical Role in Drug and Alcohol Testing

Occupational medicine physicians can obtain additional training in the areas of drug and alcohol testing, and serve as medical consultants or medical review officers (MROs). Service as a medical consultant generally requires the physician to have additional expertise in the area of toxicology. More often, occupational physicians will serve as MROs as defined in several places within federal statutes, most particularly within statutes outlining the U.S. Department of Transportation drug testing program. Additional training in these areas is sponsored by several organizations in addition to the American College of Occupational and Environmental Medicine, particularly the American Academy of Medical Review Officers and the American Society of Addiction Medicine. Most physicians who serve as MROs proceed beyond this additional training to obtain certification through the Medical Review Officers' Certification Council.

A medical review officer's primary purpose is to protect the donor from a false accusation of illicit drug use when there is alternate medical explanation for the laboratory result. Another purpose is to serve as the employer's resource to help them develop policies and deal with regulatory issues in the areas of drug and alcohol testing. Examples of alternate medical explanation for a positive laboratory result include use of prescription and nonprescription medications, as well as dietary confounding factors such as poppy seed bagels. A useful and concise textbook is available pertaining to the activities of medical review officers (Swotinsky, 1992).

Health Hazard Evaluation

Occupational physicians can provide useful input during worksite health hazard evaluations. A physician's knowledge of toxicology and pathophysiology is particularly helpful, as specific disease-related input can be provided to the health and safety team at the time of the evaluation. An occupational physician can also evaluate the worksite monitoring data provided by the industrial hygienist, and proceed with a risk assessment analysis. This may require a highly specialized knowledge of toxicology; and occupational physicians who undertake these types of evaluations often require subspecialty training in biometrics and toxicology (Figure 3-5).

General Health Promotion and Preventive Medicine

At worksites, general health promotion and preventive medicine activities often require physician input. Occupational physicians work closely with

Figure 3-5. An occupational physician can also evaluate the worksite monitoring data provided by the industrial hygienist, and proceed with a risk-assessment analysis. (Courtesy American Board of Occupational Health Nurses)

occupational health nurses in providing education and other measures aimed at general health promotion (also see chapters 4 and 9). Preventive medicine activities can include a vaccination program and monitoring a wellness center that teaches task-specific exercises. Physicians may direct a travel medicine program that provides preventive medicine directed to high-altitude risk, malaria, hepatitis, and many other potential risks to the traveling or expatriated worker (also see chapters 15 and 29).

For example, a 38-year-old geologist is traveling to Peru for a mining site visit. The occupational physician profiles the trip, and discovers that the geologist needs to fly into the Amazon side of the Andes Mountains, then travel by road up to a mine location at an elevation of 4,550 meters (14,700 feet). This geologist might reasonably require hepatitis A vaccination, malaria prophylaxis, and acetazolamide to prevent high-altitude sickness.

Corporate Medical Direction

Multistate and international corporations often require occupational medical direction. In this role, the physician does the following:
- interfaces with regional medical providers
- develops uniform treatment guidelines
- supervises medical surveillance programs
- provides risk assessment services to management.

By definition, an occupational physician's activities are generally preventive in nature.

Good examples of corporate medical direction are seen when occupational physicians provide for expatriate health services. Use the example of the geologist traveling to Peru on company business given previously; in addition to the individualized preventive medicine measures, corporate medical direction is needed. In this case, the corporate medical direction function would provide the individual with points of contact for injury and illness care along with company policy that would direct the geologist in case of a medical emergency.

Corporate medical directors often are involved with individual disability determinations and other similar evaluations directed to individual employees. Out-of-state peer review may be acceptable, but these physicians must be careful to avoid practicing medicine outside the scope of their specific state licensure. A prudent corporate medical director will not make medical determinations regarding individual evaluation and treatment options outside the protections provided under peer review provisions included in insurance programs. Provision of "telemedicine" direct patient-care services, outside of the arena of insurance peer review, to out-of-state patients is generally not covered by a physician's license and malpractice insurance.

One of the most useful functions served by a corporate medical director is direct communication with regional medical providers. All physicians are trained to be receptive to legitimate peer review input from other physicians. Specific medical information may be communicated without burdensome medical releases and an individual's treatment and prognosis may frankly be discussed. This type of corporate medical direction may provide valuable information to company representatives. It will also serve to direct the activities of regional medical directors and make them more sensitive to the company's needs.

ROLE AS MEMBER OF THE HEALTH AND SAFETY TEAM

The occupational physician plays an important role on the occupational health and safety team, offering the following:
- inputs medical information to the team
- advocates for patients in the realms of employment and management
- coordinates with outside medical providers.

These roles are explored in the following sections.

Physician Input to the Health and Safety Team

Occupational physicians provide the health and safety team with a physician's perspective on routes of exposure, ergonomic factors, pathophysiology of specific disease, and risk assessment. The physician provides medical direction in individual cases in which medical evaluation treatment is needed. Otherwise, the physician serves as an equal member to other members of the health and safety team in a true interdisciplinary environment.

Patient Advocacy in the Realm of Employment and Management

A physician is trained to be a *patient advocate*. However, a physician's advocacy is different than the type of advocacy practiced by attorneys. A physician is trained to advocate for his patient's health, while an attorney is trained to advocate for his client's interests. This principle is illustrated in the following example: If an employee asks an occupational physician to help him or her with a disability application, the role of the physician would be to advocate for appropriate medical evaluations in order to accurately diagnose and estimate the disability implications of the employee's condition. The physician is ethically obligated to be objective in this determination. If the same employee asks an attorney to help him or her with a disability application, the attorney's advocacy would include selecting particular types of evaluations that would cast the employee in a favorable light for award of his disability insurance. Thus, an occupational physician's advocacy should not be seen as providing support for the employee in a conflict with an insurance company over disability benefits. On the other hand, provision of this type of support (or advocacy) is precisely what the employee's attorney is ethically bound to provide.

Coordination with Outside Medical Providers

An occupational medicine physician's role can include coordination of care with outside medical providers. This may occur in the following situations:
- preplacement evaluations
- an employee's return to work after a leave-of-absence period
- during the coordination of care directed to a worker's compensation case.

This type of coordination is often provided in conjunction with occupational health nurses.

SUMMARY

Physicians are trained to evaluate and treat disease, and their primary focus is on the health of individuals and groups. Occupational medicine is a specialty, and occupational physicians are particularly trained to recognize and treat occupational injury and illness. Occupational medicine physicians also have an expertise in preventive medicine that is particularly valuable to the occupational health and safety team. Physicians who are board-certified in occupational medicine are best qualified to provide occupational medicine services in most instances.

Organizations like the American College of Occupational and Environmental Medicine are valuable resources for information pertaining to occupational medicine. Ethical guidelines have been developed by ACOEM, and these are helpful in defining the physician's role. Other specialty organizations may also have subspecialists who are involved in occupational medicine, and these organizations can provide additional information pertaining to occupational medical practice, e.g., American Board of Preventive Medicine, American Board of Medical Specialties.

Occupational physicians have a wide array of roles. The scope of an occupational medicine physician's direct patient care is broad, encompassing preplacement evaluations, periodic medical evaluations, evaluation of fitness for duty, initial and follow-up care of occupationally injured and ill employees, determination of impairment, and evaluation of disability. Many of the most valuable activities of the occupational physician are nonclinical, including performance of risk assessment, provision of medical review officer duties, and corporate medical direction activities.

REFERENCES

American Medical Association. *AMA Guides to the Evaluation of Permanent Impairment,* 4th ed. Chicago: American Medical Assn, 1993.

Castleman BI. Asbestos: Medical and legal aspects. In Clifton NM (ed). *Law and Business.* New York: Harcourt Brace Jovanovich, 1984.

Hales TR in Wald PH, Stave GM (eds). *Physical and Biological Hazards of the Workplace.* New York: Van Nostrand Reinhold, 1994.

Hall DN: *Tales of Pioneer Practice.* Denver: Carson Printing Co, 1937.

Hunter D. *The Diseases of the Occupations.* Boston: Little, Brown, 1955.

Packard FR: *History of Medicine in the United States.* New York: Paul Hoeber, 1931.

Selikoff IJ, in Rom WN (ed). *Environmental and Occupational Medicine,* 2nd ed. Boston: Little, Brown, 1992, xxiii–xxiv.

Silverstein BA, Fine LJ, Armstrong TJ. Hand and wrist cumulative trauma disorders in industry. *Br J Ind Med* 43:779–784, 1986.

Swotinsky RB (ed). *The Medical Officer's Guide to Drug Testing.* New York: Van Nostrand Reinhold, 1992.

U.S. Equal Employment Opportunity Commission. *Americans with Disabilities Act Technical Assistance Manual.* Pittsburgh: Superintendent of Documents, 1992.

Chapter 4

Occupational Health Nursing Profession

by Sharon D. Kemerer, RN, MSN, COHN-S
Mary Amann, COHN-S

61 **Overview**
62 **Role Definition**
62 **Placement in the Organization**
Within human resources ■ Within environmental affairs ■ Within production
63 **Training and Educational Requirements**
64 **Certification/Licensure**
69 **Code of Ethics**
69 **Professional Organizations**
American Association of Occupational Health Nurses, Inc. ■ Association of Occupational Health Professionals in Healthcare ■ Case Management Society of America
71 **Scope of Practice**
72 **Roles**
Direct care ■ Manager/coordinator/director ■ Educator/advisor ■ Case manager ■ Consultant
74 **Summary Matrix**
74 **Emerging Roles**
74 **Role as Member of the Health and Safety Team**
76 **Summary**
76 **References**

OVERVIEW

The occupational health nurse (OHN) is a licensed registered nurse who specializes in the delivery of health care services, either directly or indirectly, to workers and worker populations. The occupational health nurse can act as a tremendous positive force for health and safety within an organization. By applying the nursing process to a working population, the nurse is in an ideal position to identify and manage employees' health risks related to the workplace, and the employer's liability related to the provision of health care.

This chapter describes the role that an occupational health nurse can play for a company. Topics covered include the following: a basic definition of the various roles; training and educational requirements; professional activities; and the scope of practice of the occupational health nurse.

This chapter will assist the reader in being able to accomplish the following:

1. define the various roles that an occupational health nurse can perform within or for an organization, an employer and/or multidisciplinary workplace health teams
2. understand the educational, regulatory and professional preparation necessary to specialize in occupational health nursing
3. discuss the roles of the membership organization—American Association of Occupational

Health Nurses, Inc. (AAOHN)—and the certifying body—American Board for Occupational Health Nurses, Inc. (ABOHN)—within the occupational health nursing specialty
4. describe the scope, the code of ethics and standards of practice which constitute the framework for occupational health nursing.

ROLE DEFINITION
The definition of the occupational health nurse's role has evolved along with advances and changes in the healthcare delivery system and dynamic relationships with the client population of employees. Increasing independence and accountability for professional actions and judgements are a hallmark of that evolutionary process.

With its roots in public health, the primary goal of the occupational health nurse is to assist employees to achieve and maintain optimal physical and psychological functioning (Figure 4-1). The client is not an isolated patient, but rather an employee who interacts with his or her work environment—an environment that may be hazardous to health.

The following official definition of occupational health nursing as provided by AAOHN was last revised in 1994.

> *Occupational health nursing is the specialty practice that provides for and delivers healthcare services to workers and worker populations. The practice focuses on promotion, protection, and restoration of workers' health within the context of a safe and healthy work environment. Occupational health nursing practice is autonomous, and occupational health nurses make independent nursing judgments in providing occupational health services.*
>
> *The foundation for occupational health nursing practice is research-based with an emphasis on optimizing health, preventing illness and injury, and reducing health hazards. This specialty practice derives its theoretical, conceptual, and factual framework from a multidisciplinary base including, but not limited to: nursing science; medical science; public health sciences such as epidemiology and environmental health; occupational health sciences such as toxicology, safety, industrial hygiene, and ergonomics; social and behavioral sciences; and management and administration principles.*
>
> *Guided by an ethical framework made explicit in the AAOHN Code of Ethics, occupational health nurses encourage and enable individuals to make informed decisions about health care concerns. Confidentiality of health information is integral and central to the practice base. Occupational health nurses are advocates for workers, fostering equitable and quality health care services and safe and healthy work environments.*
>
> *Occupational health nurses collaborate with workers, employers, and other professionals to: identify health needs; prioritize interventions; develop and implement interventions and programs; and evaluate care and service delivery.*

PLACEMENT IN THE ORGANIZATION
As a licensed professional, the occupational health nurse functions at a management level within an organization, and is responsible for the management of risk related to employee health. The occupational health nurse can interact with employees at many points within the employment cycle. Therefore, the needs of the organization often will dictate the best placement of this position. For example, a chemical industrial facility with high risk for hazardous exposures will have very different needs than a high-tech industry for which employee stress and burnout are more critical issues. The focus and structure of the organization often will dictate the best reporting structure, among a variety of models, including the following.

Within Human Resources
In some companies, the occupational health nurse is a primary resource regarding the selection, structuring, and administration of benefit plans to address employee and dependent health needs. The ability to interpret healthcare utilization data and to apply principles of public health makes the occupational health nurse a strong resource in this area. Also, in situations where employee-hiring rates are significant, and

Figure 4-1. With its roots in public health, the primary goal of the occupational health nurse is to assist employees to achieve and maintain optimal physical and psychological functioning. (Courtesy American Board for Occupational Health Nurses, Inc.)

appropriate job placement is a key factor, placement within the human resources area may be advantageous. Occupational health nurses who focus on case management of employee health issues often are placed within the human resources area as well. Management of workers' compensation programs, job accommodation/job placement, and compliance with such U.S. federal regulations as the Family Medical Leave Act (FMLA) and the Americans with Disabilities Act (ADA) often fall within this area.

Within Environmental Affairs

In company situations where chemical use and potential exposure is a factor, such as in chemical manufacturing, the occupational health nurse may function within the environmental affairs organizational structure. Increasingly, companies are combining this function with health and safety, and thus are making health, safety, and environmental departments a growing trend. Because of his or her knowledge of the health effects of chemical exposure, effective protective strategies, and the administration of appropriate medical monitoring programs to evaluate employee exposure, placement within the health, safety, and environmental group can be a wise choice for an organization. Since compliance with regulatory issues often falls within this area of a company, the occupational health nurse's role in OSHA regulatory compliance also makes this a good "fit." For example, compliance with hearing conservation, respiratory protection, hazard communication, OSHA record keeping (OSHA 200 log), access to medical records, and mandated medical testing for specific chemicals can be effectively managed by a professional occupational health nurse as a program leader or member of an interdisciplinary team. Another example is the critical role of healthcare personnel in the accurate reporting of TSCA 8C requirements.

Within Production

Placing the occupational health nursing position within line management for production recognizes the important role that the occupational health nurse plays in supporting a strong productive work force. This is not an unusual organizational configuration to find in manufacturing facilities. End product cannot be realized without an effective work force in place. The nurse's role in management of programs, such as ergonomics, early intervention for identified safety issues based on health effects and/or injuries, and the occupational health nurse's role in training, make this a viable choice for some organizations.

TRAINING AND EDUCATIONAL REQUIREMENTS

Occupational health nurses are registered nurses holding an active license in the state where they are working. Basic nursing preparation can take a number of pathways, including associate degree programs, diplomas from a hospital-based school, or a baccalaureate program within a university setting. In fact since the early 1980s, the professional association for occupational health nurses (AAOHN) has taken the position that the preferred educational preparation for the occupational health nurse is at the baccalaureate level. The AAOHN took this position because of the strong management component within the occupational health nursing role, and because of the frequent need for autonomous functioning.

Nurses typically enter occupational health positions after working in traditional health delivery systems. Because of the demands for clinical judgement, decision-making, and priority-setting within the occupational health nursing role, an experienced professional is preferred. Background and clinical experience within ambulatory care settings, emergency rooms or adult acute care units are often helpful in direct-care roles. Oversight and coordination of multifaceted programs requires management experience.

Occupational health positions that require a significant amount of direct care, such as immediate care for illness and injury and a high volume of physical examinations, benefit from an occupational health nurse who is also a nurse practitioner. A nurse practitioner is a professional who has completed an approved nurse-practitioner program and has graduated with a master's degree in nursing. In many states, this role has been recognized for expanded practice, including the ability to write prescriptions and manage treatment regimens for selected diagnoses. Adult or family nurse practitioners ideally are suited to working with employee populations.

A number of masters degree programs around the country are also available to prepare nurses in the area of occupational health nursing. These programs are typically associated with NIOSH Education and Resource Centers (ERCs). Since these centers also house programs in industrial hygiene, occupational medicine, epidemiology, and safety, they afford nurses a multidisciplinary educational experience. The current NIOSH ERCs are listed in Table 4-A.

The American Association of Occupational Health Nurses (AAOHN) is the membership organization for occupational health nurses in the United States. They have recently identified basic competencies important for the safe practice of occupational health nursing and have defined occupational health nurse activities at the competent, proficient and expert levels for each of those competencies. Categories of competencies are as follows:

- clinical and primary care
- program management
- health and hazard surveillance
- regulatory compliance
- disability/case management
- work force, workplace, and environmental issues
- regulatory/legislative issues
- management
- health promotion and disease prevention
- occupational and environmental health and safety education and training
- research
- professionalism.

Some occupational health settings with multiple professional personnel employ licensed practical nurses (LPNs) or licensed vocational nurses (LVNs) as part of the staffing mix. It is important to note that nurse practice acts do not allow LPNs or LVNs to practice unless they are directly supervised by a registered nurse (RN) or physician. Therefore, it is inappropriate to utilize licensed practical nurses in single-nurse units within industry.

CERTIFICATION/LICENSURE

As mentioned previously, occupational health nurses should hold licensure as registered nurses (RNs) within their states of practice. Unlike licensure, which assures the public of safe practice at a beginning level and is required for basic practice, certification reflects specialty practice and knowledge at a mastery level. Certification is a voluntary mechanism for validating a professional's knowledge and expertise in a specialty. The occupational health nurse who is certified has made a commitment to the specialty and to continued professional growth and development by successfully completing eligibility and examination requirements.

The American Board for Occupational Health Nurses, Inc. (ABOHN) is the sole certifying body for occupational health nurses in the United States. It was established in 1972 following the recommendations of a multidisciplinary committee to provide a certification program for occupational health nurses.

Table 4-A. Occupational Health Nursing Programs.

A. NIOSH Education and Research Centers (ERCs)

Program Director	Phone/Fax/E-Mail	Degrees	Program Type
University of Alabama at Birmingham School of Nursing 1701 University Blvd. Birmingham, AL 35294-1210 Kathleen Brown, R.N., Ph.D. Director, OHN	(W) 205/934-6858 (F) 205/975-6142 E-mail: brownk@admin.son.uab.edu	MSN PhD	Nurse Practitioner OHN Researcher
University of California, San Francisco School of Nursing, Box 0608 San Francisco, CA 94143-0608 Julia Faucett, R.N., Ph.D. Director, OHN Program	(W) 415/476-3221 (F) 415/476-6042 E-mail: jaf@itsa.ucsf.edu http://nurseweb.ucsf.edu/www/coeh.htm	MS PhD	Occupational Health Adult Nurse Practitioner Occupational Health Nurse Administrator Occupational Health Nurse Researcher
University of California, Los Angeles School of Nursing Louis Factor Building, Box 956919 700 Tiverton Ave. Los Angeles, CA 90095-6919 Wendie Robbins, R.N., Ph.D. Director, OHN	(W) 310/825-8949 (F) 310/206-3241 E-mail: wrobbins@sonner.ucla.edu	MSN MSN/MBA	Occupational Health Adult Nurse Practitioner Occupational Health Nurse Administrator Occupational Health Nurse Researcher
University of Cincinnati College of Nursing P.O. Box 210038 Cincinnati, OH 45221-0038 L. Sue Davis, R.N., Ph.D. Director, OHN	(W) 513/558-5280 (F) 513/558-7523 E-mail: davisls@email.uc.edu or l.sue.davis@uc.edu	MSN PhD	Advanced Practice Specialists, Management, Research
Harvard School of Public Health and Simmons College Graduate School of Health Studies 300 The Feaway Boston, MA 02115 Carol Love, Ph.D., R.N. Director, OHN Carol Summers (Harvard) Co-Director	(W) 617/521-2135 (F) 617/521-3045 E-mail: clove@simmons.edu HSPH: (W) 617/432-3327 (F) 617/432-0219	MSN MSN/SM	Occupational Health Nurse Practitioner Dual Degree Occupational Env. Health

Table 4-A (continued)

Table 4-A. Occupational Health Nursing Programs. (continued)

A. NIOSH Education and Research Centers (ERC's)

Program Director	Phone/Fax/E-Mail	Degrees	Program Type
University of Illinois at Chicago College of Nursing (M/C 802) 845 S. Damen Ave., Rm. 942 Chicago, IL 60612 Karen Conrad, Ph.D., R.N. Director, OHN	(W) 312/996-7974 (F) 312/996-7725 E-mail: konconrad@uic.edu	MS PhD MS/MPH Post Masters Cert.	Occupational Health Nurse Practitioner Occupational Health Nurse Manager
John Hopkins School of Hygiene and Public Health 615 N. Wolfe St. Rm. 7503 Baltimore, MD 21205 Jacqueline Agnew, R.N., Ph.D., F.A.A.N. Director, OHN	(W) 410/955-4082 (F) 410/955-1811 E-mail: jagnew@hsph.edu	MPH MSN/MPH PhD DrPH	Occupational Health Nurse Manager Occupational & Env. Health Researcher
University of Michigan School of Nursing Division Health Promotion Risk Reduction 400 N. Ingalls, Rm. 3182 Ann Arbor, MI 48109-0482 Sally L. Lush, Ph.D., R.N., F.A.A.N. Director, OHN Program	(W) 734/647-0347 (F) 734/647-0351 E-mail: lusk@umich.edu Website: http://www-personal.umich.edu/-lusk	MS PhD	Occupational Health Nurse Mgmt. Clinical Nursing Research (PhD) On Job-On Campus for MS degree Post-doctoral fellowships for Health Promotion/Risk Reduction WHO Collaborating Center
The University of Minnesota School of Public Health Mayo Bldg., Box 807 Minneapolis, MN 55455 Patricia McGovern, R.N., Ph.D. Director, OHN	(W) 612/625-7429 (F) 612/626-0650 E-mail: pmg@cccs.umn.edu	MPH MPH/MSN PhD	Management Program
University of Medicine and Dentistry of NJ BOHSI 170 Frelinghuysen Rd. Piscataway, NJ 05854 Gail Buckler, R.N., M.P.H., COHN-S Program Director	(W) 732/445-0126 (F) 732/445-0130 E-mail: gbuckler@eohsi.rutgers.edu	MSN MPH	Adult Nurse Practitioner Management

Table 4-A (continued)

Table 4-A. Occupational Health Nursing Programs. (continued)

A. NIOSH Education and Research Centers (ERC's)

Program Director	Phone/Fax/E-Mail	Degrees	Program Type
University of North Carolina at Chapel Hill School of Public Health Rosenau Hall - CB 7400 Chapel Hill, NC 27599-7400 Bonnie Rogers, Dr. PH, COHN-S, F.A.A.N. Program Director	(W) 919/966-1030 (F) 919/966-0981 E-mail: brogers@sphvax.sph.umc.edu	MPH MS PhD	Management Program Planning
University of South Florida College of Nursing Health Sciences Center - MDC22 12901 Bruce B. Downs Blvd. Tampa, FL 33612-1700 Candace M. Burns, Ph.D., A.R.N.P. Director, OHN	(W) 813/974-9160 (F) 813/974-5418 E-mail: cburns@coml.med.usf.edu	MS/MPH	Occupational Health Nurse Practitioner
University of Texas, Houston Health Science Center 7000 Fannis #1620 Houston, TX 77030 Thomas A. Mackey, Ph.D. N.P. Director, OHN	(W) 713/500-3250 (F) 713/500-3263 E-mail: tmackey@admin.hsc.uth.tme.edu	MPH MSN/MPH	Management
University of Utah RMCOEH Bldg. 512 Salt Lake City, UT 84112 Dewis Ondrejka, Ph.D. Program Director, OHN	(W) 801/581-8719 (F) 801/581-7224 801/585-4800 E-mail: donrejka@dfpm.utah.edu	MPH MSPH MS PhD Post Masters Cert.	Occupational Health Nurse Practitioner Management/Clinical OHN/CHN Research Bunphasis, OHN/Informatics
University of Washington School of Nursing Box 357262 Seattle, WA 98195 Mary Salazar, Ed.D., COHN-S Director, OHN	(W) 206/685-0857 (F) 206/543-6656 E-mail: msalazar@u.washington.edu	MN PhD	Administration Nurse Practitioner

Table 4-A (continued)

Table 4-A. Occupational Health Nursing Programs. (continued)

B. NIOSH Training Projects Grants

Program Director	Phone/Fax/E-Mail	Degrees	Program Type
University of Pennsylvania Occup. Health Program School of Nursing 420 Guardian Dr. Philadelphia, PA 19104 Karen Buhler-Wilkerson Director, OHN Kay M. Arendasky Associate Director	(W) 215/898-2194 E-mail: arendask@pobox.upenn.edu	MSN MSN/MBA	Options: Occup. Health Adult Nurse Practitioner Family Nurse Practitioner Occup. Health Administration/Consultation Administration/Consultation & Wharton MBA

C. List of NIOSH - ERC Program Directors (April, 1999)

Jacqueline Agnew, RN, PhD John Hopkins School of Hygiene & Public Health 615 N. Wolfe St. Rm. 7503 Baltimore, MD 21205 jagnew@jbsch.edu	University of Cincinnati College of Nursing PO Box 210038 Cincinnati, OH 45221 davisls@email.uc.edu	Gail Buckler, RN, MPH, COHN-S Program Director UMDNJ-RWJMS EOHSI 170 Frelinghuysen Rd. Piscataway, NJ 08854 gbuckler@eohsi.rutgers.edu	Sally Lusk, PhD, RN, FAAN Professor & Director University of Michigan School of Nursing 400 N. Ingalls, Rm. 3182 Ann Arbor, MI 48109 lusk@umich.edu	Dennis Ondrejka, PhD, COHN-S University of Utah Rocky Mt. Center for Occup. & Env. Health 75 S. 2000 E. Salt Lake City, UT 84112 donrejka@dfom.utah.edu
Kathleen Browne, RN, PhD OHN Program Director University of Alabama at Birmingham School of Nursing 1701 University Blvd. Birmingham, AL 35294 brownk@admin.son.uab.edu	Carol Love, RN, PhD Director, OHN Program Harvard ERC/Simmons College 300 The Fenway Boston, MA 02115 clove@simmons.edu	Karen Conrad, PhD, RN Director, OHN University of Illinois at Chicago College of Nursing 845 S. Damen Ave. Rm. 942 MC 802 Chicago, IL 60612 krpconrad@uie.edu	Thomas A. Mackey, PhD, NP Director, OHN University of Texas, Houston Health Science Center 7000 Famin, #1620 Houston, TX 77030 tmackey@edmin4.hsc.uth.toc.edu	Wendie Robbins, PhD UCLA Center for Occup & Env. Health OHN Program Director 10833 LeCorte Ave. Los Angeles, CA 90095 wrobbins@ucla.edu
Candace M. Burns, PhD, ARNP Director, OHN Program College of Nursing University of South Florida Sunshine HCS 22, 1290 Bruce B. Downs Blvd. Tampa, FL 33612 cburns@coml.med.usf.edu	Patricia McGovern, RN, PhD OHN Program Director The University of MN School of Public Health Mayo Bldg. Box 807 420 Delaware St. SE Minneapolis, MN 55455 pmg@cces.umn.edu	Julia Faucett, PhD, RN Director, OHN Program Univ. of Calf, San Francisco School of Nursing PO Box 0608 San Francisco, CA 94143 jaf@irsa.ucsf.edu	Bonnie Rogers, RN, DrPH OHN Program Director Univ. of North Carolina School of Public Health CB 7400, Rm. 263 Chapel Hill, NC 27599 brogers@sphynx.sph.umc.edu	Mary Salazar, EdD, COHN-S Director, OHN University of Washington School of Nursing Box 357262 Seattle, WA 98195 msalazar@washington.edu
L. Sue Davis, RN, PhD Associate Professor, OHN Director		Kay Arendasky, RN, PhD Associate Director University of Pennsylvania 420 Guardian Dr. Philadelphia, PA 19104 arendask@pobox.upenn.edu		

ABOHN's goals are to:
- establish standards and examinations for professional nurses in occupational health
- elevate and maintain the quality of occupational health nursing services
- stimulate the development of improved educational standards and programs in the field of occupational health nursing
- encourage occupational health nurses to continue their professional education.

ABOHN's program includes two basic certification types: the Certified Occupational Health Nurses (COHN) and the Certified Occupational Health Nurse Specialist (COHN-S). Candidates achieve certification by first meeting eligibility requirements, completing an application process, and successfully writing a national examination. Certification is maintained by demonstrating continuing education in the area of occupational health and by continued professional practice. Over 9,200 nurses have been certified by ABOHN since its founding, with over 6,600 nurses holding active certification at the time of this publication. (www.abohn.org)

Both the COHN and COHN-S certifications require that nurses work full time in occupational health for at least 2.5 years, hold an active registered nurse license, and have attended at least 75 contact hours of continuing education in the specialty area. In addition, the COHN-S certification requires a baccalaureate degree or higher for educational preparation. No certification is available for licensed practical nurses.

A number of functional certifications are also available for specific areas within occupational health. They are often acquired by occupational health nurses but do not necessarily require that the holder of the certification be a nurse. Here are some examples:

Certified Occupational Hearing Conservationist (COHC): Holders of this certificate have completed a weeklong course in occupational hearing conservation approved by the Council for Accreditation in Occupational Hearing Conservation (CAOHC). This is recommended when managing an occupational noise and hearing conservation program in industry.

NIOSH Approved Spirometry Certification: This certificate indicates completion of a multi-day course in occupational spirometry that meets the guidelines and has been approved by the National Institute of Occupational Safety and Health (NIOSH). This is extremely important when managing cotton dust and asbestos programs, and is recommended for anyone managing an occupational respiratory protection program.

Case Management: Certified occupational health nurses can obtain specific occupational health case management certification through ABOHN, after obtaining the basic professional credential (COHN/CM or COHN-S/CM). Case management certification can also be obtained from the Commission for Case Management Certification (CCM), and the American Nurses Credentialing Center also offers a general nursing case management credential (Cm). The CCM and the Cm designations are not focused in occupational health. Case management certification is often useful for those actively managing occupational injury cases and dealing with the variety of issues impacting attendance and performance of employees.

CODE OF ETHICS

A code of ethics for occupational health nurses has been developed and is updated periodically by the membership organization, American Association of Occupational Health Nurses, Inc. (AAOHN). The purpose of the "code" is to serve as a "guide for registered professional nurses to maintain and pursue professional recognized ethical behavior in providing occupational and environmental health services." The code of ethics provides a framework of universal moral principles as they are utilized within the occupational health setting, to be used for decision making and evaluation of nursing actions. These include autonomy (including the right to self-determination and confidentiality), beneficence (doing good), nonmaleficence (avoiding harm), and justice (fair treatment).

PROFESSIONAL ORGANIZATIONS
American Association of Occupational Health Nurses, Inc.

The American Association of Occupational Health Nurses, Inc. (AAOHN) is the largest professional association for registered nurses who provide on-the-job health care for the nation's workers. The AAOHN has approximately 13,000 members and accomplishes its goals via more than 180 local, state, and regional chapters throughout the United

> ### AAOHN Code of Ethics
>
> i. Occupational and environmental health nurses provide healthcare in the work environment with **regard for human dignity and client rights,** unrestricted by considerations of social or economic status, personal attributes, or the nature of the health status.
>
> ii. Occupational and environmental health nurses **promote collaboration** with other health professionals and community health agencies in order to meet the health needs of the work force.
>
> iii. Occupational and environmental health nurses strive to **safeguard employees' rights to privacy** by protecting confidential information and releasing information only upon written consent of the employee or as required or permitted by law.
>
> iv. Occupational and environmental health nurses strive to **provide quality care** and to safeguard clients from unethical and illegal actions.
>
> v. Occupational and environmental health nurses, licensed to provide health care services, **accept obligations to society** as professional and responsible members of the community.
>
> vi. Occupational and environmental health nurses maintain individual competence in health nursing practice, based on scientific knowledge, and recognize and **accept responsibility for individual judgements and actions,** while complying with appropriate laws and regulations (local, state, and federal) that impact the delivery of occupational and environmental health services.
>
> vii. Occupational and environmental health nurses participate, as appropriate, in activities such as research that contributes to the ongoing **development of the profession's body of knowledge** while protecting the rights of subjects.

States. The AAOHN speaks for the profession on public policy and public communication issues, and actively works to promote and advance the specialty. Founded in 1942, the AAOHN advances the profession by:

- promoting professional excellence through education and research
- establishing professional standards of practice and code of ethics
- influencing legislative, regulatory, and policy issues
- promoting internal and external communication
- establishing strategic alliances and partnerships.

AAOHN accomplishes these goals through the following four program areas:

Professional Affairs

Professional Affairs promulgates standards of practice and a code of ethics for the specialty, promotes preparation and maintenance of professional competency through continuing education and publication of practice-related resources, and seeks to advance the profession by expanding the knowledge base through research. The AAOHN provides more than 100 educational classes and workshops in its co-sponsorship of the annual American Occupational Health Conference with the American College of Occupational & Environmental Medicine. The AAOHN also supports quality education by maintaining an approval process for continuing education by the America Nurses Credentialing Center's Commission on Accreditation.

AAOHN's monthly publication, the *AAOHN Journal,* is professionally juried, and contains continuing education modules as well as research and general professional interest articles. The Professional Affairs arm was also responsible for the publication of *AAOHN's Core Curriculum for Occupational Health Nursing,* which provides a framework for practice.

Governmental Affairs

This portion of the program provides interface with the legislative process. AAOHN represents the profession on issues that impact workplace health and safety through regular contact with legislators and regulatory agencies.

Member Services and Public Affairs

AAOHN as a membership organization provides a variety of services to its members, including

Figure 4-2. An occupational health nurse can perform health hazard assessment and surveillance of the worker population and workplace. (Courtesy American Board for Occupational Health Nurses, Inc.)

employment listings, information clearinghouses (www.aaohn.org), a monthly newsletter, and networking through local constituencies.

Association of Occupational Health Professionals in Healthcare

The Association of Occupational Health Professionals in Healthcare (AOHP) is a smaller organization designed to meet the needs of occupational health professionals working in health care institutions. The organization was founded in 1981, and has more than 1,200 members. The organization functions through local and national activities, and publishes the *Journal of the Association of Occupational Health Professionals in Healthcare* on a quarterly basis. Their website is www.aohp.org/aohp/.

Case Management Society of America

Occupational health nurses whose practice is heavily concentrated in case management may also choose to join the Case Management Society of America (CMSA). Their website is www.cmsa.org.

SCOPE OF PRACTICE

The role of the occupational health nurse (OHN) can be as variable and diverse as the nation's industries and work environments. The scope of practice depends on the needs of the client. These are predicated on workplace exposures, job requirements, task performance, tools employed, the work environment, and the physical capabilities of the worker.

In addition to the specialized knowledge and skills required to promote and protect the health of workers in specific industry segments and work settings, the occupational health nurse possesses an understanding of a core of fundamental occupational health principles. These basic principles are derived from a framework that includes the following:

- nursing science
- medical science
- public health sciences such as epidemiology and environmental health
- occupational health sciences such as toxicology, safety, industrial hygiene, and ergonomics
- social and behavioral sciences
- management and administration principles.

The occupational health nurse functions independently or as part of multidisciplinary teams to insure the delivery of high quality, comprehensive occupational health services that include the following:

- health promotion and primary, secondary, and tertiary prevention strategies
- health hazard assessment and surveillance of the worker population and workplace (Figure 4-2)
- investigation, monitoring, and analysis of illness and injury episodes, and trends and methods to promote and protect worker health and safety
- primary care (clinical nursing diagnosis and management of occupational and nonoccupa-

> **AAOHN Standards of Practice (1994)**
>
> 1. accurate *assessment* of the worker, the work force, and the environment
>
> 2. appropriate *diagnosis* based on professional observation, research, and communication with the worker, healthcare providers, and others
>
> 3. identification of *outcomes* that serve as markers of effectiveness
>
> 4. development of *plans of action* that are directed toward achievement of outcomes
>
> 5. *implementation* of interventions to promote health, prevent illness and injury, and facilitate rehabilitation
>
> 6. systematic and continuous *evaluation* of services
>
> 7. effective management of *resources* to support occupational health services
>
> 8. continuous *professional development* and professional education
>
> 9. *collaboration* with other disciplines
>
> 10. participation in and utilization of *research*
>
> 11. use of an *ethical framework* for decision-making.

develop, and implement interventions and programs; and (3) evaluate care and service delivery. The American Association of Occupational Health Nurses has established Standards of Practice (1994) that describe and define the elements of activities performed by and expected of a professional occupational health nurse. They are as follows:

ROLES

Because occupational health nursing is a diverse specialty, the occupational health nurse may perform multiple roles consecutively or even simultaneously. The roles within occupational health nursing fall into the following general categories:

Direct Care

In this role, the occupational health nurse provides health care services to individual employees, groups and often to families of employees. Directly delivered services are based on the systematic assessment and accurate nursing diagnosis of the client(s), the work environment and/or the expectations of the job (Figure 4-3). Typical occupational health nurse functions in the direct-care role would include, but are not limited to, performing or administering the following:
- physical examinations
- health-risk appraisals
- tests that screen for early disease
- treatment for work-related injuries
- selection of personal protective equipment
- application and management of work restrictions, immunizations, and job/task analyses.

Manager/Coordinator/Director

As a manager/coordinator/director, the occupational health nurse directs the operation, staff, resource utilization, and evaluation of services of an occupational health and/or safety department within a larger employee health and safety structure. The occupational health nurse is well suited to the role of manager/coordinator/director and applies the standards of assessment, planning, implementation, collaboration, and evaluation to functions such as the following:
- collecting and analyzing data to guide decision-making
- establishing the mission, goals and objectives of an occupational health service
- designing and managing a budget, monitoring changes in regulatory or legislative requirements

tional health services, including program planning, policy development and analysis, and cost-containment measures)
- compliance with regulations and laws governing safety and health for workers and the work environment. (See chapter 10, Safety Programs for more information.)

Regardless of the specific role or type of practice, the occupational health nurse collaborates with workers, employers, and other professionals to (1) identify health needs; (2) prioritize interventions;

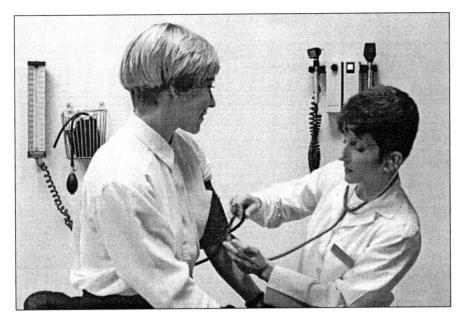

Figure 4-3. Occupational health nurses perform tests that screen for early disease. (Courtesy American Board for Occupational Health Nurses, Inc.)

- facilitating or leading multidisciplinary occupational health teams
- identifying and managing vendors
- preparing and presenting reports of occupational health activity and outcome achievement
- overseeing the selection and professional development of health and safety personnel.

Educator/Advisor

Regardless of the primary role of the occupational health nurse, most will dedicate a significant amount of time, resources, and skill to the role of educator/advisor. In this capacity, the occupational health nurse applies the principles of adult learning theory combined with the standards of nursing practice to educate and heighten awareness in groups and individual employees. Based on the desired outcomes of an educational or training intervention, the occupational health nurse does the following:

- develops mechanisms that are appropriate to the learner(s)
- incorporates techniques to facilitate behavior change
- utilizes local data and current research findings to address relevant health issues.

In an academic setting, occupational health nurse educators prepare nurses who wish to specialize in occupational health. Many occupational health nurse educators conduct research vital to the advancement of the specialty and necessary for promoting and protecting the health of workers. Typical functions of the occupational health nurse in the role of educator/advisor in the occupational health setting are as follows:

- conducting a learning needs assessment
- selecting materials to address a specific workplace hazard
- developing a teaching program to meet regulatory requirements
- conducting community education
- providing in-service education to health and safety staff
- educating managers about the health effects of specific tasks or work.

Case Manager

The occupational health nurse who performs case management as a primary role or as a function within another role is responsible for coordinating services that assist an employee in returning to pre-illness or pre-injury function or to the highest level of functioning achievable after a health challenge. The occupational health nurse case manager does the following

- establishes networks of quality providers
- recommends appropriate, cost-effective treatment
- monitors progress toward expected outcomes
- maintains a strong communication link among all parties (AAOHN, 1994).

As a vital member of the human resource, environmental safety, and management teams, the occupational health nurse insures the safe placement of workers on their return and the appropriate modifications for workers who continue to work while experiencing illness or injury.

Consultant

In this role, the occupational health nurse provides professional assessment and recommendations to groups of managers, workers, the public, and/or clients who require support in addressing workplace health and safety issues. Although all occupational health nurses perform some component of consultation in the course of fulfilling other roles, many occupational health nurses specialize in this area. Whether the occupational health nurse is employed to provide a wide range of services—including consultation—to the employer, is self-employed and provides consultative services to private clients, or is employed by a firm specializing in the provision of occupational health consultation, the role has similarities. Following a systematic evaluation of current conditions, regulatory requirements, the occupational health literature, expert data sources, and client/corporate resources, the occupational health nurse develops and presents an appropriate plan of intervention.

SUMMARY MATRIX

Because the occupational health nurse can fulfill multiple roles within an organization, depending on the needs and dynamics of the population, it is difficult to provide one universal or general job description. Table 4-B delineates functions, qualifications, and typical titles occurring within the five major roles of occupational health nursing.

EMERGING ROLES

As technology and the business practices of client populations change and evolve, the role of the occupational health nurse will adapt and expand. Occupational health nurses will learn and apply new and more advanced mechanisms of minimizing the effect of yet unknown exposures in the workplace. Changes in the models of health-care delivery will afford the occupational health nurse opportunities to apply knowledge and skills to processes such as *managed care, risk management, and 24-hour models*. Employers who are burdened by increasing costs of health care to workers regularly consult with occupational health nurses as they design and deploy flexible benefit programs. Occupational health nurses are important contributors to the development of information systems that not only manage employee health data but also work toward integrating workplace health information with the *complete computerized patient record*. And finally, as health-related legislation and regulations increase, the occupational health nurse will be critical to the development and implementation of programs to insure employer compliance and worker health and safety.

ROLE AS MEMBER OF THE HEALTH AND SAFETY TEAM

Managing occupational health risks is accomplished by first recognizing potential liabilities, then taking a *multidisciplinary* approach to reduce that risk. Effective, comprehensive health and safety programs can only result from collaborative multidisciplinary efforts.

Descriptors used to explain successful collaborative relationships include such terms as *co-equality, mutual trust and respect, complementary contributions, joint problem-solving, collegiality,* and *interdependence*. Collaborative practice enables the knowledge and skills of two or more professionals to synergistically influence an outcome.

Occupational health nursing is the application of nursing principles, particularly the nursing process of assessment, planning, intervention, and evaluation, to help workers achieve and maintain a high level of health and safety. Through the various roles addressed earlier in this chapter, the occupational health nurse serves to limit employer liability and costs in the area of worker health. Depending on the size of the company, the occupational health nurse may be the only occupational health professional on-site in the workplace on a daily basis. This fact necessitates the broadening of the occupational health nurse's role to include more responsibility for environmental assessment and safety. While continuing to apply the nursing process to individual employees, nurses increasingly view the company and its work environment as their client. However, the primary focus of occupational health nursing remains on the human health impact of working conditions.

Key elements needed to successfully achieve collaborative practice include the following:

Table 4-B. Summary Matrix-Role of Occupational Health Nurse.

Role	Direct Care	Manager/Coordinator	Educator/Advisor	Case Manager	Consultant
Job Summary:	A registered nurse who practices in Occupational Health settings and is primarily responsible for providing direct care and assists with other occupational and environmental health services as needed.	A registered nurse who is responsible for setting policy, directing, administering and evaluating health and safety services consistent with organizational goals and objectives.	A registered nurse who has responsibility for developing and delivering occupational and environmental education and training to professional practitioners as well as to individual and groups of employees in the workplace.	A registered nurse who coordinates and monitors services to assist ill and injured workers to reach maximum health and productivity.	A registered nurse who serves as an advisor for developing, selecting, implementing and evaluating occupational and environmental health and safety services to a single employer or to a group of client companies.
Functions:	Maintains medical records. Assesses job related risks. Assesses, delivers and/or coordinates treatment for illness and injury at work. Provides health screenings and other health promotion activities. Administers health and safety program using protocols. Provides surveillance testing, analyzes and reports results.	Develops, administers and maintains policy, budget, resource allocation, information systems, and organizational vision and mission. Evaluates and reports aggregate health and safety data to management. Works on inter-disciplinary teams to develop programs to meet organizational needs. Conducts research and utilizes findings to design, implement and monitor interventions.	Develops, implements and evaluates curricula to various levels of educational preparation using a variety of methods and technologies. Serves as an expert in the field of Occupational Health and Safety. Conducts research and publishes findings to advance OH knowledge.	Assesses client and family physical and psychosocial needs using data from multiple sources. Establishes comprehensive care plans to achieve mutually set goals and objectives. Monitors benefit systems and conducts cost benefit analysis to assure appropriate utilization and services delivered to client. Assists supervisors and workers in assuring safe return to work.	Performs needs assessment and analysis of occupational and environmental health and safety programs. Evaluates applicability of interventions to a specific work environment. Advises clients regarding the structure, delivery and resources required to implement appropriate interventions. Develops training and education programs in accord with current regulations.
Qualifications:	Current license to practice in state(s) of employment. 3 or more years of nursing experience in community, ambulatory or critical care settings. Physical Assessment skills. Good communication skills. BSN and certification as COHN or COHN-S preferred.	Current license to practice in state(s) of employment. 5 years experience in OH nursing. Current knowledge of all regulations and principles related to occupational and environmental health and safety. Ability to plan, organize and manage multiple projects. Ability to use technology, statistics and epidemiology to manage the collection and interpretation of health results. Good communication skills. BSN preferred, MS and certification recommended.	Current license to practice in state(s) of employment. 2 years OH nursing experience. Knowledge of theory of adult learning. Course work in specialized fields such as industrial hygiene, toxicology, ergonomics, etc. Excellent communication skills. BS in Nursing and MS required for academic settings and preferred for continuing education, staff development and worksite training settings. Certification in OH preferred.	Current license to practice in state(s) of employment. 3 or more years Nursing experience. Experience in OH and case management preferred. Current knowledge of laws and regulations governing work and worksite health and safety. Ability to organize work and manage time effectively. Excellent communication skills. BS in Nursing and certification in OH preferred.	Current license to practice in state(s) of employment. 3 years supervisory or administrative experience in OH nursing. Ability to plan, organize, work independently and manage multiple projects. Excellent oral, written and presentation skills. Current knowledge of business management principles and outcome analysis using variety of technologies. BSN required. MS and Certification in OH Nursing preferred.
Sample Titles:	Occupational Health Nurse OHN specialist Occupational and Environmental Health Nurse	OHN Manager OH Program Director Occupational Health and Safety Services Coordinator Corporate OH Director	OHN Instructor Occupational Health and Safety Advisor Director of Occupational Health and Safety Training Educational Director	OH Case Manager Occupational and Environmental Health Nurse Case Manager	Occupational Health Nurse Consultant Occupational Health and Safety Consultant

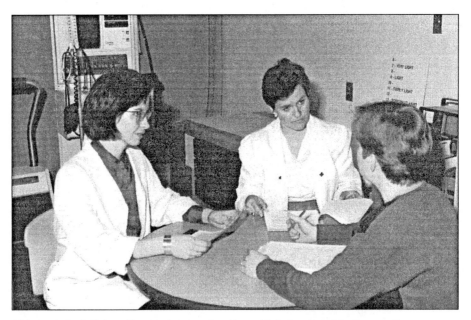

Figure 4-4. Collaboration is characterized by a high degree of both assertiveness and cooperation. (Courtesy American Board for Occupational Health Nurses, Inc.)

- ongoing dialogue conducted on a professional level, and a willingness to move beyond basic information exchange
- mutual understanding of each other's professional domains
- sharing and respecting each other's opinions
- an atmosphere of inquiry that emphasizes research and dissemination of knowledge
- an atmosphere of professionalism.

Collaboration is characterized by a high degree of both assertiveness and cooperation (Figure 4-4). Mutual understanding and respect supporting an environment of collaborative practice results in the most effective use of the expertise of the various disciplines within the occupational health and safety team.

SUMMARY

This chapter explores the profession of the occupational health nurse. The professional roles have evolved over time to enable occupational health nursing professionals to play an increasing part on an organization's health, safety, and environmental team. Occupational health nurses are guided by an ethical framework, and no matter where they are placed in an organization, occupational health nurses can accurately assess the worker, work force, and environment; appropriately diagnose, identify outcomes, develop plans of action, implement interventions, evaluate services, effectively manage resources, collaborate with members from other professional disciplines, etc. Whether on staff or in a consulting capacity, the occupational health nurse is an integral member of the health, safety, and environmental team.

REFERENCES

American Association of Occupational Health Nurses, Inc. (AAOHN). *Guidelines for Developing Job Descriptions in Occupational & Environmental Health Nursing.* Atlanta: AAOHN, 1997.

AAOHN. *Competencies and Performance Criteria in Occupational and Environmental Health Nursing.* Atlanta: AAOHN, 1999.

AAOHN. *AAOHN Code of Ethics and Interpretive Statements.* Atlanta: AAOHN, 1998.

ABOHN. *Candidate Handbook for Examination.* Hinsdale Il: ABOHN, 1998.

Bradford R. Obstacles to collaborative practice. *Nursing Management* 20: 721–72P, 1989.

Brown ML. *Occupational Health Nursing.* New York: Springer, 1981.

McLain B. Collaborative practice: A critical theory perspective. *Research in Nursing and Health* 11:391–398, 1988.

Rogers B. *Occupational Health Nursing Concepts and Practice.* Philadelphia: WB Saunders, 1994.

Salazar Mary K., Ed. *AAOHN Core Curriculum for Occupational Health Nursing.* Philadelphia: WB Saunders, 1997.

chapter 5

Safety Profession

by Richard Lack, PE, CSP, CPP, CHCM, RSP

77 **Overview**
78 **Overall Goals**
   Role definition ■ Placement in the organization ■ Training/educational requirements ■ Certification/licensure
82 **Code of Ethics**
83 **Scope of Practice**
   Administrative elements ■ Action elements
85 **Summary**
85 **References**
85 **Appendix. Sample Position Description**
   5177 Safety officer ■ Role as member of the safety and health team

OVERVIEW

Since the occupational health (medical) professional must adequately understand the role and function of safety management, this chapter presents an overview of the safety profession. The profession of safety management can be compared to the profession of management in that both are more of an art than an exact science. For this reason, there are a number of different definitions for the practice of safety.

The Board of Certified Safety Professionals in their *Candidate Handbook* has the following definition for safety professional: A *safety professional* is one who applies the expertise gained from a study of safety science, principles and practices and other subjects and from other professional safety experience to create or develop procedures, processes, standards, specifications and systems to achieve optimal control or reduction of the hazards and exposures which may harm people, property and/or the environment.

The American Society of Safety Engineers in their *Dictionary of Terms Used in the Safety Profession,* 3rd edition, defines *safety* as a general term denoting an acceptable level of risk of, relative freedom from, and low probability of harm. This dictionary also defines *safety professional* as an individual who, by virtue of specialized knowledge and skill and/or educational accomplishments, has achieved professional status in the safety field. He or she may also have earned the status of Certified Safety Professional (CSP) from the Board of Certified Safety Professionals.

Frank Bird's definition of safety in his book *Practical Loss Control Leadership* is "Safety is the control of accidental loss." According to Homer K.

Lambie, "*safety is the control of hazards*". Following on from this, a simplified definition of a hazard is "the potential for harm. Hazards include the characteristics of things and the actions or inactions of people" (Manuele, 1997).

To summarize, a safety professional is a person who has both technical knowledge and skills in the identification and control of hazards combined with managerial skills in the development and administration of proven safety management systems.

OVERALL GOALS
Role Definition

Two trend-setting pieces of legislation impacting the role of safety professionals are the Cal/OSHA Injury and Illness Prevention Program 1991 and the UK Management of Health and Safety at Work Approved Code of Practice 1992.

The California Code of Regulations Title 8 Section 3203 Injury and Illness Prevention Program states (in part):
(a) Effective July 1, 1991 every employer shall establish, implement and maintain an effective Injury and Illness Prevention Program. The Program shall be in writing and shall, at a minimum:
 (1) *Identify the person or persons with authority and responsibility for implementing the Program.*

The UK Health and Safety Commission on 3 September 1992 approved the Code of Practice entitled *Management of Health and Safety at Work* to go into effect 1 January 1993. Regulation 6 of this Code of Practice states (in part):
 (1) Every employer shall, subject to paragraph (6) and (7) appoint one or more competent persons to assist him in undertaking the measures he needs to take to comply with the requirements and prohibitions imposed on him by or under the relevant statutory provisions.
 (5) A person shall be regarded as competent for the purposes of paragraph (1) where he has sufficient training and experience or knowledge and other qualities in undertaking the measures referred to in that paragraph.

Further developments in the United Kingdom were the publishing in 1995 of National Vocational Qualifications for Occupational Health and Safety Practice Levels 3 and 4. These documents were developed by the Occupational Health and Safety Lead Body (OHSLB).

The OSHLB is an independent group of professionals representing the Confederation of British Industry, the Trades Union Congress, the Health and Safety Executive, the Employment Department and Local Authorities, assisted by the National Council for Vocational Qualifications and City and Guilds of London Institute.

To give readers an idea of the scope of these OHSLB documents, the following are the key items listed in the section *Domain of Knowledge and Understanding*:
- management of risk
- perceptions of, and responses to risk
- risk control methods
- assessment of success/failure
- the judicial system
- civil liability
- statute law and European Community legislation
- the health and safety at work, etc. act 1974 and associated legislation
- other health and safety legislation
- other legislation bearing on the work of health and safety professionals
- legislation about provision of information case law
- general aspects of occupational health
- toxicology, epidemiology, and pathology
- environmental and biological monitoring
- environmental engineering
- personal protective equipment
- maintaining integrity of machinery, plant, and structures
- safe use of machinery
- safety in hazardous processes
- safety in movement of people and materials
- safe use of electricity
- fire prevention and precautions
- oral communications
- effective listening
- written communications
- meeting skills
- interpersonal relations.

This trend can be expected to expand both in the United States and internationally.

The American Society of Safety Engineers (ASSE) has conducted ongoing research on this subject for many years, and has published two

guideline documents: *The Scope and Functions of the Safety Profession* (1966 with later amendments). In this document, the four key areas of the safety position are:
1. Anticipate, identify and evaluate hazardous conditions and practices.
2. Develop hazard control designs, methods, procedures and programs.
3. Implement, administer and advise others n hazard controls and hazard control programs.
4. Measure, audit and evaluate the effectiveness of hazard controls and hazard control programs.

More recently (1992), a position paper was prepared by the ASSE Governmental Affairs Committee. The paper was titled *Management Responsibilities of the Safety Professional*. These responsibilities were outlined as follows:
1. Assist the line leadership in assessing the effectiveness of the unit safety and health programs.
2. Provide guidance to line leadership and employees so they understand the programs and how to implement them.
3. Assist line leadership in the identification and evaluation of high risk hazards and develop measures for their control.
4. Provide staff engineering services on the safety and health aspects of engineering.
5. Maintain working relations with regulatory agencies.
6. Attain and maintain a high level of competence on all related aspects of the profession.
7. Represent the organization in the community on matters of safety and health.

Placement in the Organization

The placement of the safety function can be at a variety of levels and departments. The safety function can report to any of the following areas:
- human resources director
- risk manager
- vice president and general manager
- environmental and safety director
- legal
- finance
- engineering
- maintenance
- operations
- technical services
- security.

The ideal placement is where the safety professional has a direct reporting relationship to top management. Nevertheless, a safety professional with good influencing and communications skills can perform effectively wherever he or she is placed in the organization.

The current trend toward consolidation at the facility level indicates that the functions concerned with asset protection are moving closer together. For example there could be a human and environmental affairs unit that includes the functions of human resources, training and education, risk management, safety, industrial hygiene, medical, environmental, security, fire protection, related engineering professions, and legal.

Training/Educational Requirements

The American Society of Safety Engineers' document *Scope and Functions of the Professional Safety Position* outlines the education, training, and experience needed to competently perform in this position.

To perform their professional functions, safety professionals must have education, training and experience in a common body of knowledge. Safety professionals need to have a fundamental knowledge of physics, chemistry, biology, physiology, statistics, mathematics, computer science, engineering mechanics, industrial processes, business, communication and psychology. Professional safety studies include industrial hygiene and toxicology, design of engineering hazard controls, fire protection, ergonomics, system and process safety, safety and health program management, accident investigation and analysis, product safety, construction safety, education and training methods, measurement of safety performance, human behavior, environmental safety and health, and safety, health, and environmental laws, regulations and standards. Many safety professionals have backgrounds or advanced study in other disciplines, such as management and business administration, engineering, education, physical and social sciences and other fields. Others have advanced study in safety. This extends their expertise beyond the basics of the safety profession.

Because safety is an element in all human endeavors, safety professionals perform their functions in a variety of contexts in both public and private sectors, often employing specialized knowledge and skills. Typical settings are manufacturing, insurance, risk management, government, education, consulting, construction, health care, engineering and design, waste management, petroleum, facilities management, retail, transportation, and utilities. Within these contexts, safety professionals must adapt their functions to fit the mission, operations and climate of their employer.

Not only must safety professionals acquire the knowledge and skill to perform their functions effectively in their employment context, through continuing education and training they stay current with new technologies, changes in laws and regulations, and changes in the workforce, workplace and world business, political and social climate.

As part of their positions, safety professionals must plan for and manage resources and funds related to their functions. They may be responsible for supervising a diverse staff of professionals.

By acquiring the knowledge and skills of the profession, developing the mind set and wisdom to act responsibly in the employment context, and keeping up with changes that affect the safety profession, the safety professional is able to perform required safety professional functions with confidence, competence, and respected authority.

The Board of Certified Safety Professionals (BCSP) has five specific requirements in order to qualify for the Certified Safety Professional title:
1. Complete an application to BCSP.
2. Meet an academic requirement.
3. Meet a professional safety experience requirement (four years with substitutions for master's and doctoral degrees).
4. Pass the Safety Fundamentals Examination.
5. Pass the Comprehensive Practice Examination.

In the case of the academic requirements, candidates must have a minimum of an acceptable associate degree in safety and health or a bachelor's degree in any field.

The BCSP *Candidate Handbook,* 3rd ed., October 1996, lists the schools with accredited bachelor's (BS) or master's (MS) degrees in safety.
- Illinois State University (BS)
- Indiana State University (BS, MS)
- Millersville UPA (BS)
- Murray State University (BS, MS)
- Indiana University of Pennsylvania (BS, MS)
- Oregon State University (BS)
- West Virginia University (MS)

Now that the BCSP requires a degree to qualify for their certification, the number of schools offering a safety degree may increase in the future. More information on this trend can be found in the literature (Schattel, 1990; Kohn et al, 1991; Frederick et al, 1999; Blair, 1999; Manuele, 1993).

Some highlights of these articles include the following:
- Schattel (1990) lists the following types of courses that should be included in the educational requirements for attaining a professional level of competence in safety: management, risk management, psychology of human behavior, consultant skills, engineering, industrial health and hygiene, regulations, standards and codes. He stresses that safety professionals must have knowledge and training in management techniques because they are key skills for a successful safety career.
- Kohn et al (1991) reported on a survey of safety professionals and participants were asked to rate areas of proficiency most important to health and safety success. When asked to list the specialized training programs necessary for career advancement, participants ranked the following areas in terms of importance:
  1. management
  2. computer science
  3. industrial hygiene
  4. ergonomics
  5. hazardous materials
  6. fire science.
- In Blair's survey of 450 CSPs (1999), participants identified the following top 10 competencies needed for successful safety professionals:
  1. communicating effectively
  2. actively listening

3. accepting responsibility
4. motivating others
5. maintaining flexibility
6. maintaining a vision
7. thinking creatively
8. auditing and analyzing.
9. obtaining participation
10. translating solutions into reality

They recommend that students prepare themselves for the safety profession by including communication skills, psychological aspects, business management, accounting, and marketing.

- Frederick et al (1999) conducted a study of the key safety responsibilities and skills. The top 10 identified are as follows:
  1. safety, health, and environmental management
  2. safety, health, and environmental regulations
  3. verbal or written communications
  4. training
  5. program management
  6. workers' compensation/interpersonal skills
  7. loss control
  8. compliance audits/leadership skills
  9. computer skills/fire
  10. statistics and measurement.

The role of the safety professional continues to expand. More research and activity regarding safety curricula are needed; and it should involve all stakeholders—including industry, government, accrediting bodies, and educators. Other significant published reports contain much useful advice and information for those entering the safety profession (Graham, 1997; Kedjidjian, 1998). As quoted in these articles, Nancy J. McWilliams, president of McWilliams Risk Management, Inc., agrees that, "understanding the technical side is fine, but it doesn't help you when you're presenting your program to senior management and asking for a piece of the budget. . ."

East Carolina University's James P. Kohn agrees "Everybody demands that if you come into business, you've got to be able to communicate effectively." . . . Fred Manuele, president of Hazards Limited in Arlington Heights, Ill., and author of *On the Practice of Safety* believes that students who graduate with degrees in safety can be valuable safety professionals, but their learning has just begun. "Safety is a changing, growing, and diverse field, and university programs cannot cover every topic. If there's ever a profession that is in transition, it is safety," says Manuele. "The safety professional is forever in a continuing education program. . ."

Many now employed in the safety profession came from different backgrounds, education, and experience. As noted, future successful safety professionals will need technical, management, and human relations skills.

Certification/Licensure

The number and type of certifications offered by numerous organizations is beyond the scope of this short chapter. This chapter will focus on those certifications recognized as the leading ones in the safety profession. Numerous certifications in related fields that safety professionals should consider will also be described.

Safety and Health Certifications

The following organizations offer safety and health certifications in various categories:

Board of Certified Safety Professionals (BSCP)

The BCSP offers certification as a Certified Safety Professional (CSP) or Associate Safety Professional (ASP). Candidates must possess a degree, have a minimum of four years experience and satisfactorily pass an examination. There are substitutions for experience with masters or doctorate degrees.

Board of Hazard Control Management

The Board of Hazard Control Management offers a Certified Hazard Control Manager (CHCM) certification. Candidates may satisfy the Board with their degree/s and experience or may take the CHCM examination.

National Safety Management Society (NSMS)

Recently, NSMS established the Institute for Safety and Health Management, that offers a Certified Safety and Health Manager (CSHM) and Associate Safety and Health Manager (ASHM) certifications. Candidates must possess demonstrated knowledge of health and safety management skills and techniques and satisfactorily pass an examination.

American Society of Safety Engineers (ASSE)

The ASSE has recently established a National Registry of Safety Professionals. Safety professionals desiring to be listed in the Registry must cite their position status, employment history and professional activities.

Safety Organization (WSO)
The WSO offers the following professional certification programs:
- WSO—Professional Certifications
- WSO—Certified Safety Executive
- WSO—Certified Safety Manager
- WSO—Certified Safety Specialist
- WSO—Certified Safety/Security Director
- WSO—Registration Program
- WSO—Registered Safety Director.

National Safety Council (NSC)
The National Safety Council and its regional offices offer a wide range of training courses on safety and health topics. Their *Advanced Safety Certificate* is an internationally recognized certificate and the study curriculum includes Principles of Occupational Safety and Health, and two of the following three courses:
- Safety Training
- Fundamentals of Industrial Hygiene
- Safety Management Techniques.

The Institution of Occupational Safety and Health (IOSH)
The IOSH, with headquarters in the United Kingdom, offers a *Register of Safety Practitioners*. Candidates must complete a very detailed application outlining their professional experience and evidence how their knowledge and skills were gained: by professional conferences, seminars, training courses, etc. Those that are accepted by IOSH are entitled to the designation Registered Safety Professional (RSP).

Other Certifications and Licenses
There are engineering licensures in related fields that are offered by most states, such as Fire Protection Engineers, Traffic Engineers, Quality Engineers, Nuclear Engineers, Industrial Engineers, Corrosion Engineers, and Chemical Engineers.

California may be one of the last states that has an examination and registration for safety engineers. This registration is reportedly entering a "sunset" process that will cause the registration to be terminated at a date to be determined in the next century.

Certifications in related professional fields again are too numerous to develop a complete listing. Two of the most significant certifications are *Certified Industrial Hygienist* (*CIH*) offered by the American Board of Industrial Hygiene and *Certified Protection Professional* (*CPP*) offered by the American Society for Industrial Security.

Certification Maintenance
All these certification boards have requirements for certification maintenance. Holders are required to evidence their continuing professional activity and development. Increasingly, employers require professional certifications for their in-house safety positions. For example, 10 years ago, the City and County of San Francisco Class 5177 Safety Officer position required candidates to be registered in California as a PE (professional engineer) in Safety Engineering. This has now been changed to Certified Safety Professional (CSP).

CODE OF ETHICS
In the *Oxford Dictionary*, *ethics* are defined as follows:
1. the science of morals in human conduct
2a. moral principles; rule of conduct
2b. a set of these (medical ethics).

In *The Power of Ethical Management* by Kenneth Blanchard and Norman Vincent Peale (1988), the authors describe an *ethics check* for any action a professional manager is proposing or planning. The ethics check questions are as follows:

1. *Is it legal? Will I be violating either civil law or company policies?*
2. *Is it balanced? Is it fair to all concerned in the short term as well as the long term? Does it promote win-win relationships?*
3. *How will it make me feel about myself? Will it make me proud? Would I feel good if my decision was published in a newspaper? Would I feel good if my family knew about it?*
4. *Every professional society or organization has a code of ethics, or sometimes called code of professional conduct.*

A review of the codes of several professional societies reveals several commonalities.
- Members must abide by the legal requirements relating to their practice.
- Members shall observe the precepts of truthfulness, honesty and integrity.
- Members shall be competent in discharging professional responsibilities.

- Members shall not undertake responsibilities as safety and health professionals which they do not believe themselves competent to discharge. Members shall acknowledge any limitations in their own competence.
- Members shall safeguard confidential information and exercise due care to prevent its improper disclosure.
- Members shall not maliciously injure the professional reputation or practice of colleagues, clients or employers.
- Members shall not improperly use their membership of the society for commercial or personal gain.
- Members shall seek to avoid their professional judgment being influenced by any conflict of interest and shall inform their employer or client of any conflict between the members' personal interest and service to the relevant party.

SCOPE OF PRACTICE

The broad scope of practice for safety professionals was outlined in the *Role Definition* section of this chapter. The technical aspects that the professional may need to understand and provide guidance as to safe operating procedures vary with the nature of the organization and the level of the safety position in that organization.

Examples of typical hazards confronting safety professionals include the following:
- *work areas*—work surfaces, illumination, ladders, scaffolds, platforms
- *machines*—mechanical power and transmission, point of operation, production machines, punch presses and metalworking machines, woodworking machines, saws
- *equipment*—compressors, cranes, hoists and slings, conveyors, gad welding and burning, pumps, pipes, valves and tanks
- *portable tools*—hand tools manual, hand tools power
- *hazardous materials, health-related*—skin irritants, toxic (systemic poisons)
- *hazardous materials, traumatic*—compressed gases and liquids, molten metal, hot materials, acids and caustics, flammables and explosives, flying particles
- *electrical*—motors and generators, transformers and rectifiers, wiring-conductors, control devices, electrical hand tools, electrical welding, high voltage switching.

Figure 5-1. Safety professionals also train employees. (Courtesy American Board for Occupational Health Nurses, Inc.)

- *vehicles*—power trucks, forklifts, fleet safety, railroads, aircraft, marine.
- *multicategory*—lockout/tagout, confined space entry, excavations.

Topics a safety professional should be familiar with include these:
- hazard communication program
- process system safety management
- ergonomics
- fall protection
- hearing conservation
- respiratory protection program
- exposure control program for bloodborne pathogens
- indoor air quality
- asbestos and lead management program
- medical surveillance
- fire prevention plan
- emergency procedures.

Safety professionals also train employees (Figure 5-1). The scope of training responsibilities varies with the nature and size of the organization. Typical categories of safety training include conducting new employee safety orientation, management/ supervisory training, and training on hazard communication, and emergency procedures.

Figure 5-2. Typically a safety professional conducts research and provides professional consultation. (Courtesy American Board for Occupational Health Nurses, Inc.)

Other types of specific training can include asbestos awareness, electrical safety, ergonomics, excavation safety, exposure control, fall protection, hazardous waste operations, hearing conservation, lockout/tagout, confined space entry and rescue, defensive driving, first aid/CPR, forklift operation, crane operation, and aerial lift operation.

Most safety professionals are able to conduct some of this training. But the highly specialized technical training can be conducted by in-house specialists or outsourced to qualified consultants or training providers.

A chief function of an effective safety professional is to develop information for managers and the occupational safety and health team to use in decision-making. In a typical day a full-time safety professional might include some or all of the following activities:

- meeting with one manager or supervisor to discuss the status and progress of his/her safety and health systems
- conducting an audit or survey with or without members of management to assess the effectiveness of the unit's safety and health systems
- reviewing any loss event that may have occurred and assisting line management with the investigation of those considered significant
- working up programs such as training presentations, new or revised procedures, committee meetings and assignments
- conducting training
- attending committee or other meetings
- conducting research and providing professional consultation (Figure 5-2)
- developing and guiding staff members, if in a supervisory role.

An effective safety professional can be a system administrator whose major duties and responsibilities include the following:

- Recommend to management for the continuous improvement of the unit's safety (health, security, fire protection, etc.) systems.
- Explain the principles and procedures of the systems to the various management levels.
- Provide consulting service on safety (health, security, fire) regulations and standards.
- Help develop good relationship with government agencies.
- Effectively manage the safety department.
- Take appropriate steps to ensure continuing professional development, through attending conferences, seminars and professional meetings.

In his or her capacity as a system administrator, the safety professional is responsible for helping to establish and maintain effective safety management systems.

Administrative Elements

Primary elements of this system are as follows:
- safety manual (procedures and guidelines)
- safety committees and coordinators
- safety training, interest and motivation.

Action Elements
Primary elements of this system are as follows:
- safety inspections
- hazard control
- job safety analysis
- safety meetings
- accident investigation.

SUMMARY
As part of the occupational safety and health team, the safety function is broad and overlaps with that of industrial hygiene and occupational medicine. Thus adequate time and attention must be given to team-building and careful delineation of responsibilities. This will result in development and implementation of an appropriate organizational structure for all sizes of operations. Many facilities have a full-time safety professional—others used consultants. Either way, the safety function and professional is a major part of the occupational safety and health team.

REFERENCES
Kohn JP, Timmons DL, Bisesi M. Occupational health and safety professionals: Who are we? What do we do? *Professional Safety* 36(1):24–28, 1991.

Blair EH. Which competencies are most important for safety managers? *Professional Safety* 44(1):28–32, *1999.*

Frederick LJ, Winn GL, Hungate AC. Characteristics employers are seeking in today's safety professionals. *Professional Safety* 44(2):27–31, 1999.

Graham S. Job outlook '98. *Safety + Health* 156(3):40–42, 1997.

Kedjidjian C. Safety 101—Today's students, tomorrow's safety pros. *Safety + Health* 157(4):38–42, 1998.

Manuele FA. On becoming a profession. *Professional Safety* 38(10):22–27, 1993.

National Safety Council. *Accident Prevention Manual for Business & Industry,* 11th ed. Itasca IL: National Safety Council, 1997.

Schattel JL. Needs of the safety professional in the 1990s. *Professional Safety* 35(6):39–42, 1990.

APPENDIX. SAMPLE POSITION DESCRIPTION:
The following position description is a composite of an actual civil service position for a Class 5177 Safety Officer for the City and County of San Francisco Airport Commission, together with major responsibilities for the position that were taken from a performance appraisal document.

5177 Safety Officer
Duties

This position is responsible for the development and implementation of a comprehensive health and safety program which meets the needs of a major operating department. This includes advanced professional level work in the areas of safety, accident prevention, emergency response, hazardous waste management, industrial hygiene, loss control, risk management, accident investigation, safety training, development of safety and training materials and programs, procedures analysis, trend analysis, and equipment modification studies. The Safety Officer works with considerable independence and supervises subordinate safety/industrial hygiene staff.

Minimum Requirements

1. Requires possession of an accredited baccalaureate degree in Safety Management, Industrial Hygiene, Safety or Environmental Engineering, or a related field, and
2. Four years of increasingly responsible professional health and safety experience (as described above) which must have included safety program development, safety management, hazardous materials management, and industrial hygiene management. This must also have included one year of supervisory experience, and
3. Possession of valid driver's license to be presented at time of examination.

Note: Additional years of qualifying experience in safety management as described above may substitute for the degree on a year for year basis. Additional related education may substitute for the experience on a year to year basis up to one year. A CSP may substitute for one year of supervisory experience. Certain positions may require certified Safety Professional Certification.

Major Responsibilities of this Position

1. Assesses the effectiveness of the Airport safety and health program and recommends for its continuous improvement.
2. Provides training for supervisors so they understand the Airport program and implement it effectively.
3. Manages hazardous material and industrial hygiene program.
4. Conduct and/or coordinate safety and health training for all levels of employees.
5. Serves as advisor to the Executive Safety Advisory Committee and Industrial Safety Committee.
6. Assists management and supervision in investigation of significant incidents and serious accidents.
7. Maintains all safety and health reports and records.
8. Keeps current on all related legislation and standards.
9. Provides expertise on safety engineering aspects.
10. Works towards maintaining good relations with regulatory agencies.

Role as Member of the Safety and Health Team

The safety professional must work closely with other unit asset protection professionals including system safety engineers, industrial hygienists, ergonomist and ergonomic specialists, environmental specialists, fire protection and security specialists, occupational medicine physician and nurses, and workers' compensation adjuster. The safety professional must:

- cooperate in establishing and maintaining joint loss control.
- provide constant resource information and advice to these functions.
- Regularly review plans and activities of all these functions to assume coordinated action and maximum efficiency.

In some cases, the safety professional will function as manager of the entire asset protection department. Outside of departmental functions, the safety professional must interact with all other departments in the enterprise. Other than top management, there are certain departments and positions that will be especially important for the safety professional to be directly involved with and develop a fully trusting relationship. These are the chief engineer, maintenance manager, planning and scheduling section, human resources director, legal counsel, controller or finance director, purchasing manager, operations/facilities manager/s.

These positions have the most direct influence over the success of the enterprise's safety and health mission, goals and strategies.

The chief engineer and the technical services group provide vital control over safe design and construction. The maintenance manager and staff, including planning and scheduling, are also a critical group in safely maintaining the unit's equipment and systems, and also carrying out necessary repairs and modifications where hazards may have been identified.

The human resource director and staff are a key group in ensuring that the workforce are competent and that controls are established to minimize performance problems such as those arising from drug and alcohol abuse or other criminal activities.

The safety professional must maintain regular dialogue with legal counsel and seek their guidance on all matters where legal liability may be a concern.

The controller and accounting staff and purchasing and warehousing staffs are important groups for the safety professional to work with on such issues as purchase and storage of chemicals and personal protective equipment.

Operations/facilities managers and their staff will need special assistance and support depending on the nature of the process and type of buildings involved. Emergency and evacuation procedures, and health and security concerns are typical issues requiring the safety professional and other specialists input and guidance.

chapter 6

Industrial Hygiene Profession

by Douglas P. Fowler, PhD, CIH

OVERVIEW

This overview of the industrial hygiene profession covers definitions, code of ethics, professional organizations, professional education, and certification and licensure practices in the profession. Each topic is explored from a practical perspective.

Definitions

The American Industrial Hygiene Association (AIHA) defines *industrial hygiene* as:

> *the science and practice devoted to the anticipation, recognition, evaluation, and control of those environmental factors and stresses arising in or from the workplace, that may cause sickness, impaired health and well being, or significant discomfort and inefficiency among workers and may also impact the general community.*

The AIHA defines an *industrial hygienist* as:

> *a person having a baccalaureate or graduate degree from an accredited college or university in industrial hygiene, biology, chemistry, engineering physics or a closely related physical or biological science who, by virtue of special studies and training, has acquired competence in industrial hygiene. Such special studies and training must have been sufficient in the above cognate sciences to provide the ability and competency: (1) to anticipate and recognize*

- 87 **Overview**
  Definitions ■ Code of ethics ■ Professional organizations ■ Professional education ■ Certification and licensure
- 90 **The Industrial Hygienist on the Health and Safety Team**
  In-house industrial hygiene services ■ Industrial hygiene consulting services
- 92 **Practicing Industrial Hygiene**
  Anticipating occupational health hazards ■ Recognizing occupational health hazards ■ Evaluating occupational health hazards ■ Controlling occupational health hazards
- 105 **Summary**
- 105 **References**

the environmental factors and stresses associated with work and work operations and to understand their effect on people and their well-being; (2) to evaluate, on the basis of training and experience and with the aid of quantitative measurement techniques, the magnitude of these factors and stresses in terms of their ability to impair human health and well-being, and (3) to prescribe methods to prevent, eliminate, control, or reduce such factors and stresses and their effects.

Code of Ethics

The professional industrial hygienist is bound by a *Code of Ethics for the Practice of Industrial Hygiene:* (AIHA, 1998)

Objective

These canons provide standards of ethical conduct for industrial hygienists as they practice their profession and exercise their primary mission, to protect the health and well-being of working people and the public from chemical, microbiological and physical health hazards present at, or emanating from, the workplace.

Canons of Ethical Conduct

Industrial hygienists shall:
1. *Practice their profession following recognized scientific principles with the realization that the lives, health and well-being of people may depend on their professional judgement and that they are obligated to protect the health and well-being of people.*
2. *Counsel affected parties factually regarding potential health risks and precautions necessary to avoid adverse health effects.*
3. *Keep confidential personal and business information obtained during the exercise of industrial hygiene activities, except when required by law or overriding health and safety considerations.*
4. *Avoid circumstances where a compromise of professional judgement or conflict of interest may arise.*
5. *Perform services only in the areas of their competence.*
6. *Act responsibly to uphold the integrity of the profession.*

Professional Organizations

The American Conference of Governmental Industrial Hygienists (ACGIH) was the first (1938) U.S. industrial hygiene organization, and currently has approximately 5,400 members. The ACGIH limits full membership to health and safety professionals practicing in government agencies and educational institutions, but offers affiliate membership to others in the health and safety field. Two services offered by the ACGIH are of particular value to the profession. One of these is the manual, *Industrial Ventilation—A Manual of Recommended Practice*, by the Industrial Ventilation Committee of ACGIH. This manual gives guidance in design of ventilation systems for the control of airborne health hazards, and is a standard reference source for engineers and industrial hygienists who design or evaluate ventilation systems.

The other is the annual publication of the TLV-BEI listings. The ACGIH, through its Threshold Limit Values Committee, publishes its listing of *Threshold Limit Values (TLVs) and Biological Exposure Indices (BEIs)* annually. The TLVs and BEIs are widely used as measures of allowable exposure limits; in fact, the 1968 TLVs were adopted into law as the Permissible Exposure Limits in the OSHAct. Both of these documents are discussed elsewhere in this chapter.

The American Industrial Hygiene Association (AIHA) is the predominant U.S. industrial hygiene organization. The AIHA was formed in 1939 by a small group of industrial hygienists, and has grown in the intervening years to more than 13,000 members. The only restrictions on membership are those of educational qualifications and practice of industrial hygiene—a full member must have an appropriate degree (usually in one of the physical or biological sciences), and have practiced industrial hygiene full-time for three years. The AIHA provides many services to the profession, including the accreditation of laboratories offering analytical services. There are local (regional) sections of the AIHA in all areas of the United States, and many foreign occupational health professionals are members, as well.

Many industrial hygienists are also members of organizations representing major and minor industries. The American Petroleum Institute, the Chemical Manufacturer's Association, and the National Agricultural Chemical Association are examples of industry organizations with active committees on various aspects of occupational health on which industrial hygienists serve.

Some industrial hygienists also maintain membership in organizations representing the disciplines of their original training. The American Chemical Society and the American Institute of Chemical Engineers are examples of societies claiming many industrial hygienists among their members. Industrial hygienists also belong to organizations with peripheral interests in industrial hygiene, or representing fields within which industrial hygienists have responsibilities. Such organizations as the National Safety Council, the Society for Epidemiological Research, the Air and Waste Management Association, the American Management Association, the American Society of Safety Engineers, the American Public Health Association, the American Society for Testing and Materials, and the Genetic and Environmental Toxicology Association are examples. Finally, general scientific societies (such as the American Association for the Advancement of Science) also claim their share of industrial hygienists as members.

Professional Education

As noted earlier, designation as a professional industrial hygienist requires at least a college degree in one of the sciences and some specialized training and/or experience. In past years, the industrial hygienist was usually trained as a chemist or engineer and came to the field of industrial hygiene as the result of employer assignment to a position with industrial hygiene responsibilities. Such assignment might have been accompanied by short course training by a government agency or university school of public health. After a few years of experience the practicing industrial hygienist may have returned to a school of public health to obtain a master's degree in the field. A few intent on a teaching career obtained doctoral degrees in the field at the few universities offering them. Several prominent early industrial hygienists were originally trained as physicians.

As the field has grown in the past 30 years, it has become more common for industrial hygienists to be introduced to the field as undergraduates, and to go directly into graduate programs for master's degrees in industrial hygiene before entering the workplace. The typical science background of industrial hygienists has also changed. Many of the entrants to the field in the 1980s and 1990s came from a background in the biological sciences, rather than the physical sciences or engineering. This sort of background gives them greater facility in dealing with the ever more complex developments in such fields as genetic toxicology; but they can be less well equipped for such tasks as design of local exhaust systems.

A recent trend has been for industrial hygienists to have increasingly important management responsibilities, and to supplement their scientific backgrounds with management training either at the graduate level or through short courses and summer institutes and the like.

Certification and Licensure

Certification of industrial hygienists began in 1960, when the AIHA and ACGIH established the American Board of Industrial Hygiene (ABIH), to set up certification requirements. Three classes of certification are currently recognized:

- Diplomate (permitted to use the designation Certified Industrial Hygienist—CIH)
- Industrial Hygienist in Training (IHIT) and Occupational Health and Safety Technologist (OHST)
- Construction Health and Safety Technicians (CHST).

CIH Certification is currently offered in comprehensive practice, and in the specialty area of chemical aspects—focusing on analytical chemistry. Beginning in 1993, specialized competence of diplomates in the subspecialty of indoor air quality was formally recognized by the board. Before 1992, certification in the specialized fields of acoustical, air pollution, engineering, radiological, and toxicological aspects of industrial hygiene was offered, and those certifications remain valid, although no new certificates will be offered. Certification as a CIH is granted when education in the sciences equivalent to college graduation, successful performance in the field for a minimum of five years, recommendation by two practicing industrial hygienists, and completion of two written (daylong) examinations are shown. The first of these examinations (the core examination) can be taken after one year of work in the field, and successful performance allows use of the IHIT designation. The second examination (covering comprehensive practice or a specialty) cannot be taken until the CIH candidate has completed a total of at least five years of acceptable experience. Maintaining certification requires documenting continued activity in certain defined professional activities and continuing education. The subspecialty recognition is granted after successful completion of a written one-half day examination. Diplomates are eligible for membership in the American Academy of Industrial Hygiene. The ABIH publishes a directory each year in which are listed all CIHs, IHITs, OHSTs, and CHSTs.

Although many industrial hygienists are becoming interested in seeking some form of government licensure, no state or federal licensure requirements had been established by mid-1999. In a few states, such as Illinois and California, state *title protection* laws have been passed, which make use of the title *Certified Industrial Hygienist* or *CIH* illegal for non-CIHs. However, the CIH designation is often required for industrial hygiene practice on projects or jobs supported by government funds. The CIH is increasingly recognized as a minimum requirement to assure that industrial hygiene work is being performed according to professional standards, especially when litigation is anticipated or feared.

THE INDUSTRIAL HYGIENIST ON THE HEALTH AND SAFETY TEAM

The industrial hygienist specializes in the anticipation, recognition, evaluation, and control of health hazards. In most organizations, the industrial hygienist is principally responsible for chemical agent hazards, but he or she will often both be qualified to exercise, and have responsibility for control of physical agents, including noise and vibration, ionizing radiation, and nonionizing radiation. (In larger and more complex organizations, the control of radiation can devolve upon the radiation specialist, the health physicist.) As the need for control of biological agents (especially for indoor air quality, such as prevention of legionnaires' disease or other communicable diseases) has become more widely recognized, many industrial hygienists have assumed this responsibility as well. Depending on his or her expertise, the industrial hygienist may also have responsibility for ergonomics. The industrial hygienist maintains lists of chemical, biological, and physical agents found within the facility, evaluates exposures to those agents, and institutes controls to ensure that exposures to those agents are within tolerable limits.

In-House Industrial Hygiene Services

The industrial hygienist must have a good relationship with the engineering department, since many of the most effective controls on exposure will involve modifying the physical facility. The environmental staff—those responsible for control of emissions outside of the facility—also should be part of the industrial hygienist's major contacts. Controls to reduce exposures within the facility can lead to increased emissions; a unified approach to reducing exposures is essential.

Two of the most important allied departments for the industrial hygienist are the safety and medical departments. The safety department often has complementary or overlapping functions—particularly for record keeping and regulatory compliance—and the professional expertise of the safety professional may be needed to ensure that equipment or process modifications made to reduce exposures do not lead to traumatic injury hazards.

The relationship between the medical and industrial hygiene departments is a traditional source of strength to both. The industrial hygienist provides information to the physicians and nurses in the medical department on workplace conditions, the introduction of new chemicals, and changes in use of familiar chemicals. The medical staff provide information on changing patterns of injury and illness within the facility, and on the special needs of workers who may require accommodation. As noted in the following, increasing numbers of small- to mid-size companies are using industrial hygiene and medical consultants, rather than full-time industrial hygienists or medical directors.

In some such cases, the industrial hygienist is called in only to monitor for specific chemicals, or in the event of a problem, and the physician is viewed only as a source of medical examinations and treatment for workplace injuries. The physician usually does not have the time to visit the facility and observe operations with sufficient frequency to give him or her an adequate picture of what normal operations are like, or to fully understand the interactions of the various occupational categories within the corporation or other workplace setting. Diagnoses of occupational diseases might be made without reference to the body of knowledge regarding potential and actual exposures that the industrial hygienist may have available to assist the physician.

Similarly, without frequent interaction with the medical personnel the industrial hygienist might begin an evaluation unaware of clinical effects that can guide him or her in developing a rational evaluation protocol. An example is the existence of significant skin disease among a worker population that should lead the industrial hygienist to more carefully evaluate the work practices involved in handling irritating or otherwise skin-active compounds, and to pay careful attention to use of appropriate personal protective equipment. The industrial hygienist and occupational physician should always work in cooperation, particularly in those settings where many environmental factors are present, and their evaluation is complex.

The industrial hygienist should have access to the facility management, since effective controls can be expensive, and the need for these controls must be made clear to the decision makers within the company. To properly convey the impact of health and safety issues on the company, the industrial hygienist must speak the language of the managers who will make the financial decisions. Thus, many industrial hygienists are developing competence in management and fiscal topics.

Finally, and perhaps most importantly, the industrial hygienist must be recognized by the workers within the facility as reliable, believable, and honest, so that recommendations for changes in work practices are more diligently implemented and assurances that conditions are acceptable are more likely to be believed.

The industrial hygiene function can be placed in a variety of possible locations within the company. Most commonly, the industrial hygiene function is placed in the human resources department (traditionally with the medical function). Some companies combine the health and safety function, and others combine the environmental and industrial hygiene departments. Occasionally, the industrial hygiene function is found in the engineering department, or the security department.

For many small and mid-size companies, a full-time industrial hygienist cannot be justified economically. In these cases, a consulting agreement with a qualified local consulting firm or individual consultant can be the most appropriate way for the company to obtain essential services. Ordinarily, the consultant reports to the safety director of the company. There are numerous qualified industrial hygienist consultants, and many companies are taking this approach, as described in the following section.

Finally, the capabilities of insurance companies and of the consulting services offered by OSHA should not be neglected. The insurance company (usually the workers' compensation carrier) has a vested interest in improving the workplace environment, to reduce the potential for occupational illness. Many insurance companies have professional industrial hygiene staff members, who can provide knowledgeable assistance to the company. The difficulty in relying upon this source of assistance is that the service provided by the industrial hygiene consultant from an insurance company is necessarily limited. Similarly, the consulting services offered by OSHA in many states cannot fully duplicate the dedication and familiarity with operations offered by either a full-time or consulting industrial hygienist within the company.

Industrial Hygiene Consulting Services

The ideal circumstance for most companies is to have a full-time industrial hygienist on staff. However, because of economic factors, and the lack of fully qualified personnel, industrial hygienists can be employed in a consulting capacity.

There are drawbacks to this approach. The consultant industrial hygienist is ordinarily called in only when problems arise, and may not be as familiar with the facility and company personnel as a full-time industrial hygienist would be. The extent of this problem can be lessened by scheduling regular industrial hygiene consulting visits to discuss policy issues and inspect the facility during normal conditions. Some of these routine visits should coincide with visits of the medical consultant (many companies with industrial hygiene consulting services also have medical consulting services).

Several sources are available to assist in the quest for a competent and qualified industrial hygiene consultant. First is the semiannual listing of industrial hygiene consultants published by the AIHA. The AIHA is happy to provide this list to inquirers. Presence of a name on the list does not guarantee competence, however. For an industrial hygiene consultant to be considered, the minimum requirement should be certification (i.e., designation as CIH) and familiarity with the industry or other occupational setting of interest. As with other professional services, personal recommendations from satisfied users of the consulting services are often the best sources for information—enabling a differentiation between otherwise apparently equivalently qualified consultants.

In addition to the qualifications of the person being considered as a consultant, resources of his or her firm can also be important. If the topic of interest includes process technology, or other technical issues such as toxicology, then the consultant should have quick access to other specially qualified consultants. When quick turnaround of analytical results from air monitoring or other workplace monitoring is required, the availability of an in-house laboratory within the consultant's firm can be important. In any case, the laboratory considered for use should be accredited by the AIHA. This accreditation should be required regardless of whether the analysis required is one for which specific accreditation is offered. The AIHA accreditation includes evaluation of such general areas as quality control and record keeping, in addition to performance on specific analytes.

In some cases, trade associations have experience with consulting firms and can recommend those that are familiar with the industry of interest. Formal consulting agreements with the industrial hygiene consultant should include adequate time to discuss problems of mutual interest with the medical director and engineering staff.

PRACTICING INDUSTRIAL HYGIENE

Anticipating, recognizing, evaluating, and controlling occupational health hazards are the four definitive elements of industrial hygiene. Anticipating and recognizing health hazards are most important since they must take place before proper evaluation, or (if needed) control can take place. Upon anticipating or recognizing a health hazard, the industrial hygienist should be able to identify the set of measures necessary for proper evaluation. When the evaluation is completed, the industrial hygienist then is in a position, in consultation with other members of the occupational health team, to implement controls needed to reduce exposures to tolerable limits.

Anticipating Occupational Health Hazards

It is sometimes difficult to differentiate between *anticipation* and *recognition* of health hazards. The usual differentiation is between an existing facility or process, where hazards may be *recognized*, and a planned facility or process, where hazards may only be *anticipated*. Similarly, the line between *evaluation* and *control* may be blurred if the facility or process is not yet in existence, or if new technology is being evaluated, as discussed in the following section.

Recognizing Occupational Health Hazards

Recognizing occupational health hazards in an existing facility or process is the first step in the process leading to *evaluation* and *control,* and entails identifying potentially harmful materials and processes. This identification can be based on general knowledge of the characteristics of materials and processes, on clinical findings that disease or discomfort is present in the exposed population, on reports from others in the scientific literature, in bulletins from trade associations or governmental agencies, in conversations with peers, or from reports from workers.

However, the principal tool of the industrial hygienist for *recognition* of health hazards is the observation of the workplace. There is no real substitute for competent observation of the following:
- work practices used
- the extent of use of chemical and physical agents
- apparent effectiveness of control measures.

A health hazard does not exist in isolation from the workplace—a chemical or physical agent is necessarily only a danger if workers can be exposed. The industrial hygienist must use all of his or her senses to evaluate the workplace. In addition, discussions with workers and managers are extremely helpful in evaluating the potential for exposure.

Walk-Through Survey

The *walk-through survey* is the first and most important technique to recognize potential workplace hazards. A warning is in order here. A competently done walk-through survey, while it appears to be simple and straightforward, can be the most difficult of all industrial hygiene functions. The industrial hygienist must draw upon a substantial background of knowledge to evaluate observations. Any "survey" done without this background is more properly characterized as a *tour*.

The survey begins with a proper introduction to facility management, and discussion of the purpose of the survey. A clear understanding of the labor-management relationships within the facility is important, so that the industrial hygienist does not inadvertently make comments, or enter into discussions that will interfere with the orderly progress of any negotiations that may be going forward. The initial discussions should include a review of the history of the facility and of the processes used there, as well as a thorough discussion of any complaints that have arisen in the workplace. If appropriate, a simplified process flow diagram should be obtained (or prepared) at this time.

Following the process flow through the facility is often the most productive way to evaluate the workplace. The survey proper begins at the loading dock, to observe materials entering the facility. Warning labels, descriptive material on chemical composition of materials, and packaging of incoming materials should be noted. Questions regarding the handling of unknown materials, or materials about which insufficient information is available should then be answered. The pathway of the incoming materials should then be followed into the process flow stream, and each of the processes of interest in the facility should be visited. Of interest throughout the survey are the methods used for materials handling, as well as the effective labeling of materials, particularly at points where they are transferred from original manufacturers' containers into vessels for use within the facility.

At each point in the process, the industrial hygienist should observe the materials handling procedures in use, as well as any protective measures used. The controls that may be appropriate are listed in the discussion of control of health hazards. However, use of respiratory protection and protective clothing, and the apparent effectiveness of engineering controls as indicated by absence of characteristic odors of the chemicals in use, absence of visible dust accumulations, absence of loud noise, and other common sense observations should be recorded. The survey should continue through to the final product produced by the facility and its packaging. At each point, notes should be made regarding monitoring to be recommended or required.

In addition, the industrial hygienist should follow the pathways of any waste materials and determine their disposal sites. In addition to observing the nature of the processes in the facility, the industrial hygienist should also carefully note (where appropriate) the periodicity of processes. That is, whether the processes are *batch* or *continuous* is important in determining the opportunity for exposure to short-term but high concentration peaks at intervals throughout the process.

The number of employees at each step of the process should be noted, as well as any relevant data on their sex, ethnicity, or age that might affect their sensitivity to chemicals in the workplace. Look for obvious stigmata such as reddening or roughening of skin (as might be expected where inappropriate exposure to solvents has occurred), although the clinical observations of the physician or nurse on site are usually far more important in this regard. Where the situation permits, discuss work practices with the employees directly involved, since the perception of those practices is often very different on the shop floor than it is in the executive offices. Throughout the survey the industrial hygienist should convey a friendly and open demeanor so that workers feel comfortable discussing any concerns with him or her.

At the end of the walk-through survey, the industrial hygienist ordinarily has a closing conference with the facility management, where any obvious concerns are discussed, and any recommended follow-up measures are identified. Where the industrial hygienist is a regulatory agency representative, follow-up surveys can require special notices, and interaction with agency officials, as well as facility officials. In any case, a report on the walk-through survey, together with any conclusions and recommendations, should be completed for the record.

Data Review

An important part of the industrial hygienist's role in recognizing health hazards in the workplace (and anticipating health hazards from new processes) is review of the data. Such data can include reports from physicians on clinical findings that can be related to exposures in the workplace, as well as review of company records on materials coming into the workplace, which can have some significant health hazards. The current OSHA workers' right-to-know (Hazard Communication Standard) regulation requires the employer to inform workers of the nature and hazards of the materials to which they can be occupationally exposed. Where exposures are to materials purchased from a supplier, the data on materials and their hazards most often comes from the suppliers' Material Safety Data Sheets (MSDSs).

The industrial hygiene review of MSDSs and other forms coming from manufacturers of materials should include particular attention to any identifiable health hazards, as well as any recommended control measures. While recent MSDSs are informative, there will be substantial differences for the information provided by different manufacturers for the same (generic) materials. In addition, the MSDSs provided often are prepared by personnel without substantial health science background, and can represent merely a regurgitation of data from conventional sources, often outdated. The industrial hygienist will thus find it necessary to compare and to balance the recommendations made by manufacturers, to provide a unified program for control of similar materials regardless of their commercial source(s).

Currently, chemical manufacturers tend to stress protective measures more completely than previously. This effort by manufacturers is based on the technical concern that their materials were, in some cases, being misused, and from fear of litigation if they fail to recommend the most effective protective measures. In some cases, the recommended personal protective measures are unnecessarily complex—particularly where the chemicals are used in very small quantities. The industrial hygienist can recommend less restrictive personal protection than that recommended on the MSDS, if the combination of quantities used, inherent toxicity, process controls, and other engineering control measures combine to reduce exposures to acceptable levels. However; such reduction of protection below the levels recommended by the MSDS must be undertaken with great caution, to protect both the worker from unacceptable exposure and the industrial hygienist—or his or her employer—from unnecessary liability.

The industrial hygienist also should be aware of the commercially available MSDS compilations, which are often a useful supplement to the MSDSs provided by chemical manufacturers selling to the industrial hygienists' firms. The industrial hygienist should also carefully and routinely review current industrial hygiene, medical, and toxicological literature to determine if previously unsuspected hazards can attach to chemicals in use within the facility. Use of the commercial databases and literature files available through computer searching may be appropriate. Some industrial hygienists have routine standing orders for literature appearing on these data files for all chemicals to which workers in their facility could be exposed.

In some cases, the industrial hygienist will be faced with chemicals for which no reliable toxicological data are available. Extreme caution must be exercised in determining whether a hazard exists from the use of this material in the specific occupational setting of interest. An important consideration is that the worker must be appropriately protected, even in the face of uncertainty. If uncertainty exists, it should be resolved in favor of a greater standard of care. However; it is frequently extremely difficult to predict the potential human health effects of exposure to chemicals for which data exist only for nonhuman species. In such cases, the industrial hygienist must consult with qualified toxicologists and physicians for assistance in setting appropriate exposure guidelines or ensuring adequate levels of personal protective equipment are available and used.

The industrial hygienist is also concerned with process data from the facility, especially where chemical manufacturing or formulation are carried out. One of his or her main goals will be to determine what trace contaminants and by-products can be produced or exist during the processes used, and the hazards associated with those chemicals. This necessitates a complete understanding of the chemical processes being used, and frequent interaction with process personnel. The industrial hygienist should participate with the medical department in review of medical reports, with particular attention to such items as occupational diseases that may have arisen from occupational exposures.

Evaluating Occupational Health Hazards

Evaluating health hazards is the next function of the industrial hygienist. The evaluation phase is often the phase that requires the most time and careful attention to detail for the industrial hygienist. The Industrial Hygiene Survey Checklist summarizes the steps that an industrial hygienist takes when physically evaluating a workplace with a survey (Figure 6-1). Evaluation includes measuring exposures (and potential exposures), comparing those exposures to existing standards, and recommending controls, if needed.

An Industrial Hygiene Survey Checklist.

- ❏ Determine purpose and scope of study.
  - Comprehensive industrial hygiene survey?
  - Evaluate exposures of limited group of workers to specific agent(s)?
  - Determine compliance with specific recognized standards?
  - Evaluate effectiveness of engineering controls?
  - Response to specific complaint?
- ❏ Discuss purpose of study with appropriate representatives of management and labor.
- ❏ Familiarize yourself with facility operations.
  - Obtain and study process flow sheets and facility layout.
  - Compile an inventory of raw materials, intermediates, byproducts and products.
  - Review relevant toxicological information.
  - Obtain a list of job classifications and environmental stresses to which workers are potentially exposed.
  - Observe the activities associated with job classifications.
- ❏ Review the status of workers' health with medical personnel.
- ❏ Observe and review administrative and engineering control measures used.
- ❏ Review reports of previous studies.
- ❏ Determine subjectively the potential health hazards associated with facility operations.
- ❏ Review adequacy of labeling and warning.
- ❏ Prepare for field study.
  - Determine which chemical and physical agents are to be evaluated.
  - Estimate, if possible, ranges of contaminant concentrations.
  - Review, or develop if necessary, sampling and analytical methods, paying particular attention to the limitations of the methods.
  - Calibrate field equipment as necessary.
  - Assemble all field equipment.
  - Obtain personal protective equipment as required (hard hat, safety glasses, goggles, hearing protection, respiratory protection, safety shoes, coveralls, gloves, etc.).
- ❏ Prepare a tentative sampling schedule.
- ❏ Review specific applicable OSHA regulations.
- ❏ Conduct field study.
  - Confirm process operating schedule with supervisory personnel.
  - Advise representatives of management and labor of your presence in the area.
  - Deploy personal monitoring or general area sampling units.
  - For each sample, record the following data:
    1. Sample identification number.
    2. Description of sample (as detailed as possible).
    3. Time sampling began.
    4. Flow-rate of sampled air (check frequently).
    5. Time sampling ended.
    6. Any other information or observation which might be significant (e.g., process upsets, ventilation system not operating, use of personal protection).
  - Dismantle sampling units.
  - Seal and label adequately all samples (filters, liquid solutions, charcoal or silica gel tubes, etc.), which require subsequent laboratory analyses.
- ❏ Interpret results of sampling program.
  - Obtain results of all analyses.
  - Determine time-weighted average exposures of job classifications needed.
  - Determine peak exposures of workers.
  - Determine statistical reliability of data, e.g., estimate probable error in determination of average exposures.
  - Compare sampling results with applicable industrial hygiene standards and regulations.
- ❏ Discuss survey results with appropriate representatives of management and labor.
- ❏ Implement corrective action comprised of, as appropriate:
  - Substitution
  - Engineering controls (isolation, ventilation, etc.).
  - Administrative controls (job rotation, reduced work time, etc.).
  - Personal protection.
  - Biological sampling program.
  - Medical surveillance.
  - Education and training.
- ❏ Determine whether other occupational or environmental safety and health considerations warrant further evaluation:
  - Air pollution?
  - Water pollution?
  - Solid waste disposal?
  - Safety?
  - Health physics?
- ❏ Schedule return visit(s) to evaluate effectiveness of controls.

Figure 6-1. Industrial Hygiene Survey Checklist.

As noted previously, if the process or facility does not exist, direct physical evaluation cannot be performed. Thus, if anticipation, not recognition, precedes the evaluative phase, controls may be designed based on estimates of exposures derived from similar processes, or similar chemical unit operations, rather than from actual workplace measurements.

Exposure measurements are intended to be determinations of dose to the worker exposed. The mere existence of chemicals in the workplace, or even in the workplace atmosphere, does not necessarily imply that the chemicals are being delivered to a sensitive organ system at a dose sufficient to cause harm. Hence, the notion that exposure does not equal dose. The effective dose will be affected by such things as particle size of dusts in the air; the use of protective devices (such as respirators or protective clothing), and the existence of other co-contaminants in the workplace. The dose delivered to the worker can be further complicated by multiple pathways of absorption. Such contaminants as lead are absorbed readily both through inhalation and ingestion; and both routes of intake must be considered in evaluating the potential for harm. Similarly, many solvents are readily absorbed through the skin, and mere measurement of airborne levels is not sufficient to determine the complete range of risks associated with their use.

Sampling and Analysis of Airborne Contaminants

Although skin disorders via direct contact and subsequent irritation, represent the majority of reported occupational diseases, inhalation of airborne contaminants is still considered the major route of entry for systemic intoxicants. Passive dermal exposures to airborne contaminants are usually quite small in comparison to the potential dose received via inhalation. Thus, evaluating and controlling airborne contaminants is always an important part of any occupational health program.

Sampling and analyzing airborne contaminants is the function that is perhaps most well known as the definitive function of the industrial hygienist. While it is the responsibility of the hygienist to interpret the results of such measurements, measurement alone contributes to the awareness of hazards, as well as to their evaluation. Recent developments in instrumentation enable field measurements to be made of extraordinarily low concentrations of airborne chemicals, such that previously unsuspected contamination is being discovered. Examples include the discovery of measurable amounts of organic solvent vapors in the air of *clean* office buildings, where no significant chemical use takes place.

In some cases, these measurements, coupled with evaluations of health status of those exposed, have led to discoveries of connections between relatively low levels of airborne contaminants and health effects. The field of indoor air quality is one such general case. The determination of exposures to occupants of buildings (office workers) usually has not received substantial attention by industrial hygienists in the past, but apparently real health effects have been found at concentrations of contaminants well below those for which occupational standards have been established.

In addition, standards typically have been lowered, as both our ability to discern clinical effects and our expectations of no detectable health effects have been heightened. An example is found in the recent concerns regarding asbestos in buildings. It is generally accepted (from the standpoint of avoiding "fault" in litigation) that no avoidable exposure to asbestos should be tolerated, since it cannot be shown with certainty that a threshold exists below which all harmful effects of asbestos exposure can be ruled out. Further, background measurements of asbestos are common as part of asbestos abatement projects. Thus, measurements of asbestos concentrations down to and including ambient levels have become relatively commonplace.

There are two major general sampling approaches to determining airborne contaminant levels. In the first (personal, or breathing zone sampling), the industrial hygienist places a collection device near the breathing zone of the worker whose exposure is to be evaluated. The collection device can either be *active*—requiring that air is drawn through the collection device (Figure 6-2), or *passive*, requiring no pump or other suction source. The second approach (area sampling) employs sampling stations in the work area.

Personal breathing zone sampling is ordinarily the more desirable of the two approaches, since exposures are measured at the point nearest to the actual entry of airborne contaminants, and the sampling system moves with the worker. Thus, the measurements made are more likely to represent actual exposures. In addition, breathing zone sam-

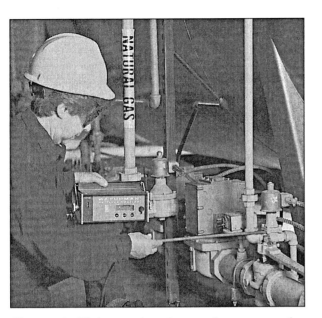

Figure 6-2. Worker monitors the area for presence of various gases in the atmosphere. (Courtesy Mine Safety Appliances)

pling is ordinarily required to show compliance with the OSHA *Permissible Exposure Limits* (PELs).

There are disadvantages to this approach, however. First, the volume of air sampled is limited by the capacity of the battery-operated pumps used (or the diffusion coefficient of a passive collection device). Thus, trace contaminants can be difficult to measure. Secondly, when complex evaluations are required, the number of collection devices may be too cumbersome for practical installation in the worker's breathing zone. In these circumstances, or when direct-reading instruments (usually larger and often requiring line power) are to be used, area monitoring can be used. Fixed monitoring stations can also be used to measure sources, background concentrations, or concentrations in several areas simultaneously to evaluate the effectiveness of controls.

Determining the *time-weighted-average* (TWA) exposure levels is most important. A thorough understanding of the process is important; and the potential variability of concentrations with time and different operating conditions should be identified before beginning the sampling. This ensures that all periods when employees can be exposed to a hazardous substance will be appropriately sampled for concentrations of the substances.

Time-weighted average exposure determinations should be made for the entire period of work to be evaluated. (OSHA requires measurement for at least seven hours out of an eight-hour workday if the potential exposure lasts the entire day.) In a continuous process environment or an assembly line type process, the period of exposure will usually be the entire workshift. The TWA exposure level is usually required to determine compliance with relevant standards (see following sections) and also can be useful for comparing exposures at various points within the facility.

In addition to TWA exposure determinations, a thorough understanding of the time-course of exposures is often needed. Although long-term, chronic diseases are more often than not the result of long-continued exposure, peak exposures can be important in acute effects, and can also be more directly important in long-term exposure than their relative contribution to a TWA would indicate. Further, in some cases exposures only occur for a relatively short period of time within the overall workshift. Such cases arise when a job is done intermittently or when a specific operation is only required at intervals during the workday or at periods of one or several days. Peak exposures can occur at times of maximal exertion (and respiration) and thus lead to unusually high inhalation of airborne contaminants. Peak exposures can be determined by taking an integrated sample for a relatively short time (the period of performance of a specific operation, or for 10–30 minutes at a time when maximum exposure is expected), or by using direct-reading instruments for real-time measurements.

A thorough discussion of sampling strategy (Hawkins et al, 1998) was presented by the AIHA Exposure Assessment Strategies Committee. The ACGIH (1995) has collected data on most of the available instruments and techniques for evaluating airborne contaminants, and NIOSH (1994) and OSHA (1994) have compiled generally accepted sampling/analytical methods for many common contaminants. These detailed references should be consulted for details on specific contaminants, and for guidance on setting up a professional air sampling program. The following material is introductory in nature.

Gas and Vapor Sampling
Gas and vapor sampling is accomplished by any of several methods. These can be divided into five major categories. The first is active collection by drawing a measured volume of air through a collection system, which is subsequently analyzed.

Second, air can be pulled through a device in which a color change in the collection medium is proportional to concentration of the contaminant, and which can be read directly. A third type of device is a passive dosimeter; which collects gas or vapor molecules by diffusion from the atmosphere, with the collection device subsequently being analyzed. A fourth type of collector is an evacuated container used to carry a sample of air to a convenient site for analysis. Finally, direct-reading instruments sensitive to one or several atmospheric gases or vapors can be used. Direct-reading instruments are particularly useful to evaluate gases and vapors, especially when the consequence of overexposure can be death—such as entry into confined spaces—and when the delay required for laboratory analysis is not tolerable (Figure 6-3).

Particulate Material Sampling

Airborne particulate material can be either solid (dusts or fumes), liquid (mists, fogs, or droplets), or biological (bacteria or fungi). The particles of health concern are usually those that are small enough to be inhaled (less than 10 microns), and are thus usually both small enough to remain airborne for long periods of time, and too small to be readily seen with the unaided eye.

Airborne particulate contamination can be measured either by collecting integrated samples with subsequent analysis, or by direct-reading instruments. Integrated sample collection and analysis is by far the more common modality of evaluation, both because of certain inherent difficulties associated with direct-reading instruments for particles, and because of the greater precision associated with laboratory analysis.

Modern particle sampling for solid or liquid particles is ordinarily done with filters that are subsequently analyzed for the concentration(s) of specific analytes. In the case of materials, such as asbestos, where the number concentration of asbestos fibers is the most important factor affecting the toxicity of the environment, the number of fibers on the filter is counted by microscopic techniques. In some cases, sampling is done with the aid of a size-selective sampling device (examples are *cyclones* and *impactors*) preceding the filter upon which the material is to be collected for analysis. Only those particles small enough to penetrate to and be retained within the respiratory space of interest (an example is the use of certain cyclones to discriminate against those particles too large to penetrate to the deep lung space) will pass through the selective device and be captured on the filter for analysis. Sampling for microorganisms is a specialized area, and requires specialized techniques and instruments (ACGIH, 1998).

In some complex environments, it is appropriate to use combined particulate and gaseous (or vapor) collection devices. This can be the case where a material exists in particulate form in the atmosphere, but has an appreciable vapor pressure so that substantial amounts may evaporate following collection on a filter. In this case, a vapor-sorbing material would be used following the filter to assure complete collection. Such a combined sampling is often used for effective collection of pesticides.

Surface Evaluation (Wipe Sampling)

Evaluating surface contamination can be a useful supplementary technique to evaluate exposure potential and the effectiveness of control measures. The approach also is useful in identifying contaminated areas, where a spill of a particularly toxic material has occurred. As an example, wipe sampling routinely is used to evaluate the extent of contamination resulting from spills of such materials as pesticides, and other materials for which absorption through the skin can be an important route of entry. However, surface sampling is ordinarily an indirect way to evaluate the potential risk of exposure by inhalation, and is relatively imprecise.

Surface sampling can also be a useful adjunct to programs evaluating effectiveness of housekeeping measures, particularly in manufacturing facilities where separation of manufacturing areas from cafeterias, offices, or dressing rooms is important. In such a setting, a diary of contamination levels can be kept, so that effective cleanup and engineering control can be evaluated on a routine basis. A typical program calls for the wipe sampling of identical areas on a monthly or quarterly basis. Surface sampling must be done according to a well-defined protocol if it is to be useful for long-term evaluations—isolated measurements are of little use in defining health risk. Most commonly, a template of a defined size (usually 10 cm x 10 cm) is prepared, and wiping or other surface sampling is done within the exposed area of the template, for the sake of uniformity.

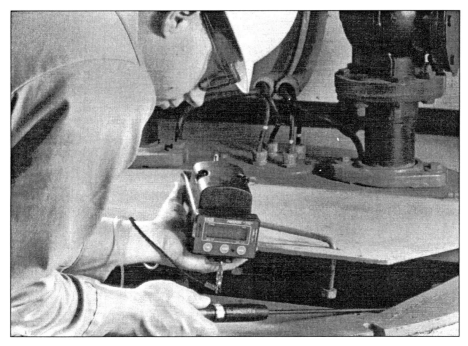

Figure 6-3. Direct-reading instruments are particularly useful to evaluate gases and vapors, especially when the consequence of overexposure can be death, such as entry into confined spaces—and when the delay required for laboratory analysis is not tolerable.

Physical Agent Evaluation

Evaluating physical agents requires specialized equipment that is often not routinely available, except sound-level meters required to evaluate noise exposures. In general, evaluating either ionizing or nonionizing radiation requires specialized training, and may be best left to a trained health physicist. However; many industrial hygienists have expertise in these evaluations and can make very accurate evaluations.

Evaluating exposures to noise is a traditional industrial hygiene function, and any trained industrial hygienist should be able to perform such evaluations. The equipment used for evaluations will be of two principal types. First (and most common), sound-level meters can be used. These devices consist of a microphone and associated electronic circuitry with a meter; which gives a readout in decibels. The circuitry typically contains filtering circuits, which permit evaluation of exposures to components of the noise spectrum weighted in accordance with their effects upon hearing. The *A-weighting network* has been functionally adopted as the standard for determining occupational noise exposure. In this weighting scheme, the very low frequencies and very high frequencies are suppressed, and the mid-range of frequencies (in the range of 1,000–6,000 Hz) is slightly accentuated. This gives primacy to the *speech frequencies* and reflects the importance of these sound frequencies in speech communication.

Sound-level meters can also be fitted with filtering circuits that enable determination of noise levels within discrete bandwidths. Often, one-octave or (less commonly) one-third octave bandwidth circuits are used. Such a device can isolate and identify the specific frequencies of occurrence of the noise. This enables identification of sources, and is essential to effective control in complex noise environments.

Another approach is use of noise dosimeters. In these devices, which employ a recording circuit, a small microphone is placed near to the ear of the worker to be evaluated, and his or her noise exposure during the day or period of exposure is recorded. The devices can either give an overall integrated average exposure for the course of the measurement period, or (as with the direct-reading instruments for gases and vapors and in combination with a data logger), can give a readout showing exposure as a function of time. Dosimetry is generally preferred to sound level meter monitoring since the exposures measured are specific and unique to the individual and offer the same advantage over area sampling as indicated above for breathing zone sampling for airborne contaminants.

Observing Work Practices and Process Variables

The quantitative measurement techniques discussed previously are of little use unless the industrial hygienist is able to put them into context of the actual work environment. Exposures often vary substantially from time to time during a day, week, month, or year. These differences often are due to changes in process configuration and through-put rates, as well as to individual differences among and between workers. The work practices used by those workers for whom exposures are measured should be observed during the monitoring period, and reported. The description of the workplace must include an adequate description of personal protective devices being used by the workers during the sampling period, so that an estimation of *true exposure* (that is, the actual intake of chemical into the worker's body) can be provided.

Operating conditions of ventilation equipment and other engineering controls must also be evaluated and stated clearly, so that the sampling results are placed into a sensible context. The workers and management in the workplace will be able to provide a reasonable estimate as to how closely conditions during the survey period approximate the usual conditions. Use caution with these estimates, since it is often difficult to relate true exposure to observed environmental conditions. Those contaminants most readily visible, e.g., dust or apparent odors, may or may not be related to exposures to other contaminants within the workplace.

The general conditions in the workplace, including such things as whether windows and doors are open or closed, must also be evaluated and recorded. The ideal industrial hygiene report will be sufficiently detailed so that another industrial hygienist entering the workplace several months or a year later will be able to determine whether conditions are the same or different than those that existed during the period previously surveyed.

Comparison to Standards

After the first phase of evaluation (the determination of levels of environmental contamination and conditions of exposure to chemical or physical agents), the industrial hygienist must determine whether the exposures measured are likely to cause harm to those exposed. If such harm seems likely, then action must be taken to reduce exposures to tolerable levels. (See the section on control.) In most cases, the industrial hygienist refers to a set of standards for various individual chemical contaminants or physical agents. He or she will conclude that exposures are acceptable if the measured concentrations are less than the allowable upper limit, and if exposures are unlikely to rise above that allowable upper limit.

The most important set of U.S. standards for industrial hygienists is the table of *Threshold Limit Values* (TLVs), which has been published annually by the Threshold Limit Values Committee of the ACGIH since the late 1940s. The TLV Committee reviews and updates the specific recommended exposure limits for each substance as new information is obtained. The TLV list is not only used in the United States; it is often adopted verbatim in other countries. In the OSHAct of 1970, the 1968 TLVs were adopted and given the status of law. They were named *Permissible Exposure Limits* (PELs). OSHA has made relatively slow progress in updating the PELs over the three decades since the original passage of the OSHAct. In 1990, OSHA attempted a wholesale upgrading of the PELs to reflect the accumulation of knowledge since 1968, by adopting revised standards recommended by the ACGIH, NIOSH, and others for over 400 chemicals. That attempt was declared illegal by a federal court and has produced substantial criteria regarding the legally applicable standards since the 1990 upgrade does not have the force of law. Therefore, the reader should carefully check and ascertain the *current status* of published standards. The OSHA website (www.OSHA.gov) is an excellent source of information.

The ACGIH also publishes a looseleaf binder (updated periodically) that states the basis for the specific TLVs (ACGIH, *Documentation for the TLVs and BEIs,* annual). The TLVs include values for chemical substances and physical agents (heat, ionizing radiation, lasers, noise and vibration, radio frequency and microwave radiation, ultraviolet and infrared radiation, and visible light). In addition, the book now contains a section on *Biological Exposure Indices* (BEIs) for a few chemicals for which well-established acceptable levels of the parent chemical or its metabolites in body fluids have been documented.

Despite the ACGIH warnings clearly enunciated in the beginning of the TLV booklet, some people consider the TLVs (and the PELs) as being levels at which all exposed are free of any risk. Nothing could be further from the truth. The TLVs have always been intended as guidelines for control of the workplace atmosphere to offer protection to *most of those exposed,* by health professionals with adequate training and experience in the field of industrial hygiene.

Too many personnel interpreting occupational exposure measurements (both industrial hygienists and others) have implied that exposures that are barely below the TLVs are acceptable. It has always been considered good practice to hold exposures to the minimum practically possible. No unnecessary exposure to any toxic material should be tolerated, unless it can be shown that no harm will arise from such an exposure. In some cases it is necessary, because of economic or engineering factors, to expose workers to levels greater than zero (ambient) levels. In such cases, the TLVs should be used as guides to the tolerable exposure levels. A competent industrial hygienist will attempt to hold exposures to the lowest level practically possible, or to a level at which risk is acceptable.

In recognition that a nonzero fraction of those exposed may develop disease as a consequence of their exposures at the TLV level, many organizations, both corporate and government, have established a policy of setting standards at some fraction of the TLV. In some cases, 10%, 25%, or 50% of the TLVs will be established as the internal control level. Some companies have gone so far as to attempt to remove all contamination from workplace atmospheres. Any detectable odor or irritation is considered to be unacceptable, and control measures are inaugurated to reduce exposures when any process effluvia are detected.

Where no established standards are available for guidance, a substantial amount of in-house research may be necessary to establish reasonable guidelines in the workplace. Where a chemical not previously used is being widely adopted in a particular industry, a trade association study of the effects of that chemical can be an appropriate venue for such research. Because of the potential risks associated with subtle health effects not easily foreseen, such control limits should be established only with great caution.

Several other sources of recommended exposure limits are available to an industrial hygienist. Among these are the Workplace Environmental Exposure Limits promulgated by the American Industrial Hygiene Association for several chemicals not listed by the TLV Committee. The various *Criteria Documents* of NIOSH are another set of valuable references in determining allowable exposures. In this set of documents, NIOSH has provided an evaluation of the existing literature, recommended control measures, and a recommended limit for exposure (REL). Because of budget constraints, NIOSH has not recently published any *Criteria Documents*.

Although many countries outside the United States adopt the ACGIH TLVs without substantial modification, several foreign countries have active committees evaluating allowable exposure limits. The International Labor Office (1991) has published, in tabular form, the occupational exposure limits for airborne toxic substances from all countries. This tabulation is very useful in identifying substances for which exposure limits lower than the TLVs might reasonably be established. Given this large body of published exposure limits, the reader must remember that in the United States, only published OSHA PELs are legally binding.

Occupational Epidemiology

Conducting occupational epidemiological studies is an increasingly important aspect of the industrial hygienist's involvement in evaluating health effects. While most industrial hygienists in industry may never be directly involved in an epidemiological study, the exposure data generated today may be used to evaluate the effects of such exposure many years in the future. The subject is complex, and has been the subject of substantial recent work. An introduction to the topic is given in Rappaport and Smith (1991) and other standard textbooks of epidemiology.

The industrial hygienist's roles in this effort are to record the processes and materials to which the population at risk has been exposed, and to record the exposures of the population. An important aspect of the industrial hygienist's task is to record, with adequate reference to published methods, the techniques used to make measurements, so that the exposures measured today can be compared with those in the past and future.

Controlling Occupational Health Hazards

Upon completing the evaluation, the industrial hygienist should be in a position to recommend appropriate controls, if needed. Recommendations on controls should take into account not only the conditions found during the survey, but also conditions that can reasonably be expected to occur in the future. Controls should be adequate to prevent unnecessary exposure during conditions of upset processes, as well as during normal operating conditions. Planned process modifications should be taken into account, and the controls recommended made as adaptable as possible to future process needs.

Consideration must be given to *failsafe* operation of controls. That is, recommended controls should always operate to protect workers, regardless of process fluctuations. There is a natural hierarchy of control approaches and the control approach within a given occupational situation should generally reflect this hierarchy.

Substitution

First, all possible avenues of *substitution* for the particular toxic material or agent under consideration should be explored. The reasons for this are obvious—toxic material can only cause harm if it is present and available for exposure to workers. However; use of this approach can lead to other hazards since there are several instances where an apparently "harmless" substitute for an obvious hazard was later found to be harmful itself. Nonetheless, substitution should be the first alternative to be explored. A useful substitute is one that is suitable for existing processes, or for which the processes can be relatively simply adapted.

Engineering Controls

Second in order of desirability as general control strategies are the *engineering controls*. Engineering controls include the general principles of *enclosure, isolation,* and *ventilation*. Engineering controls are desirable because they are built into the process, and can be made part of process equipment, thus reducing the need for human action in order to assure safety. Clearly, such controls as enclosure (building structures around the sources of emissions) and isolation (placing hazardous process components in areas with limited human contact) are desirable.

The use of *ventilation* for control of airborne contaminants is often a desirable approach. Both local exhaust ventilation and general ventilation are sometimes satisfactory approaches, but local ventilation is the more desirable of the two methods. The use of local exhaust ventilation control (that is application of the local exhaust near to the source) follows the general rule that control should be implemented as near to the source as is practically possible. Thus, application of a local exhaust inlet on a specific tool would be more desirable than performing the grinding operation in a ventilated hood. In turn, the ventilated hood would be more desirable than installing general ventilation in the room where the grinding is performed.

In a situation where a particularly toxic material is being manipulated in such a way that exposure is possible, all three ventilation systems might be reasonable to use. Thus the operator would be protected by the ventilation of the specific tool, nearby workers (as well as the operator) would be protected by the hood, which would capture any materials escaping the tool ventilation control and the remainder of the building would be protected by the general ventilation system to restrict passage of air from the workroom to nearby offices, lunchrooms, etc.

Design of ventilation systems for contamination control should be left to engineers with specific background or experience in this field. Unfortunately, some companies rely upon local sheet metal shops to design such systems, and the consequences are often improper design. An industrial hygienist without an extensive background of engineering training and experience in the process to be controlled can also produce an unsatisfactory design. An engineer without industrial hygiene support will often produce a system inadequate for control, while an industrial hygienist without process engineering experience will often produce a design adequate for control, but nearly impossible to use in the process flow.

As mentioned earlier; the ACGIH produces *Industrial Ventilation: A Manual of Recommended Practice,* a document on industrial ventilation. This document provides guidance to the industrial hygiene engineer on approaches to use in specific circumstances, as well as general guidance on ventilation control principles.

In addition to the general engineering approaches of ventilation, enclosure, and isolation, some very specific engineering controls can be appropriate in the specific process environment. For example, it is often necessary to design process pipelines and valves such that splashes and ejection of toxic chemicals are made unlikely. Control systems that will enable a safe and orderly shutdown of the process to avoid such disasters as runaway reactions can also be a substantial benefit to the overall occupational health and safety status of the facility. These elements of safe process design can only be accomplished with the wholehearted cooperation and collaboration of process design and operation engineers.

Controls Over Human Behavior
Third in the hierarchy of desirable control approaches is the category of controls involving human behavior. These can be subdivided into the general categories of *administrative controls* and *work practice controls*.

Administrative controls. Administrative controls imply management control of behavior patterns within the process environment, including such things as establishing prohibited areas, areas where smoking and eating are either prohibited or allowed, and safe pathways through the work environment. Administrative controls also include such control measures as work scheduling. Work can be scheduled in such a way that particularly dangerous operations are carried out when the minimum number of workers in nearby areas are present. This is clearly a sensible procedure to follow.

The practice of scheduling individual workers to perform tasks for short periods, where excessive exposures would be incurred over an extended period of time, is less desirable, at least where exposure to carcinogenic agents or agents with reproductive effects is concerned. While the individual risk may be relatively low, the notion of distributing an exposure with potential genetic effects to many members of the population is inherently unsound.

However, administrative controls including scheduling are usually essential to control of the work environment. As an example, employees who do not have adequate training should be prohibited from entering spaces where health or safety hazards exist. Administrative controls are necessary to assure that they do not do so.

Work practice controls. The second behavioral control approach, *control of work practices,* implies control over the behavior patterns of individual workers in the performance of their job(s). Such details as regulating the handling of contaminated tools and appliances are included. Education (on the hazards to be avoided) and training (on the desired practices) are always required in order to make work practice controls effective in the workplace. Substantial first-line supervision of workers is needed to enforce satisfactory compliance with desired work practices. Controls on work practices are particularly important where engineering controls are neither adequate nor possible and where airborne contaminants can be generated outside of controlled spaces.

These general approaches to control workplace exposures rank third (below the more generally applicable substitution and engineering controls), because of the need to regulate human behavior. This can be a difficult problem—constant management attention, education of managers and workers, and specific training in techniques and approaches are necessary if these general approaches are used. The rules to follow are often contained in health and safety handbooks; on posters in work areas, change rooms, and cafeterias; and in employee training booklets, which can be handed out at the end of safety meetings. Video-based packaged training is another effective method.

Personal Protective Equipment
Fourth (and last) in the hierarchy of general approaches to control is the use of *personal protective equipment*. Personal protective equipment (although often essential) is less desirable than the aforementioned approaches. It is difficult to ensure that it is properly fitted, used, and effective. Examples of personal protective equipment that are commonly used are the basic protective devices that are seen (as an example) on construction sites—hard hats and safety shoes. In laboratory environments, the use of protective eyewear is common, as is the use of protective garments, such as laboratory coats. The obvious function of these garments and devices makes their use fairly straightforward. Respirators are commonly used personal protective devices in a variety of occupational settings. Figure 6-4 shows the application of a self-contained breathing apparatus (SCBA) in the field.

However, a simplistic approach to using personal protection can result in unsuspected exposures. This is because there are significant complexities in both design and function of the protective devices used to reduce exposures.

Figure 6-4. Respirators are commonly used personal protective equipment in a variety of occupational settings. This photo shows a chemist using a half-mask respirator as she analyzes materials in a hazardous environment. (Courtesy Mine Safety Appliances)

In the area of respiratory protection, a worker who is issued and wears a respirator may feel that he or she is now adequately protected from all potential hazards in the workplace. This employee may therefore not use engineering controls, violate administrative control guidelines, and ignore required work practices. In fact, without substantial attention to selecting, fitting, training, and maintaining respirators, exposures during their use can be nearly as high as those of "unprotected" workers.

In too many cases, respirators are handed out without adequate attention to any of these items. It is common to see workers with substantial facial hair wearing air-purifying respirators (i.e., those using a cartridge through which air is drawn on inspiration) and working in environments where elevated concentrations of contaminants are found in the air. If the wearer has significant facial hair; it is nearly impossible to obtain an adequate fit for such a respirator.

Similarly, using protective gloves to avoid exposure to solvents must be guided by a thorough understanding that many common solvents readily penetrate many commonly used glove materials, particularly after prolonged contact. Therefore, carefully selecting glove materials is essential. In addition, prolonged wearing of gloves into which materials hazardous to the skin have either leached or leaked through holes can result in a substantial exposure to the worker (sometimes higher than would be found without the gloves), because the material is held in close proximity to the skin for long periods of time.

Relying on personal protective equipment alone, without an adequate overall control program is very expensive, if that personal protection is to be adequate. The goal of a control program is to reduce true exposures of workers to tolerable levels. The expense of using personal protective equipment is less clearly visible than the (sometimes considerable) expense for designing and installing engineering controls. But it arises in the substantial management time and effort required to both design and maintain the personal protection program and in the (usually) reduced efficiency and productivity of workers encumbered by personal protection. Other important aspects of control are adequate housekeeping and proper disposal of waste materials. The simplest forms of control can often be the best.

A well-regulated control program in a company with diverse operations nearly always uses all of the previously mentioned methods of control. Substitution should be the first consideration. When substitution cannot be rationally adopted, isolate workers from exposure and enclose hazardous sources. If no suitable substitute material is available, and if complete isolation and enclosure are not possible, consider local exhaust ventilation. General exhaust ventilation is a useful supplement to local exhaust ventilation and ordinarily should be part of the ventilation design.

When none of these engineering controls can completely abate the hazard, then administrative controls, work practice controls, and personal protection may be necessary. In most circumstances, several of these approaches to control will be appropriate.

The control process must be viewed as a continuing one, in which existing controls are continually evaluated for their effectiveness. Equipment ages, workers change, processes evolve, and level of management attention to control varies with time. All of these forces change the effectiveness of a given control. The evaluation of effectiveness is the province of the industrial hygienist, but he or she must involve managers, engineers, and workers in the evaluation.

SUMMARY

This chapter was a brief introduction to a complex and fascinating field. The industrial hygienist must be acquainted with engineering principles and practice, analytical chemistry, toxicology, clinical medicine, epidemiology, statistics, and law, as well as topics specific to industrial hygiene. The field of industrial hygiene has grown significantly during the past 20 years, and will continue to grow in the future, since there is no sign that all occupational health problems are likely to be solved anytime soon. There are few fields in which you can find an equally satisfying blend of technical challenge and satisfaction that you have helped people to avoid harm.

The National Safety Council text *Fundamentals of Industrial Hygiene* (Plog et al, 1996) is a useful, more detailed summary of the fundamentals of this field. Interested readers can obtain information on additional educational opportunities in the field of occupational health in general, from the Educational Resource Centers funded by NIOSH. These are located at universities across the country. Education, in the form of short courses as well as undergraduate and graduate education, is available from these centers. The AIHA, ACGIH, and NIOSH also offer various short courses in topics of interest at various nationwide locations.

REFERENCES

American Conference of Governmental Industrial Hygienists (ACGIH). *Industrial Ventilation: A Manual of Recommended Practice*, Cincinnati: ACGIH, published biannually in even-numbered years.

ACGIH. *Air Sampling Instruments for Evaluation of Atmospheric Contaminants,* 8th ed. Cincinnati: ACGIH, 1995.

ACGIH. *Threshold Limit Values for Chemical Substances and Physical Agents and Biological Exposure Indices.* Cincinnati: ACGIH, issued annually.

ACGIH. *Documentation of TLVs.* Cincinnati: ACGIH, published/updated periodically.

ACGIH. *Bio Aerosol Sampling.* Cincinnati: ACGIH, 1998.

American Industrial Hygiene Association (AIHA). *Workplace Environmental Exposure Limits.* Fairfax VA: AIHA, published/updated periodically.

Hawkins NC, Norwood SK, Rock JC (eds). *A Strategy for Occupational Exposure Assessment,* 2nd ed. Fairfax VA: AIHA, 1998.

International Labour Office (ILO). *Occupational Exposure Limits for Airborne Toxic Substances,* 3rd rev. ed., Occupational Safety and Health Series No.37, Geneva: International Labour Office, 1991.

National Institute for Occupational Safety and Health (NIOSH). *NIOSH Manual of Analytical Methods,* 4th ed. NIOSH Publication No. 84-100. Washington DC: U.S. Government Printing Office, 1997.

Occupational Safety and Health Administration (OSHA). *OSHA Analytical Methods Manual.* Salt Lake City: OSHA Analytical Laboratory, 1991 & 1993. (Available through U.S. Government Printing Office, Washington DC.)

Plog BA, Quinlan P, Niland J (eds). *Fundamentals of Industrial Hygiene,* 4th ed. Chicago: National Safety Council, 1996.

Rappaport SM, Smith TJ (eds). *Exposure Assessment for Epidemiology and Hazard Control.* Cincinnati: ACGIH, 1991.

Part 3

Occupational Health and Safety Programs

chapter 7

Occupational Health and Safety Program Design

by Michael Larsen

109 **Objectives**
110 **Introduction**
Role of health and safety in the workplace ■ Role of the health and safety professional
110 **Trends in Workplace Demographics**
Aging workforce ■ Workplace diversity ■ Language ■ Culture ■ Our litigious society ■ Pace of change ■ Managing stress ■ Role of technology
112 **Program Considerations**
Designing a proactive program ■ Acceptable level of risk ■ Communication strategies ■ Driving the program to the individual level ■ Fitting the program to the organization ■ Assessing program needs ■ Setting clear and consistent goals ■ Setting the metrics ■ Protecting the bottom line ■ Managing productivity ■ Records-retention strategy ■ Continuous education programs
115 **Integrating Quality into Program Function**
Quality management requirements
115 **Accountability**
Operational accountability ■ Program accountability ■ Design considerations ■ Managing audit results ■ Legal documents
116 **Summary**

OBJECTIVES

Companies in every line of business have an inherent responsibility to their employees. This responsibility is stated clearly in the enabling language of the Occupational Safety and Health Act Section 5(A)(1). It states:

> 5. Duties (a) Each employer -
> (1) shall furnish to each of his employees employment and a place of employment which are free from recognized hazards that are causing or are likely to cause death or serious physical harm to his employees;

It is under this sweeping mandate that individual companies must design and operate their occupational safety and health programs.

The purpose of this section is to provide guidance to the architect of these programs, to assure the program rests on a solid foundation. This chapter introduces the role of the safety and health program and the individuals responsible for carrying out program functions.

Trends in workplace demographics illustrate the ever-changing picture of the U.S. work force. The health and safety program designed even a few years ago may no longer meet the needs of the organization today. A rapidly growing organization may need to reconsider program aspects frequently, as conditions fluctuate. An old management rule-of-thumb states that every time a company grows by more than 40% a new administrative structure is required. Continuous improvement will be a theme throughout this chapter.

The section on program considerations asks some questions, though difficult to answer, will be critical in the overall design of the program. Some of the issues discussed may suggest a paradigm shift will be required from old style (command and control) management thinking.

Quality control issues are discussed in relationship to program credibility. And for those companies where the paperwork documentation level is high, alternatives are discussed so the manager is assured that program objectives are being met.

This chapter provides guidance to companies generally smaller then 500 employees, in all Standard Industrial Classification codes. In some cases the information may be too general and in others too detailed. However, the objective is to provide enough information to all sizes and types of companies that will assure the right questions are asked; and that the answers to those questions can lead to the formation of proper program elements and a system in which those elements can successfully operate.

INTRODUCTION
The roles of the health and safety function as well as those who design and effect the program are discussed in the following sections.

Role of Health and Safety in the Workplace
The health and safety of employees is a vitally important function within an organization. It is sometimes difficult for safety and health professionals to keep a perspective while tightly focused on their jobs.

Depending on the organization's management style, the health and safety function can be seen from several points of view. One viewpoint is that health and safety is a *necessary evil* that must be tolerated while struggling to meet production goals. The other end of the spectrum is equally distorted in that a safe workplace is the only goal (while we produce a product). While the first viewpoint may be more prevalent, the issue is generally distorted in some manner.

Some observe that health and safety programs are equally as important to the well-being of an organization as the accounting function. But a comprehensive health and safety program has dual objectives. The first objective is to contain costs related to injury and illness. The second is to maintain productivity. Employees cannot perform at peak efficiency if they are uncomfortable, in pain, or if they are worried about illness or physical injury.

Like the health and safety program itself, employers also have dual objectives. Employers have a legal responsibility to their employees through the OSHAct; and they hope to limit lost incidents, injuries and the potential of lost workdays. The cost of accidents, illness, and loss of productivity will jeopardize the company's fiduciary responsibility to its stakeholders.

Role of the Health and Safety Professional
Today's health and safety professional wears many hats. Not only are they expected to be a coach, psychologist, lawyer, engineer, investigator, and scientist; but often they are responsible for unrelated collateral duties. In the area of health and safety, the requirements are clear. They must assist management in understanding the nature of the risks involved in the business process, and craft and execute a program that will mitigate those risks in a cost-effective manner acceptable to management and staff. While this can seem a Herculean task, it is composed of several manageable building blocks. In crafting solutions for the challenges on some of these blocks, here are some design considerations.

TRENDS IN WORKPLACE DEMOGRAPHICS
It does not matter how long the company has been in business. Changes in the work force, technology and other demographic factors have worked together to create a different organization than even a few years ago. Consider the following factors.

Aging Workforce
Our work force is growing older. The Baby Boom Bubble has now passed and smaller numbers of young people are entering the job market. Finding qualified workers is becoming more difficult and competitive, as large numbers of new companies actively seek the same workers that the more estab-

lished companies seek. With smaller numbers of younger workers, many companies find the average age of the work force is increasing. This demographic factor has both positive and negative aspects that need to be considered in a health and safety program.

The positive aspects include the following:
- Older more experienced workers tend to have fewer accidents, a fact demonstrated repeatedly in studies.
- More experienced workers carry institutional knowledge of solutions that have worked in the past and those that failed.
- More experienced workers tend to work within established policies and procedures.

The negative aspects include the following:
- Older workers tend to have more chronic disease processes, which could be adversely affected by a manufacturing process or process chemical.
- With age, strength, reflexes, and hand-eye coordination tend to decrease.

What this information suggests is that the older worker is an increasingly valuable resource that may require additional ergonomic considerations to function fully within an organization.

Workplace Diversity

As the number of native-born North Americans entering the job market decreases, the number of immigrant workers is increasing dramatically. Today we see fewer European immigrants and more from the Pacific Rim and Latin American countries. While these immigrants bring a wealth of skill and talent, they also bring issues that must be addressed in the context of a health and safety program. The most important issue is language.

Language

Regardless of the skill or talent of the non-English-speaking worker, health and safety requirements must be conveyed in a clear and unambiguous manner. Training classes must be structured so all workers clearly understand the rights and responsibilities required in their jobs. Even when using a translator, do not assume all workers from Latin American or Asian countries will understand the specific dialect of your translator. Dramatic differences in meaning exist across similar Hispanic and Asian languages and, therefore, leave much room for miscommunication to occur.

Culture

With the passage of the Religious Freedom Restoration Act in 1993, some companies find the need to accommodate religious needs of some workers. The Occupational Safety and Health Administration has made one exemption for Amish workers to wear hard hats, other less formal exemptions are frequently made. Cultural issues also arise in the context of diet, holidays, working hours, and the manner in which the worker relates to management and workers of other gender.

Our Litigious Society

In the past, workers' compensation was the only recourse an employee had, by law, to an occupational injury or illness. The workers' compensation system was intended to shield companies from legal action by their employees regarding injury and illness. This legal requirement exists today in approximately the same form as when it was passed by the U.S. Congress and various states. Recently, however, the shield has been eroding. In some states, employees have been permitted to sue the employer, if some level of management in the organization knew of a hazard, and took no steps to correct the hazard. The Occupational Safety and Health Administration defines this as a *willful violation*. Other ways companies can be sued over health and safety conditions include third-party suits (by spouses or relatives) or the widely publicized mass action suits such as those involving companies that worked with asbestos.

Pace of Change

One thing noticeable to everyone is the pace at which our workplaces change. This is brought about by the increasing reliance on technology. New means of communication, including e-mail, pagers, cell phones, and group communication software such as Lotus Notes have forever changed traditional work practices. The Internet, company networks, and enterprise software have increased productivity dramatically, in some cases. Dramatic restructuring of organizations has taken place after reengineering and right-sizing. In many cases organizations are left with completely new operating structures and the same employees as before. By the nature of what has happened, these employees may be under extreme stress. Gains achieved through technology can be lost quickly by stressed employees.

Managing Stress

Stress, according to Hans Selye, a Canadian considered by many the father of stress theory, is the nonspecific reaction people have to change. The more change that occurs, the more stress is experienced. While a debate goes on about "good" stress or "bad" stress, one thing is clear; stress is linked in some way to most major disease processes. The best example may be the link between heart disease and the intense type A personality. On-the-job stress is now recognized as an illness covered by workers' compensation in many states.

Why is stress bad? Stress creates a flood of body hormones that are intended to do a number of things. First, it alters the entire body chemistry and invokes the *fight-or-flight response* (to one degree or another). In this type of stress response, the body releases adrenaline, levels of insulin fall, bloodflow diverts to extremities, hormones that speed blood clotting are produced, and other steroids that suppress inflammation are produced. In short, stress leaves a distinct mark on the body each time it occurs. An objective of each health and safety program should be to manage stress within a working population just as carefully as a chemical exposure would be managed.

Role of Technology

Computer technology can make today's health and safety practitioner more effective and productive. In the past, it was not unusual for a health and safety professional to spend 50% of his or her time searching for information. Locating items such as previous reports or test results, regulatory requirements, correspondence, or previous requests would consume a large amount of time because of the volumes of paper involved. This created stress on the practitioner because while looking for information, more pressing issues were not being addressed.

Today the technology with its modern databases, scanners, contact information, and program management software, holds the promise of easier access to information. Off-the-shelf program management software and easily customizable information management software are now available for small programs at modest prices.

PROGRAM CONSIDERATIONS

The goal in program design is to create a *proactive* occupational safety and health program, while avoiding a *reactive* program design. This is described in the following material.

Designing a Proactive Program

What is a *reactive program*? This type of program waits and compiles statistics on the effectiveness of the program. When the statistics suggest something is not working optimally, the program reacts to make a correction. This is also referred to as the *body-count method,* because no improvement occurs until something happens. At that point a valuable member of the team can be injured and unavailable to participate, damaging overall productivity. Workers' compensation and insurance premiums also can be adversely impacted. It is a costly way to manage a program in terms of money and human well-being.

The *proactive program* looks for every means available to prevent an incident. This type of program seeks out advice on better and safer means of performing the work. Every organization has experts on their work right there in the building. These are the employees who do the work everyday. Generally speaking, the people who perform the work daily know best how to do it and how to make improvements.

Note that every program needs a mechanism to acquire this input. These three levels of input are as follows:

- We can call this first level *general health and safety consciousness*. Employees need to be aware that performing their work safely and efficiently is their top priority.
- The second level of input is the *need to capture and understand unplanned events,* regardless of whether an injury or illness occurs. Often referred to as *near misses* (or should they be called near hits?), these events can provide critical information on the effectiveness of discrete program elements. Assessing the potential for these unplanned events to cause injury or illness is central to improvement of a program.
- This leads to the third level of input, which is *the results of a thorough and complete investigation of each unplanned event.* An investigation that delves deeply into such an event should have as an objective, the *root cause.* Finding the root cause of an event can lead to systematic changes in management and program structure, not just a superficial change in operations.

Acceptable Level of Risk

An important question that is critical to the overall design of a program is often never asked. How much risk is the organization willing to accept? The OSHA standards, which often are viewed as minimum levels of acceptability, generally look at risk levels of one event in one thousand. The EPA and other organizations involved in public health look at more conservative levels of risk, one event in ten thousand or one million. The more conservative the level of risk tolerance, the more highly engineered and (often) costly is the solution. With a payoff of lower insurance premiums and (sometimes) higher productivity, a balance is needed in program design. Many risk-assessment models are available to assist in determining the level of risk inherent in the specific design of high-risk operations.

Communication Strategies

An effective communication strategy is absolutely critical to the operation of an effective health and safety program. In all human endeavors, it is natural to expect differing opinions in the creation of any program. These differing opinions are normal and should be looked at not as a source of dissention but a potential treasure. Individuals with differing opinions merely mean they perceive the problem from different perspectives, which is valuable to the program planning phase. The health and safety program should be designed in such a manner to capture these differences of opinion at the earliest possible moment. If not caught early, a risk is run of escalation to disagreement and on to dispute, from which negative consequences may flow. If a system is in place to take a difference of opinion and explore each perspective in a nonthreatening manner, the best of each perspective can be derived with the active participation of all parties, before a dispute elevates emotional levels.

Driving the Program to the Individual Level

The active participation of each individual in the business is important. The answer to the most vexing health and safety problem in the facility can just as easily be held by the cleaner who sees all parts of the facility as the general manager. It should be the responsibility of each individual to be as fully engaged in the health and safety program as possible. When each individual employee feels responsibility for his or her co-workers' safety, it can create a powerfully motivated work force. Effective safety programs do not happen in the front office. They happen when individuals are empowered to make decisions and are permitted to make a difference.

Fitting the Program to the Organization

All companies vary dramatically in customer base, cash flow, profit margins, and growth. This, in turn, requires a wide variety of methods with which to meet requirements and provide a safe and healthy workplace.

Assessing Program Needs

A complete inventory of environmental, health, and safety risks is necessary to determine the essential program elements. This inventory must include the following:

- chemical exposure through inhalation, skin contact, and ingestion potential
- safety risks including walking/working surfaces, fire and life safety, material handling, machine guarding, safety of tools, and welding and electrical safety
- biological risks including bloodborne pathogens, mold and bacteria, and other infectious diseases
- ergonomic factors, fitting the worker to the work.

Understanding the risks presented by these potential hazards is the most important question to be answered. The presence of a hazardous material or a specific machine in a facility and the risk posed by the work practices are two different issues. Determining the risks from hazardous materials or specific work practices can require the input from occupational physicians, industrial hygienists, and safety professionals. This input also will assist in determining staffing needs on a day-to-day basis.

A board-certified occupational health physician or a physician familiar with the hazards associated with your business should review the corporate program. The physician's opinion will be valuable in determining if your program would benefit from a review by an industrial hygienist or safety professional.

Setting Clear and Consistent Goals

The design and operation of a program should be guided by goals determined by your management team. General goals will assist in the selection and design of overall program elements. Specific goals can do the same for various aspects of program operation. Only simple rules apply to goal setting. The goals can be changed as needed, but the description should always be stated clearly and consistently within the context of the program. Everyone should be informed of the program goals, how they effect them, and the role each person plays in the achievement of that goal. Constant feedback is critical to maintain forward momentum.

Setting the Metrics

"If it cannot be measured, it cannot be managed" is another old management axiom. In order to manage your program, *accurate metrics* must be established and communicated to appropriate management. Some words of caution are important. Metrics must be selected carefully, keeping in mind the *"Law of Unintended Consequences."* Experienced managers are skilled at managing to numbers. Other metrics may suffer if the right parameters are not selected. It must also be determined how the metrics will be measured, and how the manager will be held accountable. Unless senior management of an organization takes seriously the health and safety of employees, lower-level management will not. Consequently, the evaluation of managers' health and safety metrics must have a meaningful affect on their performance evaluation. It will have an even greater effect if it is tied to a performance bonus or merit pay increase.

Protecting the Bottom Line

Keep in mind the function of the health and safety program. It protects the workers who are the lifeblood of the organization. In many knowledge-based industries, the workers are the intellectual capital of the organization. In every organization, each employee contributes in his or her own important way to the health and well being of the organization. The organization has a responsibility to protect the health and well being of its workers.

Many costs are associated with the establishment of a work force. Some of those costs are directly related to the effectiveness of the health and safety program. Medical insurance costs (that part paid by the company, if any), workers' compensation, and general liability insurance premiums are some of the costs directly related to the program. Indirectly related costs include loss of productivity from increased sick time, and increased employee turnover. Productivity plummets when workers are in pain from ergonomic factors, such as poorly set up computer workstations or excessive repetitive upper body motions. The function of a health and safety program is to protect the organization's human capital, intellectual capital, and financial capital.

Managing Productivity

Every company has units of work to sell to its customers. These units will vary widely in nature but workers are required to produce all output. The fewer workers required per unit, the more competitive the company becomes. This is one of the strongest measures of productivity in the workplace. Maintaining high productivity is one of the functions of a comprehensive health and safety program. Productivity is the first victim of high stress in the workplace. Some of the factors that affect stress include:

- physical comfort
  - thermal (heat and cold) stressors
  - poorly fitted tools (ergonomic stressors)
  - strong odors, nuisance dust,
  - high noise levels.
- psychological factors
  - understanding what is required
  - understanding the level of quality of the output
  - has the proper tools to do the work
  - has the time to do the work properly
  - has an opportunity to have his or her views heard
  - is clear on what others expect.

While psychological factors clearly are a function of management, comfort factors fall under the health and safety program. A well-managed, comfortable worker can work faster.

Records-Retention Strategy

One aspect of a health and safety program that is often overlooked is the plan for record retention. If one understands that all documentation derived from the operations of the program may someday find its way into a court of law, it is clear that these records should be treated as legal documents. In that light it is important to document compliance with your organization's overall programs goals and objectives as well as regulatory requirements.

The types of records your organization retains (e.g., full reports, summary, field notes) should be discussed as a legal issue affecting the organization.

Certain chemicals have prescriptive record-retention requirements pertaining to industrial hygiene and medical records. States operating their own OSHA plans can have additional requirements.

In addition to considering the regulatory requirements, your organization should decide what records to retain, in what format, and for how long. Those decisions should be incorporated into a policy statement within the context of the health and safety program.

Electronic (paperless) filing systems and the old standby, the filing cabinet, will mean nothing unless a sound system is in place for filing and retrieval of reports. While this may sound obvious, many companies spend vast sums of money to test, evaluate, and inspect their environment, and then are unable to locate these inspection reports when asked. When the reports are lost, all money spent on the inspections, analyses, and report production has been wasted. It is as if no test or evaluation has ever been performed.

Continuous Education Programs

It is a common view that things today change at a faster rate than ever before. The rate of change has become so great that some scientists estimate that 20% of everything we know becomes obsolete every year. From a practical side, conservative politicians will quickly tell you how many pages are added to the *Federal Register* each year in the form of additional regulatory requirements. Pressure from the government and the public is increasing the need to reduce pollutants to the air, water, and landfill. In addition to that, technology churns out faster, (perhaps) better, and (hardly ever) cheaper ways to produce your units of output. This creates the need for a very flexible safety, health, and environmental program, able to respond to a rapidly changing workplace environment.

The need to adapt is becoming a highly desired skill. Managers of health and safety programs need to be constantly educated about issues involving every process change and capital "improvement." (Stories abound about companies switching to a chemical or process they thought was more safe, less toxic, or better in some manner, only to find out they have introduced a far more dangerous operation into the facility.) When speaking from a base of knowledge and fact, health and safety concerns will generally be well received.

INTEGRATING QUALITY INTO PROGRAM FUNCTION

Quality management means nothing more than a means of assurance that the program is doing what it is intended to do. It is a critical means of providing assurance the reports generated are reliable. It is felt by some to be so important that one of our major automobile manufacturers has stated that "Quality is Job 1." Indeed, without a quality assurance program in place, your organization's reports will be meaningless in court. Without assurance that the report findings have been generated in a standardized manner, with calibrated instruments, the report will be dismissed as worthless.

Quality Management Requirements

Quality assurance management is part of OSHA's prescriptive requirements for specific hazardous chemicals. In standards regulating use of the chemical formaldehyde, for instance, methods used for employee exposure evaluation must meet specific statistical criteria. Highly hazardous chemicals, falling under the Process Safety Management Standard (29 *CFR* 1910.119) have extensive requirements for quality control.

In addition to these specific requirements, other data collection efforts must have similar quality control measures incorporated to establish credibility for the results. One model to consider is the criteria used by OSHA compliance officers to collect data for compliance purposes. Detailed instruction on these procedures can be obtained from the *Technical Manual* at the OSHA website.

ACCOUNTABILITY

Every plan must assign responsibilities for the operational aspects of the program. The determination of how well these responsibilities are carried out is one aspect of accountability.

It becomes even more important in this period of rapidly changing technology, to determine if all conditions of the program still are pertinent. Two levels of accountability will be discussed: the operational accountability in an organization and the health and safety program.

Operational Accountability

The operation of a health and safety program can range from simple to complex, with supervision ranging from simple administrative oversight to several layers of management. Periodically, it must be determined if the program is functioning as intended. This is usually accomplished through a program review or audit.

The audit is intended to answer, at a minimum, the following questions:
1. Is the organization in compliance with regulatory requirements?
2. Is the organization following company policy and procedure?
3. Are improvements in the program or administration possible?

Program Accountability
Periodically, the structure and operation of the health and safety program should be reviewed. Again, the purpose is to determine if the goals of the program are being met. At a minimum, the audit should be designed to answer the following questions:
1. Does the structure and operations of the program assure regulatory compliance?
2. Are the program goals and objectives still valid and do they meet the requirements of the company?
3. Are program costs in line with the goals and objectives?

Design Considerations
The design of an audit program should include some basic considerations of human nature. They include:
- The auditor(s) should be unbiased and detached. The auditor (or team) should not consist of any person with operational responsibility (or ties to a person with operational responsibility) to the program. The auditor(s) should be experienced, and if possible, a subject matter expert in the field.
- The audit will likely generate highly sensitive information about the organization and its operation. It should be structured so that the results remain confidential from regulators and outside legal processes. Several approaches are possible, but can vary with jurisdiction. Consult legal counsel before proceeding. However, the results of the audit should only be shared with those individuals with a need to know.
- Senior management should review the report.

Managing Audit Results
The audit will have limited use unless assurance that corrective action will be taken is built into the process. Depending on the goal and structure of the audit, voluminous reports are often common. Tracking progress in correcting the findings has been made easier with the introduction of simplified database software. This becomes especially important if the organization has multiple sites. As discussed previously, the database should be arranged to give only limited access to those individuals with a need to know. Confidentially of the data should be a high priority.

Legal Documents
The program administrator should review, at some level, all documents produced by the health and safety program. Keeping in mind that the documents may find their way into a court, all documents should be accurate and complete, reflecting conditions that were present at the time of the observation. The quality assurance program should give the findings a high degree of reliability, well into the future. The documents should reflect a well-designed and well-executed program that assures a safe and healthy worksite for all employees.

SUMMARY
Health and safety is a critically important function in most organizations. This becomes obvious when you read newspaper accounts of companies that have experienced a catastrophic failure in the health and safety program. Industries that have learned from a history of disastrous loss, such as the petrochemical industry, now operate highly sophisticated health and safety programs. Do not sit on a reactive program. Create a proactive, self-corrective program that will serve your organization for many years.

chapter 8

Occupational Medicine Programs

by Charles E. Becker, MD
revised by Gary R. Krieger, MD, MPH, DABT

- 117 Introduction
- 117 Association/Causation
  Dose-response relationship ■ Sensitivity and specificity ■ Significance of animal models
- 120 Basic Principles of Toxicology
- 120 Factors Influencing Toxic Effect
- 121 Toxicity Ratings
  Lethal dose 50 ■ Toxicokinetics
- 122 Risk Assessment
- 124 Epidemiology
- 126 Medical Surveillance
- 126 Summary
- 126 References

INTRODUCTION

As occupational health and safety professionals, our goal is to identify hazards—the specific causes of safety and health problems in the workplace. Studying the occupational environment can provide us with this information; but the diversity of modern work environments and the number of factors interacting continuously to affect health and safety complicates this method of workplace hazard assessment. Scientists are often asked to ascertain if conditions in the workplace can cause or exacerbate a safety or health problem. This chapter gives a brief overview of the various disciplines that relate to the study of occupational health and safety.

ASSOCIATION/CAUSATION

There is a relationship or association between hazards and occupational illness and injury. Several methods are used to examine the strength of any association in causing certain problems. All require an understanding of the dose-response relationship, sensitivity and specificity, and the significance of animal models.

Dose-Response Relationship

The dose-response relationship is the term used when an increasing effect can be identified with an increasing dose. It is likely that the association is more than coincidental and can be a cause. Dose-response relationships hold for most body systems except the immune system where even small doses can have profound effects. The fact that an increased dose has an increased effect also suggests that the agent being studied is reaching a specific location. It is critical to select a specific effect (e.g., the risk of causing cancer), but that is not often easy. Often the specific effect chosen is the morbidity (illness) or mortality (death) of animals or humans exposed to the dose. Thus, specialized knowledge of dose-response allows us to tentatively establish thresholds for given specific effects, including cancer.

The cause is not always studied on the basis of life and death. Rather, specialized knowledge of causation can be aimed at more subtle health and safety consequences, such as worker satisfaction or days missed from work.

Obviously, the cause precedes the effect, but a suspected cause is just the first manifestation and not necessarily a basic cause. Therefore, the true cause cannot be determined until it has been proved to precede the effect.

The best example of making sure that a cause clearly preceded the event is with a tumor known as *mesothelioma* malignant tumor of the peritoneum, pleura, and pericardium. This type of tumor occurs infrequently, but was first seen in persons living near facilities that manufacture asbestos products. Because of the infrequent occurrence of mesothelioma, studies established that this type of tumor is caused by prior significant exposure to asbestos.

Sensitivity and Specificity

Sensitivity and specificity are methods of quantitatively estimating errors in predicting association and cause. Sensitivity is the proportion of truly caused events (diseased) that are tested positive; specificity is the proportion of noncaused events (nondiseased) that are tested negative (Figure 8-1). These two quantitative measures are inversely related to one another. Although they are not absolutely reciprocal, if we enhance the sensitivity of a given scientific method, specificity will suffer. Thus, as investigators we must constantly adjust the sensitivity and specificity of our methods because of the importance of assigning quantitative values to causes and dealing with the problems of misidentifications, i.e., false-positive and false-negative results. If a health study is designed to identify the true proportion of sick persons in a population with a given exposure, the testing method has to be very sensitive so as not to miss identifying any truly sick individual. However, we must also use a specific method to identify correctly all of those workers who are not ill. By altering sensitivity and specificity for a given health problem, we can balance the two extremes and thus get closer to identifying a true cause.

A good example of the importance of sensitivity and specificity for occupational health involves screening of persons for heart diseases. A routine electrocardiogram, a stress test on a treadmill, or a stress test on a treadmill when the subject was given special radioisotopes have all been used to identify cardiovascular diseases in persons who do not show any of the usual symptoms.

As screening medical techniques in asymptomatic persons, the electrocardiogram and the stress-test electrocardiogram have such a high false-positive rate (low sensitivity) that their routine use can cause many persons to have unnecessary expensive medical evaluation. If, however, the electrocardiogram and the stress test are applied to patients with heart disease, such as chest pain, the test is specific and sensitive enough to provide useful information. If the patient shows symptoms, for the very best sensitivity and specificity to correctly identify heart disease, the radioisotope stress test is currently the best method.

Significance of Animal Models

Laboratory animals help establish causes of occupational diseases. An animal model is any condition found in an animal that is of value in studying a biological phenomenon, e.g., a pathological mechanism of an animal disorder useful in studying human disease. Results of certain animal tests can predict how humans will respond to health factors. To ensure that tests done with laboratory animals are valid, researchers must: (1) select appropriate animal species; (2) use a sufficient number of animals; (3) administer appropriate dosage; (4) maintain animals so that the exposure is the only difference between exposure and control groups; (5) make careful pathological observations; and (6) administer exposure materials throughout the expected natural life span of the animals.

An experience with methanol poisoning is an example of the assets and liabilities of how findings from animal experiments relate to humans. Many cases of methanol poisoning in humans were reported, but when animals were exposed, no

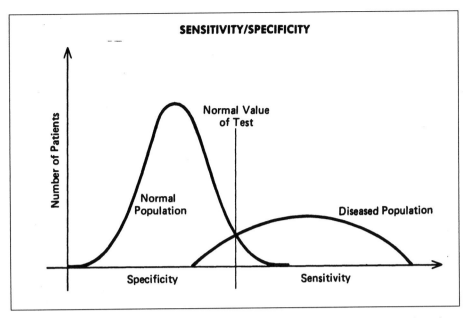

Figure 8-1. Sensitivity is the proportion of truly caused events that are tested positive. Specificity is the proportion of noncaused events that are tested negative. They are inversely related to each other.

species developed poisoning when exposed to methanol. It was then recognized that only those species of animals deficient in the vitamin folic acid had methanol poisoning. Further research studies showed that most animals cannot be methanol poisoned, unless they are made deficient in the vitamin folic acid.

This example illustrates that the dose-response relationships in animals must be related to the basic biochemistry of the hazardous substance, in order to gain certainty as to cause. Dose-response relationships studied in sensitive and specific manner and verified by results of laboratory animal studies can help establish cause.

Other questions must be answered when establishing the cause of a specific problem:
- Is the observed association reproducible?
- Is it sufficiently strong?
- Does it support the findings of other scientific studies?

These types of questions have been expanded and are known as the *Hill Criteria,* after Bradford Hill, the British epidemiologist who performed seminar work on the relationship between smoking and cancer.

For example, in an investigation to find information that would protect workers from a possible carcinogen (cancer-causing agent), identifying the required strength of the carcinogenic agent is the goal of the investigation.

Do short-term cancer tests provide reliable information about the carcinogenic effects of given chemical substances? In determining cancer risk, a single negative result does not provide sufficient evidence that a chemical cannot cause cancer. Even if a study establishes appropriate dose-response relationships, and controlled specificity and sensitivity testing offers sufficient evidence of cancer-causing effect, the findings must still be confirmed by additional study.

The most frequent short-term cancer test used in common occupational health practice is the Ames test. This is a specialized urine test in which bacteria are exposed to agents that alter their basic molecular structure. Common carcinogenic agents alter these urinary tests and become a useful monitor for persons exposed to carcinogens. It is now recognized that common foods, such as vegetables and cooked meats, cause these tests to become positive in persons in whom there is no evidence of cancer risk. Therefore, it is important to realize that although these tests are useful for research purposes, the results must be carefully analyzed for clinical significance. Otherwise, the early results can cause undue concern.

In the field of occupational health, sound conclusions must be established before suggesting that certain actions are caused by a given agent or process. If the biochemical and physiological background and findings suggest an association, the investigator is on the right track. The studies of industrial hygiene, toxicology, and epidemiology are used to identify such causes and effects.

- Industrial hygiene is used to study the environment.
- Toxicology is used to study the effect on the human body.
- Epidemiology assesses the work force for an outcome.

BASIC PRINCIPLES OF TOXICOLOGY

Toxicology is the study of dangerous or poisonous physical and chemical agents and their injury potential for living cells. This injury can be reversible or irreversible and affects a single cell, an organ system, or an entire animal. Toxicologists gather data on the toxicity of materials to predict their impact on human populations and the environment. These qualitative and quantitative studies can be related to the fields of clinical, occupational, and veterinary medicine.

By attempting to predict injury to living cells, toxicologists work directly with patients, indirectly in industry, or with the government. Because human experimentation in the field of toxicology is not always feasible, it is often necessary to base health decisions on biochemistry, interpretation of animal studies, and careful assessment of environmental monitoring results.

The word toxic indicates that an agent can cause reversible or irreversible injury to living cells. Hazard, on the other hand, involves the probability that injury will result from the agent under specific conditions of use. A toxicologist can describe a chemical or physical agent as producing a reversible or irreversible injury to a living cell, but the important question is whether or not the living cell will ever be exposed to the toxic effect. Thus, knowing the toxicity of the given physical or chemical agent is not sufficient. It also must be studied during the conditions of use.

Risk is another term sometimes used in toxicology. Risk, like hazard, is the probability that a substance will produce a toxic effect under specific conditions of use. Safety is the reciprocal of risk—the probability that no toxic effect will occur under specific conditions of use. To ensure standardized communication, these terms must be used precisely and uniformly by occupational health and safety professionals.

FACTORS INFLUENCING TOXIC EFFECT

Many factors must be considered in determining a toxic effect. Dose is the single most important factor, but the dose must be delivered to a molecular target or specific cell to have its effect. That is, if a toxic chemical is placed on the skin but does not penetrate the skin to gain access to the circulation, no toxic effect will occur. Hence, exposure does not equal dose.

The spectrum of undesired effects is so broad that the nature of the injury induced by a toxic chemical must be carefully defined. For example, a small quantity of nickel can induce a toxic effect by an allergic reaction, which actually is the result of a previous sensitization. This reaction is not a toxic effect from the nickel, but from an earlier alteration of the body's immunological system. The dose-response relationship of a chemical causing an allergic effect would look very different if the same chemical caused a direct toxic effect on a given cell or organ. In the basic pathology and mechanism of injury, these reactions are quite different and their distinction is crucial in understanding cause and effect.

A toxic agent has chemical and physical characteristics, as well as impurities. Impurities can also account for the observed effect. For example, 1,3 butadiene, a chemical widely used in the rubber industry, may produce its carcinogenic toxicity due to a specific impurity, dithiocarbamate. One of the major problems in toxicology is identifying toxic effects of chemical agents because impurities are not always obvious. For example, it can be as long as 30 years before the adverse effects of exposure are recognized. This delay is known as latency, and is a particular problem in identifying causes of cancers—many do not develop until 10–40 years after exposure. Further, the observed toxic effect may not always occur in an adult animal exposed to the toxic agent, but can be transmitted to the next generation. An example of this type of transgenerational toxicity is laboratory animals, which procreate after being exposed to a certain dosage of lead. The lead exposure of the parents passes to the next generation and results in an alteration in the sperm of the male laboratory animal children.

Toxicity also depends on exposure factors. The rate of absorption of the toxin and duration of the effect can be critical factors. For example, the body normally is able to metabolize a small quantity of cyanide, but with a large cyanide exposure, the human body's ability to metabolize is exceeded.

Thus, the dose and rate of absorption become critical factors.

Toxicity is affected by the presence or absence of special warning properties. Certain toxins have pungent odors that provide warning. Some toxins are acids (e.g., sulfuric acid) and some are alkaline materials (e.g., sodium hydroxide). When either of these is placed on the skin or inhaled into the lungs, it will cause irritation. Such local effects as skin irritation provide some degree of warning; but acid or alkaline substances can injure the protective barrier of the skin and thereby allow the toxic agent to penetrate. These toxins then injure living cells directly (like hydrofluoric acid does).

After entering the body, most toxins do not affect organs uniformly. Usually a specific molecular target or organ receives the primary toxic effect. These targets differ according to the specific agent. For example, the peripheral nervous system is the target of arsenic. Thus, toxicological study requires specialized knowledge of the most likely cells to be injured by a given toxin. Inorganic lead damages the nervous system, bone marrow, testes, and kidney. Thus, if the toxicology of lead was examined in terms of its effect on the liver function, its primary toxic effects would be entirely overlooked.

Another factor in understanding a toxic effect is the degree of protection afforded by a given cell. First, the cell is shielded from the delivery of the toxic chemical by exogenous means. The best example of this would be respiratory protection worn by fire fighters. Second, vital tissues can be protected endogenously by certain cofactors that protect cells from injury.

The most important of these endogenous cofactors that protect persons from injury is the sulfa-containing amino acid, glutathione. This material is derived from dietary sources that are high in sulfa. The body takes the dietary supplement and converts it to glutathione, which protects cells from injury from exogenous chemicals.

Moreover, there are certain inherent factors in cells that cause toxic effects to vary. Species, age, sex, and nutritional factors can also contribute to observable effects. Thus, in assessing factors that influence a toxic effect, it is important to control all variables other than the toxic agent. Increasingly, scientists are recognizing that there are three levels of study when analyzing exposed individuals, as follows:

1. susceptibility
2. sensitive measures of exposures, e.g., blood, urine, exhaled breath assays
3. early measures of adverse effects, e.g., low molecular weight proteins, found in urine and produced by kidney damage from exposures to heavy metal.

Finally, environmental factors, such as temperature, humidity, and pressure, influence occupational health and safety because they can influence and alter the potential injury to a living cell from a toxic exposure.

Because of the many factors that influence a toxic effect, interactions of several toxins should be considered. Toxins can interact by enhancing absorption, altering protein-binding, or rendering the molecular targets more or less susceptible to a toxic effect. It is not enough to say that two given compounds produce an individual response and, when given together, will produce a greater response. In biological systems, toxins added together not only can produce an additive effect, but also can create antagonistic effects, e.g., benzene and toluene compete for the same enzyme systems. Benzene only produces its toxicity via its metabolites; therefore, the simultaneous presence of both toluene and benzene tend to lower the toxicity of the absorbed benzene dose.

TOXICITY RATINGS

Because of the many factors that determine a toxic effect, it is sometimes useful to look first at the inherent toxicity of a given chemical. Table 8-A lists an approximation (mg/kg) basis of the comparison of dose and average response in rats, rabbits, and humans. The system presented here is very similar to many rating systems in toxicology textbooks. Although this method is useful for predicting toxicity, it is oversimplified and inadequate for determining actual toxicity.

Lethal Dose 50

In the field of toxicology, the most important dose-response curve is usually the lethal response known as an LD_{50} (lethal dose). The LD_{50} is the dose that causes death for 50% of the test subjects. Since the measurement of death is precise, the LD_{50} provides useful information in a given species as to the potential toxicity of a given agent. However, animals may not die but instead develop severe toxicity. When testing a new chemical agent, that caused

Table 8-A. Animal and Human Oral Toxicology Ratings.

Toxicity Rating	Commonly Used Term	LD_{50} Single Oral Dose Rats, mg/kg	LD_{50} Single Oral Dose Rabbits, mg/kg	Probable Lethal Oral Dose for Humans
1	Extremely toxic	1 mg or less	20 mg or less	Taste, 5–7 drops
2	Highly toxic	1–50	0–200	Pinch, 7 drops–1 tsp
3	Moderately toxic	50–500	200–1,000	1 tsp–2 tbsp (1 oz)
4	Slightly toxic	500–5,000	1,000–2,000	1 oz–1 pt (1 pound)
5	Practically nontoxic	5,000–15,000	2,000–20,000	1 pt–1 qt
6	Relatively harmless	15,000	20,000	1 qt

death, pathological assessment must be performed to determine the cause of the lethal event.

A typical LD_{50} curve produces a shape as depicted in Figure 8-2. It is evident that two different compounds could have the same LD_{50} and have widely differing toxicities, depending on the steepness of the curve. Furthermore, the lethality of a given agent may not relate to the more important levels of clinical toxicity that it can cause. A good example is those organophosphate pesticides which, upon exposure, inhibit the enzyme acetylcholinesterase. This can be lethal. However, some organophosphates, which do not completely inhibit acetylcholinesterase, can inhibit other esterases. This effect accounts for clinical toxicities such as delayed neurotoxicity.

Data at the lowest end of the LD_{50} curve are much less reliable than the data presented at the steepest part of the curve. It is dangerous to extrapolate from a given animal LD_{50} curve to low-dose human exposure:

- Animal species LD_{50} levels vary in relation to human levels.
- Extrapolation of a no-dose effect obtained on a relatively insensitive portion of an LD_{50} curve can provide faulty scientific information.

Toxicokinetics

Given basic toxicological information and knowledge of the importance of dose-response relationships, the key variable left is the mechanism of toxin movement throughout the body. The study of this movement is called toxicokinetics. To injure a specific molecular target, a given toxin must have gained entrance to that target. The physical and chemical properties of the toxic agent determine its absorption by the gastrointestinal tract, lung, or skin. Once the toxin is absorbed, the body has the ability to metabolize and excrete it via urine, bile, breath, sweat, or stool. It is only when the rate of absorption exceeds the rate of excretion that the toxic compound accumulates in sufficient quantity and for a sufficient duration to injure the living cell.

Thus, the concentration of toxin at the molecular target, being a function of the amount of the toxic agent that is available for absorption, depends on:

- rate of absorption
- distribution of the agent within the body
- rate of biochemical transformation
- rate of excretion.

A special problem in toxicokinetics occurs when the injury to the living cell occurs at a primary organ of elimination. If a toxin injures the lung, the liver, or the kidney, which are the primary organs of elimination, the body's ability to rid itself of the toxic agent will be retarded. Consequently, higher concentrations of the toxic agent will be delivered to the molecular target, causing enhanced injury.

An example of a toxin that injures its organ of elimination is mercury. Mercury directly injures the kidney, which is its primary route of elimination. In sufficient doses, the mercury, which is ingested into the body, damages the kidney and causes even higher levels to remain in the body, further damaging the kidney and other important organs, such as the brain.

RISK ASSESSMENT

Given a basic knowledge of toxic agents, a clear understanding of the factors influencing type and level of toxic effect, and reliable information concerning toxicokinetics, the toxicologist is then able to make a risk assessment (Table 8-A). That is, the toxicologist is often called on to determine how to extrapolate from known effects of toxic agents at high doses to risks at lower levels, given basic information concerning exposure and doses and a recognition of the danger of toxic effects with long latency. It is clear that new toxicological tools are

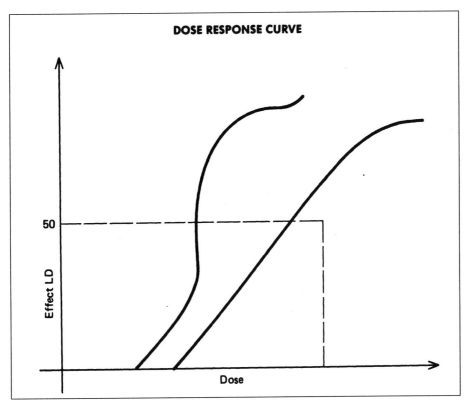

Figure 8-2. LD_{50} curves of two compounds having the same LD_{50} value but widely different toxicities (shown by the steepness of the curves).

necessary to refine this process. Since no human action or exposure is without some risk, the investigator must be careful to quantitate scientifically all of the parameters of the risk and then to ask who is at risk. Is a worker at risk? The employer? Society?

The process of risk assessment in toxicology is plagued with uncertainty and always is inherently a statistical exercise in which estimates must be made. Since absolute safety is the goal of society in most occupational health and safety endeavors, we must use this imperfect system to our best interests; but we must also strengthen the validity of these estimates by constantly updating our scientific information through experience and research.

Scientific evaluation of occupational health risks can lead, on the one hand, to needlessly prohibiting important action or, on the other, to permitting an unacceptable risk to a working individual. Since risk assessment is always an inherently statistical exercise, the analyst must constantly update animal models, in particular, and their associations to humans. In risk assessment, which is the crucial factor in industrial hygiene studies of exposure levels, investigators must apply epidemiological studies of populations of workers and toxicological information. For instance, predicting a carcinogenic, mutagenic, or teratogenic risk requires all of the following:

- epidemiological studies of human populations
- bioassays of experimental animals
- short-term tests of isolated cells or bacteria
- analysis of the common structural chemistry of the agent.

None of these factors alone is sufficient, but, when taken together, they can provide a scientifically sound risk assessment.

Risk assessment has become its own specialty area, complete with dedicated scientific journals and professional societies. Risk assessment is an integral part of environmental evaluations and investigations. The U.S. EPA has issued detailed guidance covering the practice of risk assessment for environmental problems.

In a workplace setting, the risk assessment process is less well-defined, although some substances, e.g., asbestos, had risk-based exposure levels. Internationally, the term *risk assessment* has a very different meaning and application than as used in the United States. Qualitative or semiquantitative risk-assessment evaluations are commonly performed as part of the standard health and safety practice in the United Kingdom. In the United States workplace, quantitative risk assessments tend to be part of process safety evaluations.

EPIDEMIOLOGY

Epidemiology is the scientific study of how, when, where, and why an illness distributes itself over time in a population. The acute infectious and transmissible diseases were first to be recognized as stemming from environmental agents. Initial epidemiological logic came from the interaction of three factors:
- infectious agent
- host
- environment of both.

The first studies of epidemiology involved the infectious agents of plague and cholera. The most recent example of infectious agents for which there is important epidemiological data is AIDS (acquired immune deficiency syndrome). Epidemiological logic clearly identified the infectious nature of AIDS and that many life-style factors were associated with its transmission.

Epidemiology is both science and medical philosophy. It is the logical process for examining and understanding the interplay between the physical environment, biological effects, and social phenomena that relate to human health and illness in the workplace. Epidemiology is increasingly sophisticated and interacts with many other scientific disciplines to explain health problems.

There are basically two types of epidemiological studies—descriptive and experimental. Descriptive epidemiology is an attempt to explain and thereby prevent human disease by observing outcomes among naturally occurring study groups—those not created experimentally. It involves examining dose-response relationships in terms of time, place, and the person of exposed individuals. In experimental epidemiology, a hypothetical situation is constructed and controlled by the investigator to gain scientific data that will aid in testing hypotheses relating to nonexperimental situations.

There are several investigative epidemiological methods. The first is the formulation of hypotheses. This is a never-ending task of epidemiology in which (a) a clinical or pathological observation is made; (b) a hypothesis is expressed; and (c) new observations are gathered and tested to refine the hypothesis. A given hypothesis is tested by statistical analysis, dose-response relationships, time sequence, and basic knowledge of science. Often this testing is merged with basic experimental laboratory information. This type of testing is often carried out in animal studies in which a controlled environment is deliberately manipulated in an attempt to avoid biases inherent in nonexperimental studies.

The three primary methods or types of epidemiological study are:
- prevalence studies, or cross-sectional studies
- retrospective studies, or case control studies
- prospective studies, or cohort or incident studies.

Each of these basic studies has a number of advantages and disadvantages that are briefly discussed (Table 8-B).

Prevalence or cross-sectional studies attempt to study the relationship between a cause and effect at a given point in time. This type of study is based on an existing problem and investigates only present-time cases. As an example, a good cross-sectional investigation might study all the new and existing cases of influenza infection in a given population. It is called a prevalence study because all persons who are currently affected will be considered.

Prospective or cohort or incident studies involve all persons who are developing a given condition. From this distinction, it is clear that the cross-sectional study cannot establish causation, is most useful for common diseases, and gives very rapid results. This type of study is relatively simple, economical, and rapid because the basic approach is to examine an unselected population, categorize it with regard to presence or absence of disease, and then associate the condition with an exposure.

The most scientifically useful epidemiological study is the prospective study, or cohort or incident study. As was just discussed, the basic approach is to select two similar groups of persons whose only difference is that one group is exposed to the factor under study and the other is unexposed, a control group. With this selection, the investigator goes forward in time to determine the development of the disease, thus providing incidence rates. The primary bias in this type of study is in quantifying the exposure. Its major advantage is that multiple disease outcomes can be studied; its disadvantages are expense, the long time delay (especially when studying rare diseases), and the large numbers of subjects whose conditions require a long follow-up.

Table 8-B. Three Basic Epidemiology Study Methods.

Prevalence (Cross-Sectional)	Case Control	Incident/Cohort
Studies existing living cases	Studies disease/nondisease	Studies exposure/ nonexposure
Present time	Retrospective	Prospective
Quick and inexpensive	Moderate time, cost commitment	Expensive and prolonged
Poor for rare disorder	Good for rare disease	Poor for rare disease
Subject to bias in identifying active case	Subject to bias in non-disease group	No bias of exposure
Cannot establish cause	Gives only relative risk but not cause	Gives attributable risk getting at cause

The attrition rate is high, and changing medical knowledge can change the criteria of study.

Retrospective or case control studies are similar to prevalence studies. In this approach, two groups of people are selected who are similar in all relevant aspects except that one group has the disease and the other does not. The investigator then goes back in time to determine the presence or absence of an exposure. Retrospective studies are relatively simple to perform, relatively inexpensive, faster than prospective studies, and are feasible for studying rare diseases.

Unfortunately, case control studies suffer from biases and difficulty in selecting controls and matching a section of controls. Moreover, this type of study cannot be used to establish cause or determine risk directly, since it begins with a defined disease. Many sources of bias in this type of study occur with respect to selection of cases and controls and the determination of exposure.

Bias can occur when any epidemiological study is undertaken. The following are some potential biases of scientific studies of occupational health and safety.
1. Selection of a study group that does not have the proper characteristics will lead to inadequate or inappropriate sampling for study.
2. In classifying subjects as to exposure of disease, bias can result from poor memory, leading questions, and an inappropriate time frame.
3. Use of observers with varying levels of experience, competence, or training can introduce bias when collaborative studies are performed in different locations.
4. Bias can be introduced by weak classifications of borderline cases caused by poorly designed questionnaires or inconsistent use of medical information by observers.
5. The final bias in epidemiology occurs when the results of the study can apply only to the study group or another identical group. Extrapolation must be done with extreme caution and is valid only to the extent that the sample study group is adequately shown to be representative of the larger group.

To prevent bias, the sample group must be well defined, precisely described, and strictly followed. The comparability of all study subjects must be established by checking not only for characteristics required in the sampling plan, but also for additional characteristics not included in the plan. The subjects and investigators should be blind as to the outcome of the study, and observer variations must be minimized by recording and adjusting for interinvestigator differences. Despite best efforts, bias is always a danger in epidemiological studies and, when present, it must be quantitated.

In the analysis of epidemiological studies, all study methods require comparison with a control group. A careful selection of a sample group is the most important way to control epidemiological bias. Here, the adjustment for age comes from recognizing that age is a significant factor in most illnesses. A standard mortality ratio (SMR) can be applied to correct for age in any index of health in the population. Yet, the problem of adjusting epidemiological data for age is not always resolved with these methods.

For example, all 50-year-old persons would not be expected to have the same smoking history or the same exposure to environmental pollutants. In fact, populations with different life expectancies can have identical SMRs. Thus, the adjustment for an SMR is merely accounting for the single variable of age in a situation where there can be other confounding variables. A useful method of dealing with these epidemiological problems especially in incidence/cohort studies is use of so-called life table analysis. A life table is a statistical instrument expressing the mortality of a cohort studied during a definite interval of time. This type of analysis is probably the most precise of occupational health and safety procedures.

Epidemiology attempts to place a disease in proper perspective so that its importance can be assessed in the population and its causes identified and separated from coincidental associations. Logic of study design and the rigorous prevention of bias are essential. Cohort studies combined with life table analyses are the most effective epidemiological methods for occupational safety and health problems. However, epidemiological studies will always involve a compromise with perfection in that anticipated biases can destroy certainty.

MEDICAL SURVEILLANCE

The principles of toxicology and epidemiology can be used by the facility's health professional to rationally design a medical surveillance program for exposed individuals. At its first level, any surveillance program must be compatible with existing legal requirements, e.g., OSHA standards or other similar binding regulations.

Assuming this threshold is met, the program must evaluate three areas that can be investigated currently: (1) susceptibility, (2) sensitive indicators of exposure, e.g., biological exposure indices, and (3) early measures of adverse effect. There are substantial ethical and legal issues surrounding determinations of susceptibility. However, the rapid expansion of genetic analysis at a molecular level is going to increasingly confront health professionals with how best to balance legal and ethical concerns with knowledge that regulatory exposure levels are probably underprotective for a select few individuals.

Determinations of sensitivity, at a molecular level, are active research problems and have not systematically affected general workplace practice. However, all indications are that this situation is likely to change within the next 5–10 years.

The ability to measure biologic exposure via blood, urine, stool, or exhaled breath is well-established. In its *Threshold Limit Values and Biological Exposure Indices* publication, the ACGIH gives its biological exposure indices for a variety of common workplace chemicals. At present, the limiting problem with this approach is the sheer number of workplace chemicals versus the available biological assays.

Finally, systematic analysis for early measures of adverse effects must be part of every workplace surveillance program. The OSHA standards tend to substantially lag behind the most current scientific practice. For example, sensitive measurements of kidney damage are now available for workers with cadmium exposure. The use of tests with low sensitivity and specificity will produce uninterpretable results and needless worry and confusion among workers. Since there will always be problems with adequate availability of sensitive and specific tests, the workplace medical professional should always emphasize primary prevention through engineering controls of potential exposures.

SUMMARY

The goal in occupational health and safety is to prevent injury and illness in the workplace. No human experience is risk-free. No absolute scientific principle is any more appropriate than the most recent reproducible data. Because the scientific approach to occupational health and safety is invariably a process of compromise, scientific decisions often occur in a polarized environment requiring rigorous scientific review to avoid bias and conflict. A rigorous scientific approach to occupational safety and health is a constant educational process in which we progressively discover our own ignorance by creating more problems to solve.

REFERENCES

Clayton GD, Clayton FE (eds). *Patty's Industrial Hygiene and Toxicology,* 3rd ed. (vols. I, II, and III). Somerset NJ: Wiley, 2000.

Klaassen CD, et al (eds). *Casarett and Doull's Toxicology: The Basic Science of Poisons.* New York: Macmillan, 1998.

Sullivan JB, Krieger GR (eds). *Hazardous Material Toxicology,* 2nd ed. Baltimore: Lippincott Wiliams Wilkins, 1999.

chapter 9

Occupational Health Nursing Programs

by Connie Lawson, MS, COHN-S
Nita Drolet, RN, MPH, CIC
Regina M. Cambridge, RN, COHN-S
Kenneth R. Pelletier, PhD
Thomas J. Coates, PhD
Edwin B. Fisher, Jr., PhD
Joan M. Heins

- 127 Overview
- 128 Objectives
- 128 Medical Screening/Surveillance
  Definitions ■ Purpose
- 128 Physical Examinations
- 129 Preventing Occupational Injuries and Illnesses
- 130 Health Promotion/Adult Education
- 132 Fitness Programs
  Benefits ■ What benefits and costs weigh ■ Evaluations of worksite health-promotion programs ■ Long-term changes in behavior ■ Community/corporate organization approaches to reaching hard-to-reach workers
- 135 Employee Assistance Programs
- 136 Work/Life Strategies
- 136 Case Management
- 137 Hearing Conservation Program
- 137 Stress Management
  Preventing job stress ■ Steps toward prevention
- 139 Hazard Communication (HAZCOM) Program
  Federal standard ■ Chemical inventory ■ Signs and labels ■ Material safety data sheet ■ Hazard communication training ■ Hazard communication program
- 140 Drug and Alcohol Testing Program
  Types of testing ■ Policy ■ Supervisory training ■ Regulations
- 142 Department of Transportation Guidelines
- 143 Occupational Exposure to Bloodborne Pathogens
  Virus information ■ OSHA bloodborne pathogens standard ■ Universal precautions ■ Engineering controls and housekeeping ■ Personal protective equipment ■ Immunizations ■ Post-exposure follow-up ■ Record keeping
- 144 Occupational Exposure to Tuberculosis
  Background ■ Tuberculosis bacteria ■ TB standard
- 146 Workplace Violence Prevention and Critical Incident Stress Debriefing
  Measures an organization can take ■ Profile ■ Conducting a critical incident stress debriefing
- 149 Summary
- 150 References

OVERVIEW

Occupational health and safety professionals strive to optimize worker health and safety by minimizing preventable disease and injury while maintaining regulatory compliance. Traditionally, health and safety programs operated independently from one another. Today we know a multidisciplinary model is needed in order to better focus on interventions and understand near-miss incidents.

Chapter 4 discussed the occupational health nursing profession and the role occupational health nurses play within various sectors of an organization. As licensed professionals, manager/coordinator/director, occupational health nursing professionals assume a leadership role within an organization for many programs.

This chapter describes a sampling of the many occupational health nursing programs including the following:
- medical screening/surveillance
- physical examinations
- occupational injuries and illnesses prevention

- health promotion/adult education
- employee assistance programs/work/life strategies
- case management
- hearing conservation
- stress management
- hazard communication (hazcom) program
- drug and alcohol testing program
- department of transportation guidelines
- occupational exposure to bloodborne pathogens
- occupational exposure to tuberculosis
- workplace violence prevention/critical incidence stress debriefing.

OBJECTIVES

This chapter will assist the reader in being able to accomplish the following:
1. define the various programs an occupational health nurse can manage, coordinate or direct
2. discuss the basics of the various programs that fall under the standards of practice for occupational health nursing
3. understand how each of these programs requires a multidisciplinary effort between organizations to positively impact worker health and safety.

MEDICAL SCREENING/SURVEILLANCE

Medical screening and medical surveillance are two fundamental strategies for optimizing employee health. Although the terms are often used interchangeably, they are quite distinct concepts. Medical screening is, in essence, only one component of a comprehensive medical surveillance program. Both can contribute significantly to the success of worksite health and safety programs. The following definitions will help in understanding the terminology used in this section.

Definitions

Screening—A method for detecting disease or body dysfunction before an individual would normally seek medical care. Screening tests are usually administered to individuals without current symptoms, but who may be at high-risk for certain adverse health outcomes.

Surveillance—The analysis of health information to look for problems that may be occurring in the workplace that require targeted prevention, and thus serves as a feedback loop to the employer. Surveillance may be based on a single case or sentinel event, but more typically uses screening results from the group of employees being evaluated to look for abnormal trends in health status. Surveillance can also be conducted on a single employee over time. Review of group results helps to identify potential problem areas and the effectiveness of existing worksite preventive strategies.

Purpose

The fundamental purpose of screening is early diagnosis and treatment of the individual and thus has a clinical focus. The fundamental purpose of surveillance is to detect and eliminate the underlying causes (i.e., hazards/exposures) of any discovered trends and thus has a prevention focus.

The Occupational Safety and Health Administration uses the term *medical surveillance* to describe protocols for monitoring specific exposures. However, the processes and personnel involved in this monitoring extend beyond medicine; hence it is actually a type of *health surveillance*. See Table 9-A for examples of the kinds of surveillance required by some OSHA standards.

PHYSICAL EXAMINATIONS

There are a variety of examinations used in the workplace. Each of these assessments should include a thorough interview of the worker and be objective, nonjudgmental, and based on data. The types of occupational examinations given include the following:
- *baseline examination*—findings provide documentation of a current illness or injury
- *preplacement examination*—focuses on the ability to perform the essential functions of the job
- *Americans with Disabilities Act (ADA) examination*—This examination is done to remove barriers against potential workers with known disabilities (Pruitt, 1995); to assess the need for reasonable accommodation for the worker with disabilities; and to assess the condition of the worker at the time of the examination without considering the possibility that a disability could develop in the future
- *periodic examination*—can supplement a baseline and provide ongoing health status information
- *exit examination*—provided when an employee leaves the company, usually based on mandated programs or exposure history

Table 9-A. Examples of OSHA Standards Requiring Medical Surveillance

Standard No.	Substance
29 CFR 1910.1000	Hazardous materials (hazmat)
29 CFR 1910.1001	Asbestos
29 CFR 1910.1002	Coal tar pitch volatiles
29 CFR 1910.1003	4-nitrobiphenyl
29 CFR 1910.1004	alpha naphthylamine
29 CFR 1910.1006	methyl chloromethyl ether
29 CFR 1910.1007	3,3'-dichlorobenzidine and its salts
29 CFR 1910.1008	bis-chloromethyl ether
29 CFR 1910.1009	beta-naphthylamine
29 CFR 1910.1010	benzidine
29 CFR 1910.1011	4-aminodiphenyl
29 CFR 1910.1012	ethyleneimine
29 CFR 1910.1013	beta-propiolactone
29 CFR 1910.1014	2-acetylaminofluorene
29 CFR 1910.1015	4-dimethylaminoazbenzene
29 CFR 1910.1016	n-nitrosodimethylamine
29 CFR 1910.1017	vinyl chloride
29 CFR 1910.1018	inorganic arsenic
29 CFR 1910.1025	lead
29 CFR 1910.1027	cadmium
29 CFR 1910.1028	benzene
29 CFR 1910.1029	coke oven emissions
29 CFR 1910.1043	cotton dust
29 CFR 1910.1044	1,2-dibromo-3-chloropropane
29 CFR 1910.1045	acrylonitrile
29 CFR 1910.1047	ethyl oxide
29 CFR 1910.1048	formaldehyde
29 CFR 1910.1050	4,4'-methylenedianiline

Examples of OSHA Standards That Require Medical Clearance

Standard No.	Specific Activity
29 CFR 1910.134	Respiratory protection
29 CFR 1910.156	Fire Fighters

Post Exposure Surveillance

Standard No.	
29 CFR 1910.1030	Bloodborne Pathogens
29 CFR 1910.95	Noise exposure

Source: Occupational Safety and Health Standards for General Industry

- *fitness-for-duty examination*—its purposes are to (1) assure that the worker is physically and mentally capable of performing job functions; and (2) assure safety of the worker and the public (Pransky et al, 1988)
- *internal job-transfer evaluation*—done to determine if the job transfer will place the worker at increased risk for injury or aggravation of an existing problem
- *return-to-work evaluation*—assess the worker's ability to resume work after time away from work; the goal is to return the worker to productive employment.

PREVENTING OCCUPATIONAL INJURIES AND ILLNESSES

Occupational injury and illness prevention programs must educate employees to be diligent about their safety. Safety diligence can be defined as "the approach to a task that assures that the task can be successfully performed with manageable risk to the individual, client, property and company assets." It focuses employees' attention on the vital role safety plays in their day-to-day lives. Safety diligence is also a constant state of awareness and realization that shortcuts can be expensive in terms of time, injury, and increased risk. Tying this to a corporation's quality-improvement process, safety diligence identifies four *quality tools* that are necessary to do a job safely. They include the following:

- skills
- materials
- correct equipment
- ability to do the job safely.

The goal should be for employees to take just a minute to ask themselves if they have these four tools before they proceed to perform their jobs.

To design programs and anticipate occupational health and safety hazards, an occupational health nurse must know the workplace and the work being performed, appreciating unique attributes, including risk factors that may characterize a worker population.

Two critical steps in any worksite program include the following:

- recognizing existing hazards
- identifying potential hazards

Recognition is the process of identifying and describing existing workplace hazards. *Hazard* is "the potential for harm or damage to people, property, or the environment." *Anticipation* is the foresight to recognize hazards in equipment and processes during the planning stages so they can be eliminated from the design (Manuele, 1994).

One method of hazard identification is a site survey or walk-through, which is a worksite inspection not related to any particular incident or concentrated area or piece of equipment (Travers & McDougall, 1995).

Job safety analysis, accident investigations, exposure monitoring, and process safety reviews are covered in the following chapters: 10, Safety Programs; 11, Industrial Hygiene Programs; 12, Ergonomics Programs; and 16, Radiation Safety Programs.

HEALTH PROMOTION/ADULT EDUCATION

Occupational health professionals must take a leadership role in health promotion and adult education. We must use multiple methods (e.g., Web technology) to provide tools and resources that maximize effectiveness during the learning process. Adult employees should be enabled to access knowledge, attain skills, and accept accountability for their health and safety. The desire to have all employees obtain and maintain an optimal level of health, is truly a noble universal goal. In these days of increasing concern with return on investment however, health promotion must be quantified as a benefit to the organization, not just a benefit to the employees. There is a logical link between healthy employees and increased benefits to the company. Organizations with health promotion programs find they have lower medical and workers compensation costs, better attendance, and happier more productive employees. This is due to the fact that employees with healthy lifestyles have fewer illnesses and injuries, and recover from illness and injury faster. Employees who participate in regular physical activity have increased cardiovascular endurance. Those employees who eat low fat diets, refrain from smoking, and get adequate rest add years to their lives. Employees who do all of the above are more alert, have a better mental attitude, and deal more effectively with stress, rapid change and the permanent white water conditions present in companies today.

Healthy People 2010: National Health Promotion and Disease Prevention Objectives describe the national objectives related to major chronic illness, injuries, and infectious diseases; its major emphasis is health promotion and disease prevention. Health promotion can take many forms, and be as simple or as complex as a company desires. Basic health promotion should include employee education about diet and nutrition. This information can be provided by health resources within the organization or obtained using community health departments and nonprofit organizations. Covering the importance of good nutrition first is a cost-effective means of health improvement because changing dietary habits of employees doesn't require expensive equipment or health club memberships. Employees will be purchasing and preparing food regardless, and teaching them what food to buy and how to prepare it is cost effective. It also doesn't require an employee to make major lifestyle changes. Large companies that have their own on-site cafeterias can start by providing a healthy alternative with a low fat, high fiber menu. Employers should at least consider offering a healthy alternative menu, in addition to the regular cafeteria menu. Another key component of health promotion is regular physical exercise. Walking or other low impact aerobic exercise programs can be organized at the work site, before and after work, or during meal breaks. Walking provides an opportunity to warm up before work, or relax and reduce stress after work and during breaks. These are low-cost alternatives to having an on-site health club, or paying for membership to a fitness facility.

Employers should also consider starting with a *health risk appraisal (HRA)* or health screening. The health risk appraisal is a health education tool, easy to administer, that's used to compare an employees health-related behaviors and characteristics with those of the general population. A health screening is usually done to measure an employee's pulse, blood pressure, cholesterol, and body fat. Additionally some test of aerobic capacity is a good measure of overall fitness. The medical departments in those larger organizations that have them could provide health screenings. Those organizations that do not have access to medical facilities could arrange for screenings with their health care provider or through any number of companies now in the wellness business.

If you choose to have an on-site health club, you should consider the factors involved. This is an excellent employee benefit, for those employees that are interested (Figures 9-1 and 9-2). The employer must consider the potential costs and liability issues. There will be a significant overhead expense for the facility and equipment, and if space is added there will be construction or remodeling costs. It is important to have quality equipment. Fitness club quality aerobic equipment is expensive

Figure 9-1. Among the reported indirect benefits of workplace health-management programs are enhanced employee morale, consistency of a corporate product with the image of a healthy company, and availability of the program as a perk. (Courtesy Johnson & Johnson Health Management, Inc.)

but necessary to prevent injury and provide for a long service life. Trained staff will need to be available to instruct employees on the proper exercise techniques, since improper use of equipment can lead to injury to muscles and joints. On-site childcare may be needed if not already available in the organization. Employees that have not seen their children all day may not feel comfortable staying after work to exercise. In addition, liability concerns arise in the form of workers compensation for employees injured while engaged in on-site aerobics or resistance training. Some employees prefer exercising in the privacy of their own homes or with friends at a neighborhood health club. Paying an employee's membership in fitness facilities, or reimbursing their membership costs may be a more prudent approach. There would be no exposure to liability (employees would not be directed into these programs) on the company's part. Employees will likely have the ability to participate in a greater variety of aerobic programs, have a facility with childcare, and more flexible days and hours of operation.

Employee Assistance Programs (EAPs) are also an important part of employee health promotion. The EAPs can provide employee education on wellness issues in addition to information on stress reduction and smoking cessation programs. Company EAPs are commonly thought of as the first point of referral when an employee's personal problems begin to impact work performance. Self esteem and personal image issues that manifest themselves as health and fitness problems are often closely related to emotional problems that can be

Figure 9-2. Certified aerobics instructors work with groups of employees in the worksite wellness program offered by Johnson & Johnson Health Management, Inc. (Courtesy Johnson & Johnson Health Management, Inc.)

addressed by EAPs. It is common knowledge that aerobic exercise is effective in reducing stress and improving one's overall sense of well being. Tying exercise and mental health together can be an important component of your wellness program.

Employee participation in health promotion must be voluntary, but incentives can be provided. Incentives can take the form of lower or subsidized insurance premiums and or deductibles, company logo health and fitness gifts, or subsidized memberships to gyms and fitness centers. Some organizations provide gift certificates to local stores or malls. This way employees can collect a larger monetary incentive over time and purchase a major gift or large piece of fitness equipment such as a treadmill.

FITNESS PROGRAMS

Employers have a strong motivation to prevent or reduce the occurrence of acute and chronic health conditions among their employees (Fielding, 1984). Employer-sponsored programs in disease prevention and health promotion are becoming an integral part of an overall strategy to improve the management of medical care costs and to minimize the occurrence of productivity-reducing conditions (Jones & Dosedel, 1986; Felton & Cole, 1963). Beyond involving employees in such programs, employers are extending their efforts to dependents and the burgeoning retiree population.

Health-promotion programs educate employees on how to maintain or enhance psychological and physical health. These programs can be comprehensive or focused on a single risk factor, such as smoking or hypertension. Comprehensive health-promotion programs show greater health and cost efficacy than single-factor programs. The comprehensive programs identify organizational high-risk factors, such as environmental toxins, in addition to individual risk factors. Problems are addressed through policy-level organizational changes, and personal treatments and referrals. Disease-prevention programs focus on the early detection of and behavioral intervention into one or more detectable risk factors.

The workplace is an excellent setting for health-promotion programs (Pelletier et al, 1988a; Pelletier et al, 1988b). Workplace programs have access to large numbers of people, and the programs are relatively easy and efficient to implement. Most people spend the majority of their waking hours at work, so there can be a relatively large "dose" of the program. The workplace offers psychosocial supports, including peer support and peer pressure. Time and travel barriers to employee participation are less than at other locations. The target population is relatively stable, which facilitates follow-up. There also can be related programs and existing facilities in place, such as an employee medical department or an employee assistance program (EAP) with on-site health staff. Finally, many employers are willing to sponsor and fund some or all program costs. Of the private workplaces in the United States with 50 or more employees, approximately 66% sponsor one or more types of health-promotion activities, and the employer pays the costs in the majority of cases (Fielding & Piserchia, 1989).

Benefits

Employers have many direct and indirect reasons for undertaking health-promotion and disease-prevention programs. These benefits include improving the health of employees, providing additional employee benefits, and controlling the costs of health care, accidents, and absenteeism (Mullen, 1988). Employer-sponsored health-promotion programs can satisfy several employee health needs, such as promoting health and reducing risk, managing chronic illness, rehabilitation, and improving participation in the clinical process (Fielding & Piserchia, 1989). Reports from the Bureau of National Affairs indicate that about 50% of worker absenteeism was avoidable through appropriate attention to the physical and emotional needs of employees.

To date, health-promotion programs focus on a single risk factor, such as hypertension, smoking, stress, low-back disability, and cholesterol screening. There is increasing interest in evaluating the health-effectiveness and/or cost-effectiveness of these single-factor programs. At the same time, the overall field of health-promotion programs in the workplace has not been evaluated adequately by rigorous design and appropriate data analysis. Given the lack of such analytic work, it is not surprising that no specific area has been evaluated adequately. Numerous workplace health-management programs are justified by the sponsoring corporation based solely on indirect benefits. In these instances, the direct cost-effectiveness and/or cost-benefit are not important.

Among the reported indirect benefits are enhanced employee morale, improved corporate image, ability to attract and retain key personnel, consistency of a corporate product with the image of a healthy company, and availability of the program as a perk for key executives. These are legitimate reasons, but in a decade of increasing competition, mergers, takeovers, and concern for the "bottom line," it is more likely than not that such good intentions can initiate a program but not sustain it. Overall, the trend is to seek objective evaluations of both health-effectiveness and cost-effectiveness in all aspects of medical benefits and health promotion. Health-effectiveness is evident in specific areas such as smoking cessation and hypertension, but the cost-effectiveness remains equivocal.

What Benefits and Costs Weigh

The demand to show the "cost-effectiveness" of health promotion is often confusing. It sounds, on its face, reasonable and prudent; however, this demand may be too broad. Consider a company that is losing large amounts of money because of pilfering and petty theft after hours. Management might review security procedures and devise a new strategy for after-hours security and consider switching from internal security staff to vendors or vice versa. The company might decide to increase emphasis on electronic surveillance or on the presence of visible security officers. No doubt, there are many other possibilities. However, it is extremely doubtful that the company would consider eliminating security services altogether. Questioning the cost-effectiveness of health promotion in general is like considering the use of security because losses are resulting from pilfering. The meaningful question is not whether health promotion is effective or needed, but what approach to health promotion is most effective:

- What health needs most require attention?
- What health programs can effectively address those needs?
- How can choices of needs or programs be influenced by the nature of the company, by the nature of the work force, or by other local conditions?

As another analogy, asking whether health promotion is effective is similar to asking whether health care is effective. Obviously, some health care strategies are more effective than others.

Effectiveness needs to be considered in terms of specific procedures and circumstances, rather than globally.

Putting these points into concrete terms, measures to alter risk factors related to life-style are adequately justified by the demonstrated costs attributable to diseases associated with these risk factors, and by the fact that conventional health care does not alter these risk factors effectively. Then, the general commitment to health promotion is validated by the cost of unhealthy life-styles to companies and employees.

Cost-effectiveness or cost-benefit analysis can, however, indicate results of a company's general commitment to health promotion. Evaluating the effects and costs of specific approaches to health promotion can help the company focus and refine those approaches. Is the cost of ongoing smoking cessation activities within a workplace more effective than simple screening, self-help manuals for quitting smoking and implementation of a policy to limit smoking at the worksite? Are weight-loss clinics more cost-effective than weight-loss competitions backed up by self-help manuals on losing weight?

The costs attributable to a problem in a specific setting can guide decisions about whether or not to address that problem. For instance, a low prevalence of obesity in a work force can obviate a weight-loss clinic; however, improved nutrition might still be a worthwhile goal. Questions about programming need to reflect local circumstances, including the nature of the work force. Such specific, adequately focused questions are important when weighing the benefits and costs of a health-promotion program.

Evaluations of Worksite Health-Promotion Programs

The majority of evaluations of health-promotion programs are focused on health outcomes, such as changes in relatively objective health risk factors (e.g., smoking, blood pressure, fitness) and changes in subjective indicators (e.g., stress, feeling of support for health change, self-efficacy). Based on such measures, there is growing evidence that workplace health-promotion programs can improve employee health behavior and health status. For example, projects at the workplace achieve much higher rates of hypertension control than the national average of other types of programs (Leviton, 1987). Workplace outcomes for smoking cessation and cholesterol reduction are on a par with what can be expected in clinics (Fisher et al, 1988). Physical activity and fitness levels also have increased through worksite programs. Evidence of enduring changes in other outcomes, such as weight control, is less well documented but positive outcomes are reported (Fielding, 1982). There is growing emphasis on evaluation of stress management programs in small-to-large corporations, but outcome data are limited (Leviton, 1987).

Many experts believe that two critical ingredients in a successful comprehensive program are targeting the entire workplace or work force for the program and achieving high rates of active participation. Potentially, the use of peer influence in the workplace, and an environment that supports health activities, influence increasing rates of participation and minimize recidivism. Thus, developing a reproducible process that can account for site-specific differences and still produce high participation is an important prerequisite to effective programming and improved return on investment. In an ideal situation, because of environmental and peer cues and supports, a health promotion program should cause changes in health behavior among much of the work force, not only those who participate in specific program components (Shephard et al, 1982). By focusing on all the employees at a workplace, it is possible to assess the impact of the program on passive participants, and avoid some of the evaluation problems related to participant self-selection.

Long-Term Changes in Behavior

Finally, monitor the effects of a program over time. To date, the majority of evaluations do not exceed one year. There are few evaluations of the long-term effects of either individual or comprehensive programs. Since the trend among the general population toward improved health habits is part of the rationale for employer sponsorship, the ability to show continued improvement over time is an important issue.

Community/Corporate Organization Approaches to Reaching "Hard-to-Reach" Workers

Extending programs to "hard-to-reach" employees or those at heightened risk can be accomplished through communication networks. Often, these workers can be remote from formal communication channels offering health-promotion programs or encouraging healthier life-styles. While isolated from or unresponsive to such formal channels, these employees may be linked by strong informal channels such as discussions among co-workers, educational television programs on site, or noon lecture series. Thus, one way to reach these employees is via an informal network. For example, informal networks are important in low-income and minority communities in supporting health activities and providing connections to the more formal services (Pilisuk et al, 1982).

One way to enlist informal networks is to increase links between informal networks of co-workers talking among themselves and formal program networks, which provide specific classes or interventions. A program can include formal roles for members of hard-to-reach, informal networks, thereby recruiting those networks to carry the program's message to other members of the networks. Asking representatives of such groups to join a program steering committee is a widely applicable way of linking informal networks to a health promotion program. Another approach is to make sure that programs include activities that are highly valued by hard-to-reach groups. Basketball or bowling (although not the best aerobic activity) can recruit into a fitness program those who initially reject jogging suits and exercise bicycles. To enhance the linkage, ask the leaders of the basketball or bowling league to join the program committee.

The stakes are very high in developing workplace life-style interventions to prevent chronic disease and disability. The Carter Center report "Closing the Gap" (Gibbs et al, 1985) indicates that, given our present medical knowledge and capabilities, approximately 66% of all deaths in this country are premature and, further, about 66% of all years of life lost before age 65 years are savable. Given the 1992 medical budget of more than $850 billion, or 14% of the GNP, it is tragic and ironic that less than 5% of that budget was expended in primary prevention or health-promotion programs.

The problem of preventing premature morbidity and mortality before age 65 and reallocating billions of dollars spent is overwhelming. The cornerstone of such efforts should be programs concerned with life-style intervention. However, the resolution of such an undertaking depends upon the effective coordination of three basic elements of a true health care system: (1) quality, cost-effective medical care; (2) procedures for cost containment and utilization review to guard against overutilization or underutilization; and (3) quality, cost-effective programs in health promotion and behavioral medicine. At present, there are a few promising instances of addressing these three elements in health care and occupational settings.

Businesses are beginning to see that, instead of being reactive and spending great sums of money only after employees become ill, it makes good sense to institute programs designed to prevent employees from becoming ill in the first place. With such programs, corporations not only improve the health of their employees (a worthy objective in its own right), but they contribute to containing medical costs at the same time—clearly a "win-win" situation for all concerned. In short, businesses are finding that an ongoing, active interest in the health of their employees has a direct bearing on their own long-range health as a business.

EMPLOYEE ASSISTANCE PROGRAMS

Each of us is touched by personal problems at some point in our lives. And when we are, we find that the supportive help given by family and friends will often see us through a time of personal challenge. There may be times though when our problems are more than we can manage alone. Without effective help, these problems can interfere with our physical and emotional health, and lead to unhealthy changes in our lifestyles at home and at work. The pressures of unresolved personal problems can rob us of our peace of mind, our energy, and even our physical health.

As mentioned previously, *EAPs* are confidential professional counseling services for employees. Many companies also cover immediate family members and other eligible dependents covered under company sponsored medical plans that request help with personal problems, large or small. The goal of an EAP is to prevent life's problems from turning into more serious difficulties by identifying problems early and helping employees arrive at a resolution.

Services can be provided through EAP providers or in-house professionals to deal with a variety of problems. Counselors can provide services many

times to employees and their families on issues like the following examples.
- personal problem assessment
- brief counseling (usually a predetermined number of sessions)
- referral for additional professional help, if needed
- postreferral follow-up.

Counselors can help with problems such as these:
- family management
- stress management
- depression
- problem drinking
- drug abuse
- psychiatric disorders
- eating disorders
- referral for financial counseling
- referral for legal assistance.

Counseling services are usually provided at private and convenient locations and times. If a referral is necessary, counselors will try to recommend treatment options that are covered under employee insurance plans, and help them arrange the first appointment. Counselors also schedule follow-up contacts to monitor the success of the referral.

Employers generally encourage the use of EAP programs because healthy, productive employees mean a healthy, productive organization.

WORK/LIFE STRATEGIES

Today more than ever, many employees face the difficult challenge of balancing family, work, and personal life. To help employees succeed; many employers provide work/life programs that help offset competing work and life demands by providing employees with an array of programs and information to make educated decisions for themselves and their family at various points in their life. Balancing childcare and elder care responsibilities with a career can be difficult at times. Many employers realize this and provide work/life programs to address employee concerns, whether it is locating child care or elder care facilities and services or providing employees with educational materials about parenting and caregiving issues. Work/life programs can generally help or refer employees to all types of care including the following:
- family day-care homes
- child care centers
- preschools
- school-age programs
- special needs programs
- adult day-care centers
- home health workers
- nursing home facilities
- housing alternatives for the elderly
- transportation services for the elderly
- home meal delivery.

In addition to referrals, consultants can help with parenting and caregiving issues such as the following:
- understanding why your toddler bites
- getting ready for parent-teacher conferences
- communicating with your teenager
- dealing with the stress of coping with a dependent adult child who is mentally challenged
- understanding your role as a caregiver of an older relative
- becoming more knowledgeable about Medicare.

Occupational health professionals managing work/life programs generally ensure that a wide variety of guidebooks, checklists, information fact sheets, and other educational materials are offered to help employees become more knowledgeable about the day-to-day concerns of caregiving. Work/life strategies are an important way for companies to invest in employees and their families.

CASE MANAGEMENT

Case Management is a process of coordinating an individual client's health care services to achieve optimal, quality care delivered in a cost-effective manner (AAOHN, 1994). Case management can take place on site, by telephone, or off-site by a provider. Employers decide which type of case management and the type of cases they wish to manage. Case management could include the following:
- high-cost catastrophic cases
- work-related injury or injury cases
- nonwork related injury and illnesses for workers and their dependents.

Comprehensive case management is ideally provided by occupational health nurses; because of the emphasis on return to pre-injury function, occupational health nurses are well positioned to support an appropriate return-to-work plan as a case management goal (AAOHN, 1994; ABOHN, 1994).

HEARING CONSERVATION PROGRAM

The passage of the Hearing Conservation Amendment to the Occupational Safety and Health Act in 1983 provided the thrust toward the development of hearing conservation in industry. The purposes of a *hearing conservation program (HCP)* are to accomplish the following:

- prevent the hearing loss of workers
- identify the progression of hearing loss so that preventive measures can be taken
- identify temporary hearing loss before it becomes permanent
- comply with federal regulations or OSHA-approved state plans (OSHA Noise Standard *CFR* 1910.95).

If reliable information indicates noise exposure in the worksite, noise measurements need to be conducted. These can be performed by a (an):

- acoustical engineer
- industrial hygienist
- occupational audiologist, or
- professional proficient in noise-level measurement.

After measurements are conducted the results:

- identify areas of the worksite where hazardous noise levels exist
- identify workers to be included in a hearing conservation program (HCP)
- classify workers' noise exposure in order to define hearing protection device (HPD) policies and prioritization of areas for noise control efforts
- identify safety hazards in terms of interference with speech communication and warning-signal detection
- evaluate noise source for noise control purposes
- document noise levels for legal purposes.

Worker training and education, audiometric testing, and hearing protection devices need to be supplied as appropriate. Testing is performed to determine baseline hearing and to monitor effects of noise exposure. The test environment and equipment must meet criteria set by the American National Standards Institute (ANSI).

An audiometric evaluation can reveal the following:

- *normal hearing*—in general, falls within hearing threshold levels between 0 and 25 db. It may vary slightly from left ear to right ear and may be age-dependent.
- a *standard threshold shift (STS) (also referred to as significant threshold shift) is an average shift in either ear of 10 db or more at 2,000, 3,000, and 4,000 Hertz compared to baseline audiogram.* Workers must be notified in writing within 21 days of determination of the STS.
- a *temporary threshold shift (TTS) occurs shortly after exposure and improves gradually if the noise has not been too loud or the exposure too long.* The greatest recovery takes place in 1 to 24 hours if the worker is removed from exposure. Adequacy of HPD should be checked.
- a *permanent threshold shift (PTS) occurs when hearing loss persists after removal from exposure.* It is associated with damage to the delicate sensory hair cells in the inner ear. If there is no improvement within one week, the loss is usually permanent.

There are specific record-keeping requirements for audiometric testing records, noise survey forms, audiograms, exposure histories, etc. These must be kept for duration of employment (more than 30 years is often suggested). Periodic self or external audit of the overall HCP should be performed to ensure compliance with all relevant regulatory requirements.

STRESS MANAGEMENT

Stress is a contributing risk factor to cardiovascular disease and more than half of the visits to health care professionals are for stress-related disorders. Also modifiable lifestyles have been identified as being a major cause or premature deaths in the United States.

Preventing Job Stress

So, while no standardized approaches or simple "how to" manuals exist for developing a stress prevention program, program design and appropriate solutions will be influenced by several factors:

- the size and complexity of the organization
- available resources
- the unique types of stress problems faced by the organization.

In one organization, for example, the main problem is work overload. In another, employees may be bothered by difficult interactions with the public and an inflexible work schedule.

Although it is not possible to give a universal prescription for preventing stress at work, it is possible to offer guidelines on the process of stress prevention in organizations. In all situations, the process for stress prevention programs involves three distinct steps: problem identification, intervention, and evaluation. For this process to succeed, organizations need to be adequately prepared. At a minimum, preparation for a stress prevention program should include the following:

- building general awareness about job stress (causes, costs, and control)
- securing top management commitment and support for the program
- incorporating employee input and involvement in all phases of the program
- establishing the technical capacity to conduct the program (e.g., specialized training for in-house staff or use of job stress consultants).

Bringing workers or workers and managers together in a committee or problem-solving group may be an especially useful approach for developing a stress prevention program. Research has shown these participatory efforts to be effective in dealing with ergonomic problems in the workplace, partly because they capitalize on workers' firsthand knowledge of hazards encountered in their jobs. However, when forming such working groups, care must be taken to be sure that they are in compliance with current labor laws. (The National Labor Relations Act may limit the form and structure of employee involvement in worker-management teams or groups. Employers should seek legal assistance if they are unsure of their responsibilities or obligations under the National Labor Relations Act.)

Steps Toward Prevention

Low morale, health and job complaints, and employee turnover often provide the first signs of job stress. But sometimes there are no clues, especially if employees fear losing their jobs. Lack of obvious or widespread signs is not a good reason to dismiss concerns about job stress or minimize the importance of a prevention program.

Step 1. Identify the Problem

The best method to explore the scope and source of a suspected stress problem in an organization depends partly on the size of the organization and the available resources. Group discussions among managers, labor representatives, and employees can provide rich sources of information. Such discussions may be all that is needed to track down and remedy stress problems in a small organization. In a larger organization, such discussions can be used to help design formal surveys for gathering input about stressful job conditions from large numbers of employees.

Regardless of the method used to collect data, information should be obtained about employee perceptions of their job conditions and perceived levels of stress, health, and satisfaction. The list of job conditions that may lead to stress and the warning signs and effects of stress provide good starting points for deciding what information to collect.

Objective measures such as absenteeism, illness and turnover rates, or performance problems can also be examined to gauge the presence and scope of job stress.

However, these measures are only rough indicators of job stress—at best.

Data from discussions, surveys, and other sources should be summarized and analyzed to answer questions about the location of a stress problem and job conditions that may be responsible. For example, are problems present throughout the organization or confined to single departments or specific jobs?

Survey design, data analysis, and other aspects of a stress prevention program may require the help of experts from a local university or consulting firm. However, overall authority for the prevention program should remain in the organization.

Step 2. Design and Implement Interventions

Once the sources of stress at work have been identified and the scope of the problem is understood the stage is set for design and implementation of an intervention strategy.

In small organizations, the informal discussions that helped identify stress problems may also produce fruitful ideas for prevention. In large organizations, a more formal process may be needed. Frequently, a team is asked to develop recommendations based on analysis of data from Step 1 and consultation with outside experts.

Certain problems, such as a hostile work environment, may be pervasive in the organization and require company-wide interventions. Other problems such as excessive workload may exist only in some departments and thus require more narrow solutions such as redesign of the way a job is performed. Still other problems may be specific to certain employees and resistant to any kind of organizational change, calling instead for stress management or employee assistance interventions.

Some interventions might be implemented rapidly (e.g., improved communication, stress management training), but others may require additional time to put into place (e.g., redesign of a manufacturing process).

Before any intervention occurs, employees should be informed about actions that will be taken and when they will occur. A kickoff event, such as an all-hands meeting, is often useful for this purpose.

Step 3. Evaluate the Interventions
Evaluation is an essential step in the intervention process. Evaluation is necessary to determine whether the intervention is producing desired effects and whether changes in direction are needed.

Timeframes for evaluating interventions should be established. Interventions involving organizational change should receive both short- and long-term scrutiny. Short-term evaluations might be done quarterly to provide an early indication of program effectiveness or possible need for redirection. Many interventions produce initial effects that do not persist. Long-term evaluations are often conducted annually and are necessary to determine whether interventions produce lasting effects.

Evaluations should focus on the same types of information collected during the problem identification phase of the intervention, including information from employees about working conditions, levels of perceived stress, health problems, and satisfaction. Employee perceptions are usually the most sensitive measure of stressful working conditions and often provide the first indication of intervention effectiveness. Adding objective measures such as absenteeism and health care costs may also be useful.

However, the effects of job stress interventions on such measures tend to be less clear-cut and can take a long time to appear.

The job stress prevention process does not end with evaluation. Rather, job stress prevention should be seen as a continuous process that uses evaluation data to refine or redirect the intervention strategy.

HAZARD COMMUNICATION (HAZCOM) PROGRAM
The hazard communication standard (HCS) was promulgated by OSHA in 1983 as a means of reducing risks related to chemical exposure in the American workforce. Hazcom programs in the workplace need to be tailored to comply with the law and to protect workers against exposures specific to their work settings because no single written program will work for all worksites. The law, requires employers using hazardous chemicals in their operations to: prepare a formal inventory of those chemicals; inform and train employees on the nature of the chemicals through signs, labels, and material safety data sheets; and prepare a written program summarizing the employer's hazard communication program.

This federal standard has turned out to be the most sweeping approach to allowing employees the "right-to-know" the hazards of the chemicals they work with and precautions to take in working with those chemicals.

The standard is generic and performance-oriented so that an employer can adapt the broad provisions of the standards to the specific chemical hazards encountered in a facility. The standard does not apply to hazardous waste, which is covered by separate federal regulations.

This broad-based federal law has wide implications for many of the millions of facilities that come under OSHA enforcement control for safety and health. Since 1983 it has become the federal safety standard most frequently cited for violations by compliance officers.

A number of states have developed their own state-level "right-to-know" law, some of which have more stringent provisions than the federal standards. Occupational safety and health professionals who are analyzing implications of the federal standards should also check to see if their state has hazard communication requirements.

Federal Standard
The OSHA standards can be summarized into the following seven key elements:
1. Prepare an inventory of chemicals used in the facility.
2. Drums and containers of chemicals must be identified with a sign and labels.
3. Material safety data sheets on each chemical must be available.
4. Hazard communication training is to be provided to employees.
5. A written hazard communication program must be prepared.

6. A spill or emergency plan must be developed.
7. Methods of informing outside contractors of the hazards that they will be exposed to in your facility must be developed.

Chemical Inventory

Employers are required to survey the facility and list every chemical in use in the process, in maintenance and cleaning activities, excluding only packaging materials such as corrugated and folding boxes, tobacco and cosmetic products, and "articles" this is, your company's finished product. The basic inventory should be revised when a new chemical is added or one is dropped from use.

Signs and Labels

Manufacturers and distributors of hazardous chemicals must clearly identify incoming supplies of chemicals with identification labels and warnings as to the content of the chemicals. Portable, in-facility containers (transfer containers) used by individual employees do not require labeling if the material has been transferred from a labeled container and the materials is to be used up during the employee's workshift.

Material Safety Data Sheet

The employer is required to obtain from suppliers a material safety data sheet (MSDS) for each chemical used in the facility. These MSDS must be readily available to supervisors or employees who may have questions about the chemicals with which they work. A MSDS must be on hand for every chemical listed on the chemical inventory. Employers should follow-up to obtain MSDSs from "upstream" suppliers and advise the supplier they will reject any new chemicals coming in which do not have such labels affixed. Suppliers "trade secrets" are protected under the law. The MSDS can be retained in a local computer system or a binder prepared for the MSDS at each location with MSDS made readily available to all supervisors and all employees at all times.

Hazard Communication Training

Employees who work with chemicals must be trained in the nature of the chemicals they work with, precautions in using the chemicals, first aid and emergency steps to take, and the existence of the MSDS file or binder where this information on the chemicals is kept. Where specific procedures such as engineering controls or work practice controls are used to lessen employee exposure to the chemicals, these steps must be outlined in the written program. Training must take place at the time of initial assignment or when an employee is transferred to work involving the chemicals.

Hazard Communication Program

The steps listed previously are to be summarized in a written Hazard Communication Program that is available to employees, their union representatives or OSHA compliance officers for review.

Each year the OSHA Hazard Communication Standard leads the list of violations of OSHA standards. Citations are issued for such basics as lack of a written hazard communication program, lack of training of employees in the hazards of the chemicals with which they work, availability of MSDSs at the worksite, and labeling of containers of chemicals.

It is not difficult to comply with the Hazard Communication Standard. The investment of a few hours of staff time to review the standard, develop training, and a written program is a ready trade-off to the waste of management time in handling OSHA inspections, responding to citations, meeting in information conferences, and paying OSHA fines.

A copy of the federal Hazard Communication Standard 1900.1200 is available from the nearest OSHA area office (under U.S. Department of Labor in the blue pages of the telephone book) or state OSHA offices.

DRUG AND ALCOHOL TESTING PROGRAM

It is estimated that substance abuse costs employers billions of dollars each year as a result of increased injuries, fatalities, absenteeism, and excessive use of health care benefits, decreased productivity, theft, and alcoholism. The worksite is a strategic place for preventing and identifying early substance abuse and instituting drug-free worksite policies and programs. The purpose of a drug and alcohol testing program is to avoid hiring workers who use illegal drugs, to deter workers from abusing drugs and alcohol, and to identify and refer to treatment those workers who are presently abusing drugs and alcohol.

Types of Testing

As most people are aware, there are several types of substance testing, often distinguished by timing rather then type of examination. Some organizations use all of the forms of test schedules, and others use only one or two. The decision should be made individually based on the needs of the organization and the goals of substance testing.

Preemployment Testing

Preemployment substance testing should be used to screen all applicants for particular positions or all positions. Organizations may want to test all new hires, particularly if they have a drug free workplace or have another type of test schedule in place. They may only want to test those employees in hazardous or high-risk positions. Organizations need to remember that any time a group is singled out for any reason, particularly a reason as controversial as substance testing, concerns about discriminatory practices can arise.

Random Testing

Random substance testing is testing of employees on an irregular unannounced schedule. The use of random testing is critical if employers are serious about addressing the illegal use of drugs and alcohol in the workplace. This method is the only way to apply non-discriminatory, nonselective substance testing, and it is the only way to detect if there are casual substance abusers in the organization. If a company is going to have a meaningful substance-testing program, a random test is the most critical component of that program. Organizations can mitigate risk if they ensure that the tests are truly random. This can be done by using a random number generating computer program so the selection of employees for testing is not influenced by human decision.

Scheduled Annual or Semiannual Testing

These tests are most effective when used as part of an ongoing or completed rehabilitation program. They have severe limitations if they are the only tests in organizations that are serious about controlling substance abuse. The only persons who test positive in an annual test are those who have reached an addictive state that preclude them from abstaining prior to the test.

For-Cause/Reasonable Suspicion Testing

Testing employees suspected of substance abuse on the job can take several forms. Organizations can test employees who behave erratically, smell of alcohol, or are the subject of complaints from the public or other employees. Employers can also routinely substance test after any accident.

Policy

Having a written policy is critical any time a substance abuse testing program is initiated. There are several important issues to address as part of a substance abuse detection and testing program. The first part of the policy should make a clear statement of the reasons for the program. For example, it is required by law, part of a zero tolerance/drug-free workplace program, or a safety and security issue, etc.

The second part of the policy must state the method used for detecting substance abuse. As previously mentioned, there are several categories of tests that can be given. There should be an initial test of all employees, followed by random, post accident and for-cause tests. It is preferred that all employees be tested as part of the program. This sends a message of fairness and equality, and shows that there is a top down commitment to stopping a growing problem. Some employers choose to test only those employees who operate equipment, work with hazardous materials, work with or around minor children, or handle large sums of money.

Employers that have collective bargaining agreements must remember that drug testing is a mandatory subject of bargaining. An employer that begins a substance abuse testing program during a contract period may be charged with making unlawful changes to a collective bargaining agreement. Employers also must remember to take care to follow procedures outlined in the bargaining agreement when discussing potential discipline against employees.

Supervisory Training

Supervisors play an important role in the day-to-day administration of a substance abuse detection program. Supervisors must be trained on the effects of substance abuse both at work and at home. They need to understand that absence or performance problems at work and financial or relationship problems at home may be related to substance abuse.

Supervisors need to know what types of substances are being tested for and the type of test to be used. Supervisors should refer employees to the human resources/benefits area as the focal point for rehabilitative processes. Benefits available through the organization's insurance company or EAP, and community-based resources can provide the first steps in the rehabilitative process.

Regulations

Certain employers, due to the nature of their business, are covered by regulations that involve some level of substance abuse testing. If you are in an industry that is impacted by the Department of Transportation requirements or the Drug Free Workplace Act, you have specific guidelines for your business. These are not the only regulations to consider, but they are the most recognizable. The DOT regulations cover millions of employees, and most of the general public have no objection to these employees being tested, given the number of lives at stake in the operations of public transportation.

A variety of workers are covered by the DOT Regulations, including mass transit workers (Federal Transit Administration); motor carrier workers, primarily truck drivers with a commercial drivers license (Federal Highway Administration); aviation workers (Federal Aviation Administration); railroad workers (Federal Railroad Administration); marine workers (Merchant Marine covered by Coast Guard regulations); and natural gas and pipeline workers (Research and Special Programs Administration); the regulations also include maintenance personnel, dispatchers, and security guards who carry firearms in these operations.

Employees covered by these regulations must be tested for alcohol, marijuana, cocaine, opiates, phencyclidine, and amphetamines. The regulations require uniform testing standards. They include a very complex series of testing and documenting procedures, along with employee notice, retest and confidentiality requirements. The procedures outlined in the DOT regulations can be used as a good framework to establish a drug-testing program.

The Drug Free Workplace Act of 1988 applies to all federal contractors or federal grant recipients with contracts in excess of $25,000. The act requires a written policy statement that prohibits the manufacture, possession, distribution and consumption of controlled substances at the workplace. The act includes provisions for pre-employment certification that the employee will not use drugs in the workplace, drug prevention training during orientation and continued drug prevention training for all employees.

Title VII of the Civil Rights Act permits employers to refuse to employ people who engage in unlawful drug use as long as it is not a pretext for race, color, religion, national origin or sex discrimination, even if it has a disparate impact on classes under Title VII.

The Americans with Disabilities Act permits organizations to have policies that prohibit the possession of drugs and alcohol in workplace. The act also allows the prohibition of on-duty drug or alcohol use, or being under the influence of drugs or alcohol at work. The key part of the ADA to remember is that individuals who have been disabled by the use of alcoholism are protected, while those actively engaged in current alcohol or illegal drug use are not protected. The ADA permits employers to have a substance abuse testing policy, when such testing complies with all laws regarding test sample quality, test confidentiality, and employee rehabilitation, which should already be part of your program. The ADA does not cover applicants who test positive for illegal drugs. Under the ADA a drug test is not considered a medical examination. An organization may drug test to determine that an employee is no longer engaging in drug use without violating the ADA. Employees with a disability are held to the same performance standards as other employees, to the extent they can perform the essential functions of their duties, and do not pose a safety threat to others.

DEPARTMENT OF TRANSPORTATION GUIDELINES

Various federal statutes and regulations dictate what employment testing must be done before applicants qualify for certain jobs. The Department of Transportation (DOT) has authority over motor vehicle drivers under 49 *CFR* 391.40.

Before starting employment the applicant must have a Commercial Drivers License (CDL) and a physical examination. The examination is extensive and must be performed by a licensed physician. The purpose is to detect the presence of physical, mental or organic defects that would affect the applicant's ability to operate a commercial vehicle. Drug and alcohol testing is included. Physicians can grant waivers for certain physical conditions after review.

If the employer has an on-site clinic, the examination can be done there. If not, local medical clinics employ physicians and nurses certified in occupational medicine that are qualified to do the examination and testing needed.

Random drug testing is required periodically for all drivers licensed under DOT. Licensure and the physical examination are good for two (2) years, then must be renewed.

OCCUPATIONAL EXPOSURE TO BLOODBORNE PATHOGENS

In the late 1970s and early 1980s, a new workers' compensation claim was filed. An estimated 12,000 health care workers a year became infected with the hepatitis B virus after exposure to hospitalized, ill patients. Of this number, approximately 600 developed acute illness requiring hospitalization and lost work time; approximately 1,000 became carriers; and approximately 250 died.

The Centers for Disease Control and Prevention (CDC) published their *Recommendations for Protection Against Viral Hepatitis* in 1985 to give guidance to medical providers to prevent their acquiring the virus. Drug companies prioritized monies for the development of a vaccine. A vaccine was released for public use, but the high cost was to be borne by the employees themselves, with a focus on "safe work practices" instead of vaccination protection.

In the mid-1980s, another virus caused sensational worldwide response. The HIV (human immunodeficiency virus) not only made news headlines, but the medical experts needed to respond to an illness that resulted in death within six weeks of diagnosis. Although exposure and infection to medical workers was not evident, the CDC responded with a publication entitled *Recommendations for Prevention of HIV Transmission in Health-Care Settings* in 1987.

In 1988, the CDC updated their recommendations for medical worker safety in *Universal Precautions for Prevention and Transmission of Human Immunodeficiency Virus, Hepatitis B Virus, and Other Bloodborne Pathogens in Health-Care Settings*. And, because of the impact of worker safety, lost work time, and worker deaths (attributed to hepatitis B virus), OSHA was preparing their federal regulations to address the issue.

Virus Information

A virus is a living protein that attacks healthy cells to change or disrupt cell function or replication. The viruses HIV and hepatitis B (and others) are called *bloodborne pathogens* because they have their highest concentration (viral load) in the blood of infected people. Contact with viral-infected blood (or fluids that contain blood) present a high risk of infection to non-infected people. Body fluids that transmit these viral infections include (listed in order of viral load): blood, any fluid containing blood, vaginal secretions, semen, spinal fluid, abdominal cavity fluid, chest cavity fluid, joint fluid, amniotic fluid, breast milk (also tissue or cell cultures). Table 9-B is an easy, concise reference for information about those viruses.

OSHA Bloodborne Pathogens Standard

The Bloodborne Pathogens Standard (29 *CFR* 1910.1030) was released in December 1991, with a compliance date of July 1992. The standard addresses safety in the workplace by providing employers and employees strategies to reduce potential exposures to hepatitis B and HIV. Employers are required to provide a safe working environment for their employees. Employees are required to work safely, using universal precautions, PPE, and engineering controls.

The standard mandates that employers develop a written exposure control plan, specific to their facility. The plan consists of risk classification, task identification, universal precautions, work practice, engineering controls, housekeeping, personal protective equipment, immunization, exposure follow-up, and record-keeping.

Risk classification/task identification are shown in Table 9-C.

Universal Precautions

Unfortunately, not all people or blood or body fluids can be identified as virus-positive. Therefore, to treat *all* people, *all* blood and *all* fluids *as if* they were virus-infected, or to treat "universally", is good work practice. Handwashing (friction with soap for at least 10 seconds) is recommended before and after tasks, especially with blood or body fluid contact.

Engineering Controls and Housekeeping

Employers are required to provide safe working conditions for employees to reduce the potential for viral exposure. These include "safety" devices,

Table 9-B. Virus Information.

VIRUS	SYMPTOMS	MODE OF TRANSMISSION
HEPATITIS B Liver inflammation	■ Nausea, vomiting, stomach cramping, diarrhea, clay colored stools ■ Yellow tinge to skin, eyes (jaundice) ■ Dark, cloudy urine ■ Fatigue, aching muscles	■ Blood-to-blood contact ■ Blood-to-abraded, chapped, or nonintact skin ■ Blood-to-mucous membranes (eyes, nose, mouth)
HIV Immune system	■ Swollen lymph glands ■ Night sweats, recurrent fever ■ Fatigue ■ Rapid weight loss, diarrhea, loss of appetite ■ Yeast infections, pneumonia, other illnesses	■ Blood-to-blood contact ■ Blood-to-abraded, chapped, or nonintact skin ■ Blood-to-mucous membranes (eyes, nose, mouth)

such as self-sheathing needles and syringes, scalpel shields, puncture-resistant sharps disposal containers, biohazard cabinets, biohazard bags for waste, and EPA-approved disinfectants for cleaning. Training in the proper use of these items is also required.

Personal Protective Equipment

Personal protective equipment (PPE) places a barrier between blood and body fluids and the employee's mucous membranes, skin, and clothing. These include gloves, masks, goggles, face shields, gowns, and aprons that are fluid resistant. The employer must provide and have these items available to the employee. The employee must be trained in the selection of the barrier that best fits the task. For example, handling soiled linens would require gloves and aprons or gowns, and possibly masks. Handling tissue cultures would require gloves and fluid-resistant lab coats.

Immunizations

Hepatitis B vaccine must be offered free of charge within 30 days of hire to all employees who have the potential for exposure to blood or body fluids. Employees who refuse the vaccine (a series of three) must sign a refusal waiver, which is kept in their medical records.

Post-Exposure Follow-up

Although every effort is made to prevent viral exposure, the exposure plan must contain employee counseling, exposure occurrence evaluation, treatment, and referral. Recent information concerning viral transmission recommends immediate reporting of the incident to begin treatment within hours of the exposure. Employers should contract with emergency care facilities to expedite care delivery to the employee. Acute care delivery facilities frequently use an algorithm to determine treatment modalities.

Record Keeping

All documented exposures are kept for the duration of employment, plus 30 years. Access is limited to the employee, or their designee (with written consent). All training records must include the training dates, the contents, and the names and qualifications of the trainer. Training records are maintained for three years.

OCCUPATIONAL EXPOSURE TO TUBERCULOSIS

Background

Healthy People, the U.S. national prevention initiative, is a compilation of health promotion and disease prevention objectives to improve the nation's health. One of the objectives of *Healthy People 2000* (1980) has been the elimination of tuberculosis. Improved housing and living conditions, and advances in medical care have reduced the incidence of tuberculosis in the United States, so the goal of elimination by the year 2000 was feasible.

However, alarming statistics in 1990 identified an increase in tuberculosis cases (approximately 20% since 1985), especially among the HIV-infected populace. Also, some health care providers, those caring for hospitalized HIV-infected tuberculosis patients, contracted tuberculosis and died.

In a response to provide guidance for employers and employees, the Centers for Disease Control published *Guidelines for Preventing the Transmission of Tuberculosis in Health Care Settings, With Special Focus on HIV-Related Isues in 1990*, updating and expanding the guidelines in 1994 to include all issues.

Table 9-C. Risk Classification/Task Identification.

Job Classification	Task	Mode of transmission
Medical staff (doctors, nurses, technicians)	■ First aid response ■ Specimen collection/handling ■ Using needles, syringes ■ Dirty equipment handling ■ Pulmonary function testing	■ Blood/body fluids ■ Blood/body fluids ■ Blood/blood; needlesticks ■ Blood/body fluids ■ Open mouth wounds
Fire, police	■ First aid response ■ Contact with illegal drug syringes ■ Contact with guns, knives ■ Contact with unruly suspects ■ Crime scene investigations	■ Blood/body fluids ■ Open mouth wounds ■ Blood/blood; needlesticks ■ Blood/body fluids ■ Blood/body fluids ■ Open mouth wounds ■ Blood/body fluids
EMTs, paramedics	■ First aid response ■ Specimen collection/handling ■ Using needles, syringes ■ Dirty equipment handling ■ Contact with illegal drug syringes ■ Contact with guns, knives ■ Contact with unruly suspects	■ Blood/body fluids ■ Open mouth wounds ■ Blood/body fluids ■ Blood/blood; needlesticks ■ Blood/body fluids ■ Blood/blood; needlesticks ■ Blood/body fluids ■ Blood/body fluids ■ Open mouth wounds
Custodians, housekeepers	■ Cleaning bathroom facilities ■ Clean-up (blood, vomit, etc.) ■ Waste removal (feminine hygiene items, others) ■ General cleaning	■ Blood/body fluids ■ Blood/body fluids ■ Blood/body fluids ■ Blood/body fluids; needlesticks
Corrections institutions employees	■ First aid response ■ Contact with illegal drug syringes ■ Contact with guns, knives ■ Contact with unruly suspects/ searches	■ Blood/body fluids ■ Open mouth wounds ■ Blood/blood; needlesticks ■ Blood/body fluids ■ Blood/body fluids ■ Open mouth wounds
Laundry employees	■ Handling soiled linens	■ Blood/body fluids; needlesticks
Trash removal employees	■ Handling Biohazard waste ■ Handling general waste	■ Blood/body fluids; needlesticks ■ Blood/body fluids; needlesticks
Coroners, morticians, etc.	■ Incisions, syringe handling, sample collection and preparation, examination and equipment handling	■ Blood/body fluids; needlesticks

Increased TB infections, workplace exposures, and employee deaths encouraged OSHA to evaluate employer provisions for employee protection from tuberculosis exposure.

Tuberculosis Bacteria

Mycobacterium tuberculosis is the bacterium that causes tuberculosis (TB). The bacteria generally attack the lungs, causing chronic or acute infection. Bacteria is carried in airborne particles (from speech, sneeze, and cough) of those infected, and spread to others who inhale the bacteria-infected air. Symptoms appear after a number of weeks to months. They include productive coughing, night sweats, unexplained weight loss, loss of appetite, and fever. The TB infection is verified by x-rays, skin tests, and sputum cultures.

Although there is evidence of TB infection occurring since 2000 BC, the TB of the 1990s is very worrisome. Prior to this decade, inexpensive antibiotics successfully treated TB infection. However, the emerging multiantibiotic-resistant strains of bacteria have changed that scenario. Presently, a five-drug regimen is recommended for TB therapy. Once TB cultures are obtained and the antibiotic sensitivities identified, the drug treatment therapy can be re-evaluated.

TB Standard

Respiratory Protection for M. Tuberculosis (29 *CFR* 1910.139) was published in January 1998. The standard incorporated the 1994 CDC Guidelines, and the OSHA Standard Respiratory Protection (29 *CFR* 1910.134). The purpose of the standard is to reduce health care workers' risk for exposure to TB. This is accomplished by identifying infectious persons, isolating those persons, providing respiratory protection for health care workers, providing medical surveillance and treatment, and maintaining documentation. Administrative measures, engineering controls, personal protective equipment, and education accomplish these.

Respiratory protection for tuberculosis requires diligence on the parts of employers and employees. *Best work practice,* which includes protective equipment, environmental controls, and administrative measures will reduce the incidence of infectious tuberculosis as a work-related injury (see Table 9-D).

WORKPLACE VIOLENCE PREVENTION AND CRITICAL INCIDENT STRESS DEBRIEFING

Because of its epidemic growth, workplace violence is creating a new form of job hazard and an increased sense of worker vulnerability. Almost daily the subject of workplace violence is captured in a newspaper headline sending shock and alarm through many business communities. Although workplace violence entering the work environment logically follows workplace problems such as theft and substance abuse, workplace violence inspires fear in a way that neither of the other criminal behaviors has done. This may be true because it appears to be produced by the *irrational and enraged employee,* one of the most frightening elements threatening our culture.

It also appears that no business is immune to workplace violence. Once a risk limited to gas stations, liquor stores, convenience stores, taxi drivers, law enforcement, etc., workplace violence has infiltrated every commercial arena. The contributors to aggressive behavior are manifest everywhere: financial insecurity, layoffs, job stress and work related conflicts, the cultural glamorizing of violence, and the accessibility of weapons. The deadly results of domestic disputes are increasingly enacted in the workplace.

Table 9-D. Complying With the OSHA TB Respiratory Protection Standard.

Measures	Outcomes	Achieved by:
Administrative	■ Identify infectious patients	■ Review medical history, TB contacts, lifestyle, etc.
		■ Observe signs & symptoms
		■ Confirm with chest x-ray, skin test, sputum cultures
	■ Medical surveillance for health care workers	■ Identify "at risk" tasks & jobs
		■ Annual skin testing
		■ Post-exposure medical management
	■ Program effectiveness	■ Designate person or committee for plan oversight & evaluation
	■ Continuity of care	■ Policies/procedures
Engineering controls	■ Prevent spread of TB-infected droplets	■ Local exhaust ventilation (isolation room)
		■ Directionally controlled airflow (negative pressure)
		■ Air filtration and/or ultraviolet irradiation
Personal protective equipment	■ Reduce/ prevent exposure	■ NIOSH-approved filtration masks
		■ Fluid-resistant barrier gowns
		■ Gloves
Education	■ Safe work practice	■ Medical evaluation & NIOSH mask fit-testing
		■ Protocol for mask storage, cleaning, and disposal
		■ Formal & informal information sessions about TB, treatment, policies
		■ Hazard communication (labels, signs, chemicals)
		■ Equipment handling, cleaning, storage

Measures an Organization Can Take

There are measures organizations and particularly health and safety professionals along with human resource departments, can take to reduce the risk of employee-committed violence that go beyond physical security precautions. Creation of a workplace culture that provides the following can contribute to the reduction of workplace violence:
- encourages open communication
- empowers employees
- provides support systems
- treats all fairly
- fosters an atmosphere less susceptible to aggressive behavior
- peer review and team-building programs
- employee hotlines
- anonymous tip programs
- EAPs.

According to the National Safe Workplace Institute, characteristics of a high-risk workplace, include: chronic labor/management disputes, frequent grievances filed by employees, a disproportionate number of injury claims, especially psychological; understaffing or excessive demands for overtime and an authoritarian management style.

Occupational health and safety professionals working with human resource staff can update their recruitment policies to include in-depth reference checks, interview techniques, and psychological tests designed to reveal those applicants prone to violent behavior. There is some controversy around the use of psychological tests for fear that their validity will be tested legally; however, many nationally known companies in the business of psychological testing will stand behind their instruments and support any employer in a legal challenge.

Profile

Data collected to-date provides a profile of the perpetrator of workplace violence. He is male, 35 years of age or older, has a previous history of violence toward women, children, or animals, owns a weapon, and reflects self-esteem externally connected with his job, concurrent with minimal outside interests. He is likely to be withdrawn, or a loner, and is characterized by a tendency to externalize blame for his life disappointments. There is a high probability of military history; and substance abuse and/or mental health concerns may be prominent, although often identified only after the fact.

Occupational health and safety professionals should train managers to recognize and defuse potentially violent situations. Many perpetrators signal their intent and could be thwarted. Aggressive behavior can be defused. There are communication skills useful in de-escalating a workplace confrontation that can be taught in a classroom. Most of these verbal skills create and sustain conversation—the longer the dialogue the less the probability of physical violence. There are many good training sources for the management of aggressive behavior.

Sometimes a violent incident will erupt before there is an opportunity to intercede. It is smart to be prepared to deal with the worst possible scenario. A crisis management or threat management team can be prepared to respond to any predictable disaster—no matter how unlikely. Evacuation plans, communication networks, community resources, etc. should be garnered and formalized before a crisis takes place. In the event of workplace violence, immediate on-site counseling and intervention is critical. It makes a tremendous difference to the mental health and recovery time of the people involved. Employee assistance professionals are the best resource for those companies that employ them. For those who do not, most towns and cities are now served by community-based counseling organizations. Occupational health and safety professionals should make contact with those professionals and ask for their help in the planning so that if needed they will know you and your organization and be ready to respond in an emergency.

The creation of a healthy work environment prepared to respond to a workplace crisis requires the employment of sound occupational health and safety as well as human resource practices, management training, and the formation of an emergency response team and plan. (For tips on writing a plan for your facility, see chapter 18, Emergency Response Programs in this book and the *On-Site Emergency Response Planning Guide,* available from the National Safety Council.)

Conducting a Critical Incident Stress Debriefing

A *critical incident* can be defined as a workplace event that is extraordinary in nature and that could be expected to produce significant reactions on the part of victims or those otherwise either directly or indirectly impacted, such as witnesses, colleagues, and/or family members. *Critical incident stress* is often described as the natural reaction of a normal person to an extremely abnormal situation. It can manifest itself as a physical, cognitive, and/or emotional response that may be experienced almost immediately or may be delayed days, weeks or even months.

While the primary internal organizational response to such an event is usually coordinated by appropriately trained members of the organization's occupational health and safety team or human resources department, it is usually in the best interest of both the organization and its staff to involve employee assistance professionals trained in *critical incident stress debriefing* immediately following any significant incident, or as soon as possible thereafter. Such professionals are specifically trained to deal with the effects of crisis in the workplace and can provide many valuable services to an organization. Some of the objectives met by using such professionals are listed:

- *Conduct a debriefing immediately following an incident.* Those affected may feel stunned and/or extremely emotionally volatile. A trained critical incident consultant can conduct a debriefing that provides a structured setting for victims to talk about the event and begin to process it in their minds.
- *Provide reassurance.* Knowing that the organization is taking measures to help, as witnessed by calling in a professional consultant, can make everyone feel a little less traumatized and bolster morale and confidence.
- *Ask appropriate questions and/or de-escalate.* For people who are not used to routinely dealing with the immediate effects of trauma, normal reactions to a critical incident can seem frightening and/or out of control. A trained professional knows the appropriate questions to ask to determine whether a response is normal or if additional help should be sought. He or she can also be instrumental in de-escalating a highly emotional situation and/or individual(s).
- *Provide support to management.* A trained professional can help management to structure their response to the crisis as the situation evolves and to evaluate the need for particular services on an ongoing basis. He or she may also be able to provide the personal and objective support needed by management in a time of great stress.
- *Assess the emotional damage.* A trained professional can help identify how individuals and/or groups are responding to the crisis and intervene to help prevent escalating and/or long-term problems.
- *Link employees with additional help if and when needed.* A trained professional is able to make referrals for any outside counseling or additional help that is clinically indicated.

In almost all instances following the occurrence of a critical incident where a crisis or EAP professional is called into an organization, one of the first steps is to conduct a *debriefing*. A discussion qualifies as a formal debriefing if it meets the following criteria: it is a scheduled meeting, it is structured with a specific purpose and direction, it includes all members of a designated group, and it is facilitated by an outside professional who was not involved in the incident. In almost all cases, a debriefing is most effective when it is conducted within three days of the precipitating incident. Formal debriefings usually comprise the following five elements and basically follow the format briefly outlined as follows.

Introduction
It is advisable for an executive with the organization to introduce the meeting by stating why the meeting is being held. That individual should make a strong statement regarding confidentiality. Although the debriefing facilitator will reiterate this information, it is important for the affected staff to hear it from a well respected and a concerned internal source. The executive should then introduce the facilitator and leave the room, unless that individual was directly involved with the incident in which case they should remain a part of the group. All individuals attending should be made to feel as comfortable as possible, being free to move about the room as necessary. Simple refreshments should be made available.

The facilitator should then restate the purpose of the meeting by letting those attending know that everyone is present because all of them experienced a very dramatic event. It should be discussed that experience has shown that, in similar situations, it has been beneficial to those involved to discuss what happened. Even though some people may feel that it is unnecessary to do so, some of what they might have to say may be very useful to someone else who is having a more difficult time dealing with the experience. It should be pointed out that participation, while voluntary, will be appreciated and that what is talked about during the meeting(s) must remain confidential.

Describing the Event
As the facilitator neither witnessed nor was involved in the incident, it is important that those who were present and/or involved talk about it and creates as vivid and complete a picture as possible. Also it is common for people who are directly involved in such an incident to develop tunnel vision and become very focused on their own functioning and lose a sense of the big picture. Therefore, this phase can be especially helpful in putting everything in perspective for all who were involved. During this phase, a discussion should be conducted through which the incident is recreated by having each individual attempt to factually describe in detail what he or she did as part of the critical incident. This is usually most helpful if done chronologically.

Reactions
Following the chronological recreation of the critical incident it is important to allow those involved to share their reactions, those reactions experienced both during and following the incident. Some of the reactions reported may include any one or several of the following symptoms: time distortion, depersonalization of victims, intrusive thoughts or visions and/or physical symptoms such as difficulty breathing, shakiness, or nausea.

Education
Information should be imparted throughout the debriefing process. It is important to normalize reactions and to be reassuring and supportive. Sufficient time should be allowed for questions and answers.

Follow-up
Prior to the end of the debriefing session, materials should be distributed which describe some of the most common responses to traumatic events, the primary stages of the healing process, common stress reducers, and relevant support information for family members/significant others. Those individuals participating in the group meeting or debriefing should also know how to reach the facilitator should they wish to talk further. It is also helpful for the participants to sign a list providing the facilitator with their name and phone number so that a follow up call might be initiated, if appropriate.

It should be noted that this model for debriefing strategy would apply to occurrences such as natural disaster, serious accident/death in the workplace, violent crime in the workplace, death of a co-worker outside of work, serious illness of a co-worker, arrest of or charges pressed against a co-worker, suicide attempt/completion, or downsizing and/or layoff of personnel. In the case of victimization by crime in the workplace or as a function of their job, it is best to intervene with the victim individually and to formally debrief the other affected individuals as a group. At a later time, another debriefing may be conducted with the victim and others together if the victim feels that it would benefit him or her in some fashion.

The purpose of any *critical incident stress debriefing* is to assess, stabilize, and support those individuals impacted as a result of a critical incident. It is designed to help minimize, to the greatest degree possible, negative disruption to the lives of individual employees and to the work life of the organization. While many feel that the initial debriefing can best be accomplished by an externally trained mental health/EAP professional who has been identified and screened by the organization long before a need for their services is identified, the ongoing strategic organizational support role following a critical incident is most effectively accomplished by members of management inclusive of occupational health and safety professionals along with the human resources team. For more information on this subject, see chapter 17, Employee Safety and Security Programs.

SUMMARY
Occupational health nursing (OHN) requires a broad base of knowledge in nursing, public health, and human relations. As market driven health care reform continues, the occupational health nurse is an organization's primary source regarding health care issues and the delivery of occupational health services. Occupational health nurses provide value-added knowledge and expertise to case management services, return-to-work planning, and management of integrated disability management and workers' compensation services. Occupational health nurses are also adept at being the liaison

between employer benefit plans and organizations providing health care delivery. There is no doubt that occupational health nurses will continue their phenomenal growth, and in collaboration with other professionals be part of an interdisciplinary approach to providing expertise in occupational health and safety for the purpose of improving workers' health, safety and well-being.

REFERENCES

AAOHN. Position statement: The occupational health nurse as a case manager. *AAOHN Journal,* 42 (4), 1994.

American Board for Occupational Health Nurses. Position statement: The certified occupational health nurse as a case manager. Hinsdale, IL:ABOHN, February 1994.

CDC. Guidelines for preventing the transmission of *Mycobacterium tuberculosis* in health-care facilities. *MMWR* 43 (RR13):1–32, 1994.

CDC. Guidelines for preventing the transmission of tuberculosis in health-care facilities, with special focus on HIV-related issues. *MMWR* 39 (No. RR17), 1990.

CDC. Recommendations for prevention of HIV transmission in health-care settings. *MMWR* 36 (suppl. 2S), 1987.

CDC. Recommendations for protection against viral hepatitis. *MMWR* 34:313–324, 329–335, 1985.

CDC. Update: Universal precautions for prevention of transmission of human immunodeficiency virus, hepatitis B virus, and other bloodborne pathogens in health-care settings. *MMWR* 37:377–382, 387–88, 1988.

Code of Federal Regulations 29 CFR 1910.139. Washington DC: U.S. Government Printing Office, 1997.

Code of Federal Regulations 49 CFR 391.40. Washington DC: U.S. Government Printing Office.

Code of Federal Regulations 29 CFR 1910.1030. Washington DC: U.S. Government Printing Office, 1988.

Dishman RK et al. Worksite physical activity interventions. *Am J Prev Med* 15(4):344–361, 1998.

Felton J, Cole R. The high cost of heart disease. *Circulation* 27:957–962, 1963.

Fielding JE. Effectiveness of employee health improvement programs. *J Occ Med* 24:907–916, 1982.

Fielding JE. Health promotion and disease prevention at the worksite. *Ann Rev Pub Health* 5:237–265, 1984.

Fielding JE, Piserchia PV. Frequency of worksite health promotion activities. *Am J Pub Health* 79:16–20, 1989.

Fisher EB, Bishop DB, Mayer J, et al. The physician's contribution to smoking cessation in the workplace. *Chest* 93(2):56S–65S, 1988a.

Fisher EB, Nord W, Warren-Boulton E. Organizational factors in implementing patient education. *Diabetes* 37:175A, 1988.

Gibbs J, Mulvaney D, Henes C, et al. Worksite health promotion: Five-year trend in employee health care costs. *J Occ Med* 27:826–830, 1985.

Jones J, Dosedel J. The impact of corporate stress management on insurance losses. *Legal Insight* 1(4):24–27, 1986.

Lawson C. An interdisciplinary approach to safety and health promotion: The "whole person" model. *The Occupational and Environmental Medicine Report* 13 (2), 1999.

Lawson C. Bell Atlantic benefits from whole person model. *Health Promotion Practitioner* 6 (4), May 1997.

Lawson C. Bell Atlantic calls on safety diligence. *Occupational Hazards,* July 1996.

Lechner L, DeVries H. Participation in an employee fitness program: Determinants of high adherence, low adherence, and dropout. *J Occup Environ Med* 37(4):429–436, 1995.

Leviton LC. The yield from work site cardiovascular risk reduction. *J Occ Med* 29:931–936, 1987.

Mullen PD. Health promotion and patient education benefits for employees. *Ann Rev Pub Health* 9:305–332, 1988.

Manuele FA. Learn to distinguish between hazards and risk. *Safety + Health* 150 (5): 70–74, 1994.

Marcus BH et al. Physical activity interventions using mass media, print media, and information technology. *Am J. Prev Med* 15(4):362–378, 1998.

Pelletier KR, Klehr NL, McPhee SJ. Town and gown: A lesson in collaboration. *Business & Health* February 1988a, pp 34–39.

Pelletier KR, Klehr NL, McPhee SJ. Developing workplace health-promotion programs through university and corporate collaboration. *Am J Health Promotion* (2)75–81, 1988b.

Pilisuk M, Parks SH, Kelly J, et al. The helping network approach: Community promotion of mental health. *J Primary Prev* 3(2):237–242, 1982.

Pransky GS, Frumkin H, Himmelstein JS. Decision making in worker fitness and risk evaluation. In J.S. Himmelstein, G.S. Pransky (eds). *Occupational Medicine: State of the Art Review* 3(2):179–191, 1988.

Pruitt RH. Pre-placement evaluation: Thriving within the A.D.A. guidelines. *AAOHN Journal* 43(3):124–130, 1995.

Salazar MK (ed). *AAOHN Core Curriculum for Occupational Health Nursing.* Philadelphia: WB Saunders, 1997.

Shephard R, Corey P, Ruezland P, et al. The influence of an employee fitness program and lifestyle modification program upon medical care costs. *Can J Public Health* 73:259–263, 1982.

Shirasaya K et al. New approach in the evaluation of a fitness program at a worksite. *J Occup Environ Med* 4(3):195–201, 1999.

Travers PH, McDougall C. *Guidelines for an occupational health & safety service.* Atlanta: AAOHN Publications, 1995.

chapter 10

Safety Programs

by Dan Petersen, PE, CSP

153 Introduction
153 Principles of Safety Management
    System failure ■ Human error
155 Safety Program Requirements
155 Evaluating Current Safety Systems
156 The Packaged Audit Controversy
156 The Self-Built Audit
156 Alternatives to Audits
157 Surveys
157 Company Safety Policy
158 Safety Rules and Regulations
    The safety organization ■ Responsibilities ■ Assigning responsibility ■ Accountability
160 The Worker and Safety
    Training and motivation ■ Safety committees pro and con ■ Safety inspections ■ Incident investigation ■ Job safety analysis
165 Behavior-Based Safety
166 Summary
166 References

INTRODUCTION

The criteria for a successful safety process are consistent, even though companies differ from one another and no single program can be recommended for all. A minimum of these six criteria must be met if any program is to be successful:

- Senior management must visibly lead the way.
- Middle management must be involved.
- The process must force supervisory performance.
- Employees (workers) must be involved meaningfully.
- The system must be flexible enough to change when change is indicated and flexible enough to allow individual decisions and ownership.
- The safety system must be perceived as positive, not negative and boring, by the employees.

PRINCIPLES OF SAFETY MANAGEMENT

Current thinking on safety management and its principles is based on a number of assumptions. First, incidents are symptoms that the management system is at fault. Experience shows that most severe injuries happen as the result of activities considered nonroutine or nonproduction, activities involving high energy sources, certain construction situations, etc. Safety should be managed like any other company function. Achievable goals should be set, and a process of planning, organization objective setting, and control should be established to meet those goals. An important part is the determination of persons who are to be held accountable for performance. Operational errors that result in incidents should be analyzed carefully to determine why they occurred and whether they could have been prevented with more effective controls. Incidents, injuries, and other losses result from both system failure and human error. These two must be taken into account by management.

System Failure
Management must determine policy for the organization, defining responsibility, authority, and accountability at each level. Before an incident occurs, it must be clear who is accountable for what, and how those persons can fulfill their responsibility for safety. In setting up the safety process, the following questions might be considered.
- What types of safety inspections should be used? By whom? How often, etc.?
- How are chemical hazards to be identified? What systems are needed?
- What should be done when unsafe conditions are found?
- How are employees who are involved in incidents identified and what approaches might be used?
- How should new employees be trained with regard to general safety and safety for specific jobs?
- How are employees to be selected for particularly hazardous jobs?
- Is the company's medical program adequate for the most severe problem that can be anticipated?
- What types of records should be kept on incidents, and how should they be used?

Answers to these and hundreds of other questions should provide an outline for developing a company's policy for handling system failures.

Human Error
Although an incident occurs because of failures in the system, it also occurs because of human error. Human error can result from worker overload (too much work, too little time, etc.), an individual decision to err, problems in the workplace, or a combination of these and more.

Overload (also discussed in other chapters) is a significant cause of human error. The cause can be physical (work that is too difficult manually); physiological (work that overstresses the human system); or psychological (too much work, too little work, and other stressful conditions). Incidents in which overload can be a factor can be dealt with on an individual basis. Each person involved must be studied with regard to his or her current workload, physiological capacity, and psychological state. Has the individual's workload been increased recently? Has body strength been weakened by recent illness? Does he or she face psychological problems either at work or at home? Similarly, pressure, fatigue, and drug or alcohol abuse must also be considered. Practically every aspect of the worker's life, both at work and in the home, can contribute to human error. Thus, the problem of determining whether overload is the cause of an incident is complex and requires considerable time and judgment to resolve.

Human error can also stem from an unconscious or conscious decision of the worker to err. If pressured by supervisors or peers to improve production, the worker usually chooses an unsafe rather than a safe approach to save time and get the job done. The decision by the worker to work safely or unsafely is determined by a number of factors. Consider these just for starters:
- What has the worker learned in terms of the payoffs for safe or unsafe behavior? Has the worker been reinforced for working safely? Or has the worker gotten a message from the boss and co-workers that taking chances and getting the job done is better?
- What does the worker's peer group endorse as the more acceptable behavior? What are the peer group norms? That is what the worker will do.
- How has work performance been measured by the worker's supervisor in the past for safety? For production? Which measure has been more precise? Usually we measure productivity (number of pieces completed) much more accurately than safe behaviors.
- How has the worker been rewarded for safety? For productivity? Which reward happens most often? Which is larger?

We know today that the worker who chooses to work unsafely does so because those behaviors are completely logical to that person. The worker has learned that unsafe behavior (taking chances) pays off to his or her boss, to his or her peers, to the organization.

To change worker behavior, you must first change the psychological environment that makes unsafe behavior normal and logical to most workers.

Some workers simply do not believe that anything negative will happen to them in any aspect of their lives. They also will take chances (work

unsafely) when working. Taking chances happens for many reasons.
- It makes sense in the psychological environment in which a worker finds himself or herself.
- The worker may not know any better (lack of training).
- The worker doesn't care (lack of motivation).
- The worker is physically or psychologically tired (e.g., believing that as a result of corporate downsizing, he or she is being asked to do too much each day; or is being asked to work too long, etc.).
- The worker may have had to cope with too much, both at work and at home.
- The worker's personality may be such that he or she will do whatever is necessary to get the job done, even if it means taking risks.

The design, maintenance, housekeeping, etc., of the workplace can cause human errors. The workplace can be incompatible because there is too little light or too much noise, or for many other reasons. Today we know that an unsuitable work environment can cause unsafe behavior (human error). People work best and most safely when in their comfort range on environmental variables. Work performance deteriorates when people are forced to work outside of their comfort ranges.

Designing workplaces to reduce system-caused human errors and workstations that are compatible with the human body is a facet of ergonomics that is an integral part of safety management. This ergonomic approach to design of the workplace today is a departure from the workplaces of the past, which may have lacked workstations that were appropriate for the worker and the job functions the worker was required to perform.

Lack of adequate ergonomically correct design leads to hazardous workplace conditions. When we design a workstation in which a person cannot reach the needed controls, we design to ensure error; or when we design a machine that operates in a way that is opposite to what we expect, we ensure error. For example, demanding employees perform repetitive motions for long periods without providing adequate protective equipment and techniques often results in the employees complaining of repetitive motion problems, or cumulative trauma disorders (CTDs). These have ended up costing companies millions of dollars and prompted the drafting of new federal guidelines to cover this type of situation. Before an accident occurs, management should reevaluate the condition of the workplace, to determine whether it is properly designed and if controls are adequate.

SAFETY PROGRAM REQUIREMENTS

The requirements of a company safety process vary according to its type of operation, size, equipment and processes, geographic location, and organizational climate. Hazards vary by type of industry and between companies in the same industry, because no two companies have identical design and layout.

If the process is to be successful, all employees of the company, from the president to the line employees, must be committed to safety. Most important, management cannot view safety as separate and distinct from other concerns. Safety must be an integral part of production at all stages if the company is to be efficient and profitable. Although safety should be the highest priority of each employee, the direction of and driving force behind the safety process must rest with management. Safety performance should become such an integral part of every management job that levels of performance can affect chances for advancement, salary increases, etc.

EVALUATING CURRENT SAFETY SYSTEMS

There are numerous tools for evaluating the effectiveness of a safety process. One of the older and least useful methods compares the incident rate in a company with its experience in previous years and with incident rates in other companies. To determine the incidence rate, we need to know (1) the number of injuries and illnesses recorded during the year; and (2) the total number of hours actually worked by all employees during the year. With these numbers and the following formula, the incidence rate can be obtained.

$$\frac{\text{Number of injuries and illnesses} \times 200{,}000}{\text{Total employee hours worked}}$$

The incidence rate does not indicate how effective or ineffective a safety program is in all cases. For larger companies, it could be a useful indicator, but for smaller ones, it only indicates that the company has been lucky. The incidence rate cannot be used to indicate the safety performance of individual managers or supervisors.

Other methods can be used to evaluate a safety program. One is the audit, used by many companies to assess the effectiveness of its safety system. An example is the OSHA-recommended safety audit, which provides insight into the effectiveness of the program by showing where changes are needed. Better, audits should be self-built within the organization.

THE PACKAGED AUDIT CONTROVERSY

As the audit concept became popular, the idea of the packaged (purchased from outside the company or from OSHA) audit also became popular. Believed to have started in Canada, the concept evolved throughout the world.

The whole concept of the packaged audit is that there are certain defined things that must be included in a safety system to get a high rating or the biggest number of stars.

How does this thinking jibe with the research? Not too well:

- A NIOSH study in 1978 identified seven crucial areas needed for safety performance. Most are not included in the previously described packaged programs. A Michigan study had similar results.
- Foster Rhinefort's doctoral dissertation at Texas A&M University suggested there was no one right set of elements.
- A major study done by the Association of American Railroads conclusively showed that the elements in most packaged programs were not correlated with bottom-line results.

Clearly the research questions the validity of the concept of packaged safety audits. It does not question the value of the audit concept only of packaged audits.

In the light of today's management thinking and research, the packaged audit concept is suspect. When packaged audits became popular, many questions were left unanswered. In most systems a number of elements were defined; and in most, all elements were equally weighted. Thus, having the right books in the corporate safety library counted as much as whether supervisors were held accountable for doing anything about safety. Most of us never questioned this nor what components were included in the safety program. Thus, some systems had five important components, some 19, some 21, some 30. It all depended on who made up the system.

In addition, there seemed to be little effort made to correlate the audit results to the incident record (safety practitioners were buying into an unproved concept). When some correlational studies were run, the results were surprising:

- One Canadian oil company location consciously chose to lower their audit score and found their frequency rate significantly improved.
- One chain of U.S. department stores found no correlation between the audit scores and workers compensation losses. They found a negative correlation between the audit scores and public liability loss payout.

THE SELF-BUILT AUDIT

The self-built audit could be the answer: to construct an audit that accurately measures the performance of the safety system. The process of audit construction consists of the following:

1. defining the safety system elements
2. defining the relative importance of them (weighting)
3. defining the questions to find out what is happening.

Some painful lessons are learned in the audit construction process. In one organization, an audit that looked great ended up with no correlation to results.

The audit must be tested against reality. In one organization, the accountability section was heavily weighted, but not enough attention was paid to the quality of performance. The result was a paper program only.

ALTERNATIVES TO AUDITS

Audits have some severe drawbacks for instance, they
- stifle autonomy.
- stifle creativity.
- force uniformity.
- tend to be based on one person's opinion as to what is right and wrong.
- are subjective.

They also have some real advantages; for instance, they
- get attention.
- force performance.

How can we get the advantages without the drawbacks? One way to do this is to audit against agreed-to criteria instead of company-dictated (or outsider-dictated) standards. This allows total flexibility in programs as long as the system meets certain criteria.

What are the criteria? You'll have to decide for yourself based upon what you believe causes injuries and what constitutes an effective safety system. Then, based upon what you believe, identify your criteria against which to judge the safety system.

These are my criteria.
- Does your safety system force supervisory performance?
- Does your safety system involve middle management?
- Does your safety system get visible management commitment?
- Does your safety system get total employee participation?
- Is your safety system flexible?
- Is your safety system seen as positive?

SURVEYS

One excellent way of assessing the safety system is through the use of a perception survey. Some organizations are currently marketing safety climate surveys, which might provide a dipstick of current thinking. A perception survey can define clearly and validly the strengths and weaknesses of many of the elements of your safety system.

There seems little question that the way the worker sees the company safety program strongly influences not only his or her behavior on the job but the ability to learn from, and to respond to safety materials. Workers in different companies characterize their safety programs quite consistently in any given company. Research shows that a properly constructed perception survey of employees, properly administered, seems to be the most valid indicator of safety program effectiveness.

Comparison of data from the survey with other indicators of safety performance confirmed that units with the highest positive response to survey questions were generally those with the best performance as measured by other indicators of safety performance.

The analyses of survey data proved to be extremely useful to management. For instance, on one railroad the survey clearly identified the weaker management systems affecting safety performance as:
- *recognition for good safety performance*—with a low 48% favorable response from hourly rated employees
- *inspections*—with only 51% favorable response
- *supervisor training*—49% favorable response
- *quality of supervision*—50% favorable response.

This cluster clearly indicated a major problem in perception of supervisory performance and suggested that management needed to immediately train supervisors and hold them accountable for performance in all aspects of safety, emphasizing observations of people and conditions and positive reinforcement techniques.

In this same company, the widest difference in perception of program effectiveness between hourly rated employees and management was in these same four categories. Not only did they have a serious problem, but they were largely unaware the problem even existed.

One of the locations in a company was consistently lower in positive hourly rated response in nearly every category than the other regions. Management in this region clearly had a severe credibility problem with employees in nearly every phase of their safety efforts.

In one company, a single location's scores on recognition for good safety performance were significantly lower than those for any other location. When hourly rated employee perception is at the 40% level, it means that 6 out of 10 workers believe they are not receiving the kind of recognition due them for the job they are doing.

COMPANY SAFETY POLICY

The company's policy on safety should be made known to all levels of the organization. Such a document might indicate the following:
- The process requires employee awareness and total involvement.
- Safety is a primary company goal and has top priority.
- Every effort must be made to reduce the possibility of incidents or any other undesirable occurrence.
- The company intends to comply with all safety laws and do more than meet minimal safety requirements.

In addition, company procedures could also include the following:
- how to develop and implement safety standards for equipment, working methods, and product processes

- a mechanism to conduct safety inspections to identify potential hazards in all areas of the company
- how to investigate incidents that occur, as a means of developing future preventive actions.
- how to analyze incident records and hazards to determine trends and determine targets for protective action
- methods to educate and train all employees in safety principles
- when necessary, methods to provide and fit employees with the most advanced technological protective equipment to prevent injury on the job
- how to promote and encourage employee interest and commitment to the safety program
- methods to train employees in off-the-job incident prevention

Figure 10-1 is an example of a company safety policy statement.

SAFETY RULES AND REGULATIONS

Management should develop general safety rules for the company that will help avoid incidents (as well as minimize health hazards) on the job. Since most industrial incidents consist of eye injuries, foot injuries from dropping items, and back strains resulting from moving items that are too heavy or that have been moved incorrectly. The most important safety rule is to engineer out safety hazards. When this is not feasible, management must require that employees must wear personal protective equipment. Employees must also be trained in proper methods of lifting and moving heavy items.

The Safety Organization

Safety is a line responsibility and must be approached in the same way as other line responsibilities, such as production, cost, and quality control. Consequently, duties should be assigned in the same way. Each supervisor should be responsible not only for output, but for doing some things that will ensure the safety of those being supervised. In the past, industrial incidents were attributed to unsafe acts, unsafe conditions, or both. But present thinking assumes management and supervisory staff are partly responsible if they did nothing to eliminate unsafe conditions when they may have known about them.

Note: This sample safety policy was excerpted from the Columbia River Paper Company Safety Policy.

SAFETY POLICY: Columbia River Paper Company

Introduction
A. In order to clarify the safety activities of Columbia River Paper Company, division of Boise Cascade Corporation, the following is set forth as a basic program to clearly establish its existence. It consists essentially of an outline of the relationships of management and employee responsibilities, which are necessary for an effective safety program.
B. We wish it to be known that this program will become a basic part of our management policy and will govern our judgment on matters of operation equally with considerations of quality, quantity, personnel relations, and other phases of our management policy.
C. The program is developed and administered, upon approval, by the safety director. In doing so, the safety director is performing a function of the resident manager. His/her primary purpose is to assist management by encouraging safety consciousness and employee participation. This is accomplished by the development of promotional material, program planning, motivation incentives, safety meetings, inspections, etc.

Purpose
The management of Columbia River Paper Company holds in high regard the safety, welfare, and health of its employees. We believe that Production is not so urgent that we cannot take time to do our work safely. In recognition of this and in the interest of modern management practice we will constantly work toward:
A. The maintenance of safe and healthful working conditions.
B. Consistent adherence to proper operating practices and procedures designed to prevent injury and illness.
C. Conscientious observance of all federal, state, and company safety regulations.

Responsibility
A. The resident manager has taken the responsibility to develop an effective program of accident prevention.
B. Plant superintendents are responsible for maintaining safe working conditions and practices in the areas under the jurisdiction.
C. Department heads and supervisors are responsible for the prevention of accidents in their departments.
D. Foremen are responsible for the prevention of accidents in their crews.
E. The safety director is delegated by the resident manager and has the responsibility to provide advice, guidance, and any such aid as may be needed by supervisors in preventing accidents.
F. The safety director also keeps adequate records and coordinates safety programs and other educational activities.
G. Employees are responsible for exercising maximum care and good judgment in preventing accidents.

Figure 10-1. Example of a safety policy statement. (Adapted with permission from Petersen D. *Techniques of Safety Management: A Systems Approach,* 3rd ed. Des Plaines, IL: American Society of Safety Engineers, 1998.)

If a company is large enough, management can employ a safety officer or manager, either part-time or full-time, whose sole responsibility is to advise on matters of safety and to assist line managers in controlling hazards and preventing incidents. Large companies may opt to employ a full-time safety professional and an industrial hygienist and, perhaps, even a physician and nurse. For smaller companies, this approach may not be financially feasible, but they can obtain advice and assistance from federal and state agencies, insurance companies, and health and safety consultants.

Responsibilities

The responsibilities of the safety and health personnel are as follows:

- assisting line management to fulfill its safety responsibilities through teaching, counseling, and advising.
- providing for periodic safety appraisals, particularly where new machinery has been installed or existing equipment relocated.
- promoting and maintaining an ongoing safety education program to create an awareness of and commitment to safety at all levels of the organization.
- maintaining and interpreting incident and injury statistics and data.
- developing safety standards and measurements for adequately evaluating each responsible supervisor and manager in terms of incident prevention and loss control.
- preparing instructional materials on safety that will inform supervisors and employees about the process.
- conducting research into new methods of safety and incident prevention, as well as controlling health hazards in the workplace.
- maintaining liaisons with safety organizations and participating in their activities to keep abreast of recent findings in the field.

Assigning Responsibility

All companies, large and small, whether they have full-time safety and health staffs or depend on outside consultants, must clearly define the safety accountability of employees throughout the organization.

The most important areas of accountability are those of senior and middle management since they must make sure that supervisors perform their safety functions properly. They must visibly demonstrate to all employees that safety is one of the highest priorities of the company. Supervisory staff is the next most important group. Their tasks in this regard include indoctrinating new workers in safety requirements and developing safe working habits, continuously training and observing both new and old employees, encouraging a commitment to safety among workers, inspecting the workplace for hazards, investigating and reporting on incidents, etc. If a company has a permanent safety and health staff, they constitute the third important area of responsibility for the program. They are responsible for advising and assisting line managers, primarily senior and middle management.

Accountability

Unless managers and supervisors are aware that they are accountable and that their careers will be affected by a lack of performance in safety, the process will not succeed.

Accountability can be assigned by a number of means. The simplest but least effective is to assign all incidents to the department in which they occur. If a worker suffers an injury, that fact appears on the department manager's or supervisor's work record. Or the costs may be attributed to and charged against the line manager's performance and reflected in his or her budget.

Line managers and supervisors are the people who are most responsible and thus accountable for safety performance. Their safety performance should be evaluated periodically, in terms of the following:

- number and content of safety meetings held
- number of toolbox meetings (informal talks with workers on the job)
- activity reports on safety
- inspection results
- incident investigation results
- incidence reports
- job hazard analyses
- or any other defined tasks.

In evaluating safety as in other job performance appraisals, the most effective standard for evaluating performance is comparison to objectives set. Managers and supervisors are more likely to understand and accept this approach and could resent comparison with other, competing groups, or being measured by luck.

THE WORKER AND SAFETY

Safety of the worker is the purpose of any safety process. Efforts at this level must include training and motivation of both new and current employees, safety and health involvement teams in which workers participate, safety inspections of the workplace, incident investigations, job safety analysis, and many more.

Training and Motivation

All new employees and employees changing jobs should be immediately instructed in safety procedures. It is important to establish immediately the importance of safety procedures and personal protective equipment and to motivate employees to observe safety requirements both for their own well-being and the good of the company. If employees are properly instructed in safety at the beginning, they will be more likely to observe good work habits. If safety orientation is delayed, improper work habits can develop and will be difficult to overcome later. Before any employee is assigned to a job, or current workers are reassigned to new jobs or different equipment, they should be determined to be physically and mentally capable of doing the work and be properly trained.

Safety training is based on the assumption that the development of a positive attitude predisposes an individual to safe work habits and conduct. Thus, those doing the training and motivating should use a positive approach. First, the instructor must explain why safe work practices are required for each job in terms of the worker's safety. Specific examples should be provided to show why safety is necessary. Throughout the session, every effort should be made to motivate employees to do their jobs properly.

During the training session, the instructor should
- make the employee feel at ease.
- explain that he or she should not be afraid to ask questions, especially when new situations arise.
- explain the job in detail in terms employees can easily understand.
- indicate why protective equipment, if appropriate, must be worn, and point out consequences of not wearing it.
- demonstrate at each job step, the precautions and safe work practices that must be followed and explain their importance.
- question employees to determine whether they understand all instructions and, if so, allow employees to demonstrate or practice what they have learned.
- motivate employees by demonstrating the company's commitment to safety and safe practices.

Training sessions can include first aid courses, cardiopulmonary resuscitation, fire extinguisher use, lifting and moving of heavy objects, emergency procedures, etc.

Follow-up training might include toolbox sessions brief, informative get-togethers of supervisors and workers. These informal sessions foster open communication and stimulate safety consciousness. The topic should be planned in advance, and the session should be held at the job site and last no longer than 15 minutes. Suggested topics include the following:
- previous incidents and corrective action
- job safety in individual tasks and operations
- housekeeping methods
- techniques of manual lifting, individually or in groups
- personal protective equipment hearing protection, when to wear protective goggles, and types used for different procedures
- lockout/tagout cleaning, oiling, repairing,
- materials handling safety (cranes, hoists, forklifts)
- preventing falls (safety belts, lanyards, tying off).

The National Safety Council publishes training and support materials, including books, booklets, packaged training modules, videos, and posters on various employee safety issues. Such training tools support previous training, such as safety meetings, etc., and give ideas for future activities.

Additional training sessions may be indicated if there is a high incidence of injury and illness, an increase in the number of incidents that could have resulted in incidents, excessive waste or scrap after production, a high rate of labor turnover, a change in processes or introduction of a new process, or any sharp, unexplained increase in accidents.

Training provides workers with the knowledge and skill to accomplish the work safely, but unless they are motivated, employees will not normally work that way. There are various factors that influence worker motivation together, these indicate whether an employee will perform in a safe or unsafe manner. However, motivation to safety is not that simple. It is also influenced by pressures in daily life, whether a person likes or dislikes the job, whether the job is meaningful, the personality of the worker, and the job climate. Finally, does the worker get a reward for performing in a safe manner? In other words, does he or she feel satisfied with what was accomplished?

In motivating for safety, the program must take into account the current needs of the workers. Most people want jobs that are meaningful and to feel that they are part of things. They may have many more needs. If those needs are met, they will very likely be motivated to do what you want. Thus, the job must offer a chance for achievement and recognition, it must be fun, it must offer the opportunity for responsibility, and it must permit personal growth. These factors can be applied to safe working performance: Will the employee be recognized for his or her safety record? Will this person be given additional responsibility because of that record? Will the record contribute to career growth? Management and the safety programs that take these factors into account will succeed in motivating employees.

Workers can also be motivated by having satisfactory interpersonal relationships with managers and supervisors. Studies have identified two types of supervisors: the employee-centered supervisor who achieves high production because he or she is primarily interested in his workers, and the job-centered supervisor, who does not achieve high production because he or she emphasizes it more than an interest in the workers. Thus, the most effective manager is one who is interested in both people and production, combining high performance goals with the good of his or her people. The key to employee motivation to safety, then, is a management that cares about and rewards its employees for safe conduct.

Safety Committees Pro and Con

Some companies establish safety committees in which supervisors and workers participate. Such committees are only a part of the safety program, but if properly organized and used, they augment the safety program. However, they are not essential and can have some inherent limitations. First, only a few workers can participate in a committee's work at any given time. The safety committee concept also has the following additional limitations:

- Busy supervisors can refer safety problems to the committee, problems that they should be handling themselves.
- Only a small percentage of the work force will have a chance to serve on the committee; often the same members are appointed year after year.
- Meetings and committee attendance can be irregular, sometimes allowing too much time to elapse between a problem and its solution.
- Committee members often are not qualified to handle complicated safety problems and consequently tend to concentrate on superficial problems.
- Committee recommendations, particularly if they require large expenditures, can be bypassed by management.
- Most organizations today are forming improvement teams in lieu of the old safety committee. In an improvement team, members select who they will work with and what problems they will work on. The process is under their control, not that of management.

Safety Inspections

Safety inspections are important to identify potential hazards before incidents occur. In general, hazards can be categorized as the results of unsafe practices or unsafe conditions; however, incidents seldom result from only one cause but usually from a combination of causes.

Unsafe practices include operating equipment without authority or training; using equipment, tools, materials, or vehicles unsafely; operating at unsafe speeds; operating defective equipment; making safety devices inoperative; failing to use personal protective equipment; unsafe lifting; taking an unsafe position during procedures; or adjusting, cleaning, or oiling machinery that is in motion; etc. Unsafe conditions include unsafe design or construction of equipment or the facility; defective tools or equipment; poor housekeeping in the workplace; inadequately safeguarded or unguarded equipment; inadequate illumination or ventilation; etc.

Supervisors should regularly ensure that safety inspections are conducted, and middle and senior managers should periodically conduct unannounced spot-checks for hazards. Before inspections, review previous incident reports and job safety analyses to identify problem and high-risk areas. Facility and individual checklists also can prove helpful.

Those conducting the inspection should communicate with workers in the area and encourage their assistance and suggestions during the inspection, because the person at a machine is more familiar with it than anyone else. Workers in certain areas are generally more aware of the problems or hazards. Moreover, if a worker's suggestion is accepted, there is a positive reaction. If it is not accepted, the reasons should be explained.

By law, certain inspections will require special expertise and must be conducted by safety personnel or industrial hygienists. These include (1) inspections of such equipment as cranes, pressure vessels, fire extinguishers, chains, and ropes; and (2) industrial hygiene surveys to identify health hazards, etc.

Periodic inspection by supervisors and managers not only reduces the potential for incidents in the workplace, but also demonstrates the company's commitment to safety.

Incident Investigation

The investigation of incidents is usually a part of the incident prevention program. It identifies hazards so that future similar incidents can be prevented by mechanical improvements, better supervision, improved employee training, etc. It determines the causes of human errors. An investigation makes supervisors and employees more aware of particular hazards in the workplace, and it directs attention to safety and incident prevention, in general. Minor injuries and near-incidents should be reported as well as lost-time incidents, because nothing is learned from unreported incidents. Near-incidents are often predictors of future serious incidents; potential hazards can be corrected, and the need for medical treatments of involved workers may not surface until later.

All levels of management are responsible for incident investigation, but the first-line supervisor is best qualified for the following reasons:

1. Supervisors are closest to and most knowledgeable about jobs, working conditions, and workers. They know the details of jobs and job procedures, hazardous environmental conditions in their areas, and any unusual circumstances that must be taken into account. They also know their workers and their personal characteristics, job experience, and job language.
2. A supervisor's sense of responsibility is reinforced by incident investigations. After each investigation, supervisors become more aware of their responsibility for safe procedures and incident prevention.
3. Supervisors learn more about hazards by investigating incidents. By such investigations, they learn the unsafe practices and conditions to look for, how best to train new employees in safety considerations, what hazards to caution employees about, and generally how to prevent incidents.
4. Supervisors are the primary instrument of incident prevention. They can help to prevent the recurrence of incidents as well as prevent minor injury incidents because they are closest to the scene. Information that they obtain during incident investigations will show them whether they should change their instructions on safe job procedures, eliminate unsafe conditions, or correct the sources of such conditions.

Of course, supervisors know how to recognize hazards, but they need support. Supervisors may require assistance from the safety staff, equipment suppliers, etc., when investigating incidents. Management should see to it that supervisors are given the training and assistance that they need.

Workplace hazards are many and varied; consequently, there will be many types of corrective actions some already in effect, some pending but not yet implemented, and some still not identified. The incident investigation report must specify which of these applies and, in the case of the last, must offer recommendations for remedial action. These need not be detailed technical descriptions, but preferably should be general descriptions with an explanation of the purpose. If the recommendation is technically detailed, it may be rejected because of some minor technical point. Figure 10-2 is an example of an incident investigation report form.

Job Safety Analysis

Job safety analysis (JSA) identifies the hazards or potential hazards associated with each step of a particular job, determines the severity of the hazard, attempts to establish a probability of incident occurrence, and attempts to reduce or eliminate hazards. To be effective, the procedure should include input from workers on how each task is performed and the hazards that they perceive. A description and examples of job safety analysis are given in the latest edition of the *Accident Prevention Manual for Business & Industry: Engineering & Technology*, available from the National Safety Council.

ACCIDENT INVESTIGATION REPORT

Case Number _____

Company _____ Address _____
Department _____ Location (if different from mailing address) _____

1. Name of injured

2. Social Security Number

3. Sex ☐ M ☐ F

4. Age

5. Date of accident

6. Home address

7. Employee's usual occupation

8. Occupation at time of accident

9. Length of employment
☐ Less than 1 mo.   ☐ 1-5 mo.
☐ 6 mo. - 5 yr   ☐ More than 5 yr.

10. Time in occup. at time of accident
☐ Less than 1 mo.   ☐ 1-5 mo.
☐ 6 mo. - 5 yr   ☐ More than 5 yr.

11. Employment category
☐ Regular, full-time   ☐ Regular, part-time
☐ Temporary   ☐ Seasonal   ☐ Non-Employee

12. Case numbers and names of others injured in same accident

13. Nature of injury and part of body

14. Name and address of physician

15. Name and address of hospital

16. Time of injury
A. _____ a.m.
         p.m.
B. Time within shift
C. Type of shift

17. Severity of injury
☐ Fatality
☐ Lost workdays—days away from work
☐ Lost workdays—days of restricted activity
☐ Medical treatment
☐ First aid
☐ Other, specify _____

18. Specific location of accident _____

On employer's premises?   ☐ Yes   ☐ No

19. Phase of employee's workday at time of injury
☐ During rest period   ☐ Entering or leaving plant
☐ During meal period   ☐ Performing work duties
☐ Working overtime   ☐ Other _____

20. Describe how the accident occurred
_____
_____
_____
_____

21. Accident sequence. Describe in reverse order of occurrence events preceding the injury and accident. Starting with the injury and moving backward in time, reconstruct the sequence of events that led to the injury.

A. Injury event _____

B. Accident event _____

C. Preceding event #1 _____

D. Preceding event #2, 3, etc. _____

Figure 10-2 (continued)

Figure 10-2. Example of an incident investigation report. (Copyright 1995 National Safety Council)

22. Task and activity at time of accident	23. Posture of employee
General type of task	
Specific activity	
Employee was working: ☐ Alone ☐ With crew or fellow worker ☐ Other, specify	**24. Supervision at time of accident** ☐ Directly supervised ☐ Indirectly supervised ☐ Not supervised ☐ Supervision not feasible

25. Causal factors. Events and conditions that contributed to the accident.
Include those identified by use of the Guide for Identifying Causal Factors and Corrective Actions.

26. Corrective actions. Those that have been, or will be, taken to prevent recurrence.
Include those identified by use of the Guide for Identifying Causal Factors and Corrective Actions.

Prepared by _____ Approved _____

Title _____ Title _____ Date _____

Department _____ Date _____ Approved _____

Developed by the National Safety Council
©1995 National Safety Council

Title _____ Date _____

Figure 10-2. Continued

BEHAVIOR-BASED SAFETY

The safety world in this country and others has been informed in recent years that our future progress will come from *behavior-based* approaches. The difficulty to many of us in the field is that while we tend to agree, we're not quite sure what *behavior-based* means.

We've believed since the 1930s that most incidents are caused by human behavior (unsafe acts); and that the preponderance of our efforts therefore should be to do things to lessen the probability of unsafe behavior. Does this mean that most of what we have been doing traditionally can be called behavior-based approaches? In the 1970s our government regulations and fear of fines forced us to concentrate on the physical conditions in a facility, worker health, record keeping, machine guarding, occupational diseases, and ergonomics, depending upon what year it was, and what was the current flavor of the year from the government. These thrusts were regulatory/compliance-driven; however, training, awareness campaigns, and use of multimedia continued during these years in most organizations. These techniques all aim at influencing the worker's behavior.

Perhaps the definitions we hear most about today are those that define behavior-based safety as a process of involving workers in defining the ways they are most likely to get hurt. Thus, workers are asked to become involved and "buy-in" to the overall approach. In addition, the workers are asked to observe other workers in order to determine progress in the reduction of unsafe behaviors.

A behavioral safety program begins with identifying and defining desired behaviors, setting goals and standards, and providing instruction. The consequence of safe behavior is a positive reward for safe acts. The emphasis is on observable behavior that can be measured (i.e., performed in either a *safe* or an *unsafe manner*).

The positive program involves providing employees with predetermined consequences for behavior. Safe practices are rewarded with tangible benefits, social reinforcement (e.g., praise), or information (feedback) about performance.

Subsequent maintenance of the desired safe behaviors depends on continued but less frequent reward than does establishment of these behaviors. The efficiency of the reward system in maintaining behavior depends on the frequency of reward (schedule of reinforcement).

Thus, regular reward is important in establishing a desired level of safe practice, but its maintenance can be achieved through less frequent and intermittent rewards. Should reward no longer be offered, however, the level of safety behaviors will return to its original level.

The overwhelming evidence collected from actual worksite investigations to-date suggests that positive reinforcement programs are effective in increasing the level of safe performance; and that this level can be maintained over the long term. Unlike most training programs, a direct link can be demonstrated between the program and actual performance on the job.

A true behavior-based approach to safety requires behavioral measurements. As an upstream measure it requires a statistically valid observation/sampling process (are people behaving more safely today than last week/month?). To achieve statistical validity, a number of observations are needed, with consistency over time (one sampler) looking at *all* unsafe behaviors, not just a selected few.

For downstream measures, statistical validity is also needed. For most of us that means shedding the old measures (frequency rates, incidence rates, severity rates, etc.) and replacing them with valid, meaningful measures (process improvements achieved, perception surveys, etc.).

Behavioral-based safety is the building of a management system in safety that each level of the organization defines precisely what behaviors are required, measures whether or not these behaviors are, in fact, present an reinforces regularly (daily, hourly) for the behaviors desired. All levels of the organization, from the shop floor to the executive suite must be involved.

Behavior-based safety is one of the primary answers to future improvement, but only if and when it does the following:

1. defines the behaviors needed at each level of the organization, from the bottom to top
2. ensures that each person clearly understands the required behaviors
3. measures whether the behaviors are, in fact, there
4. rewards (reinforces) for the behaviors on a regular (daily, hourly) basis.

Behavior-based safety systems that are real management systems do not turn over safety to any one level of the organization. It is great to get worker involvement; in fact, it is crucial. But to eliminate and isolate management from the process is unproductive.

One of the fundamental tenets of safety is that safety systems can be flexible and company-specific, as long as these systems meet two key criteria:
1. There is a system of accountability in place that defines roles, ensures knowledge to fulfill the roles at each and every level from work to CEO, measures role fulfillment; and has rewards contingent upon role fulfillment.
2. The system asks for, allows for, requires, and ensures participation at each level.

SUMMARY

An occupational health and safety program is a shared responsibility, from top management down. All employees must be committed to safety as a part of the job. Anyone responsible for an occupational health and safety program should evaluate the current safety policy and program; compare current incident rates with those of previous years; analyze each job to determine the potential hazards associated with each job; and conduct safety program audits and perception surveys. He or she should stress the importance of training both for the job and on how to use required personal protective equipment. Working with appropriate employees and the safety committee, he or she should routinely inspect the workplace and investigate all incidents. Workers who have good safety records should be recognized and rewarded.

REFERENCES

Bird F. *Loss Control Management.* Loganville GA: Institute Press, 1976.

Diekemper R, Sparz D. A quantitative and qualitative measurement of industrial safety activities. *Professional Safety,* December 1970.

Heinrich H, Petersen D, Roos N. *Industrial Accident Prevention,* 5th ed. New York: McGraw-Hill, 1980.

National Safety Council. *Accident Prevention Manual for Business & Industry,* 11th ed. Itasca IL: National Safety Council, 1997.

NIOSH. *Safety Program Practices in Award-Winning Companies.* Cincinnati: NIOSH, 1970.

Petersen D. *Safe Behavior Reinforcement.* Goshen NY: Aloray, 1988.

Petersen D. *Safety Management: A Human Approach,* 2nd ed. Des Plaines IL: American Society of Safety Engineers (ASSE), 1998.

Petersen D. *Techniques of Safety Management,* 3rd ed. Des Plaines IL: ASSE, 1998.

Petersen D. *Analyzing Safety Performance,* 3rd ed. New York: Van Nostrand Reinhold, 1996.

Petersen D. *Human Error Reduction and Safety Management,* 3rd ed. New York: Van Nostrand Reinhold, 1996.

Petersen D. *Safety Supervision,* 3rd ed. Des Plaines IL: ASSE, 1999.

Petersen D. SBO: *Safety by Objectives.* New York: Van Nostrand Reinhold, 1996.

Planek TW, Driessen G, Vilardo FJ. Industrial Safety Study. *National Safety News,* August 1967.

Planek TW, Fearn KT. Reevaluating safety priorities: 1967 to 1992. *Professional Safety,* October 1993.

chapter 11

Industrial Hygiene Programs

by Sara Joswiak, MPH

167 **Overview**
168 **Sampling**
Personal sampling ■ Area sampling ■ Wipe sampling
173 **Personal Protective Equipment**
Skin protection ■ Respiratory protection ■ Hearing protection and the Hearing Conservation Amendment
183 **Hazard Communication**
Hazard communication program ■ Written HCP
■ Employee training
186 **Laboratory Safety**
186 **Substance-Specific Standards**
187 **Employee Training**
187 **Indoor Air Quality**
188 **Summary**
188 **References**
189 **Appendix**

OVERVIEW

The Occupational Safety and Health Act (OSHAct) specifies the employers' obligation to furnish to each employee a place of employment free from recognized hazards that are causing or likely to cause death or serious physical harm, and to comply with the Occupational Safety and Health Administration (OSHA) standards. Responsibilities of an employer include determination of whether a hazardous condition exists in a workplace, evaluation of the degree of hazard, and where necessary, the control needed to prevent occupational injury and illness.

The industrial hygiene program is vital to an organization's overall health and safety program. Without an industrial hygiene program the organization would not be able to prevent occupational disease and injury, and provide a safe and healthful workplace.

There are many objectives of an industrial hygiene program. Some of these objectives are as follows:
- Determine the need and effectiveness of ventilation.
- Determine the need and level of personal protective equipment (PPE) required.
- Delineate the work areas or work tasks where protection is needed.
- Compliment a medical monitoring program to assess the potential health effects of exposure.
- Develop a periodic schedule of monitoring.
- Establish that exposures to toxic materials are within acceptable limits.

A comprehensive industrial hygiene program will include monitoring workplace conditions. This includes personal and area air sampling, surface sampling, indoor air quality investigations, noise surveys, ergonomic studies, employee training, and determination of PPE. For more detailed information on industrial hygiene and industrial hygiene programs, consult *The Fundamentals of Industrial Hygiene*, 4th edition (Plog, 1996).

SAMPLING

Inhalation hazards are evaluated through the collection of air samples. Typically, the measured concentration of an airborne contaminant is compared to a recognized exposure limit. The two most common limits used are the Permissible Exposure Limit (PEL) set by OSHA and the Threshold Limit Value (TLV) set by the American Conference of Governmental Industrial Hygienists (ACGIH). Specific sampling methods for collection of many airborne contaminants have been developed. These standardized methods provide all the information needed to sample for a specific air contaminant and ensure that the information collected is accurate and meaningful. In addition, the analytical method contains information for the laboratory, which is needed for the analysis (Figure 11-1).

Prior to air sampling a sampling strategy is developed. A walkthrough survey in the workplace is performed first to identify potential hazards. The strategy is then prepared to reflect the hazard sources, degree of hazard, and adequacy of controls (if present). This strategy will also outline what, how, where, whom, when, and how long to sample. The analytical method is consulted for this information.

There are two basic types of air monitoring sampling techniques. The sampling strategy will outline what, how, where, whom, when, and how long to sample. The objectives of the industrial hygiene survey dictate whether either or both sampling methods may be used. Some sampling objectives are as follows:
- Evaluate employee exposure.
- Assist in the design or evaluation of control measures.
- Document compliance with government regulations.

Personal Sampling

Personal sampling is the preferred method of evaluating worker exposure to air contaminants because the data collected closely approximates the concentration inhaled. Personal air sampling is accomplished by attaching an air-sampling pump to the employee and placing the sampling media in the employee's breathing zone. The breathing zone is defined as "the imaginary globe of two foot radius surrounding the head" (Plog, 1996). In some circumstances, passive dosimeter badges, which are attached to the employee's collar, may also be used to collect air contaminant samples.

Area Sampling

Area samples are taken at fixed locations in the workplace. In general, this type of sampling does not provide a good estimate of worker exposure. It is useful to evaluate background concentrations, locate sources of exposure, evaluate the effectiveness of control measures, locate high exposure areas, indicate flammable or explosive concentrations, or to determine if an area should be isolated or restricted to prevent employees from entering. Area sampling for airborne contaminants is performed by placing an air-sampling pump with its respective media at a fixed location within the work area.

Wipe Sampling

Another sampling technique available to evaluate exposure to contaminants is *wipe sampling*. This method however is used to evaluate *surface contamination* instead of *airborne contamination*. Wipe samples are taken by using a section of a surface to be sampled using a filter with an open area exactly 100 square centimeters (Appendix).

The goals of wipe sampling are as follows:
- to assess the potential for ingestion of a chemical
- to assess the potential for skin absorption of a chemical
- to assess housekeeping
- to assess the effectiveness of PPE
- to assess worker decontamination
- to assess migration of a contaminant out of an isolated area into a clean area
- to assess the potential for take-home exposure of a chemical.

Thousands of chemicals pose the risk of ingestion, absorption, skin irritation and occupational skin diseases. In the workplace, materials can be transferred to other surfaces inadvertantly. This transfer can happen through poor decontamination procedures or lack of hygiene practices. Some examples are: wearing a lab coat into a lunch room, drinking a beverage while at a workstation, or tak-

METHYLENE CHLORIDE 1005

CH_2Cl_2 MW: 84.94 CAS: 75-09-2 RTECS: PA8050000

METHOD: 1005, Issue 3	EVALUATION: FULL	Issue 1: 15 August 1984 Issue 3: 15 January 1998

OSHA: 25 ppm; STEL 125 ppm
NIOSH: lowest feasible; carcinogen
ACGIH: 50 ppm; suspect carcinogen
(1 ppm = 3.47 mg/m³)

PROPERTIES: liquids, d 1.323 g/mL @ 20 °C; BP 40 °C; MP -95 °C; VP 47 kPa (349 mm Hg) @ 25 °C, nonflammable

SYNONYMS: dichloromethane, methylene dichloride

SAMPLING		MEASUREMENT	
SAMPLER:	SOLID SORBENT (2 coconut shell charcoal tubes, 100/50 mg)	TECHNIQUE:	GAS CHROMATOGRAPHY, FID
		ANALYTE:	methylene chloride
FLOW RATE:	0.01 to 0.2 L/min	DESORPTION:	1 mL CS_2
VOL-MIN: -MAX:	0.5 L @ 500 ppm 2.5 L	INJECTION VOLUME:	1 µL
SHIPMENT:	separate front and back tubes	TEMPERATURE-INJECTION: 250 °C -DETECTOR: 300 °C -COLUMN: 80 to 150 °C at 10 °C/min	
SAMPLE STABILITY:	ca. 30 days @ 5 °C [1]	CARRIER GAS:	Helium, 2.4 mL/min
BLANKS:	2 to 10 field blanks per set	COLUMN:	capillary, 30 m x 0.32-mm ID, 0.25-µm film polyethylene glycol, Stabilwax, or equivalent
ACCURACY		CALIBRATION:	solutions of methylene chloride in CS_2
RANGE STUDIED:	1700 to 7097 mg/m³ [2]	RANGE:	1.4 to 2600 µg per sample [1]
BIAS:	- 4.1%	ESTIMATED LOD:	0.4 µg per sample [1]
OVERALL PRECISION ($\hat{S}_{rT}$):	0.076 [1, 2]		
ACCURACY:	±15.8%	PRECISION (S_r):	0.026 [1]

APPLICABILITY: The working range for GC-FID analysis is 0.4 to 749 ppm (1.4 to 2600 mg/m³) for a 1-L air sample [1]. An electron capture detector (ECD) also may be used to obtain lowest feasible level of detection and quantitation [3]. Conditions for using an ECD are listed in the APPENDIX.

INTERFERENCES: No specific interferences were identified. However, any compound with a similar retention time may interfere. Alternate chromatographic columns are Carbowax-PEG and DB-1 fused silica capillary columns. The capacity of the charcoal is greatly reduced when sampling in high humidity [4].

OTHER METHODS: This revises NMAM 1005 (dated 8/15/94) [5]. If sampling in an atmosphere with high relative humidity (≥ 80%), a tube with larger bed of charcoal is recommended. OSHA Method 59 uses a sampler with three sorbent sections, each containing 350 mg of charcoal, and has been evaluated for a 10-L air sample at 1 ppm of methylene chloride with 80% relative humidity [4]. OSHA Method 80 uses a carbon molecular sieve sampler and GC-FID analysis, and has been evaluated at 10 ppm and 500 ppm [6].

(continued)

Figure 11-1. An analytical method for sampling methylene chloride. (From the NIOSH *Manual of Analytical Methods*)

METHYLENE CHLORIDE: METHOD 1005, Issue 3, dated 15 January 1998 - Page 2 of 4

REAGENTS:

1. Methylene chloride, chromatographic grade.*
2. Carbon disulfide, chromatographic grade. *
3. Helium, purified.
4. Hydrogen, prepurified.
5. Air, filtered.

* See SPECIAL PRECAUTIONS

EQUIPMENT:

1. Sampler: glass tube, 7 cm long, 6-mm OD, 4-mm ID, flame-sealed ends with plastic caps, containing two sections of activated (600 °C coconut shell charcoal (front = 100 mg, back = 50 mg) separated by a 2-mm urethane foam plug. A silylated glass wool plug precedes the front section and a 3-mm urethane plug follows the back section. Tubes are commercially available (SKC #226-01, Supelco ORBO-32s, or equivalent).
2. Personal sampling pump, 0.01 to 0.2 L/min, with flexible connecting tubing.
3. Gas chromatograph, flame ionization detector, integrator, and column (page 1005-1).
4. Vials, autosampler, with PTFE-lined caps.
5. Microliter syringes, 10-µL and other sizes as needed, readable to 0.1 µL.
6. Flasks, volumetric, various sizes.
7. Pipets, various sizes.

SPECIAL PRECAUTIONS: Carbon disulfide is toxic and a serious fire and explosion hazard (flash point = -30 °C). Methylene chloride is a suspected carcinogen. Wear appropriate protective clothing and work with these compounds in a well ventilated hood.

SAMPLING:

1. Calibrate each personal sampling pump with a representative sampler in line.
2. Break ends of sampling tubes immediately before sampling. Attach sampler train to personal sampling pump with flexible tubing.
3. Sample at an accurately known flow rate between 0.01 and 0.2 L/min for a total sample size of 0.5 to 2.5 L.
4. Separate the front and backup tubes to prevent migration of methylene chloride between tubes. Cap the samplers and pack securely for shipment.

SAMPLE PREPARATION:

5. Place front (along with first glass wool plug) and back sorbent sections of each tube in separate vials. Discard the foam plugs.
6. Add 1 mL of desorption solvent, CS_2, to each vial and attach crimp caps.
   NOTE: Significant heat is generated when the desorbing solution is added to the charcoal. To minimize sample loss, chill the desorbing solution before adding to the charcoal [4].
7. Allow to stand 30 min with occasional agitation to aid desorption.

CALIBRATION AND QUALITY CONTROL:

8. Calibrate daily with at least six working standards over the range of interest. Three standards (in duplicate) should cover the range from the LOD to LOQ.

Figure 11-1. Continued

METHYLENE CHLORIDE: METHOD 1005, Issue 3, dated 15 January 1998 - Page 3 of 4

   a. Add known amounts of methylene chloride to CS_2 in 10-mL volumetric flasks and dilute to the mark.
      NOTE: An internal standard such as 2-butanol or hexane can be added to the CS_2.
   b. Analyze together with samples and blanks (steps 11 and 12).
   c. Prepare calibration graph (peak area or height vs. µg methylene chloride).
9. Determine desorption efficiency (DE) at least once for each lot of charcoal tubes used for sampling in the calibration range (step 8). Prepare three samplers at each of 6 levels plus three media blanks.
   a. Remove and discard back sorbent sections of sampler and media blanks.
   b. Inject a known amount of methylene chloride directly onto the front sorbent bed of each charcoal tube.
   c. Allow the tubes to air equilibrate for several minutes, then cap the ends of the tubes and allow to stand overnight.
   d. Desorb the samples (steps 5 through 7) and analyze together with standards and blanks (steps 11 and 12).
   e. Prepare a graph of DE vs. µg methylene chloride recovered.
10. Analyze three quality control blind spikes and three analyst spikes to ensure that the calibration graph and DE graphs are in control.

MEASUREMENT:

11. Set gas chromatograph according to manufacturer's recommendations and to conditions given on page 1005-1. Inject a 1-µL sample aliquot manually using solvent flush technique or with an autosampler.
    NOTE: If peak area is above the linear range of the working standards, dilute with appropriate solvent, reanalyze and apply the appropriate dilution factor in the calculations.
12. Measure peak areas.

CALCULATIONS:

13. Determine the mass, µg (corrected for DE), for methylene chloride found in the sample front (W_f) and back (W_b) sorbent sections, and in the average media blank front (B_f) and back (B_b) sorbent sections.
    NOTE: If $W_b > W_f/10$, report breakthrough and possible sample loss.
14. Calculate concentration, C, of methylene chloride in the air volume sampled, V (L):

$$C = \frac{(W_f + W_b - B_f - B_b)}{V}, \text{mg/m}^3$$

NOTE: µg/mL ≡ mg/m³

EVALUATION OF METHOD:

The method was originally evaluated for methylene chloride (Method S329) and validated over the range 1700 to 7100 mg/m³ at 25 °C and 763 mm Hg using a 1-L sample [2, 7]. Overall precision ($\hat{S}_{rT}$) using flame ionization detection was 0.073 with a recovery of 95.3% (non-significant bias). The concentration of methylene chloride was independently verified by calibrated syringe pump. Desorption efficiency was 0.97 over the range of 1.3 mg to 5.3 mg methylene chloride per sample. Breakthrough (5% on backup section) occurred at 18.5 min when sampling an atmosphere containing 6726 mg/m³ methylene chloride at 0.187 L/min at 0% RH. The stability of methylene chloride on charcoal was not determined in the original method development.

Method 1005 was evaluated for methylene chloride using electron capture detection (ECD)[3]. Toluene was determined to be the best desorption solvent for use with ECD. Desorption efficiency was determined to be 75% at a 230 ng spiking level. After 30 days storage at 5 °C, the stability of 320 ng methylene chloride samples was determined to be 87.3% [3]. The NIOSH storage stability criterion of ≥90% recovery was within

Figure 11-1. Continued

METHYLENE CHLORIDE: METHOD 1005, Issue 3, dated 15 January 1998 - Page 4 of 4

the 95% confidence interval of 80.9 to 93.7%.

REFERENCES:

[1] Pendergrass SM [1997]. Backup Data Report for Methylene Chloride by ECD. Cincinnati, OH: National Institute for Occupational and Health, DPSE (NIOSH)(unpublished report).
[2] NIOSH [1977]. Documentation of the NIOSH Validation Tests. S329. Cincinnati, OH: National Institute for Occupational Safety and Health, DHEW (NIOSH) Publication No. 77-185.
[3] NIOSH [1997]. Analytical Report for Methylene Chloride, Data Chem, NIOSH Sequence #8799-L (unpublished report).
[4] OSHA [1986]. Methylene Chloride: Method 59. OSHA Analytical Methods Manual. Salt Lake City, Utah: DOL, Occupational Safety and Health Administration.
[5] NIOSH [1994]. Methylene Chloride: Method 1005. In: Eller PM, Cassinelli ME, eds. NIOSH Manual of Analytical Methods, 4th ed. Cincinnati, OH: National Institute for Occupational Safety and Health, DHHS (NIOSH) Publication No. 94-113.
[6] OSHA [1990]. Methylene Chloride: Method 80. OSHA Analytical Methods Manual. Salt Lake City, Utah: DOL, Occupational Safety and Health Administration.
[7] NIOSH [1977]. Method S329: Methylene Chloride. In: Taylor DG, ed. NIOSH Manual of Analytical Methods, 2nd ed., V3. Cincinnati, OH: National Institute for Occupational Safety and Health, DHEW (NIOSH) Publication No. 77-157-A.

METHOD REVISED BY:

Stephanie M. Pendergrass, NIOSH/DPSE

APPENDIX. MEASUREMENT CONDITIONS FOR METHYLENE CHLORIDE BY GC-ECD.

Desorption: 1 mL toluene, 30 min with agitation
Calibration: Solutions of methylene chloride in toluene
Range: 0.23 to 1.0 µg per sample (instrumental) [1]
Estimated LOD: 0.002 µg per sample [1]
Precision ($\tilde{S}_r$): 0.050 at 320 µg

Analytical column, GC temperature program, and injection volume are the same as on page 1005-1.

Figure 11-1. Continued

ing a smoking break prior to washing hands. For this reason, the use of wipe sampling for the presence of contaminants in various areas in the workplace is an additional type of sampling that should be considered.

PERSONAL PROTECTIVE EQUIPMENT

Personal Protective Equipment (PPE) is one of the three methods used to control health hazards in the work environment. These controls, typically referred to as the "hierarchy of controls" are used in the workplace in the following order:

1. engineering
2. administrative
3. PPE

When engineering or administrative controls do not work or are not feasible, then the use of PPE should be considered for use in the workplace.

On March 25, 1994, an OSHA amendment to 29 *CFR* 1910, Subpart I - PPE was authorized and signed by OSHA secretary Joseph A. Dear. This amendment added new paragraphs to 1910.132 - General Requirements; revised information in 1910.133 - eye and face, 1910.135 - head protection and 1910.136 - foot protection; and added new sections to Subpart I as follows:

- 1910.38 - Hand Protection
- Appendix A - References for further information
- Appendix B - Nonmandatory Compliance Guidelines for Hazard Assessments and PPE Equipment Selection

The new amendment requires employers to:

- assess the workplace for hazards which would necessitate the use of PPE.
- select appropriate PPE that would protect the employee from the hazard identified.
- communicate PPE selection decisions to affected employees.
- remove defected or damaged PPE from service.
- provide training to employees required to wear PPE.
- ensure PPE meets the American National Standards Institute (ANSI) standards.
- provide hand protection based on the hazard assessment.

Skin Protection

Skin protection is one kind of PPE that is commonly chosen for protection. This may be related to the fact that, one of the top 10 work-related diseases-injuries in the United States according to NIOSH is dermatological in nature. This should come as no surprise as the skin is the largest organ of the body, and with the large number of substances and conditions capable of causing a skin disorder in the workplace. We know that direct dermal exposure can under some circumstances exceed inhalation doses. However, usually dermal absorption associated with vapor exposure is significantly less than via the direct inhalation route. There are five classifications of the causes of skin disorders. These are:

- chemical
- mechanical
- physical
- biological
- botanical.

The number of chemicals found and used in the workplace from day to day are endless. Each year additional chemicals are introduced in the marketplace. All of these chemicals can be separated into two groups: *primary irritants* and *sensitizers.*

A primary irritant is a substance that will produce an irritating effect when it comes in contact with the skin. A sensitizer is a material that can cause an allergic reaction of the skin. Primary irritants affect most people and include effects like burns, dermatitis, dry skin, and oil acne. Sensitizers sensitize a person so that initial contact may not produce a reaction, but prolonged or repeated exposure may result in an allergic contact dermatitis, e.g., nickel chromium.

Chemicals can be irritants and sensitizers. Some substances that are examples of both are turpentine, formaldehyde, chromic acid, and epoxy resins. Mechanical agents like friction, pressure, and sharp objects can result in skin disorders like blisters, calluses, and lacerations. Physical agents like heat, cold, and radiation can also cause occupational skin disorders. Biological agents like bacteria, viruses, fungi, and parasites can produce a variety of diseases and infections. Examples include "anthrax in hide processors, yeast infections of the nail in dishwashers and bartenders, animal ringworm in veterinarians, and HIV or hepatitis B in health care workers" (Plog, 1996). The most common botanical agents that cause contact irritant dermatitis are poison ivy, poison oak, and poison sumac.

Skin disorders caused by specific substances and conditions in the workplace can be prevented. As discussed earlier, PPE is one way to reduce these skin disorders. Clothing is one type of PPE available for the skin. Clean clothing should be provided in those environments where chemical contact is likely. Clothing worn on the job should not be worn at home due to the likelihood of the worker's family developing illnesses from exposure to the clothes. Contaminated clothing should be changed at the end of the workshift, and laundered thoroughly before reuse. If clothing becomes grossly contaminated as a result of a spill then it should be removed at once. Many organizations have found the use of uniforms to be a cost-effective way to provide employees with clothing.

Protective clothing is a barrier used to protect against exposure to many substances. This type of PPE is designed to be worn over the employees work clothes. This type of protection is available in many different materials. For example, protective clothing is available in rubber, plastic, leather, and cotton or synthetic fiber. Protective clothing is offered in many styles such as aprons, jackets, pants, or whole-body suits.

Hands are constantly bombarded with exposures. To protect hands from injury, handwashing is the first line of defense. Gloves are used to provide extra protection. Gloves are developed for all types of uses. There are gloves for laboratories, cut-resistance, heat-resistance, and heavy-duty industrial use. Gloves are available in the following types of materials (Plog, 1996):

- natural rubber latex
- synthetic rubber latex and synthetic rubber: butyl, Neoprene®, Viton®, nitrile, styrene-butadiene block polymer, and styrene-ethylene-butadiene.
- plastic polymers: ethylene-methyl methacrylate, polyethylene (PE), polyvinylalcohol (PVA), polyvinylchloride (PVC)
- laminated plastic polymers: a laminate of polyethylene-ethylenevinylalcohol copolymer-polyethylene (Silver Shield® and 4-H gloves®)
- leather
- textiles: natural or synthetic; woven from fabric, knit, or terry cloth; also coated with rubber or plastic; fabric such as cotton or nylon
- other: wire cloth made of stainless steel or nickel.

When selecting gloves or other protective clothing there are a number of factors to be considered. What are the employees handling? What hazard are you protecting them from? If chemicals are present, are they mixtures? What is the temperature of the hazard? Does the PPE create a new hazard? Is there machinery being used where gloves can get caught? These and other questions will ensure the PPE you select will adequately perform the job.

Respiratory Protection

Another type of PPE commonly chosen in the workplace is Respiratory Protection. As stated earlier, the main objective of an industrial hygiene program is to control exposures through engineering controls or administrative control measures. When this is not feasible, or while engineering controls are being implemented, respirators may be necessary. The Occupational Safety and Health Administration has a standard established for respiratory protection, 29 *CFR* 1910.134. On Jan. 8, 1998 OSHA issued a final rule for respiratory protection. This rule replaces the previous respiratory standard which was adopted in 1971. The new respiratory standard applies to all air contaminants including biohazards. Tuberculosis is not covered under the current standard.

The OSHA standard outlines the minimum requirements of the respiratory protection program. These requirements are as follows:

- written respiratory protection program with worksite specific procedures
- proper respiratory protection selection procedures
- medical evaluations of employees required to use respirators
- employee training of respiratory hazards in routine and emergency situations
- employee fit-testing for tight fitting respirators
- routine use and emergency use procedures
- respirator cleaning, disinfection, storage, inspection, repairing, discarding and other maintenance procedures
- atmosphere supplying respirator procedures to ensure adequate air quality, quantity and flow of breathing air
- employee training of proper use and limitations of respirators
- program evaluating procedures
- air sampling.

Written Procedures

Employers are required to develop a written respiratory protection program with worksite specific procedures. In addition, the employer is required to select a program administrator. A program administrator is a person qualified through training or experience, who will oversee the program, evaluate its effectiveness, and make changes as necessary. Written procedures must be made available to employees and revised and reviewed as conditions change.

Selection Procedures

In order for the employees to select the proper respirator, the workplace must be evaluated for respiratory hazards. In addition to workplace evaluations, there are a variety of factors to consider when selecting a respirator. Those selection factors are:
- chemical and physical properties of the contaminant
- toxicity and concentration of the hazardous material
- amount of oxygen present
- the nature and extent of the hazard
- work rate
- area to be covered
- mobility
- work requirements
- limitations and characteristics of the available respirators
- other PPE to be worn.

If an employer cannot reasonably estimate the exposure then the atmosphere must be considered immediately dangerous to life and health (IDLH). Respirators for use in IDLH atmospheres are the self-contained breathing apparatus (SCBA) (Figure 11-2) and the combination type pressure demand air-line respirators with auxiliary self-contained air supply. All selected respirators must be approved by the National Institute for Occupational Safety and Health (NIOSH) and the Mine Safety and Health Administration (MSHA).

Finally, the employer must "select a respirator from a sufficient number of respirator models and sizes so that the respirator is acceptable to and correctly fits the user." (29 *CFR* 1910.134, OSHA Standards for General Industry). *Practices for Respiratory Protection,* ANSI Z88.2, can also help with selection of respirators.

Medical Evaluations

For employees assigned to tasks that require the use of a respirator, a medical evaluation must be performed to ensure the employee is physically able to perform the work. Wearing respirators can

Figure 11-2. Respirators for use in IDLH atmospheres include the self-contained breathing apparatus (SCBA) type. (Courtesy Mine Safety Appliances)

result in a variety of physiological responses of the body. These responses will vary based on the type of respirator worn, the job performed, the existing workplace conditions, and the medical status of the employee.

In order to determine the employee's medical status, a medical evaluation must be conducted prior to the employee being fit-tested and given a respirator to use in the workplace. However, an annual medical evaluation is not required.

Medical evaluations must be performed by a physician or other licensed health care professional (PLHCP). A PLHCP may be a nurse practitioner, physician assistant, or occupational health nurse provided he or she is authorized by a state license, certification, or registration. Part of the evaluation involves a medical questionnaire or exam which includes the same information on the questionnaire (Figure 11-3).

Follow-up medical examinations and pulmonary function tests may be required if the PLHCP deems them necessary. The employer must provide additional medical evaluations if
- an employee has symptoms from respirator use.
- PLHCP, supervisor or program administrator determines the employee needs to be reevaluated.
- a change in workplace conditions occurs (PPE, temperature).

- fit-testing and program evaluation determine a need for employee reevaluation.

Employee Training

Employees required to use respirators must be given training on the proper use and limitations of the equipment. In addition, employees must be trained on the respiratory hazards encountered in their workplace in routine and emergency situations. The training must be comprehensive, understandable, and given at least annually. The following items should be covered in the training program.

- the nature of the respiratory hazard, and what can happen if the respirator is not used properly
- engineering and administrative controls being used and the need for the respirator as added protection
- reason(s) for the selection of a particular type of respirator
- limitations of the selected respirator
- methods of donning the respirator and checking its fit and operation
- proper wear of the respirator
- proper procedure for handling emergency situations
- respirator cleaning procedures.

In addition to these items, the training must include practice time for employees to put on the respirator, handle it, adjust it, perform a positive- and negative-user seal check to determine if it fits properly, and wear it in a test atmosphere (fit-test).

Videotapes can be used to supplement a training program, but should not be used in place of one. Training of employees should be documented with the date, name, and contents of the program.

Fit-Testing and Respirator Fit

An employee required to wear a negative- or positive-pressure tight-fitting facepiece respirator must be fit-tested prior to being issued the respirator. The respirator used for the fit-test must be the same make, model, style, and size of the respirator that will be used. If the respirator used changes, the fit-test must be repeated. All fit-tests must be conducted on an annual basis thereafter. There are three fit-test methods outlined in the OSHA standard that can be performed.

The employer also must prevent any conditions or situations that may result in facepiece seal leakage. Respirators equipped with a tight- or loose-fit-

---

Appendix C to § 1910.134: OSHA Respirator Medical Evaluation Questionnaire (Mandatory)

To the employer: Answers to questions in Section 1, and to question 9 in Section 2 of Part A, do not require a medical examination.

To the employee:

Can you read (circle one): Yes/No

Your employer must allow you to answer this questionnaire during normal working hours, or at a time and place that is convenient to you. To maintain your confidentiality, your employer or supervisor must not look at or review your answers, and your employer must tell you how to deliver or send this questionnaire to the health care professional who will review it.

Part A. Section 1. (Mandatory) The following information must be provided by every employee who has been selected to use any type of respirator (please print).

1. Today's date: _____
2. Your name: _____
3. Your age (to nearest year): _____
4. Sex (circle one): Male/Female
5. Your height: _____ ft. _____ in.
6. Your weight: _____ lbs.
7. Your job title: _____
8. A phone number where you can be reached by the health care professional who reviews this questionnaire (include the Area Code): _____
9. The best time to phone you at this number: _____
10. Has your employer told you how to contact the health care professional who will review this questionnaire (circle one): Yes/No
11. Check the type of respirator you will use (you can check more than one category):
    a. _____ N, R, or P disposable respirator (filter-mask, non-cartridge type only).
    b. _____ Other type (for example, half- or full-facepiece type, powered-air purifying, supplied-air, self-contained breathing apparatus).
12. Have you worn a respirator (circle one): Yes/No
    If "yes," what type(s): _____

Part A. Section 2. (Mandatory) Questions 1 through 9 below must be answered by every employee who has been selected to use any type of respirator (please circle "yes" or "no").

1. Do you *currently* smoke tobacco, or have you smoked tobacco in the last month: Yes/No
2. Have you *ever had* any of the following conditions?
    a. Seizures (fits): Yes/No
    b. Diabetes (sugar disease): Yes/No
    c. Allergic reactions that interfere with your breathing: Yes/No
    d. Claustrophobia (fear of closed-in places): Yes/No
    e. Trouble smelling odors: Yes/No
3. Have you *ever had* any of the following pulmonary or lung problems?
    a. Asbestosis: Yes/No
    b. Asthma: Yes/No

(continued)

Figure 11-3. OSHA Respirator Medical Evaluation Questionnaire - 29 *CFR* 1910.134, Appendix C.

c. Chronic bronchitis: Yes/No
d. Emphysema: Yes/No
e. Pneumonia: Yes/No
f. Tuberculosis: Yes/No
g. Silicosis: Yes/No
h. Pneumothorax (collapsed lung): Yes/No
i. Lung cancer: Yes/No
j. Broken ribs: Yes/No
k. Any chest injuries or surgeries: Yes/No
l. Any other lung problem that you've been told about: Yes/No
4. Do you *currently* have any of the following symptoms of pulmonary or lung illness?
   a. Shortness of breath: Yes/No
   b. Shortness of breath when walking fast on level ground or walking up a slight hill or incline: Yes/No
   c. Shortness of breath when walking with other people at an ordinary pace on level ground: Yes/No
   d. Have to stop for breath when walking at your own pace on level ground: Yes/No
   e. Shortness of breath when washing or dressing yourself: Yes/No
   f. Shortness of breath that interferes with your job: Yes/No
   g. Coughing that produces phlegm (thick sputum): Yes/No
   h. Coughing that wakes you early in the morning: Yes/No
   i. Coughing that occurs mostly when you are lying down: Yes/No
   j. Coughing up blood in the last month: Yes/No
   k. Wheezing: Yes/No
   l. Wheezing that interferes with your job: Yes/No
   m. Chest pain when you breathe deeply: Yes/No
   n. Any other symptoms that you think may be related to lung problems: Yes/No
5. Have you *ever had* any of the following cardiovascular or heart problems?
   a. Heart attack: Yes/No
   b. Stroke: Yes/No
   c. Angina: Yes/No
   d. Heart failure: Yes/No
   e. Swelling in your legs or feet (not caused by walking): Yes/No
   f. Heart arrhythmia (heart beating irregularly): Yes/No
   g. High blood pressure: Yes/No
   h. Any other heart problem that you've been told about: Yes/No
6. Have you *ever had* any of the following cardiovascular or heart symptoms?
   a. Frequent pain or tightness in your chest: Yes/No
   b. Pain or tightness in your chest during physical activity: Yes/No
   c. Pain or tightness in your chest that interferes with your job: Yes/No
   d. In the past two years, have you noticed your heart skipping or missing a beat: Yes/No
   e. Heartburn or indigestion that is not related to eating: Yes/No
   f. Any other symptoms that you think may be related to heart or circulation problems: Yes/No
7. Do you *currently* take medication for any of the following problems?
   a. Breathing or lung problems: Yes/No
   b. Heart trouble: Yes/No
   c. Blood pressure: Yes/No

d. Seizures (fits): Yes/No
8. If you've used a respirator, have you *ever had* any of the following problems? (If you've never used a respirator, check the following space and go to question 9:)
   a. Eye irritation: Yes/No
   b. Skin allergies or rashes: Yes/No
   c. Anxiety: Yes/No
   d. General weakness or fatigue: Yes/No
   e. Any other problem that interferes with your use of a respirator: Yes/No
9. Would you like to talk to the health care professional who will review this questionnaire about your answers to this questionnaire: Yes/No

Questions 10 to 15 below must be answered by every employee who has been selected to use either a full-facepiece respirator or a self-contained breathing apparatus (SCBA). For employees who have been selected to use other types of respirators, answering these questions is voluntary.

10. Have you *ever lost* vision in either eye (temporarily or permanently): Yes/No
11. Do you *currently* have any of the following vision problems?
    a. Wear contact lenses: Yes/No
    b. Wear glasses: Yes/No
    c. Color blind: Yes/No
    e. Any other eye or vision problem: Yes/No
12. Have you *ever had* an injury to your ears, including a broken ear drum: Yes/No
13. Do you *currently* have any of the following hearing problems?
    a. Difficulty hearing: Yes/No
    b. Wear a hearing aid: Yes/No
    c. Any other hearing or ear problem: Yes/No
14. Have you *ever had* a back injury: Yes/No
15. Do you *currently* have any of the following musculoskeletal problems?
    a. Weakness in any of your arms, hands, legs, or feet: Yes/No
    b. Back pain: Yes/No
    c. Difficulty fully moving your arms and legs: Yes/No
    d. Pain or stiffness when you lean forward or backward at the waist: Yes/No
    e. Difficulty fully moving your head up or down: Yes/No
    f. Difficulty fully moving your head side to side: Yes/No
    g. Difficulty bending at your knees: Yes/No
    h. Difficulty squatting to the ground: Yes/No
    i. Climbing a flight of stairs or a ladder carrying more than 25 lbs: Yes/No
    j. Any other muscle or skeletal problem that interferes with using a respirator: Yes/No

Part B Any of the following questions, and other questions not listed, may be added to the questionnaire at the discretion of the health care professional who will review the questionnaire.
1. In your present job, are you working at high altitudes (over 5,000 feet) or in a place that has lower than normal amounts of oxygen: Yes/No
   If "yes," do you have feelings of dizziness, shortness of breath, pounding in your chest, or other symptoms when you're working under these conditions: Yes/No
2. At work or at home, have you ever been exposed to hazardous solvents, hazardous airborne chemicals (e.g., gases, fumes, or dust), or have you come into skin contact with hazardous chemicals: Yes/No

   If "yes," name the chemicals if you know them: _____

3. Have you ever worked with any of the materials, or under any of the conditions, listed below:
   a. Asbestos: Yes/No
   b. Silica (e.g., in sandblasting): Yes/No
   c. Tungsten/cobalt (e.g., grinding or welding this material): Yes/No
   d. Beryllium: Yes/No
   e. Aluminum: Yes/No
   f. Coal (for example, mining): Yes/No
   g. Iron: Yes/No
   h. Tin: Yes/No
   i. Dusty environments: Yes/No
   j. Any other hazardous exposures: Yes/No
   If "yes," describe these exposures: _____

4. List any second jobs or side businesses you have: _____

5. List your previous occupations: _____

6. List your current and previous hobbies: _____

7. Have you been in the military services? Yes/No
   If "yes," were you exposed to biological or chemical agents (either in training or combat): Yes/No

8. Have you ever worked on a HAZMAT team? Yes/No

9. Other than medications for breathing and lung problems, heart trouble, blood pressure, and seizures mentioned earlier in this questionnaire, are you taking any other medications for any reason (including over-the-counter medications): Yes/No
   If "yes," name the medications if you know them: _____

10. Will you be using any of the following items with your respirator(s)?
    a. HEPA Filters: Yes/No
    b. Canisters (for example, gas masks): Yes/No
    c. Cartridges: Yes/No

11. How often are you expected to use the respirator(s) (circle "yes" or "no" for all answers that apply to you)?:
    a. Escape only (no rescue): Yes/No
    b. Emergency rescue only: Yes/No
    c. Less than 5 hours *per week*: Yes/No
    d. Less than 2 hours *per day*: Yes/No
    e. 2 to 4 hours *per day*: Yes/No

(continued)

Figure 11-3. Continued

f. Over 4 hours *per day:* Yes/No

12. During the period you are using the respirator(s), is your work effort:
    a. *Light* (less than 200 kcal per hour): Yes/No

    If "yes," how long does this period last during the average shift:_____hrs._____mins.

    Examples of a light work effort are *sitting* while writing, typing, drafting, or performing light assembly work; or *standing* while operating a drill press (1–3 lbs.) or controlling machines.

    b. *Moderate* (200 to 350 kcal per hour): Yes/No

    If "yes," how long does this period last during the average shift:_____hrs._____mins.

    Examples of moderate work effort are *sitting* while nailing or filing; *driving* a truck or bus in urban traffic; *standing* while drilling, nailing, performing assembly work, or transferring a moderate load (about 35 lbs.) at trunk level; *walking* on a level surface about 2 mph or down a 5-degree grade about 3 mph; or *pushing* a wheelbarrow with a heavy load (about 100 lbs.) on a level surface.

    c. *Heavy* (above 350 kcal per hour): Yes/No

    If "yes," how long does this period last during the average shift:_____hrs._____mins.

    Examples of heavy work are *lifting* a heavy load (about 50 lbs.) from the floor to your waist or shoulder; *working* on a loading dock; *shoveling; standing* while bricklaying or chipping castings; *walking* up an 8-degree grade about 2 mph; *climbing* stairs with a heavy load (about 50 lbs.).

13. Will you be wearing protective clothing and/or equipment (other than the respirator) when you're using your respirator: Yes/No

    If "yes," describe this protective clothing and/or equipment: _____

14. Will you be working under hot conditions (temperature exceeding 77° F): Yes/No

15. Will you be working under humid conditions: Yes/No

16. Describe the work you'll be doing while you're using your respirator(s):
    _____

17. Describe any special or hazardous conditions you might encounter when you're using your respirator(s) (for example, confined spaces, life-threatening gases):
    _____

18. Provide the following information, if you know it, for each toxic substance that you'll be exposed to when you're using your respirator(s):

    Name of the first toxic substance: _____
    Estimated maximum exposure level per shift: _____
    Duration of exposure per shift: _____
    Name of the second toxic substance: _____
    Estimated maximum exposure level per shift: _____
    Duration of exposure per shift: _____
    Name of the third toxic substance: _____
    Estimated maximum exposure level per shift: _____
    Duration of exposure per shift: _____
    The name of any other toxic substances that you'll be exposed to while using your respirator:
    _____

19. Describe any special responsibilities you'll have while using your respirator(s) that may affect the safety and well-being of others (for example, rescue, security):
    _____

Appendix D to § 1910.134 (Non-Mandatory) Information for Employees Using Respirators When Not Required Under the Standard

Respirators are an effective method of protection against designated hazards when properly selected and worn. Respirator use is encouraged, even when exposures are below the exposure limit, to provide an additional level of comfort and protection for workers. However, if a respirator is used improperly or not kept clean, the respirator itself can become a hazard to the worker. Sometimes, workers may wear respirators to avoid exposures to hazards, even if the amount of hazardous substance does not exceed the limits set by OSHA standards. If your employer provides respirators for your voluntary use, of if you provide your own respirator, you need to take certain precautions to be sure that the respirator itself does not present a hazard.

You should do the following:

1. Read and heed all instructions provided by the manufacturer on use, maintenance, cleaning and care, and warnings regarding the respirators limitations.

2. Choose respirators certified for use to protect against the contaminant of concern. NIOSH, the National Institute for Occupational Safety and Health of the U.S. Department of Health and Human Services, certifies respirators. A label or statement of certification should appear on the respirator or respirator packaging. It will tell you what the respirator is designed for and how much it will protect you.

3. Do not wear your respirator into atmospheres containing contaminants for which your respirator is not designed to protect against. For example, a respirator designed to filter dust particles will not protect you against gases, vapors, or very small solid particles of fumes or smoke.

4. Keep track of your respirator so that you do not mistakenly use someone else's respirator.

Figure 11-3. Continued

ting facepiece must not be worn if facial hair interferes with the valve function of the face-to-facepiece seal. Goggles, glasses, and faceshields must also be worn without interfering with the face-to-facepiece seal. Some respirators accommodate prescription glasses without adversely affecting the facepiece seal.

Respirator Maintenance and Care
Respirators used by employees must be maintained by the employee. However, the employer must provide procedures for the cleaning, disinfection, storage, inspection, and repair of this equipment.

Cleaning
A used respirator must be cleaned and disinfected before it is reissued. Personally assigned respirators must be cleaned and disinfected on a regular basis. Emergency use and shared respirators must be cleaned and disinfected after each use. Respirators may be washed in a detergent solution and disinfected by immersion in a sanitizing solution. Dishwashing liquid, or commercial cleaner sanitizers are available. Alcohol-free wipes can also be used to clean a respirator. When cleaning respiratory equipment, caution should be taken when using strong cleaning agents and solvents. These products may damage rubber or elastomeric respirator parts, thereby shortening the life of the respirator.

Inspection
All respirators must be inspected for wear and deterioration of their components before each use. Also, a respirator kept for emergency use must be inspected at least monthly. SCBAs must be inspected monthly. Inspection records must be kept for emergency-use respirators that include the inspection date and findings (Figure 11-4). Respirators that do not pass inspections must be immediately removed from service to be repaired or replaced.

Storage
Respirators must be stored to protect against dust, sunlight, heat, extreme cold, excessive moisture, or damaging chemicals. Respirators should also be protected from mechanical damage. Respirators should be stored so that the facepiece exhalation valves rest in normal positions to prevent the rubber or plastic from reforming into abnormal shape. A good way to store a respirator is in a sealed plas-

Figure 11-4. A respirator kept for emergency use must be inspected at least monthly. Inspection records must be kept for emergency-use respirators that include the inspection date and findings. Respirators that do not pass inspections must be immediately removed from service to be repaired or replaced. (Courtesy Mine Safety Appliances)

tic bag in a respirator pouch which is kept in a location away from contamination. A paint spray booth, tool box, or peg board in the maintenance shop are not good storage locations.

Air Quality
The employer is to provide employees using atmosphere-supplying respirators with breathing air of high quality. High quality air must meet the requirements of grade D as outlined in American National Standards Institute (ANSI)/Compressed Gas Association (CGA) standard G-7.1-1989.

Program Evaluation
Employers are required to evaluate the workplace to ensure that the respiratory protection program is being implemented properly. Interviewing employees using respirators and observing employees wearing them are prudent ways to conduct this evaluation. In addition, air monitoring should be conducted to ensure the chosen respirator is adequate protection and that the program is still effective.

Record Keeping

As with most OSHA standards, employers are required to keep written documentation. Written documentation of medical evaluations should be retained and made available according to 29 *CFR* 1910.1020. Fit-testing records should include the following:
- employee name
- type of fit-test performed
- specific make, model, style, and size of respirator
- test date
- pass/fail results for qualitative fit-test
- fit factor and strip chart recording for quantitative fit-test

A written copy of the respirator program should be kept and users of respirators should be able to gain access to it and be familiar with its contents.

Voluntary Use

One thing to note in the new OSHA 1910.134 standard is *voluntary use*. In certain cases, an employee may want to wear a respirator where it is not required by OSHA or the employer. In these cases the employer may allow the respirator use as long as no hazard is created as a result. The employer must provide the employee with basic information in Appendix D of the standard, (written or oral), ensure the employee is medically able, and ensure that the respirator is clean, stored, and maintained so it does not present a health hazard. Voluntary use of a filtering facepiece (dust mask) does not require that the employee needs to be included in the respiratory protection program.

Respirators

There are two classes of respirators:
1. air-purifying
2. atmosphere-supplying

Air-purifying respirators use filters or sorbents to remove harmful substances from the air. Mechanical filters remove particles and chemical cartridges remove gases and vapors from the air. These types of respirators do not supply oxygen, may not be used in oxygen-deficient atmospheres (<19.5%), and may not be used in IDLH atmospheres. Some examples of air-purifying respirators include the following:
- mechanical filter cartridge
- chemical cartridge
- combination mechanical filter/chemical cartridge
- gas masks
- power air-purifying

Atmosphere-supplying respirators provide breathable air from a clean air source outside of the contaminated work area. Some examples of atmosphere-supplying respirators are:
- supplied air
- self-contained breathing apparatus (SCBA)
- combination SCBA and supplied air

More detailed information on respirators and their use can be found in the *Fundamentals of Industrial Hygiene,* 4th ed. (Plog, 1996), ANSI Z88.2, 29 CFR 1910.134, and the *3M Respirator Selection Guide.* Also see Figure 11-5, the OSHA Respirator Use Requirements Flow Chart from the *Major Requirements of OSHA Respiratory Protection Standard* 29 CFR 1910.134, March 1998.

Hearing Protection and the Hearing Conservation Amendment

The OSHA Hearing Conservation Amendment (HCA) requires employers to monitor employee noise exposure levels to determine their levels of exposure. Those employees with exposures equal or exceeding 85 dBA as an eight-hour Time Weighted Average (TWA) must be included in the company's hearing conservation program. This is referred to as the Action Level (AL).

For employees exposed to levels equal to or exceeding an eight-hour TWA of 90 dBA, hearing protection and engineering controls are mandatory requirements to reduce levels to below the 90 dBA level. The 90-dBA level is referred to as the Permissible Exposure Limit (PEL).

Employees are entitled to observe monitoring procedures and must be notified of the results. Letters to the employee with a copy for the employers file is a good method to notify employees of exposure results.

Employees typically are not static during the day. Many workers move about to several workstations and departments in order to perform their job duties. For this reason personal noise dosimetry is the preferred type of monitoring for determining a noise dose. A noise-level survey should preceed the noise dosimetry and will assist in locating the areas and operations where workers may be exposed to hazardous noise levels where dosimetry is needed. Instruments used for monitoring must be calibrated to ensure the measurements are accurate. Manufacturers' instructions are a good guide to use for calibration frequencies and procedures. Typically, instruments are calibrated before and after use and

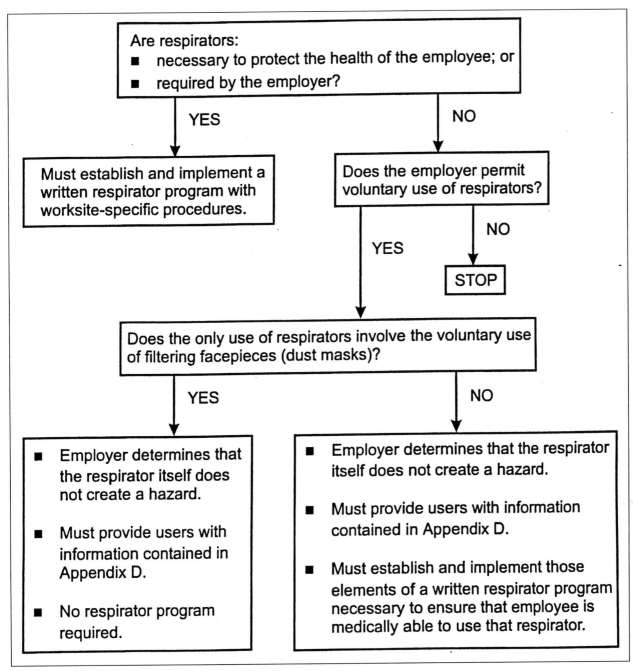

Figure 11-5. The OSHA Respirator Use Requirements Flow Chart from the *Major Requirements of OSHA Respiratory Protection Standard* 29 CFR 1910.134, March 1998.

factory calibration is done annually. In addition, the instrument measurement range of the instrument must include all noise within an 80–130-dBA range. Monitoring should be repeated when changes in production, process, or controls change noise exposure potential.

Audiometric Testing

The audiometric testing program includes a baseline audiogram, annual audiograms, and audiogram evaluations. The purpose of the audiograms is to determine whether hearing loss is being prevented by the hearing conservation program. The baseline is the reference audiogram against which future annual audiograms are compared. The baseline should be conducted within six months of exposure to noise at the action level. An annual audiogram must be conducted within one year of the baseline. Action-level audiograms should be preceded by at least 14 hours without exposure to workplace noise. All employees exposed to noise levels at or above the action level of 85 dBA are required to get audiograms.

A professional (audiologist, occupational health nurse, otolaryngologist or physician) must be responsible for the audiometric testing program, but need not be present when a qualified occupational hearing conservationist conducts the testing. Another responsibility of the occupational hearing conservationist is to ensure the tests are conducted in an appropriate test environment, making sure the audiometer works correctly, reviewing audiograms for standard threshold shifts (STS) and identifying audiograms that require further evaluation by a professional. A STS is a measured average hearing loss in either ear of 10 dB or more at the 2000, 3000 and 4000 Hertz frequencies.

Hearing Protection

All workers exposed to noise exposures at or above the action level must have hearing protection made available. This ensures that employees will have access to protection before they experience a loss in hearing.

Hearing protection must be worn by employees who have a STS, employees exposed to noise at or above 90 dBA as an 8-hr TWA. Employees who are exposed to an 8-hr TWA at or above the action level of 85 dBA and have not received a baseline audiogram are required to wear hearing protection until the baseline is performed.

Hearing protection must be selected by the employee from a variety of choices supplied by the employer (typically three types). The hearing protection must reduce employee exposures to at least 90 dBA. If an employee has an STS, then the hearing protection must reduce noise levels to 85 dBA.

Employee Training

Employee training for the hearing conservation amendment must be conducted on an annual basis. The following information should be included in the training program:
- effects of noise
- the purpose, advantages, and disadvantages of various types of hearing protection
- the selection, fit, and care of hearing protection, and
- the purpose and procedure of audiometric testing.

Record Keeping

As with most OSHA standards record keeping is a part of compliance. Noise exposure measurement records must be kept for two years. Records of audiometric test results must be kept for the duration of the employees employment. The audiometric test results must include the following information:
- employee name and job classification
- the date
- the examiner's name
- the date of the last acoustical calibration
- measurements of background sound pressure levels in audiometric test rooms
- the employees most recent noise exposure measurement.

Hearing Protectors

Hearing protection devices fall into four categories.
- enclosures
- aural inserts or earplugs
- superaural protectors or canal caps
- circumaural protectors or earmuffs.

Enclosure-type hearing protectors do just what the name implies, enclose the head. The attenuation is achieved by the material the enclosure is made out of. A helmet is an example of this type of protection. Helmets typically are used in high noise areas where there is an additional hazard of head injury. Another use of the helmet would be as a second form of hearing protection to further attenuate noise. Cost and bulkiness however make this type of protection uncommon in a typical industrial environment.

Earplugs are probably the most common type of hearing protection. Earplugs are made from rubber, plastic, fine glass down, foam and wax impregnated cotton. They can be formable, custom molded and premolded. The cost of this type of protection is very reasonable. However, the earplug has a limited life of one use to several months.

Canal caps attenuate sound by sealing the external opening of the ear. Canal caps are made of soft rubber and held in place with a headband.

Earmuffs fit over the entire ear. Earmuffs are made of plastic and lined with foam and held in place with a headband.

In order to select the proper hearing protection for use in the workplace, there are a variety of factors to consider.
- characteristics of noise levels in the area
- type of work area where using protection
- other PPE worn (safety glasses, hard hats)
- frequency of noise exposure.

After weighing all the options by using the factors above, the most appropriate type of hearing protection can be chosen. For example, if hard hats and safety glasses are also worn, ear muff protection attached to the employees hard hat may be the best choice. If hearing protection use is rare, an insert or plug may be the best choice.

HAZARD COMMUNICATION

Hazard Communication (HC) is the OSHA standard that was promulgated on Nov. 25, 1983 and is located in 29 *CFR* 1910.1200. After more than 10 years it remains the most frequently cited OSHA standard. The objective of this standard is to provide employees with information about the hazards and identities of the chemicals they are exposed to in the workplace. In addition, they are entitled to know the protective measures available to protect them from the hazards of these chemicals.

The HC covers physical hazards such as flammability, as well as health hazards like irritation, lung damage, and cancer. Carbon monoxide exhaust from a forklift truck and metal fumes from welding would also be covered by the HC standard.

Chemical manufacturers, importers, and distributors of hazardous chemicals are required to evaluate the hazards of the chemicals they produce, import, or distribute. Using the information from the evaluation process they must prepare labels for containers and technical health and safety bulletins called material safety data sheets (MSDS). The two most common styles of MSDS are the OSHA 174 form (Figure 11-6) and the ANSI Z400.1 Hazardous Industrial Chemicals MSDS Preparation, 1998.

Hazard Communication Program

Employers who use hazardous chemicals must have a program to ensure that the labels and MSDS information is communicated to their employees. This program is referred to as a Hazard Communication Program (HCP). There are four main components to the HCP. These are:

- written hazard communication program
- labels and other forms of warning
- MSDS
- employee information and training.

Written HCP

The written program is the blueprint that describes how you intend to comply with the HC standard. There are six points that must be addressed in the written program. These six points are as follows:
1. MSDS
2. labels
3. hazards in unlabeled pipes
4. nonroutine tasks
5. information exchange with contractor employees
6. employee training and information.

An MSDS must be maintained for each hazardous chemical in the workplace. This is referred to as a chemical inventory. A facility wide walk through is a prudent way to develop a chemical inventory. Once developed, it should be kept on file and updated periodically.

After determining the inventory, an indexing system must be developed. Index MSDSs by the type of chemical, area in which it is used, or alphabetically. The purpose of the index is to allow the user to easily locate a MSDS. This is especially important during an emergency situation. The MSDSs may also be maintained electronically as long as it does not interfere with the employee gaining access to the information.

The employer must develop a *labeling system*. The employer must explain how to inform employees about contents in containers. The label must contain an "appropriate hazard warning" and the contents of the container. Appropriate hazard warning means the label will provide a specific target organ with the hazard. For example "may cause central nervous system damage if inhaled." The written program should outline who will check if containers are labeled, what action is taken if containers are not labeled, what will be done to replace worn and damaged labels, and who is responsible to label secondary containers. A secondary container is a smaller container in which material from a larger container is transferred into. If the material is transferred back into the original container by the same person within the same workshift, it does not require a label.

Although pipes are not required to be labeled there must be a method in place to inform employees of their contents. It is especially important if the same pipes carry different materials at different times.

As the workplace becomes more dynamic, more companies have begun to use contractors more frequently. This is referred to as a *multi-employer worksite*. Painters, carpenters, electricians, and HVAC contractors are all covered by the multi-

Material Safety Data Sheet
May be used to comply with
OSHA's Hazard Communication Standard,
29 CFR 1910.1200. Standard must be
consulted for specific requirements.

U.S. Department of Labor
Occupational Safety and Health Administration
(Non-Mandatory Form)
Form Approved
OMB No. 1218-0072

IDENTITY *(As Used on Label and List)*

Note: Blank spaces are not permitted. If any item is not applicable, or no information is available, the space must be marked to indicate that.

Section I

Manufacturer's Name	Emergency Telephone Number
Address *(Number, Street, City, State, and ZIP Code)*	Telephone Number for Information
	Date Prepared
	Signature of Preparer *(optional)*

Section II — Hazardous Ingredients/Identity Information

Hazardous Components (Specific Chemical Identity; Common Name(s))	OSHA PEL	ACGIH TLV	Other Limits Recommended	% *(optional)*

Section III — Physical/Chemical Characteristics

Boiling Point		Specific Gravity (H_2O = 1)	
Vapor Pressure (mm Hg.)		Melting Point	
Vapor Density (AIR = 1)		Evaporation Rate (Butyl Acetate = 1)	

Solubility in Water

Appearance and Odor

Section IV — Fire and Explosion Hazard Data

Flash Point (Method Used)	Flammable Limits	LEL	UEL

Extinguishing Media

Special Fire Fighting Procedures

Unusual Fire and Explosion Hazards

(Reproduce locally) OSHA 174, Sept 1985

Figure 11-6. OSHA 174 MSDS form.

employer worksite. The written HC plan must explain how information on hazardous chemicals will be exchanged with these other employees. This exchange of information includes the contractors, who are responsible for informing the worksite of the hazardous chemicals they use and

Section V — Reactivity Data

Stability	Unstable		Conditions to Avoid
	Stable		

Incompatibility (*Materials to Avoid*)

Hazardous Decomposition or Byproducts

Hazardous Polymerization	May Occur		Conditions to Avoid
	Will Not Occur		

Section VI — Health Hazard Data

Route(s) of Entry:	Inhalation?	Skin?	Ingestion?

Health Hazards (*Acute and Chronic*)

Carcinogenicity:	NTP?	IARC Monographs?	OSHA Regulated?

Signs and Symptoms of Exposure

Medical Conditions
Generally Aggravated by Exposure

Emergency and First Aid Procedures

Section VII — Precautions for Safe Handling and Use

Steps to Be Taken in Case Material Is Released or Spilled

Waste Disposal Method

Precautions to Be Taken in Handling and Storing

Other Precautions

Section VIII — Control Measures

Respiratory Protection (*Specify Type*)

Ventilation	Local Exhaust	Special
	Mechanical (*General*)	Other

Protective Gloves	Eye Protection

Other Protective Clothing or Equipment

Work/Hygienic Practices

Page 2

☆ USGPO 1986-491-529/45775

Figure 11-6. Continued

bring on-site.

The HCP must address how the facility intends to handle these nonroutine tasks. Some of the most hazardous jobs we do are those that are not done regularly. The jobs we only do once or twice per year may become more hazardous due to the fewer opportunities we have to recognize controls. A confined space entry is an example of this type of job.

Employee Training

The last and most important element to include in the written program is employee training. In this section you should describe how you will train your employees. This includes, routine and non-routine tasks. Training can be done for each material, or by classes of hazards (acids, solvents), depending on how many materials the employees work with. Training must be given at the time of initial assignment and whenever a new hazard is introduced. Prudent practice would include annual training as a minimum. The program should cover the following information.

- the nature of hazards posed by the chemicals in the workplace
- measures that employees can take to protect themselves from these hazards
- instructions on work practices, PPE and any special procedures to be followed in an emergency
- an explanation of your HC program, including information on labeling and MSDSs.

For more information on what the written program should contain see the OSHA compliance directive CPL 2.2-38.

LABORATORY SAFETY

Laboratories are workplaces that contain a much wider array of hazards than most workplaces. In laboratories employees encounter hazards of chemicals used, hazards of new substances created, and hazards of new types of experiments. In contrast to manufacturing facilities, laboratories typically use smaller volumes of chemicals and have shorter time periods where exposure occurs.

It can be difficult to manage safety and health in the laboratory workplace due to the mobility of employees, variability of activities performed, wide range of chemicals used, and variety of equipment worked with. For that reason the OSHA standard 29 *CFR* 1910.1450, which regulates exposure to hazardous chemicals in laboratories, is a performance-oriented standard. That means that instead of listing individual requirements to maintain compliance, OSHA requires compliance through performance measures. Although the laboratory standard provides guidelines, some specific requirements are also included. For example, the employer must assign a chemical hygiene officer (CHO), develop a chemical hygiene plan (CHP) and evaluate and control hazards as necessary in order to protect the employees' health.

There are 11 components that should be included in the CHP. These are as follows:
1. standard operating procedures relevant to all laboratory operations
2. chemical procurement, distribution and storage
3. environmental monitoring
4. housekeeping and maintenance inspections
5. medical program
6. PPE and equipment
7. record keeping
8. signs and labels
9. spills and accidents
10. training and information
11. waste disposal.

The objectives for the CHP should be to guide all laboratory workers to perform their work safely, to protect others exposed to hazards in the laboratory, and to protect the environment.

For more information on the CHP see the following sources: *Prudent Practices for Handling Hazardous Chemicals in Laboratories* from the National Research Council, and *Developing a Chemical Hygiene Plan* from the American Chemical Society.

SUBSTANCE-SPECIFIC STANDARDS

In addition to all these standards, OSHA has substance-specific standards for 29 chemicals. These standards require initial employee monitoring, employee training and have set PELs, Action Levels (AL) and Short-Term Exposure Limits (STELs). For those employers that exceed the action level and PEL, some additional requirements need to be met. These are *medical monitoring, expanded employee training, PPE,* and *written compliance programs*. Some of the more commonly found chemicals in the workplace that have substance specific standards in 29 *CFR* are the following:
- Asbestos *CFR* 1910.1001
- Lead *CFR* 1910.1025
- Cadmium *CFR* 1910.1027
- Formaldehyde *CFR* 1910.1048
- Methylene Chloride *CFR* 1910.1052
- Benzene *CFR* 1910.1028.

It is important that the employer refer to the MSDSs on all products used in their facility to determine if any of these products contain one or more of the 29 substance-specific chemicals. If you determine that you have these substance-specific chemicals in your workplace, become familiar with where and how they are used.

Asbestos can be found in floor tile, spray-on insulation and thermal-pipe insulation. Lead is in solder, paints (typically reds and yellows), plating operations and welding rods. Cadmium is found in inks, brazing and plating operations, welding rods, and paints. Formaldehyde can be found in funeral homes, medical facilities, veterinary clinics, metal-working fluids, paints, biocides, and injection-molding operations. Methylene chloride can be found in paint strippers, printing inks, printing cleaners, foam adhesives, injection-mold releases, and cleaners.

In addition to the substance-specific standards, OSHA has three categories of air contaminants with PELs located in 29 *CFR* 1910.1000. Table Z-1 is for general contaminants that were adopted from ACGIH standards. This table has approximately 500 contaminants that cover things like solvents, acids, and metals. Table Z-2 is for select air contaminants. This table has approximately 21 contaminants, and covers hydrogen sulfide, toluene, and trichloroethylene. All chemicals in Table Z-2 have a ceiling limit in addition to a PEL. Table Z-3 is for mineral dusts like silica, graphite, and coal dust. There are about nine contaminants in Table Z-3.

Like the substance-specific standards, it is important to check the MSDSs for chemical components in products used in the workplace, to determine if and to what extent air monitoring is required.

EMPLOYEE TRAINING

Training is becoming more important today because employees work with a wide array of products and processes. Training has also become a standard component in many of the OSHA regulations. In addition, for employees to understand how to protect their health in the workplace, the information provided in training programs is vital.

There are many sources available today to enhance and aid in the development of training programs. When developing a training program, the best way to begin is to refer to the OSHA requirements for each of the various standards. A good way to review requirements is through *Training Requirements in OSHA Standards and Training Guidelines* (1998), a reference book that outlines OSHA training requirements for the employer. This guidebook details eight steps that comprise the training program process. These steps are as follows:
1. Determine if training is needed.
2. Identify training needs.
3. Identify goals and objectives.
4. Develop learning activities.
5. Conduct training.
6. Evaluate the program.
7. Improve the program.
8. Document the training.

It is important to remember that all training is to be site-specific. Every workplace has different hazards, and it is the employer's responsibility to instruct each employee in the recognition and avoidance of hazards or unsafe conditions and the ways to control or eliminate any hazard that may result in illness or injury.

INDOOR AIR QUALITY (IAQ)

Many industrial hygiene professionals find themselves working in the area of indoor air quality today. However, unlike the other issues mentioned in this chapter, IAQ has no OSHA standard available for compliance. Instead, three consensus standards that affect the IAQ issue have been developed by the American Society of Heating, Refrigerating and Air Conditioning Engineers (ASHRAE). These are the following:
1. ASHRAE 62 on ventilation for IAQ.
2. ASHRAE 55 on thermal comfort.
3. ASHRAE 52 on air filtration.

Note that the ASHRAE IAQ standards are under review and revision. The ASHRAE website, www.ashrae.org has the most current status of these standards.

The IAQ problem seems to have developed as a result of the energy crisis in the 1970s. During that time period, in order for employers to reduce energy costs, outside air brought into a building was reduced and new building materials were introduced. The result has been buildings that are under-ventilated and materials that generate new air contaminants. These things in combination have caused the occupants of these buildings to complain about air quality.

According to NIOSH, there are four major sources of the IAQ problem.
1. deficiencies in ventilation of the building
   - lack of outside air
   - poor air distribution
   - uncomfortable temperatures and humidity
   - sources of contaminants in the system
2. indoor air contaminants
   - formaldehyde
   - solvent vapors
   - dust
3. outdoor contaminant source
   - motor vehicle exhaust
   - pollen
   - fungi
   - smoke
   - construction dust
4. no observable cause

Deficiencies in the ventilation of the building represent 50% of the cases. These cases have a few other common traits: forced-air ventilation, energy-efficient buildings, central thermostats, and high-population densities. Whatever the problem, the industrial hygienist should try to work with the facility to investigate, correct, and control it to maintain quality indoor air.

SUMMARY

An industrial hygiene program is necessary to provide a place of employment free from recognized hazards that may cause death or serious physical harm to workers. This chapter was written to provide those not trained in industrial hygiene with enough information to enable them to implement an industrial hygiene program in their workplaces. The major components of an industrial hygiene program were identified, including personal and area air sampling; surface sampling; indoor air quality investigations; noise surveys; ergonomic studies; employee training; and determination of personal protective equipment, including respiratory protection, skin protection and hearing protection; hazard communication; laboratory safety; substance-specific standards; employee training; and indoor air quality. The importance of the industrial hygiene program was explained; and we explored the various relevant guidelines and regulations.

REFERENCES:

ASHRAE. www.ashrae.org

Developing a Chemical Hygiene Plan. Washington DC: American Chemical Society, 1990.

Ladou J. *Occupational Medicine.* East Norwalk CT: Appleton & Lange, 1990.

National Research Council. *Prudent Practices for Handling Chemicals in Laboratories.* Washington DC: National Academy Press, 1981.

Centers for Disease Control (CDC)/NIOSH. *Manual of Analytical Methods,* 4th ed. Cinncinati: NIOSH, 1994.

Plog BA. *Fundamentals of Industrial Hygiene,* 4th ed. Itasca IL: National Safety Council, 1996.

3M. *Respirator Selection Guide.* St Paul MN 3M, 1998.

U.S. Dept. of Labor (OSHA). *Major Requirements of OSHA Respiratory Protection Standard.* 29 CFR 1910.134, March 1998. Washington DC: U.S. Government Printing Office, 1998.

U.S. Dept. of Labor (OSHA). *Occupational Safety and Health Standards for General Industry.* Riverwoods IL: Commerce Clearing House, 1998.

U.S. Dept. of Labor (OSHA). *Training Requirements in OSHA Standards and Training Guidelines,* OSHA No. 2254. Washington DC: U.S. Government Printing Office, 1998.

Vincoli JW. *Basic Guide to Industrial Hygiene.* New York: Van Nostrand Reinhold, 1995.

APPENDIX

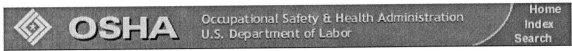

| TECHNICAL LINKS | OSHA TECHNICAL MANUAL |

OSHA Technical Manual

SECTION II: CHAPTER 2 [Extensively Revised and Featuring New Additions]

SAMPLING FOR SURFACE CONTAMINATION

Contents:
I. Introduction
II. The Use of Surface Contamination Sampling in Evaluating Safety and Health Programs
III. Media and Techniques for Wipe Sampling
IV. Bibliography

Appendix II:2-1. Substances Listed With a Skin Notation by the ACGIH TLV's and/or the OSHA PEL's
Appendix II:2-2. Example Procedures for Isocyanates
Appendix II:2-3. Example Procedures for Aromatic Amines

I. **INTRODUCTION.**

A. Worksite analysis (i.e., hazard assessment) is a basic component of an effective safety and health program. A complete worksite analysis requires the assessment of surface contamination since workers may be exposed to these contaminants directly through dermal and ingestive routes (e.g., isocyanates, pesticides), and indirectly through inhalation of contaminants that become re-entrained in the air (e.g., asbestos, lead).

B. Dermal and ingestive routes of entry are much more significant than inhalation for a large number of chemicals. For example, a fifteen-minute exposure of the hands and forearms to liquid glycol ethers [2-methoxy-ethanol (ME) and 2-ethoxy-ethanol (EE)] will result in a dose to the body well in excess of the eight-hour inhalation dose at their recommended air exposure limits. (Biological monitoring for the urinary metabolites methoxyacetic acid and ethoxyacetic acid was used to estimate the absorption via skin and lung.) Unfortunately, many industrial hygienists are only familiar with air sampling and fail to evaluate significant exposures caused by surface contamination.

C. Wipe sampling is an important tool of worksite analysis for both identifying hazardous conditions, and in evaluating the effectiveness of personal protective equipment, housekeeping, and decontamination programs. As described below, wipe sampling is an important tool for assessing compliance with certain OSHA requirements even though there are few specific criteria for acceptable surface contamination amounts.

D. The terms wipe sampling, swipe sampling, and smear sampling are synonyms that describe the techniques used to assess surface contamination on the skin, work surfaces, and PPE surfaces (e.g., gloves, respirators, aprons, etc.). The term "wipe sampling" will be used in this document.

II. THE USE OF SURFACE CONTAMINATION SAMPLING IN EVALUATING SAFETY AND HEALTH PROGRAMS.

29 CFR 1910.132 requires employers to "assess the workplace to determine if hazards are present, or are likely to be present, which necessitate the use of personal protective equipment (PPE)." To this purpose, wipe sampling can be useful in categorizing work areas for certain types of controls, such as PPE and/or special cleaning and decontamination. It is also useful in assessing the effectiveness of these controls, including proper work practices. Examples are provided below for three generalized work areas: controlled areas that require the use of PPE, controlled areas that require the use of special cleaning and/or decontamination, and non-controlled work areas that require neither PPE or special cleaning.

A. CONTROLLED WORK AREAS REQUIRING PPE.

1. These are areas where it has been determined (e.g., from an employer's hazard assessment) that PPE is necessary to prevent dermal exposures to a surface contaminant in spite of an aggressive, yet feasible cleaning regimen. Many production areas and specific job tasks fall into this category.

2. Wipe sampling can be used in assessing the effectiveness of the PPE program. Many elements of PPE programs are intended to prevent contamination to certain locations, such as the use of gloves to prevent contamination to the skin of the hands. Surface contamination found in those "protected" locations usually indicates a problem with the program. For example, the presence of surface contamination inside a glove is normally the result of either PPE failure (e.g., the contaminate soaked through the glove material or a tear in the glove), and/or an improper work practice for using the PPE, such as the worker inserting a contaminated hand inside the glove. Additional sampling and observation can be used to determine the specific source of the program failure and possible abatement (e.g., changing gloves more often, checking for tears before donning, cleaning hands before donning, etc.). Sampling after abatement measures are implemented can be used to show the effectiveness of the abatement.

3. It is important to recognize that this sampling is not attempting to assess the health risk resulting from the contamination inside the glove. Rather, it is to identify failures in the PPE program. Therefore, the criteria for concluding that contamination exists does not need to be quantitative. Criteria and reproducible procedures should be selected that provide confidence that contamination has not been adequately controlled (i.e., contaminant levels are above background). The use of wipe pads that change color upon contact with the contaminant is ideal both in locating contamination and as a visual tool in training workers on the consequences of poor work practices.

B. **CONTROLLED WORK AREAS REQUIRING SPECIAL CLEANING.**

   1. Wipe sampling in these areas can show that a feasible and practical regimen of special cleaning and/or decontamination precludes the need for PPE or additional cleaning. The cleaning of lunch room tables, and the decontamination of equipment before being removed from a restricted area are examples of this category. Other examples include cleaning surfaces to reduce accumulation of toxic materials (e.g., asbestos, lead, beryllium) that may become re-suspended in air and thus contribute to airborne exposures.

   2. Wipe sampling is used in these areas as a quality control test of the specialized cleaning (or decontamination) regimen. Therefore, samples are taken to assess contamination levels of those surfaces for which the special cleaning is required. Samples found in excess of an acceptable, task-specific, surface contamination limit (see below) indicate a failure in the cleaning or decontamination program. More aggressive training and supervision of the cleaning procedures and/or scheduling may need to be implemented.

   3. Again, it is important to recognize that this sampling is not attempting to assess the health risk resulting from the contamination. Rather, it is to ensure that the cleaning and decontamination regimen is being effectively implemented. Establishing an acceptable contamination limit will depend on the purpose of the cleaning, and what is feasible for the procedures utilized. For example, periodic vacuuming of floor surfaces in a lead production area may be used to reduce the amount of lead dust available for re-entrainment, but significant lead contamination of the floor would still be expected. An acceptable surface contamination limit for this type of cleaning would be set much higher than a limit used to evaluate cleaning of tables in the break room.

   4. A few surface contamination concentration guidelines have been published, but typically concentration limits must be established by an employer for a specific task. The limits should be based on sufficient initial sampling to determine a "normal" range of contamination that can be expected after utilizing prescribed cleaning procedures. It would be appropriate to include documentation for the limits and their purpose, in the worksite Safety and Health Program.

C. **NON-CONTROLLED WORK AREAS.**

   1. These are work areas for which no special cleaning or PPE are required by the Safety and Health Program. Examples of this category are office areas that are physically separated from the production areas. These areas are often "assumed" to have no significant contamination. Wipe sampling is useful in demonstrating the lack of contamination. If samples do show contamination, further investigation would be needed to determine the cause. Consistent positive results would require a re-assessment of whether the area requires controls.

2. As with sampling to evaluate PPE programs described above, procedures and criteria for sampling non-controlled areas need to provide confidence that contamination has not occurred (i.e., surface concentrations are not above background). Again, the use of wipe pads which change color upon contact with the contaminant is ideal. The "direct reading" capability makes it possible to quickly screen an entire work area (and a single pad may be used for multiple locations within the area).

3. Sample those locations within the non-controlled area that accumulate dust (e.g., tops of filing cabinets), and surfaces that have potential for contamination from production areas (e.g., paper work brought in from the production areas).

4. Additional surfaces to consider for sampling include those that may come into contact with food and other materials that are ingested or placed in the mouth (e.g., chewing tobacco, gum, cigarettes).

   Contaminated smoking materials may allow toxic materials or their combustion products (e.g., lead, mercury) to enter the body via the lungs. Wiping of surfaces that smoking materials may touch, including the hands, may be useful in evaluating this possible route of exposure.

D. **EVALUATION OF SAMPLING RESULTS.**

1. The investigator must use professional judgment on a case-by-case basis when evaluating the significance of wipe-sampling results. As described above, acceptable surface contamination amounts will vary widely for the same toxic agent depending on the purpose and location of the sample. Any concentration above background is sufficient to identify a problem with the PPE program for some sample locations.

2. When evaluating results, consider the toxicity and the contribution of skin absorption and/or gastrointestinal absorption to the total dose. Additional factors are the ambient-air concentrations, skin irritation, etc.

3. The OSHA Technical Links Internet site includes Chemical Sampling Information which lists substances that have a potential for ingestion toxicity, skin absorption, and/or a hazardous effect on skin. This information may be found under the "Health" notation. Additional toxicological information concerning chronic skin absorption, dermatitis, etc. should be used in determining if the resulting exposure presents a potential employee hazard (see Bibliography and other references in Technical Links).

III. MEDIA AND TECHNIQUES FOR WIPE SAMPLING.

A. SAMPLING SURFACES FOR CONTAMINATION.

1. Techniques and media for collection of wipe samples from surfaces vary with the agent and purpose of the sample. It is recommended that the Technical Links Chemical Sampling Information, and/or the Salt Lake Technical Center's Applied Industrial Hygiene and Chemical Procedures Group be consulted when selecting a sampling procedure for a specific chemical or contaminant.

2. Classic wipe sampling techniques involve wiping a surface with a filter, which is then submitted to the SLTC for chemical analysis.

    a. Glass fiber filters (GFF), 37 mm in diameter as used in air sampling, are recommended for many chemicals that are analyzed by gas chromatography (GC), or high-performance liquid chromatography (HPLC). These may be used dry, or wetted with an appropriate solvent as specified in the Technical Links Chemical Sampling Information file.

    b. Paper filters are generally used for collection of metals. Mixed cellulose ester filter discs (AA filters) or smear tabs, or their equivalent, are most often recommended.

    c. Polyvinyl chloride filters are available for substances which are unstable on paper-type filters.

    d. Squares of a gauze material, available from the Salt Lake Technical Center upon request, may be used for many organic substances, and have the advantage of being more durable than filter media, especially when wiping rough surfaces. They may be used dry; or wetted with water or solvent to enhance collection efficiency.

    e. Charcoal-impregnated pads may be useful for collection of volatile solvents from surfaces. They work by trapping the solvent on activated charcoal, similar to air sampling charcoal tubes.

    f. In certain specialized cases, such as isocyanates and aromatic amines, highly reactive and unstable compounds must be collected on a filter medium that has been treated with a derivatizing reagent. These are available from the SLTC quality control department.

3. For a limited number of chemicals, direct-reading colorimetric wipe sampling procedures are available for qualitative or semi-quantitative detection of surface contaminants. These can be used for acids and bases, isocyanates, aromatic amines, organic solvents (not solvent specific), lead, platinum salts, explosives and hydrazine. Contact the Salt Lake Technical Center for more information.

4. For a variety of pesticides and certain other toxic chemicals, immunoassay kits can provide qualitative or semi-quantitative information on-site, and within about an hour. Some wet chemistry is required. Contact the <u>Salt Lake Technical Center</u> for more information.

B. <u>SAMPLING SKIN FOR CONTAMINATION</u>.

1. Techniques and media for wipe sampling of skin contamination vary with the agent and purpose of the sample. It is recommended that the Technical Links Chemical Sampling Information, and/or the Salt Lake Technical Center's Applied Industrial Hygiene and Chemical Procedures Group be consulted when selecting a sampling procedure for a specific chemical or contaminant.

2. There are concerns related to direct wipe sampling of the skin, including the possibility of promoting skin absorption with the use of certain solvents. Contact the Salt Lake Technical Center prior to taking wipe samples directly on the skin to receive agent specific procedures and precautions. Where feasible, biological monitoring is often the most effective means of assessing overall absorption of a contaminant, including through the skin.

3. Before any skin wipe is taken, explain why you want the sample and ask the employee about possible skin allergies to the chemicals in the sampling medium or wetting solution. Employees may elect not to allow sampling of their skin.

4. As an alternative to direct skin sampling, an indirect measurement of skin contamination (as well as PPE failure) can be assessed by wipe sampling surfaces that workers can touch (e.g., table tops, handles, control knobs, inside surfaces of protective equipment).

5. Classic wipe sampling techniques as described earlier, employing glass-fiber filters, mixed cellulose ester filters or smear tabs, or gauze squares, charcoal impregnated pads, may be used for sampling contaminants on the skin. If it is deemed desirable to moisten the collecting medium to improve collection efficiency, procedures will normally utilize distilled or de-ionized water, or a 50% solution of isopropyl alcohol in water.

6. Hand washes may be appropriate in some cases. Twenty ml of distilled or de-ionized water, or a dilute solution of mild soap may be added to a zipper-style sandwich bag. The hand to be sampled is inserted, and the bag held tightly closed around the wrist. After a few seconds of agitation, the hand is carefully removed, and the wash solution is poured back into a scintillation vial for shipment to the laboratory.

7. For a limited number of chemicals, direct-reading colorimetric wipe sampling procedures are available for qualitative or semi-quantitative detection of contaminants. These can be used for acids and bases, isocyanates, aromatic amines, organic solvents (not solvent-specific), platinum salts, and hydrazine. The technique differs from that used for surface wipes. Contact the Salt Lake Technical Center for more information.

8. The same technology employed in the colorimetric wipe sampling procedures described above has been applied to a band-aid-type format. These can be applied to the hands inside gloves to demonstrate glove permeability or breakthrough. They can serve as an excellent tool in employee training.

C. PROCEDURES FOR COLLECTING WIPE SAMPLES.

1. Preloading a group of vials with appropriate filters is a convenient method to carry the sample media to the worksite. (The smear tabs should be inserted with the tab end out.) Clean plastic gloves should be worn when handling the filters. The gloves should not be powdered.

2. The following are general procedures for taking wipe samples:

    a. If multiple samples are to be taken at the worksite, prepare a rough sketch of the area to be wipe sampled.

    b. A new set of clean, impervious gloves should be used for each sample to avoid contamination of the filter by previous samples (and the possibility of false positives) and to prevent contact with the substance.

    c. Withdraw the filter from the vial with your fingers or clean tweezers. If a damp wipe sample is desired, moisten the filter with distilled water or other solvent as recommended.

    d. Depending on the purpose of the sample, it may be useful to determine the concentration of contamination (e.g., in micrograms of agent per area). For these samples, it is necessary to record the area of the surface wiped (e.g., 100 cm^2). This would normally not be necessary for samples taken to simply show the presence of the contaminant.

    e. Firm pressure should be applied when wiping.

    f. Start at the outside edge and progress toward the center of the surface area by wiping in concentric squares of decreasing size.

    g. Without allowing the filter to come into contact with any other surface, fold the filter with the exposed side in, then fold it over again. Place the filter in a sample vial, cap and number it, and note the number at the sample location on the sketch. Include notes with the sketch giving any further description of the sample (e.g., "Fred Employee's respirator, inside"; "Lunch table").

    h. At least one blank filter treated in the same fashion, but without wiping, should be submitted for each sampled area.

    i. Some substances should have solvent added to the vial as soon as the wipe sample is placed in the vial (e.g., benzidine). These substances are indicated with an "X" next to the solvent notation in the Technical Links Chemical Sampling Information File.

j. Submit the samples to the Salt Lake Technical Center (SLTC) with an OSHA 91 form.

IV. BIBLIOGRAPHY.

Adams, R.M. 1983. *Occupational Skin Disease.* New York: Grune and Stratton.

Benezra, C. et al. 1982. *Occupational Contact Dermatitis.* Clinical and Chemical Aspect. Philadelphia; Saunders. 1st ed.

Chaiyuth, C. and L. Levin. "A laboratory evaluation of wipe testing base on lead oxide surface contamination." Am. Ind. Hyg. Assoc. J. 45:311-317, 1984.

Clayton, G.D. and F.E. Clayton. 1981. Patty's Industrial Hygiene and Toxicology. New York: John Wiley and Sons. Vol. II.

Fisher, A.A. 1986. *Contact Dermatitis.* Philadelphia: Lea and Febriger. 3rd ed.

Gellin, G. and H.I. Malbach. 1982. *Occupational and Industrial Dermatology.* Chicago: Year Book Medical Publisher.

Lees, P.S.J. et al. "Evidence for dermal absorption as the major route of body entry during exposure of transformer maintenance and repairman to PCBs." Am. Ind. Hyg Assoc. J. 48:257-264, 1987.

Occupational Safety and Health Administration (OSHA), U.S. Dept. of Labor. 1995. "Chemical Sampling Information." Washington, D.C. Available on the OSHA Internet site.

Kezic, Sanja; et al.; *Dermal absorption of vaporous and liquid 2-methoxyethanol and 2-ethoxyethanol in volunteers*, Occup. Environ. Med., 54, 1997, pp. 38-43.

APPENDIX II:2-1. SUBSTANCES LISTED WITH A SKIN NOTATION BY THE ACGIH TLV's AND/OR THE OSHA PEL's

NAME:	CAS # :	NAME:	CAS # :
B) Acrylamide	79-06-1	T) Hexachloroethane	67-72-1
T) Acrylic acid	79-10-7	B) Hexachloronaphthalene	1335-87-1
T) Acrylonitrile	107-13-1	T) Hexafluoroacetone	684-16-2
B) Aldrin	309-00-2	T) Hexamethyl phosphoramide	680-31-9
B) Allyl Alcohol	107-18-6	T) 2-Hexanone	591-78-6
T) 4-Aminodiphenyl	92-67-1	B) Hydrazine	302-01-2
B) Aniline	62-53-3	B) Hydrogen Cyanide	74-90-8
B) Anisidine (o,p-Isomers)	29191-52-4	B) Lindane	58-89-9
B) Azinphos-Methyl	86-50-0	B) Malathion (Total Dust)	121-75-5
T) Benzidine	92-87-5	T) Manganese Cyclopentadienyl Tricarbonyl (as Mn)	12079-65-1
B) Bromoform	75-25-2		
B) 2-Butoxyethanol	111-76-2	T) Mercury (organo) Alkyl Compounds (as Hg)	
T) n-Butyl Alcohol	71-36-3		
		T) Mercury (Vapor) (as Hg)	7439-97-6

B) Butylamine	109-73-9	B) Methyl Acrylate	96-33-3
B) tert-Butyl Chromate (as CrO_3)	1189-85-1	T) Methyl Acrylonitrile	126-98-7
		T) Methyl Alcohol	67-56-1
T) o-sec-Butylphenol	89-72-5	B) Methyl Bromide	74-83-9
T) Captafol	2425-06-1	B) Methyl Cellosolve	109-86-4
T) Carbon Disulfide	75-15-0	B) Methyl Cellosolve Acetate	110-49-6
T) Carbon Tetrachloride	56-23-5	T) Methyl Chloride	74-87-3
T) Catechol	120-80-9	T) o-Methylcyclohexanone	583-60-8
B) Chlordane	57-74-9	T) Methylcyclopentadienyl	12108-13-3
B) Chlorinated Camphene	8001-35-2	T) Methyl Demeton	8022-00-2
T) Chloroacetyl Chloride	79-04-9	T) 4,4'-Methylene-bis (2-Chloroaniline)	101-14-4
T) o-Chlorobenzylidene Malononitrile	2698-41-1	T) 4,4'-Methylenedianiline	101-77-9
B) Chlorodiphenyl (42% Cl)	53469-21-9	B) Methyl Iodide	74-88-4
B) Chlorodiphenyl (54% Cl)	11097-69-1	B) Methyl Isobutyl Carbinol	108-11-2
B) Chloroprene	126-99-8	B) Methyl Isocyanate	624-83-9
T) Chlorpyrifos	2921-88-2	T) Methyl Parathion	298-00-0
B) Cresol (All Isomers)	1319-77-3	T) Monocrotophos	6923-22-4
B) Cumene	98-82-8	T) Monomethyl Aniline	100-61-8
B) Cyanide (as Cn)	57-12-5	B) Monomethyl Hydrazine	60-34-4
T) Cyclohexanol	108-93-0	B) Morpholine	110-91-8
T) Cyclohexanone	108-94-1	B) Nicotine	54-11-5
T) Cyclonite	121-82-4	B) p-Nitroaniline	100-01-6
P) DDT	50-29-3	B) Nitrobenzene	98-95-3
B) Decaborane	17702-41-9	T) 4-Nitrodiphenyl	92-93-3
B) Demeton	8065-48-3	B) p-Nitrochlorobenzene	100-00-5
T) Diazinon	333-41-5	B) Nitroglycerin	55-63-0
T) 2-n-Dibutylaminoethanol	102-81-8	T) N-Nitrosodimethylamine	62-75-9
T) 3,3'-Dichlorobenzidine	91-94-1	B) Nitrotoluene	99-08-1
B) Dichloroethyl Ether	111-44-4	B) Octachloronaphthalene	2234-13-1
T) 1,3-Dichloropropene	542-75-6	P) Paraquat, respirable dust	1910-42-5
B) Dichlorvos (DDVP)	62-73-7	B) Parathion	56-38-2
T) Dicrotophos	141-66-2	T) Pentachloronaphthalene	1321-64-8
B) Dieldrin	60-57-1	B) Pentachlorophenol	87-86-5
T) Diethanolamine	111-42-2	B) Phenol	108-95-2
T) Diethylamine	109-89-7	T) Phenothiazine	92-84-2
B) Diethylamino ethanol	100-37-8	P) p-Phenylene Diamine	106-50-3
T) Diethylenetriamine	111-40-0	T) Phenyl Glycidyl Ether	122-60-1
B) Diisopropylamine	108-18-9	B) Phenylhydrazine	100-63-0
B) Dimethyl Acetamide	127-19-5	T) Phorate	298-02-2
B) Dimethylaniline (N,N-Dimethylaniline)	121-69-7	B) Phosdrin	7786-34-7
		P) Picric Acid	88-89-1
T) Dimethyl-1,2-dibromo-2, 2-dichloroethyl phosphate	300-76-5	T) Propargyl Alcohol	107-19-7
		T) Propyl Alcohol	71-23-8
		T) 1,2-Propylene Glycol Dinitrate	6423-43-4

B) Dimethylformamide	68-12-2	B) Tetramethyl Succinonitrile	3333-52-6
B) 1,1-Dimethylhydrazine	57-14-7	P) Tetryl	479-45-8
B) Dimethyl Sulfate	77-78-1	B) Thallium, soluble	7440-28-0
B) Dinitrobenzene (All isomers)	25154-54-5	T) Thioglycolic Acid	68-11-1
B) Dinitro-o-cresol	534-52-1	T) Tin, organic compounds	1983-10-4
B) Dinitrotoluene	25321-14-6	T) o-Tolidine	119-93-7
B) Dioxane	123-91-1	T) Toluene	108-88-3
T) Dioxathion	78-34-2	T) m-Toluidine	108-44-1
B) Dipropylene Glycol Methyl Ether	34590-94-8	B) o-Toluidine	95-53-4
		T) p-Toluidine	106-49-0
T) Diquat	85-00-7	T) 1,1,2-Trichloroethane	79-00-5
T) Disulfoton	298-04-4	B) Trichloronaphthalene	1321-65-9
B) EPN	2104-64-5	T) 1,2,3-Trichloropropane	96-18-4
T) Endosulfan	115-29-7	T) Triethylamine	121-44-8
B) Endrin	72-20-8	B) 2,4,6-Trinitrotoluene	118-96-7
B) Epichlorohydrin	106-89-8	T) Triorthocresyl Phosphate	78-30-8
T) Ethion	563-12-2	T) Vinyl Cyclohexene Dioxide	106-87-6
B) 2-Ethoxyethanol	110-80-5	T) m-Xylene-alpha,alpha'-diamine	1477-55-0
B) 2-Ethoxyethyl Acetate	111-15-9		
P) Ethyl Acrylate	140-88-5	T) Xylidine	1300-73-8
T) Ethylamine	75-04-7		
T) Ethyl Bromide	74-96-4		
B) Ethylene Chlorohydrin	107-07-3	Note: P = Skin notation in the PEL	
T) Ethylenediamine	107-15-3	T = Skin notation in the TLV	
T) Ethylene Dibromide	106-93-4	B = Skin notation in both references	
B) Ethylene Glycol Dinitrate	628-96-6		
B) Ethyleneimine	151-56-4		
B) N-Ethylmorpholine	100-74-3		
T) Fenamiphos	22224-92-6		
T) Fenthion	55-38-9		
T) Fonofos	944-22-9		
T) Formamide	75-12-7		
T) Heptachlor	76-44-8		
T) Hexachlorobenzene	118-74-1		
T) Hexachlorobutadiene	87-68-3		
B) Propylene Imine	75-55-8		
B) Sodium Fluoroacetate	62-74-8		
T) Styrene	100-42-5		
B) TEDP	3689-24-5		
B) TEPP	107-49-3		
T) 1,1,2,2-Tetrachloroethane	79-34-5		
P) Tetrachloronaphthalene	1335-88-2		
B) Tetraethyl Lead (as Pb)	78-00-2		
B) Tetramethyl Lead (as Pb)	75-74-1		

APPENDIX II:2-2. EXAMPLE PROCEDURES FOR ISOCYANATES

This example is provided for information only and should not be taken as the basis for OSHA policy.

Prepared by: Richard Lawrence, Chemist   Date: August 21, 1997

Aromatic Isocyanate Surface Contamination Sampling and Evaluation Techniques

The OSHA Salt Lake Technical Center received a request for assistance at a facility that uses methylene bisphenyl isocyanate (MDI, CAS no. 101-68-8) in their production process. It was decided that an inspection would include screening and sampling for possible MDI surface contamination (as well as air sampling for MDI). The following surface monitoring techniques were evaluated in preparation for that inspection.

A route of workplace exposure to chemicals with low vapor pressures, such as aromatic isocyanates, may be through skin contact with contaminated solvents or surfaces (1,2). Aromatic isocyanates present a respiratory sensitization hazard (1,2). Laboratory studies with animals have indicated that respiratory sensitization to both TDI (3,4) and MDI (5) can be induced by dermal contact alone. The ability to determine surface contamination may be useful in evaluating the effectiveness of housekeeping, decontamination and chemical protective equipment.

Direct reading indicators, such as commercially prepared pads that change color when they come in contact with specific chemicals (or classes of chemicals), are available for aromatic isocyanates. These types of indicators may be used as a screening tool, when assessing the extent of surface contamination, because they are inexpensive and the results are immediate. If an indicator wipe yields a positive result, a wipe sample can then be taken and sent to a laboratory for confirmation.

The mention of the commercial products Swypes® and Permea-Tec® does not constitute an endorsement but is inteded solely as an identification of the type of product deemed suitable for the use intended.

The use of direct reading instruments and indicators can be effective in helping employers to comply with the OSHA personal protective equipment standard, 29 CFR <u>1910.132</u> (d)(1)(I) and (f)(1)(iii) and (iv).

Wipe Sampling for Screening

Materials: At the time of the inspection, the OSHA Salt Lake Technical Center had an aromatic isocyanate indicator wipe kit in stock called a Swype® kit (6). This kit consists of indicator wipes called Swype®, a spray bottle of a developer solution that contains a chemical which activates the Swype®, and indicator patches that are worn under PPE, such as gloves, to test the effectiveness. The kit also contains a spray bottle of decontamination solution and a skin cleaner. The effectiveness of the skin cleaner and decontamination solution was not investigated.

   1. Gloves and other personal protective equipment must be worn during testing, as appropriate. "Best" style 727 nitrile gloves should provide protection to the hygienist's hands for the time required to perform the screening.

2. Gloves must be changed after a positive result to avoid cross contamination of any subsequent tests.

3. Spray the area to be sampled lightly with the Developer Solution. Use as little as needed to ensure that the surface is wet. Excess solution will dilute contamination, possibly below the detection limit. When testing a vertical surface or knob, some of the solution may begin to "run-off" or drip. This "run-off" should be captured onto the pad to ensure that any contaminant present has not been lost. The Developer Solution contains a proprietary component that activates the Swype® pad.

4. Wait approximately 30 seconds for any aromatic isocyanate to dissolve, then wipe the surface with a surface Swype® pad.

5. Allow 2 to 3 minutes for the color to develop. A pastel red-orange or pink color indicates aromatic isocyanate contamination. The color varies depending on the type of isocyanate present. The surface Swype® detection limit is approximately 3-5 μg.

6. Record appropriate information as needed.

7. If the surface Swype® tests indicate a positive for contamination, the hygienist may want to take corresponding surface wipe samples for laboratory quantitation and confirmation in key samples.

Wipe Sampling for Laboratory Analysis

Materials: Glass fiber filters, scintillation vials and a derivatizing solution are required for this sampling procedure.

1. Gloves and other personal protective equipment must be worn during sampling, as appropriate. "Best" style 878 butyl gloves provide protection to the hygienist's hands for the time required to perform the sampling.

2. Gloves must be changed after obtaining each sample to avoid cross contamination of any subsequent samples.

3. A solution of 10.0 mg 1-(2-pyridyl) piperazine per milliliter acetonitrile is prepared. The 1-(2-pyridyl) piperazine is a derivatizing agent that stabilizes the isocyanate. The derivatives also allow a greater sensitivity during the analysis.

4. Estimate the number of samples that will be taken. An overestimate might be better. Prepare vials by adding 1.0 milliliter of the 1-(2-pyridyl) piperazine solution to each vial. It is recommended that the solution be pipetted into the vials in a controlled environment, before sampling, to eliminate any chance for contamination.

5. Untreated glass fiber filters are the appropriate wipe media.

6. Reagent grade acetonitrile is used to wet the filter. The acetonitrile acts as a solvent to dissolve and extract any contaminant present.

7. Using a dropper, wet the filter until almost dripping.

8. Select an area immediately adjacent to the area where the Swype® yielded a positive test result.

9. Using the filter, wipe an area about 100 cm², rubbing the entire area side to side, then up and down. In many cases (such as knobs and levers) it may not be possible to wipe 100 cm².

10. Place the filter in a scintillation vial containing the derivatizing solution, label the vial, and record appropriate information.

11. The samples are ready to be analyzed by a laboratory. The OSHA Salt Lake Technical Center uses OSHA method 47 for analysis of MDI, and OSHA method 42 for analysis of 2,6-TDI and 2,4-TDI.

Monitoring Inside Protective Equipment (gloves, suits)

Studies have shown that solvents containing chemicals may act as a vehicle allowing the chemicals to permeate gloves and protective clothing (7). The Permea-Tec® aromatic isocyanates detection system may be used for worksite evaluation of chemical protective PPE. For example, a negative result (no color change) of the Permea-Tec®, after being worn under a glove for a time period, in a work environment known to have contaminants present, demonstrates that the glove protection was effective for that time period, in those working conditions. The Permea-Tec® is an indicating pad that is attached to a band-aid-like adhesive strip.

1. Place one or more Permea-Tec® patches (pad side out) on the fingers, palms, wrist, lower arm (near cuff of glove), wherever there is likely to be permeation or contamination.

2. Workers should then don their PPE and work for a time period as they normally would. (If the workers normally change gloves every two hours, for example, then the time period is two hours.)

3. After the shift, allow the workers to doff the gloves as they normally would, then collect, identify, and note the color of the pads.

4. In most cases the Permea-Tec® pad should not need any further treatment. If solvent (containing isocyanates) permeation has occurred, then this solvent should be sufficient to activate the pad.

5. If permeation or penetration of the PPE by the solvent containing aromatic isocyanates has occurred, a reaction turns the pad a pastel red-orange to pink. It is a reliable indicator to a detection limit of 3-5 µg aromatic isocyanates.

Evaluating PPE for Dry Chemicals

1. There may be situations where the Industrial Hygienist may want to use Permea-Tec® pads for dry chemicals. (For example: unprotected handling of components that are assumed to be totally cured.) In this case, after the pads have been collected, 3 drops of solvent (methanol works well) needs to be placed directly onto the pad. Again gloves must be worn during sampling and solvent dispensing.

2. The methanol (or other solvent) wicks into the pad and enables a reaction that turns the pad a pastel red-orange to pink if aromatic isocyanate contamination is present. It is a reliable indicator to a detection limit of 3-5 µg.

References

1. "Preventing Asthma and Death from Diisocyanate Exposure", Department of Health, Education and Welfare, NIOSH, ALERT, Cincinnati, OH, March, 1996, DHHS (NIOSH) Publication No. 96-111, 2-3.

2. Material Safety Data Sheet for MDI, DOW Chemical Company, Midland, MI, 1995, MSD002334, 1.

3. Bickis, U. Investigation of dermally induced airway hyperreactivity to toluene diisocyanate in guinea pigs. Ph.D. thesis, Department of Pharmacology and Toxicology, Queen's University, Kingston, Canada. November, 1994.

4. Karol, M.H., Hauth, B.A., Riley, E.J., and Magreni, C.M. Dermal contact with toluene diisocyanate (TDI) produced respiratory tract hypersensitivity in guinea pigs. Toxicol. Appl. Pharmacol, 1981, 58, 221-230.

5. Rattray, N.J., Botham, P.A., Hext, P.M., Woodcock, D.R., Fielding, I., Dearman, R.J., and Kimber, I. Induction of respiratory hypersensitivity to diphenylmethane-4,4'-diisocyanate (MDI) in guinea pigs. Influence of route of exposure. Toxicology, 1994, 88, 15-30.

6. CLI, Colormetric Laboratories, Inc. 1261A Rand Road, Des Plaines, IL, 60016-3402, Telephone: (847) 696-3036.

7. Gunderson, E. C., Kingsley, B. A., Witham, C. L. and Bromberg, D.C. A Practical Study in Laboratory and Workplace Permeation Testing. Appl. Ind. Hyg., 1989, Vol. 4, 12, 324-329.

APPENDIX II:2-3. EXAMPLE PROCEDURES FOR AROMATIC AMINES

This example is provided for information only and should not be taken as a basis for OSHA policy.

<u>Prepared by: Richard Lawrence, Chemist</u>   <u>Date: August 21, 1997</u>

Aromatic Amine Surface Contamination Sampling and Evaluation Techniques

Air sampling by itself may be an inadequate technique to evaluate potential worker exposure to compounds with a low vapor pressure, such as aromatic amines (e.g., MOCA, MDA). A major route of workplace exposure to these chemicals is through skin contact with contaminated surfaces. Aromatic amines are persistent chemicals and once released into the work environment they may remain on surfaces for months and even years. When contaminated surfaces are contacted by unprotected skin, significant exposures are possible. This screening and sampling technique can be useful in evaluating the effectiveness of chemical protective equipment, housekeeping, and decontamination.

Direct reading indicators, such as commercially prepared pads that change color when they come in contact with specific chemicals (or classes of chemicals), are available for aromatic amines. These types of indicators may be used as a screening tool, when assessing the extent of surface contamination, because they are inexpensive and the results are immediate. If an indicator wipe yields a positive result, a wipe sample can then be taken and sent to a laboratory for confirmation.

The mention of the commercial products Swypes® and Permea-Tec® does not constitute an endorsement but is intended solely as an identification of the type of product deemed suitable for the use intended.

The use of direct reading instruments and indicators can be effective in helping employers to comply with the OSHA personal protective equipment standard, 29 CFR 1910.132 (d)(1)(I) and (f)(1)(iii) and (iv).

-Using the surface contamination Swype®*.

1. Gloves and other personal protective equipment must be worn during testing, as appropriate.

2. Gloves must be changed after a positive result to avoid cross contamination of any subsequent tests.

3. Spray the area to be sampled lightly with the Cleaning Solution. Use as little as needed to ensure that the surface is wet. Excess solution may dilute any contamination, possibly to less than the detection limit. When testing a vertical surface or knob, some of the solution may begin to "run-off" or drip. This "run-off" should be captured onto the pad to ensure that any contaminant present has not been lost. The Cleaning Solution contains a proprietary component which activates the Swype® pad.

4. Wait approximately 30 seconds for any aromatic amine to dissolve, than wipe the surface with a surface Swype® pad.

5. Allow 2 to 3 minutes for the color to develop. A red color indicates aromatic amine contamination. The surface Swype® detection limit is approximately 3-5 μg.

6. Record appropriate information as needed.

7. If the surface Swype® tests indicate a positive for contamination, the hygienist may want to take corresponding surface wipe samples for laboratory quantitation. The laboratory may also confirm the presence of the contaminant in key samples.

-Taking a surface wipe sample.

1. Gloves and other personal protective equipment must be worn during sampling, as appropriate.

2. Gloves must be changed after obtaining each sample to avoid cross contamination of any subsequent samples.

3. Glass fiber filters that have been prepared for air sampling for MOCA or MDA are the appropriate wipe media. These filters have been prepared by soaking each filter with 0.5 ml of 0.26 N sulfuric acid and drying them in an oven. The sulfuric acid converts the amine to a more stable amine salt.

4. A solution of 60/40 v/v methanol/water is used to wet the filter. The methanol acts as a solvent to dissolve and extract any contaminant present. The water activates the sulfuric acid already in the filter.

5. Using a dropper, wet the filter until it is almost dripping.

6. Select an area immediately adjacent to the one where the Swype® yielded a positive test result.

7. Using the filter, wipe an area about 100 cm^2, rubbing the entire area from side to side, then up and down. In many cases, such as knobs and levers, it may not be possible to wipe 100 cm^2.

8. Place the filter in a scintillation vial, label the vial, and record appropriate information.

9. The samples are ready to be analyzed by a laboratory. The OSHA Salt Lake Technical Center uses OSHA method 71 for analysis of MOCA, and OSHA method 57 for analysis of MDA.

-Monitoring Inside Protective Equipment (gloves, suits).

Field and laboratory studies have shown that aromatic amines may permeate gloves and protective clothing. The Permea-Tec® aromatic amine detection system may be used for work site evaluation of chemical protective PPE. For example, a negative result (no color change) of the Permea-Tec®, after being worn under a glove for a time period, in a work environment known to have contaminants present, demonstrates that the glove protection was effective for that time period, in those working conditions. The Permea-Tec® is an indicating pad similar to the Swype® attached to a band-aid like adhesive strip.

1. Have the selected workers wear one or more Permea-Tec® patches (pad side out) on the fingers, palms, wrist, lower arm (near cuff of glove), wherever there is likely to be permeation or contamination.

2. After the shift, allow the workers to doff the gloves as they normally would, then collect, ID, and develop the pads.

3. To develop, ten drops of water are applied slowly to the white strip. The white strip contains a component that activates the Permea-Tec® pad.

4. The water wicks up into the pad and the pad turns red if aromatic amine contamination is present. It is a reliable indicator to a detection limit of 3-5 µg

The following compounds are among the suspected agents that can be detected through this procedure:

- methylene dianiline (MDA)
- 4,4'-methylene bis(2-chloroaniline)
- benzidine
- α-napthylamine
- β-napthylamine
- 4-aminobiphenyl
- o-toluidine
- aniline
- 2,4-toluenediamine
- 1,3-phenylenediamine
- napthylenediamine
- 2,4-xylidine
- o-chloroaniline
- 3,4-dichloroaniline
- p-nitroaniline

---

* Mention of the Swypes® and Permea-Tec® brand names does not constitute an endorsement of those products.

chapter 12

Environmental Regulations

by Mardy Kazarians, PhD
Kimberly Doherty Bradley

- 207 Introduction
- 208 National Environmental Policy Act
- 209 Clean Air Act
- 211 Clean Water Act
- 213 Toxic Substances Control Act
- 214 Resource Conservation and Recovery Act
- 215 Underground Storage Tanks
- 216 Comprehensive Environmental Response Compensation and Liability Act
- 217 Emergency Planning and Community Right-To-Know Act
- 218 Pollution Prevention Act
- 219 Summary
- 219 References

INTRODUCTION

This chapter provides a brief overview of the wide range of regulations that address environmental issues and focuses on a select set of regulations that general health professionals working in an occupational setting should understand.

Since the late 19th century, laws have been passed in the United States of America that were intended to give some protection to an environmental element. However, environmental laws, as we now know them, were promulgated starting from the late 1960s, and have since evolved into a very complex collection of statutes, regulations, guidelines, and case interpretations called the environmental law system. Thomas F. P. Sullivan provides the following definition in the *Environmental Law Handbook* (1997). "The environmental law system is an organized way of using all of the laws in our legal system to minimize, prevent, punish or remedy the consequences of actions which damage or threaten the environment, public health and safety." In addition to the regulations at the federal level, several states have promulgated laws that are specific to their states.

The Environmental Protection Agency (EPA) was created in the early 1970s to administer the environmental laws at the federal level. The main function of EPA is to interpret the environmental laws and develop guidelines or specific regulations for the implementation of the laws. The EPA also conducts research and enforces the implementation of its own regulations.

The environmental laws and regulations are promulgated through the legislative process at federal, state and local levels. A bill is introduced in the U.S.

House of Representatives or the U.S. Senate. The bill is studied by committees of the legislative body and is recommended for a vote. When a bill passes, it becomes an act. Most environmental regulations are based on acts. An act becomes law upon the President's signature. At that point, generally the EPA gets the responsibility to write the regulations to implement the requirements of the act. Also, often the EPA publishes guidelines to provide clarification on how to implement certain elements of the law. Clarification is also provided, in some cases, by state level organization and even court cases.

Depending on how the act is written, in some cases the states are mandated to develop their own program for implementing an environmental law. In other cases, the states have a choice to adopt the law or to leave it to the federal EPA to implement it within the state. For example, the state of California often chooses to adopt federal regulations and in most cases rewrites the laws with more stringent requirements. It is very common for local governments to implement portions of the laws and regulations per local needs and conditions.

Environmental laws touch a large spectrum of legal, political, financial, scientific, engineering, and even management issues. Given the breadth and depth of the regulations, professionals now specialize in only a few aspects of these laws. For example, there are professionals who only work with air emissions-related issues, others in water-related issues, etc. For a government body to implement a specific regulation and likewise for a business to comply with those regulations requires knowledge from many disciplines and a wide spectrum of resources. For example, for compliance with air-related regulations, the business needs to gain an intimate knowledge of the materials that are released from its machinery and use thermodynamic, and heat and mass transfer models to estimate the extent of dispersion of those chemicals. This also requires knowledge of local meteorology and biological effects of the chemicals.

Since the late 1960s a large number of laws have been promulgated and amended. The following is a select list of those laws:

- National Environmental Policy Act (NEPA)
- Clean Air Act (CAA)
- Clean Water Act (CWA)
- Toxic Substances Control Act (TSCA)
- Resource Conservation and Recovery Act (RCRA)
- Underground Storage Tanks (UST)
- Comprehensive Environmental Response, Compensation and Liability Act (CERCLA)
- Emergency Planning and Community Right-To-Know Act (EPCRA)
- Pollution Prevention Act (PPA)
- Oil Pollution Act
- Safe Drinking Water Act
- Pesticides
- Federal Facility Compliance Act
- Occupational Safety and Health Act (OSHAct)
- Endangered Species Act.

As mentioned, this chapter provides a brief overview of the wide range of regulations that address environmental issues and focuses on a select set of regulations that have a broad impact.

NATIONAL ENVIRONMENTAL POLICY ACT

The *National Environmental Policy Act (NEPA)* initiated the environmental regulatory movement of the nation. It was enacted in 1970 and declared the national environmental policies and goals. It created the Council on Environmental Quality and set the requirement for Environmental Impact Statement (EIS) for all the activities of the federal government. The main purpose of the National Environmental Policy Act was to ensure that environmental factors are given the same level of importance as other factors in the federal agencies. Therefore, it was intended to be all encompassing and affect all federal agencies and practically all the actions taken by those agencies.

The National Environmental Policy Act specifies for public involvement and has been interpreted as cascading down to state, local, and facility levels. All governmental decisions, regardless of the level (i.e., federal, state or local) are interpreted as linked to an action of a federal agency. Because of this interpretation, the National Environmental Policy Act has affected practically all aspects of the lives of the people in the United States of America. It is now common for a business to submit an environmental impact statement to a local permitting agency as part of obtaining permits to construct or to operate.

Compliance with the National Environmental Policy Act requires a multistage environmental impact assessment with increasing level of detail and sophistication at different stages of approval.

An important goal of the multistage approach is an attempt at minimizing the paperwork. The National Environmental Policy Act requires an *Environmental Assessment* (*EA*) to be submitted to assist in the decision whether or not to request an *Environmental Impact Statement* (*EIS*). The Environmental Assessment is a simple checklist that includes the following:

- a brief description of the need for the proposed action
- potential alternatives to the proposed action
- environmental impact of the proposed action and the alternatives.

The responsible agency, upon the review of the Environmental Assessment issues a Finding of No Significant Impact (FONSI) On Human Environment that documents the decision. If a Finding of No Significant Impact cannot be issued, an EIS will be required.

The Environmental Impact Statement preparation and final decision require the involvement of a multidisciplinary team and several interest groups of the proposed action. For a majority of facilities, several different permitting agencies are involved. In addition, the Environmental Impact Statement must address issues of interest to those agencies. Therefore, a lead agency has to be identified to prevent duplication and minimize paperwork. The involved agencies negotiate among themselves and appoint a lead agency to be the focal point for all Environmental Impact Statement-related activities. The lead agency takes on the responsibility of preparing the Environmental Impact Statement. However, it is a common practice for the lead agency to ask the business to prepare the Environmental Impact Statement on behalf of the agency. Typically, the agency selects a contractor to prepare the Environmental Impact Statement. The contractor, then, works closely with the business in putting together the Environmental Impact Statement. The business funds the contractor's efforts. The Environmental Impact Statement is then submitted to the lead agency, who distributes it among the interested parties.

The lead agency is responsible for identifying different interest groups (e.g., nearby residents and the labor force) and coordinating the interested agencies. In the early stages of Environmental Impact Statement development, the scope of the study is determined by the interest groups and the permitting agencies. Typically an Environmental Impact Statement contains the following:

- adverse environmental impacts that cannot be avoided
- reasonable alternatives
- commitment of resources
- possible mitigation measures
- comparison among alternatives and proposed action
- demographic information and effects on the local scene
- detrimental and beneficial effects
- cumulative impacts.

The Environmental Impact Statement is distributed among the interested parties who can comment on it. There are set meetings where these comments are aired and the business is given the opportunity to resubmit the Environmental Impact Statement with modifications reflecting the concerns of the various groups.

CLEAN AIR ACT

The Clean Air Act (*CAA*) (40 *CFR* Parts 50 through 95) was enacted in 1967 to set a standard for air quality and to control air pollution. The Clean Air Act has since been amended three times in 1970, 1977, and 1990. The Clean Air Act works through the states. Each state is required to submit a *State Implementation Plan* (*SIP*) describing the state of the air quality in the state and the programs in place to meet Clean Air Act requirements.

The Clean Air Act addresses air pollution by prohibiting emission of pollutants that exceed the ambient air quality, by putting stringent standards on new sources, by mandating a permitting process, and by setting a chemical specific standard known as the National Ambient Air Quality Standard (NAAQS).

The National Ambient Air Quality Standard is a central element of the Clean Air Act and addresses pervasive pollution associated with six pollutants, known as *criteria pollutants*. The criteria pollutants are as follows:

- sulfur dioxide (SO_2)
- nitrogen oxides (NO_x)
- particulate matter, known as PM-10
- carbon monoxide (CO)
- ozone
- lead.

Current attainment designations for all areas of the United States are listed in the most recent version of 40 *CFR* 81.301-356. The Clean Air Act requires EPA to revisit the standard once every five years and revise them if deemed appropriate. However, since its first publication, the standard has not been changed. An attempt was made to reduce the standard for particulate matter to 2.5 micron (PM2.5) but it proved to be very difficult to implement.

States have the primary responsibility to ensure that the air quality within their borders are maintained at or below that of the National Ambient Air Quality Standard. As mentioned above, the states are required to submit State Implementation Plans (SIPs) to the federal EPA to show how they meet the National Ambient Air Quality Standard. A State Implementation Plan is required to include the following:

- emission limitations imposed on sources
- air quality data at different parts of the state and a display of areas that meet the standards and those that do not
- emission data from various sources
- enforcement activities of air quality regulations
- interstate pollution across state boundaries
- funding and authority of various state bodies responsible for air quality
- contingency plans for dealing with special conditions (e.g., curbing emissions during peak concentrations)
- prevention of significant deterioration (PSD) by controlling new sources
- permit fees.

It is required that the State Implementation Plan be submitted to EPA for approval. The federal highway fund is tied to the State Implementation Plan. The EPA uses it as leverage to force those states that have not met the federal standards to act to develop strict programs for curbing air emissions.

If an area does not meet the federal standards (i.e., the area is in nonattainment), the state and local governments are expected to take drastic measures. The local government is required to obtain from all sources, on a periodic basis, the emissions inventory for those pollutants that are in nonattainment, and keep that data up-to-date. All existing sources must apply a reasonably available control technology (RACT) to reduce their emissions inventory. New or modified sources must be subjected to a strict permitting program that uses emissions budgeting to control the added emissions into a nonattainment area.

Source specific programs should be developed for nonattainment areas. EPA has defined levels of nonattainment, based on which source specific program is implemented. For example, ozone five severity levels have been defined as marginal, moderate, serious, severe, and extreme. Marginal areas were required to meet the National Ambient Air Quality Standard by November 1993, moderate areas by 1996, serious by 1999, etc. Also, approaches are required for these levels. For example, areas that are in the serious category of nonattainment, reasonably available control technology (RACT) should be used. Two severity levels are defined for carbon monoxide (moderate and serious) and just one for PM-10 that should be achieved by 2001.

There are specific requirements for new sources (i.e., businesses who are trying to build a new facility or bring in new processes into their existing processes). Since upgrading existing sources to emit less can prove to be economically prohibitive, new sources are subjected to more stringent levels of control than existing sources. The Clean Air Act defines New Source Performance Standards (NSPS) in terms of 60 categories of sources, types and sizes of process. The categories are by industry and includes further breakdown within each industry. The New Source Performance Standards calls for specific design features, equipment type, work practices and many other specific requirements for various sources. The New Source Performance Standards is regarded as the minimum level of control.

New sources are subjected to preconstruction review and permitting process. Attained areas fall under Prevention of Significant Retention (PSD) program that tries to prevent new sources from severely offending the existing margin in meeting the National Ambient Air Quality Standard. For these areas, a major new source is defined as one that emits more than 250 tons per year (tpy) or 100 tpy for 28 specific pollutants. Businesses that plan to install a major source are required to use *Best Available Control Technology (BACT)*. The Best Available Control Technology applies to those technologies that are available in the market and offer the most reduction in emissions taking into account economic and other factors. The Best Available Control Technologies are defined by EPA or state regulatory bodies in terms of specific technologies for specific applications.

For nonattainment areas the definition of major source is more stringent than attainment areas and it varies depending on nonattainment category level. For areas other than extreme, 100 tpy is considered as a major source and 10 tpy for extreme nonattainment areas. Lowest Achievable Emissions Rate (LAER) is required for new sources or obtaining emission credits from other facilities nearby. If a business chooses the latter approach, the local regulatory bodies require better than one-to-one ratio of emissions exchange.

Reconstruction cases, that is when a facility is repaired or refurbished to extend its useful life, for attained areas fall under the New Source Performance Standards program and not the PSD. However, a modification is defined as practically all activities that involves a source that may increase the emission rate. Routine maintenance activities or replacement of equipment is not considered as a modification. The PSD program applies to this case.

In addition to the six pollutants in the National Ambient Air Quality Standard, the Clean Air Act also addresses the following set of specific pollution problems:

- air toxics
- acid rain
- visibility
- stratospheric ozone protection
- fuels for vehicles.

For air toxics, 189 substances (organics, metals, etc.) have been identified. A major source for these materials is defined at 10 tpy per substance or 25 tpy when more than three air toxic material is considered. The law also defines 174 source categories (refineries, chemical plants, etc.) that must control air toxic emissions. Maximum Achievable Control Technology (MACT), a technology-based standard, is required for these materials. For air toxics technology specific control requirements are defined and exemptions are offered to those facilities that can show voluntary reduction of 90% or better. An important part of the air toxics part of the law deals with accidental release of highly hazardous materials (e.g., chlorine and ammonia). This part of the law resembles OSHA's process safety management requirements and is know as the Risk Management Plan or Accidental Release Prevention Program.

Acid rain-related issues deal with the control of SO_2 and NO_x (i.e., NO_2, NO_3, etc.) emissions. This part of the law affects mainly electric utilities. Title IV of Clean Air Act requires emissions allowance and trading program for SO_2. An "allowance" is defined as one ton of SO_2, which is auctioned by EPA on an annual basis. The law includes severe penalties if annual allowances are exceeded

The issue of visibility was introduced to protect the air quality of the national parks. The Federal Land Manager is required to get involved in these cases and participate in the permit process. For this Best Available Retrofit Technology (BART) is required and so far it has had very limited impact.

The well-known Montreal Protocol has been addressed in the Clean Air Act that deals with stratospheric ozone protection. It may be noted here that ozone is a concern at the ground level and it is a by-product of certain organic materials when they are released into the air. However, at stratospheric level ozone is needed to provide protection of the earth's surface from the penetration of ultraviolet rays. This part of the law has stringent requirements on the phase out of ozone depleting substances and use of such materials. Chlorinated organics, which includes freon (a widely used refrigerant) have been most severely affected by this part of the law.

Reduction in tailpipe emissions of hydrocarbons and NO_x from vehicles has been the focus of the Clean Air Act for a long time and has had a profound effect on the design of internal combustion engines used in automobiles.

Another important part of the Clean Air Act is its operating permit program. Title V of the 1990 Clean Air Act Amendments requires a comprehensive operating permit program for most sources of air pollution. One of the key aspects of Title V is the consolidation of various permitting activities and requirements. Under Title V a major source is defined as 100 tpy. The source categories are consolidated from various parts of the Clean Air Act and other regulations. Nonmajor sources are allowed to defer compliance for five years. A permit application must be submitted by businesses that includes information on major emissions and regulated air pollutants, identification of points of emission, emission rates, description of air pollution control equipment, monitoring and measurement techniques, schedule for compliance (if not in compliance). The application provides a shield from any enforcement actions until it is reviewed by the state regulatory body.

CLEAN WATER ACT

The Clean Water Act (CWA) is a comprehensive program for protecting the nation's waters from human-made pollutants. The federal EPA administers numerous programs under the Clean Water Act including: the National Pollutant Discharge Elimination System (NPDES); a dredge and fill permitting process; and a municipal wastewater treatment program.

The earliest federal law pertaining to the preservation of water was the Refuse Act of 1899 and focused protection on the navigation of bodies of water rather on the water quality. In the 1960s, federal regulators used the broad sense of the Refuse Act to create the Federal Water Pollution Control Act of 1972. This act also addressed water quality issues and is the foundation for today's Clean Water Act. In 1976 a lawsuit forced the federal EPA to develop effluent water quality standards for industry on an industry-by-industry basis. These standards were derived based on the pollution control technologies available. These standards were the framework for the 1977 Amendments to the Federal Water Pollution Control Act. By December 1979, the EPA had created a list of 65 toxic pollutants in 21 major industrial categories.

The act was amended again in 1987 to focus the program on toxic control by establishing a Toxic Hot Spots program to identify and improve waterways that were expected to remain polluted with toxins even after the most stringent technology based requirements were met. In 1990, after the Exxon Valdez oil spill, the oil spill provisions of the act were overhauled. The Oil Pollution Act of 1990 was created.

The objectives of the Clean Water Act are to restore and maintain the chemical, physical and biological integrity of the nation's waters. To achieve these objectives; the Clean Water Act established the following two goals: (1) eliminate the discharge of pollutants into surface waters; and (2) achieve water quality levels in order to protect fish, shellfish, and wildlife and for recreation in and on the water. The Clean Water Act also prohibits the discharge of toxic pollutants in toxic amounts.

The Clean Water Act includes the following elements in order to attain its objectives and goals. The primary elements are as follows:
- prohibition of discharge except as in compliance with the Clean Water Act
- a permit program to allow for authorized discharges
- discharge limitations to be imposed on regulated discharges
- a federal and state implementation process
- a system for preventing, reporting and responding to spills
- a dredge and fill material permit program
- an enforcement program.

The Clean Water Act terms defining the prohibition of discharge of pollutants are interpreted very broadly. The discharge of a pollutant is defined as any addition of any pollutant or material to navigable waters from any point source. Pollutant and/or materials are defined as chemicals, biological materials, wastes, rocks, sand, sludge, dirt, heat, and dredged spoil. Navigable waters are considered to be practically all bodies of water, including; oceans, lakes, streams, wetlands, some local springs and even ponds for salt production. In general, groundwater is not included unless the groundwater has a hydrological connection to the surface water. Groundwater is included in many states' definitions of "Waters of the State," and in these cases, *point source discharges* to groundwater are covered by the state's interpretation of the Clean Water Act. A *point source* is any conveyance from which a pollutant may be discharged such as, pipes, ditches, erosion channels, and gullies. Based on the wide reach these definitions allow, industrial facilities must pay very close attention to the requirements of the Clean Water Act.

The Clean Water Act is implemented through a permit program called the National Pollutant Discharge Elimination System (NPDES) and requires a permit for every discharge of a pollutant from a point source. The primary purpose is to establish enforceable effluent limitations. Either the EPA or an authorized state issues these permits. Currently only 10 states are not authorized, but additional states can be authorized at any time. To determine if your state or the EPA governs your program, call your local EPA Regional office. The NPDES permits have a five-year limit, and allow sources to discharge, within a set limit, specified pollutants into specified waters. Permits are required for facilities with industrial activities and specifically for process water, cooling water, and storm water runoff. Therefore facilities must have a NPDES permit for discharging offsite as well as for discharging to a municipal sewer treatment plant. Permits issued by a state authority, or a local treatment and may be reviewed and possibly reissued by EPA.

The permit application, whether for EPA or a state authority, requires extensive information

about the facility and the nature of the discharge(s). All application forms, drafts, and final plans must be signed and certified by a responsible corporate officer. It is important to note that applications must be submitted 180 days before commencement or renewal of a discharge or operations cannot start. The permit establishes
1. enforceable discharge limits
2. monitoring, reporting, and record-keeping requirements
3. duty to properly operate and maintain equipment
4. best management practices (i.e.; basic housekeeping)
5. reporting of upset and unusual conditions
6. inspection and entry requirements
7. a schedule for compliance.

It is also important to keep in mind that the terms of the permit are negotiable with the agency permit writer.

The Clean Water Act also mandates establishing effluent limitations. All dischargers must meet treatment requirements based on EPA's assessment of methods that are technologically and economically achievable in the discharger's industry. In addition, there are more stringent water quality goals set by EPA. The water quality goals are based on one or a combination of the following: chemical specific limitations, effluent toxicity control, and biological criteria. All new discharge sources must conform to specific required treatment technologies known as New Source Performance Standards (NSPS).

The Clean Water Act has many requirements for industrial and treatment facilities. It is important to keep abreast of the continuous changes in the Clean Water Act requirements prior to modifying a facility or constructing a new facility.

TOXIC SUBSTANCES CONTROL ACT

The Toxic Substances Control Act (TSCA) (40 *CFR* Parts 700 through 799) gives EPA the authority to regulate the manufacture, use, distribution and disposal of chemicals and requires the manufacturers to provide data on the health and environmental effects of chemical substances and mixtures. The objective of TSCA is to understand the risks of new and existing chemicals and regulate those that pose unreasonable risk.

The Toxic Substances Control Act was enacted in 1976 and has since has been amended three times to include additional titles that address specific toxics. Title I addresses testing requirements, pre-manufacture review of new chemicals, defines the authority to limit and prohibit the manufacture, use, distribution and disposal of existing chemicals, requires record keeping, and defines export and import provisions. Title I also promulgated the prohibition of PCB manufacture and control of its usage. Title II addresses the asbestos hazards; Title III, indoor radon, and Title IV, lead-based paint exposure. The activities that are regulated by the Toxic Substances Control Act are vaguely defined in the law. It is interpreted to include manufacture, production, extraction, mixing, and importation. Processing is also addressed. However, not all "processors" were intended to be touched by this law.

The EPA is required to compile, keep current and publish a list of substances, as the *TSCA Inventory,* which establishes the existing chemicals. These are chemicals that are either manufactured or processed in the United States for commercial purposes. The list excludes several categories of chemicals. Mixtures, chemicals manufactured for a noncommercial purpose, pesticides (they are addressed under a different regulation), polymers, impurities, by-products, chemicals from incidental reactions, and nonisolated intermediates are not in the Toxic Substances Control Act list. Toxic Substances Control Act defines the methods for adding and delisting of chemicals. Manufacturers are required to report production volume and plant sites once every four years. It is interesting to note that EPA regards genetically engineered microorganisms covered under Toxic Substances Control Act.

Manufacturers and importers of new chemicals (i.e., chemicals that are yet not part of the Toxic Substances Control Act Inventory) bear the responsibility for developing adequate data regarding the effects of the chemical on human health and the environment. The business is required to file a Pre-manufacture Notice (PMN) where the chemical identity, usage, anticipated amounts to be manufactured, by-products, and disposal method (if known) are indicated. The notice should also include any test or other data on the effects of the new chemical on human health and the environment. Upon receipt of a PMN, the EPA has 90 days to review and respond. At this point, EPA has three options (1) either request an additional 90 days; (2) limit or prohibit production if insufficient information is provided if there is reasonable basis that the chemical poses an unreasonable risk; or (3) no action. If EPA does not take any action within this time period, the manufacturer or importer may proceed with the production and distribution of the material and send a Notice of Commencement (NOC) to EPA.

Several federal, state, and local level regulations address classes of substances that may overlap with the Toxic Substances Control Act. Because of the Federal Insecticide, Fungicide, and Rodenticide Act (FIFRA), pesticides are exempted from the Toxic Substances Control Act. However, the Toxic Substances Control Act applies during the research and development phase of these materials. Substances regulated under the Food, Drug and Cosmetics Act are exempt from the Toxic Substances Control Act. This includes foods, food additives, drugs, and cosmetics.

RESOURCE CONSERVATION AND RECOVERY ACT

The Resource Conservation and Recovery Act (RCRA) addresses the hazardous waste problem of the nation. The magnitude of this problem is easy to envision, especially when observing the activities of large communities. For example, 275 million tons of hazardous waste were generated in 1994. The objective of the Resource Conservation and Recovery Act is to reduce the generation of or to eliminate hazardous waste as expeditiously as possible. The Resource Conservation and Recovery Act tries to address materials control from cradle to grave. It touches generators; transporters; and treatment, storage, and disposal (TSD) sites. In addition, many recycling and reclamation activities are considered as part of the Resource Conservation and Recovery Act. This law is focused on solid waste and active disposal sites. Nonactive sites are addressed under CERCLA (see later sections of this chapter). The Resource Conservation and Recovery Act encourages waste minimization and puts a ban on land disposal. Underground storage tanks are also addressed under the Resource Conservation and Recovery Act and because it is a major issue for many businesses, it is discussed as a separate topic in the following material.

Waste is defined as any discarded material from industrial, commercial, mining, agricultural, and community activities. It includes garbage, refuse, and sludge. It includes solid, liquid, semisolid, and even gaseous materials (in a vessel). It also includes reclaimed and recycled materials. *Hazardous waste* is defined as a waste material which because of its quantity, concentration, or physical, chemical or infectious characteristics may cause or contribute to an increase in mortality or serious illness or pose a hazard to human health or the environment when improperly treated, stored, transported or disposed of.

The regulations, court judgment, and EPA have identified certain wastes as exempt from the Resource Conservation and Recovery Act. Domestic sewage and waste are not addressed under the Resource Conservation and Recovery Act. Solid or dissolved materials in irrigation return flows are not covered either. Liquid point sources are addressed in the Clean Water Act. The materials that are passed through Publicly Owned Treatment Works (POTW) are part of the regulations that govern the effluents from those types of facilities. Mining overburden that is returned to the mine are not part of the Resource Conservation and Recovery Act. Also, the wastes from coal combustion for power generation is not covered. Oil and natural gas exploration drilling waste, wastes from processing ores and minerals, cement kiln dust wastes, arsenic-treated wood wastes, certain chromium bearing wastes, petroleum-contaminated media and debris, and universal wastes (e.g., cadmium batteries, recalled pesticides, and mercury thermostats) are other categories of materials that are specifically not addressed under the Resource Conservation and Recovery Act.

The EPA has established three hazardous waste lists to identify a waste. The first list covers wastes from nonspecific sources. These materials are identified by an "F" number. For example, F001 is assigned to various spent solvents. The second list addresses wastes from specific sources. These materials are identified by a "K" number. For example, K048 is a certain oil refinery waste. The third list covers pure chemicals. This list includes off-spec chemicals, spill residues, etc. These materials are identified with either a "P" number or a "U" number. The "P" numbers are used for acutely hazardous materials. For example, P076 denotes nitric acid. All other pure chemical wastes are identified with a "U" number. For example, U002 is acetone. If a waste is not listed as hazardous, the waste is covered by RCRA if it is ignitable, corrosive, reactive, or toxic. These wastes get a D number designation. D001 is used for ignitable wastes, D002 for corrosive wastes, D003 for reactive wastes, and D004 through D0043 for different various toxic wastes.

The EPA has provided an extensive guideline for characterizing hazardous waste. The *mixture-rule* addresses mixtures of hazardous waste and other wastes or effluents in terms of concentration of hazardous waste in the mixture and quantity of hazardous waste (discontinued or discarded material). The *derived-from rule* defines those materials that are derived from treating a hazardous waste. This rule includes provisions for de-listing and defines reclaimed materials as not a waste.

The cradle-to-grave process is initiated by the identification of wastes by the generator. The generator has many responsibilities under the Resource Conservation and Recovery Act. The EPA identification number must be assigned to all wastes. No waste can be shipped prior to obtaining this number. The companies in the transportation business and the owners and operators of treatment, storage, and disposal sites cannot receive a waste without an EPA identification number. A manifest must be issued for every waste shipment. The treatment, storage, and disposal site is responsible to return a copy of the manifest which will indicate that the waste was received by the treatment, storage, and disposal site in quantity and condition defined in the manifest. Neither transporter nor the treatment, storage, and disposal site can accept a waste without the manifest. The law specifies that the manifest should be retained in the files for three years, but because of court cases and other regulatory actions, many businesses archive them for longer retention.

Compliance with DOT regulations is required for preparing waste for transportation. On-site accumulation of waste, up to 90 days, is allowed without a RCRA permit. Special regulations exist for small quantity generators (100 to 1,000 kg per month). They may store waste on site up to 270 days. Generators of less than 100 kg per month are exempt from the Resource Conservation and Recovery Act.

Transporters of hazardous waste are also affected by the Resource Conservation and Recovery Act. As mentioned previously, they are required to obtain an EPA identification number and must follow DOT standards for hazardous materials. They are responsible for the waste, including clean-up in case of a spill, until it is delivered to a treatment, storage, and disposal site. They may hold hazardous waste at their own sites for up to ten days without obtaining a Resource Conservation and Recovery Act permit and may mix compatible wastes if deemed economically advantageous.

TSD facilities must comply with strict requirements defined under the Resource Conservation and Recovery Act. There are several types of treatment, storage, and disposal sites—storage tanks, landfills, incinerators, etc. Specific requirements from treatment, storage, and disposal sites include testing of wastes received, preparation of Waste Analysis Plan (WAP), installation of a security system, provisions for protection from known natural hazards, formal training of personnel, formal accident prevention program, institution of mitigative features, establishment of an emergency/contingency plan, maintaining liability insurance and financial responsibilities, record keeping, and, finally closure requirements. These issues are addressed in what is known as Part A and Part B applications that are submitted to local authorities and EPA.

A number of different treatment, storage, and disposal sites are exempt from the Resource Conservation and Recovery Act or are covered under a different regulation. Ocean disposal is covered by Marine Protection, Research and Sanctuaries Act. Underground injection is covered by Safe Drinking Water Act. The POTWs, farmers disposing waste pesticides, totally enclosed treatment facilities, neutralization and wastewater treatment units, and spill containment activities are covered by other regulations.

UNDERGROUND STORAGE TANKS

Underground storage tanks (*USTs*) (40 *CFR* Part 280) are widely recognized as a major source of soil and groundwater pollution. It is estimated that there are over two million tank systems in the United States in over 700,000 facilities. Underground storage tanks are known to have leaks caused by corrosion of the tanks and the associated piping. It is also estimated that 75% of the tank systems may be leaking or posing a threat, primarily because they have historically been constructed of steel, or have been subject to poor operating practices resulting in spills and overflows during filling. Since 1984, the government has spent billions of dollars to clean up underground storage tank-related leaks, and yet much more has to be spent to complete the cleanup process.

The U.S. Congress had several basic objectives when it enacted the Underground Storage Tank provisions of The Resource Conservation and Recovery Act in 1984. The statute addressed both issues related to the environmental problems caused by existing underground storage tank systems and to the design of new underground storage tank systems to eliminate the past problems. The primary objective of the Underground Storage Tank Program is to bring all existing tank systems to a certain design and operating standard, clean up the contamination caused by leaking tank systems, and to impose strict design, operating, and reporting requirements on all new underground storage tank systems. This applies to partially buried tanks and associated piping, tanks that contain crude oil and substances derived from crude oil, tanks that contain regulated substances per CERCLA definition. All existing tank systems were to be upgraded, replaced, or closed by Dec. 22, 1998.

An underground storage tank system should have been be upgraded, replaced, or closed if the system has any of the following:
- single-wall steel tanks including waste oil
- tanks that don't have spill and overflow protection
- any piping without corrosion protection
- single-wall tanks and piping that hold hazardous substances other than motor fuel.

If a tank system has one or more of the above characteristics, the underground storage tank system is out of compliance under the Underground Storage Tank Program and should be modified or closed immediately.

The requirements for new underground storage tanks containing hazardous substances are as follows:
- All tanks must have secondary containment.
- All tanks must be monitored by an interstitial system or be monitored visually if you can see under and around the tank.
- All piping must have a lined trench, a vault, or a secondary piping around it.

All piping must be monitored by one of the following methods:
1. continuous monitoring system with audible and visual alarm, via interstitial space
2. leakline detector system
3. annual piping integrity testing
4. continuous monitoring system with pump shutdown if the monitoring system fails or disconnects.

All underground storage tanks are required to be designed and operated according to a strict design and performance standard in order to prevent releases. The appropriate design and construction codes and standards are addressed in 40 *CFR* Section 280.20. It is important to note that either EPA implements the underground storage tank requirements or authorizes states to do so. Each authorized state has the authority to make the federal Underground Storage Tank requirements more stringent and, if this is the case, the underground storage tank owner must comply. Under a state program, the Underground Storage Tank regulations may be under the oversight and enforcement of the fire authorities, the county health departments or another local agency.

The EPA has identified the following facility types that are either directly exempt from Underground Storage Tank regulations or are covered under other environmental laws and therefore, not addressed as part of this regulation.
- farm or residential tanks of 1,100 gallons or smaller
- heating oil storage tanks servicing buildings
- septic tanks
- surface impoundments, pits, ponds, and lagoons
- stormwater or wastewater collection systems
- flow-through process tanks
- liquid traps
- storage tanks in an open area below grade level
- pipeline facilities subject to other regulations
- hazardous waste storage tanks subject to other RCRA rules
- wastewater treatment tanks regulated under the Clean Water Act
- equipment and machinery
- tanks less than 110 gallons
- tanks with low concentration of regulated substances
- emergency spill or overflow containment tanks.

Underground storage tank regulations require a considerable amount of reporting and record keeping per the Resource Conservation and Recovery Act Subtitle I. The following is a list and brief description of the recording and record-keeping obligations:
- Initial notification of each system via registered mail must be made within 30 days of use with the local, state or federal authority.

- Suspected releases must be reported to the local implementing agency.
- Spills and overfills must be contained, immediately cleaned up and documented.
- Confirmed releases must be reported to the implementing agency within 24 hours or reasonable time period determined by the agency. Also, any reportable quantity per the Comprehensive Environmental Response Compensation and Liability Act must be reported immediately to the National Response Center.
- Corrective action to address site assessment and cleanup of a leak or spill must be reported and documented during each step of the process.
- Permanent closure or change-in-service must be reported at least 30 days in advance to the administering agency, unless action is in response to corrective action.
- Financial responsibility must be demonstrated and documented to provide assurance that corrective action can be financed if needed.
- UST inspection, testing and repair records must be documented and maintained.
- Release detection performance documentation must be maintained.
- Permanent closure or change-in-service records must be maintained for at least three years.

For more detail on each of the discussed requirements refer to 40 *CFR* Part 280.

Identification and cleanup of existing chemical spills and the prevention of future leaks and spills from underground storage tanks and enforcing strict standards for new underground storage tanks provide many challenges for the government and private industry. The regulations and responsibilities that underground storage tanks owners and operators face are numerous and confusing.

COMPREHENSIVE ENVIRONMENTAL RESPONSE COMPENSATION AND LIABILITY ACT

The *Comprehensive Environmental Response Compensation and Liability Act (CERCLA)* was enacted by Congress to deal with inactive hazardous waste sites. This law complements RCRA. It was enacted in 1980 and was amended in 1986. The amendment is known as *Superfund Amendments and Reauthorization Act (SARA)*. The Comprehensive Environmental Response Compensation and Liability Act has become a very complex set of rules that has been criticized heavily by both sides of the environmental debate.

It references a large list of substances as candidates for clean-up activities. In 40 *CFR* Part 302 there is a summary list of the substances, which includes hazardous waste covered under the Resource Conservation and Recovery Act, substances and toxic pollutants covered under Clean Water Act, air pollutants of Clean Air Act, and imminently hazardous chemical substances identified under the Toxic Substances Control Act. The Comprehensive Environmental Response Compensation and Liability Act also includes benzene, toluene, and xylene (BTX).

The Comprehensive Environmental Response Compensation and Liability Act gives the EPA the authority to respond to a release or substantial threat of a release of any "pollutant or contaminant" which may present an imminent and substantial threat to public health or welfare. For example, corroding or abandoned tanks have been interpreted as a substantial threat to release. This provision of the law gives a wide authority to EPA and encompasses practically everything. The EPA can respond to practically every release deemed to include a pollutant, and charge release recovery and cleanup costs to the companies and persons deemed to be responsible. The EPA also has the authority to force the liable party to cleanup. Routine workplace releases or Clean Water Act NPDES releases are excluded from the Comprehensive Environmental Response Compensation and Liability Act.

Under the Comprehensive Environmental Response Compensation and Liability Act an elaborate mechanism for collecting and distributing funds has been created. The fund is known as Hazardous Substance Superfund and is intended for long term remedial actions. Taxes are imposed on petroleum and chemical companies for this purpose and cleanup costs have been charged to liable parties. It should be noted that federal facilities cannot use Superfund money.

EMERGENCY PLANNING AND COMMUNITY RIGHT-TO-KNOW ACT

The Emergency Planning and Community Right-to-Know Act (EPCRA) requires states to establish a process for developing local chemical emergency preparedness programs and to receive and disseminate information on hazardous chemicals present at facilities within a community. The Emergency Planning and Community Right-to-Know Act was enacted in 1986 as part of Superfund Amendments and Reauthorization Act (SARA), and is known as SARA Title III. It requires emergency planning, emergency release notification, community right-to-know reporting and toxic chemical release inventory reporting.

The Emergency Planning and Community Right-to-Know Act requires for each state to organize itself and develop a community level emergency response plan, and collect and maintain the necessary data from each community. It defines a State Emergency Response Commission (SERC) and requires for the states to be divided into Emergency Planning Districts. It requires formation of Local Emergency Planning Committees (LEPCs) within each district and provides the List of Extremely Hazardous Substances.

A local emergency response plan must be developed for each district that covers the following issues:
- Identify all the facilities within the district.
- Identify all the routes used for transporting extremely hazardous substances.
- Identify sensitive receptors.
- Identify risk-related facilities (e.g. natural gas pipeline, and high-tension transmission lines).
- Provide a description of response methods.
- Designate a community emergency response coordinator.
- Identify all the facility emergency response coordinators.
- Describe the emergency notification procedure and evacuation plans.
- Describe the method for identifying a chemical release and the method for determining potential area and people who may be affected by the effects of the spill.
- List the emergency equipment, its locations, and responsible persons of the community.
- Describe the training program of the community emergency-response crew.
- Schedule for drills and exercises.

The Emergency Planning and Community Right-to-Know Act requires all owners and operators of facilities that produce, store or use a hazardous chemical to notify the SERC and LEPC of any releases of a listed hazardous substance. The Emergency Planning and Community Right-to-Know Act requires facilities to submit MSDSs, inventory, location, and other information about the hazardous chemicals on-site. Businesses must submit whether the substances that they handle are considered as an acute hazard, chronic hazard, mechanical hazard, or reactive hazard. Hazardous properties of a mixture can be reported in terms of either the properties of the mixture or of individual components.

The Emergency Planning and Community Right-to-Know Act requires owners or operators of facilities to report their annual releases (routine and accidental) into the environment via air, water, land, POTWs and transfer to other locations for treatment, storage, or disposal. This requirement applies to facilities in SIC codes 20 through 39, that have 10 or more full-time employees or handle one of the chemicals listed in Section 313. Janitorial and grounds maintenance products, building and structural materials, personal use materials and foods, motor vehicle related fluids and chemicals in intake water or air are exempt from Emergency Planning and Community Right-to-Know Act reporting.

The Pollution Prevention Act of 1990 mandates collection of data on releases, recycling, and pollution prevention to support the EPA in its biennial report to congress on the source reduction programs. The report should include: amount entering a wastestream before recycling; treatment or disposal, amount recycled; change from previous year; amount treated on-site and off-site; estimates for next year; change from previous year; source reduction practices used (e.g., equipment, technology, etc.); methods used to identify source reduction opportunities (e.g., audit program, employee suggestion, etc.); ratio of production change; and accidental releases. Part of this information is collected from businesses on what is known as Form R.

POLLUTION PREVENTION ACT

The latest development in environmental control and regulations is the *Pollution Prevention Act (PPA)* of 1990 that departs from the traditional prescriptive method of regulating environmental issues discussed in the preceding sections of this chapter. The Pollution Prevention Act requires the EPA to develop and implement a strategy to promote source reduction. The EPA is to come up with measurable goals for source reduction. Also, EPA is to submit biennial reports to the U.S. Congress on this source reduction program. In doing this the Pollution Prevention Act is attempting to change the national approach and focus.

The EPA has modified its approach from directing industries to encouraging them in achieving a reduction in pollution. The pollutant-by-pollutant approach of the past is now replaced with a facility-wide integrated program where the facility owner tries to reduce overall emissions and discharges. Now, a facility owner may chose to use a moderate polluting machinery at one end of the operation, and yet severely reduce emissions at other places of the facility that in the balance provides a reduction in emissions and discharges. This approach provides the facility owners the flexibility to use various equipment based on their economic value and yet achieve overall environmental goals.

SUMMARY

The amount of environmental regulations and requirements is very great and oftentimes overwhelming and confusing to private industry as well as government. It can involve diverse elements and multidisciplined activities of a facility. Various agencies at federal, state, and local levels are mandated with the implementation of various aspects of the environmental regulations. It is not only important to understand the current requirements that a facility must meet, but also imperative to understand that the requirements are fast changing. Great consideration of these requirements, and the underlying forces driving the changes in environmental law, must be taken into account during facility planning, construction, and operation. Although history presents the government as offering prescriptive forms of environmental compliance, the new trend is for the government to offer more user-friendly and self-directed solutions to industry.

REFERENCES

Sullivan TFP. *Environmental Law Handbook.* Rockville MD: Government Institutes, 1997.

U.S. Code of Federal Regulations: 40 *CFR* Parts 50 through 95, 40 *CFR* 122, 123, and 405 through 471, 40 *CFR* 81.301-356, 40 *CFR* Parts 700 through 799, 40 *CFR* Part 280, 40 *CFR* Section 280.20, 40 *CFR* Part 302. Washington DC: Available through the U.S. Government Printing Office.

chapter 13

Radiation Safety Programs

by Ara Tamassian, PhD
revised by Keith Keller

- 221 **Introduction**
  Uses of radioactive materials and radiation ■ Natural sources ■ Biological effects of radiation
- 222 **Regulatory Controls**
  Licensing ■ Radiation safety programs
- 224 **Protecting Against Radiation Hazards**
  Internal contamination hazards ■ External radiation hazards ■ Administrative controls
- 229 **Hazard Monitoring Program**
  Environmental ■ Personnel ■ Instrumentation
- 231 **ALARA Concept**
- 232 **Facility and Equipment Design**
- 233 **Summary**
- 233 **References**

INTRODUCTION

The harmful effects of radiation have been known from the early days of its use. A few months after Roentgen's discovery of x-rays in 1895, the first case of human injury was reported in the literature. The first case of high-level radiation-induced cancer was reported seven years later. Early evidence of the harmful effects of radiation exposure in humans became available in the 1920s and 1930s through the exposure experiences of radiologists, miners exposed to airborne activity from radon gases, and workers in the radium industry. The long-term biological effects of low-level radiation, however, were not appreciated until relatively recently. With the increased uses of nonionizing radiation and ionizing radiation in medicine (for treatment of cancer or diagnosis of bone disease), industry (density gauges, industrial radiography for weld inspection), and research (as labels attached to drugs or genes for tracing them in the cells or animals), comprehensive radiation safety programs are important.

Uses of Radioactive Materials and Radiation

Radioactive materials and radiation are used in a range of medical research and industrial applications. Here are some of the most common uses:

- *Medical*—Radioactive materials and ionizing radiation are widely used in nuclear medicine, radiology, and radiation oncology, for diagnostic and therapeutic purposes.
- *Utilities*—Nuclear fission (splitting of the atom) is used to generate heat, which in turn generates steam to operate electricity-producing turbines.

- *Industrial radiography*—Intense sources of radiation are used to produce images of weld joints, castings, pipe connections, and other structures. The images are analyzed for signs of fractures or other defects.
- *Industrial*—Sources of radiation are used in general industrial applications for measuring the thickness of paper, the thickness of gold plating, the level of liquid in cans, and the soil moisture in well-drilling operations. Radiation is also a source of ionization in fire or smoke detectors, static eliminators in photographic industry, and baggage inspection devices in airports and other areas for security.
- *Biomedical research*—Radioactive materials are also widely used as tracers in biomedical research applications, where small quantities are incorporated into biological samples.

Radioactive materials are also used as sources of power in nuclear batteries, in cardiac pacemakers used in humans, and in satellites. Radioactive materials are also used in consumer products, such as smoke alarms and emergency exit signs that contain a radioactive gas that produces luminescence in the powder coating inside the light tube. These lights are used in public areas to illuminate exit signs should the power supply fail.

Natural Sources

It is important to note that there are several sources of radiation that occur naturally. The radiation from these sources is identical to the radiation that results from human-made sources. The four major sources of naturally occurring radiation exposures are as follows:
- cosmic
- sources in the earth's crust, referred to as *terrestrial radiation*
- sources in the human body, referred to as internal sources
- radon.

Biological Effects of Radiation

For ease of discussion, we can divide the biological effects of radiation into two groups:
1. *Genetic*—Effects that occur in the reproductive cells and can be inherited.
2. *Somatic*—Effects that result from damage to the cells in the body and are observed in the person who has been exposed.

Genetic effects are long-term, since they are manifested in the future generation(s). The *somatic effects*, on the other hand, can be either early (acute) or late (chronic). The terms *acute* and *chronic* are also used to describe the exposure period. An *acute exposure* takes place within seconds, minutes, or hours. The early, or acute, effects develop from within minutes to a few weeks after the exposure. A *chronic exposure*, on the other hand, can extend over weeks, months, or years. It may not be continuous, and the late, or chronic, effects can be produced during or after the exposure.

With some irradiation, increased dosage is associated with a more severe biological response. For example, skin may show only a slight reddening at low doses, but severe gross tissue damage at higher doses. This response is termed *nonstochastic* and usually has a threshold below which the response is not detected. In other instances, irradiation can produce leukemia where the severity is independent of dose; the effect is either manifested or not. This response is termed *stochastic*. In stochastic response, the probability of the response depends on the dose—that is, the higher the dose, the greater the probability of the response.

Understanding the early and late effects of ionizing radiation is important for establishing safety guidelines designed to minimize the risk inherent in the uses of this radiation. The first standards for radiation protection were developed to protect workers from the acute effects. The current standards, which were originally recommended by the International Commission on Radiological Protection (ICRP) and later adopted by most regulatory agencies, are mainly based on protecting workers from late stochastic effects, such as cancer and genetic effects.

A comprehensive radiation safety program should take all of the following variables into account:
- tissue or organ exposed
- age of exposed person
- amount and duration of exposure
- type of radiation (alpha, beta, x-ray).

For example, because the effects of exposure depend on the age of the exposed person, a fetus is highly radiation-sensitive. Therefore, companies may need separate and more stringent guidelines to protect employees of reproductive age (the fetus) from potential damage.

REGULATORY CONTROLS

Because radioactive materials and radiation are used in a range of applications, their use has to be subject to a strict regulatory control to prevent unnecessary exposure to the workers or the public. In the United States, the use of ionizing radiation, such as x-rays, is controlled by the state agencies. The use of radioactive materials is controlled by the U.S. Nuclear Regulatory Commission (NRC) or by state agencies in agreement states—(states that have agreed to set up their own agencies to regulate the uses of radioactive materials). At a minimum, agreement states must follow the NRC guidelines, and most agreement states tend to be more restrictive than the NRC. In other countries, a national agency regulates the uses of radiation and radioactive materials.

Licensing

Before a facility can purchase or possess any radioactive material, it must obtain a license from the appropriate agency. Depending on the scope of the operations and the proposed uses, the applicant must provide detailed information on items such as the following:

- types and quantities of radioactive materials to be used
- locations of use, including details of any engineering controls used to prevent unnecessary exposures or airborne activity
- information on qualifications and experiences of users
- detailed procedures outlining methods of contamination control and surveys for detecting contamination
- instrumentation used for detection and analysis
- monitoring of personnel and environmental exposure
- training program for users
- policy and procedures manuals
- quality control procedures
- the appointment of a radiation safety officer (RSO), who is responsible for the implementation of the commitments made.

Once the regulatory agency is satisfied with the application, it issues a license. The license contains a number of general and specific conditions that must be followed. Regulatory agencies routinely inspect facilities to verify compliance with these conditions. Licensing plays an important role in the overall safety scheme by ensuring that the applicant has enough knowledge, as well as the appropriate facilities and instrumentation, to maintain a safe work environment.

Radiation Safety Programs

The overall organization and management of an adequate radiation safety program is not much different from that of any other safety program. The company needs a strong commitment from management combined with a comprehensive worker-training program. The program must be designed as a collaborative venture between the users and the safety staff, rather than purely requiring compliance.

The federal and state regulations require that any company using radioactive materials or radiation establish a radiation safety committee to oversee their safe use. The committee must consist of representatives from management and various user groups from within the facility. Its role is to routinely review policies and procedures, accidents, exposures, and operations and to make the changes necessary to ensure safety. The committee must meet at least quarterly and maintain written minutes of these meetings. Another legally mandated link in the radiation safety program is the *radiation safety officer* (*RSO*). That person's role is to educate and inform the users about safety procedures and to monitor the implementation of the safety requirements. The latter task involves developing a comprehensive safety inspection program.

The inspection program is important in identifying potential hazards before they evolve into accidents. A radiation safety inspection program is similar in scope to other safety inspections in that it seeks to identify unsafe situations or procedures. These include:

- unauthorized uses or users of radioactive materials
- using materials in unapproved areas
- improperly storing or transporting materials
- inadequate shielding
- failing to use appropriate safety clothing
- using volatile materials in areas other than approved hoods
- bypassing interlocks; consuming food in areas where radioactive materials are used or stored
- failing to label the equipment or "use" areas clearly to warn others of the presence of radioactive materials or radiation.

The results of the inspection process are important in analyzing the overall effectiveness of the program as well as identifying trends that might be developing. The RSO and the radiation safety committee should review the results of each inspection and amplify their efforts in appropriate problem areas. The committee also should present the results of the inspections to workers and their supervisor with recommendations on how to correct problems. To monitor the progress of the corrective actions taken, the committee should institute a follow-up inspection program.

PROTECTING AGAINST RADIATION HAZARDS

As described earlier, working with ionizing radiation or radioactive materials, like any other hazardous material, entails some potential risk. The hazards of radioactive materials and ionizing radiation fall into two major classes: (1) *internal* contamination hazards and (2) *external* radiation hazards.

Internal Contamination Hazards

Contamination is defined as the unwanted presence of radioactive materials. The possibility of a radioactive material entering the body inadvertently represents an internal contamination hazard. Generally, once this occurs, the remedial actions are somewhat limited. Therefore, the facility must emphasize preventing the entry. Prevention is usually achieved through a combination of engineering design, education of the users, and development of sound experimental techniques and administrative controls.

Inhalation

Airborne radioactive materials can enter the body by being inhaled. In that case, the radioactive material will directly expose the lung tissue, and if in soluble form, it could enter the bloodstream and expose other organs.

To eliminate the potential for breathing in radioactive materials, workers should perform all operations involving the use of volatile materials, or those that can produce dust or mists, in a properly designed and functioning enclosure such as a fume hood or a glove box. If the operation is such that full enclosure is not possible, then the facility must institute a respiratory protection program involving the use of appropriate personal protective respiratory equipment. Note that respiratory protection should be the final option; the facility should emphasize facility design, engineering out the radiation exposure hazard, as the primary protective measure.

Of course, management should consider the exhausted air from the enclosure. If necessary, the facility should have proper filtration systems to protect the environment and the public from exposure to effluents.

Ingestion

A major hazard with all uses of open (unsealed) radioactive material is that it can be ingested. Either direct consumption of a radionuclide or consumption of contaminated food can constitute *ingestion*. The food can be contaminated by coming in contact with either the radionuclide itself, other contaminated items such as plates or utensils, or hands.

To eliminate this potential hazard, impose administrative controls such as these:
- Prohibit storage and consumption of food or beverages where radioactive materials are used or stored.
- Require that workers use disposable gloves when handling radioactive materials.
- Instruct workers to wash hands thoroughly before eating or drinking.
- Label containers used for radioactive materials.
- Designate clean areas for storage or consumption of food in a location away from the radioactive use area.

Absorption

Some radioactive materials are absorbed through the skin. To eliminate this hazard, the facility can require workers to use disposable impervious (rubber, vinyl) gloves.

Puncture

Radioactive materials can also enter the body through punctures in the skin. Therefore, it is important to cover all wounds and skin punctures with dressings and to use disposable gloves. Also, all sharp objects, broken glassware, needles, and syringes that could be contaminated with radioactive materials should be placed in hard-sided containers and handled with utmost care. Any cuts received while working with radioactive materials must be checked immediately for contamination. If contamination is found, the worker should obtain immediate medical care.

External Radiation Hazards

Radioactive materials are sources of radiation emission; this property is often used for various applications. In order to use the radiation emissions, and at the same time eliminate the potential for contamination, the radioactive material is placed in a closed container. This is known as a sealed source. Use of a sealed source, however, could result in high radiation doses to the users. The three basic principles of protection against external radiation are *time, distance,* and *shielding*.

Time

The radiation dose from a given radioactive source is delivered at a specific rate, which depends on the radionuclide and its quantity. Therefore, for a given quantity of the material, this rate is constant as expressed by this equation:

$$\text{Dose} = \text{Time} \times \text{Dose Rate}$$

Thus, if the time during which a person is exposed to radiation is reduced, the total exposure will decrease in direct proportion. For example, reducing the time to one-half will cut the exposure to one-half.

These time reduction concepts suggest some principles to include in a safety program.

- Preplan and discuss the task before entering the area (e.g., walkthrough procedures). Discuss engineering and design concerns.
- Use only the number of workers required to do the job.
- Have all necessary tools and materials before entering the area.
- Use mockups and practice runs that duplicate work conditions.
- Take the most direct route to the job site.
- Never loiter in an area controlled for radiological purposes.
- Work efficiently, but swiftly.
- Do the job right the first time.
- Perform as much of the work as possible in lower dose-rate areas.
- Use good communication between workgroups and radiological control personnel. This will help to ensure timely job completion.
- Do not exceed stay times.

Radiological control personnel may have to limit the amount of time a worker can stay in an area. A *stay time* can be assigned when basic exposure reduction techniques are not practical or when a worker may be approaching administrative control levels.

Calculations:

$$\text{Stay Time} = \frac{\text{Allowable Dose}}{\text{Area Dose Rate}}$$

$$\text{Dose} = \text{Area Dose Rate} \times \text{Time}$$

Distance

The intensity of radiation decreases with an increase in the distance between the source of radiation and the person. This reduction follows what is known as the "inverse square law": The intensity of radiation decreases as an inverse function of the distance squared. Mathematically this can be represented as follows:

$$D1 \times X1^2 = D2 \times X2^2$$

where *D = Dose, Exposure, Intensity* and *X = Distance*

Distance can thus be effectively used as a method of reducing the dose to workers and the public. In practice, workers can take the following steps to achieve this:

- Use remote handling tools.
- Stand away from the source of radiation.
- Store sources as far as possible from the occupied areas of the facility.
- Move your work to a lower dose-rate area away from the radiological source, if possible.

Shielding

As radiation travels through matter, it interacts with the matter and loses some of its energy to the matter. As this process continues, the radiation eventually will lose all its energy. Therefore, if enough matter is placed in the beam's path, the radiation can be absorbed to a level that is indistinguishable from natural background levels of radiation.

Shielding refers to placing absorbers between the source of radiation and the person to reduce the intensity of radiation. The absorption ability of shielding not only depends on the nature and thickness of the matter used for this purpose, but also on the type and quantity of the radiation. For example, beta particles require material with a low atomic number, such as plastic, lucite, or aluminum. Gamma rays, in contrast, require materials with a high atomic number, such as lead.

The walls of rooms used for x-ray machines or other high-radiation-exposure units should include adequate shielding to protect the public from exposure. The type of shield used will depend on the type and intensity of radiation, desired exposure outside the shielded area, cost-benefit ratio, and physical constraints of the area (Figure 13-1).

Shielding should be well designed and carefully implemented to ensure that it does not create other problems. For example, large amounts of uncovered lead shielding could result in other undesired

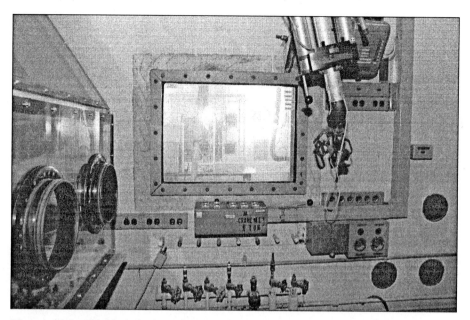

Figure 13-1. Shielded viewing window and manipulator arms on a sealed "cave" used to handle radioactive materials safely. (Courtesy Argonne National Laboratory)

health effects. Shielding one type of radiation could result in the production of another. For example, when electrons (beta particles) are decelerated, part of their energy is converted into *bremsstrahlung*, or braking radiation. The shielding design should account for this radiation production and provide for absorption. Using shielding may actually increase the total dose for a job. The human-hours involved with placing and removing of the shielding may be more than doing the job using other basic protective measures.

Administrative Controls

Besides specific steps to protect against the various types of hazards, a radiation safety program must include a series of operational procedures and guidelines. These should not only address the criteria for facility and experimental design, but also place detailed administrative controls over the entire operation (Figure 13-2).

Procurement and disposal

The federal and state regulations require a detailed accounting of all radioactive materials from the time they enter the facility to the time they are disposed of as radioactive waste. Therefore, the facility should establish a central receiving area where all incoming shipments are logged in and checked for presence of contamination on the package. The package can then be released for use by the person who ordered it. There also should be a central radioactive waste facility where all waste is collected and disposed of. This will allow for proper disposal, inventory control, and documentation (Figure 13-3).

Usage

The facility should develop procedures outlining the proper handling techniques and should make them available to all workers using radioactive materials or radiation-producing equipment. The procedures covered should include instructions on the proper preparation of the work area, posting of appropriate warning signs in all use and storage areas, segregation of contaminated and noncontaminated waste, and use of safety equipment. The procedures should also identify and address special hazards such as volatile materials or work activities with high-exposure equipment. Personnel should maintain detailed records of each use.

Waste

Proper disposal of radioactive waste is important for complying with the law as well as for preventing contamination of the environment. The way radioactive waste is segregated, packaged, labeled, and disposed of will depend on the category of the waste generated. In general, dry, biological, aqueous liquid, organic liquid, and special nuclear waste must be packaged separately. All packaging, labeling, and manifests must comply with regulations of the NRC, the state, and the U.S. Department of Transportation.

GENERAL SAFETY PRECAUTIONS FOR RADIOACTIVE WORK

1. Always keep radioactive and nonradioactive work separated as far as possible. Remember separating work spaces will decrease your exposure by both increasing the distance from radioactive sources and potential contaminations and by decreasing the time spent in the area of radioactivity.
2. Always wear lab coats and disposable gloves when handling radioactivity. Change gloves frequently, especially when moving from radioactive to nonradioactive areas. When iodinating wear two pairs of gloves. There is some passage of vapor through vinyl gloves.
3. Plastic safety glasses will reduce eye exposure from high-energy beta-particles such as 32-P.
4. Always cover work surfaces with absorbent paper. Change the paper frequently and use a spill tray lined with absorbent paper if possible. It is easier to clean a radioactive contamination by rolling up the paper and disposing of it in the radioactive trash then it is to scrub the counter.
5. Wear film badges and/or finger rings, except for the low-energy beta emitters such as 3-H, 14-C, and 35-S.
6. Use automatic pipets. **NEVER MOUTH PIPET.**
7. Cover containers (vials, test tubes, etc.), which hold volatile and air-reactive radioactive compounds, such as radioiodine, borotritides, tritiated water, labeled methylhalides. Use covered tubes when centrifuging radioactivity. Also cover tubes with parafilm or tops when vortexing is required.
8. Never work with radioactive materials when open cuts may be contaminated.
9. Eating, drinking, smoking, and applying cosmetics in areas where radionuclides are used or stored is expressly forbidden. Food and drink containers or wrappers may not be placed in the trash cans in a radioactive area.
10. Handle gamma and energetic beta-emitting sources and stock bottles using tongs or forceps. Crucible tongs with rubber tubing on the top to increase the gripping effectiveness are usually quite good to use.
11. Make sure a functioning survey meter is available in the work area when handling radionuclides, especially 125-I and 32-P. Use the appropriate probe for the radionuclide, i.e., a scintillation probe for 125-I and a G-M probe for 32-P. Survey meters must be calibrated yearly. Check the meter before each use to verify operations.
12. Survey glassware and apparatus used in radioactive experiments for contamination prior to releasing them to the general dishwashing services or releasing them for general usage.
13. Be informed. Know the mechanical, chemical, and radiation hazards of an experiment that is to be performed. Frequently it is useful to do a "cold-run" to see if an experiment is feasible.
14. Monitor working areas regularly for possible contamination.
15. Use a fume hood whenever airborne radioactivity is handled, e.g., fine powders, volatile liquids, tritiated water. Any radioisotopes, which are very hazardous or high-activity (greater than 1 mCi) must be handled in a hood, particularly radioiodine.
16. Handling **RADIOIODINE** presents a volatility hazard to personnel. I_2 and I are highly reactive and readily absorbed through the skin and vinyl gloves. Significant thyroid burdens have been observed when inappropriate handling techniques have been used.
    a. NaI should be kept at alkaline pH (pH 7.8–11.0). Avoid acidic solutions that result in volatile iodine. Store NaI at room temperature. Freezing results in instability of the compound and volatilization.
    b. Always work in a fume hood with a minimum face velocity of 100 linear feet per minute. The industrial hygiene program can be called to check your hood.
    c. Use two pairs of gloves. Volatile iodine compounds can penetrate each layer of glove within 10 minutes. Always dispose of the gloves in the radioactive waste.
    d. Less volatilization will occur if 125-I is taken from the stock vial via the rubber septum using a syringe or needle, rather than pipeting from an open container.
    e. Waste should be placed in plastic bags and stored in the hood until pickup. Liquid waste should be alkalinized with NaOH and stored in the hood.
    f. Finger rings must be worn by all personnel handling stock solutions containing 1 mCi or more of 125-I. Significant extremity exposure can occur if a stock vial of 125-I is handled directly.
17. Always wash your hands and monitor yourself before leaving a radioactive area.
18. Have radioactive trash removed promptly. Package it before your pickup date.
19. Wash and monitor your hands before leaving the work area and especially before eating or drinking.

Figure 13-2. General Safety Precautions for Radioactive Work—a radiation safety program must include a series of operational procedures and guidelines.

RADIOISOTOPE USAGE FORM

PRINCIPAL INVESTIGATOR: _____ BLDG./RM. # _____ DEPARTMENT: _____

1. RADIOACTIVE MATERIALS RECEIVED

DATE	ISOTOPE	CHEMICAL COMPOUND	ACT (μCi)	VENDOR	LOT. OR SERIAL #
				NEN/ICN/AM/____	

2. SURVEY RESULTS

Wipe Test (cpm), Bkgd. _____, surface _____, vial _____. Exp. Rate (mR/hr): Bkgd: _____, Surface ____, Im _____

Tech. Name: _____ Lab. Receipt Date: _____. Received By: _____

Equipment Used: Type _____, Model No. _____ Last Calibration. _____.

3. USAGE

DATE	ACT. USED μCi	INIT.

4. METHOD OF DISPOSAL

DATE	ACT. μCi	L/D/V/B/ABS	TRANSFER μCi	BALANCE on hand μCi	REMARKS	INIT.

L=Liquid; D=Dry; V=LSC Vials; B=Biological; ABS=Absorbed Liquid

Figure 13-3. The Radioisotope Usage Form allows for documentation of inventory control and proper disposal of radioactive materials.

Because the availability of disposal facilities is limited and the cost to dispose of radioactive waste is escalating, the facility should make every effort to reduce the total volume of waste produced. These efforts should include substituting "nonradioactive" procedures for radioactive ones when available, as well as setting up procedures for reducing the contaminated material to a minimum.

Training
Regulations require that all personnel working with radioactive materials receive adequate training before they use radionuclides, as well as routine refresher courses. The scope of a particular training program will depend on specific uses of the radioactive materials in the facility and will often be a condition of the license.

Incident reporting and investigation
All incidents involving radioactive materials must be fully investigated and properly documented. The investigation should include the nature and causes of the accident and specify a corrective action the facility will take to prevent recurrence of the accident. Certain accidents involving loss of radioactive materials, exposure to personnel, or damage to property above a certain value must immediately be reported to the licensing agency.

Inspection program
As discussed earlier, the inspection program is a tool to monitor the effectiveness of the control programs as well as a way to identify trends and ensure compliance. The inspection should be comprehensive and follow a structured format. The use of a checklist is highly advisable.

The inspection program should include a means of enforcement. This will allow the facility to take appropriate action against anyone who disregards the established guidelines (Figure 13-4).

HAZARD MONITORING PROGRAM
Establishing a program to monitor hazards is an integral part of comprehensive safety management. The program can be divided into two main categories: environmental monitoring and personnel monitoring.

Environmental
Environmental monitoring is used to evaluate the efficacy of the control mechanisms put in place to protect the work area and the public. This effort might include surveys conducted to determine the presence of contamination in the work area (air, soil, water, and vegetation sampling for evaluation of potential releases). In addition, environmental monitoring often includes taking measurements outside of a high-exposure area to determine the existence of excessive levels of contamination.

Personnel
Personnel monitoring determines workers' potential exposure to radioactive substances. The scope of the monitoring program will depend on the types and quantities of radioactive materials workers use as well as the actual use procedures. The monitoring program should be designed to detect both internal and external exposure.

External exposure
The primary way to detect and quantify external exposure is with personnel dosimeters such as film badges or finger rings. These devices contain a sensitive receptor (a photographic film) that produces a reaction (darkening of the film) when exposed to radiation. This degree of darkening can be correlated to a standard for measurement of exposure. Workers wear these devices for a period of time, usually one month, after which the devices are sent to a certified laboratory for processing. Any person over the age of 18 years, who has the potential to receive more than 10% of the allowable whole-body dose limit, must wear an external dosimeter. If under age 18 years, their exposure must be kept under 10% of the allowable dose limits (i.e., the same as the general public). The results of this monitoring constitute legal documents and must be maintained indefinitely. Note that external radiation-monitoring devices (film badges) cannot detect low-energy radiation emitters such as tritium (radioactive hydrogen).

Internal exposure
Internal radiation exposure can result if radioactive material enters the body. To determine the extent of a person's exposure to airborne radioactive contamination, the company should develop and implement a bioassay program. The results of bioassays have several applications:

- estimation of internal organ doses
- determination of the presence of airborne radioactive materials, which could indicate a failure in, or inadequacy of, engineering controls

RADIOISOTOPE LABORATORY SAFETY AUDIT

DATE _____

USER _____ PERMIT # _____ MAIL CODE _____ PHONE _____

DEPARTMENT _____ PERSON INTERVIEWED _____

CAMPUS _____ BLDG _____ ROOM _____

I. BEHAVIORAL
1. Smoking
2. Eating
3. Drinking
4. Storage of food
5. Use of cosmetics
6. Mouth pipeting
7. Housekeeping

II. EDUCATION AND INFORMATION
8. Users authorized
9. Storage area posting
10. Training and experience forms
11. Radiation safety manual
12. Work area posting
13. Legal requirements
14. Containers and/or equipment labeled
15. User training frequency met
16. Decontamination procedures known
17. Radiation
18. Knowledge

III. PERSONNEL SAFETY
19. Protective clothing
20. Protective gloves
21. Fume hood evaluation
    Date: _____

IV. PERSONNEL MONITORING
22. Film badges
23. Finger rings
24. Proper use and storage of dosimeters
25. Dosimeters report availability
26. Bioassay frequency

V. AUDIT FREQUENCY STATUS
27. Bioassay performance
28. Appropriate survey meter present
29. Appropriate survey meter use
30. Survey meter calibration
31. Area monitoring frequency
32. Wipe test frequency
33. Wipe test locations

VI. INVENTORY AND ADMINISTRATIVE
34. Record keeping (survey, decontamination, usage, transfer, bioassay, calibration, incidents, dosimetry, audit, quarterly inventory)
35. Sealed sources
    a. Leak test
    b. Inventory (quarterly)
36. License limits exceeded?
    Isotope(s) _____ Quantity _____
37. Currently used isotopes _____
38. Unauthorized isotopes _____
39. Use locations

VII. STORAGE
40. Proper storage
41. Proper disposal
42. Radioactive material

VIII. RADIATION SAFETY MONITORING
43. Attach diagram with location and map

IX. MONITORING EQUIPMENT
44. Badges
45. Monitor/detectors
46. Other

Figure 13-4. This type of Radioisotope Laboratory Safety Audit can be used as part of the facility's inspection program.

- evaluation of work habits and experimental designs.

Bioassays are performed according to one of the following methods:
- analyzing samples of blood, tissue, or urine
- monitoring the organ of interest to determine the presence of radioactive materials.

Which method to use depends upon the type of radioactive material involved (e.g., urine analysis for tritium and thyroid monitoring for radioiodine).

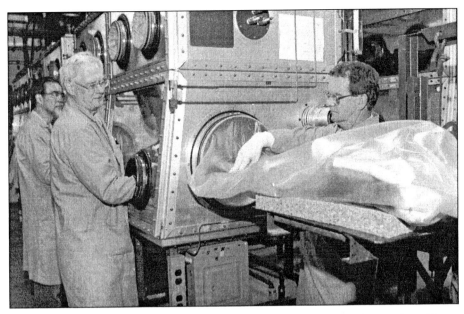

Figure 13-5. Transferring a part from a glovebox hood into a containment bag. The bag and tube from the hood are closed by heat sealing. (Courtesy Argonne National Laboratory)

Ideally, bioassays should be performed routinely, either quarterly, semi-annually, or annually or when anyone suspects that an internal contamination might have occurred, and after an accident.

Instrumentation

Radiation-monitoring programs require the use of adequate and proper instrumentation. The RSO must ensure that the available instruments can detect all radioactive materials used in the facility. Furthermore, these instruments must have a low limit of detection, and they must be able to measure at ideally below the thresholds set by the regulatory agencies. Often, employees fail to detect low levels of radioactive contamination because they use inadequate analytical equipment. Also, to provide adequate measuring capabilities in the event of an accident, the instruments should also be capable of measuring the highest exposure rates possible.

ALARA CONCEPT

The ALARA concept or *As Low As is Reasonably Achievable exposure* is the goal for all occupational exposures. Here, occupational exposure includes all the external exposures and radioactive material intakes incurred by a worker during periods of work, but excludes medical and natural radiation, unless the latter is enhanced as a result of a particular working environment (Figure 13-5). The limitation of occupational exposures should be at the source of radiation. This should be achieved through design and engineering criteria as much as possible, so that the use of personal protective equipment, in general, is supplemental to the more fundamental provisions. Access to controlled areas should be restricted and subject to local operating instructions (a radiation safety manual outlining the details must be developed and provided for the employees).

External exposure can be reduced by the use of shielding, distance, and reduction of exposure time. The potential for contamination can be reduced, or eliminated, by proper containment and cleanliness. A combination of these precautions and good work habits should greatly reduce the occupational exposures. Paramount in achieving these goals are good worker training and educational programs. Institutions should develop basic programs that clearly inform the workers of the potential hazards from the radioactive materials and radiation sources, as well as proper safety precautions (Figure 13-6).

The federal and state agencies have set occupational exposure limits, as well as maximum permissible airborne contaminant concentration levels, which must not be exceeded under normal operational conditions. Even though these limits are set at levels that provide a very low risk of injury, it is prudent to avoid unnecessary exposures. A good radiation safety program should set internal limits

Figure 13-6. Some operations require complete enclosure. Here, at a glove box, a technician works with aluminum powder, used in atomic reactor fuel elements. He is wearing a film badge and air sample on his lapel.

ranging from 10%–25% of these limits. The program should include an investigation of any exposures exceeding these internal limits. Once the causes have been identified, the action should be taken to reduce these as much as possible. This objective of reducing occupational levels by means of good radiation protection planning and practice is the concept of maintaining the occupational exposure to radiation to As Low As is Reasonably Achievable.

FACILITY AND EQUIPMENT DESIGN

Substantial exposure reduction can be achieved by proper design and use of facilities and equipment required for the uses of radioactive materials, or ionizing radiation. The details and particulars of each would depend on the types and quantities of radioactive materials, as well as the intended operations. However, there are some fundamental guidelines that apply to all cases. These include the following:

- Providing optimum distance between areas of frequent occupancy and radioactive usage areas.
- Placing fume hoods used for volatile radioactive materials in remote areas of the laboratory. The airflow through the fume hood must be sufficient to remove all volatile materials and provide adequate dilution factors.
- Designating clearly marked sinks for rinsing and disposing of radioactive contaminated materials.
- Designating special, and when necessary, shielded areas for storage of radioactive materials.
- Designating special waste containers. These should be clearly marked to avoid improper disposal.
- Laboratory floors should be covered with vinyl type materials, which can be easily decontaminated.

- Providing for appropriate placement of radiation and contamination monitoring equipment.
- Limiting the use of radioactive materials to properly designed and designated laboratories.
- Instrumentation used for detection of the contamination must be appropriate for the radioactive materials used. These should also be capable of detecting the low levels set by federal and state regulatory agencies.

SUMMARY

A radiation safety program is basically similar to any other safety program. While the specifics are different, it follows the same principles of facility design, user training, detailed procedures, administrative controls, and accident prevention. The company should apply to radiation safety the detailed approaches to administering a safety program that are outlined in other chapters in this book.

A comprehensive radiation safety program requires that management provide both moral support and financial commitment. As is the case with other safety programs, providing a safe work environment can be expensive. But considering the potential hazards associated with the use of radioactive materials, the investment is small and will return dividends far into the future.

REFERENCES

Brodsky A (ed). *Handbook of Management of Radiation Protection Programs,* 2nd ed. Boca Raton FL: CRC, 1992.

Brodsky A (ed). *Handbook of Radiation Measurement and Protection.* Boca Raton FL: CRC, 1979.

Code of Federal Regulations, 10, (10 *CFR*) Energy, Parts 0-50, United States Nuclear Regulatory Commission.

International Commission on Radiological Protection (ICRP) Reports. This series is available from Pergamon Press, Tarrytown NY.

Murphy BL, Traub RJ, Gilchrist RL, et al. *Fundamentals of Health Physics for the Radiation Protection Officer.* Olney MD: Lectern Assoc., Inc. Available through National Technical Information Service (NTIS), U.S. Department of Commerce, Springfield VA 22161.

National Council on Radiation Protection and Measurement (NCRP). *Recommendations on Limits for Exposure to Ionizing Radiation.* NCRP Report 91. Bethesda MD: NCRP, 1987.

NCRP. *Recommendations on Operational Radiation Safety Program,* NCRP Report 59. Bethesda MD: NCRP, 1978.

NCRP. *Recommendations on Exposure of the U.S. Population from Occupational Radiation,* NCRP Report 101. Bethesda MD: NCRP, 1989.

NCRP. *Recommendations on Implementation of the Principle of as Low as Reasonably Achievable (ALARA) for Medical and Dental Personnel,* NCRP Report 107. Bethesda MD: NCRP, 1990.

National Research Council Committee on Biological Effects of Ionizing Radiation. *Health Effects of Exposure to Low Levels of Ionizing Radiation* (BEIR V). Washington DC: National Academy Press, 1990.

Wang Y. *Handbook of Radioactive Nuclides.* Boca Raton FL: CRC, 1969.

chapter 14

Workers' Compensation Management Programs

by Bertram Cohen, JD

235 **Overview**
236 **Cost of the System**
237 **Federal Workers' Compensation Acts**
Federal employees ■ District of Columbia Workers' Compensation Act ■ U.S. Longshoremen's and Harbor Workers' Compensation Act
238 **State Funds**
238 **Self-Insurance**
239 **Coverage-Statutory Scope of the Laws**
Who is an employee? ■ AOE/COE
240 **Occupational Disease**
240 **Policy Exclusions**
240 **Benefits**
241 **Disability Classifications**
Permanent total disability ■ Temporary total disability ■ Permanent partial disability ■ Temporary partial disability ■ Death benefits or survivors' benefits ■ Social Security disability benefits ■ Subsequent (second) injury funds
243 **Compulsory or Elective Coverage**
Employment of minors ■ Injury outside the jurisdiction ■ Failure to insure
243 **Administration**
Industrial/medical facilities ■ Reporting of injuries ■ Discrimination charges
244 **The Claims Process**
245 **Financial Incentives**
Cost incentives of self-insureds ■ Cost incentives of insureds ■ Cost relationships
247 **Summary**
247 **References**

OVERVIEW

Although the concept of a master's responsibility for the consequences of injuries befalling a servant is thousands of years old, workers' compensation as a form of social insurance is a relatively modern idea, first formalized in Europe in the 1880s and introduced to the United States around the turn of the 20th century. Before then, compensation for job-related injuries or death was governed by principles of English common law that allowed recovery only where the employer was solely negligent, and only after suits were brought for damages. The mechanism was slow, cumbersome, expensive, and inadequate, with overwhelming difficulties of proof. A recognition that society as a whole has a valid interest in the health and safety of those involved in producing goods and services led to demands for major reform.

By the early 1900s, many states passed laws that tended to restrict or modify common law defenses, but employer negligence was still the key to recovery; and the system remained unwieldy for workers and costly for employers. By 1911, the first comprehensive workers' compensation laws passed, in some instances allowing the worker a choice between suing for civil damages and accepting workers' compensation benefits of a lesser but fixed amount, without regard to negligence or fault and without the delays of formal litigation. Over the years, most states removed the civil suit option, and by now workers' compensation is the exclusive remedy for job-related injuries. Only a few instances remain where a choice of civil suit still exists, generally where an employer illegally does not carry insurance or has committed intentional torts. Likewise, only very few areas still exist in which the fault concept is relevant to work injuries. One situation is where the employer has brought about the injury by its own serious and willful misconduct and becomes liable for penalties punitive in nature and generally uninsurable. Such penalties are usually fixed as a percentage of compensation paid by the employer or insurer. For the vast majority of situations, workers' compensation today in the United States is a true no-fault system requiring only a recognized injury to trigger automatic entitlement to statutory benefits.

All 50 states, U.S. territories, and the federal government now have workers' compensation laws, which vary from venue to venue, but have many common features based on the same overall principles:

- Victims of work-related injuries receive prompt, reasonable, and certain income and medical benefits, delivered without formal litigation. In case of industrial death, dependents receive income and burial benefits.
- Negligence and fault are largely immaterial and do not affect a workers' right of recovery. However, the worker gives up the right to sue for damages in civil court.
- An appeals process exists to resolve disputed claims, with as little complexity and delay as possible.
- Fees to lawyers and expert witnesses are minimized, and costs of litigation are reduced or are reimbursable to the worker regardless of outcome.
- The amount of benefits is predetermined and fixed by statute, with but few exceptions as noted above.
- The system includes mechanisms to return the worker to the labor market through rehabilitation if possible, to minimize losses to the worker and to the labor force.
- Enforcement penalties are included to provide an economic incentive to pay benefits.
- Special funds and provisions exist to compensate special situations, such as injuries while working for uninsured employers, cases of latent disability like asbestosis, and injuries sustained by workers already partially disabled.

The Occupational Safety and Health Act of 1970 (OSHAct) required creation of a national commission on state workers' compensation laws. The commission's landmark 1972 report recommended 84 improvements, 19 of which were deemed essential for modernizing the various systems and meeting minimum standards nationwide. It was recommended that if such legislation did not occur, the federal government should move to preempt state laws.

In the years since, a flurry of state statutory reform resulted in increased benefits and a degree of uniformity throughout the United States. But there is still a wide disparity in benefits from state to state and in many jurisdictions the level of benefits lags far behind the economics of the workplace. The changes are ongoing, and the threat of federal preemption continues to galvanize reform, although during the 1980s there was a significant decline of congressional interest in federal preemption, as the states made at least some efforts to enhance their systems. During the 1990s a number of states engaged in further comprehensive reform of their workers' compensation delivery systems, attempting with varying degrees of success to eliminate the potential for fraudulent claims.

Moreover, the changing face of industry and advances in medical science, refining definitions of occupational disease, continue to bring about alterations in the philosophy and goals of the compensation delivery system. At the same time, however, the mushrooming costs of medical care and monetary benefits continue to result in constant pressure for cost containment from employer and insurance carrier interests. As a result of all these factors, it is unlikely that stability and total uniformity in the compensation programs nationwide will be seen in the near future.

COST OF THE SYSTEM

Most costs of workers' compensation programs are borne by the employer community, either directly in cases of self-insurance or through premiums to insurance carriers, and passed on to the consumer as a cost of doing business. In some cases, primarily where public agencies that do not sell their goods and services for a price are concerned, this is impossible. In such instances costs are defrayed directly through taxation. In either event, employees do not contribute to the payment of insurance premiums or the costs of their benefits; in most states contribution by the worker is forbidden by law.

Whether administered directly by the states, the federal government, or through competitive insurance carriers, workers' compensation is big business. After rising rapidly in the 1980s, benefits and costs deceased during the early 1990s, but even after the decline the national total of benefits paid in 1995 was $43.5 billion. Employer costs, including insurance premiums and self-insurance administrative costs were $57 billion in 1995.

In 1988, the average cost per $100 of payroll was $2.36, up from $2.13 in 1979. Recent increases in insurance rates were caused by benefit increases, also by inflation. Rates are usually adjusted annually on the basis of past industry experience and projected benefit levels. In recent years, there has been a trend toward rate reductions. These reductions frequently were coupled with benefit increases, causing net increases in insurance rates. All of this means that injured employees are receiving greater benefits and employers are paying more for coverage.

Rates charged by insurance carriers generally anticipate that about 65% pays for claims. The balance is used for operating expenses, including claims handling, profits, and dividends. Most states allow an up-front premium discount based on the size of the account and historical profit picture of the carrier.

Rates are promulgated by rating organizations; some states have their own. The National Council on Compensation Insurance, an association of workers' compensation insurance companies, is the licensed advisory organization. The main functions of this and similar organizations are to collect statistics and calculate proposed rates, establish policy, develop experience and retrospective rating plans, and serve as the filing organization with the various insurance commissioners for its member companies.

Some states allow carriers to set and file their own rates, which is referred to as open rating. The major problem with open rating is that individual carriers do not have adequate and complete historical data on which to base future projections. Other states allow carriers to deviate from published rates, and still others publish minimum rates that all carriers must charge.

All rates and rating programs are subject to state insurance department scrutiny regardless of the pricing mechanism. One of the major functions of such insurance departments is to determine that carriers remain solvent. Otherwise, the injured employee might become a victim of an inadequate social system and a burden to the taxpayer because someone will have to compensate the injured.

The net result of increased benefits and liberalization of statutes is increased product cost to consumers. Many employers chose to insure themselves in an attempt to reduce costs and improve their cash flow. Inflation had some impact, but most of the increased costs are associated with increased benefit levels. Many states now automatically increase maximum indemnity benefits annually, based on average weekly wages. Most set the maximum benefit at two-thirds of the state average wage; however, some are at 100%—and that benefit is tax-free income. Not all workers receive the maximum benefit, and for some, the benefit is less than two-thirds of their preinjury wages.

FEDERAL WORKERS' COMPENSATION ACTS

The federal government enacted several laws to cover its own employees, as well as railroad, maritime and admiralty employees, and longshore and harbor workers.

Federal Employees

The Federal Employers' Liability Act covers employees of interstate rail carriers only. Workers are compensated for injury or death on a comparative negligence basis. Juries generally grant substantial award, because injuries are usually serious.

The Federal Employees Compensation Act (FECA) governs the workers' compensation system for civilian employees of the U.S. government. This act does not apply to military personnel.

District of Columbia Workers' Compensation Act

Enacted by the District of Columbia City Council in July 1982, this act replaced the Longshoremen's Act that was previously in effect in the District of Columbia. The act excludes military and federal government employees.

U.S. Longshoremen's and Harbor Workers' Compensation Act

In March 1927, Congress passed U.S. Longshoremen's and Harbor Workers' Compensation Act

(L&HWCA) to close the coverage gap for employees engaged in maritime employment, including longshoring, stevedoring, ship repairing and building—those workers who were outside the scope of state compensation laws. The act, however, excludes members of the crew of any vessel. There have been many legal disputes, mostly jurisdictional in nature, between the scope of the federal act and the state acts; and many precedents set that have added to the confusion. Further, the U.S. Supreme Court granted jurisdiction to either state or federal law, depending on which of the two was sought for remedy. Furthermore, although the L&HWCA was thought to be the exclusive remedy, the Court did not always view the act in that light.

In 1972, Congress amended the L&HWCA as follows:

- Hold-harmless agreements between stevedore companies and vessel owners were held contrary to public policy and unenforceable.
- The use of the unseaworthiness doctrine was disallowed as basis of recovery against third parties by those who came under the act.
- L&HWCA was the exclusive remedy to injured employees coming within its scope.
- Benefits increased substantially. The L&HWCA benefits are tied to changes in the national average weekly wage, and have increased substantially over the years. As of 1999, the maximum weekly indemnity payment was $871.76, well above the compensation benefits available in most states.
- The jurisdictional scope of the act expanded onto land and to employees engaged in maritime employment.

Since the law extended coverage onto land, it became widely subject to interpretation. Moreover, the amendment did not define the term *maritime employment*, which caused further confusion. Among the employers hit hardest were proprietors of small boat repair yards, who could no longer find carriers to provide coverage, even at the high rates dictated by such increased exposure. In response the act was amended effective 1986 to exclude such employers from L&HWCA jurisdiction.

STATE FUNDS

Simply put, state funds are insurance companies owned and operated by the state. There are two types of state funds—*monopolistic* and *competitive*. In a monopolistic fund, the state is the only carrier authorized to write workers' compensation insurance. Some monopolistic states, however, do allow self-insurance. Competitive state funds, in contrast, compete for business along with private carriers and those who self insure.

Initially, many of these state funds were created to fill an important need. Private carriers sometimes refused to cover certain classes of risks to which workers were subject. Since coverage was mandated by federal statute and had to be provided, certain states went into the business themselves. Originally, many funds did not compete for business but wrote only those risks that could not find coverage at a reasonable price in the private marketplace.

SELF-INSURANCE

Employer self-insurance has become an attractive alternative to conventional forms of insurance with private carriers or state funds. This trend has now leveled off and may be reversing itself. Currently, about 10% of the nation's workers' compensation premiums have been diverted to self-insurance.

The switch to self-insurance is a logical reaction for businesses to reduce costs and maximize cash flow. In recent years, benefits, claim reserves, litigation, and associated administrative costs have risen dramatically. In general, so have interest rates and rates of return on invested income. Risk managers for the large sophisticated accounts perceived that they could do as well as the insurance industry by handling their own claims, while enjoying the advantage of the investment income and cash flow themselves.

Workers' compensation claims have a relatively long pay-out period. The more serious claims tend to be paid over a long period, sometimes five years or more. Insurance carriers tend to reserve for the ultimate potential of each case. Self-insureds, on the other hand, tend to be less conservative in reserving and increase reserves as the need becomes apparent; some even operate on a pay-as-you-go basis, unless the jurisdiction is one where state requirements mandate maintenance of reserves for all self-insureds. The difference is the cash that the self-insured account can apply directly to operations. The cash flow advantage is usually lost after about five years, however. At that point, claims payments for the current and prior years, plus reserve increases, excess insurance, and administrative costs, are about the same as an insured program.

Few companies are truly self-insured. Most purchase excess insurance above a self-insured retention limit, and these policies usually are written on a per-injury or per-occurrence basis. Although some states allow aggregate excess policies for the policy term, others do not. The ultimate exposure could be nearly unlimited. For example, if a company purchased an excess policy for $500,000 to $10 million and had 10 of $400,000 claims each, the excess policy would not come into play and the cost to the self-insured account would be $4 million. When administrative services, claims, loss control, and the like are purchased, the lowest price is not always the best bargain.

Experience modifications can usually be maintained if an administrator or self-insured continues to make unit statistical filings with the jurisdictional rating bureau. This can make a significant difference in cost.

Buying back into a standard insurance program can be expensive. The state agency having jurisdiction will require that securities or a bond be posted to cover liabilities for outstanding claims. Consequently, those claims will have to be paid, claims-handling expense will be incurred, and new premiums will have to be paid, all at the same time.

COVERAGE—STATUTORY SCOPE OF THE LAWS

Workers' compensation policies provide two basic coverages: workers' compensation and employers' liability.

- Workers' compensation provides benefits to injured workers and their dependents for injuries and illnesses related directly to their employment. Coverage and benefits are required by law, and there are no limits on this coverage.
- Employers' liability protects employers from suits in tort. But employees or their dependents can sue the employer if their injuries are such that the Workers' Compensation Act is not applicable, or if there are special exceptions. This provision also provides employers with defense coverage if the allegation is within the scope of coverage. Some examples might be:
  1. Suits by third parties, such as dependents of the injured.
  2. In action-over cases, when the injured brings suit against a third party and that third party in turn brings an action against the employer.
  3. Dual capacity issues in which, basically, the employee having collected workers' compensation benefits, also brings an action against the employer as the manufacturer of a product or provider of a service. This situation can be covered under employers' liability in the future, although it is debatable, since such coverage was never intended by the system. Coverage provided for employers' liability is limited by the policy. Policy limits generally vary from $100,000 to approximately $1 million. They can be increased at the carrier's option for an additional premium.

Since dual-capacity suits are directly contrary to the exclusive remedy rule, in many jurisdictions appellate courts have been reluctant to expand situations in which such actions can be brought. In fact, the recent trend is to limit dual-capacity alternatives, either by court decision or by statute.

Who Is an Employee?

The test for employee status varies by state, indeed, in some cases by the circumstances of an individual claim. As a general rule, an employee is defined as ". . . every person who performs a service for an employer expressed or implied, oral or written, whether lawfully or unlawfully employed including aliens and minors." Generally, officers and general partners are considered employees. In some states and under certain circumstances, however, they can elect to be excluded from coverage. The overall national trend has been to interpret the employment relationship broadly, in the interest of bringing as many workers as possible under the workers' compensation umbrella.

The employee must receive some form of remuneration and the employer some benefit from the services performed. The employer must possess the right to control and direct the employee's activities, not only as to result but also as to method (what they want done and how to do it).

Bona fide independent contractors are not employees. There are many tests for independent contractor status, but the primary test is control. The surest method to protect against this exposure is to require certificates of insurance from all independent contractors. Usually, individual independent contractors cannot purchase coverage for themselves.

When the carrier cannot determine independent contractor status, it will probably charge a premium for this exposure, usually a percentage of the contract cost.

AOE/COE

Workers' compensation laws apply only to injuries or illnesses that arise out of employment (AOE) and, in many states, occur in the course of employment (COE). The AOE concept relates to the causal factors, e.g., whether the risk of injury arose from the employment itself rather than nonemployment factors personal to the worker. The COE concept relates to the time, place, and circumstances in which the injury occurred. These concepts are far broader in scope than those of civil tort law, which requires foreseeability and proximate causation; the result is that most injuries and illnesses that have some direct or indirect relationship to the workplace are compensable.

Other forms of insurance provide coverage for off-the-job injuries and illnesses, for example, accident and health, state disability, income protection plans, private short-term and long-term disability plans, unemployment insurance, and Social Security.

OCCUPATIONAL DISEASE

Occupational disease is rapidly becoming a major coverage and benefit concern to employees and employers. Most states allow for this coverage, provided the disease was caused or aggravated by the employment. Some states have specific lists of diseases that are covered but most include diseases that are related to employment.

Occupational disease can be caused by toxic materials, radiation, heat, vibration, noise, air contaminants, and skin irritants, all of which can result in adverse effects on the human body. Because of the significant latency period for some occupational diseases, they may not be detected for 20–30 years after exposure.

Most workers' compensation carriers provide to policyholders occupational health services that test, monitor, and advise on control of hazards and exposures. The concept of cumulative injury is recognized by many jurisdictions. Similar to occupational disease in that it occurs over time, a cumulative injury results from repetitive activities, the ultimate effect of which is a condition causing disability and need for treatment. As with occupational diseases, cumulative injury liability may be spread among two or more employers and carriers.

POLICY EXCLUSIONS

Most policies have standard exclusions, which generally pertain to job type, multiple policies, contractual liability, punitive damages, statute of limitations, and workers' compensation claims.

- *Employment*—Some jobs are excluded, and some are covered on a voluntary basis. These exclusions vary by state but frequently include domestic workers, children employed by parents, voluntary service, ski patrols, and, in a few states, farm employees.
- *Other insurance*—To avoid duplicate coverage, carriers exclude coverage when there are multiple locations and separate policies were purchased for the other locations or when other locations are self-insured.
- *Contractual liability*—Contractual liability exclusions apply solely to employers' liability. The policy states that liability assumed by the employer under any contract is not covered. This coverage can be specifically provided under the contractual liability part of a general liability policy.
- *Punitive damages*—Any penalty or punitive damage assessed against an employer for violations of law is not insured under the workers' compensation policy. Examples would be (a) fines and penalties for illegal employment of a minor; (b) failure to insure; (c) failure to report an injury; or (d) serious and willful misconduct.
- *Statute of limitations*—No coverage is provided for claims presented after the statute of limitations has expired—usually a period of five years after an injury was discovered and three years for employers' liability.
- *Workers' compensation exclusion*—Under employers' liability, coverage is excluded for awards obtained under workers' compensation or occupational disease laws, unemployment insurance, disability benefits laws, or any other similar law.

BENEFITS

The level and scope of benefits vary significantly from state to state, with each legislature determining the appropriate limits. Many factors contribute to the calculation of benefits, including the quality and makeup of the labor force, wage levels, and often most importantly the relative bargaining power of important interest groups, such as organized labor, employers, medical providers, insurance carriers, and attorneys. It is a sad but undeniable fact of life that benefit levels and scope are frequently determined more by the political influence of various lobbying interests than by the real needs of the injured worker.

Workers' compensation laws place a definite but limited liability on the employer for employee disabilities caused in the workplace. Benefits payable to the disabled workers, under the law, attempt to compensate for part of the economic loss of earnings and extra expenses associated with the injury. The types of benefits provided are income or cash benefits, medical benefits, rehabilitation benefits, and death benefits.

Income or cash benefits replace an injured employee's loss of income or supplement his or her reduced earning capacity. These benefits are provided for both physical impairments and disability. Disability benefits are available whenever there is an impairment and a wage loss.

Medical benefits are provided for injured workers without dollar or time limitation. Most states use a medical fee schedule, which is somewhat less than the customary charge for compensating industrial injuries and illnesses. Physicians and pharmacies are obligated to accept the fee schedule as payment in full. Insurance carriers have been known to make exceptions in the case of specialists and pay them their usual and customary fees. Medical benefits include all medical, surgical, and hospital treatment, including medicine, prescription drugs, medical and surgical supplies, crutches, and other apparatus including orthopedic replacements.

In some states, the injured has a free choice of physician; in others the employer can designate the physician. Some states allow a selection from a panel, and some allow the employee or the employer to request or honor a change of physician.

Vocational rehabilitation benefits are provided for those cases involving severe disabilities and include both medical rehabilitation and vocational rehabilitation. This distinction was not made until vocational rehabilitation came on the scene. As a matter of practice, they are provided in all states even when not specified by statute. The provisions of the benefits vary significantly between states. As a general rule, benefits are available to a worker whose industrial injury or illness permanently precludes him or her from engaging in the former occupation or job in which he or she was injured. Benefits can include temporary disability while undergoing rehabilitation, medical treatment, living expense, vocational and psychological counseling, or retraining and job placement assistance. The goal of a rehabilitation plan is to return the injured to suitable, gainful employment, consistent with physical limitations, and with the minimum possible wage loss. Death benefits are provided to dependents or, in some cases, to the state if no dependents exist.

DISABILITY CLASSIFICATIONS

There are different classifications of disability: *permanent total disability; temporary total disability; permanent partial disability;* and *temporary partial disability*. These classifications are covered in the following sections.

Permanent Total Disability

When the injured is *totally disabled* (*not able to do any kind of work*) and *will not improve regardless of treatment,* he or she is classified as totally disabled. For this category, the benefit in most states is two-thirds of average wages, subject to a minimum and a maximum. In most states, the benefit is for the duration of the injured person's life. Some states provide additional amounts for dependents. A few provide permanent total disability benefits for a maximum number of weeks. For example, Indiana provides a benefit for 500 weeks with an additional 150-week benefit from a second injury fund. An example of a permanent total disability would be loss of sight in both eyes. In certain states, such as California, once the facts are established giving rise to permanent total disability, the benefits are payable accordingly, even though the worker is able to do some work, on the theory that the employee has suffered total inability to compete for employment in an open labor market, i.e., against workers similarly situated but without the impairment.

Temporary Total Disability

Most injuries fall into the *temporary disability category*. In this case, the injured is expected to recover but is temporarily unable to work. The benefit is payable, subject to a minimum and a maximum and based on the average wage, until the person is able to return to work. Benefits begin after a waiting period, and the waiting period is retroactively covered after a certain number of days off work have elapsed. The major distinction between temporary and permanent disability is that the former benefit is payable during the period of convalescence, and represents compensation for wage loss during that time, while entitlement to the latter only arises after the worker's condition becomes relatively stable and the extent of his or her long-term impairment is ascertainable.

Permanent Partial Disability

A *permanent partial disability* causes only a partial reduction in the ability to compete for employment. The injured may even be able to perform the customary job, but earning capacity and ability to compete in the open labor market are impaired. These benefits are designed to compensate for the effect of the injury on the future earning capacity and are calculated on a scheduled or nonscheduled basis.

- Scheduled injuries are compensated for, according to a specific schedule. The schedule is fixed by statute and the benefit is paid whether the injured party is working or not. Most schedules provide payment over a set number of weeks (based on the body part affected), multiplied by the weekly benefit amount (based on a percentage of earnings at the time of injury), and subject to a minimum and a maximum.
- Nonscheduled injuries are compensated for on the basis of a replacement of a percentage of the wage loss. The percentage used is the difference between wages earned before and after the injury.

Temporary Partial Disability

Temporary partial disability means the worker is convalescing, but still capable of working. However, the employee can perform only part of the usual duties of the job until recovery is complete. This type of injury is not common. The injured would be compensated a percentage of the difference between the previous and current wages.

Death Benefits or Survivors' Benefits

Benefits are paid to the dependent spouse or children if an employee is killed on the job. If there are no immediate total dependents, benefits can be paid to partial dependents such as mother, father, grandchildren, brothers, sisters, or others who were receiving support from the deceased employee. Currently, there are significant differences in the amounts and methods of payment among states.

Social Security Disability Benefits

The Social Security Act provides monthly benefits that can supplement workers' compensation, to those under age 65 years, who are disabled beyond the required waiting period, which is five months after an application has been filed. Benefits are calculated as though the injured had reached the age of 62 years. Under the act, the term disability means inability to engage in any form of gainful employment, and the disability is expected to last for a period of not less than 12 months or to result in premature death.

Combined Social Security disability and workers' compensation benefits cannot exceed 80% of the *average current earnings* before the disability. Social Security benefits are reduced from what they would otherwise be so that the combined workers' compensation benefit and the reduced Social Security benefit equal 80%. In some states, workers' compensation benefits are reduced by all or part of Social Security benefits. This is known as a reverse offset.

Subsequent (Second) Injury Funds

A preexisting injury can combine with a second injury to produce a disability greater than the second injury alone. For example, a worker loses a hand in an industrial accident and later suffers loss of the other hand, becoming totally disabled.

The amount of workers' compensation benefits is considerably increased when the injured receives benefits for the combined disabilities. This potential increase in the cost of compensation benefits can be a financial burden for an employer. The subsequent (second) injury funds were established to solve this problem. The employer or insurance carrier pays compensation for the second injury alone, even though the employee receives benefits relating to the combined disability. The difference is made up from the subsequent (second) injury fund of the state. Most states, but not all, currently have subsequent (second) injury funds. In the states that do not have such funds, employers would be liable for the combined benefits.

COMPULSORY OR ELECTIVE COVERAGE

When workers' compensation laws first came into being, the employer had the choice of coming under the act or not. By purchasing coverage, the employer would elect to come within the law. The advantage was that workers' compensation then became the exclusive remedy for compensation of the injured with benefits set by statute. But the employer gave up many of the common law defenses that were used in tort laws at that time.

Similarly, in some states, employees could reject the act by filing a rejection at the time of hire or before an accident. Since workers' compensation is a no-fault system, the employee was certain to collect benefits if injured. Therefore it was generally to his or her advantage to accept the provisions of the act. If the employee rejected the act, the employer retained the common law defense. The basic reason for elective coverage was to avoid the constitutional issue that the employer might be deprived of the right of due process.

Under compulsory laws, the employer must provide coverage and will be penalized for failing to do so. Moreover, the employee cannot reject benefits and sue in tort.

Employment of Minors

Because of possible penalties and criminal sanctions, every employer should know the child labor laws in his or her state. Moreover, all employers should verify the age of any minors hired. The hiring of minors is affected both by labor laws and workers' compensation law. Labor laws spell out the type of work that minors can perform. None of the states allow minors to work on machinery with dangerous moving parts or in hazardous occupations. Employers who do not abide by these laws are subject to fine and possible imprisonment. If a minor is injured on the job, compensation laws provide for uninsurable penalties against the employer. Some states provide unlimited double indemnity, which means the employer must match what is paid by the carrier. Others, such as New Jersey, give the injured minor the right to sue for civil damage in addition to collecting workers' compensation benefits.

Injury Outside the Jurisdiction

As a general rule, when an employee is hired in one state and generally works in that state, but is injured in another, he or she has a choice of where to file a claim. Employees usually select the state that provides the greatest benefits. Although some states will not normally accept jurisdiction, unless some work is performed in the state where the contract of hire was made, in others the making of the employment contract in that state is enough to confer jurisdiction, regardless of where the work is performed.

When the contract of hire is in one jurisdiction and the employee is injured in an area of federal jurisdiction, the employer may have a liability under either the state or federal jurisdiction. An injured employee cannot make a double recovery by filing more than one jurisdiction. However, the employer, not the carrier, can be liable for benefits in excess of those provided in the state of hire, when coverage was not purchased in the state of injury.

Failure to Insure

In those states where workers' compensation is compulsory, failure to insure, or to qualify as a self-insured, will result in an uninsurable penalty (fine) and possible criminal charges. Moreover, the employers' operations can be stopped until coverage is secured. Some states provide an Uninsured Employers' Fund from which the injured can collect benefits and still bring an action against the employer for penalties and attorney fees. Additionally, the Uninsured Employers' Fund has a right to recover benefits paid the injured from the uninsured employer.

ADMINISTRATION

Administration of a program encompasses many aspects of an employer's responsibilities. These include arranging for and providing initial medical treatment for any work-related injury. Employer control over medical treatment varies from state to state. Injuries must be reported accurately and in a timely manner; and employees who have been injured and have filed a claim must be protected from any sort of discrimination. These topics are covered in the next sections.

Industrial/Medical Facilities

It is the primary responsibility of the employer to arrange for and provide initial medical treatment for any work-related injury. This duty includes all medical, surgical, nursing, and hospital care reasonably required to cure or relieve the effects of the injury. This responsibility can also include therapy and various other healing techniques. The employee is not required to ask for medical treatment, but to notify the employer of the injury. It is the employer's responsibility to provide or offer treatment. Given this responsibility, the employer would be well advised to make arrangements with a quality medical facility, in advance, to accept injured employees. The facility should be within a reasonable distance from the worksite and should be able to accommodate emergencies.

The employer should meet with the administrator of the facility to discuss the types of work done by employees (skills, physical requirements), the types of injuries expected, and the opportunities for light-duty work following an injury. Job descriptions and duties also should be provided to the facility. The physician who will treat injured employees should tour the work location to become familiar with the exact job requirements.

Initial and long-term control over medical treatment varies from state to state. Some jurisdictions give the employer or carrier complete say over who shall treat, subject only to a "reasonableness" standard. In other jurisdictions, the worker is given free choice as to the treating entity, generally after an initial brief period of employer control. In the latter states it is more difficult for the employer or carrier to conduct an ongoing dialogue with the provider of medical services; but since there remains some measure of control through the employer's payment of the costs, it is still worthwhile and often effective for the employer or carrier to acquaint the health care provider with facts about the workplace. Whatever the local law may be regarding control, it is always desirable and frequently essential for full and accurate job descriptions to be made available to the medical provider. A physician or facility unfamiliar with the needs of the particular job cannot comment intelligently on whether vocational rehabilitation is needed.

Reporting of Injuries

The employer must report injuries (within certain time limits) to the carrier and usually to the state agency having jurisdiction. The time limits for reporting vary from immediately to 30 days. Failure to report can result in fines and, in some cases, imprisonment. The carrier will usually file on behalf of the employer with the required state agency.

In some states, it is not necessary to report first aid cases treated in-house or by a physician, provided that the injury did not result in lost time. Not all state compensation laws require record keeping on injuries. OSHA, however, does require record keeping on all work injuries. The same form used to report the injury to the carrier will suffice for OSHA.

The employer's first report to the carrier is confidential and in most states is not admissible in the appeals process. The treating physician is also required to make an initial report; however, that report is admissible.

Discrimination Charges

Most states have a provision to protect the injured employee from discharge or harassment because of filing a claim or testifying in support of another employee filing a claim. Discharge of an employee for these causes will almost always result in a litigated claim that will probably cost the employer more than an unlitigated claim. In some cases, uninsurable criminal penalties and fines can be charged against the employer, and the employee can also bring a wrongful discharge suit. If a terminated employee's case is upheld, the employee will most likely be reinstated with back wages.

The law is not well settled as to whether an injury (generally psychiatric) caused solely by job termination is itself compensable under the workers' compensation laws, since arguably such trauma does not arise during the employment's existence and therefore does not meet the COE test. The prevailing view is that the worker's remedy in this kind of case must be sought through grievance and labor arbitration procedures and civil suits for unlawful termination, rather than through workers' compensation.

THE CLAIMS PROCESS

Unlike nonindustrial personal injury claims, which usually are resolved after filing of lawsuits, the workers' compensation program is designed to be self-executing, reporting of the injury serving as the only trigger for payment of benefits. Once an injury report is filed with the carrier, it will either accept the claim, reject the claim, or investigate to make a determination. If the claim is accepted, medical payments begin and are usually paid 100% by the carrier with no limits as to total payments paid. In some cases, fee schedules are set by state legislatures. Disability benefits are paid by a scheduled figure based on earnings at the time of injury after the required waiting period.

Permanent disability is determined by medical evaluation without a pain and suffering concept, but rather with a medical determination of the known and/or subjective disability that affects the injured's work performance. The commission, board, or agency having jurisdiction makes a final decision, and payment is made either as a weekly benefit for a prescribed period of time or in a lump sum settlement, usually indicating a final conclusion to the claim and eliminating the possibility of further recourse on the part of the injured worker. If a claim is rejected, usually over an issue of AOE/COE (previously described), the injury can be covered by some other form of insurance. If the injured believes that he or she has a right to workers' compensation benefits, he or she can appeal to the board, commission, or agency having jurisdiction. When a claim is investigated by a carrier, it is usually to obtain enough data to make a determination or to evaluate the possibility of a third-party subrogation action.

As a general rule, the rank-and-file worker does not understand the workers' compensation system. When injured, employees frequently think that they must sue their employer to recover benefits. They do not understand that workers' compensation is a no-fault system and that they are due benefits, but they generally have no idea what benefits are due them. Usually, they think the employer is upset with them because they had an injury, and they feel guilty, unsure, and worried about financial matters. A visit by an employer or his or her representative to the hospital, to the home, or even a phone call to explain that the employer is not upset and wants the worker back as soon as possible and to explain benefits that the injured can expect will frequently accelerate the recovery, reduce the cost of the claim, and minimize litigation.

In its *Report to the Industry,* the California Workers' Compensation Institute concluded that most people seek legal assistance because they do not understand the system and the benefits due them. It would seem that by simple communication, employee needs would be better served and the cost of workers' compensation would be reduced.

Each year following the expiration of a policy, carriers must file with the treating organization a report of payroll and claims costs by classification. This information is used to promulgate industry rates and the experience modification (credit or debit charged against the premium) for the individual account. Prior to filing that report, the employer may wish to review all open claims with the carrier as to status and reserves because, once those figures are filed, they are frequently irreversible.

FINANCIAL INCENTIVES

One of the basic objectives of the workers' compensation system is to encourage safety in the workplace. This objective can be achieved by a number of financial incentives (ability to reduce the ultimate cost), primarily by preventing and controlling occupational injuries and illnesses. For self-insureds, the incentive is reduced claim payments and costs and, for insureds, it is reduced net premiums. To the extent that the cost of prevention is perceived as cost-effective, management will usually respond positively. The larger the potential savings, the greater the potential preventive effort.

Cost Incentives of Self-Insureds

There is a direct one-to-one relationship between the reduction of costs and preventive effort. For every injury prevented, the self-insured employer saves the costs of such benefits as medical, indemnity, claims handling and legal fees, permanent disability, and rehabilitation, plus the loss of use of reserves and the tax credit on those reserves.

Cost Incentives of Insureds

Experience rating, retrospective rating, dividend plan, and cash-flow plans are among the cost incentives for insureds. These are covered in the following sections.

Experience rating

Experience rating is one of the most influential cost incentives for insureds. It can be an up-front discount or surcharge for the individual employer based on past loss performance. Losses for an entire category of insureds can be predicted with a reasonable degree of accuracy. However, it is more difficult to forecast or predict losses for the individual risk.

Experience rating attempts to predict the losses of the individual risk by comparing the individual insured to a homogeneous group on a statewide basis. If the experience of the individual risk is better than average for a group, insureds will receive a credit rating, or if worse, a debit rating. The rating, expressed as a percentage, is applied to the "manual premium." Some modifications can be as low as 30%, some as high as 300%.

Not all insureds qualify for experience rating. Loss experience for smaller firms is not statistically reliable for forecasting purposes. There is usually a set premium level to qualify and not a sufficient premium to cause a financial incentive for smaller insureds. These insureds use manual rates without a modification applied.

Because of the formulas used to calculate the modification, the very large account can become self-rating. Their losses are not discounted in the calculation, and the effect is similar to that of self-insurance. In effect, they pay 100% of their losses in their future projections. For those insureds, loss experience has a significant impact on premiums and that becomes a strong financial incentive. As a general rule, for any experience-rated insured, lower losses will ultimately mean lower premiums. Three years of past payrolls and losses are used in the calculation, excluding the most recent expired policy year. The net result is that good or bad year(s) will have an impact on the calculations for three future years. The financial incentive of experience rating often exceeds the financial incentive of self-insurance, and self-insureds should study each carefully.

Retrospective rating

There are a number of different retrospective rating plans approved by the various insurance departments. Basically, a retrospective rating plan is a contractual agreement whereby the insured's premium includes carrier expense and the insured's own losses and loss adjustment expenses and is subject to a minimum and maximum premium. With excellent experience, the minimum will be less than the standard premium. However, with poor experience, the maximum is charged and the premium will be higher than the standard premium. Standard premium is calculated as the manual premium multiplied by the experience modification factor.

Again, the financial incentive is greater for the larger insured, and to the extent that this insured can reduce injuries, it can save money. With a retrospective rating plan, premiums usually must be adjusted several times after the policy has expired before the true cost can be determined.

Dividend plan

A dividend plan is basically a statement of intent by the insurance carrier to allow the insured to participate in a refund of excess profits, if earned, from the premium they have paid the carrier based on that insured's own losses. Dividends are paid from accumulated earned surplus (profits), must be declared by the board of directors, and cannot be guaranteed. To a large extent, dividends are declared at the discretion of the carrier and based to some extent on the carrier's overall profitability as well. It is hard to justify dividends when a carrier is losing money. Dividends can be significant financial incentive, and are materially affected by the insured's own loss results.

Cash-flow plans

There are a number of cash-flow plans in existence today. Some are directly related to paid losses, and others to deferred premium payments. All have built-in financial incentives to reduce injuries and illnesses (previously discussed).

Cost Relationships

Historically, management made its business decisions in terms of "the bottom line," in other words, the effect on its profits. There are some truly significant relationships between workers' compensation costs and profits. Once some of these relationships are realized, management will tend to be more responsive to its safety priorities. The Table shows two identical examples in terms of manual premium, gross sales, and planned net profits to illustrate some of these relationships.

Table. Impact of Experience Modification on Calculation of Insurance Premiums.

Manual premium		$100,000
Annual sales		$5,000,000
Net profit (planned), percent		5
Hours worked		200,000
	Company A	Company B
Experience modification	0.50	1.50
Standard premium	$50,000	$150,000
Variance	$ (50,000)	$+50,000
Percent of sales	1.0	3.0
Required sales	$1,000,000	$3,000,000

For this example, the same net profit percentage was used for both companies. Actually, the net profit for Company B would have been reduced by the standard premium differential. It would also have increased the allocation to sales, the cost per hour, and required sales to support that cost. The required sales are calculated by dividing the cost of insurance by the net profits (Cost/Profit = Required Sales). From required sales, the number of units, given the cost per unit, and/or the months of production required to support the cost of the insurance can be projected.

SUMMARY

Workers' compensation represents a significant business expense for employers, in some cases, 20%–30% of payroll. The system, however, is critical to the economic well-being of the injured worker. There is a clear trend of increased benefits to the injured and increased cost to the employer, which are passed on to consumers, in terms of higher priced products. Only one thing is certain: workers' compensation will cost more, not less, in the future.

To a large extent, employers can influence their net costs for this coverage, positively or negatively, through their own efforts. There are sufficient cost incentives in the system to motivate management positively in that direction. The enormity of the cost and the means of controlling these costs must be understood, accepted, and applied. Management is the key. When health and safety becomes a high enough priority for top managers to pay attention to it, they, their employees, and the public will realize the direct benefits.

REFERENCES

Analysis of Workers' Compensation Laws, 1982. Washington DC: U.S. Chamber of Commerce, 1982.

Grimaldi JV, Simmons RH. *Safety Management,* 3rd ed. Homewood IL: Irwin, 1975.

Hanna W. *California Law of Employee Injuries and Worker's Compensation,* rev. 2nd ed. New York: Matthew Bender, 1999.

Malecki D, Donaldson J, Horn R. *Commercial Liability Risk Management and Insurance,* vol. 1. Malvern PA: American Institute for Property and Liability Underwriters, 1978.

The Producer's Guide to Workers' Compensation. Encino CA: Zenith Insurance Co., 1980.

Tarrants WE. *The Measurement of Safety Performance.* New York: Garland STPM, 1980.

Witt. *California Workers' Compensation Reporter.* Berkeley, February 1998.

Zenz C. *Occupational Medicine: Principles and Practical Applications,* 2nd ed. Chicago: Yearbook, 1986.

chapter 15

Travel Health and Remote Work Programs

by Stephen Burastero, MD, MPH

249 **Introduction**
251 **Trip Preparation**
251 **Individual Pretravel Arrangements**
Consultation ■ Traveler's health history ■ Traveler's medical kit ■ Insurance
252 **Immunizations**
Required immunizations ■ Recommended immunizations
260 **Motion Sickness**
261 **Jet Lag**
Symptoms ■ Treatment of jet lag
262 **Traveler's Diarrhea**
Preventing traveler's diarrhea ■ Treating traveler's diarrhea
265 **Malaria**
Preventing malaria ■ Prophylactic medications ■ Protection against mosquitoes
269 **Sexually Transmitted and Bloodborne Diseases**
AIDS, HIV infection, and bloodborne hepatitis ■ HIV-infected travelers
271 **Skin Conditions**
Sunburn and other problems of sun exposure ■ Fungal infections ■ Other skin conditions
272 **Environmental Hazards**
Air pollution ■ Heat illness and its prevention ■ Altitude sickness ■ Toxic hazards ■ Swimming and bathing hazards ■ Injuries ■ Crime
274 **Mental Health and Substance Abuse Abroad**
Alcohol hazards ■ Consequences of illicit drug use abroad ■ The recovering addict and travel
275 **International Air Evacuation**
275 **More Information on Travel Health**
275 **Summary**
276 **References**

INTRODUCTION

Today, international travel is as common as it is important. Estimates indicate that in the late 1990s more than 500 million persons traveled abroad. This figure will increase as corporations and other institutions continue to globalize their operations. Even now, a large segment of U.S. business involves frequent and/or extensive overseas travel.

As common as foreign travel is, it also involves significant health risks. One person in four traveling abroad experiences at least one illness or injury that will interfere with his or her itinerary (Steffen, 1988). Some risks are especially high in remote, tropical, or developing countries, where currently more than 20 million U.S. residents travel annually (Jong & McMullen, 1992). The relative frequency of different kinds of health problems among visitors to the developing world is summarized in Figure 15-1.

Most travelers' diseases result from the immersion of a poorly prepared foreign visitor into an environment where inadequate public health and sanitation are the norm and infectious diseases are prevalent. Travelers may not be prepared for the extremes of heat and sunlight common in the Tropics. In many areas, rapid population growth and haphazard development have resulted in severe environmental pollution, traffic congestion, crime, and political turbulence; these conditions too can present hazards to foreign visitors.

The science that developed to address such risks is *emporiatrics* (derived from the Greek word *emporos*, meaning shipboard passenger).

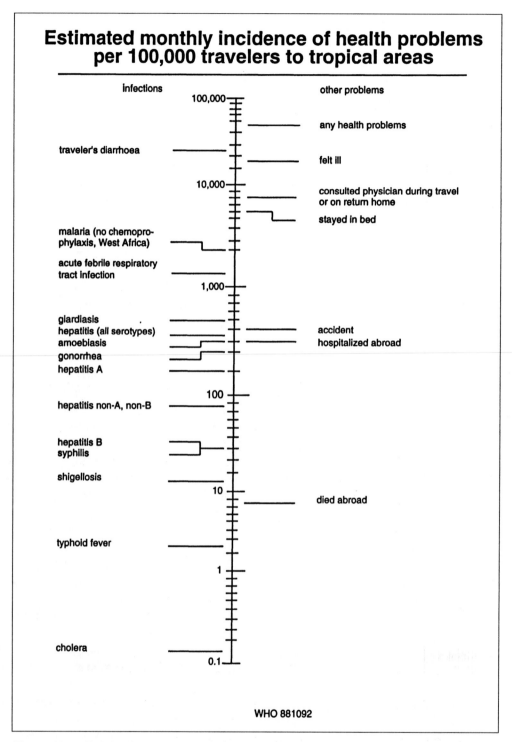

Figure 15-1. Relative frequency of different kinds of health problems among visitors to the Tropics. (Reprinted with permission from Steffen R. *Proceedings of the Conference on International Travel Medicine, Zurich 1988.* Berlin: Springer-Verlag, 1988.)

This is also known as travel medicine. Emporiatrics originated with early attempts by European nation states to control outbreaks of tropical diseases in their colonial empires. Since that time, national governments and public organizations such as the World Health Organization (WHO) and the U.S. Centers for Disease Control and Prevention (CDC) have made great strides in controlling the importation and spread of infectious diseases.

During the global travel boom of recent years, *emporiatrics* has become increasingly important. Specialized travel clinics, often university-affiliated, have proliferated throughout the United States. These clinics offer a range of preventive health advice to prospective overseas travelers. Many of these clinics rely on computer links that provide up-to-date news of disease outbreaks and changing health conditions around the world.

Developments in emporiatrics have made most travelers' diseases well understood and preventable. Unfortunately, travelers often neglect to take the precautionary health measures available to them before, during, and after they travel (Lobel et al, 1987). These travelers are most likely to fall ill during a trip.

Not only business travelers, but also the managers who plan or oversee the travel of others, need to be well informed of the risks and precautions involved in travel to developing countries. This knowledge will ensure the most fulfilling and productive work abroad.

TRIP PREPARATION

When planning health needs for an overseas project, the project manager should give serious consideration to the following topics:

- geographic location of the site(s)
- the mental and physical job demands of the assignment(s).

A rigorous, remote or long-term assignment may necessitate a fitness-for-duty examination by a physician to qualify the candidate for an assignment. A written description of job demands will make the decision to qualify a candidate for overseas work. This examination may include, but not be limited to, a comprehensive physical examination, laboratory tests, a dental evaluation, TB skin testing, etc. Chronic medical conditions should be under optimal medical control. A history of uncontrolled mental illness or substance abuse may also be a relative contraindication to approval. Multiple risk factors for coronary artery disease (e.g., diabetes, hypertension, smoking, etc.) should be reviewed by the evaluation physician, and a treadmill ECG considered.

INDIVIDUAL PRETRAVEL ARRANGEMENTS

Pretrip arrangements are necessary to ensure the traveler's safety and health. Travelers must determine which inoculations, if any, are required and should have thorough physical and dental examinations before departing on an extended trip. Bringing a complete health history and traveler's medical kit as well as checking on health insurance coverage while traveling are also recommended.

Consultation

Begin planning a trip about six to eight weeks before departure. Consult with a private physician, health department, or travel clinic to obtain the most recent information available on the health hazards and precautions peculiar to the area to which you or others are traveling. Do not rely on health advice from travel agents. Travelers who rely solely on health information from travel agents are often misinformed about serious health hazards such as malaria.

If you plan an extended stay overseas (e.g., greater than four to six months), have a physical and dental examination, along with laboratory screening tests, blood-typing, and TB skin tests. Insurance plans or corporate policy can dictate the nature of these examinations. To obtain the best possible advice, tell your health practitioner where you are going, how long you will be there, and what kind of work you will be doing. Women who are pregnant or who plan to become pregnant should convey this information to their physician.

Traveler's Health History

The traveler's health history is highly recommended for any visit to the developing world (Table 15-A). Health professionals can help you organize key health information clearly and concisely. A summary of up-to-date immunization records, current medications and medical problems, and known drug allergies can be compiled on one easily accessible page. A traveler's health history should also include the name and telephone number of the person's regular physician and a person to notify in the United States in case of serious illness or injury. Female travelers should also include their most recent menstrual history and indicate if they are pregnant.

Traveler's Medical Kit

Many experts recommend a traveler's medical kit for trips overseas. The kit contains medications for traveler's diarrhea and malaria, as well as over-the-counter remedies that can be useful abroad. The traveler should include an ample supply of any other prescription medications currently prescribed

Table 15-A. Traveler's Health History*

International travelers should assemble the following information in a concise and clearly written form to carry with them:

1. An up-to-date immunization record (preferably the *International Certificates of Vaccination*).
2. A list of current medications by both trade name and generic name as well as the actual dose.
3. A list of all medical problems, such as hypertension, diabetes, and heart disease (cardiac patients should carry a copy of the most recent electrocardiogram).
4. A list of known drug allergies.
5. ABO blood type and Rh factor type.
6. Name and telephone number (and telefax number, if available) of his or her regular doctor.
7. Name and telephone number of the closest relative or friend in the United States who might assist if the traveler incurs serious illness while out of the country.

* Reprinted with permission from *The Travel & Tropical Medicine Manual*, 2nd edition, edited by Elaine C. Jong, MD, and Russell McMullen, MD. Philadelphia: WB Saunders Company, Copyright 1995.

for chronic conditions, along with a doctor's note stating their intended use. This will prevent unexpected delays when passing through customs. See Table 15-B for a comprehensive list of drugs that may be included in the traveler's medical kit.

Many pharmacies in the developing world do not carry the medications and/or sundries that U.S. citizens are used to; for example, contact lens solution, contraceptives, sanitary napkins, dental floss, insect repellent, or sunscreen. Familiar over-the-counter drugs may be unavailable. Sometimes toxic or narcotic drugs that have been banned in the United States are dispensed freely by developing-world pharmacists without a prescription. Medications in the developing world have not been subject to quality controls like those applied in the United States and should be avoided.

Insurance

Carefully check insurance plans for coverage of sickness or injuries suffered while abroad. Some health insurance companies will ask you to take along special forms to make it easier for you to be reimbursed for health care you might receive abroad.

American embassies or organizations like the International Association for Medical Assistance to Travelers (736 Center Street, Lewiston, NY 14092; 716/754-4883) keep lists of well-qualified English-speaking doctors abroad. Some companies offer direct insurance payments or air ambulance services, or both (e.g., International SOS, Box 11568, Philadelphia, PA 19616; 800/523-8930, which caters to both corporate clients and individuals).

IMMUNIZATIONS

Immunizations are absolutely essential in preparing for world travel. With the aid of health professionals (e.g., travel clinic or health department), carefully plan the vaccination schedule at least six to eight weeks before departure. Do not rely on information from travel agents or foreign embassies about immunization requirements. These sources often quote only compulsory vaccine requirements from outdated travel guides.

No vaccine is 100% effective. Therefore, vaccines are not a substitute for proper attention to food, water, and insect control. These topics will be addressed later in the chapter. Vaccines work by introducing a small amount of bacteria or virus into the body through injection. This antigenic material stimulates the immune system to produce protective antibodies that make the person less susceptible to infection. Occasionally, boosters can be administered to bolster preexisting vaccine-induced immunity.

Live vaccines (e.g., measles and yellow fever vaccines) contain live microorganisms able to stimulate an immune response, but the microorganisms are altered so that they are not virulent enough to cause disease in normal persons. Because live vaccines can cause illness in those who have debilitated immune systems (e.g., persons on immunosuppressive drug therapy, radiation therapy, or persons with the HIV virus), they should not be given routinely to these individuals without careful medical consultation. Pregnant and nursing women also should avoid live vaccines because of the risk of infecting the fetus or infant. Travelers with a history of skin allergies, egg intolerance, or prior severe reactions to immunizations are at higher risk of adverse effects and should seek medical advice before receiving any vaccines.

It is important to distinguish between required and recommended vaccines. Required vaccines are mandatory for legal entry into some nations. Recommended vaccines, in contrast, are medically indicated, based on itinerary, prior individual immune status, and reproductive status. Anticipated occupational exposures (e.g., contact with mammals, insect vectors, or blood products) are crucial in determining recommended vaccinations. From a health perspective, recommended vaccines

Table 15-B. The Traveler's Personal Medical Kit.*

It is wise traveler who is prepared ahead of time for unexpected emergencies that can arise. Below are listed some of the suggested items for the traveler's personal medical kit. Not all of these items are necessary or appropriate for every traveler: items should be selected based on the style of travel and destination(s).

PRESCRIPTION ITEMS

Antibiotics, General: Antibiotics may be useful for travelers at risk for skin infections, upper respiratory infections (URIs), and/or urinary tract infections (UTIs).

Skin infections:
- Dicloxacillin, 250-mg capsule. Two PO q 6 h x 7 days; or
- Cephalexin, (or cephradine) 500-mg capsule. One PO q 6 h x 7 days
- Mupirocin 2% topical antibiotic ointment. 15-gm tube or 1-gm foil packets. Apply to infected skin lesions three times a day.

Upper respiratory tract infections:
- Erythromycin, 250-mg tablet. One PO q 6 h x 7 days; or
- Trimethoprim 160-mg/sulfamethoxazole 800-mg double-strength tablet (TMP/SMX DS). One PO q 12 hr x 7 days; or
- Doxycycline, 100-mg capsule. One PO q 12 hr x 7 days; or
- Azithromycin, 250-mg tablet. Two PO first dose, followed by one PO q 24 h x 4 additional days.

Urinary tract infections (uncomplicated):
- TMP/SMX DS tablet: One PO q 12 h x 3 days; or
- Norfloxacin, 400-mg tablet: One PO q 12 h x 3 days.

Multipurpose Antibiotics: These would provide empiric coverage for skin infections, URI, and UTI. (Note that ciproflaxacin and ofloxacin are used for treatment of traveler's diarrhea as well):
- Amoxicillin, 500-mg plus clavulanate, 125-mg tablet (Augmentin, 500 mg). One PO q 8 h x 3–7 days; or
- Ciprofloxacin, 500-mg tablet. One PO q 12 h x 3–7 days; or
- Ofloxacin, 300-mg tablet. One PO q 12 h x 3–7 days.

Allergic Reactions: (To bee, wasp, yellow jacket, or hornet stings; food, etc).

Epi Pen emergency injection of epinephrine:
Use according to package directions for severe reaction to bee sting or for other allergic reaction causing shortness of breath or wheezing; or swelling of the lips, eyes, throat, or severe hives. This will give short acting relief. As soon as the afflicted person can swallow, give them Benadryl tablets as directed below.

Benadryl (diphenhydramine), 25-mg tablet:
Take two tablets by mouth immediately, then one to two tabs q 6 h x 2 days following a allergic reaction. Use Benadryl alone for mild to moderate allergic skin reactions and itching, and take Benadryl following the use of the EpiPen.

Medrol (methylprednisolone) DosePack:
For use with severe and persistent allergic reactions or skin rashes. Follow the instructions for the tapering dose schedule in the packet. May be required in addition to Benadryl for severe allergic reactions.

Ventolin (albuterol) inhaler (MDI, multidose inhaler):
Use for asthma attacks or for allergic reactions that cause persistent wheezing. Two puffs 2 minutes apart, each inhaled as deeply as possible into the lungs. Do this four times a day.

Cough Suppressant: A small bottle of prescription cough syrup or a few tablets of codeine-containing medication (Tylenol #3 tablets will serve this purpose and that of medication for severe headache or pain).

Diarrhea Treatment: An antimotility drug (Lomotil or Imodium) plus an antibiotic (e.g., trimethoprim/sulfamethoxazole, ciprofloxacin, ofloxacin, tetracycline) may be prescribed for self-treatment.

High-Altitude Illness: Acetazolamide (Diamox) may be prescribed for prophylaxis of high-altitude illness for high-altitude destinations.

Jet Lag: In some cases, a short-acting sleeping medication is helpful in treating sleeping problems associated with jet lag.

Malaria Pills: As the malaria situation in many countries continues to change, malaria chemoprophylaxis changes as well. Updated information on the malaria situation for specific destination(s) needs to be carefully reviewed, and appropriate medications prescribed.

Motion Sickness: Travelers who experience motion sickness may be prescribed a medication for this.

Nausea and Vomiting: Compazine (prochloperazine), 25-mg rectal suppository. This may be helpful when oral medications cannot be tolerated and an injectable antiemetic is not available.

Pain Relief: A modest supply of prescriptive pain medication may be needed for headache, toothache, or musculoskeletal injury.
- Tylenol #3 (Tylenol with 30 mg codeine) tablets: one to two tablets PO q 4–6 h prn severe headache or pain; or
- Dilaudid (hydromorphone), 2-mg tablets: one to two tablets PO q 4 h for relief of severe pain (useful for people allergic or intolerant to codeine).

---

* Reprinted with permission from *The Travel & Tropical Medicine Manual,* 2nd edition, edited by Elaine C. Jong, MD, and Russell McMullen, MD. Philadelphia: WB Saunders Company, Copyright 1995.

Table 15-B (continued)

Table 15-B. The Traveler's Personal Medical Kit. (concluded)

NONPRESCRIPTION ITEMS
Aspirin or Tylenol or Ibuprofen (Advil, Nuprin): For general relief of minor aches and pains or headache.
Antibiotic Ointment: (Neosporin or Bacitracin) for topical application on minor cuts and abrasions.
Antifungal Powder or Cream: For travelers prone to athletes' foot and/or other fungal skin problems.
Antifungal Vaginal Cream or Troches: For women prone to yeast vaginitis associated with changes in climate or following antibiotic use.
Decongestant Tablets (Actifed, Sudafed, Contac, etc.): For nasal congestion due to colds, allergies, or water sports.
Diarrhea Prevention: Bismuth subsalicylate (Pepto-Bismol) tablets may be taken, two tablets qid every day of the trip to prevent traveler's diarrhea.
Hydrocortisone Cream: For topical relief of itching due to insect bites or sunburn.
Laxative: For relief of "traveler's constipation" due to changes in diet and schedule. Patients with a history of this problem may need to take a fiber supplement and/or a prescription stool softener.
Oral Rehydration Salt Packets: WHO-ORS (Jianas Brothers, Kansas City, MO), IAMAT Oral Rehydration Salts (IAMAT address in the Appendix): to be mixed in purified water safe for drinking for fluid replacement and rehydration during severe diarrhea.
Throat Lozenges: For relief of throat irritation due to air pollution or upper respiratory infection.

GENERAL HEALTH AND FIRST AID SUPPLIES
Antiseptic Solution: Topical solution for cleansing of minor cuts and abrasions (Hibiclens).
Bandages: Bandaids, 4 x 4-inch sterile gauze pads, 2-inch roll gauze dressing.
Elastic Bandage: For minor sprains (Ace wrap).
Eyeglasses: If corrective lenses are used, bring an extra pair of eyeglasses along.

---

are often more important than required vaccines. No vaccinations are required or recommended for travel between the United States, Canada, Western Europe, and Japan. Table 15-C lists potential immunizations for overseas travel.

Required Immunizations

As noted, required vaccines are mandatory for legal entry into certain countries. The most commonly required mandatory vaccine is the yellow fever immunization.

Yellow Fever

Yellow fever historically has posed hazards to workers in the Tropics (Figure 15-2). Completion of the Panama Canal was seriously hampered by epidemics of yellow fever among American and French construction crews. This viral disease is spread by the bite of Aedes and jungle mosquitoes. The live vaccine is highly effective, and immunity lasts for about 10 years. Certificates of vaccination are compulsory for legal entry into several countries, especially for travelers arriving from an endemic region. The yellow fever vaccine can be administered only at specially designated vaccination centers in the United States.

Recommended Immunizations

Immunizations are recommended for the following diseases: tetanus and diphtheria; measles, mumps, and rubella; polio; typhoid; hepatitis A and B; rabies; Japanese encephalitis; meningococcal disease; tuberculosis; plague; and pneumococcal disease and influenza. Vaccines for cholera, tuberculosis, and plague are of limited or doubtful efficacy. Recommended vaccines are prescribed based on the traveler's age, medical history, prior immunization status, and exposure risk. Exposure risk depends on detailed itinerary, duration of stay and the occupational risks of the traveler.

Tetanus and Diphtheria

Tetanus is contracted through wounds infected with ubiquitous *Clostridium tetani* spores. Tetanus is a life-threatening illness but is easily preventable. All persons should update their tetanus immunity for life in the United States, as well as when traveling abroad.

Diphtheria occurs worldwide. Adult travelers at high occupational risk of infection are those who work closely with children (e.g., teachers and health care workers). All adults should have been immunized during childhood and should receive booster injections of tetanus-diphtheria toxoid every 10 years.

Measles, Mumps, and Rubella

Widespread vaccination efforts reduced the incidence of *measles* in the United States. However, risk of measles remains high in parts of the developing world where immunization programs have

Table 15-C. Dosing Schedules for Commonly Used Vaccines for Travel.*

Vaccine	Primary Series	Booster Interval
Cholera, parenteral	2 doses 1 wk or more apart	6 mo
Cholera, live oral CVD 103-HgR	1 dose PO on an empty stomach for 2 yr old and above	6 mo (optimum booster schedule not yet determined)
Hepatitis A (Havrix) (1440 EL.UJmL, adult) (USA)	1.0 mL IM into the deltoid muscle	Booster dose at 6–12 mo predicted to confer immune protection for 10 yr or more
Hepatitis A (VAOTA)	1.0 mL (50 U) IM into the deltoid muscle (adults); 0.5 mL (25 U) IM into the deltoid muscle for children 2–17 yr	Booster dose at 6–18 mo predicted to confer immune protection for 10 yr or more
Hepatitis B (Engerix B) (accelerated schedule)	3 doses at 0, 30, and 60 d	A fourth dose is recommended at 12 mo to boost immunity
Hepatitis B (Engerix B) (standard schedule)	3 doses at 0, 1, and 6 mo	Need for booster not determined
Hepatitis B (Recombivax) (standard schedule)	3 doses at 0, 1, and 6 mo	Need for booster not determined
Immune globulin (Hepatitis A protection)	1 dose IM in the gluteus muscle	Boost at 3- to 5-mo intervals depending on initial dose received
Japanese encephalitis (JEV) (Japanese manufacture, Biken Brand)	2 or 3 doses given 1 wk apart	Booster dose of 1.0 mL at 12–18 mo, then at 4 yr intervals
Meningococcus Quadrivalent (A/C/Y/W-135)	1 dose* SC	None (variable immunogenic response in children < 4 yr of age: revaccination for this group recommended after 2–3 yr who continue to be at high risk)
Plague	1st dose; 2nd dose 4 wk later; dose 3 3–6 mo after dose 2	Boost if the risk of exposure persists: give the first 2 booster doses 6 mo apart; then give 1 booster dose at 1–2 yr intervals as needed
Poliomyelitis, enhanced inactivated (eIPV) (killed vaccine) (safe for all ages)	Give doses* 1 and 2 SC or IM 4–8 wk apart; give dose 3 at 6–12 mo after dose 2	Give a dose once to people prior to travel in areas at risk
Poliomyelitis, oral (OPV) (attenuated live virus)**	Give doses* 1 and 2 orally, 6–8 wk apart; give dose 3 at 6 wk after dose 2	Give a dose once to people less than 18 yr prior to travel in areas of risk
Rabies, human diploid cell vaccine (HDCV)	3 doses (0.1 mL ID) on days 0, 7, and 21 or 28	Boost after 2 yr or test serum for antibody level. Must not use chloroquine prophylaxis until 3 wk after completion of vaccine series
Rabies (HDCV) or rabies vaccine absorbed (RVA)	3 doses (1 mL IM in the deltoid area) on days 0, 7, and 21 or 28	Boost after 2 yr or test serum for antibody level
Typhoid, heat-phenol-inactivated parenteral vaccine	2 doses 4 or more wk apart	Boost after 3 yr for continued risk of exposure
Typhoid Ty21a, oral live-attenuated (Vivotif)**	1 capsule PO every 2 d for 4 doses (> 6 yr old)	5 yr
Typhoid Vi capsular polysaccharide, injectable (Typhim VI)	1 dose	Boost after 2 yr for continued risk of exposure (booster interval 3 yr in Canada)
Yellow fever**	1 dose	10 yr

* See manufactuer's package insert tor recommendations on dosage.
** Caution, may be contraindicated in patients with any of the following conditions: pregnancy, leukemia, lymphoma, generalized malignancy, immunosuppression caused by HIV infection or treatment with corticosteroids alkylating drugs, antimetabolites, or radiation therapy.
Adapted from Jong EC: Immunizations. In Jong EC, McMullen R (eds): *The Travel & Tropical Medicine Manual,* 2nd Edition, Philadelphia: WB Saunders Company, Copyright 1995; with permission.

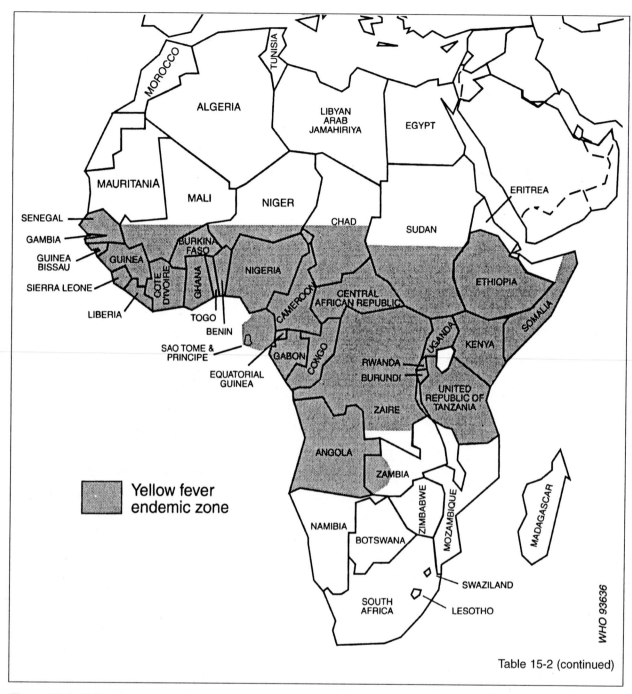

Figure 15-2. Yellow fever endemic zones in Africa and the Americas. Note that the "yellow fever endemic zones" are areas where there is a potential risk of infection on account of the presence of vectors and animal reservoirs. Some countries consider these zones as "infected" areas, and require an international certificate of vaccination against yellow fever from all travelers arriving from these areas. (Reprinted from CDC. *Health Information for International Travel, 1999–2000.* Washington DC: U.S. Government Printing Office, 1999.)

not been implemented. As a result, measles is predominantly a disease of the developing world. Approximately 20% of the measles cases reported in the United States between 1981 and 1987 were directly or indirectly related to international travel and immigration. Measles is a far more serious disease in adults than in children. Complications of measles can include pneumonia, infection of the central nervous system, and, rarely, death. Young adults born in 1957 or later who travel overseas should be considered for vaccination against measles. Those who were born before 1957 have probably already been infected, and thus have acquired natural immunity.

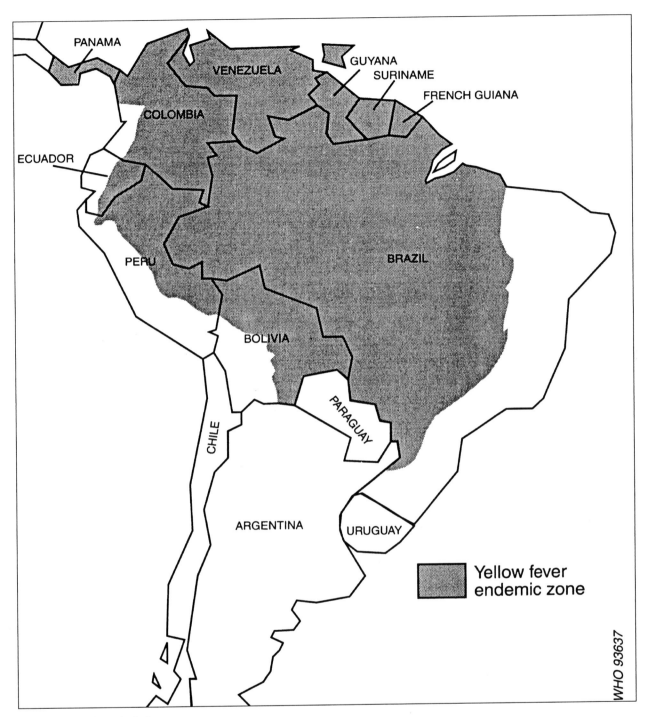

Figure 15-2. (concluded)

Nonimmune adults, especially women of childbearing age, should also have a vaccine for *rubella* (*German measles*). Rubella is still a common cause of congenital malformations in the developing world.

Mumps vaccine also is suggested for nonimmune young adults, as well as adolescents and children. For greater convenience, a combined measles-mumps-rubella vaccine is often administered.

Poliomyelitis

Poliomyelitis, formerly known as infantile paralysis, is a viral illness of the nervous system, spread by contaminated food and water. It is still common in developing countries. The risks of polio to short-term travelers are low. However, travelers planning to spend longer than three weeks in areas of poor sanitation are at higher risk and should update their primary series with a booster dose of oral polio vaccine (OPV).

Typhoid

Typhoid fever is a systemic bacterial illness caused by *Salmonella typhi*. The disease is spread via contaminated food and water, or by direct contact with infected persons. It occurs worldwide and is especially problematic in areas with poor sanitation (e.g., rural India and Pakistan). The risk to travelers is approximately 100 per 1 million travelers to India, Pakistan, or Peru. Travelers planning to spend longer than three weeks in areas with these conditions should be vaccinated. An oral vaccine is available, which requires doses of four capsules on alternate days. Side effects with this vaccine are minimal. There is also an intramuscular vaccine that confers two to three years of immunity.

Viral Hepatitis A

Hepatitis A is a common and disabling viral disease of the liver, characterized by fever, malaise, fatigue, and jaundice. The disease typically begins three to five weeks after the ingestion of fecally contaminated food and water. Afflicted persons usually recover without complications, but they are frequently unable to return to work for one month or longer after contracting hepatitis A.

The development of a safe and effective hepatitis A vaccine has been a great breakthrough in travel medicine. This vaccine provides several years of immunity.

Viral Hepatitis B

Hepatitis B is also known as serum or transfusion hepatitis, because it is transmitted via blood and bodily secretions. It can also be spread through sexual contact with infected persons. Hepatitis B usually causes more severe illness than hepatitis A; it can progress to chronic liver disease and cirrhosis. Hepatitis B has also been epidemiologically linked to liver cancer in the developing world (Figure 15-3).

The vaccine is recommended for health care workers, laboratory technicians, and emergency responders. It is also recommended for those who anticipate sexual contact with the local population or who are likely to seek local medical or dental care. The series of three immunizations is safe and virtually free of side effects. It provides long-lasting immunity.

Rabies

Rabies is a viral infection of the brain, transmitted by the bite of infected animals, especially dogs, foxes, skunks, and bats. The risk of rabies is quite low for routine travel, but the disease is invariably fatal.

Rabies immunization is usually recommended for travelers planning to spend longer than one month in areas of high incidence. In addition, all veterinarians, field biologists, and other animal handlers should be immunized. Antibody testing and/or boosters are recommended every two years.

If an animal should bite, wash the affected area thoroughly with soap and water for at least 10 minutes. If the animal is potentially rabid, consult a physician for further treatment.

Japanese Encephalitis

The mosquito-borne viral illness known as *Japanese encephalitis* occurs sporadically throughout areas of China, Japan, Korea, and Eastern Russia, especially in rural areas where rice and pig farming are common. Outbreaks occur during the rainy season (May through October). Illness is often asymptomatic, but it can be severe, resulting in neurological damage or death. A vaccine (JE-Vax) has been developed in Japan and is available worldwide. It is a series of three injections.

Travelers to Asia who are planning excursions into rural areas or long-term stays should get vaccinated. Travelers should also protect themselves from mosquito bites (see the discussion of insect control later in this chapter).

Meningococcal Diseases

Meningococcal disease is present worldwide. Its most feared manifestation is meningitis (infection of tissues surrounding the brain and spinal cord), a life-threatening illness. The best-known endemic area is the *meningitis belt* in sub-Saharan Africa (Figure 15-4), where epidemics occur regularly during the rainy season. Sporadic outbreaks have also been reported recently in Mongolia and Saudi Arabia. Meningococcal infections are uncommon among Western travelers, even during epidemics, but vaccination is currently recommended for visitors to Mongolia, Kenya, and Tanzania, and some regions of India, Saudi Arabia, or sub-Saharan Africa (CDC, 1999).

The meningococcal vaccine is quite safe and is sometimes recommended for persons traveling to high-risk areas. Recommendations for vaccination will depend on up-to-date reports of epidemics around the world. Travel clinics and local health

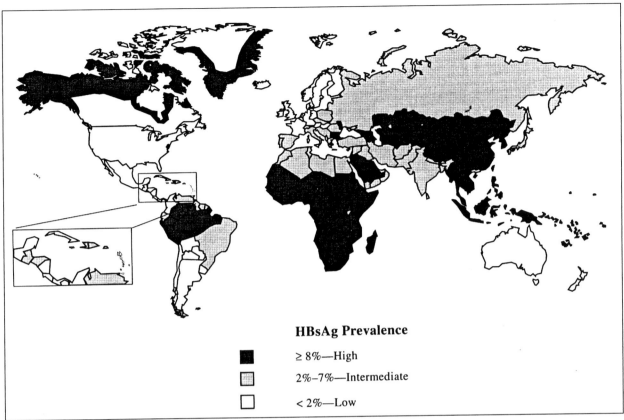

Figure 15-3. Hepatitis B has been epidemiologically linked to liver cancer in the developing world. (Reprinted from CDC. *Health Information for International Travel, 1999–2000*. Washington DC: U.S. Government Printing Office, 1999.)

departments usually have access to this information. Information can also be obtained via the CDC hotline at 404/332-4559.

Pneumococcal Disease and Influenza

Pneumococcal and *influenza* vaccines are recommended for persons 65 years and older and some chronically ill persons. Influenza vaccines are recommended for similar groups as well as health care workers. Check with health authorities about eligibility requirements.

Tuberculosis

During the industrial revolution, *tuberculosis (TB)* was known as the white plague and was the leading cause of death among young people. Tuberculosis is still prevalent worldwide, especially in areas with widespread poverty, overcrowding, and malnutrition. Tuberculosis is usually transmitted through inhalation of infectious aerosols coughed up by infected persons. Another way to contract TB is to ingest unpasteurized milk or dairy products. Therefore, tuberculosis is not a serious hazard to travelers unless they anticipate prolonged indoor exposure to infected persons (as health care workers might). These persons should undergo periodic diagnostic TB skin testing (i.e., pretravel and post-travel).

There is an immunization against tuberculosis, called the *bacille Calmette-Guerin (BCG)* vaccine. It is not usually recommended for routine travels. However, long-term residents of highly endemic areas may consider the vaccine if normal skin testing and chest x-ray facilities are unavailable. Check with a physician; the vaccine is rarely medically indicated and often difficult to obtain in the United States.

Cholera

Cholera has long been known as a hazard to travelers and overseas residents. Vasco da Gama's expedition probably suffered from cholera outbreaks in 1490. Toward the end of the late 18th century, thousands of English colonial soldiers in India died from cholera.

Cholera is an acute diarrheal illness caused by the bacteria *Vibrio cholera*. The infection is acquired by ingesting contaminated food and water. Cholera out-

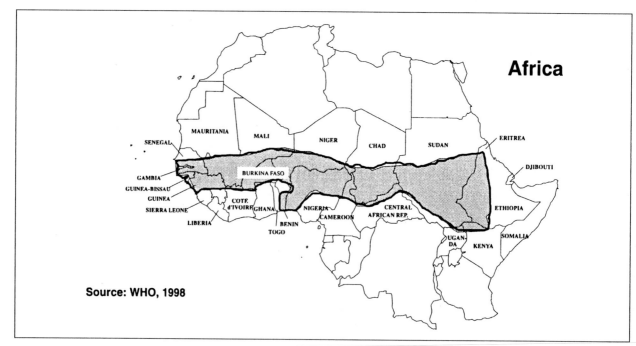

Figure 15-4. Areas with frequent epidemics of meningococcal meningitis. (Reprinted from CDC. *Health Information for International Travel, 1999–2000*. Washington DC: U.S. Government Printing Office, 1999.)

breaks occur sporadically in areas of Africa, the Middle East, and Asia. In 1991, an extensive epidemic began in Peru and spread to neighboring countries in the Western Hemisphere. In 1997, nearly 150,000 cases of cholera from 65 countries were reported to the World Health Organization (CDC, 1999).

The risk of cholera to American travelers is extremely low (Figure 15-1). The cholera vaccine has been removed from the World Health Organization's list of required vaccines. The vaccine is only about 50% effective in preventing the disease, and this partial immunity lasts for only three to six months after vaccination. Thus, the immunization is rarely recommended except for travelers to rural areas of extremely high endemicity. However, new oral cholera vaccines look promising and may become routinely recommended in U.S. travel clinics.

Plague

Plague is a bacterial infection primarily associated with rodents. It was an important traveler's disease during the Middle Ages. In Venice, Italy, in a.d. 1348, docking ships and their passengers were quarantined for 40 days to prevent its spread.

Despite its colorful history, plague is no longer a scourge, except in certain remote areas, including rural Southeast Asia. The vaccine is recommended only to travelers who anticipate frequent occupational contact with rodents for prolonged periods, as might be the case with laboratory researchers and field biologists. Antibiotic prophylaxis is also available with tetracycline.

Other Travel Infections

There are many serious infectious risks for which vaccines are not routinely available. Take special care to avoid modes of transmission. *Anthrax* is a serious cutaneous bacterial infection transmitted via direct contact with livestock in the developing world. Travelers should avoid handling contaminated fur and wood products, especially those made from goat skins. There is a human killed vaccine against anthrax that is available for persons at occupational risk (animal handlers, butchers, etc.). This vaccine also has been used to protect soldiers at risk from biological attack.

Dengue fever is an extremely serious illness spread by the bits of the *Aecles* mosquito in tropical regions. It can progress to potentially fatal dengue shock syndrome or hemorrhagic fever. No vaccine is currently available, so personal protection against mosquito bites is vital for those traveling in endemic areas.

MOTION SICKNESS

Motion sickness affects about 1% of commercial airline passengers during routine flights. During severe turbulence, about 8% are affected. The exact mechanism of motion sickness is unknown, but it is believed to arise when sensory input from the eyes and the *vestibular (balance) system* fail to match past expectations of motion cues. This sensory "mismatching" results in increasing nausea, pallor, and sweating, which can culminate in vomiting.

Some passengers find that sitting in the front of the plane can minimize these troublesome symptoms. Lying down, keeping the eyes closed, and the head still also can help to reduce motion sickness during air travel (Kozarsky, 1998).

Two drugs found to be useful in preventing motion sickness are *meclizine (Antivert)* and *transdermal scopolamine (Transderm Scop)*. Meclizine, an oral medication, is quite effective. It can be sedating, however, and should not be taken with alcohol. Transdermal scopolamine is a medicated adhesive bandage that delivers the drug through the skin. It is effective, less sedating, and long-lasting; the effects endure up to 72 hours. It must be applied at least six to eight hours before air travel. Side effects can occur, especially in elderly travelers and/or those with cardiac, neurological, or psychiatric conditions. Thus, it is important to consult a physician before using these drugs. Table 15-D lists some contraindications to air travel.

JET LAG

Rapid travel across many time zones causes desynchronization of circadian rhythms. Researchers have identified over 50 circadian rhythms, the most important of which are involved in the regulation of sleep, wakefulness, mental, and physical performance, hunger, and defecation. Disturbing these rhythms results in the symptoms that together are known as jet lag.

Symptoms

More than 90% of air travelers on flights across time zones are estimated to suffer from jet lag. Major symptoms include, in order of frequency, daytime sleepiness, insomnia, poor concentration, slowed reflexes, indigestion, hunger at odd hours, irritability, and depression. Of these symptoms, daytime sleepiness and insomnia are the most bothersome, lasting from two to seven days. Depending on age and individual susceptibility, symptoms can last for weeks. As a result of jet lag, travelers often notice decreased efficiency and impaired ability to negotiate and make key decisions.

Table 15-D. Examples of Contraindications to Air Travel.

- Severe anemia
- Sickle-cell anemia
- Severe otitis media and sinusitis (middle-ear or sinus infections)
- Acute contagious or communicable disease
- Angina pectoris (severe)
- Recent myocardial infarction (heart attack)
- Cardiac rhythm disorders
- Congestive heart failure (severe)
- Recent cerebral infarction (stroke)
- Uncontrolled arterial hypertension (e.g., >200 mmHg systolic)
- Peptic ulceration with hemorrhage within 3 weeks
- Simple abdominal operation within 10 days
- Major chest surgery within 21 days
- Contagious or repulsive skin diseases
- Mental illness without escort and sedation
- Pregnancies after 35th week (long journeys) or 36th week (short journeys)
- Pneumothorax

Treatment of Jet Lag

As a rule, it takes approximately one day per time zone crossed to fully recover from jet lag. Adjustment seems to be far easier following westward rather than eastward flights—that is, when the travel day is lengthened. It does appear easier to delay biological rhythms than to advance them. This may be related to the fact that circadian rhythms have a natural period of slightly more than 24 hours.

Several strategies can help to minimize the fatigue and stress of air travel. While in flight, walk in the aisle, avoid excessive intake of alcohol or coffee, and drink plenty of fluids. Some travelers cope better by arranging one-day layovers to break up long flights. The additional travel time results in more enjoyment and improved mental concentration upon arrival.

Some experienced travelers believe that dietary modifications before and during air travel may be helpful in combating jet lag. However, there is little scientific data to support this assertion. Some travelers use the Argonne National Laboratory's antijet-lag diet.

During the first few days after arriving in a foreign destination, it is wise to moderate your schedule. Postpone crucial meetings and decisions until later in the stay. If important conferences must be scheduled, arrange them to coincide with morning hours back home, a time of optimal alertness.

For help in readaptation to a different time zone, use environmental cues. Immediately after arriving at the new time zone, set a routine for such activities as eating, bathing, and exercising. These cures are the key to resynchronizing circadian rhythms during the day. Circadian rhythms can be reset by exposing oneself to bright light. This exposure is best accomplished by walking outside in sunshine during specific times. For example, after flying eastward, morning exposure to bright light will advance bodily rhythms toward synchronization with the new time zone. Following a westward flight, a walk in the afternoon sun will delay circadian rhythms and thus prevent sleepiness. Some researchers are developing artificial light sources for resynchronizing body rhythms during sleep.

Melatonin is a hormone synthesized in the pineal gland and secreted at night. When taken exogenously, it may help decrease the symptoms of jet lag, especially when taken after arrival. However, significant concerns remain about its safety and efficacy, as it is not reviewed by the Food and Drug Administration (FDA). The standardization of melatonin content and potency is also not available in many over-the-counter preparations.

Insomnia brought on by jet lag can be ameliorated by short-acting *benzodiazepine* medications such as *triazolam* (*Halcion*), and *temazepam* (*Restoril*). These drugs can be effective for inducing sleep, especially following westward flights. However, do not take these drugs along with alcohol or other sedative drugs, and do not use them during air travel, because they can cause confusion and temporary amnesia. Do not use benzodiazepines chronically, as they can be habit-forming or have other significant side-effects.

TRAVELER'S DIARRHEA

Traveler's diarrhea is by far the most common health problem among international travelers; it is believed to afflict 20%–50% of visitors to the developing world (World Health Organization, 1992). It affects some 40% of American travelers to Mexico, whose nicknames for the condition, *turista* and *Montezuma's revenge,* aptly describe both the victim and the illness.

Traveler's diarrhea can significantly interfere with plans for business overseas. One study found that about 30% of those afflicted with it are bedridden. Another 40% were so sick that they had to change their travel itinerary (Barer, 1987). Fortunately, the syndrome is usually short-lived, lasting no more than three to four days.

Traveler's diarrhea typically occurs during the first week of a visit to a country that lacks adequate sanitation (Figure 15-5). It results when drastic dietary changes lead to a rapid change in the balance of microorganisms present in the gastrointestinal tract. Many different microbes can cause traveler's diarrhea.

Preventing Traveler's Diarrhea

Fortunately, traveler's diarrhea can be prevented. The key to prevention is to pay careful attention to eating and drinking habits. Prudent dietary practices can also decrease the risk of other infectious diseases, such as hepatitis A, typhoid fever, and cholera.

Dietary Precautions

Even while eating at luxury hotels, travelers need to exercise caution in their dietary habits. It is difficult at times to refuse food offered by a gracious host, especially in rural areas. Decide whether offending a host is less desirable than a bout of diarrhea.

The best rule of thumb is to eat hot, freshly cooked food. Food prepared by street vendors is often contaminated and should be avoided. Raw fruits and vegetables that can be peeled are usually safe to eat. Avoid salads, even at the most reputable restaurants. Do not consume unpasteurized milk and dairy products. In summary, "boil it, peel it, cook it, or forget it." See Tables 15-E and 15-F for more information.

Avoid eating shellfish in the developing world. In many parts of the world, shellfish are collected from areas close to the dumping of raw human sewage. Shellfish tend to harbor pathogenic organisms because of their unique feeding habits, which involve filtration of large quantities of ocean water. As a result, ingestion of raw or undercooked shellfish is a risk factor for many infectious diseases, including hepatitis A virus, *Salmonella, Shigella,* and *Vibrio parahaemolyticus.* Shellfish may also contain poison or biotoxins when consumed during certain seasons.

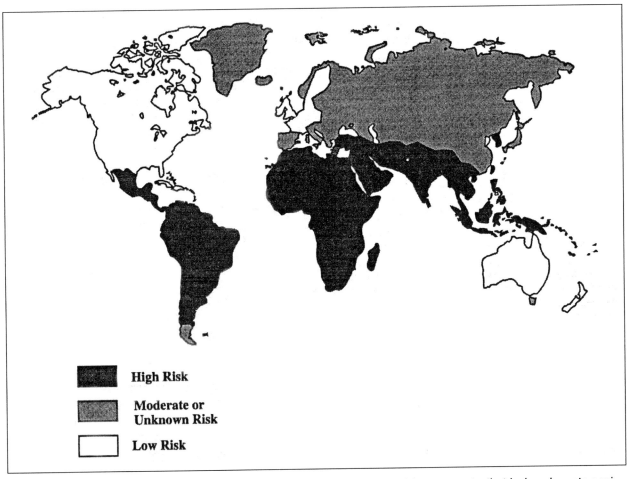

Figure 15-5. Traveler's diarrhea typically occurs during the first week of a visit to a country that lacks adequate sanitation. (Reprinted with permission from Jong EC. Travel-related infections: Prevention and treatment. *Hospital Practice,* November 1989.)

Water and Other Beverages

It is prudent to drink only bottled carbonated drinks, beer and wine, or hot drinks such as tea or coffee. Ice cubes and tap water can be contaminated and should be avoided in the Tropics. The myth that the ethanol present in liquor is sufficient to sterilize water in ice cubes is simply not true. Drinks should be chilled by putting them *on ice* rather than *on the rocks.*

Sterilize pure tap water before drinking it. The most reliable way to disinfect water is to boil it vigorously for at least three minutes. Boiling tends to take the taste out of drinking water because it reduces the content of dissolved gases. The taste usually improves, however, if the water is allowed to sit and cool for a few hours. Water for small children or infants should always be boiled or disinfected because they are highly susceptible to diarrheal diseases.

You might take along a portable electric-coil heater on the trip. The heater can be used in the hotel room to boil water. Do so carefully, however, as the heater is a potential fire hazard.

If boiling facilities are not available, an alternative is to chemically disinfect drinking water after removing suspended solids and organic matter. Chlorine is a widely used disinfectant available in the form of tablets or household chlorine bleach.

While many people find the taste of iodine to be unpleasant, it is generally a more effective disinfectant than chlorine, because it usually kills cysts of *Giardia* and amoebas. Iodine is available in tablet form or 2% tincture solution. One tablet or 10 drops are adequate to disinfect a quart (liter) of clear water. The water should then be allowed to stand for 30 minutes before drinking. The methods of chemical disinfection are summarized in Table 15-G.

If it is impossible to obtain or prepare potable drinking water, the thirsty traveler can drink scalding

Table 15-E. Relatively Safe and Unsafe Foods, Beverages, and Culinary Practices.*

	SAFE	UNSAFE
Foods	Steaming hot Packaged Dry Jelly, syrup Peeled fruit Citrus Ripened cheeses Cooked vegetables	Raw fish or meat Cold salads Cold sauces Hamburgers Unpeeled fruit Some desserts Fresh soft cheeses Strawberries
Beverages	Carbonated Bottled water Boiled water Iodized water Packaged ice Fresh fruit juice Irradiated milk	Tap water Chipped ice Unpastuerized milk
Culinary Practices	Recommended restaurants Careful Judicious alcohol Considers heat of food	Vendors Adventuresome Excessive alcohol Buffet food at room temperature

* Reprinted with permission from DuPont HL, Steffen R. *Textbook of Travel Medicine*. Hamilton Ont: B.C. Decker Inc. 1997.

Table 15-F. 10 Tips for Selection of Safe Food and Water.*

1. Drink purified water or bottled carbonated water.
2. Eat foods that are thoroughly cooked, and served piping hot.
3. Eat fruits that have thick skins (they should be peeled at the table by the traveler).
4. Avoid salads made with raw vegetables, especially leafy green vegetables.
5. Do not use ice cubes in any beverages, even beverages containing alcohol.
6. Only eat and drink dairy products made from pasteurized milk.
7. Avoid shellfish, and raw or undercooked seafood, even if "preserved" or pickled with lemon or lime juice or vinegar.
8. Do not buy and eat food sold by street vendors.
9. If canned beverages are cooled by submersion of the can in a bucket of ice water or in a stream, be sure to dry off the outside of the container before drinking the contents.
10. Use purified water for brushing teeth and for taking medications.

* Reprinted with permission from *The Travel & Tropical Medicine Manual*, 2nd edition, edited by Elaine C. Jong, MD, and Russell McMullen, MD. Philadelphia: WB Saunders Company, Copyright 1995.

hot tap water that has been allowed to cool. This water is relatively safe.

For long-term travelers, contact inactivation can be an effective alternative to boiling or halogenation. Iodine resins are a relatively new technology in water purification. They release iodine on contact with microbes and leave little aftertaste. Commercial purifiers are available from Water Technologies Corp. (800/627-0044) and Sweet Water Filter (800/444-5865), Travel Well (1 lb, 6 oz [660 mL]), and Pur (12 oz [360 mL]). These two products contain filter systems that remove *Giardia* cysts, thus making potable water available promptly after it leaves the system. The problem in using these devices is to know when the iodine resin is used up and the purification system is ineffective. Water filtration with sediment or microbial filters is advertised, but not reliable (Jong, 1995).

Prophylactic Drugs

Because travelers are not always able to control their food intake, they run a relatively constant risk of developing some traveler's diarrhea. Using medication to prevent traveler's diarrhea is not usually recommended, except when the consequences of the condition are extremely severe, for instance, during crucial business or government negotiations, military maneuvers, or international competitions. Travelers should discuss the risks and benefits of such drugs with their physician before including them in the traveler's medical kit.

Bismuth subsalicylate (Pepto-Bismol) can be helpful in preventing traveler's diarrhea. Taken orally, in liquid or tablet form, bismuth subsalicylate can reduce the incidence of traveler's diarrhea by up to 60%. However, it does have some adverse effects. A majority of users experience darkening of the tongue or blackening of the stools, and 16% of users report ringing in the ears, a symptom similar to that of aspirin toxicity. People with aspirin sensitivity, gout, bleeding disorders, or impaired kidney function, or those taking anticoagulants, probenicid, or methotrexate, should avoid this drug.

Prophylactic antibiotics can reduce the incidence of traveler's diarrhea, but they are not routinely recommended for travelers, because they are

Table 15-G. Methods for Purification of Water.*

METHOD	BRAND NAME	QUANTITY TO BE ADDED TO 1 QUART OR 1 LITER OF WATER
Iodine compound tablets†	Potable Aqua, Coughlins	Two tablets are added to water at 20°C, and the mixture is agitated every 5 minutes for a total of 30 minutes.
Chlorine solution, 2–4%	(Common laundry bleach)	Two to four drops are added to water at 20°C, and after mixing, the solution is kept for 30 minutes before drinking.
Iodine solution†	(2% tincture of iodine)	Five to ten drops of iodine are added to water at 20°C, and after mixing, the solution is kept for 30 minutes before drinking.
Heat		Water is heated to above 65°C for at least 3 minutes. (At 20,000 ft altitude or 6000 m, water boils at 70°C.)

* The methods presented here are sufficient to kill *Giardia* cysts in most situations. Heat is the best method when tested in a laboratory situation.
† Iodine-containing compounds should be used with caution during pregnancy.

commonly associated with adverse effects, such as skin reactions, vaginal yeast infections in women, and the development of drug-resistant microorganisms (Ericsson, 1998).

Treating Traveler's Diarrhea

The keys to treatment of self-limited cases of travelers diarrhea are oral rehydration and salt replacement (Table 15-G). During diarrheal episodes, consume only fluids, such as fruit juices, noncaffeinated carbonated drinks, and broths. Later, add solid foods such as salted crackers, toast, rice, or bananas to the diet. Avoid excessively hot or cold drinks, caffeinated beverages, spices, dairy products, and foods high in fat or fiber.

Many developing-world pharmacies sell World Health Organization oral rehydration solution (ORS) packets. These packets contain powders that, when mixed with water, can replace lost fluids and electrolytes. Or to make your own rehydration solutions, mix eight level teaspoons of sugar or honey and one-half teaspoon of salt in 1.8 pints (1 L) of water.

Therapy with empirical antibiotics can shorten the typical three- to four-day course of traveler's diarrhea to one to one and one-half days. *Ciprofloxacin* is highly effective therapeutically and works well as a single-dose (750 mg) treatment. Bismuth subsalicylate (Pepto-Bismol) liquid suspension also can be effective therapy. Take one teaspoonful every half hour for eight hours.

Professionals who need to keep important commitments or travel long distances may wish to reduce symptoms by taking *antimotility agents*. These include *codeine, paregoric, loperamide* (*Imodium*), and *diphenoxylate hydrochloride* (*Lomotil*). These medications contain narcotics and other agents that reduce diarrheal symptoms by decreasing the motility of the gastrointestinal tract. Small children and pregnant women should not use antimotility agents. Also, avoid antimotility agents and contact a physician if one or more of the following problems arise:

- temperature greater than 102°F (38.9°C),
- severe abdominal pain,
- blood or mucus in the stools,
- severe dehydration,
- duration of diarrheal symptoms longer than three days.

Patients on diuretic medications for high blood pressure or congestive heart failure are especially prone to dehydration and salt depletion resulting from traveler's diarrhea. If a bout of traveler's diarrhea occurs, they should stop using these diuretics and consult with a physician as soon as possible.

It is not wise to purchase over-the-counter antidiarrheal agents in the developing world, even if they are recommended by local pharmacists. One common but hazardous antidiarrheal drug is *iodochlorhydroxyquin* (*Entero-Vioforme*), which is no longer available in the United States, but remains on the shelves in many developing-world pharmacies. This drug is ineffective and associated with severe neurological side effects (CDC, 1996).

MALARIA

First described by the ancient Greeks, *malaria* has long been recognized as an important health hazard.

Table 15-H. Checklist for Travelers to Malarious Areas.*

The following is a checklist of key issues to be considered in advising travelers. The numbers in parentheses refer to pages in the text where these issues are discussed in detail.

Risk of malaria
Travelers should be informed about the risk of malaria infection and the presence of drug-resistant *P. falciparum* malaria in their areas of destination.

Anti-mosquito measures
Travelers should know how to protect themselves against mosquito bites.

Chemoprophylaxis
Travelers should be:
- Advised to start prophylaxis before travel, and to use prophylaxis continuously while in malaria-endemic areas and for 4 weeks after leaving such areas.
- Questioned about drug allergies and other contraindications for use of drugs to prevent malaria.
- Advised which drug to use for prophylaxis, and, if chloroquine is used, whether Fansidar® should be carried for presumptive self-treatment.
- Informed that antimalarial drugs can cause side effects; if these side effects are serious, medical help should be sought promptly and use of the drug discontinued.
- Warned that they may acquire malaria even if they use malaria chemoprophylaxis.

In case of illness, travelers should be:
- Informed that symptoms of malaria may be mild, and that they should suspect malaria if they experience fever or other symptoms such as persistent headaches, muscular aching and weakness, vomiting, or diarrhea.
- Informed that malaria may be fatal if treatment is delayed. Medical help should be sought promptly if malaria is suspected, and a blood sample should be taken and examined for malaria parasites on one or more occasions.
- Reminded that self-treatment should be taken only if prompt medical care is not available and that medical advice should still be sought as soon as possible after self-treatment.

Special categories
Pregnant women and young children require special attention because of the potential effects of malaria illness and inability to use some drugs (for example, doxycycline).

* * *

(Adapted from International Travel and Health, World Health Organization, Geneva, 1995)

---

* Reprinted with permission from *The Travel & Tropical Medicine Manual*, 2nd edition, edited by Elaine C. Jong, MD, and Russell McMullen, MD. Philadelphia: WB Saunders Company, Copyright 1995.

In the late 17th century, Bernardo Ramazzini, known as the father of occupational medicine, commented on the intermittent fevers of malaria and advocated treating them with Peruvian chinchona bark (from which the current antimalarial drug quinine is derived). Nowadays, malaria is probably the single most important health hazard in the Tropics (Figure 15-6). Three-hundred million people currently contract the disease each year. About 1.5–2.7 million persons annually die from malaria throughout the world. In the industrialized countries, about 10,000–30,000 persons contract malaria each year. Many of these are contracted by travelers who failed to comply with health precautions (Kain & Keystone, 1998) (see Table 15-H). The symptoms of malaria may begin to appear five to ten days following the bite of a female *Anopheles* mosquito carrying the *Plasmodium* parasite, which causes the disease. *Plasmodium falciparum* is the most deadly species. The *Plasmodium* parasite can incubate for long periods, however, causing silent infection for up to one year before symptoms occur. Symptoms of malaria can initially resemble influenza: fever, headache, malaise, and joint aches are common. These symptoms can progress to the classic malaria paroxysm, the dramatic onset of shaking chills, temperatures above 105°F (40.6°C) followed by profuse sweats. Symptoms vary significantly from case to case, so malaria should be suspected in any febrile illness. Because of this wide variation in malaria's manifestations, U.S. physicians often misdiagnose the disease. The key factors influencing survival from malaria are early diagnosis and appropriate treatment.

Preventing Malaria

Although malaria is present worldwide, the risks vary depending on the traveler's itinerary, and are constantly changing. The situation is worsening in many areas. Moreover, travel agents are notoriously ill informed on this matter. Many travel agents, for example, have been known to recommend "malaria shots" for travelers, when no such vaccine is currently available. For this reason, it is crucial to consult with informed health authorities before traveling to the Tropics.

Prophylactic Medications

Most American travelers to malarious areas should take drugs to prevent malaria (Table 15-I). No prophylactic drug gives complete protection. Travelers need to follow a comprehensive

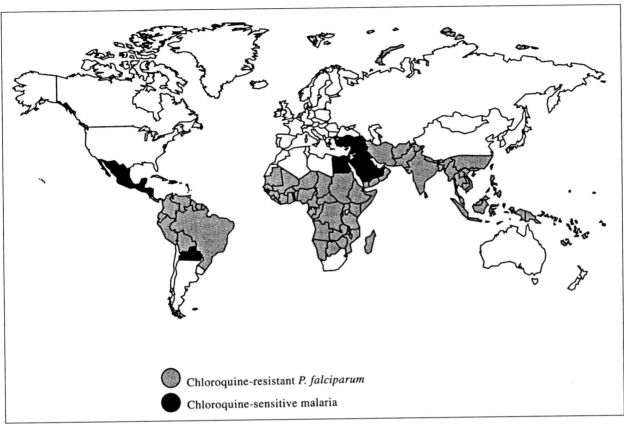

Figure 15-6. Distribution of malaria and chloroquine-resistant *Plasmodium falciparum*, 1997. (Reprinted from CDC. *Health Information for International Travel, 1999–2000*. Washington DC: U.S. Government Printing Office, 1999.)

program of malaria prevention, including antimalarial medications, avoidance of mosquito bites, and conscientious planning of potential medical care.

Travelers usually should begin taking prophylactic medicine one to two weeks before going to malarious regions. This precaution allows the traveler and physician to deal with any side effects that can result from taking the drugs. The traveler should continue taking antimalarial drugs throughout the trip and for four weeks after leaving the malarious region. The post-trip medication is intended to suppress any incubating infections that the traveler may have acquired. The specific drug regimen a physician prescribes will vary according to the traveler's itinerary, the duration of the stay, and the traveler's tolerance of side effects.

Chloroquine has been the standard prophylactic drug for low-risk malarious areas. It is taken weekly and is usually well tolerated. Minor side effects include gastrointestinal discomfort, headache, dizziness, blurred vision, or itching. These side effects can usually be minimized by taking the drug with meals or dividing the dose into two biweekly intervals. There has been some concern over the effects of chloroquine on the retina in long-term users; these effects probably occur only in people taking high-dose chloroquine for other diseases. The prescribing physician may recommend that a person taking chloroquine continuously for more than three years receive periodic retinal examinations.

For travelers who elect to use chloroquine—it is a presumptive treatment medication often included in the traveler's medical kit. This dose is to be taken at once if the traveler falls ill with a high fever and adequate medical care is not available. Travelers taking this dose should seek an immediate medical evaluation.

Mefloquine (Lariam) is the antimalarial drug of choice for travelers to high-risk areas where chloroquine-resistant *P. falciparum* exists (Figure 15-6). Adverse effects including hallucinations and convulsions have been reported. The drug should be avoided by persons in occupations involved with spatial discrimination and/or fine coordination (e.g., airline pilots), pregnant women,

Table 15-I. Drugs Used in the Prophylaxis of Malaria.*

Drug	Usage	Adult dose	Pediatric dose	Comments
Mefloquine (Lariam®)	In areas with chloroquine-resistant *Plasmodium falciparum*	228 mg base (250 mg salt) orally, once/week	<15 kg: 4.6 mg/kg base (5 mg/kg [salt]) once/week; 15–19 kg: 1/4 tab/wk 20–30 kg: 1/2 tab/wk 31–45 kg: 3/4 tab/wk > 45 kg: 1 tab/wk	Contraindicated in persons allergic to mefloquine. Not recommended for persons with epilepsy and other seizure disorders; with severe psychiatric disorders; or with cardiac conduction abnormalities.
Doxycycline	An alternative to mefloquine	100 mg orally, once/day	> 8 years of age: 2 mg/kg of body weight orally/day up to adult dose of 100 mg/day	Contraindicated in children < 8 years of age, pregnant women, and lactating women.
Chloroquine phosphate (Aralen®)	In areas with chloroquine-sensitive *Plasmodium falciparum*	300 mg base (500 mg salt) orally, once/week	5 mg/kg base (8.3 mg/kg [salt]) orally, once/week, up to maximum adult dose of 300 mg base	
Hydroxychloroquine sulfate (Plaquenil®)	An alternative to chloroquine	310 mg base (400 mg salt) orally, once/week	5 mg/kg base (6.5 mg/kg [salt]) orally, once/week, up to maximum adult dose of 310 mg base	
Chloroquine + Proguanil	A less effective alternative for use in Africa, only if mefloquine or doxycycline cannot be used	Weekly chloroquine dose as above, plus daily proguanil dose 200 mg orally, once/day	Weekly chloroquine doseas above, plus < 2 years: 50 mg/day 2–6 years: 100 mg/day 7–10 years: 150 mg/day >10 years: 200 mg/day	Proguanil is not sold in the United States, but is widely available in Canada, Europe, and many African countries.

* Reprinted with permission from *The Travel & Tropical Medicine Manual*, 2nd edition, edited by Elaine C. Jong, MD, and Russell McMullen, MD. Philadelphia: WB Saunders Company, Copyright 1995.

travelers using beta-blockers, and travelers with a history of mental illness or epilepsy.

Doxycycline and *proguanil* (*Paludrine*) are alternative prophylactic medications occasionally prescribed for travelers to areas where chloroquine-resistant strains of malaria are present or for those who tolerate mefloquine poorly. Doxycycline alone taken daily may be appropriate for those with a history of sulfa drug intolerance. Proguanil (Paludrine) is not available in the United States. For travelers to East Africa, it can be an effective prophylactic, when taken weekly with chloroquine.

Some strains of malaria can persist for years in the body after the traveler has returned home. For this reason, long-term residents of high-risk areas (e.g., Peace Corps volunteers or missionaries) can be prescribed another drug, called *Primaquine* (an 8-aminoquinoline), upon their return to the United States. This drug can eradicate latent malaria infections in the liver and prevent relapse. In any case, it is important to seek prompt medical attention for any flu-like illness accompanied by fever, even after returning from the trip. Volunteer the travel history to the treating physician, so that he or she will consider the likelihood of malaria.

Children traveling in the Tropics are also at risk for malaria. Prophylactic doses and drugs for children usually vary from adult regimens. If children are to accompany adults during the journey, seek appropriate pediatric expertise. Travelers with special needs should also seek medical advice.

Protection Against Mosquitoes

To round out a comprehensive preventive approach to malaria, take precautions to protect against mosquitoes. Such precautions are vital for workers who spend a lot of time outdoors (e.g., construction and forestry workers). Properly employed, mosquito control measures can reduce the risk of malaria by a factor of 10. Here are a few simple recommendations:

- When outside during dawn and evening hours, wear lightly colored clothing with long sleeves and long pants. Mosquitos are not attracted to light colors.
- Use insect-repellents containing DEET (N,N-diethyl-m-toluamide—see Table 15-J). Reapply the repellent to exposed skin every two hours if excessive sweating occurs.
- Sleep in rooms that are properly screened or air-conditioned, or use mosquito nets around the bed at night in unscreened rooms. Make sure that the net is not torn. Some nets are impregnated with insecticides for added protection. Clothing can be soaked in permethrin solution for added protection (see Figure 15-7).
- Use electric insecticide dispensers or burn mosquito coils. Mosquito coils are effective, but they do not usually last through the night. Pyrethrum-containing insect sprays also are safe and effective.

Precautions to avoid mosquitoes can prevent more than just malaria. They reduce the risk of other insect-borne infectious diseases like dengue fever, Japanese encephalitis, and yellow fever. Dengue fever, which is carried by mosquitoes, has been a problem in East Asia for many years and now has spread to the Caribbean and parts of Latin America. There is currently no effective vaccine or preventive drug for dengue fever. It is often present with a flu-like syndrome resembling the initial symptoms of malaria. These symptoms can progress to serious complications, including the fatal hemorrhagic shock syndrome (CDC, 1996).

SEXUALLY TRANSMITTED AND BLOODBORNE DISEASES

Sexually transmitted infections are ubiquitous throughout the developing world. Syphilis, gonorrhea, herpes simplex, and other more exotic venereal diseases also are common in the Tropics. Further complicating the problem of gonorrhea, many drug-resistant strains of the diseases have emerged, especially in West Africa and Southeast Asia. The increasing resistance can be a result of frequent self-treatment with antibiotics by prostitutes and others.

Table 15-J. Insect Repellents and Insecticides.*†

INSECT REPELLENTS CONTAINING DEET— FOR SKIN APPLICATION
Ultrathon Insect Repellent: 35% DEET in polymer formulation, up to 12 h protection against mosquitoes; also effective against ticks, biting flies, chiggers, fleas, gnats (3M, Minneapolis, MN 55144).

DEET Plus Insect Repellent: 17.5% DEET with 2.5% R—326, apply every 4 h for mosquitoes, every 8 h for biting flies (Sawyer Products, Safety Harbor, FL 34695).

Skedaddle Insect Protection for Children: 10% DEET using molecular entrapment technology (Little Point Corp, Cambridge, MA).

PERMETHRIN-CONTAINING INSECTICIDES— FOR APPLICATION TO EXTERNAL CLOTHING AND BED NETS
Permanone Tick Repellent: Contains permethrin in a pressurized spray can; repels ticks, chiggers, mosquitoes, and other bugs (Coulston International Corp, Easton, PA 18044).

Duranon Tick Repellent: Contains permethrin in a formula lasting up to 2 weeks, supplied in a pressurized spray can (Coulston International Corp, Easton, PA 18044).

PermaXill 4 Week Tick Killer: 13.3% permethrin liquid concentrate supplied in 8-oz bottle, can be diluted (1/3 oz permethrin concentrate in 16 oz water) to be used with a manual pump spray bottle; or diluted 2 oz in 1 1/2 cups of water to be used to impregnate outer clothing, bednets, and curtains (Coulston International Corp, Easton, PA 18044).

* Brand names are given for identification purposes only, and do not constitute an endorsement.

† * Reprinted with permission from *The Travel & Tropical Medicine Manual*, 2nd edition, edited by Elaine C. Jong, MD, and Russell McMullen, MD. Philadelphia: WB Saunders Company, Copyright 1995.

AIDS, HIV Infection, and Bloodborne Hepatitis

Human immunodeficiency virus (HIV), which causes AIDS, has been reported in over 140 countries worldwide. An estimated 9–11 million people carry HIV infection (World Health Organization, 1992). HIV can be acquired through sexual contact or by exposure to blood or blood products. There is no evidence that HIV is transmitted through casual, nonsexual contact or through contact with swimming pools or rest rooms. Mosquitoes and other insects have not been shown to transmit the HIV infection.

Prostitutes in the developing world are at extremely high risk of carrying HIV. For example, up to 50%–85% of urban prostitutes in Africa and

Use Permethrin (Permanone®) to treat clothing, mosquito netting, or tent fabric.
Do not apply Permethrin on your skin.

Spraying—Lay out clothing to be sprayed. For shirts and trousers, spray the front of each garment for 60–90 seconds. Use a slow sweeping motion, holding the can about 12 inches from the fabric. Spray inside cuffs. Turn the garment(s) over and repeat. For socks and bandanas, spray for 30 seconds per side. The fabric should be slightly damp. Hang up to dry. You'll need 1/2 to 2/3 of a can to treat a shirt and trousers. The sprayed garments, even after multiple launderings, will kill mosquitoes and ticks for about six weeks. This is because permethrin binds tightly to the fabric.

All fabrics, even your best silk shirt, can be treated without leaving an odor or stain. Permethrin is also safe to humans. Very little is absorbed through the skin and even these small amounts are rapidly neutralized by the body. Minor rashes only have been reported from skin contact.

Impregnating—Soaking protective clothing or mosquito netting in a permethrin solution is a method used to achieve long protection. Follow the directions below using 4 Week Tick Killer 13.3% solution. Treated fabric will kill insects for several months.

Technique for Impregnation Clothing or Mosquito Netting with Permethrin Solution

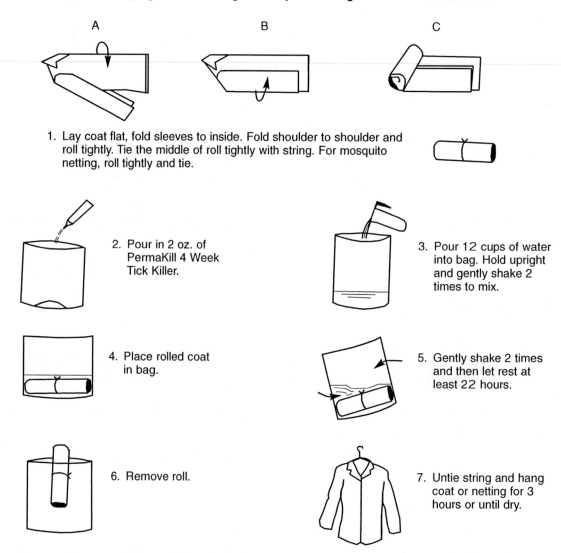

1. Lay coat flat, fold sleeves to inside. Fold shoulder to shoulder and roll tightly. Tie the middle of roll tightly with string. For mosquito netting, roll tightly and tie.
2. Pour in 2 oz. of PermaKill 4 Week Tick Killer.
3. Pour 12 cups of water into bag. Hold upright and gently shake 2 times to mix.
4. Place rolled coat in bag.
5. Gently shake 2 times and then let rest at least 22 hours.
6. Remove roll.
7. Untie string and hang coat or netting for 3 hours or until dry.

Figure 15-7. How to apply permethrin to your clothing. (Reprinted with permission from *The Travel & Tropical Medicine Manual*, 2nd edition, edited by Elaine C. Jong, MD, and Russell McMullen, MD. Philadelphia: WB Saunders Company, Copyright 1995.)

Thailand are infected with HIV (Jong & McMullen, 1995). It is wise to avoid any sexual contact with prostitutes. Furthermore, unprotected sexual contact between travelers and any new sexual partners should be strongly discouraged. Men should always use latex condoms during sexual intercourse with a new partner. Women should use diaphragms with spermicidal jelly, and should insist that their male partner wear a latex condom. Condoms and other forms of contraception are not routinely available in the developing world, so travelers should obtain them before leaving.

Transfusions of blood and blood products are well known to transmit HIV, hepatitis, and other diseases. Many developing nations do not routinely screen their donor blood pool for HIV or viral hepatitis before transfusing blood. For this reason, avoid transfusions of blood products in the developing world, except in life-threatening emergencies. In most instances, trauma victims can be resuscitated by artificial physiological plasma expanders until blood can be appropriately screened for HIV or until medical evacuation can be undertaken.

Another potential source of HIV transmission in the developing world is contaminated needles. "Disposable" needles are often reused by developing-world medical professionals. Moreover, adequate sterilization equipment is often lacking in these regions. Some Western travelers have begun to pack their own supply of disposable needles, in case injections will be needed in an emergency. Related hazards to avoid while traveling in developing countries include experimental medical treatments, dental work, acupuncture, ear piercing, tattoos, and intravenous drug abuse.

Vaccines in the developing nations, which are often derived from blood plasma products, present another potential source of contamination. It is advisable to obtain every vaccination potentially needed during a trip before departure, rather than assuming vaccinations can be added during a trip. Taking sexual precautions and avoiding contaminated blood products and needles can also protect the traveler from other bloodborne diseases, including hepatitis B and hepatitis C.

HIV-Infected Travelers

People with a preexisting HIV infection need to take extra precautions in travel planning. Live vaccines, for example, can cause serious adverse effects in HIV-positive persons. These immunosuppressed persons are also extremely susceptible to infectious diarrheas and skin conditions and other illnesses.

HIV testing or test results are necessary for entry or long-term residence in some countries (e.g., Iran and China). HIV-positive travelers can be refused entry into some countries. Before leaving, ask health authorities about current requirements.

SKIN CONDITIONS

Skin conditions can result from overexposure to the sun, fungal infections, and hot, humid weather conditions.

Sunburn and Other Problems of Sun Exposure

Ultraviolet radiation is the cause of skin problems related to excessive sun exposure. Ultraviolet B rays can cause sunburn, while ultraviolet A rays have been linked to premature aging and skin cancer (Council on Scientific Affairs, 1989).

Sunlight can be intense in the Tropics, especially at midday. It makes good sense to avoid excessive sun exposure. Plan outdoor activities for the morning or late afternoon. Wear a hat and protective clothing outdoors. If work necessitates being outdoors during periods of intense sunlight, then sunscreens are crucial.

Chemical sunscreens are graded by *sun protection factor* (*SPF*). The SPF values provide a rough estimate of the strength of protection. For example, a lotion with SPF of 10 enables the skin to endure sun without burning about 10 times longer than it could with no protection. Lotions are available with SPFs of 2 to 50. Fair-skinned persons should use a preparation with an SPF of no less than 15. Bald-headed men also should regularly apply sunscreen to their head. Reapply sunscreen after swimming or excessive sweating.

Many formulations contain *PABA* (*para-aminobenzoic acid*) or its derivatives. These additives occasionally irritate the skin of sensitive individuals. Try a sunscreen at home before taking it abroad, and bring an ample supply on the trip, because sunscreens are not routinely available for purchase in the developing world.

Reflective sunscreens contain zinc oxide, talc, or titanium dioxide in an ointment base. They can be extremely effective when applied for added protection of sun-sensitive areas like the nose, lips, and shoulders.

Fungal Infections

Travelers are prone to develop superficial fungal infections in tropical humid climates. Common fungal diseases include *athlete's foot* (*tinea pedis*) and *jock itch* (*tinea cruris*). A related problem is *truncal ringworm* (*tinea corporis*), which is characterized by itching and the formation of ring-shaped, scaly patches.

Travelers should carry topical antifungal solution such as *clotrimazole solution* (*Lotrimin 1%*), in case the need arises. A good preventive measure is to apply powders after bathing. Wearing sandals can help to prevent athlete's foot by allowing proper ventilation of the spaces between the toes.

Women are more susceptible to vaginal yeast infections in the Tropics, especially if they are pregnant, diabetic, or on antibiotic therapy. To prevent this condition, women should keep the genital area cool and dry, wear cotton underpants, and avoid synthetic fabrics. Should the problem occur, a daily douche with vinegar and water can be effective. The application of yogurt can be helpful, as can vaginal antifungal creams or suppositories.

Other Skin Conditions

Travelers are prone to friction blisters and should carry tape to cover any that occur. Foot powders will help prevent athlete's foot and blisters. Always wear shoes or sandals—even at the beach—to prevent penetration of the skin by larvae of intestinal worms living in the soil. Persons suffering from acne or eczema may find that their condition is exacerbated in hot, humid climates, so they should be prepared for this. Travelers returning home with unusual skin lesions should seek medical attention. Inform the physician of recent travels.

ENVIRONMENTAL HAZARDS

Travelers are affected by environmental factors, such as air pollution, temperature extremes, altitude, toxins in the air and water, as well as hazards and crime.

Air Pollution

Rapid population growth and industrialization in many developing world cities have resulted in severe air pollution. In cities such as Mexico City, Beijing, and Sao Paulo, atmospheric concentrations of air pollutants are often several times higher than allowable levels in North American cities. High levels of ozone, sulfur dioxide, and oxides of nitrogen can irritate the eyes, throat, nasal passages, and bronchial tree. Persons with asthma or other pulmonary conditions can experience respiratory difficulties in some polluted cities.

Travelers should refrain from strenuous exercise during periods of excessive air pollution. Contact lens wearers are especially prone to eye irritation due to smog or dust. They should carry an adequate supply of cleaning, rinsing, and soaking solutions, as well as an extra pair of eyeglasses.

Heat Illness and Its Prevention

The heat and humidity of the Tropics can be oppressive and can result in *heat exhaustion* when an individual overexerts. This mild condition is characterized by elevated body temperature and dehydration. It can usually be effectively treated with rest and rehydration. However, it is also an early warning sign of *heatstroke*.

Heatstroke is a medical emergency characterized by a seriously elevated body temperature, nausea, fatigue, headache, and flushed, dry skin. Symptoms can progress to confusion, loss of consciousness, and coma. This condition is more common in elderly, debilitated persons and heavily intoxicated persons. People engaging in strenuous activity in tropical climates should rest frequently and consume alcohol only in small quantities.

Treatment of heatstroke requires emergency medical care with attention to rapid cooling (e.g., ice-water bath) and monitoring of vital signs and cardiac rhythms. Where no medical facilities exist, cool the body by applying towels soaked in cold water and/or massaging with ice bags, and arrange for emergent medical evaluation.

Acclimatization can prevent heat illness by increasing the body's ability to tolerate heat stress. A gradual increase in the duration and magnitude of physical exertion over a period of eight to ten days can prevent heat illness.

Altitude Sickness

Altitude sickness is usually thought of as a condition suffered by mountain climbers. However, many unacclimatized persons experience low-grade symptoms following sudden ascent to intermediate altitudes (e.g., cities like Denver, Mexico

City, or Nairobi). Symptoms such as headache, insomnia, nausea, and loss of appetite are common, and vary with individual susceptibility (see Table 15-K).

Symptoms can be prevented by gradual acclimatization. Ascend slowly, resting as necessary. Alternatively, a medication called *acetazolamide* (*Diamox*) can help prevent altitude sickness.

Low-grade symptoms are warnings to the traveler to rest and postpone further ascent (Table 15-L). Be aware that mild symptoms can progress to a more malignant, life-threatening acute mountain sickness when the traveler is at altitudes above 9,000 feet or is pursuing strenuous activity.

Toxic Hazards

Besides the dietary hazards reviewed in association with traveler's diarrhea, others can pose a potential risk to the traveler. To avoid sources of toxic food contamination, try to learn about local dietary habits. In many cases, local people are aware of the problem and can detoxify food by preparation. In Japan, for example, the *fugu*, or *puffer fish*, is so poisonous that chefs must be licensed in order to prepare it. The roe, liver, and skin, which contain the deadly *tetrodotoxin*, must be removed before the fish is consumed. Similarly, the unripe, raw, or undercooked ackee fruit of Jamaica is highly toxic. It contains a powerful substance that rapidly lowers blood sugar, causing vomiting, unconsciousness, and eventually death.

Mycotoxins—toxic substances produced by a fungus—are another dietary hazard. These are found on food containing molds. The best-known example is *aflatoxin*. It is a potent carcinogen found on moldy peanuts in Africa and other continents.

Many indigenous peoples commonly use traditional folk remedies and can offer them to ailing travelers. While many of these remedies are known to be medically effective, some are hazardous. For example, Mexican folk remedies *azarcon*, *greta*, and *albayalde* have been known to contain toxic quantities of lead-based salts (Baar & Ackerman, 1988). *Gordolobo* herbal tea, used in many countries to treat cough and diarrhea, may cause venoclusive liver disease (Gordon, 1988).

Commonly used substances in the developing world are harmful. For instance, lead poisoning has been reported in relation to the use of ceramic cookware. Lead-based glazes are still used on pottery in Mexico, Italy, and other countries. *Coumarin*, a potent anticoagulant used only by prescription in the United States, is used as a flavor enhancer in vanilla extract in Latin America. Excessive consumption can cause bleeding tendencies. As these examples suggest, it would be prudent to read up on indigenous medicinal and culinary customs in order to avoid toxic ingestions.

Swimming and Bathing Hazards

Exposure to freshwater rivers, canals, and lakes can be hazardous in certain developing countries because of *schistosomiasis* (*bilharziasis*), a parasitic disease that infects more than 200 million people in developing nations. *Schistosomiasis* can manifest as acute systemic disease (*Katayama disease*) or can lead to the development of chronic disease of the

Table 15-K. Tips for Prevention of High Altitude Illness.*

1. A slow ascent to high altitude is most important. Because each person adapts differently to every exposure to altitude, rather than give a formula for how slow is slow enough, travelers should monitor each other for signs of high-altitude illness, no matter what the ascent rate.

2. Climb high, but sleep low. The sleeping altitude should rise gradually. A conservative rate of ascent is prudent, taking 2 days or more to get a sleeping altitude of 10,000 ft (3050 m). Thereafter, avoid raising the sleeping altitude more than 1000 ft (300 m) a day.

3. Carry drugs for high-altitude illness.

4. Avoid sedatives, tranquilizers, and narcotic analgesics. These drugs result in hyperventilation and greater oxygen destruction during sleep. They should be avoided at altitudes above 8000 ft (2450 m).

5. Monitor food and water intake. Drinking plenty of fluids is routinely recommended. The diet should be palatable, and low in salt.

6. Monitor physical activity. Avoid strenuous overexertion for the first few days at altitude. Conscious effort to increase the depth and frequency of breathing at altitude is beneficial.

7. Wear appropriate clothing. Hypothermia is synergistic with the deleterious effects of altitude. Adequate clothing is necessary; the temperature drops 3.5°F for every 1000 ft (0.65°C for 100 m) of ascent.

8. Avoid external transport to higher altitude. When an individual is having physical problems continuing on his/her own, mechanized or animal transport to a higher altitude may lead to worsening high-altitude illness.

* Reprinted with permission from *The Travel & Tropical Medicine Manual*, 2nd edition, edited by Elaine C. Jong, MD, and Russell McMullen, MD. Philadelphia: WB Saunders Company, Copyright 1995.

Table 15-L. Recognition of Significant Altitude Illness*

Travelers should suspect significant altitude illness in themselves if they have:
- A headache and feel "hung over"
- Dyspnea and a respiratory rate above 20 at rest
- Anorexia
- Vomiting
- Ataxia
- Unusual fatigue while walking

Travelers should suspect significant altitude illness in their companions who are:
- Skipping meals
- Exhibiting antisocial behavior
- Stumbling
- Having the most difficulty with the activity
- Arriving last at daily destination and are the most fatigued.

* Reprinted with permission from *The Travel & Tropical Medicine Manual*, 2nd edition, edited by Elaine C. Jong, MD, and Russell McMullen, MD. Philadelphia: WB Saunders Company, Copyright 1995.

---

lungs, liver, intestines, and/or bladder. It is a hazard to travelers especially in parts of Brazil, Puerto Rico, Egypt and sub-Saharan Africa, southern China, Southeast Asia, and the Philippines. It is usually contracted when free-swimming larvae (*cercariae*) released by infected snails penetrate human skin during swimming or bathing in fresh water. Freshwater swimming can also pose hazards for contracting traveler's diarrhea, leptospirosis, etc. Such infection cannot be acquired in salt water or in chlorinated swimming pools.

Swimming at beaches can also pose dangers, as oceans may be contaminated by raw human sewage and/or toxic wastes, particularly near large cities. Because governments sometimes falsify statistics to promote tourism, it is wise to check with local inhabitants or environmental groups about contamination.

Persons whose occupations involve constant exposure to seawater (e.g., fishermen, shrimpers, and merchant marines) are at increased risk for infection with *Vibrio* bacteria, which can cause diarrhea or wound infections. They should also beware of corals, jellyfish, and other marine creatures known to be hazardous.

Injuries

The most common cause of death and disability among Westerners visiting the developing world is injuries, especially those sustained in motor-vehicle crashes. In fact, deaths from motor-vehicle collisions may be from seven to thirteen times higher in developing countries than in the United States.

Given the sorry state of traffic safety standards in the developing world, travelers must do what they can to protect themselves. Measures include wearing seat belts, avoiding reckless chauffeurs, and doing one's own driving if necessary. The best way to avoid an emergency blood transfusion overseas is to prevent a traffic collision.

Other major causes of injuries include drowning, carbon monoxide poisoning, electric shocks, and adverse reactions to hazardous drugs (CDC, 1996–1997). Remember not to purchase over-the-counter drugs in the developing world unless you are familiar with the product.

Crime

Crime has become increasingly common in many developing-world cities. Any traveler who has been robbed would concur with the old English proverb "The heaviest baggage for a traveler is an empty purse."

To discourage criminals, dress modestly and keep expensive jewelry or cameras hidden. To avoid physical assault, travel in groups or with trusted local inhabitants whenever possible. Americans can be targets for political terrorism or kidnapping overseas. For information on the political situation in any foreign country, call the State Department in Washington, DC (202/647-5225).

MENTAL HEALTH AND SUBSTANCE ABUSE ABROAD

During travel and overseas assignments, travelers will certainly be subjected to significant psychological demands. "Culture shock" is a well-known phenomenon. Social isolation from family an friends is common and can precipitate mental health problems like depression and substance abuse.

When traveling, try to limit alcohol intake, and avoid using illegal drugs, such as cocaine, opiate narcotics, and marijuana. The adverse health effects of these substances, as well as the legal consequences, should dissuade travelers from "overindulging" while abroad.

Alcohol Hazards

Whether for business or for pleasure, travelers tend to overindulge in alcohol when they are abroad. Heavy consumption of alcohol can predispose the traveler to heat exhaustion, heatstroke, and accidents. It can also complicate problems such as motion sickness, jet lag, traveler's diarrhea, and dehydration. For this reason, travelers should moderate their drinking, especially during times of active travel or physical weakness.

Consequences of Illicit Drug Use Abroad

Illegal drugs such as cocaine, opiate narcotics, and marijuana are frequently available to substance abusers overseas. These drugs are often for sale at low prices, and apparently many Americans do purchase illegal drugs when abroad. According to U.S. State Department reports, more than 1,200 Americans were arrested in 1987 for drug-related offenses. The majority of these arrests occurred in West Germany, Mexico, and the Caribbean Islands (Lange & McCune, 1989).

The legal consequences of drug possession can be extremely harsh in foreign countries. Many prisoners are denied prompt access to legal counsel or trial by jury. Penalties can be severe. In Saudi Arabia, alcohol is treated as an illicit drug, and foreigners have been subjected to whippings as punishment for drunken behavior. Intravenous drug abuse abroad is associated with a significant risk of bloodborne hepatitis and HIV infection, just as it is in the United States.

The Recovering Addict and Travel

Persons attempting to recover from alcoholism may find it difficult to abstain during their trip. Alcoholic beverages are dispensed freely aboard commercial air flights. Hosts in foreign lands often press their guests to drink. The lack of familiar social supports makes it difficult to resist these pressures. For alcoholics, *disulfiram (Antabuse)* can be helpful to prevent impulsive drinking. Alcoholics Anonymous also has branches overseas, should the traveler wish to attend a meeting. Of course, it would be held in the local language. In the early stages of rehabilitation, the traveler's spouse or a colleague should accompany the traveler. Should intoxication and/or withdrawal from drug or alcohol abuse occur overseas, adequate medical and rehabilitation facilities are often unavailable. Prompt repatriation may be necessary.

INTERNATIONAL AIR EVACUATION

Despite careful planning to the contrary, a significant number of persons (approximately 0.07%) will become ill enough to require air evacuation. As discussed in the pretravel planning section there are a variety of organizations that provide these services. Much of the pioneering work in this area has been done by Swiss-Air Ambulance (REGA). There are many components to in-travel emergency management, which need to be orchestrated smoothly. Of 5,366 emergency calls to REGA from all over the world, 2,105 (39%) were managed via counseling and telemedicine assistance, while 61% were repatriated by scheduled airline or by air-ambulance. The decision to air evacuate is a difficult one, based on illness severity, available medical facilities and psychosocial setting.

MORE INFORMATION ON TRAVEL HEALTH

For more information, contact the Center for Disease Control International Traveler's Hotline at 404/332-4559 or CDC Fax Information at 404/639-2888. United States citizens can call the U.S. Department of State Citizens Emergency Center at 202/647-5225. Internet resources are also useful. The CDC Travel website can be found at www.CDC.gov. The web page offers more detailed health information. The International Society of Travel Medicine website at www.istm.org has several useful links including a directory of travel clinics and educational resources.

SUMMARY

International travel can be a personally rewarding experience. Much of the enjoyment of foreign work comes from interacting with new cultures: trying new foods, meeting new people, and exploring new environments. Indeed, adopting certain local behaviors is often the key to smoothing human relations and accomplishing work goals abroad. Good health also is a prerequisite to enjoying a productive trip abroad, and effective attitudes toward prevention will develop naturally if travelers are aware of common sense precautions.

The traveler should not be dissuaded from experimenting with local customs out of fear of contracting exotic tropical diseases. Rather, it is more beneficial to understand the health hazards and know ways to prevent problems. After professional consultation, the traveler will be able to prioritize risks and to intelligently decide which behaviors are safe to practice and what adjunctive protection, if any, is necessary. This thought process is analogous to the way that occupational health and safety professionals weigh various toxic risks in the workplace. However, while many of the health hazards due to toxic exposures have not yet been scientifically determined, the vast majority of travelers' illnesses are well studied and are preventable and/or amenable to appropriate medical care.

As with any endeavor, proper planning is essential. With professional consultation, the traveler can conveniently arrange immunizations, a traveler's medical history, and a medicine kit. Effective techniques are available to reduce jet lag and motion sickness, enabling travelers to orient quickly to new surroundings upon arrival.

During the first few days of the trip, pay careful attention to eating habits. It may be necessary to refuse food or drink from a gracious host. This can be done discreetly, without offending the locals. These measures will help to prevent traveler's diarrhea and related conditions. Control of mosquitoes and other insects will reduce chances of malaria and other insect-borne diseases. Safe-sex precautions and the avoidance of blood transfusions are vital to the avoidance of AIDS and other sexually transmitted and/or bloodborne diseases. Environmental pollution is becoming severe enough in developing nations to threaten the health of foreign visitors, but advance knowledge of high-risk areas can reduce these threats.

Following the precautions outlined in this chapter will greatly decrease the chance of difficulties during a trip. With added care, employees with medical conditions, pregnant women, and rehabilitating substance abusers will be able to travel safely by following a few extra precautions.

Should an illness occur abroad, follow the directions provided for self-help in this chapter. Consult with well-qualified health care providers preselected from appropriate referral networks.

REFERENCES

Baar RD, Ackerman A. Toxic Mexican folk remedies for the treatment of empacho: The case of azarcon, greta, and albayalde. *J Ethnopharmacology* 24(1):31–39, 1988.

Centers for Disease Control. *Health Information for International Travel.* Washington DC: U.S. Government Printing Office, 1996, also 1999–2000.

Council on Scientific Affairs. Harmful effects of ultraviolet radiation. *JAMA* 262(3):380–384, 1989.

DuPont HL, Steffen R. *Textbook of Travel Medicine and Health.* Hamilton Ontario: BC Decker, 1997.

Ericsson CD. Travelers' diarrhea: Epidemiology, prevention and self-treatment. *J Inf Dis Clin N Amer* 12(2):285–303, 1998.

Hargarten SW, Baker SP. Fatalities in the Peace Corps: A retrospective study: 1962 through 1983. *JAMA* 254:1326–1329, 1985.

Jong EC. Immunizations for international travel. *J Inf Dis Clin N Amer* 12(2):249–267, 1998.

Jong EC, McMullen R. *Traveler's Health Guide,* 2nd ed. Seattle: University of Washington School of Medicine, Division of Allergies and Infectious Diseases, 1995.

Kain KC, Keystone JS. Malaria in travelers. *J Inf Dis Clin N Amer* 12(2):267–283, 1998.

Kozarsky PE. Prevention of common travel ailments. *J Inf Dis Clin N Amer* 12(2):305–321, 1998.

Lange WR, McCune BA. Substance abuse and international travel. *Adv Alcohol Substance Abuse* 8(2):37–51, 1989.

Lobel HO et al. Use of prophylaxis for malaria by American travelers to Africa and Haiti. *JAMA* 275:2626–2627, 1987.

Mann JM. Emporiatric policy and practice: Protecting the health of Americans abroad. *JAMA* 249:3323–3325, 1983.

Montgomery AB, Mills J, Luce JM. Incidence of acute mountain sickness at intermediate altitude. *JAMA* 261(5):732–733, 1989.

Parrot AC. Transdermal scopolamine: A review of its effects upon motion sickness, psychological performance, and physiological functioning. *Aviation, Space Env Med* 60(1):1–9, January 1989.

Steffen R. *Proceedings of the Conference on International Travel Medicine,* Zurich 1988. Berlin: Springer-Verlag, 1988.

World Health Organization (WHO). *Vaccination Certification Requirements and Health Advice for International Travel.* Geneva: WHO, 1992.

chapter 16

Ergonomics Programs

by Ira Janowitz, PT, CPE
David Thompson, PhD, CPE

277 **Introduction**
278 **Designing and Redesigning Jobs**
Laying out and adjusting the work space ■ Sitting vs. standing work ■ Repetitive work
281 **Materials Handling**
281 **Risk Factors**
Work-related risk factors ■ Personal risk factors
282 **Biomechanics and Biochemistry of the Spine**
Spinal degeneration ■ Sitting ■ Back school ■ Administrative controls ■ Work involving computers ■ Inspecting and monitoring jobs ■ Musculoskeletal stress
296 **Equipment Design**
User-machine communication and cooperation ■ Operating machines
297 **Hand Tool Design**
Factors causing musculoskeletal stress ■ Proper working height and location ■ Summary causes of cummulative motion trauma
300 **Evaluating the Environment**
Physical hazards ■ Safe noise levels ■ Lighting ■ Heat and humidity ■ Whole-body vibration
303 **Evaluating Job Demands**
Survey methods ■ Techniques for analyzing human motion ■ Ergonomic audits and evaluations
305 **Employee Exercise and Strengthening Programs**
Effects of exercise on musculoskeletal stress ■ Appropriate exercises ■ Effect of strength
307 **Designing Light-Duty and Modified-Duty Work**
Matching injured workers with light-duty jobs ■ Selecting alternative jobs ■ Follow-up evaluation and return to "active duty"
310 **Summary**
312 **References**

INTRODUCTION

Ergonomics (from the Greek *ergon,* "to work," and *nomos,* "study of") is literally the study of work, or the work system, including the worker, his or her tools, and his or her workplace. In modern usage, it refers to an applied science concerned with human characteristics that need to be considered in designing and arranging things that they use in order that people and things will interact most effectively and safely.

The Board of Certification in Professional Ergonomics, established in 1990, defines ergonomics as a body of knowledge about human abilities, human limitations, and other human characteristics that are relevant to design. Ergonomic design is the application of this body of knowledge to the design of tools, machines, systems, tasks, jobs, and environments for safe, comfortable, and effective human use.

Industrial ergonomics can be defined more specifically as the application of relevant areas of the life sciences, physical sciences, and engineering to the work system consisting of the interactions between workers and their occupations, tools and equipment, and working environment such as atmosphere, heat, light, vibration, and sound. Specialists in industrial ergonomics aim to evaluate and design jobs, facilities, equipment, training methods, and environments to match the capabilities of the users and workers.

Other terms that are used to define approximately the same discipline are human factors, human engineering, and engineering psychology. Until fairly recently, ergonomics was used primarily outside the United States, the other terms within the United

States. But with the internationalization of this field, ergonomics is now commonly used throughout the world and will be used in this chapter to represent all of these disciplines. The health budget is becoming very large in most organizations (Bureau of Business Practice, 1991), and an organization's health and safety committee can advise on the competent management of these funds. The occupational health and safety professionals within an organization can plan programs (including supervisor training in ergonomics) that can dramatically reduce injury costs and premiums for workers' compensation insurance. From an ergonomics standpoint, supplying employees with poorly designed tools and equipment, poor job training, and workspace layouts, that are not competently designed or are overcrowded, will lead to poor productivity, reduced employee satisfaction, and high turnover rates.

In developing a new work process, the principles and procedures of ergonomics provide useful guidelines for proper job design. For existing jobs, this information sensitizes supervisors and health care professionals to work methods, work space, tooling, equipment, and environmental problems that exist, and provides some direction for solutions. Complex problems should be handled by a professionally trained ergonomist or human factors engineer.

This chapter emphasizes particular areas of knowledge that specifically apply to workplace health and safety. For a more extensive treatment of the subject, see Bailey, 1982; Kantowitz & Sorkin, 1983; Alexander & Pulat, 1985; Eastman Kodak, 1983; Grandjean, 1987; National Safety Council, 1992; and Chaffin & Andersson, 1991.

DESIGNING AND REDESIGNING JOBS

The ideally designed job places no more physical demand on the worker, operator, user, driver, or clerk than that required to function in a manner that is efficient (in the sense of accomplishing the greatest result with the minimum amount of effort). Frequently used materials are well within the worker's reach; strength and endurance capacities are considered; and a safe and healthy work environment is provided.

The approach of the occupational health and safety professional is often two-pronged—to improve the design of jobs that appear unsafe and to ensure that hazardous job conditions are corrected. The design principles discussed here apply to both of these situations. These ergonomic principles of job design and redesign are relevant to the office as well as to the factory floor and the warehouse and/or distribution facility.

Laying Out and Adjusting the Work Space

To design workplaces of proper size, we must know the physical dimensions and the range of motion workers need in order to complete a task. Optimally, the workplace design should be flexible enough to accommodate any worker. Only from such information, together with the job specifications, will an appropriate workplace geometry emerge.

One of the primary reasons for physical stress on the job is the mismatch in size between the worker and the workplace or machine. This mismatch can result in excessive reaching for frequently used tools or objects, working bent over or with one or both arms and shoulders held high for long periods, holding a power tool at some distance for long periods, or sitting on a stool or bench that is too low, too high, or otherwise improperly fitted.

Eliminating "Waist Motion"

The most important rule of physical design for a sedentary job is that the operator should be able to reach all parts, supplies, keyboards, tools, controls, etc., without leaning, bending, or twisting at the waist. In addition, it is preferable to restrict frequent reaching to movements of only the forearm. Figure 16-1 illustrates the envelope for the *forearm-only reach* (preferable) and for the *full-arm reach* (acceptable) for men and women. Task designs requiring frequent reaches and moves outside the full-arm reach envelope are generally unsatisfactory.

These rules for reach envelopes are particularly important if the work involves high forces or weights. Thus, assembly workers who have to reach to a high shelf or reach behind themselves to get parts, or bus drivers who must lean forward across a steering wheel and pull it laterally in the direction they wish to turn, are at risk in this regard. Even the continual movement of the upper body without outside forces or weights can fatigue and injure the lower back.

These characteristics of workplace design are, of course, ideals. Some movement of the waist and

Figure 16-1. Workplace layout. All work to be performed, as well as parts, supplies, machines, keyboards, etc., should be located within the preferable work area for the hand doing the work. Less frequently accessed locations can be in the acceptable work area but these should be avoided if possible. Jobs should not be designed with motions required outside the *acceptable* work areas if at all possible.

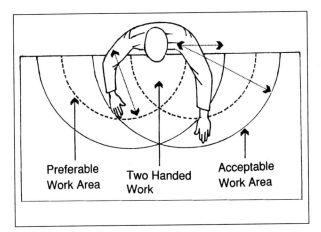

Figure 16-1a. Top view of reach envelope arcs defining the outer limits of preferable work areas (forearm movement only) and acceptable work areas (full arm movement), that do not require bending and twisting at the waist.

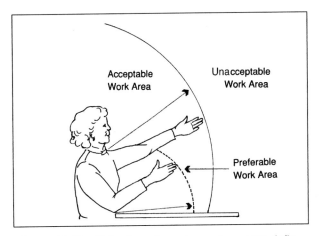

Figure 16-1b. Side view of reach envelope arcs defining the outer limits of preferable and acceptable work areas that avoid situations that propel workers to bend and twist at the waist.

shoulders can occur even in well-designed jobs. However, the design rule still applies: The less bending and twisting required of the waist and torso, the better the job design.

Example. In an electronic inspection task, the worker reached across a workbench to the conveyor belt, grasped an electronic assembly, pulled it off the belt onto the bench to modify and test, then pushed it back onto the conveyor when finished. This method required bending forward and stretching to push and pull the assembly. In the new method, the workbench was turned 90° so that the operator could sit next to the conveyor and easily reach it without bending at the waist. Assemblies were placed on nylon pads to slide more easily.

Example. In an automotive windshield plant, the worker previously clamped plastic sheets to an overhead cable by reaching over the shoulder, flexing the right hand, and compressing the clamp to open. In the new method, the cable was automatically lowered to shoulder height during clamping, eliminating all flexion of the hand. The clamp was later changed to a hook, eliminating the need to squeeze the clamp.

Avoiding Static Positions

An assembly task can involve holding an object rigidly in one hand and working on it with the other, or a maintenance job can involve holding a hand or power tool tightly in one hand for extended periods. In these cases, the continuous muscular tension is associated with local muscle fatigue and tendon strains. Static holding is usually avoided by using holding clamps and bench vises or, in the case of hand or power tools, alternating hands, using self-locking tools, and overhead suspension.

When some holding remains necessary, it is helpful to redesign the tools, tool handle, or tool trigger to increase comfort. This can be accomplished by rounding sharp edges, using padding, or substituting a handle that allows a more natural hand position (e.g., fingers not tightly curled) and better handle leverage. The goal is to grip the tool with less force and little or no wrist flexion, extension, or deviation.

Example. In a quality assurance task, the worker checked a part by picking it up and supporting it with the left hand and forearm while attaching testing clamps and making adjustments. Following job redesign, the part was placed into a waist-high rolling rack (pulled along by the quality assurance inspector), and workers made attachments and adjustments with both hands. Muscular cramping in the left wrist, forearm, and hand were eliminated, and the work was speeded up.

Static positions tend to be more pervasive among sedentary workers, and need to be compensated for. For example, when people are typing, entering data, or operating a lab machine, they tend to hold their body in a fixed position for extended periods of time to have a consistent physical relationship with the equipment to which they are relating. That is, a touch typist must maintain a fixed spatial relationship between the shoulders and the keyboard to strike the proper key each time without looking. The same is true for operators of adding machines and telephone control panels (e.g., information and PBX operators, receptionists, reservation clerks). Simple stretching exercises and frequent changes in position introduced throughout the workday can alleviate the accumulated muscle fatigue. In addition, parts of a task can be automated to allow for adequate recovery time.

A special situation arises with data-entry clerks, inspectors, word processors, microscope operators, and others who must, in addition to maintaining a fixed shoulder position, maintain the rigid neck position required by looking continuously in the same direction. In this case, the workplace can be redesigned along the lines suggested later under the discussion of stress caused by poor job design, as well as in Chaffin & Andersson, 1999, chapter 10, and in Webb, 1982. In addition, these workstations can be redesigned to allow rapid changes of the work surface to accommodate standing or sitting. This will relieve the constrained muscle and joint positions of the legs and lower back.

Sitting vs. Standing Work

If the preceding guidelines are followed, the workplace design will place tools, equipment, parts, and supplies within easy reach of the worker. However, this ideal situation is not always possible, even with a lot of careful ergonomic design, because of the constraints of the job or the machinery used. For some machinery, it is necessary to ensure that the vertical positioning of the worker is optimum. For a seated worker, the adjustable chair should be raised or lowered so that the worker is essentially doing his or her primary work near the level of the elbows, with the upper arms comfortably at the worker's sides. A worker who is standing either should be on a raised platform, or the work area should be raised in order to produce the same relative dimensions. A well-designed chair is important in helping to position the seated worker to be as comfortable and productive as possible. A *forward-tilting seat pan* or a *sit-stand chair* can help, particularly for the worker who performs extensive reaching tasks.

In the case of the seated worker, a footrest can accommodate a relatively shorter worker seated in an elevated position, but can result in increased *waist motion* when reaching is required. For the standing worker, a high stool permitting alternate standing and sitting reduces fatigue substantially. An elevated footrail, (8–10 in. [20–25 cm] high), appropriately placed, allows the worker to shift his or her weight from one leg to the other during the day. Both seated and standing workers should be working with their hands near elbow height. Optimally, this goal should be reached by adjusting the height at the work surface rather than using footstools or standing platforms, which constrain the worker's movements.

Repetitive Work

As described later in the chapter, muscular stress can be caused by the frequent, repetitive use of the same muscle group throughout the day (e.g., using the trigger finger in power tools, trimming meat with a knife, reaching to an upper shelf). If the force required exceeds the strength of the muscle group involved, significant fatigue and stress generally occur.

Motions applying relatively low levels of force, if very repetitive, also can be responsible for musculoskeletal trauma. For example, using a keyboard for data or text entry, operator entry rates can exceed 12,000 to 15,000 keystrokes per hour. Even though each keystroke may require only 9 to 12 ounces of force per stroke (2.5 to 3.4 newtons), the cumulative force exerted in the fingertips and hands can exceed 80,000 pounds or 40 tons of force (360,000 newtons) per day.

The physiological effect of this excessive use of the same muscle and tendon group over long periods of time can be inadequate removal of metabolic end-products and swelling and decreased lubrication within the synovial sheaths that support the tendons. This, in turn, causes added pressure and friction on the tendons. Moreover, persistent swelling of the synovial sheaths can lead to pressure on the median nerve in the wrist, particularly a flexed and deviated wrist, and can contribute to carpal tunnel syndrome.

It is sometimes difficult to reduce repetitive physical motions and still maintain efficient job design and work group productivity. However, because of the physical trauma and discomfort involved, the company should seriously consider such approaches as *job redesign, job rotation* (varying the task mix during the day), *job enrich-*

ment (expanding the complexity and variety of the job), and *job enlargement* (expanding the scope of the job).

MATERIALS HANDLING

Low-back pain is the most common cause of limited activity in adults between 20 and 45 years of age, and the most expensive health care problem in that age group. Back problems will affect approximately 80% of all Americans at some point in their lives, and they cause the loss of more than 93 million workdays per year in the United States. Approximately 3% of the U.S. population are temporarily impaired by low back pain at any one time, and 1% of the U.S. population is disabled by low back pain. The annual cost for this is a staggering $50 billion per year in lost productivity, sickness benefits, and treatment.

Although low back injuries account for approximately 25% of workers compensation claims, they consume between 33% and 41% of the total costs. Of particular importance is the fact that less than 10% of work-related cases of low back pain consume 75% of the costs and approximately 68% of the working days lost. But less than 10% of work-related low back pain is actually reported through the workers compensation system. Many people continue to work in pain and pay for medical expenses either out-of-pocket or through their personal health insurance.

RISK FACTORS

Risk factors are the properties of a person or the environment that increase the probability of developing an injury or disease. Prevention programs are most effective when they focus on known risk factors. As is true in the case of other occupational health problems, exposure to risk factors, both on and off the job, has a synergistic effect. This section therefore covers both work-related and personal risk factors.

Work-Related Risk Factors

The relative importance of risk factors at work will vary considerably from industry to industry. For example, slips and trips are significant contributors to low back pain in some industries such as food services and public utilities, but are much less common in the health care field. With this in mind, here are some of the most common work-related risk factors:

- frequent bending and stooping; especially below knee level
- heavy lifting (weights >25–35 lb (11.3–15.8 kg)
- frequent lifting (>8–10 lifts/min)
- pushing and pulling, especially with >50 lb (22.5 kg) of force exerted at the hands
- heavy carrying, especially >33% of body weight
- prolonged standing, especially >6 hr/shift
- prolonged sitting (>6 hr/shift), especially when combined with vibration
- vibration; especially in the range of 4.5–6.0 hz
- high levels of physical exertion, especially >33% of a person's aerobic capacity
- job dissatisfaction, especially with high levels of responsibility or job stress
- monotonous and repetitive work
- unexpected movements (slips, trips, and falls account for 9%–10% of low back pain in most industries).

One of the most pervasive risk factors to which modern workers are exposed is the vibration from the operation of motor vehicles or heavy equipment. Unfortunately, much of this vibration is in the range of 4.5–6.0 Hz, which corresponds to the resonant frequency of the human spine in a sitting position. As a result, the vibration is amplified by a factor of approximately 1.76 when applied to the buttocks of a seated person. Since this combines two risk factors, sitting and vibration, it is not surprising that operators of automobiles, trucks, helicopters, tractors, forklifts, and earth-moving equipment experience three times the rate of herniated discs of the general population.

Several of the risk factors listed are relative to the physical capacity of the worker performing the tasks. For example, workers with high levels of endurance are using a lower proportion of their aerobic capacity than their less-fit co-workers. The same phenomenon applies to risk factors such as lifting, carrying, pushing, and pulling. Nevertheless, jobs cannot often be redesigned from worker to worker as a function of each employee's physical capacity. Therefore, the organization needs guidelines to control such risk factors.

Personal Risk Factors

The age of onset of low back pain is generally between 20 and 40 years of age, although it has been found to occur earlier in workers in certain fields, such as in the building trades. Men in their first five years on the job are at especially high risk, and their injuries tend to lead to the most expensive workers' compensation claims. One possible explanation is that workers who are prone to back injuries tend to leave a trade at a relatively early age, and older workers who have survived tend to learn from their mistakes. Young men also tend to engage in more high-risk behaviors than their older or female counterparts.

For the population as a whole, the following personal risk factors are the ones most often associated with low back pain:
- previous personal history or family history of low back pain
- smoking tobacco, especially cigarettes
- decreased endurance, especially with reports of fatigue at the end of the workday
- high rate of general health problems
- social isolation at home and work, especially when accompanied by feelings of depression, worry, and tension
- low educational levels
- living in suburbs or rural areas; especially when commuting >20 miles (32 km) each way to work
- tallness (>5'7" [168 cm] for women, >5'11" [178 cm] for men)
- severe obesity
- weak trunk muscles, especially in men
- tight hip and lower-extremity muscles.

Although the vast majority of episodes of low-back pain are self-limiting, the recurrence rate is in the range of 75%–80%. This means that, by far, the most predictive personal risk factor is a previous history of low back pain. In addition, a family history of low back pain makes a person much more likely to experience that problem.

Some personal risk factors appear to be associated with the job factors listed earlier. For example, low educational levels are frequently associated with heavy physical labor. In these situations, the pay is low and the work is difficult to modify. These two factors combine to increase the probability that an injured worker will return to the same work before he or she is fully able to meet the physical demands of the job, thus increasing the risk of reinjury. In addition, such jobs tend to be monotonous and repetitive, and there may be other previously mentioned risk factors, such as job dissatisfaction.

Despite years of controversy, it appears as though minor anatomical variations do not predispose people to low back problems. Scoliosis (a right-to-left curvature of the spine) has to exceed 60° before it becomes a significant risk factor. Minor increases of lordosis ("swayback") and differences in leg length are not important factors.

BIOMECHANICS AND BIOCHEMISTRY OF THE SPINE

The spine is composed of 24 *vertebrae, the discs* and *joints* connecting them, and the nerves that descend from the brain through the spine toward the arms, trunk, and legs. A *spinal segment* consists of any two vertebrae and the structures between them. The stress on the spine is the greatest in the low back or *lumbar area,* which must support approximately 40%–50% of the body weight and which bears the brunt of many of the awkward postures and movements to which we subject the body. Note that there is a normal, inward curve in the low back.

When we stand upright, approximately 80% of our weight is borne by *intervertebral discs,* and 20% by the two facet joints. The facet joints act as doorstops, limiting motion in rotation and backward bending. Each disc has a gelatinous center (the *nucleus pulposus*) that contains 80%–88% water at age 20 years and gradually dries out as we age. Surrounding this center is a tough, fibrous covering called the *annulus fibrosis.* The combination of gelatinous center and tough covering means that the disc deforms somewhat when pressure is placed on it.

Spinal Degeneration

Most back injuries are superimposed on a gradual process of degeneration of the spine, which begins at approximately 20 years of age (Figure 16-2). Cracks and tears tend to form in the annulus fibrosis and accumulate over many years. This happens on an accelerated basis in people exposed to the risk factors discussed in the previous section (such as necessary frequent bending, twisting, and vibration exposure).

Because the annulus fibrosis has few nerve endings, the early stages of this process often occur without causing significant pain. As the annulus fibrosis weakens, the disc becomes less able to withstand pressure. Activities such as prolonged sitting, forward bending, or lifting may now cause

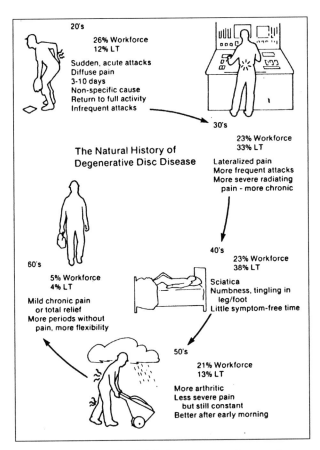

Figure 16-2. Most back injuries are superimposed on a gradual process of degeneration of the spine, which begins at about 20 years of age. For each decade between the 20s and 60s, the symptoms of low back problems are listed. The percent of the work force in those decades and the percent of lost time (%LT) from work due to low back pain are also shown. The exposure risk data are taken from the studies of Dr. Rowe (1983) over a 30-year period. (Reprinted with permission from Rodgers SH. *Working with Backache.* Fairport NY: Perinton, 1985.)

the disc to bulge backward, toward the ligaments and nerves behind the disc, initially causing pain in the low back.

Portions of the nucleus pulposus can work their way through the cracks of the annulus fibrosis. With repeated bending, twisting, or other exposure to risk factors, some of this jellylike substance may herniate, or rupture, through the annulus.

The human body does not manufacture nucleus pulposus in adulthood. Any loss of this material will tend to result in a flat tire effect, which will flatten the disc over time. This will decrease the space between the surfaces of the facet joints, increasing the pressure and friction at these joints. This process accounts for the tendency of people to shorten as they age. Backward bending, which increases the pressure on the facet joints, becomes painful as the condition of the facets worsens.

These mechanical factors do not explain all episodes of low back pain. The development of tears in the annulus fibrosis is now thought to release chemical substances from the nucleus pulposus that induce inflammation in the tissues surrounding the disc. In many people, increased pressure on the disc may not put direct pressure on nerves, but rather may squeeze out small amounts of nucleus pulposus and induce a chemical irritation.

Body Mechanics

When a person leans forward to lift an object, the back muscles contract to counteract the force of gravity on the upper body and the object being lifted. Back muscles are composed of so-called *postural fibers,* which are better designed for stabilizing the spine in slow movement than for rapid contraction. Rapid movements of the weight of the upper body and/or the load being lifted often exceed the capacity of the back muscles to respond appropriately. This accounts for the high rate of back injuries reported as a result of sudden or unexpected movements.

As these back muscles fatigue with repetitive lifting, the individual loses the ability to stabilize the spine and will tend to lose the normal inward curve in the low back. This is especially true with rapid lifting as opposed to slow, heavy lifting. As the spine sags forward into full flexion, the load is taken up by the ligaments in the spine, which can become overstretched. In addition, the intervertebral disc becomes further compressed (see Figure 16-3). If the worker twists while bending forward, the effect is like wringing out a washcloth, and increases the tendency to produce tears in the annulus fibrosis. It is these combined motions forward bending with twisting or backward bending with twisting that are most damaging to the intervertebral discs and facet joints.

Safe Practices

One way to minimize the load generated by the object being lifted is to reduce the horizontal distance between the spine and the center of gravity of the load. This decreases the torque acting on the spine by reducing the lever arm for the weight of the load, which tends to pull the upper body forward. A slow speed of lift also will minimize the reaction forces in the back muscles.

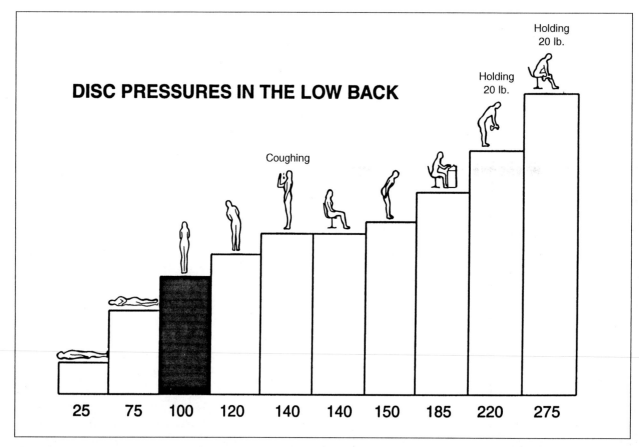

Figure 16-3. This chart shows how the pressure in the discs of the lumbar area changes as the body's position changes. If standing equals 100%, then lying on our back equals 25%; coughing in a standing position equals 140%; and sitting at a typewriter equals 185%. Lifting a 20-lb (9 kg) weight in standing and sitting positions increases the pressure to >200%. (Adapted with permission from Nachemson AL. The lumbar spine, an orthopaedic challenge. *Spine* 1(1):59–71, 1976.)

For many years, bending the knees has been advocated as a means to reduce stress on the spine. Figure 16-4 illustrates, however, that a deep squat does not necessarily reduce the factors contributing to the load on the spine when lifting. The distance in Figure 16-4 from the center of gravity of the object being lifted to the L5-S1 segment of the spine (H) is approximately 24 in. (45 cm). Objects wider than about 12 in. (30 cm) cannot easily fit between the knees of most people. In such situations, lifting from a deep squat actually increases the lever arm of the object being lifted and, therefore, increases the load on the spine as compared with a straight-knee, bent-back lift.

Ayoub (1982a) has outlined the dubious assumptions underlying the traditional deep-squat lifting pattern:
- The worker will always be in a position to maintain a straight back and bent knees.
- The object to be handled is free of obstacles or barriers.
- The shape of the object or container is assumed to be a box.
- The container can be handled without use of special coupling devices.
- The object can be placed between the knees.
- The object is free to be moved in a trajectory determined by the worker (i.e., it is unattached at each end).
- The lift can be accomplished slowly and deliberately.
- The weight of the object can be determined before the lift, and the worker will be able to assess the inertia of the object as it moves.

However, a third alternative, the *power lift*, substitutes hip flexion for both lumbar and knee flexion, so the spine can be stabilized as it is tilted forward (Figure 16-5). This lifting position is especially useful when there is an obstruction to knee flexion, as in reaching into a bin, or the trunk of a car. Another strategy allowing for stability of the spine is the *diagonal lift*, with one foot in front of the other.

The deep squat is an unpopular lift method because it requires more energy, is hard on the

Figure 16-4. A deep squat position does not necessarily reduce the factors contributing to the load on the spine when lifting. Note the 24-in. (60 cm) horizontal distance from the lumbar spine to the load's center of gravity. The spine is poorly stabilized.

knees, and takes more time. In addition, it allows for significant forward bending of the back, and often keeps the load far away from the spine. In contrast, the straight-back bend and diagonal lifts are more likely to keep the spine stable, afford better balance with a wider stance, and decrease the horizontal distance between the center of gravity of the load and the lower lumbar spine.

Different objects require different lifting techniques. The best way to lift bulky objects is to straddle them with the feet offset, often with a two-stage lift. In overhead lifting, the abdominal muscles take on increased importance in allowing the worker to avoid excessive backward bending, which puts increased pressure on the facet joints. To avoid twisting, the worker can pivot on the foot toward which he or she is about to move, as in passing a basketball. While no one lifting strategy will cover all situations, the following points summarize the most important principles:

- Plan the lift so as to avoid unexpected loads or motions.
- Lift slowly if possible.
- Avoid twisting and side bending.
- Contract abdominal and back muscles to stabilize the spine. Tuck in the chin gently if lifting overhead.
- Keep loads as close to the body as possible.
- Avoid lifting from below knee level, over shoulder height.
- Store frequently used items between waist and shoulder height.
- Get help if needed!

Sitting

When a person is sitting, the existence and position of a visual target unconsciously determines posture. As you read this book, its location relative to your eyes will determine the position of your neck and back.

Sitting Postures

Analyses of posture in sitting show that there are actually three distinct types of sitting. These are upright or erect sitting, forward sitting, and reclining.

Upright or erect sitting. Although sitting erect has traditionally been advocated as "good posture," studies of *intradiscal pressure* show that pressure on the lumbar discs is 40%–50% greater in that position than when standing. This is because, in a straight chair, the pelvis rotates backward (approximately 38°) when sitting, and the normal inward curve of the low back tends to flatten. A way to mitigate the increase of pressure is to use a lumbar cushion or support, which can restore some of the lumbar lordosis. An appropriately shaped backrest will reduce lumbar disc pressure by approximately 30%.

Traditional advice on proper sitting posture holds that it is somehow natural to have a 90^a angle between the trunk and the thigh. However, the hamstrings and trunk extensors exert a force toward a wider trunk-to-thigh angle. Numerous studies have shown that most people are more comfortable with a hip angle of 120°–130°. Such an angle will allow a lumbar support to be more effective, because the hamstrings will be less stretched.

Within the first few minutes of sitting at a desk to do paperwork virtually everyone slouches forward. This results from a conflict between the need to focus on a visual target and the posture of upright sitting. Human beings tend to orient their line of sight almost perpendicular to the surface at

Figure 16-5. Note that with a *power lift* in which the back is stabilized while it is tilted forward, the worker's spine is closer to the load, and his balance and base of support are better.

which they are looking. If the visual target is a book on the surface of a standard desk (height = 30 in. [76 cm]), the spine has to bend forward for the person to see the target. In addition, for most people, the distance from the eyes to the target will be much more than the usual adult focal length for reading of approximately 16 in. (40 cm), resulting in further forward bending. Upright sitting is most appropriate for tasks such as using a computer or driving.

Forward sitting. When reading, writing, and various workbench tasks draw the spine forward, the result is pressure on the lumbar discs approximately 90% greater than that of standing (see Figure 16-3). Yet this is the position in which most people who sit for a living work, because it accommodates their line of vision and focal distance to fine or detailed work.

Unconsciously, people try to "unweight" their spines by leaning on the workbench or desk in front of them. Alternative strategies to accommodate to the demands of forward sitting are the use of a *sitting wedge* or *forward-sloping seat* to tilt the pelvis forward so as to reduce lumbar flexion. Since kneeling chairs shift a significant portion of one's body weight onto the knees, they are not appropriate for long work periods or for people with significant knee problems.

A chest pad to help support the weight of the upper body may be helpful for situations, such as dentistry, in which forward leaning cannot be avoided. However, in most situations, it is better to raise the visual target of the worker on a slanted surface or document holder.

Reclining. The reclined position allows a proportion of body weight to rest on the back of the chair and, along with a lumbar support, reduces the pressure on the lumbar discs to approximately 25% that of standing. A problem with the reclined position arises if the visual target is too low or far away. To compensate, people may tend to increase neck flexion, increasing the stress on the neck. Reclining is most appropriate for situations in which the worker does not have to focus on small details or perform fine motor tasks.

Workplace Design

The problems that arise in the most common sitting postures have led to various attempts in the last several years to develop new chair and workstation designs that can support the worker in a variety of positions. The critical features of a good chair design are a built-in lumbar support and a hinge mechanism below the seat pan that allows for forward, horizontal, and backward tilt of the seat pan. Ergonomic chairs are highly adjustable, and some

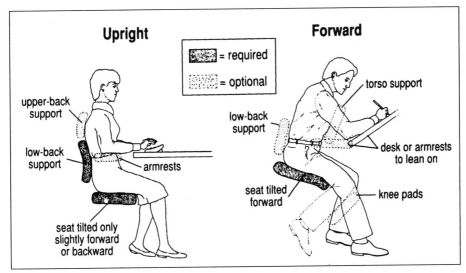

Figure 16-6. Different uses for an ergonomic chair in forward, upright, and reclined sitting positions. (Courtesy Back Designs)

come in sizes. Arm support can be provided by armrests on the chair or the work surface. In either case care must be taken to ensure that supports do not interfere with the worker's ability to move within 12–15 in. (30–38 cm) of the point of operation (hand location) needed to perform the task. Figure 16-6 illustrates some adjustments of a chair in forward and upright sitting positions.

For highly static work situations, such as light assembly or call centers, a workplace design that accommodates both sitting and standing allows for frequent changes in position. The angle between the trunk and the thigh of the person using this workplace can approximate 135° in either sitting or standing with a footrest.

Surveys of major U.S. employers have indicated that many computer users work in situations where workplace design is unsatisfactory. Good ergonomic design can increase productivity by up to 25% in workers whose employers have upgraded the furniture and equipment to better fit the workers and the tasks they are performing.

Complaints of low back pain with sitting are certainly not limited to office situations. In surveys of coalface miners, for example, the activities likely to lead to low back pain include sitting on personnel transports and automatic mining machines without low back support. Lumbar supports welded onto this equipment have significantly improved the situation.

Motor vehicles also should have lumbar support, which should be adjustable for a range of operator heights. Motor vehicle seats should be adjustable so that the backrests can be reclined to at least 20° beyond the upright. Sliding the seat forward toward the foot controls will allow operators who wish to do so to raise their knees. Easily adjustable mirrors will reduce the amount of neck and back rotation required in the act of driving. In heavy vehicles, vibration can be dampened using air-suspended seats or cushions with visco-elastic polymers.

Designing and Adjusting Seating

The primary purpose of a chair is to support the weight of the body comfortably with minimal restriction of circulation. Changes in posture permit different portions of the anatomy to support the body's weight, while other parts enjoy increased circulation. Leaning to one side, crossing the legs, slouching, and other ways of shifting in the chair all keep sufficient circulation in the legs and posterior. Thus, chair design needs to accommodate these postural variations. (For the effect of seated posture on spinal disk pressure, see Chaffin & Andersson, 1984, chapter 9.)

Chair design itself also affects blood circulation in the legs. If the seat pan is too long (more than 16 in. [40 cm]), it cuts off circulation at the popliteus (under the knees), particularly for short women. It is also helpful if the edge of the seat under the knees is smoothly curved (so-called waterfall design) to eliminate pressure points. The chair seat should be soft enough to be comfortable but not so soft that changing posture or standing up is difficult. The seat pan should be relatively flat, rather than concave, so as not to roll the thighs of heavier people inward.

The second most important aspect of seat design is lumbar support for maintaining a comfortable degree of lordosis of the spine and for helping to support the weight of the back. A chair should be easily adjustable from a sitting position to conform to the full range of back curvatures, lumbar heights, and buttock sizes. Without such support, general fatigue is much more likely, added muscular stress in the upper back tends to occur, and back pain can result.

Example. A very short data-entry clerk used a chair with nonadjustable lumbar support. The clerk's lordosis was not matched by the shape or position of the chair's back, resulting in discomfort. The employer obtained a chair for the clerk with a vertically and horizontally adjustable lumbar support, and a seat that was adjustable down to 15 in. in height.

Example. An office secretary in a large purchasing department could not find any comfortable chair to use. She was given a small, curved pillow to use as a supplementary back support for her existing chair. It was helpful, but she later bought an inflatable air pillow that was sold for the same purpose and was even more comfortable. The latter pillow conformed to her back better, and (when deflated) was quite portable.

A safe and comfortable chair design should have five legs (to reduce the likelihood of tipping over if the person using it leans backwards); and the texture of the back and seat material should be rough or nubby to allow a small amount of air circulation between the material and the body. If armrests are used, the arm height should be adjustable so as not to strike the work table or bench during normal chair movements. A person seated at a fixed-height table may have to adjust the chair to a height at which his or her feet do not touch the ground. In this case, the company must provide footrests to support the legs. (See ANSI/HFS Standard 100-1988 for VDT workplace dimension recommendations.)

Seated posture. Workers need to be instructed in the proper use of seats. Poor posture can lead to stress on the upper back and neck. Once the employer has determined that poor job factors (e.g., poor visibility, no copy holder) are eliminated, users with poor posture should be reminded to keep the upper back and neck straight.

Chairs with forward-sloping seats. Forward-sloping seats are especially advantageous when the user has to be very close to, or up over, the work, as is the case for some repair persons, typists and word processors, air traffic controllers, dentists, and surgeons. The degree of slope should be easily adjustable (in addition to adjustable height and back position) and should provide for some additional support to prevent slipping out of the chair—for example, seat-pan contouring or "saddle" or "tractor seat" shaping (Kantowitz & Sorkin, 1983; Grandjean, 1987; Kvalseth, 1983; Eastman Kodak, 1983; Mandal, 1985). Forward-tilting seat pans also preserve the spinal curve in the small of the back (lumbar lordosis) and can assist those with low back pain.

Selecting Seating for Comfort and Performance

Many types and models of chairs available today meet all or most of the requirements discussed earlier in this section. A reasonable way to select the "best" of those available is to obtain a sample of two or three appropriate chairs (with or without arms, etc.) that meet the requirements and to let the members of the employee group who will be using the chairs try them out for at least a week.

The employer should carefully consider the opinions of the employee group when ordering a supply of new chairs. If the decision is split, order some of each. A brief chair-testing trial (e.g., less than 15 minutes) is usually insufficient. Initial impressions are often different from long-term (e.g., all day) test opinions.

Seating Summary

Musculoskeletal complaints from a seated worker can be related to chair characteristics. To the extent that they are, the company might investigate the causes. Discomfort attributed to chair variables may instead be due to other factors such as the lack of armrests (to take weight off of the shoulders), visual discomfort (radiating down the neck to the back), lack of exercise, or supervisory stress. Other sections of this chapter will address these other factors.

Back School

Back school is an organized means of transmitting information and techniques designed to reduce the risk of future back injuries or to help a person cope with a present injury. As such, it is not merely a discussion, but includes the practice of techniques and exercises.

Back school has become a controversial strategy due in part to the simplistic approach to proper lifting taught in the so-called *safe lifting method*. As shown in Figure 16-3, deep-squat lifting does not necessarily reduce the horizontal distance from the center of gravity of the load to the lumbar spine. From the discussion accompanying that figure, it is easy to see why the traditional approach to safe lifting is ineffective at reducing the incidence of back injuries. On the other hand in the last 10 years, more sophisticated back-training programs have been developed.

Effectiveness of Back Schools

Perhaps nowhere in the field of back-injury prevention is there more disagreement than with regard to the issue of the effectiveness of industrial back schools. Several studies have pointed out the apparent ineffectiveness of body mechanics training for nursing personnel, 50% of whom typically have low back pain. Careful analysis has revealed, however, that patient-handling tasks in hospitals and nursing homes frequently exceed safe biomechanical limits. In such a situation, good body mechanics alone cannot be expected to protect the work force. Also, programs that omit follow-up and fail to reinforce the material learned had little success in affecting the body mechanics of the participants. On the other hand, well-designed back school programs can significantly reduce the incidence and costs associated with back injuries in some industries.

Criteria for an effective back school. Based on the experience of back schools in industry, several strategies stand out as important for effectiveness:

- Integrate back school within the organization's overall program for injury prevention. Obtain the input of management and labor from such sources as the safety or ergonomics committees.
- Select instructors who are regarded as leaders and who can motivate the work force effectively. Include pictures or demonstrations of actual workers in realistic situations. These will both entertain and challenge the audience to apply what is being taught.
- Avoid lengthy and complex descriptions of anatomy or other technical material.
- Test the participants by having them show proper body mechanics in realistic situations. Written or verbal tests can supplement this approach.
- Where possible, introduce a simple exercise program that will address the particular strengthening, endurance, and stretching needs of the worker population. Progress should be monitored. Outstanding accomplishments in this area can be rewarded and publicized.
- Encourage back school to become a forum for workers to discuss ergonomic problems. Management must be prepared to seriously consider ideas generated in these problem-solving sessions. Communication with facility safety or ergonomics committees is essential.
- A comprehensive back school cannot last for only one session. Explanation and demonstration of these concepts will usually take at least three to four hour-long sessions. The practice of good body mechanics will need to be encouraged and rewarded in the workplace on an ongoing basis.
- Periodic brush-ups will be needed at least annually. Back school must also be a part of every new employee's orientation, along with other health and safety information.
- Back school can be particularly effective as a means of secondary prevention in the event of a low back injury. Once the worker is experiencing pain, he or she is more likely to use good posture and body mechanics.
- Since low back pain is a multifactorial phenomenon, it is important to include aspects of the nonwork environment. Traveling to one's job in a slouched position or spending a Sunday afternoon gardening with poor body mechanics will increase the risk of a future back injury on or off the job.

Engineering Controls

Job surveillance is the analysis of a set of tasks in terms of the risk factors outlined earlier in this chapter. The employer should control these exposures, much as it tries to control exposures to dangerous materials or extreme temperatures. For example, exposure to vibration in the range of 4.5–6.0 Hz, can be dampened by redesigning the equipment producing the vibration or inserting a dampening system between the equipment and the operator. Forklift trucks are now available with such systems designed in. High-density visco-elastic cushions or suspended seats can be retrofitted. To reduce frequent bending or stooping in some situations, place tools and materials 20–55 in. (50–138 cm) from the floor or the surface on which the worker is standing. In addition, horizontal reach distances should be limited to 15 in. (38 cm) in both sitting and standing positions. This section will concentrate on how to develop safe limits on exposures related to manual materials handling.

Developing Safe Limits

Earlier in this century, productivity and lifting standards were developed by testing young, fit men instead of broader samples of the population. In addition, researchers rarely followed workers over time to examine the health effects of maintaining a given output over a number of years. For example, we know that repeated lifting from heights below 20 in. (50 cm) tends to produce a higher rate of back problems than lifting from 30 in. (75 cm). In addition, we know that frequent lifting tends to cause fatigue in the back and leg muscles, resulting in poor body mechanics. As these muscles fatigue, the spine tends to lose its stability, and people stoop more. Therefore, developing safe limits for manual materials handling must combine what we know of human strength capabilities, epidemiology, and physiology with a biomechanical analysis of the spine.

NIOSH Lifting Guidelines

This is the approach taken by the National Institute for Occupational Safety and Health in its *Work Practices Guide for Manual Lifting* (HHS, 1981). The formulation presented in the guide considered the following factors as modifying a person's ability to lift. Each modifier is a number between 0 and 1, which, when entered into a mathematical formula, reduces the lifting weight that is acceptable under the guidelines:

- Horizontal Modifier (HM) = The horizontal location of the worker's hands as measured from the midpoint of a line between the ankles. Used as an approximation of the horizontal distance from the lumbar spine to the center of gravity of the load, it should be measured at both the origin and destination of the lift. Higher horizontal distances reduce the weight which would be safe to lift.
- Vertical Modifier (VM) = The vertical location of the worker's hands from the floor or standing surface at the origin of the lift. Lifts that originate below or above knuckle height (30 in. from the floor) are more difficult, so the recommended weight would be reduced accordingly.
- Distance Modifier (DM) = Considered the vertical travel distance from the origin to the destination of the lift.
- Frequency Modifier (FM) = The average frequency of the lift in lifts per minute was used to take fatigue into account.

The 1981 guide (HHS, 1981) assumed no twisting of the body at any point in the lift and a good grip on the object being moved. Recognizing these limitations, NIOSH has undertaken a revision to the 1981 guide, which adds the following factors:

- Asymmetry Modifier (AM) = This takes into account the angle (A) of rotation of the body needed to grasp the object, up to a maximum of 135° of twisting. The greater the amount of twisting, of course, the higher the probability of injury. This modifier should be calculated at the beginning and at the end of the lift.
- Coupling Modifier (CM) = This factor characterizes the grip as good, fair, or poor. A poor coupling, for example, would result in a modifier of 0.90, which would reduce the acceptable weight of the lift by 10%.

In 1991, NIOSH began presenting a proposed formula for a Recommended Weight Limit (RWL) incorporating all the above modifiers. The Load Constant represents the weight that would be acceptable under optimal conditions, and is multiplied by each modifier to reduce it by an appropriate amount (Table 16-A).

RWL=(Load Constant) (HM) (VM) (DM) (AM) (FM) (CM)

The frequency multiplier divides the work duration into three categories: up to one hour per shift, up to two hours per shift, and up to eight hours per shift (Table 16-B).

Other approaches to guidelines for lifting include consideration of awkward and unusual postures. *A Guide to Manual Materials Handling* (Mital et al, 1993) presents recommended limits for a variety of lifting, pushing, and pulling tasks. Table 16-C compiled by Suzanne H. Rodgers, PhD, summarizes maximum force recommendations in various positions.

Manual materials handling that exceeds guidelines or leads workers to complain of musculoskeletal symptoms should be subjected to analysis and job redesign. The options range from

Table 16-A. Coupling Modifier Table.

Couplings	V<75 cm (30 in.)	V ≥75 cm (30 in.)
Good	1.00	1.00
Fair	0.95	1.00
Poor	0.90	0.90

Good	Fair	Poor
For containers of optimal design, such as some boxes, crates, etc., a "good" hand-to-object coupling would be defined as handles or hand-hold cut-outs of optimal design (See notes 1 to 3 below). For loose parts or irregular objects, which are not usually containerized, such as castings, stock, supply materials, etc., a "good" hand-to-object coupling would be defined as a comfortable grip in which the hand can be easily wrapped around the object.	For containers of optimal design, a "fair" hand-to-object coupling would be defined as handles or hand-hold cut-outs of less than optimal design (See notes 1 to 4 below). For containers of optimal design with no handles or hand-hold cut-outs or for loose parts or irregular objects, a "fair" hand-to-object coupling is defined as a grip in which the hand can be flexed about 90° (See note 4 below).	Containers of of less than optimal design with no handles or hand-hold cut-outs or loose parts or irregular objects that are bulky or hard to handle (See note 5 below).

1. An optimal handle design has .75 –1.5 in. (1.9–3.8 cm) diameter, ≥ 4.5 in. (11.5 cm) length. 2 in. (5 cm) clearance, cylindrical shape, and a smooth, nonslip surface.
2. An optimal hand-hold cut-out has 3 in. (2.5 cm) height, 4.5 in. (11.5 cm) length, semi-oval shape. ≥ 2 in. (5 cm) clearance, smooth, nonslip surface, and ≥ 0.43 in. (1.1 cm) container thickness.
3. An optimal container design has ≤ 16 in. (40 cm) frontal length, ≤ 12 in. (30 cm) height, and a smooth, nonslip surface.
4. A worker should be capable of clamping the fingers at nearly 90° under the container, such as required when lifting a cardboard box from the floor.
5. A less-than-optimal container has a frontal length ≥ 16 in. (40 cm) height ≥ 12 in. (30 cm) rough or slippery surface, sharp edges, asymmetric center of mass, unstable contents; gloves are required when handling these.

the use of mechanical assists to the elimination of the need for any manual handling of the load. For more information, see the references at the end of this chapter.

Administrative Controls

In situations where engineering controls are not practical, administrative strategies can be used to reduce the job risk factors for low back pain.

- Staffing patterns can be changed to convert one-person lifts to multiperson lifts.
- Continuous lifting can be subdivided among several workers, so that each is functioning within safe limits.
- Training can be expanded to cover body mechanics or the use of existing assistive devices and ergonomic furniture.

Personnel Practices

When evaluating administrative controls, consider various personnel practices. The limits for manual materials handling, for example, assume an eight-hour shift. Overtime increases the exposure to these risk factors, and limits should be adjusted downward, just as in the case of exposures to dangerous or toxic materials. Some employers have divided departments into work teams and established incentive programs to reward teams according to the number of workdays without injury. Of course, such programs also tend to pressure employees to continue working in pain and avoid filing workers' compensation claims. In our experience, the more positive reward systems are more effective in motivating workers. Other, more negative forms of incentive programs can feed into a "blame-the-victim" psychology within the work group, inducing employees to be angry with an injured co-worker.

Preplacement Screening

Another form of administrative control is worker selection techniques. These can take the form of preplacement screening of new workers and those seeking a transfer to a different job within the company. The goal of such a screening program is assurance that the applicant's physical tolerances meet or exceed the demands of a given job. During most of this century, techniques such as preplacement medical examinations, including x-ray films,

Table 16-B. Frequency Multiplier Table.

FREQUENCY	WORK DURATION (Continuous)					
	≤ 8 HRS.		≤ 2 HRS.		≤ 1 HR.	
Lifts/min	V<75 cm V<30 in.	V ≥75 cm V ≥30 in.	V<75 cm V<30 in.	V ≥75 cm V ≥30 in.	V<57 cm V<30 in.	V ≥75 cm V ≥30 in.
0.2	0.85	0.85	0.95	0.95	1.00	1.00
0.5	0.81	0.81	0.92	0.92	0.97	0.97
1	0.75	0.75	0.88	0.88	0.94	0.94
2	0.65	0.65	0.84	0.84	0.91	0.91
3	0.55	0.55	0.79	0.79	0.88	0.88
4	0.45	0.45	0.72	0.72	0.84	0.84
5	0.35	0.35	0.60	0.60	0.80	0.80
6	0.27	0.27	0.50	0.50	0.75	0.75
7	0.22	0.22	0.42	0.42	0.70	0.70
8	0.18	0.18	0.35	0.35	0.60	0.60
9	0.00	0.15	0.30	0.30	0.52	0.52
10	0.00	0.13	0.26	0.26	0.45	0.45
11	0.00	0.00	0.00	0.23	0.41	0.41
12	0.00	0.00	0.00	0.21	0.37	0.37
13	0.00	0.00	0.00	0.00	0.00	0.34
14	0.00	0.00	0.00	0.00	0.00	0.31
15	0.00	0.00	0.00	0.00	0.00	0.28
>15	0.00	0.00	0.00	0.00	0.00	0.00

were used for this purpose. Few risk factors that can be uncovered as part of this process can, in fact, help a physician determine that an individual has a high or low risk of future injury. A history of back injury, if disclosed, is the most revealing information likely to be uncovered in such an examination.

However, screening practices should not violate the requirements of the Americans with Disabilities Act (1990) or Equal Employment Opportunity Commission (EEOC) guidelines. Under the ADA, employers can only use those selection criteria that are job-related for the position in question and consistent with business necessity. All persons in that job category must undergo the same screening procedures regardless of disability, except that those with a disability (including a previous back injury) can request a reasonable accommodation, or modification, of either the preplacement test or of nonessential job functions.

The *Guidelines for Use of Routine X-ray Examinations in Occupational Medicine,* issued in 1979 by the American Occupational Medicine Association, recommended that lumbar radiographs should not be ordered routinely. Other previously used criteria for accepting or rejecting job applicants, such as body weight or height, have been held invalid under current ADA and EEOC standards.

The most direct strategy to meet ADA requirements is to set up work samples that represent the essential functions of the job as disclosed in a job analysis. For example, candidates for a job in a warehouse can be tested by lifting boxes of gradually increasing weight and size until they reach a sample representative of boxes actually handled on the job. This is patterned after the psychophysical approach, where the person being tested is instructed to stop when he or she feels unable to proceed comfortably or safely (Snook, 1978). In addition, the examiner can stop the test if unsafe actions or conditions are observed, such as poor balance. In situations where endurance is likely to be the limiting factor, the applicant's heart rate can be monitored to see that it does not exceed recommended safe limits.

Variables to Test

Evaluations of lifting, of course, do not cover the range of abilities required by most jobs. Pushing, pulling, carrying, walking, and many other activities can constitute essential job functions. In addition, certain types of employment can require sitting, driving, reaching, and standing. Evaluations of these should be part of the back injury control program for new applicants, injured workers about to return to work, and for those seeking transfer to another position. However, none of these assessments violate the requirements of job-relatedness specified by the ADA. For more information on preplacement screening, see chapter 21, Preplacement Testing.

Work Involving Computers

Computers are found in most workplaces. Operators frequently must remain at their workstations

Table 16-C. Maximum Force Recommendations.

Force Conditions	Maximum Force for Design	
	Pounds	Newtons
Whole Body, Standing		
Forward Push, Truck Handling		
Initial Force	50	220
Sustained for 1 Minute	25	110
Emergency Stop	80	355
Pull In, Waist Level	55	245
Pull Up from Floor Level	125	555
Pull Up from 20 Inches (51 cm) Height	70	310
Kneeling	40	180
Upper Body Standing		
Pull Up, Waist Height	55	245
Pull Up, Shoulder Height, Arms Extended	30	135
Boost Up, Shoulder Height	60	265
Pull Down from Overhead	100	445
Push Down, Waist Level	75	335
Lateral Push Across Body	15	65
Seated		
Forward Push, Waist Height		
Near	30	135
Arms Extended	25	110
Pull Upward, Elbow Height	25	110
Pull In, Waist Height, Near	20	90
Lateral Push Overhead	10	45
Lateral or Transverse Push Across Body	20	90
Foot Pedal Activation	90	400

Maximum Force Application Recommendations. Data on the maximum static force levels for design of handling tasks are summarized for different muscle groups. The values shown accommodate half of the female work force and most of the male work force. It is assumed that these forces will be exerted for only a few seconds unless otherwise noted. (Reprinted with permission from Rodgers SH. *Working with Backache*. Fairport, NY: Perinton, 1985.)

for the entire workday. The following provides information on the optimal ergonomic design of the computer workstations.

Layout of the Work Space

Workplaces using computers should generally follow the control-display guidelines discussed later in this chapter. Usually the monitor should be in front of the operator, with the top of the display screen below eye level and the keyboard directly in front within easy reach. Bifocal users will often need the screen to be much lower. This arrangement works well for data acquisition and editing/programming tasks where the monitor is the primary source of information. Note, however, that in the case of data entry, the monitor may not be the primary display. Word processors can require primary visual access to their copy, and data-entry clerks may look almost exclusively at the original data records (invoices, checks, etc.) that they are recording. In this situation, the copy stand or the data records should be in front of the operator, and the monitor should be positioned sufficiently off to one side or the other to accommodate this. Continual looking to one side or the other to view copy can result in muscular stress and pain in the neck and upper back.

For word processors, the copy stand should be directly adjacent to the monitor, at about the same height or between the screen and keyboard, and at about the same viewing distance. This will reduce head rotation (side to side and up and down) and eliminate refocusing the eyes when the word processor refers to both of these displays while editing documents.

For data-entry clerks, manual handling is often required to turn pages, lay aside checks or invoices, or otherwise manipulate the data records. In this case, it is necessary to compromise between the optimal handling location and the optimal viewing area (Scalet, 1987; Grandjean, 1987; Cohen, 1984).

Example. In a data-entry office, workers flipped checks one by one with the left hand, and keyed the amounts into the keyboard with the right hand. The file of checks was arranged in front of the operator (to avoid twisting the neck to the left to read them) and generally near the screen. However, the operators had to reach beyond the keyboard to manipulate

Figure 16-7. An example of an alternative keyboard with an opening angle between the two halves of the alphabetic keys allowing for neutral wrist postures for each hand when keyboarding. The keyboard also slopes laterally somewhat to allow for a degree of relaxation of forearm pronation. A sloped wrist rest is attached to the front of the keyboard, and a leveling bar under the front surface can be adjusted to produce a negative keyboard slope.

the checks, which involved suspending their arms in space and caused shoulder discomfort. Moving the checks closer to the operator would have meant more twisting to the left. Instead, padded armrests were provided to support the left forearm, taking the load off the shoulder. Separate alpha and numeric keypads allowed the checks to be placed in an optimum location between the keypads.

Keyboard Design

Most VDTs now have detachable keyboards, which allow more flexibility in the workplace arrangement. Most keyboards are modeled after the electric typewriter and arranged in straight rows. They use the Scholes keyboard layout (sometimes referred to as the QWERTY keyboard, after the keys in the row above the home row for the left hand).

This linear keyboard layout results in the typist maintaining ulnar deviation of the wrists (hands angled outward) for extended periods of time. The ulnar deviation is caused by the fact that the wrists are much closer to each other when typing than are the elbows, but the hands must be kept parallel. A solution for this problem is to arrange the keyboard in a V-shape so that the keys operated by each hand are in rows perpendicular to the long axis of each forearm, eliminating wrist deviation (Grandjean, 1987). Another new keyboard design allows rotating hands from being palm down to a more natural tilt, where the hands are rotated somewhat outward (See Figure 16-7).

When extensive numerical entry is involved, the keyboard should include a 10-key numeric keypad, arranged in an adding-machine format, to the right of the alphabetic keys. The keypad can be part of the primary keyboard, or on a separate keyboard directly in front of the shoulder. To eliminate radial deviation of the wrist (Ericsson, 1983; Eastman Kodak, 1983).

If the worker places many telephone calls using the keyboard, a 10-key telephone keypad also can be included. However, since the adding-machine layout is different from the telephone layout, the keypad chosen will probably depend on the relative frequency of data entry and telephone calls, as well as operator preference.

A wrist rest for operators who rest their wrist or hand on the edge of the keyboard or work surfaces is beneficial. The wrist rest should have an appropriate depth (1.5–2.5 in. [4–6 cm]), and be constructed of a soft rounded material. The height of the rest should support the wrist but not interfere with the keying. The use of wrist rests should be avoided while keying; for prolonged periods forearms should be supported by well-placed and padded armrests on the chair or work surface.

Visual Considerations

Vision is the primary input channel for information concerning the work being performed. This is particularly true for so-called information operators (word processors, computer programmers, data entry clerks, machine monitors, quality control inspectors, security guards, etc.). In designing these types of jobs, it is very important to take into consideration the performance characteristics of human vision.

Visual fatigue refers to the phenomena that are related to intensive use of the eyes in one's work. It can be more useful to refer instead to specific phenomena such as performance degradation, oculomotor changes, complaints of visual or ocular pain, itching, tearing, or other clinical symptoms.

Inspecting and Monitoring Jobs

Two critical jobs found in industry are those of the quality control inspector and the security guard. The inspector must carefully examine items on the assembly line or at a warehouse receiving station. Security guards may be responsible for remote monitoring of large areas of a facility on a bank of television screens in relatively unsafe or unsecured areas of the facility. Workers performing such inspection and monitoring tasks tend to have a disproportionate load on their visual systems. In addition, the levels of illumination tend to be relatively high in order to increase the likelihood of detecting the defects in objects being inspected or monitored. As a result of this added load of visual intensity and enhanced brightness, visual fatigue and ocular discomfort tend to occur relatively often among these workers.

These tasks also tend to be cognitively boring and physiologically hard on the musculoskeletal system, since they involve making repetitive observations over extended periods of time while assuming a fixed posture. Imagine observing an endless sequence of windshields or small electrical pumps or loaves of bread, or monitoring a television view of a parking lot all day or all year.

Some degree of visual, cognitive, and physical variety should be introduced into these jobs. For example, the quality control inspector could be responsible for sorting and manipulating parts inspected and for sampling plan decisions, whenever possible. The security monitor could personally check the remote areas every 15–30 minutes.

Musculoskeletal Stress

In designing and analyzing jobs, it is most important to avoid musculoskeletal stress on the worker, particularly those areas most vulnerable to injury, the back and the upper extremities. Significant increases in occupational injuries in the United States will continue to expand, reducing productivity and profitability (Muggleton et al, 1999).

The process of musculoskeletal stress typically starts with muscle fatigue, cascading into discomfort, pain, and injury in the form of cumulative trauma disorders. Increased physical stress exposure is associated with increased risk of discomfort, pain and injury. The muscle chemistry triggering fatigue in the muscle fibers enrolled in work is the build up of lactic acid and CO_2 locally faster than they can be removed by the cardiovascular system.

The levels of tolerable muscle activity may be determined by measuring a muscle group's Maximum Voluntary Contraction (MVC), that highest muscle force tolerable for a short time. Normally static muscle contraction during continuous work should not exceed 2%–5% MVC, and the mean force of contraction in dynamic work should not exceed 10%–14% MVC (Jonsson, 1978).

The symptoms of musculoskeletal stress include specific muscle weakness, fatigue, and pain or cramping of muscle groups. Musculoskeletal stress can be caused by two very different job conditions: (1) excessive load on the muscles, tendons, ligaments, and bone; and (2) insufficient circulation to the musculoskeletal system.

In the first case, the work of handling heavy products and cartons is the most obvious. Less obvious is manual work that requires disproportionate activity of a small group of relatively weak muscles, such as the continued wrist flexion required in weaving the rapid deviation of the wrist involved in carpentry, or the continuous flexion of the fingers of the dominant hand required in data entry.

An alert observer of physical effort may note the following visual cues of the onset of the fatigue process (Rodgers, 1997):

1. changes in the pattern of muscle use within active muscle groups, resulting in changes in reach techniques, switching hands, alternating task sequence, etc.
2. increasing pauses in the workflow to stretch, rub, or "shake out" the affected body parts.
3. increasing tendency to rest a hand, arm, or leg on an available support temporarily to reduce the static loading on the shoulder, back, or legs.

When observing these work characteristics, important corrective measures in the short run include allowing sufficient break time to overcome fatigue, cramping, and massage of the affected body part to improve circulation. If sufficient recovery time is not permitted as part of a physically stressful work regimen the trauma that occurs to muscle fibers accumulates over time, hence the term *cumulative trauma disorder* (CTD). Job redesign should be considered to reduce or eliminate the relevant musculoskeletal stress.

Muscular effort and recovery time should be balanced not just over the course of the work shift, but also within each job cycle. Depending on the force required, the recovery time might well exceed the effort time. Rodgers and other have advanced protocols to compute the amount of recovery time needed for various levels of exertion (Rodgers, 1997).

Example. In an assembly plant four screws had to be tightened to secure a small cover to the case of the product. At two seconds per screw, this left only two seconds of recovery time for each 10-second job cycle. The case and cover were redesigned so that two of the screws were replaced by low-force snaps, so that a total of five seconds was now required to assemble the case. In addition the screwdriver was suspended by a cable and balancer, reducing the grip force needed throughout the job cycle, especially when the tool was not in use. Recovery time was now adequate, without decreasing product output.

EQUIPMENT DESIGN

Design of equipment in the workplace today is the most important step in offering safe and ergonomically correct workstations. Human-machine interaction is one of the primary building blocks of ergonomics.

User-Machine Communication and Cooperation

Machinery and equipment might well be graded by the quality of physical and informational interaction between the operator and the machine. When the design of the user-machine system is such that the effort is primarily or completely provided by the machine and equipment, and operating control is automated with strategic decision input from the operator, the system is generally productive and not unduly stressful. To do this, the system must provide the operator with all of the information needed to make decisions in a useful form and in time.

When this information is not properly or usefully provided, physical and mental stress builds.

The result can be a variety of errors and operator ailments. The design rules described in this section should prevent system difficulties.

Designing Operating Controls

Controls (e.g., levers, switches, joysticks, pedals) enable the operator to give a machine orders or information. They also provide feedback to the operator. All primary manual controls (those used most often or of critical importance) should be within the preferable reach area of the workplace, as illustrated in Figure 16-1, and other controls should be located in the acceptable area of reach.

Designing Information Displays

The primary information displays should be in front of the operator and below eye level (but not more than 30° below eye level). This is the most comfortable position for monitoring the displays. Positioning above or below this range tends to add tension to the muscles at the back of the neck, which exacerbates the tension in the back muscles discussed earlier. Looking to one side or the other of this range puts stress on the torsion muscles of the neck. Figure 16-8 illustrates the optimal viewing and manipulation areas in a typical console workplace. The design rules for controls and displays allow them to be placed in the most convenient position for the operator and logically integrated with each other.

Example. If a steam turbine is to be monitored and operated, the displays should be in front of the operator and below eye level, and the turbine controls will generally be in front of the operator, near the hands. But a control such as rotational speed should be adjacent to and linked logically with its speed indicator display. For example, they should both be contained in a box drawn on the panel or should have a color-coded line linking them. Also, movement of the control upward or to the right should move the speed display indicator upward or to the right. This will increase the stimulus-response compatibility of the two devices, and improve the operator's ability to control them. Such simple logical linkages suggest simple responses to the information displayed to the operator, and more sophisticated logical linkages can indicate more complex responses. In this manner, the control-display geometry and graphics can simplify the information-processing load on the operator, reducing stress and error rates (Kantowitz & Sorkin, 1983; Bailey, 1982; Eastman Kodak, 1983; Woodson, 1991).

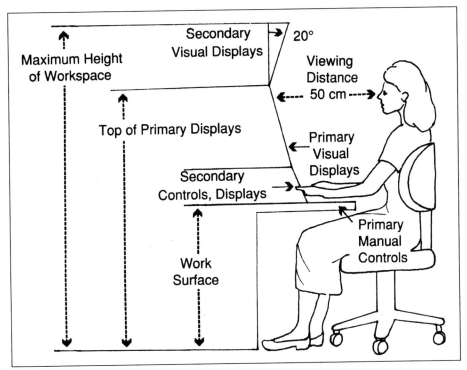

Figure 16-8. Optimal visual display and manual control areas. Controls used most often (e.g., keyboards, switches) should be placed in the *primary manual control area*. Displays viewed most often (e.g, VDTs, TV monitors, gauges) should be placed in the *primary visual display area*. Devices that also are viewed when they are operated (e.g., function buttons) should be in the secondary controls and displays area (i.e., located somewhere between the two primary areas). Generally viewing distance is controlled by the size and clarity of the display.

Operating Machines

The various machines in a factory—milling machines, bench presses, jointers, press brakes, etc.—all have specific, well-designed functions to perform. However, to complete these activities, workers are assigned to feed the appropriate material in the correct orientation and at the proper speed into the machine for processing. This is generally a difficult manipulative task and also can be boring, repetitive, and, at times, potentially unsafe. Until the necessary engineering technology is applied to automatically feed machines with the appropriate supply of raw material or in-process parts, people will have to continue operating these machines by hand.

Generally, the machines that represent a potential danger for the operator require guarding that meets OSHA standards. However, many people continue to be injured each year because the guards are not in place, are poorly designed, or were removed while the machine is operating.

The standard solution is to require that safety guards with interlocks to the machine control circuit be in place at all times where potentially unsafe conditions can develop. With a punch press, for example, where the point of operation is potentially dangerous, guarding is designed so that both hands must be safely displaced from the point of operation to operate the machine. In addition, the guards must be designed so that the interlock between the physical guards and the control circuit cannot be defeated unless the machine is locked out and energy is isolated, as during maintenance in compliance with the OSHA Control of Hazardous Energy Sources Standard (29 *CFR* 1910.147). Also, operators must be thoroughly trained in the need for guarding and operating the machine with guards in place. Finally, properly designed warnings related to operating dangers need to be clearly visible to the operator.

HAND TOOL DESIGN

The design of hand tools must address the following:
- factors causing musculoskeletal stress:
  - deviation of the wrist

- flexion of the wrist
- extension of the wrist
- grip strength
- tool handles
- use of the hand as a tool
- repetitive finger action
- two-handed tools
- hand and arm vibration
- proper working height and location
- causes of cumulative motion trauma.

These topics are detailed in the following sections.

Factors Causing Musculoskeletal Stress

Hand tools have been in use for perhaps a million years, but their design has progressed only gradually. The design of some basic tools, such as the screwdriver, hammer, and pliers, changes very slowly. This is unfortunate, since hand tools place much physical stress on the hands, wrists, and arms. Tool geometry should be determined by hand anatomy and task design. See Figure 16-9 for an explanation of hand motions.

Ulnar or Radial Deviation of Wrist

Rotation of the wrist in the plane of the palm toward the little finger, called *ulnar deviation,* is a common motion associated with tool handling or keyboard use. This motion can lead to constant loading of tendons attached to the thumb and eventually to *tenosynovitis* (inflammation of a tendon sheath) at the base of the thumb. Radial deviation of the wrist rotation toward the thumb is less common.

Palmar Flexion or Extension of Wrist

Palmar flexion of the wrist, or moving the palm toward the arm, occurs with "clothes-wringing" motions such as tightening screws, operating motorcycle-like controls, and looping wire with pliers. These motions are associated with high forces applied to the tendons in the wrist and can lead to carpal tunnel syndrome.

Wrist extension, or bending the wrist backward, can occur when scrubbing a floor with rags, or when using control knobs that are poorly placed. This motion can lead to tendonitis on the back of the wrist or at the elbow. It can also lead to carpal tunnel syndrome.

Grip Strength

In general, grip strength is reduced with any flexion or deviation of the wrist, but particularly for palmar flexion. A reduction in grip strength can impede the intended job performance and/or its speed, and can increase the likelihood that the user will lose control of the tool and drop it. This can result in injuries and damaged product. Reduced grip strength will certainly result in increased fatigue (McCormick & Sanders, 1982).

Tool Handles

To avoid tissue compression stress in the hands, handles should provide as great a force-bearing area as practicable and should be free of sharp corners and edges. This means handles should be either round or oval. A compressible gripping surface is best, but handles should at least have a high coefficient of friction in order to reduce the hand-gripping force needed for tool control. Pinch points should be eliminated or guarded.

In general, rigid, form-fitting handles with grooves for each finger do not improve the grip function unless they are sized to the individual's hand. Form-fitting handles (presumably designed for the 50th percentile hand) spread the fingers of a 5th percentile hand too far apart for efficient gripping and cause uncomfortable ridges in fingers of hands in the 95th percentile. Also, since the finger grooves are usually on both handles, the handle that fits into the palm of the hand causes added discomfort because of the ridges between the grooves.

Some tools, such as chisels and paint scrapers, transmit high force through the hand. When these forces are transmitted directly to areas overlying critical blood vessels or nerves, the user can suffer pain and chronic injury. Obstruction of blood flow, or ischemia, can lead to numbness and tingling of the hands. Thrombosis of the ulnar artery resulting from tool use has also been reported (Tichauer, 1978). Tools should be designed to not place constant pressure on these areas, particularly on the radial and ulnar arteries, such as the ulnar artery in the palm of the hand.

Example. In a maintenance department of a large plant, workers used a pair of wire clippers that pressed into the palm of the hand, resulting in pain and potential damage to the nerves and the ulnar artery. In addition, a conventional paint scraper had a handle that transmitted scraping forces by pressing into the palm of the hand. The company substituted tools with a modified handle design that distributed the work forces better thereby minimizing damage to the sensitive palm (Tichauer, 1978).

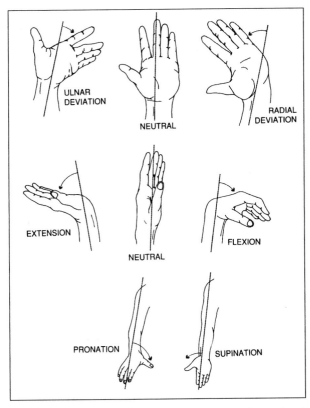

Figure 16-9. Hand tool geometry should be determined by hand anatomy and task design.

The Hand as a Tool

While the hand has obvious manipulation skills, whenever frequent finger forces exceed about 10 N (1 kgf) or frequent hand forces exceed roughly 50 N (5 kgf), physical fatigue and injury can result. The palm of the hand should never be used as a hammer. Even light tapping with the heel of the hand, done frequently on the job, can injure the nerves, arteries, and tendons of the hand and wrist. In addition, the shock waves can travel up the arm to the elbow and shoulder, causing additional physical problems (McCormick & Sanders, 1982; Kodak, 1983).

Example. In assembly work, the palm of the hand has been used for hitting electrical parts to tighten them after inserting and for hitting cabinet doors to align them after assembly. For these purposes, workers should be equipped with a mallet, which is soft enough not to damage the product but heavy enough to accomplish its intended task. Better quality control on the part of suppliers would reduce this problem further.

Repetitive Finger Action

If the index finger is used excessively for operating triggers on hand tools, excessive local fatigue can result, and a condition sometimes referred to as trigger finger develops. This condition occurs most frequently if the tool handle is so large that the tip of the finger has to be flexed while the middle of the finger must be kept straight. It may also occur with smaller handles (Tichauer, 1978).

Two-Handed Tools

Many power tools (e.g., drills, sanders, chain saws) require two hands to operate, or they are easier to operate and control with two hands. In either case, the tool generally has a primary handle with a trigger to provide for gripping by the dominant hand. If there is a secondary, stabilizing handle, it should be adjustable to either side of the tool to permit use by either left- or right-handed people. Such a design has the added advantage of permitting the user to change the trigger hand from time to time to reduce fatigue (Kodak, 1983).

Hand and Arm Vibration

Vibration of the upper limb, as in the operation of hand power tools such as handsaws, riveting hammers, sanders, pneumatic drills, and grinders, can be a continuing source of pain and physical trauma. (Whole-body vibration is discussed later in this chapter.) While not all workers exposed to vibratory hand tools experience trauma, there is a relevant set of symptoms and diseases collectively referred to as hand-arm vibration syndrome (HAVS). An example is a primary physical condition that develops called vibration-induced white finger reduced blood flow to the fingers and hand because of constricted blood vessels. It is accompanied by blanching of the hand and finger, tingling, numbness, and pain. Such vascular attacks seem to be exacerbated by cold working conditions (Sanders & McCormick, 1993; NIOSH, 1989; Wasserman, 1987).

Workers afflicted with HAVS have reduced blood flow to the skin and lowered skin temperature even when no longer working with vibratory tools. They also can have lessened touch sensitivity and fine finger dexterity, as well as a loss of grip strength. (For more information see International Standards Organization (ISO) 5343-1986; ANSI S-3.34 1986; ACGIH-TLV on Hand-Arm (Segmental) Vibration Syndrome; NIOSH 89-106; and Wasserman, 1987).

Proper Working Height and Location

The appropriate positioning of hand tools with respect to the worker can minimize muscle stress and fatigue, and improve productivity. In general, the heavier the work, the lower the work surface should be, without causing flexion of the trunk or other awkward postures.

The primary principle for positioning hand tools is that the height and location of the hand tool should be directly adjacent to the work point where it will be used. That is, if a powered screwdriver is needed to fasten parts to the right and about a foot away from the worker, then that tool should be suspended near that location from an overhead counterbalanced mechanism or spring device. Consider suspending the tool from a trolley mechanism that allows movement to the point of tool contact. Obviously, the screwdriver should be located where it doesn't interfere with other work or with the transfer of parts to and from the workstation. Locating the tool in this manner will reduce the weight and distance of moving it from a more distant location, thus reducing physical stress and speeding up the job. The suspended tool handle should be oriented to keep the wrist neutral.

Depending on the job design and the flexibility of hand tool positioning, the worker can be sitting or standing. If standing, a nearby high stool is helpful to support the worker for intermittent periods of time, reducing the stress on his or her legs and feet.

Summary Causes of Cumulative Motion Trauma

Table 16-D summarizes the occupational factors occurring in the design of jobs that are suspected of causing various cumulative trauma disorders (CTDs) of the upper extremity. Most of these job factors have to do with the use of hand tools. This checklist is a useful starting point for improving the safety of manual jobs to prevent injury, or to make the job changes necessary to accommodate a previously injured employee (Putz-Anderson, 1988; National Safety Council, 1992, Section 4).

EVALUATING THE ENVIRONMENT

The work environment affects workers performance, health, and safety in a variety of ways. The social characteristics of one's work area (e.g., isolation versus overcrowding, being driven versus being appreciated, organization flexibility versus rigidity) often play a significant role in stress-related accidents and illnesses. However, this discussion will focus primarily on the physical aspects of the environment.

Physical Hazards

Hazards come in many forms, including unguarded moving machinery or equipment, poorly designed or nonexistent railings to protect people from dangerous areas, and slippery or obstructed floors. OSHA safety and health standards outline the requirements for eliminating hazards, as do most company safety regulations. It is essential that the facility rigidly and consistently enforce OSHA and company safety standards.

Safe Noise Levels

A complaint often voiced in work environments is that there is too much noise and that it is distracting. The loudness of a sound is directly related to the mechanical pressure transmitted to the eardrum, although the sound frequency and other characteristics determine the degrading effect it has on performance. At a given intensity, lower frequencies are more likely to produce hearing impairments, while high frequencies are more apt to interfere with working concentration and thought processes. The less predictable or controllable the sound, the more annoying it is.

In quiet areas, some sound (e.g., soft music) can be preferable to mask otherwise distracting conversations nearby. Sometimes white noise (sound spread uniformly over the full hearing spectrum) is a successful alternative to music. However, workers can find it objectionable.

As sound levels rise above 50 dB, depending on their frequency and predictability, they become increasingly intrusive, objectionable, and fatiguing. Sound levels that persist above 85 dBA (decibels as recorded on a sound-level meter's A-weighted scale) for as much as eight hours, can cause hearing loss. Where noise levels routinely exceed 85 dBA, workers need to have control over sound sources or to wear hearing protection. Workers should never be exposed to sounds above 115 dBA (Kodak, 1983, Section V[B]; Bailey, 1982, chapter 22).

Lighting

Illumination, or *illuminance,* is a measure of light quantity falling on a given work area. The amount of light required for a specific task, to avoid visual fatigue, is a function of the visual difficulty of the task at the recommended work speed and quality, and the

Table 16-D. Risk Factor Checklist.

Disorder	Reported Occupational Risk Factors	Job Examples
Carpal tunnel syndrome	■ Unaccustomed, repetitive work with the hands ■ Work that involves repeated wrist flexion or extreme extension, particularly in combination with forceful pinching ■ Repeated forces on the base of the palm or wrist ■ Ulnar deviation, possibly with palmar flexion	■ New employee ■ Trainee on new task ■ Tightening bolts or screws; plucking feathers; marker; butcher ■ Bricklayer; mechanical parts assembler; rug-layer ■ Wringing out clothes; check-out cashier scanning
Tenosynovitis and peritendinitis crepitans of the abductor and extensor pollicus tendons of the radial styloid (de Quervain's disease)	■ Performance of unaccustomed repetitive work ■ Direct local blunt trauma ■ Simple, repetitive movement that is forceful and fast ■ Repeated radial or ulnar deviation, particularly with forceful exertions of the thumb	■ New employee; trainee on new task ■ Carpentry; metal forming ■ Packaging; small parts assembly ■ Key entry of data, polishing; packaging ■ Wrapping packages; meat-cutter
Tenosynovitis of finger flexor tendons	■ Exertions with a flexed wrist	■ Painting; scraping; cleaning
Lateral Epicondylitis	■ Ulnar deviation with outward rotation ■ Sudden or unaccustomed use of tendon or joint ■ Repeated manipulations with extended wrist	■ Typing; electronic assembly ■ New employee; trainee on new task ■ Wire brushing; scrubbing
Medial Epicondylitis	■ Radial deviation with inward wrist rotation ■ Repeated twisting of wrist	■ Carpentry; blacksmith; plastering; rug-laying ■ Carpentry; meat-packer
Neuritis in the fingers	■ Contact with hand tools over a nerve in the palm or sides of the fingers	■ Paint scraping; clipping wires; inserting screws

visual acuity of the worker. The demand of the visual system is typically measured by (1) the contrast between the target and its background, and (2) spatial resolution, or size of target. The worker's visual acuity, even with corrected vision, varies with age.

Recommended levels of illumination are listed Table 16-E). The selection within the range of values listed depends on the relative visual difficulty of the task, the worker's visual acuity, and the reflectance of the work area (Kodak, 1983, Section V[C]).

Workplaces involving VDTs have special lighting needs so that there is no glare from the screen. This can be accomplished by reducing the ambient illuminance to below 500 lux and using supplemental lighting for tasks and reading copy. Direct glare from windows can be avoided by correct placement of the VDT so that the window is not in the direct line of vision and by using window films and coverings. Objects with textured, dull surfaces are preferable to polished, glossy objects, which reflect more light. Even light-colored clothing can cause uncomfortable reflections on the screen.

Heat and Humidity

A rise in ambient temperature and/or humidity increases the cardiovascular load of a materials handler, and a low temperature can substantially reduce finger flexibility and accuracy. The ideal temperature and humidity for work are defined by the thermal comfort zone (Kodak, 1983, section V[D]).

The comfort zone is affected by many factors in addition to temperature and humidity. Among these are air velocity (producing a windchill effect), work load, radiant heat sources, and amount and type of clothing worn. In general, the body's core temperature should not vary by more than 1° C in either direction. The employer should adjust the factors affecting the comfort zone to accommodate this range.

Table 16-E. Illuminance Categories and Values for Various Indoor Work Activities.*

Type of Activity	Illuminance Category	Ranges of Illuminances		Reference Work Plane
		Lux	Footcandles	
Public space with dark surroundings	A	20–30–50	2–3–5	
Simple orientation for short temporary visits	B	50–75–100	5–7.5–10	General lighting throughout spaces
Working spaces where visual tasks are only occasionally performed	C	100–150–200	10–15–20	
Performance of visual tasks of high contrast or large size	D	200–300–500	20–30–50	
Performance of visual tasks of medium contrast or small size	E	500–750–1,000	50–75–100	
Performance of visual tasks of low contrast or very small size	F	1,000–1,500–2,000	100–150–200	Illuminance on task
Performance of visual tasks of low contrast and very small size over a prolonged period	G	2,000–3,000–5,000	200–300–500	
Performance of very prolonged and exacting visual tasks	H	5,000–7,500–10,000	500–750–1,000	
Performance of very special tasks of extremely low contast and small size	I	10,000–15,000–20,000	1,000–1,500–2,000	Illuminance on task, obtained by a combination of general and local (supplementary lighting)

Commercial, Institutional, Residential, and Public Assembly Interiors

Area/Activity	Illuminance Category	Area/Activity	Illuminance Category
Conference rooms		Inspection	
Conferring	D	Simple	D
Critical seeing (refer to individual task)		Moderately difficult	E
Assembly		Difficult	G
Simple	D	Very Difficult	G
Moderately difficult	E	Exacting	H
Difficult	F	Machine shops	
Very difficult	G	Rough bench or machine work	D
Exacting	H	Medium bench or machine work, ordinary automatic machines, rough grinding, medium buffing, and polishing	E
Candy-making			
Box department	D		
Chocolate department	D	Fine bench or machine work, fine automatic machines, medium grinding, fine buffing, and polishing	G
Husking, winnowing, fat extraction crushing and refining, feeding			
Bean-cleaning, sorting, dipping, packing, wrapping	D	Extra-fine bench or machine work, grinding, fine work	H
Milling	E	Paint shops	
Cream-making		Dipping, simple spraying, firing	D
Mixing, cooking, molding	D	Rubbing, ordinary hand painting and finishing art, stencil, and special spraying	
Gum drops and jellied forms	D	Fine hand painting and finishing	E
Hand decorating	D	Extra-fine hand painting and finishing	G
Hard candy		Service spaces (see also, Storage rooms)	
Mixing, cooling, molding	D	Stairways, corridors	B
Die cutting & sorting	E	Elevators, freight & passenger	B
Kiss-making and wrapping	E	Toilets & washrooms	C
Foundries		Sheet metal works	
Annealing (furnaces)	D	Miscellaneous machines, ordinary bench work	E
Cleaning	D	Presses, shears, stamps, spinning, medium bench work	E
Core making			
Fine	F	Punches	E
Medium	E	Tin place inspection, galvanized	F
Grinding and chipping	F	Scribing	F
Inspection		Storage rooms or warehouses	
Fine	G	Inactive	B
Medium	F	Active	
Molding		Rough, bulky items	C
Medium	F	Small items	D
Large	E	Welding	
Pouring	E	Orientation	D
Sorting	E	Precision manual arc-welding	H
Cupola	C		
Shakeout	D		
Garages—service			
Repairs	E		
Active traffic areas	C		
Write-up	D		

(Adapted with permission from the Illuminating Engineering Society of North America. Kaufman J E, ed. IES *Lighting Handbook: 1981 Application Volume.* Baltimore: Waverly Press, 1981)

Whole-Body Vibration

Vibration at critical frequencies and accelerations has become an important source of injury, including chest and abdominal pain, loss of equilibrium, nausea, involuntary muscle contractions, and shortness of breath. In addition, truck drivers and heavy-equipment operators have a disproportionately high rate of lumbar spinal disorders, hemorrhoids, hernias, and digestive and urinary problems. (Extended sitting and loading and unloading of trucks can contribute to these conditions [Sanders & McCormick, 1993].)

The types of vibration that most concern industrial health and safety analysts are those associated with vehicle operation (buses, forklift trucks, heavy construction equipment) and with machinery operation (large punch presses, conveyors, and furnaces) (Figure 16-10). The effect of vibration depends on its frequency, acceleration, duration, and direction (Kodak, 1983, Section V[B]).

The critical range of the torso's natural resonant frequency is 4–8 Hz, but discomfort can occur in the range of 2–11 Hz. Well-designed seats for bus and truck drivers will damp out some of the vibration in this critical frequency range, but many older seats tend to have amplification effects (seat bounces more than frame supports at the resonant frequency) of as much as 20%. Moreover, the lateral acceleration intensity can be twice the vertical intensity in some buses or trucks. Pay particular attention to the effect of seat design on driver backslap (rhythmically hitting driver's back with seat back) (Schmidtke, 1984; Sanders & McCormick, 1993; Wasserman, 1987).

Visual performance is generally impaired in the range of 10–25 Hz. Generally, truck and bus seats do not transmit vertical vibrations in this range, but other equipment including overhead cranes, lumber mill saws, and conveying machinery can.

EVALUATING JOB DEMANDS

To determine the overall impact of a set of tasks on the assigned worker, one must assess the demands of those tasks. Only then will it be possible to estimate whether or not the demands are within the means of a typical worker, or whether or not special physical or cognitive training will be required. This information will also assist in designing the workplace layout appropriate to an optimal performance of those particular tasks.

Survey Methods

There are a number of survey methods that analyze jobs in detail, assess their physiological and cognitive demands, and enhance job design. These go under the general title of *task analysis*. For information beyond the specific techniques described in this section, see Kodak, 1983, Section VI(B); and American Industrial Hygiene Association (AIHA), 1983.

Techniques for Analyzing Human Motion

Within the general area of ergonomics, a number of professional technologies have techniques for analyzing and improving job design and enhancing job health and safety. When the company needs relevant job designs and modifications, use the professional practitioners in each of these areas as resources.

The *methods engineering technique* generally relates to the design of the pattern of hand motions used by the worker, the machines and tools used on the job, and the design of the workplace layout, including location of tools, parts, jigs, and fixtures, etc. Methods engineering grew out of motion-and-time study and is a well-developed technique practiced by most industrial and production engineers.

Ergonomic Audits and Evaluations

Whenever possible, occupational health and safety professionals should tour the work areas for which they are responsible to evaluate the health and safety of the job designs, job performances, and the working conditions. During this tour, they should keep in mind the ergonomic concepts and suggestions, and should note questionable areas or activities for later study and possible job redesign. Such tours, or rounds, should occur at least quarterly, emphasizing work areas where injuries are most often reported.

In addition to redesigning unsafe and unhealthy jobs, occupational health and safety professionals might consider restructuring the job at a new skill level or new mechanization level. The techniques of job simplification (reducing physical complexity) and job enlargement (broader use of skills, more task variety) can be used. Such changes often require the aid of an ergonomist or an industrial engineer. These professionals are concerned with employee health and safety as well as productivity, since these concerns are strongly interrelated.

The occupational health and safety professional should also convene or join a health and

Table 16-F. Modified Work Opportunities: Alternative Jobs for Injured Truck Center Employees.

SEDENTARY*

Safety Trainer
- Plans training programs based on corporate materials available
- Conducts local safety training programs

Trade Skills Trainer
- As qualified, plans and conducts classes for apprentices and other junior employees in journeyman mechanical or body work skills; tests acquired knowledge
- As qualified, plans and conducts classes on new systems (e.g., diesel engines) and procedures (e.g., more visible truck scheduling) being introduced at the Center

Personal Trainee
- Attends local safety, first-aid programs, trade skills courses as available; reports back to supervision on content and quality of courses

Office Associate
- Answers telephone, responds or refers messages as appropriate
- Assists with truck rebuild schedule; calls to coordinate truck pick-up and deliveries
- As assigned, assists with document processing and handling; may tabulate data, summarize for reports
- Depending on keyboard skills, may enter or list data, or prepare simple printed reports
- Opens mail, logs or time-stamps, sorts by topic or responsible employee
- Prepares outgoing material for mailing; puts in envelope or package, weighs, computes, and applies postage

Associate Inventory Parts Clerk
- Checks inventory records for reorder point based on usage rate
- Orders supplies from best-vendor (following existing guidelines)
- Updates inventory paperwork as necessary
- Calls suppliers to inquire about parts delivery schedule; expedites when necessary
- Answers telephone and responds or refers calls

Associate Mechanic A, B, or C
- Performs bench cleaning, repair, assembly of parts (e.g., working at bench from high stool w/ back rest, all parts/tools within easy reach)

Associate Body Mechanic A or B
- Performs bench cleaning, repair, assembly of parts (e.g. working at bench from high stool w/ back rest, all parts/tools within easy reach)

LIMITED MOBILITY†

Trade Skills Trainer
- As qualified, tours work areas and observes performance of apprentice mechanics; comments on and coaches in proper mechanical or body work techniques (trainer qualified as a Mechanic A would coach Mechanic B s, etc.)
- May participate in final truck inspections, noting deficiencies to exemplify course material (any tasks listed above under *Trade Skills Trainer, Sedentary*)

Office Associate
- Files paperwork in filing cabinets; purges old files for long-term storage
- Runs off copies of office work, collates, distributes within office
- Assists secretary in routine tasks as able
- Runs errands as necessary to pick up and deliver documents within the Center and elsewhere locally
- Assists secretary in routine tasks (any tasks listed above under *Office Associate, Sedentary*)

Associate Inventory Parts Clerk
- Takes physical parts inventory of stock on hand, compares with records
- Based on records, estimates monthly stock turnover for each part in inventory; suggests shelf space and quantity (e.g., 2-mo. average supply) to keep on hand; lists parts no longer used
- Performs receiving inspection on incoming parts from suppliers, to ensure product received meets desired quantity and quality
- Fetches parts and assemblies (<10 lb [4.5 kg]) from inventory upon request from mechanics, notes withdrawals on inventory records
- Runs errands to pick up ordered parts to expedite delivery
- Reorganizes parts shelves, in response to changing stock turnover patterns and size/weight, as suggested by Inventory Parts Clerk
- Dusts and cleans inventory area; dumps packing materials received and other trash
- Performs other tasks under the direction of the Inventory Parts Clerk (any tasks listed above under *Associate Inventory Parts Clerk, Sedentary*)

(continued)

Table 16-F. Modified Work Opportunities: Alternative Jobs for Injured Truck Center Employees. (concluded)

Associate Utility
- Operates fork truck to move heavy objects or to push delivery trucks into shop
- Shuttle or relocate trucks around Center
- Does local truck pick-up and delivery
- Picks up trash, sweeps area, cleans up oil and paint spills
- Performs general area security; identifies all unknown persons in area; greets, assists visitors and delivery personnel (informal not a uniformed patrol)
- Performs minor repairs to building and equipment
- Runs general errands as requested
- Assists General Utility Person as required (any tasks listed above under *Associate Utility, Sedentary*)

Associate Mechanic A, B, or C
- Fetches small parts for mechanic from parts inventory; completes paperwork
- Keeps track of mechanical deficiencies noted in preinspection of each truck to ensure that all are corrected; checks off as each item is completed
- Keeps track of mechanical repair schedule; posts progress for each truck against schedule to inform mechanics and supervisors; reports completion of each truck
- Assists mechanics in brake, lighting, or other system tests (operation of a control such as brake pedal being tested by mechanic, or remote observation of results of tests such as checking turn signal)
- Fetches medium-sized parts or assemblies (<10 lb [4.5 kg]) for mechanic from parts inventory; may also fetch larger parts carried on rolling cart; completes paperwork
- Preinspection of truck, listing repair/replace items
- Final inspection of truck, listing exceptions; test drive
- Selected demand maintenance tasks; tune-up, check/replace electrical component operation, check fluid levels, replace oil/air filters
- Install oil pressure switch or sending unit on engine subassembly, engine warning switches, sending units and wiring; new ignition wiring, battery cables as required
- Replace windshield wiper blades, wiper arms, drive motor, washer switch, nozzles, pump as necessary
- Any mechanic duties that require only limited bending, stooping, climbing, light lifting (<10 lb [4.5 kg]) e.g., in and out of truck may be OK, but under truck may not be acceptable (any tasks listed above under *Associate Mechanic, Sedentary*)

Associate Body Mechanic A or B
- Fetches parts, assemblies for mechanic from inventory; completes paperwork
- Keeps track of body work deficiencies noted in preinspection of each truck to ensure that all are corrected; checks off as each item is completed
- Keeps track of body repair schedule; posts progress for each truck against schedule to inform body mechanics and supervisors; reports completion of each truck
- Removes/installs dashboard instruments in truck (while seated)
- Performs activities requested by Body Mechanics insofar as capable
- Preinspection of truck, listing repair/replace items
- Final inspection of truck, listing exceptions; completes paperwork
- Removes small body parts from truck (e.g., reflectors, taillights, and bulbs)
- Masks truck for painting
- Sprays primer coat on body
- Installs new small body parts after repainting (detail)
- May be able to replace larger body parts (e.g., side mirrors, window panels)
- Cleans paint spray nozzles, replaces air filter periodically (any tasks listed above under *Associate Body Mechanic, Sedentary*)

---

*Sedentary: ambulation difficult; should remain seated during workday; can communicate verbally and in writing without difficulty

† Limited Mobility: ability to stand and ambulate for limited periods (low energy level or on crutches); limited stooping, bending, climbing, carrying (<10 lb [4.5 kg]); may be able to drive delivery trucks or fork trucks for limited distances

---

safety review committee within the organization. The committee plans health and safety reviews and follow-up activities, and is a resource to management. Those who would find such activities interesting and important include industrial engineers, ergonomists, safety engineers, and human resources and risk-management personnel.

EMPLOYEE EXERCISE AND STRENGTHENING PROGRAMS

Some sedentary industrial tasks require less activity than is healthy for the body, while other tasks require greater physical demands than normal. In the former case, exercise is necessary to restore the balance between the constraining job demands and the cardiovascular needs of a healthy body. In the

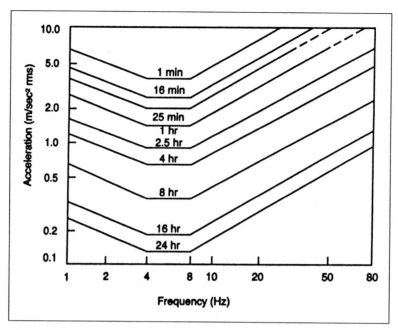

Figure 16-10. Tolerance to whole-body vertical vibration, in fatigue-decreased proficiency, as a function of vibration frequency and duration. From ANSI S3-18-1979 (ASA 38-1979, Revised 1986). (Reprinted with permission from the Acoustical Society of America.)

latter case, exercise is necessary to strengthen those muscle groups used repeatedly on the job in order to reduce the onset of musculoskeletal stress and fatigue.

Effects of Exercise on Musculoskeletal Stress

The workplace geometry of some jobs requires a rigidly held neck and shoulders. Information workers (typists and word processors, data-entry clerks, inspectors, bookkeepers, etc.) are in this category. Poor circulation and muscular fatigue in the shoulders, neck, and upper back can develop into chronic pain and cumulative trauma disorders, and should be given special attention before this occurs. Workers whose job requires a rigidly held neck and shoulders need to move those muscle groups through a full range of motion or perform exercises. For example, occasionally bending the waist to reach forward can result in useful exercise of the involved body parts. In some situations, it is preferable for these workers to get up from their chair periodically and fetch the next batch of work to be processed, rather than to have it delivered.

Moreover, because sedentary workers often get insufficient exercise and have insufficient cardiovascular activity, they need selective exercise movements (including aerobic exercises) to compensate for the muscular rigidity during work. They can use sets of in-chair exercises that move and stretch important muscle groups. Such exercise will help them maintain sufficient muscle tone and levels of alertness to remain safe, healthy, and productive (Gore & Tasker, 1986).

Example. For an office environment involving fairly continuous key entry, the employer instituted a set of simple physical exercises that would use the muscles held most rigid during keying. The exercises were mostly joint range-of-motion and stretching, and did not involve movements unsuitable for an office environment. Operators received written descriptions and sketches of the exercises, and a keying operator active in aerobics instructed workers in how to use those exercises. The employer set aside two short, paid time periods for doing these exercises daily, in addition to the regular rest breaks. Operators were also encouraged to take frequent "microbreaks" for a few quick stretches whenever a shoulder, wrist, or other body part became uncomfortable.

Appropriate Exercises

The specific set of exercises most appropriate to a particular job or job group should be selected by a registered physical therapist, an exercise physiologist, or other health professional familiar with exercise techniques. An increasing number of exercise references are becoming available to guide this process, including Gore and Tasker (1986). In addition, computer users can use computer program add-ons that have pop-up screens to remind the operator to do specific exercises on a timely basis. Principal among these are Computer Stretches (Anderson & Anderson, 1986) and Computer Health (Hoff et al, 1990). Depending on the job design, the exercise paradigm should include many microbreaks containing one or two short exercises and a few longer exercises with more extensive movement, stretching, and eye relaxation.

The appropriate exercise mix should be presented and explained to a group of workers, so that they can vary the type and frequency of exercise to suit their own particular job demands and personal needs. This allows workers to control their own levels of musculoskeletal stress and discomfort. Ideally, the exercises should be of such a type and performed often enough so that musculoskeletal tension never builds up to the discomfort level but, rather, is drained away over the day through frequent microbreaks.

Effect of Strength

Safety professionals are sometimes puzzled by the fact that a worker can be suddenly strained or injured by work that the person does routinely. However, the worker's strength and endurance may be just barely sufficient to accomplish daily tasks, with no reserves for occasional changes in job performance, product variety, a slippery floor, overtime work, or other stresses and hazards.

Results of many studies (AIHA, 1983) indicate that physical injuries are most likely to occur when a person is working near maximum strength. In contrast, if a worker's maximum lifting strength is twice the actual lifting strength required on the job, the likelihood of injury is minimized. The reason for this is that the individual has a good reserve of strength and endurance to handle any changes in job demand or work environment. Moreover, the effects of muscle fatigue, illness, and injury are cumulative, and strength and endurance reserves must be available toward the end of the day or the end of a busy week without overstressing the musculature and cardiovascular system. Consequently, materials handlers need strength and endurance training to ensure that their maximum strength and endurance far exceeds that which is required on the job. The same is true for workers who handle only light objects, because they too require extra strength and endurance if they must handle tools or materials in awkward postures, where muscles are at a mechanical disadvantage.

DESIGNING LIGHT-DUTY AND MODIFIED-DUTY WORK

Many companies make light-duty or modified-duty jobs available to previously injured or ill personnel while they are recovering. These alternative jobs contain tasks that are less physically demanding than those in the worker's regular job.

By placing injured workers in alternative jobs for which their residual physical abilities qualify them, companies keep those workers an active, productive part of the social milieu, in contact with equipment and facility changes, and in touch with other employees and supervisors. This benefits the employees physically, psychologically, and professionally, and benefits the company by keeping skilled employees in productive positions. It also improves the communication, cooperation, and understanding between the safety and health staff and the supervisory personnel.

The following example focuses primarily on physical job variables. This example is one model and should be modified for a company's specific program. It has not assessed the cognitive or experiential characteristics or the special talents of either the employees or the alternative jobs. For example, if a particular worker happens to have scheduling or typing skills or verbal presentation abilities, the employer would have to take these into account on an individual basis when assigning alternative jobs. The example also fails to take into account individual preferences and motivation. Some injured workers may feel uncomfortable working in an office environment or running simple errands. To the extent possible, the employer should consider these personal characteristics when assigning alternative jobs to injured workers.

Matching Injured Workers with Light-Duty Jobs

An ill or injured employee may be able to continue working at an alternative job. For example, the employer might consider the alternative jobs listed in Table 16-F. These alternatives are classified as sedentary and limited-mobility tasks.

Table 16-G. Physician's Restriction Guidelines for Truck Center.

To the Physician:
Please complete the following guidelines regarding the recommended physical restrictions for: _____
(Name of Employee)

ACTIVITY	NO RESTRICTIONS	RESTRICTIONS
Walking		No walking:_____ Walk <25% of the time:_____ Walk 25%–50% of the time:_____ Walk 50%–75% of the time:_____ Other:_____
Climbing, descending		No climbing or descending stairs allowed:_____ May climb/descend <25% of the time:_____ May climb/descend 25%–50% of the time:_____ May climb/descend 50%–75% of the time:_____ Other:_____
Lifting, carrying weight		No lifting or carrying allowed:_____ May lift <10 lb (4.5 kg) at a time:_____ May lift 10–25 lb (4.5–11.3 kg) at a time:_____ May lift >25 lb (11.3 kg):_____ Other:_____
range:		No reaching or lifting allowed:_____ May not reach or lift below the waist:_____ May not reach or lift above the shoulder:_____ Other:_____
frequency:		May not lift more than once per minute:_____ Other:_____
Pushing, pulling		May not push or pull more than 10 lb (4.5 kg):_____ May push or pull from 10–25 lb (4.5–11.3 kg):_____ Other:_____
Bending, stooping, squatting		No bending allowed:_____ No stooping or squatting allowed:_____ Bending/stooping/squatting with no lifting:_____ Other:_____
Twisting, turning		No turning of the head allowed:_____ No twisting at the waist allowed:_____ Other:_____
Use of the hands		May not use left hand for work:_____ May not use right hand for work:_____ Other:_____
Vision		Has the functional use of only one eye:_____ Has no effective visual abilities:_____ Other:_____

To the Physician:

1. In your opinion, does this individual take medications that would impair his or her ability to work with machinery or drive?
   Yes:____ No:____

2. Are there any other recommended restrictions? _____

3. What is the estimated length of disability? _____

Table 16-H. Minimum Physical Requirements for Alternative Jobs: Sedentary.

MINIMUM REQUIREMENTS	Safety Trainer	Trade Skills Trainer	Personal Trainee	Office Associate	Associate Inventory Parts Clerk	Associate Mechanic	Associate Body Mechanic
■ Walking, standing:							
% Expected activity							
None	X	X	X	X	X	X	X
<25%							
25%–50%							
50%–75%							
75%–100%							
■ Climbing, descending:							
% Expected activity:							
None	X	X	X	X	X	X	X
<25%–50%							
50%–75%							
75%–100%							
In/out of truck cab:							
■ Lifting, carrying:							
Weight:							
None	X	X	X	X	X	X	X
<10 lb (4.5 kg)							
10–25 lb (4.5–11.3 kg)							
25–40 lb (11.3–18.1 kg)							
Range:							
Below waist							
Above shoulder							
Frequency:							
< once/min							
>= once/min							
■ Pushing, pulling:							
None	X	X	X	X	X	X	X
<10 lb (4.5 kg)							
10–25 lb (4.5–11.3 kg)							
■ Bending, stooping, squatting:							
None	X	X	X	X	X	X	X
Without lifting							
With lifting							
■ Twisting, turning:							
None	X	X	X	X	X		
Twist at waist						X	X
Neck twisting							
■ Use of hands:							
Both hands				X	X	X	X
One hand	X	X	X				
Neither hand							
■ Vision:							
Both eyes normal				X	X		
Both eyes impaired							
One eye normal	X	X	X			X	X
■ Physical appearance	Neat	Neat	Neat	Neat	Any	Any	Any

A specific alternative job should be based on these general job descriptions, supplemented by the supervisor's knowledge of the particular employee and the nature and priorities of the work that needs to be done. Each of these alternative jobs is designed to use the worker's job knowledge and experience. The jobs are not "make-work" tasks and typically exist because of shifts in an organization's workload, staffing shortages, vacation schedules, budget priorities, etc. They are generally existing or potential jobs modified to accommodate people with physical restrictions, and consist of important tasks that will assist the organization in being more efficient and providing better service. In addition, they often result in a significant learning experience for the injured employee by exposing him or her to another side of the work environment. For example, a mechanic better appreciates the inventory parts clerk after having had to deal with vendors.

The brief job descriptions in Table 16-F are a guide to the general type of work that can be included in the alternative jobs. Mix and match among alternative job titles, adding to the listed tasks as seen fit, to create a work environment conducive to productivity and to provide a learning experience for the employee.

Selecting Alternative Jobs

To begin the procedures to select an alternative job for an injured employee, the facility's manager should have the company's occupational physician examine the employee reporting the performance-limiting illness or injury. The physician describes the employee's physical restrictions in a systematic, structured manner. One way of doing so would be to complete the guidelines in Table 16-G. (Other medical evaluations also can be required, but these guidelines are most useful in assigning an alternative job.)

Next, the facility's management compares the physician's completed restriction guidelines (Table 16-G) with the requirements for alternative jobs, as spelled out in Table 16-H. This table outlines the minimum physical requirements for alternative jobs to which an injured employee might be assigned, depending on his or her specific disability. The table is arranged by type and degree of restriction that would allow a person to perform each job. It uses the same categories of physical abilities as the physician's restriction guidelines in Table 16-G to simplify management's decisions.

Table 16-H shows sedentary jobs and Table 16-I shows limited-mobility jobs. None of the alternative jobs require extensive lifting, although the limited mobility jobs involve occasionally moving light parts or supplies (weighing less than 10 lb [4.5 kg]). Consequently, most of the jobs are suitable for recuperating from low back injuries, providing, of course, that seating has the proper back support and/or that workers use appropriate lifting techniques.

Finally, the facility's manager can assign the employee to an alternative job if, in the opinion of the physician and as reviewed by appropriate management personnel, the employee is not restricted from performing the minimum requirements for such a job. An injured worker can be considered a candidate for an alternative job if the physical requirements marked for that job on Table 16-H are no more difficult than the restrictions specified on Table 16-G. Within the range of physically acceptable alternative jobs, this assignment would also attempt to balance productive, motivated use of the employee with the facility's workload needs.

Follow-Up Evaluation and Return to "Active Duty"

After the manager has followed the procedure for assigning an alternative job, he or she should follow the employee's performance and outlook regularly. This is necessary to ensure that a reasonable, practical job decision was made initially, and that the worker is progressing toward physical recovery and a return to "active duty." A good approach is to follow up informally everyday and to prepare a written assessment at least once a week.

If the employee seems to be overcoming some of the original physical restrictions, the manager should consult the physician before considering reassignment to a more demanding job. If the employee does not seem to be accommodating physically to the alternative assignment, or if he or she reports continuing discomfort, the physician should also be informed of this. If it is deemed appropriate for the employee to return to the physician for a follow-up examination, the employee should take copies of the weekly written assessments.

SUMMARY

Successful ergonomics programs for preventing back injury combine several strategies to control a complex,

Table 16-I. Minimum Physical Requirements for Alternative Jobs: Limited Mobility.

MINIMUM REQUIREMENTS	Safety Trainer	Trade Skills Trainer	Personal Trainee	Office Associate	Associate Inventory Parts Clerk	Associate Mechanic	Associate Body Mechanic
■ Walking, standing: % Expected activity							
None							
<25%	X	X		X			
25%–50%			X		X	X	
50%–75%							
75%–100%							
■ Climbing, descending: % Expected activity:							
None	X	X					
25%–50%			X				
50%–75%							
75%–100%							
In/out of truck cab:				X	X	X	
■ Lifting, carrying: Weight:							
None							
<10 lb (4.5 kg)	X	X	X	X	X	X	
10–25 lb (4.5–11.3 kg)							
25–40 lb (11.3–18.1 kg)							
Range:							
Below waist			X	X	X	X	
Above shoulder			X			X	
Frequency:							
< once/min	X	X	X	X	X	X	
>= once/min							
■ Pushing, pulling:							
None	X	X	X	X	X	X	
<10 lb (4.5 kg)							
10–25 lb (4.5–11.3 kg)							
■ Bending, stooping, squatting:							
None	X	X	X	X	X	X	
Without lifting							
With lifting							
■ Twisting, turning:							
None	X	X	X	X	X	X	
Twist at waist							
Neck twisting							
■ Use of hands:							
Both hands		X	X	X	X	X	
One hand	X						
Neither hand							
■ Vision:							
Both eyes normal	X	X			X	X	
Both eyes impaired							
One eye normal			X	X			
■ Physical appearance	Neat	Any	Any	Any	Any	Any	

multidimensional problem. An essential component is management and labor support at the highest levels, with close coordination among divisions and departments, including production, personnel, purchasing, and planning. Key elements include the following:

- An administrative structure, including a management-labor safety committee, capable of developing a coherent strategy responsive to the particular conditions in that industry and workplace.
- Careful job analysis to assess high-risk jobs and tasks within those jobs, and to provide data to prioritize areas for intervention and to develop screening programs where needed.
- Job-specific training in body mechanics and job design for new employees and returning injured workers. Hold regular refresher classes, preferably at the work site, with actual tools and work situations encountered by employees in that department. Encourage problem-solving discussions and suggestions for improvement.
- More intensive job analysis and interventions for high-risk departments. If job redesign cannot make a substantial difference, include special training in exercise and body mechanics.
- Modified work programs that meet the specific physical tolerances of each injured worker as determined by a trained clinical specialist. Such programs help injured workers return to work sooner.
- Screening programs (matching the abilities of job applicants and injured workers with the demands of the job) that are designed to reflect the essential job functions as closely as possible.

Under the OSHAct General Duty Clause, OSHA requires that each employer shall furnish to each of his employees' employment and a place of employment that are free from recognized hazards that are causing or are likely to cause death or serious physical harm to his employees.

To this end, OSHA recognizes the need for a multifactorial program including education and training, job analysis and redesign, and early intervention/secondary prevention. We need to take a serious look at the work in our society, and continue to modify it to conform to the capabilities of people, rather than have people attempting to conform to poorly designed workplaces.

REFERENCES

Alexander DC, Pulat BM. *Industrial Ergonomics: A Practitioner's Guide.* Norcross GA: Institute of Industrial Engineers, 1985.

American Conference of Government Industrial Hygienists (ACGIH). *Ergonomic Interventions to Prevent Musculo-Skeletal Injuries in Industry.* Chelsea MI: Lewis, 1987.

(ACGIH). 1991–1992 *Threshold Limit Values for Chemical Substances and Physical Agents and Biological Exposure Indices.* Cincinnati: ACGIH, 1991.

American Industrial Hygiene Association (AIHA). *Work Practices Guide for Manual Lifting.* Akron OH: AIHA, 1983.

American National Standard for Human Factors Engineering of Visual Display Terminal Workstations (ANSI/HFS Standard No. 100-1988). Santa Monica, CA: Human Factors Society, 1988.

Anderson B, Anderson J. *Computer and Desk Stretches.* Palmer Lake CO: Stretching, Inc., 1986.

Ayoub MA. Control of manual lifting hazards: Part I: Training in safe handling. *J Occ Med* 24(8), 1982a.

Ayoub MA: Pre-employment screening programs that match job demands with workers' abilities. *Ind Eng,* March 1982b.

Ayoub MM, Mital A. *Manual Materials Handling.* New York: Taylor & Francis, 1989.

Bureau of Business Practice. *Safety Management Handbook.* Waterford CT: Prentice-Hall, 1991.

Buckle P (ed). *Musculoskeletal Disorders at Work.* New York: Taylor & Francis, 1987.

Chaffin DB. Ergonomics guide for the assessment of human static strength. *AIHA Journal* 36:505–510, 1975.

Chaffin DB, Andersson G. *Occupational Biomechanics,* 2nd ed. New York: Wiley, 1991.

Cohen B (ed). *Human Aspects in Office Automation.* New York: Elsevier, 1984.

Deyo RA (ed.). *Occupational Medicine: State of the Art Reviews.* Philadelphia: Hanley & Belfus 1(2), 1987.

Eastman Kodak Co., Human Factors Section, Health Safety and Human Factors Laboratory. *Ergonomic Design for People at Work,* vol. 1. Belmont CA: Lifetime Learning Publications, 1983.

Eastman Kodak Co., The Ergonomics Group, Health and Environment Laboratories. *Ergonomic Design for People at Work,* vol. 2. New York: Van Nostrand Reinhold, 1986.

Ericsson Information Systems. *Ergonomic Principles in Office Automation.* Stockholm: Ericsson, 1983.

Galer I (ed). *Applied Ergonomics Handbook,* 2nd ed. Boston: Butterworth, 1987.

Garg A, Herrin GD. Stoop or squat: A biomechanical and metabolic evaluation. *AIIE Tr* 11(4):293 302, 1979.

Gore A, Tasker D. *Pause Gymnastics: Improving Comfort and Health at Work.* Sydney: CCH Australia Limited, 1986.

Grandjean E. *Ergonomics in Computerized Offices.* New York: Taylor & Francis, 1987.

Griffin MJ. Levels of whole-body vibration affecting human vision. *Aviation, Space and Environmental Medicine* 46:1033–1040, 1975.

Himmelstein J, Pransky B (eds). *Worker Fitness and Risk Evaluation.* Philadelphia: Hanley & Belfus, 1988.

Hogan J, Quigley AM. Physical standards for employment and the courts. *Am Psychologist* 41(11):1193–1217, 1986.

Hoff T, Thompson D. Computer Health (software). Sunnyvale, CA: Escape Enterprises, 1990.

Human Factors Society. *American National Standard for Human Factors Engineering of Visual Display Terminal Workstations,* ANSI/HFS 100-1988. Santa Monica, CA: Human Factors Society, 1988.

International Standards Organization. *Guide for the Evaluation of Human Exposure to Whole-Body Vibration* (ISO 2631), 1978.

Isernhagen SJ. *Work Injury: Management and Prevention.* Rockville MD: Aspen, 1988.

Jonsson B. *Kinesiology, with Special Reference to Electromyographic Kinesiology, Contemporary Clinical Neurophysiology.* Amsterdam: Elsevier, 1978, pp 417–428.

Kantowitz, Sorkin. *Human Factors: Understanding People-System Relationships.* New York: Wiley, 1983.

Kvalseth TO. *Ergonomics of Workstation Design.* Boston: Butterworth, 1983.

Linton SJ, Kamwendo K. Risk factors in the psychosocial work environment for neck and shoulder pain in secretaries. *J Occ Med* 31(7):609–613, 1989.

MacKay CJ. Work with visual display terminals: Psychosocial aspects and health (report on a World Health Organization meeting). *J Occ Med* 31(12):957–968, 1989.

Mandal AC. *The Seated Man.* Denmark: Dafnia Publications, 1985.

Mandell P, Lipton MH, Bernstein J. Low Back Pain: *An Historical and Contemporary Overview of the Occupational, Medical, and Psychological Issues of Chronic Back Pain.* Thorofare NJ: Slack, 1989.

Matheson LN, Niemeyer LO. *Work Capacity Evaluation: Interdisciplinary Approach to Industrial Rehabilitation.* Anaheim CA: Employment and Rehabilitation Institute of California, 1986.

Mayer TG, Gatchel RJ. *Functional Restoration for Spinal Disorders: The Sports Medicine Approach.* Philadelphia: Lea & Febiger, 1988.

Mital A, Nicholson AS, Ayoub MM. *A Guide to Manual Materials Handling.* London: Taylor & Francis, 1993.

Muggleton JM, Allen R, Chappell PW. Hand and arm injuries associated with repetitive manual work in industry: A review of disorders, risk factors and preventive measures. *Ergonomics* 42(5):714–739, 1999.

Nachemson AL. The lumbar spine, an orthopaedic challenge. *Spine* 1(1):59–71, 1976.

National Safety Council (NSC). *Ergonomics: A Practical Guide,* 2nd ed. Itasca IL: NSC, 1992.

Oborne DJ. *Ergonomics at Work.* New York: Wiley, 1982.

Plog BA, Niland J, Quinlan P (eds). *Fundamentals of Industrial Hygiene,* 4th ed. Itasca IL: NSC Press, 1996.

Pope MH, Andersson G, Frymoyer J, et al (eds). *Occupational Low Back Pain.* St. Louis: Mosby-Year Book, 1991.

Putz-Anderson V. (ed). *Cumulative Trauma Disorders: A Manual for Musculoskeletal Diseases of the Upper Limbs.* New York: Taylor & Francis, 1988.

Rodgers SH. Work physiology-fatigue and recovery. In G. Salvendy (ed). *Handbook of Human Factors and Ergonomics,* 2nd ed. New York: Wiley, 1997, pp 36–65.

Rodgers SH. *Working with Backache.* Fairport NY: Perinton Press, 1985. (Available through SH Rodgers, PhD, 169 Huntington Hills, Rochester, NY 14622.)

Saal JA. Diagnostic studies of industrial low back pain. *Top Acute Care Trauma Rehabilitation* 2(3):31–49, 1988.

Saal JA, Saal JS. Nonoperative treatment of herniated lumbar intervertebral disc with radiculopathy: An outcome study. *Spine* 14(4):431–437, 1989.

Sanders M, McCormick E. *Human Factors in Engineering and Design,* 6th ed. New York: McGraw-Hill, 1993.

Scalet EA. VDT *Health and Safety, Issues and Solutions.* Lawrence KS: Ergosyst Associates, 1987.

Schmidtke H. *Ergonomic Data for Equipment Design.* New York: Plenum Press, 1984.

Snook S. The design of manual handling tasks. *Ergonomics* 21(12):963–985, 1978.

Tichauer ER. *The Biomechanical Basis of Ergonomics: Anatomy Applied to the Design of Work Situations.* New York: Wiley, 1978.

U.S. Department of Health and Human Services (HHS), National Institute for Occupational Safety and Health (NIOSH). *Work Practices Guide for Manual Lifting.* NIOSH Technical Report 81-122. Cincinnati: NIOSH, 1981.

Wasserman D. *Human Aspects of Occupational Vibration.* Amsterdam, Netherlands: Elsevier, 1987.

Webb RDG. *Industrial Ergonomics.* Toronto, Ontario: Industrial Accident Prevention Association, 1982.

White AH. Treatment of the industrial back, Part I: The epidemiology and diagnostics, perspective of the physician. *Top Acute Care Trauma Rehabilitation* 2(3):50–66, 1988.

White AH. Treatment of the industrial back, Part II: Clinical applications, perspective of the physician. *Top Acute Care Trauma Rehabilitation* 2(4):63–72, 1988.

Woodson WE, Tillman B, Tillman P. *Human Factors Design Handbook,* 2nd ed. New York: McGraw-Hill, 1991.

chapter 17

Employee Safety and Security Programs

by Joyce A. Simonowitz, RN, MSN

- 315 Objectives
- 315 Introduction
- 316 Workplace Violence
- 316 Cost of Violence
- 316 Causes of Violence
- 317 Types of Violence
  Type I ■ Type II ■ Type III
- 318 Predictors of Violence
- 318 Preventive Measures
  Worksite and hazard analysis ■ Hazard prevention and control ■ Post-incident response including medical and psychological services ■ Training and education ■ Record keeping and evaluation
- 322 Summary
- 322 References
- 323 Appendix

OBJECTIVES

The objectives of this chapter include providing information to accomplish the following:

- Assist managers to understand the types of violence experienced in the workplace.
- Assist managers to develop a workplace violence prevention program.
- Describe preventive actions and controls that may be needed.

INTRODUCTION

In the 1930s, President Franklin D. Roosevelt gave a "New Deal" to the people of the United States. It included labor legislation, which gave the Department of Labor a new importance. FDR then appointed Frances Perkins as Secretary of Labor, the first woman to become a member of the cabinet. Frances Perkins created the Division of Labor Standards. She integrated labor concerns of hours and wages with a new focus on health and safety conditions. She maintained that federal legislation was not the only answer to reducing health and safety hazards but that workers also had to depend on union involvement as well as a voluntary approach from state government and employers.

This voluntary approach, although adjusted, discarded, and reformed over the years, remains accurate and is a legacy to her foresight for workers' health and for all successful health and safety programs. Many safety and health hazards have improved and changed during the past 50 years, but there are new emerging hazards not recognized in Frances Perkin's time. These include biologic agents, farming health and safety issues, child labor, and workplace violence—all have moved to the forefront of concern on the part of workers, government, and employers.

WORKPLACE VIOLENCE

Workplace violence can be experienced in the commission of a robbery or other criminal acts; assaults from angry or unsatisfied customers; and from disgruntled employees, stalkers, or the spillover of domestic violence. A comprehensive safety and health program must include violence prevention programs

Although violence is pervasive in our lives and had traditionally been considered a penal code matter, the workplace generally was spared this cruel phenomenon with the exception of occasional labor strike-related violence. During the last 20 years, violence has become recognized as a workplace health and safety issue. Violence has been the subject of concern and its presence has elicited guidelines from the federal Occupational Safety and Health Administration (OSHA) and some state OSHA Plans. Violence prevention is now considered an important part of any comprehensive health and safety program.

Bulatao and VandenBos (1996) claim that workplace violence is clouded with misinformation and myth following the emotional upheaval and fear resulting from violent incidents. However, through proper planning and research, the myths are being resolved. Homicide, the most easily studied and dramatic form of workplace violence is the leading cause of workplace death for women in the United States, and is the first, second, or third leading cause of death in some states for all workers (U.S. DOL, 1997, Simonowitz, 1996).

Statistics on assaults are more elusive. Assaults have been defined by the National Crime Victimization Survey (NCVS) (Jenkins, 1996) to include "violent acts, including physical assaults and threats of assault directed toward persons at work or on duty." Assaults also can include workplace domestic violence issues, sexual assaults, and suicide. These crimes have been estimated by the NCVS to affect nearly one million U.S. residents 12 years of age and older, who are victims of rape, robbery, and assault. More than half of all victimizations are not reported to police. Many consider incidents to be minor if there is no serious injury or medical treatment needed; and some only report such incidents to a company official or security department. Unfortunately, without adequate reporting and statistics, the scope of the problem is unrecognized. Further, if not adequately recognized, measures and programs are not provided that would prevent a more serious occurrence in the future.

COST OF VIOLENCE

The cost of workplace violence is high for the victim and employer. The Bureau of Labor Statistics Survey of Occupational Injuries (1994) reported that assaults resulted in an average of five days away from work. In 1993, the federal government work force saw more than 2,000 cases of violence involving lost work time. Costs of homicide can be as high as $1 to $2 million and those costs can continue for more than a year (Wheeler & Baron, 1994). The cost to the victim may be more than financial and loss of work time. Some researchers report that cognitive emotional and physical sequelae may be present long after the victim has returned to work. Assaulted workers experience feelings of self-doubt, depression, fear, post-traumatic stress syndrome, loss of sleep, irritability, disturbed relationships with family and peers, decreased ability to function effectively at the workplace, and increased absenteeism. They may also suffer physical injury causing temporary or permanent disability or disfigurement. The mental cost to the victim of violence should be recognized and even if physical injury did not occur, professional counseling services may be required to aid in an employee's recovery.

CAUSES OF VIOLENCE

There is much speculation on the cause of workplace violence. The people of the United States appear to be simultaneously repulsed and fascinated with violence. Historically, the United States was in part settled by people escaping violence, who in turn perpetrated their own violence. The violent U.S. Revolutionary and Civil wars, slavery, and the glamorized "Wild West" are part of our heritage.

Violent crimes generate fear and at the same time curiosity and glamour, e.g., each decade seems to bring a new "trial of the century." The media thrives on pain and the unusual. On one recent day the *Los Angeles Times* had accounts of three convenience store worker slayings; however, the incident of an employee who walked into his workplace and shot two of his co-workers or supervisors was not only reported but televised. The worker who is slain during a robbery is no longer prime news.

A number of demographic and sociologic factors influence what form, distribution and intensity violent behavior takes. The tendency for people to resort to violence to resolve conflict with co-workers and others has also been identified as a cause of violence (Simonowitz, 1995). Treatment of workers as disposable goods during *downsizings* and the failures to (1) establish security systems, (2) have programs that provide employees and customers with compassionate and understanding courtesy, and (3) have programs established to deal effectively with disgruntled consumers and employees contribute to the potential for violent incidents in the workplace.

Other factors such as the influence of alcohol and use of drugs, such as crack-cocaine, racial or religious prejudice, and mental illness may be involved in violent actions. However, the majority of people who are mentally ill *do not* behave in a violent or aggressive manner and are not a danger to others. We should be aware that although the focus of the media and the public is on the worker who goes berserk and kills one or more co-workers or supervisors, these acts constitute less than 10% of workplace homicides. The killing of taxicab drivers, fast-food or convenience store workers, and public safety workers far outnumber the disgruntled worker homicides. The spillover of domestic violence in the workplace is a serious problem causing absenteeism, distraction of workers, increases in health care costs, injury and illness, and on occasion stalking and assault at work.

TYPES OF VIOLENCE

Although there may be some indicators, it is difficult to predict who will be violent. It may be easier to identify situations that may lead to or present a potential for violent acts. Simonowitz (1995) discussed three categories of workplace violence:

Type I

Type I violence is the most frequently encountered type of violence. It usually involves a person entering a small retail establishment such as a liquor store or fast-food establishment to commit robbery. The assailant has no relationship to the workplace. The security guards protecting a property are at risk as well as the store clerk. Employees who have face-to-face contact and exchange money or goods with the public and who work alone at night are at greatest risk of a Type I event.

In the health care field, reports of robbery and criminal activity have been received from hospitals where perpetrators have entered an emergency room (ER) or hospital unit and robbed, raped, and assaulted both patients and staff. Drugs are often the target that results in attacks on pharmacists.

The following are examples of Type I events. The first example illustrates subsequent *revictimization* of the employee.

Example 1. A convenience store was robbed; the manager, in an attempt to protect employees from being injured, cooperated with the robbers (which is the only type of behavior recommended by all agencies). Although no one was injured, the company subsequently fired the manager for not resisting the robbery.

Example 2. The bombing of a family planning clinic in Florida that killed a security guard and severely injured a nurse, causing her total disability, is another tragic example of a Type I event.

Type II

A *Type II* workplace violence event involves an assault by someone who is either the recipient or the object of a service provided by the affected workplace or the victim. Although Type I events are the most common workplace fatalities, Type II events include nonfatal assaults and may represent the most prevalent category of workplace violence.

Assaults on public safety and emergency personnel, bus drivers, teachers, motel/hotel attendants, and public and private sector employees who provide professional administrative or business services to the public are frequent and of serious concern.

Health care workers in acute-care hospitals, long-term care, and family planning clinics, and those who work as home health personnel, mental health and psychiatric care providers, alcohol and drug treatment staff, social welfare providers, homeless shelter staff, and emergency responders all are at greater risk of assault and fatalities. Unlike Type I events which may represent an unusual occurrence, Type II events occur on a regular or daily basis in many establishments and represent a more pervasive risk for certain providers.

Example 1. An example of a Type II event occurred in a psychiatric hospital. A patient got angry when the nurse told him he could only smoke while outside during exercise times rather than in the building. He struck the nurse breaking her jaw.

Example 2. Another example of a Type II event is seen in the attack on an employment development office where a client was refused benefits. He returned to the office with a gun and fired randomly at workers, killing five, and injuring others.

Type III

A *Type III* event is one in which an assault occurs by an individual who usually has some employment-related involvement with the workplace. The assailant may be an employee, a supervisor, or an acquaintance of a worker who may be seeking revenge for what he or she perceives as unfair treatment, intolerable, or unsatisfactory relationships. The stalker or former significant other or even an unknown person who wishes to harm an employee is an example of a perpetrator of a Type III event. These events account for a much smaller portion of injuries than those classified as Type I or II, but often attract a great deal of media attention. Because of the publicity they are often thought to be the major violence problem in the workplace. However, Kraus (1995) reports that Type III events account for only about 10% of the homicides in the workplace.

If sexual and other types of worker harassment were included in Type III events there would be a significant increase in the incidence. Bachman and Salzman (1995) report that domestic violence accounts for 5% of the homicides in the workplace; and at the same time these events account for 17% of the women killed at the workplace (U.S. DOL, 1997).

Examples of Type III events include:

Example 1. A former spouse of a receptionist in a San Francisco law firm walked into the law offices looking for his ex-wife. He did not see her immediately but began randomly firing at any worker he saw. He killed a number of workers and injured more before he shot his wife and walked out of the offices.

Example 2. In another case a worker was being disciplined for failing to perform his duties and for using aggressive behavior toward co-workers. Three supervisors and a union representative were present at the conference. The supervisors had experienced hostile encounters with this employee in the past. They had reported this to personnel and asked for assistance in dealing with this employee. They received no response or suggestions on how to handle this serious problem. They were afraid of the employee and of holding this disciplinary conference. During the conference, the employee produced a gun and shot the supervisors and the union representative, killing all four employees.

PREDICTORS OF VIOLENCE

There are no reliable predictors of who will be violent. The only reliable information is a history of violence. There are many suggested warning symptoms of potential for violent action in an employee. These include the following:

- having a fixation with weapons
- being a loner
- use of excessive profanity especially of a sexual nature
- verbalizing threats toward others
- abuse of property.

These have all been postulated as warning signs. As with any symptom, problems do not always occur. Prompt and proper attention can reduce or eliminate the undesired manifestation of violent behavior. Many believe that undesirable employees can be screened before hiring. There is no profile of the average disgruntled worker that reliably lends itself to screening before hire. Barling (1996) has proposed a combination of workplace and personal factors that may predict violence. The table has been modified to include issues of external violence, stalkers, irate customers, etc.

PREVENTIVE MEASURES

Specific preventive measures will depend on individual company circumstances, the type of service provided and the category (type) of violence being considered. Every company should expect all three types of violence. Preventive programs must consider specific methods for each type to protect staff and clients.

Managers and administrators are advised to make the provision of adequate measures to prevent violence a high priority. In California, state

Table. Personal and Workplace Factors Leading to Violence in Workplaces.

PERSONAL FACTORS
- Alcohol, drug use
- Aggressive history
- Low self-esteem
- Financial problems
- Sexual harasser
- Use of psychological aggression

WORKPLACE FACTORS
- Perceived injustice
- Working at night
- Robbery deterrents absent
- Failure to have training
- Downsizing
- Lack of security
- Angry customers

Adapted from Barling 1996.

law requires health care facilities to develop and implement security and safety programs aimed at preventing violent acts in emergency rooms and other patient care units. Some security measures may seem expensive or difficult to implement but may well be needed to adequately protect workers, clients, and physical plant and equipment. We all must recognize that the belief that certain risks are *part of the job* contributes to the continuation of the potential for violence. An attitude of postponing action until a problem manifests is foolhardy and may result in injury, death, and severe financial hardship for the company and employees.

The U.S. OSHA produced recommendations for Security for Health Care and Social Service Workers in 1996 and Workplace Violence Prevention Programs for Retail Establishments in 1998. While these guidelines are not standards, under the OSHAct the extent of an employer's obligation to address workplace violence is governed by the General Duty Clause, "Each employer shall furnish to each employee a place of employment which is free from recognized hazards that are likely to cause death or serious physical harm," (section 5(a)(1) P.L.91-596). The General Duty Clause holds an employer liable for failure to develop a plan of prevention. The guidelines provide recommendations and information about prevention strategies and are recommended resources. Simonowitz (1995) and OSHA (1998) proposed a program to prevent workplace violence that is summarized in the next sections.

Managers must be committed to develop a program leading to a *violence-proof* workplace. Management commitment and employee involvement are critical elements of all safety and health programs and should be demonstrated by implementing a written program that encourages and supports involvement in job safety, health, and security. Along with management commitment and employee involvement, a successful program must include the following program elements:
1. worksite and hazard analysis
2. hazard prevention and control
3. post-incident response including medical and psychological services
4. training and education
5. record keeping and evaluation.

Worksite and Hazard Analysis

Worksite hazard analysis is identifying the risk of having violent incidents. It consists of:

a) Review of records and past experiences. The employer should review the experience of his/her own business over the previous two or three years. This may involve examining records that can identify the company history of workplace violence. For example, injury and illness records, workers' compensation claims, or records of police department calls or robbery reports etc. However, unless reports of unusual or threatening incidents are required by the company little data may be found. This will give a false impression of security. Employees may be questioned regarding their experiences using the following examples of questions that may be helpful:
- Has the business, or employees been robbed during the past two to three years?
- Have employees been assaulted or threatened?
- Have employees been victimized by other criminal acts?
- What time of day did these acts occur?
- How many employees were on duty?
- Have any employees threatened a co-worker or supervisor?
- Have there been sexual harassment claims?
- Have security measures been instituted and were they used effectively?
(See OSHA 1998.)

Employers with more than one location should establish a history at each location. Local community experiences or similar business history or trade association can provide useful information about crime trends in the industry and in the community.

b) Perform initial and periodic worksite security inspections by conducting a walk-through survey to identify security risks and develop a plan for security measures.

- Have violent events occurred and where?
- Are there unlocked doors, windows, or unobstructed access to offices and facilities by unauthorized persons?
- Are telephone and rapid communication devices available?
- Are there lighting deficiencies in halls, stairs, or parking?
- Are doors propped open, windows broken?
- Do employees work alone or in isolation?
- Is there a security system with panic or alarm buttons?
- Is there a threat or alarm response team or system?
- Do systems in place work?
- Are offices arranged to facilitate egress?

The Appendix to this chapter contains a checklist that may be helpful for security analysis.

Hazard Prevention and Control

After assessing the physical hazards, a control method and plan should be developed and written. Physical and behavioral changes at a facility can substantially reduce the opportunity for violent incidents. For example, increased lighting and trimming plants and bushes in a parking area in addition to security escorts or employee group travel to parked cars, especially at night, may increase parking safety (Figures 17-1 through 17-5). Restricted access to treatment areas in ERs and security guard presence in the ER increases the safety of both staff and clients from intruders, would-be robbers, or clients who are under the influence of illegal drugs or alcohol or untoward reactions from prescribed medications or illnesses.

Administrative and work practice controls affect the way employees perform jobs or tasks. Integration of violence prevention activities into daily procedures such as checking security cameras, reducing the amount of cash in a register, or providing and using drop safes are all work practice controls that lead to a more secure workplace. Adopting emergency procedures in cases of robbery or security breaches, adopting safety procedures and policies for off-site work such as delivery, home health visits or other community work is essential. Such actions can include providing communication devices and a backup procedure to respond to a call for assistance. Also see chapter 18, Emergency Response Programs.

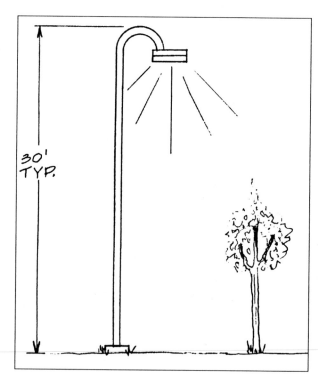

Figure 17-1. Young trees with 30-ft (9.15 m) lighting poles. (Reprinted from Lack RW (ed). *Accident Prevention Manual for Business & Industry: Security Management.* Itasca, NSC Press, 1997.)

Establish personnel policies and procedures to guide workers and supervisors in a progressive disciplinary system. Provide for and refer problem employees to an employee assistance program (EAP) or similar source of psychological counseling. Establish guidelines for handling harassment incidents of any kind, and implement programs with no exceptions for every level of employee including the CEO.

Plans and assistance for victims of sexual harassment or stalking should be made and employees encouraged to report such situations. Legal referrals, transfers, or other such assistance must be considered as the situation warrants. Open and supportive communication is essential to prevent failure to share such concerns, thereby placing the company and other employees at risk

Post-Incident Response Including Medical and Psychological Services

Once an incident has occurred, rapid recovery and corrective action is essential for the employees and the business. Standard procedures for management and employees to follow in the aftermath of a vio-

Figure 17-2. Mature 30-ft (9.15 m) trees with 30-ft (9.15 m) lighting poles. (Reprinted from Lack RW (ed). *Accident Prevention Manual for Business & Industry: Security Management.* Itasca, NSC Press, 1997.)

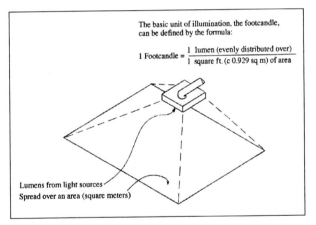

Figure 17-3. Many state and municipal security ordinances establish that the standard for parking lot security lighting should be at least 1 footcandle of light minimum maintained and evenly disbursed across the surface of the parking lot. (Reprinted from Lack RW (ed). *Accident Prevention Manual for Business & Industry: Security Management.* Itasca, NSC Press, 1997.)

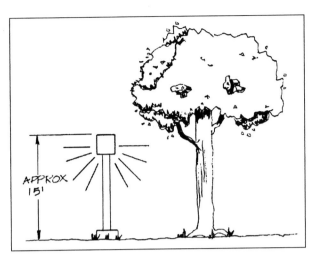

Figure 17-4. Use of many lower light fixtures with trees that canopy above them. (Reprinted from Lack RW (ed). *Accident Prevention Manual for Business & Industry: Security Management.* Itasca, NSC Press, 1997.)

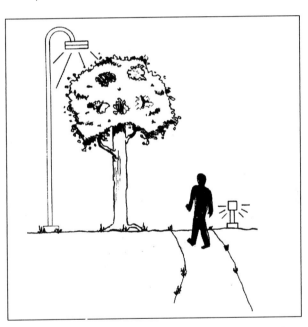

Figure 17-5. Primary and secondary lighting to create contrast. (Reprinted from Lack RW (ed). *Accident Prevention Manual for Business & Industry: Security Management.* Itasca, NSC Press, 1997.)

lent incident need to be developed and evaluated. Such procedures may include the following:

- Assure that injured employees receive prompt and appropriate medical care, including transportation to the care provider. Prompt first aid and emergency treatment are imperative but other continued assistance may be needed such as, home visits, child care, transportation to health care appointments, or other follow up depending on the extent and nature of injuries.
- Notify other authorities as required by law.
- Secure the premises to safeguard evidence and reduce distractions.
- Arrange for psychological and medical treatment for victimized employees. In addition to physical injuries, victims and witnesses may suffer mental trauma. Post-incident debriefing and counseling can reduce trauma and stress. Restoring the physical facility to its original clean and orderly state after any investigation

is completed is important to restore normal function.
- Inform management and prepare reports.

Training and Education
Training and education ensure that all staff are aware of potential security hazards and the procedures for protecting themselves and their coworkers. Training should be repeated annually or at frequencies that ensure appropriate use of techniques. There should be training for management of assaultive behavior for nurses in psychiatric situations. Persons who need to function as a team should be trained as a team. These kinds of training should be repeated every six months at a minimum. However training should never be thought of as the total answer to security problems. Security is based on using engineering measures, administrative controls, and teaching employees how to use these measures to provide a secure workplace.

Record Keeping and Evaluation
Faithful and complete record keeping helps determine the frequency and severity of the risks, evaluate methods of hazard control, and identify training needs. Employers may use the following types of records to determine the scope of the risk:
- records of employee incidents and injury
- OSHA illness and injury logs
- Incident reports of threats, abuse, or aggressive behavior are required to be completed by employees. These reports may not be used for disciplinary actions against the reporting employee.
- minutes of safety meetings where security is discussed
- procedures for and reporting of domestic-abuse issues
- records of action taken to deter violence.

Type III violence threats require action that is based on not just securing a workplace from outside threat but on establishing options for action by supervisors or employees who are threatened by co-workers or supervisors.
- Establish a true open-door policy that encourages communication among all levels of employees.
- Establish a threat-assessment team.
- Assume the validity of a threat until proven false.
- Consult human resources and security.
- Refer affected employees to EAP and place on medical leave, or suspend the offender until the threat is assessed.
- Notify targets.
- Notify law enforcement.
- Identify possible causal stresses in the offender. Such clues to stressors may be found in excessive workload or hours worked per week, shift changes, job description changes, use of grievance procedure, poor supervisor and peer relationships, absenteeism, or illnesses.

These suggestions and recommendations are not a complete discussion of all the measures that can and should be taken to prevent workplace violence incidents. It is a framework for developing a program; and assumes that the reader will adapt and develop the recommendations to fit his or her own occupational setting and types of anticipated problems.

SUMMARY
Workplace violence has emerged as an important occupational safety and health issue in many occupational settings. The pervasiveness of violence in all work settings without vigorous preventive programs makes a mockery of over a half century of struggle by state and federal governments, labor unions, and employers to secure safety and dignity for working people. It is important that companies plan preventive programs before a workplace incident occurs. Workplace violence has not yet become a subject for regulation but as Frances Perkins recommended in the 1930s, we hope to see every employer develop effective and voluntary security programs.

REFERENCES
Bachman R, Salzman L. *Violence Against Women: Estimates from the Redesigned Survey*. (pub. No. NCJ-154348). Washington DC: Bureau of Justice, Statistics Dept. of Justice, 1995.

Barling J. The prediction, experience and consequences of workplace violence. In VandenBos, & Bulata (eds). *Violence On The Job*. Washington DC: American Psychological Assn, 1996.

Bulatao E, VandenBos G. Workplace violence: Its scope and the issues. In VandenBos & Bulata (eds), *Violence On The Job*. Washington DC: American Psychological Assn, 1996.

Gates DM. Workplace violence. *AAOHN J* 43(10):536–544, 1995.

Jenkins EL. *Violence in the Workplace: Risk Factors and Preventive Strategies.* Washington DC: (DHHS [NIOSH] Publication # 96-100) 1996.

Kraus J, Blander B, McArthur D. Incidence, risk factors and prevention strategies for work-related assault injuries. *Annual Review of Public Health.* 16:355–379, 1995.

Lack RW (ed). *Accident Prevention Manual for Business & Industry: Security Management.* Itasca, NSC Press, 1997.

Rosner D, Markowitz G. Research or advocacy: Federal occupational safety and health policies during the New Deal. In Rosner & Markowitz (eds). *Dying For Work.* Bloomington IN: Indiana University Press, 1989, pp 83–102.

Simonowitz J. Violence in health care: A strategic approach. *Nurse Practitioner Forum,* 6(2): 120–129, 1995.

U.S. Dept. of Labor, Bureau of Labor Statistics (BLS). *News: Work Injuries and Illness by Selected Characteristics,* 1992. Washington, DC: BLS, 1994, pp 4–26.

U.S. Dept. of Labor, Occupational Safety and Health Administration (OSHA). *Workplace Violence,* 1997. [On line] Available at http://www.osha-sic.gov.

U.S. Dept. of Labor, OSHA. *Recommendations for Workplace Violence Prevention Programs in Late-Night Retail Establishments.* Washington DC: OSHA pub no. 3153, 1998.

Wheeler E, Baron S. *Violence In Our Schools, Hospitals, and Public Places.* Ventura CA: Pathfinder Publishing of California, 1994.

Williams ML, Robertson K. Workplace violence. Prevalence, prevention, and first-line interventions. *Crit Care Nurse Clin North Am* 9(2): 221–229, 1997.

APPENDIX: SAMPLE WORKPLACE VIOLENCE FACTORS AND CONTROLS CHECKLIST

This sample checklist can help employers identify present or potential workplace violence problems. The checklist contains various factors and controls that are commonly encountered in retail establishments. Not all of the questions listed here, however, are appropriate to all types of retail businesses, and the checklist obviously does not include all possible topics relevant to specific businesses. Employers are encouraged to expand and modify this checklist to fit their own circumstances. These factors and controls are not a new standard or regulation and the fact that a control is listed here but is not adopted by an employer is not evidence of a violation of the General Duty Clause. (Note: "N/A" stands for "not applicable.")

Yes	No	N/A	Environmental Factors
			Do employees exchange money with the public?
			Is the business open during evening or late-night hours?
			Is the site located in a high-crime area?
			Has the site experienced a robbery in the past 3 years?
			Has the site experienced other violent incidents in the past 3 years?
			Has the site experienced threats, harassment, or other abusive behavior in the past 3 years?
Yes	No	N/A	Engineering Controls
			Do employees have access to a telephone with an outside line?
			Are emergency telephone numbers for law enforcement, fire and medical services, and an internal contact person posted adjacent to the phone?
			Is the entrance to the building easily seen from the street and free of heavy shrub growth?
			Is lighting bright in parking and adjacent areas?
			Are all indoor lights working properly?
			Are windows and views outside and inside clear of advertising or other obstructions?
			Is the cash register in plain view of customers and police cruisers to deter robberies?
			Is there a working drop safe or time access safe to minimize cash on hand?
			Are security cameras and mirrors placed in locations that would deter robbers or provide greater security for employees?
			Are there height markers on exit doors to help witnesses provide more complete descriptions of assailants?
			Are employees protected through the use of bullet-resistant enclosures in locations with a history of robberies or assaults in a high-crime area?
Yes	No	N/A	Administrative/Work Practice Controls
			Are there emergency procedures in place to address robberies and other acts of potential violence?
			Have employees been instructed to report suspicious persons or activities?
			Are employees trained in emergency response procedures for robberies and other crimes that may occur on the premises?
			Are employees trained in conflict resolution and in nonviolent response to threatening situations?
			Is cash control a key element of the establishment's violence and robbery prevention program?
			Does the site have a policy limiting the number of cash registers open during late-night hours?
			Does the site have a policy to maintain less than $50 in the cash register? (This may not be possible in stores that have lottery ticket sales and payouts.)
			Are signs posted notifying the public that limited cash, no drugs, and no other valuables are kept on the premises?
			Do employees work with at least one other person throughout their shifts, or are other protective measures utilized when employees are working alone in locations with a history of robberies or assaults in a high-crime area?
			Are there procedures in place to assure the safety of employees who open and close the store?

chapter 18

Emergency Response Programs

by Richard T. Vulpitta, CEM, CUSA

325 **Introduction**
Why plan? ■ Four phases of emergency and disaster management

326 **Emergency or Disaster?**
Similarities and differences ■ Lines of defense: Local, state, and federal

327 **Roles of Government**
Federal government ■ Federal emergency management agency ■ State government ■ OSHA & SARA Title III requirements and state occupational safety and health plans ■ Emergency response consultation services ■ Local government

332 **Steps For Establishing On-Site Emergency Response Plans**
Purpose of planning ■ Planning effectiveness ■ Elements of emergency planning ■ Hazard audit ■ Risk evaluations ■ Toxic materials and material safety data sheets ■ Reviewing plan with employees ■ Chain of command ■ Assembly areas and command centers ■ Communications warning alarm system ■ Accounting for personnel ■ Special response teams

336 **Training**
337 **Drills**
338 **Personal Protective Equipment**
339 **Respiratory Protection**
341 **Confined Space Entry**
342 **Medical Assistance**
343 **Security**
343 **News Media**
343 **Business Records**
343 **Response Agreements**
343 **References**

INTRODUCTION

This chapter will detail the key elements of emergency and disaster planning, the roles of local, state and federal agencies and the essential concepts for writing an on-site emergency response plan for a facility. Safety administrators are finding more of their duties linked with emergency and disaster-preparedness issues. Examples of these responsibilities may include evacuation planning to filing the annual SARA, Title III reports. The scope of involvement is dependant upon the view of the company's management, business activities and their dedication to foster a safe work environment. There is a trend for the safety professional to function as a safety consultant. As a safety consultant, risk management options are offered to management. In disaster or emergency management planning, the consultant, is known as an emergency planner. An emergency response planner needs to be satisfied in advance that the plans in place will protect a company's employees, the public and the physical workplace. To do this one needs to understand the many roles of local, state and federal government and the emergency response planning process. To remain effective, this process will include people who will require training and practice. An emergency planner, when writing a response plan for a facility, needs to emphasize three goals. The first is to protect life, the second is to protect property, and the third is to resume normal operations. These three goals can only be achieved when the emergency planner involves all levels of management and employees during the development of the plan. This meticulous preparation will allow a company to survive a disaster or an emergency so it may resume normal operations.

Why Plan?

In a 10-minute span of time, two people will be killed, while 370 will suffer a disabling injury. On the average, 11 unintentional-injury deaths and about 2,200 disabling injuries occur every hour during the year. Unintentional injuries are the leading cause of death for all persons between the ages of 1 and 38 years old. Among people of all ages, unintentional injuries are the fifth leading cause of death. These unintentional injuries along with man-made or natural emergencies and disasters must also be considered. The potential for each type of human-made emergency and natural disaster has to be considered by an emergency planner. Every day unplanned events do occur somewhere resulting in a workplace or community disaster or emergency (Figure 18-1). The emergency planner accepts this and plans to lessen the event's effects. Emergency planning is an extension of an effective safety and health workplace program (NSC, 1997).

Four Phases of Emergency and Disaster Management

There are four phases of emergency and disaster management. These phases are an ongoing process of planning and responding to react effectively when a unplanned event occurs.

1. *preparedness*—planning for an emergency or disaster event
2. *response*—the planned response to an emergency or disaster event
3. *recovery*—the process of returning to normal operations
4. *mitigation*—steps taken to prevent the effects of an emergency or disaster event.

When properly used these phases lessen the spillover effects of an unplanned event. These disasters and emergency spillover effects disrupt the local operations and the quality of life of the persons who reside and work there. Let's use a flood as an example caused by a river that overflows its banks due to heavy rains.

- Normal lines of communications are lost.
- The flood waters disrupt transportation and the utility services to the area.
- Several of the residents and workers fleeing the rising water are injured or killed.
- Factories and residences are damaged and destroyed.
- Hazardous materials are released.
- Local medical facilities are quickly rendered inadequate.
- Looting occurs.
- Local resources are quickly exhausted, state and federal assistance are required to mitigate the event for a return to normal operations.

Advance planning and recognizing the spillover effects of a disaster would have lessened the impact on local residents and the businesses. For example, normal communications providers and other utilities are "flood-proofed" by taking the following steps:

- A new *flood early warning system* is directed to residents and business in the potential flood plain.
- Clearly marked evacuation routes along higher ground areas will save lives.
- The storage of hazardous materials by local industry is always kept at a minimum.
- Medical facilities practice for disasters.
- Building codes will prevent new companies locating in areas of potential flood waters. Evacuation routes clearly marked will aid getting residents to safety.

EMERGENCY OR DISASTER?
Similarities and Differences

To understand emergency and disaster management you need to understand their similarities and differences.

Similarities
- They begin as unexpected occurrences.
- They produce negative affects.
- They have to be dealt with immediately.

Differences
- Emergencies are handled by local resources.
- Disasters require outside resources.

How they are responded to determines the differences between emergencies and disasters. *Emergencies* are responded to locally by using local resources. *Disasters* can initially use local responders and local resources. When local resources become exhausted, outside aid in the form of responders and or resources are needed. Once these outside resources overcome the disaster, normal activities are resumed. Another difference between the two is their scope or impact. Disasters tend to

CHAPTER 18: EMERGENCY RESPONSE PROGRAMS

Figure 18-1. Disasters have an adverse impact on a larger geographic area. This photo shows hurricane damage in Florida. (Photo by Carl Griffith)

affect all normal activities in a large geographic area. Emergencies are more contained disrupting normal activities in a particular community or a single facility. There are three similar goals of emergency and disaster response.

Goals of Emergency and Disaster Response
- Protect and save people.
- Protect property.
- Resume normal activities.

Lines of Defense: Local, State, and Federal

In emergency planning/disaster preparedness governmental activities are divided into three areas or roles provided at local, state and federal levels. Each governmental area has been authorized by various laws outlining their responsibilities during an emergency or disaster event. These responsibilities become broader at the state and federal levels for emergency response planning and responding. These responsibilities can be compared as lines of defense and duties.

- local
  - responsible for local planning
  - responsible for initial responding
- state
  - plans for statewide disasters
  - allocates reserve resources.
- federal
  - responsible for national disasters
  - greater resources to back up resources of the states.

Local town/city or county police, fire, and medical personnel who respond to human-made and natural emergencies are the first line of defense (Figure 18-2). If the local responding agencies require additional assistance, the county where the event took place provides the resources and humanpower. When the emergency exhausts the resources of that county, the county's governing body requests assistance from the governor of that state. If state resources are exhausted or not adequate, the governor can request in writing disaster relief from the President of the United States. The president authorizes various federal agencies to assist.

One of these agencies is the Federal Emergency Management Agency (FEMA). At this time FEMA would become involved with recovery efforts. Besides the goal of mitigating the event, the local, state, and federal agencies share the goal of continuing governmental activities.

ROLES OF GOVERNMENT

Each level of government has characteristic resources it can bring to bear on emergency management. Simply stated, the contribution of each level can be summarized as follows:

- *federal:* legal authorities, fiscal resources, research, technical information and services, specialized personnel
- *state:* legal authorities, administrative skills, conduit between local and federal

Figure 18-2. Monticello Fire Department river rescue drill. Volunteer firefighters practice a rescue in rapids below a dam in Monticello, Indiana. Local government plans for and provides initial response for local emergencies. (Photo by R.T. Vulpitta)

- *local:* direct motivation, knowledge of the situation, personnel, proximity to both event and resources.

Federal Government

The federal government provides legislation, executive orders, and regulations that influence, more or less directly, all disaster activities. It also maintains the largest pool of fiscal resources that can be applied to emergency management. Some federal agencies are sources of specialized research, technical information, and services needed in disaster work (for example, the Nuclear Regulatory Commission). Finally, the federal government is a limited source of specialized personnel (particularly with nuclear or conventional attack preparedness and response).

Research by the National Governors' Association (NGA) has identified more than 100 federal laws containing provisions directly relating to natural, technological, peacetime, or attack-related emergencies. In fact, virtually every department and agency of the federal government has some emergency-related responsibility mandated by law. Further extending and complicating the intricate federal-level disaster authorities are a staggering variety of executive orders, regulations, and interagency agreements.

At the initiation of President Carter, Congress established FEMA in 1979 and brought a number of previously fragmented disaster programs into a coordinated structure. However, FEMA certainly does not include or direct all federal disaster efforts.

The federal government's involvement in emergency management is primarily in the areas of *assistance,* on the one hand, and *regulations and standards* on the other. Assistance can take the form of fiscal, material, personnel resources, or research and technical information.

Federal Emergency Management Agency

The Federal Emergency Management Agency (FEMA) plays coordinating and supportive/assistance roles for integrated emergency management in partnership with state and local emergency management entities. It takes a lead role in national preparedness for major peacetime or wartime crises.

FEMA's responsibilities for national preparedness include development of federal program policy guidance and plans to insure that governments at all levels can cope with and recover from emergencies. The agency administers the national civil defense program and national laws such as the Civil Defense Act of 1950. FEMA is responsible for the assessment of national mobilization capabilities and the development of concepts, plans, and systems for management of resources in a wide range of national and civil emergencies. Preparedness includes warning systems, in-place shelter planning, population protection planning, shelter identification, and disaster information such as the Emergency Broadcast System (EBS).

Another dimension of FEMA's role in emergency management is in mitigating the effects of disasters and emergencies through research and programs aimed directly at specific problem areas. FEMA's research efforts focus on increasing the nation's capability to predict, prevent, respond to, and recover from emergencies and disasters. The goal is to discover information that can help decrease loss of lives, injury, damage, and economic and social disruption from such events.

Hazardous waste disposal or transportation, especially involving nuclear waste, is increasingly seen as a threat to public health. FEMA also is helping communities in which a certain amount of acutely toxic chemicals are stored to plan for the disposal or transportation of the chemicals. This is in compliance with the Title III program, part of the 1986 Superfund Amendments and Reauthorization Act. SARA Title III mandates planning and includes right-to-know provisions similar to those found in OSHA legislation.

The agency also supports state and local governments in fulfilling their emergency mitigation, preparedness, response, and recovery responsibilities. As necessary, FEMA provides funding, technical assistance, services, supplies, equipment, and direct federal support. Among the areas in which FEMA works to aid state and local government in emergency management (in accordance with the Civil Defense Act of 1950 and NAPB-90, Nuclear Attack Planning Base-1990) are the following: civil defense, earthquakes, floods, hurricanes, tornadoes, nuclear power plant incidents, terrorist acts, dam safety, and hazardous materials incidents.

FEMA provides technical and financial assistance to state and local governments to upgrade their communications and warning systems, and operates an emergency information and coordination center that provides a central location for the collection and management of disaster and emergency information. A wide range of emergency preparedness and mobilization civil preparedness guides are available from FEMA.

One of FEMA's most visible forms of assistance is the Presidential Declaration of an Emergency or Major Disaster. Such a declaration is made when the severity of a situation cannot be adequately relieved by local and state efforts, and a request for assistance is made to the U.S. President by the governor of the affected state. The completed request, addressed to the President, is sent to the FEMA regional director, who evaluates the damage and requirements for supplemental federal assistance and makes a recommendation to the President.

Direct disaster assistance from FEMA falls into two broad categories—public assistance (aid to State and local governments) and individual assistance (aid for disaster victims and their families). Hazard mitigation efforts also are required now under disaster assistance programs to help ensure the future safety of lives and property. Federal and state funds for hazard mitigation are available to the local level.

FEMA administers the National Flood Insurance Program, which provides insurance coverage to property owners in communities with flood hazards in exchange for that community's agreement to adopt floodplain management measures to protect lives and reduce property losses. Technical assistance is provided to communities in floodplain management and postdisaster hazard mitigation activities, such as encouraging new construction away from flood-prone area. As of October 1989, there were more then 17,000 communities participating in the program and more than 2 million policies in effect.

FEMA is headquartered in Maryland and also has a variety of regional offices that can be contacted for assistance for development of emergency plans. FEMA also offers a variety of emergency response training opportunities. They offer home study courses, training through state agencies and at their training institute. The FEMA web page services offer a variety of free services. The *Emergency Management Guide For Business & Industry* can be reviewed at www.fema.gov on the Internet.

State Government

State governments have fewer public funds to allocate for disaster work, but they do have a strong public mandate (and federal encouragement) to do whatever they can to prepare for and respond to disasters. This mandate is translated into legislated authorities and extraordinary gubernatorial powers. The state, like the federal government, also is a source of laws affecting disasters (such as traffic safety codes and state fire regulations). In addition, states have responsibilities as outlined by SARA,

Title III. Under the Emergency Planning Community Right-to-Know portion of this act, states are responsible for establishing a *State Emergency Response Commission* and in turn approving districts, or areas, where *Local Emergency Planning Communities* (*LEPCs*) will be formed. The LEPCs must formulate emergency plans to be used should an incident involving the manufacture, storage, or transportation of hazardous materials occur. Finally, state government is the central point of public administration skills in emergency management (primarily in preparedness planning and long-term recovery administration).

OSHA & SARA Title III Requirements and State Occupational Safety and Health Plans

All states have offices to enforce and advise compliance on safety and environmental issues. If assistance is needed on developing an in-house safety plan call your state's OSHA compliance office. The Occupational Safety and Health Act of 1970 encouraged states to develop and operate their own state job safety and health plans under the approval and monitoring of OSHA. Eighteen states have developed "OSHA" compliance offices that are required to set effective standards, conduct inspections to enforce those standards (including inspections in response to workplace complaints), cover state and local government employees, and operate occupational safety and health training and education programs. In addition, most states provide on-site consultation to help employers to identify and correct workplace and environmental hazards. The following is a list of some of the OSHA requirements and the SARA Title III code pertaining to emergency response. (*Code of Federal Regulations, Title 29, Part 1910,* OSHA General Industry Standards.)

Subpart E - Means of Egress
1910.37	Means of egress
1910.38	Employee emergency plans and fire prevention plans

Appendix to Subpart E - Means of egress

Subpart H - Hazardous Materials
1910.120	Hazardous waste operations and emergency response. (Hazwoper)

Subpart I - Personal Protective Equipment
1910.132	General requirements - personnel protection
1910.133	Eye and face protection
1910.134	Respiratory protection
1910.135	Head protection
1910.136	Foot protection
1910.137	Electrical Protective Equipment
1910.138	Hand Protection

Subpart J - General Environmental Controls
1910.147	Control of Hazardous Energy Sources (Lockout/Tagout)

Subpart K - Medical and First Aid
1910.151	Medical services and first aid (Bloodborne Pathogen Standard 1910. 1030)

Subpart L - Fire Protection
1910.155-156	Fire protection and fire brigades
1910.157-163	Fire suppression equipment
1910.164	Fire detection systems
1910.165	Employee alarm systems

Appendix A-E of Subpart L

Subpart Z - Toxic and Hazardous Substances
1910.1200	Hazardous Communication (Hazcom)

Environmental Protection Agency Regulations
SARA Title III Regulations 40 CFR, subchapter J, pages 300-374

Emergency Response Consultation Services

State emergency management agencies can help in planning and developing programs to use when responding to occupational emergencies. Each state has an office of emergency management. Every county in a state has an office for emergency planning. These may be contacted for developmental direction and for program review of on-site response programs. State emergency response agencies are also an excellent source of emergency management planning courses. Courses offered are free and are part of the FEMA course curriculum. Expenses for some attendees from the private sector may be covered. The following are some of the courses offered.

- Introduction to Emergency Management (4-day course)
- Emergency Planning (4-day course)
- Leadership and Influence (2-day course)
- Decision Making/Problem Solving (1-day course)
- Effective Communication (3-day course)
- Developing Volunteer Resources (3-day course)
- Exercise Design (3-day course)
- Mass Fatalities Incident (3-day course)
- Incident Command System/Emergency Operations Center Interface (2-day course)

- Exercise Design (2-day course)
- Exercise Evaluation (2-day course)
- Disaster Recovery Operations (3-day course)
- Emergency Operations Center (3-day course)
- Basic Public Information (3-day course)
- Continuity of Government (2-day course)

Parallels to Federal Government

The role of state government in emergency management in many ways parallels the role of the federal sector. Legislative and executive authorities exist for state emergency programs, with a range of programs usually operating in a variety of state agencies. The state has a responsibility to develop and maintain a comprehensive program of mitigation, preparedness, response, and recovery activities. The state role is to supplement and facilitate local efforts before, during, and after emergencies. The state must be prepared to maintain or accelerate services and to provide new services to local governments when local capabilities fall short of disaster demands.

A state government is in a unique position to serve as a link between those who need assistance and those who can assist—to find out what local emergency programs need, to assess available state and federal resources, and to help the local government apply for, acquire, and use those resources effectively. The state provides direct guidance and assistance to its local jurisdictions through program development, and channels federal guidance and assistance down to the local level. In a disaster, the state office helps coordinate and integrate resources and apply them to local needs. The state's role might be best described as *pivotal*.

Gubernatorial Role

A governor of a state, responsible for the general welfare of the citizens of the state, has certain legislated powers, and resources that can be applied to *all-hazards emergency management*.

All U.S. governors (states, commonwealths, territories, and possessions) have authority and responsibility for the following:

- issuing state or area emergency declarations
- involving state response actions (personnel, material)
- activating emergency contingency funds and/or reallocating regular budgets for emergency activities
- applying for and monitoring federal assistance
- having a state emergency management agency and a preparedness plan coordinated by that agency.

Beyond these, little more can be generalized about state emergency management efforts. A closer look at individual differences and similarities among states is necessary to develop an understanding of the operation of emergency management at the state level.

Local Government

The local level—whether city, town, or other designation—is the first line of official public responsibility for emergency management activity. In an emergency, federal and state resources may not be available. Therefore, the local emergency management agency must accept responsibility to maintain an ongoing program of mitigation, preparedness, response, and recovery.

Local government is uniquely suited to do so. Locally, potential hazards are seen most clearly; resources must be most fully known; first response is made; emergency events begin. Those who know about the uniqueness of the community, where something may go wrong, where special complexities exist, and where sources of aid may be found are at the local level.

Integrated Emergency Functions

One way of summarizing the responsibilities of the local government in emergency management is to briefly review the functions of integrated emergency management. If a local jurisdiction is addressing these activities thoughtfully and effectively, it is fulfilling its important role in protecting lives and property.

This includes the following functions.

1. *emergency operations planning*—developing and maintaining emergency operation procedures appropriate to local hazards and resources
2. *direction and control*—having the ability to direct emergency response operations from an EOC or field location
3. *emergency communications*—capable of directing operating forces in an emergency
4. *alerting and warning*—able to alert public officials, response personnel, and the public that an emergency may exist.
5. *emergency public information*—distributing information on hazards relevant to the area.

6. *continuity of government*—having legally designated lines of authority and other provisions to preserve the government under emergency conditions.

Local Emergency Management Laws

The nature of local emergency management laws is guided largely by state law. The state law can be either *permissive* or *mandatory*—that is, it may allow localities either to organize and conduct emergency management systems as they see fit (permissive), or it may specify particular requirements that communities must meet (mandatory). In general terms, local laws define, with widely varying specificity and scope, who will do what in an emergency or disaster. Because local laws give the emergency management program the authority to operate, and because the local level is most directly involved with all emergencies, local laws are particularly important.

Emergency management laws ensure the legality and define the scope of the local integrated emergency management program. Authorities and responsibilities of the program should be defined and delineated clearly so that there is no question as to what is or is not included. The organization responsible for emergency management should be identified, and its functions described. However, laws also should allow some flexibility. For example, ordinances should make clear that the program encompasses all four emergency management phases—mitigation, preparedness, response, and recovery (Figure 18-3). But if they list the kinds of emergencies and disasters that come under the protection or sponsorship of the emergency program, questions of authority may arise in many unusual circumstances not listed.

Because each jurisdiction has different characteristics and requirements, local laws or ordinances must be drafted with the individual needs of the community in mind. Guidance or provisions for local authorities can be obtained from other jurisdictions with *all-hazards laws* or from higher levels of government, but each local law must be tailored to local needs.

> *Laws should provide concise definitions of vital terms. This helps to limit unnecessary controversy and misunderstanding (Introduction to Emergency Management-SM 230).*

STEPS FOR ESTABLISHING ON-SITE EMERGENCY RESPONSE PLANS

If the objective is to achieve an effective workplace emergency procedure for a facility, an effective workplace safety and health program needs to be in place. The importance of an effective workplace safety and health program cannot be stressed enough. There are many benefits from such a program including increased productivity, improved employee morale, reduced absenteeism and illness, and reduced workers' compensation rates. However, preventable injuries still occur in spite of efforts to prevent them. It is for this reason that planning for anticipated or likely emergencies is necessary to minimize employee injury and property damage.

It is hoped that businesses without safety and health plans will address this issue. The National Safety Council, a not-for-profit, nongovernmental organization, is an example of just one organization that can assist in the development of safety and health guidelines and for training of personnel. Companies that already have an effective safety and health program will find the implementation of an effective emergency procedure in place. For these companies, the following may be used as a review to assist in updating and revising the existing emergency response programs.

Purpose of Planning

The purpose of planning is to outline the basic steps needed to prepare to handle emergencies in the workplace. These emergencies can include accidental releases of toxic gases, chemical spills, fires, explosions, personal injury, and other human-made emergencies or natural disasters. Emergency response plans are not intended to serve as an all-inclusive outline but rather to provide guidelines for planning for overcoming emergencies by the people that utilize them.

Planning Effectiveness

The effectiveness of response during emergencies depends on the amount of planning and training performed. Management must show its support of plant safety programs and the importance of emergency planning. If management is not interested in employee protection and minimizing property loss, little can be done to promote a safe workplace. It is therefore management's responsibility to see that a program is instituted, and that is it frequently reviewed and updated. The input and support of all

Figure 18-3. A mobile command and communications vehicle in service at St. Joseph's County, Indiana. (Photo by R.T. Vulpitta)

employees must be obtained to ensure an effective program. The emergency response plan should be developed as a team effort involving all levels of management and employees.

Elements of Emergency Planning

The plan should be comprehensive enough to deal with all types of emergencies and written so as to be easily understood. Plans as a minimum need to include the following elements:

- Emergency escape procedures and emergency escape route assignments are detailed.
- Provide procedures to be followed by employees who remain to perform (or shut down) critical plant operations before they evacuate.
- Provide procedures to account for all employees after a completed evacuation.
- Conduct training, practice and the proper equipment for employees who are called on to perform special duties as rescue, medical duties, hazardous response, firefighting, and other responses.
- Provide the preferred means for reporting fires, medical and other emergencies that occur on-site.
- Provide a chain of command listing names or regular job titles of persons or departments who are responsible for emergency decision making and response actions to facilitate recovery.
- Identify hazards and determine the likelihood of their occurrence (*hazard audit*).
- Develop various emergency responses for the types of emergencies likely to occur.
- Train and practice on the emergency response plan by employees.
- Update the plan at least annually and communicating revisions as needed to employees.

Hazard Audit

Emergency action plans are based on the identified potential emergencies that can be reasonably expected to occur at a particular workplace. To do this, first identify all potential emergencies. Then a *risk evaluation* or a *hazard assessment* is conducted on each potential emergency. These need to be reviewed as a team project of personnel familiar with the operations of the site. For example an office building would have a maintenance engineer, office supervisor, and a safety manager. A power plant would have in addition to these a production supervisor, maintenance supervisor, and an engineer. The team of emergency planners analyze on-site emergencies by determining *what if* and *how bad will it be* if this event occurs.

The emergency planners also need to determine if other nearby businesses may also pose hazards if they have a emergency. Lines of transportation may also have to be considered as a source of a potential hazard. For example, if a truck crash on a nearby highway releases a chemical vapor into the air, this may require factories in the area to take action to protect their employees. Therefore, it is necessary to perform a hazard audit to determine the potential for one emergency to spill over to cause an emergency at another site.

Risk Evaluations

Risk evaluations include the following components:
- *critical equipment list*—Determine potential consequences or failure scenarios that may occur with critical equipment that would cause an emergency. Determine the need for a minimum of personnel to monitor and operate the equipment, in event of an emergency.
- *site utilities list*—Determine suppliers, entry points and shutoffs for on-site utilities such as electric, gas, water, and communications. Determine the need and extent of backup systems.
- *natural disasters*—Determine likelihood for tornado, blizzard, ice storm, tidal wave, hurricane, earthquake, mud slide, flood, and fire that may effect a facility and its operation.
- *human-made disturbances*—Bomb threat, arson, riot, vapor release, chemical release, terrorist attack, and possible structural failures.
- *transportation lines*—Determine if shipping, rail, air, or highway emergency events may have a carry-over effect to a facility.
- *toxic materials and or raw materials*—Determine if there a potential hazard that exists on-site.
- *other site spillovers*—Special considerations need to be determined if spill-over emergency events from other facilities need to be addressed.

Toxic Materials and Material Safety Data Sheets

Toxic materials released into the environment can cause unsafe conditions. For information on chemicals, the manufacturer or supplier of the material or substance can be contacted to obtain a Material Safety Data Sheets (MSDS). To obtain a list of the possible hazards of nearby firms, contact your local emergency response agency (LEPA) or your state's department of environmental management. They have access to the hazardous chemical lists a company must submit required by the SARA Title III Act. An MSDS includes the following sections:
- manufacturer's information
- hazard ingredients
- physical chemicals
- fire and explosive hazards data
- reactivity data
- health hazard data
- precautions for safe handling and use
- control measures.

Also see chapter 11, Industrial Hygiene Programs, for a sample MSDS.

Reviewing Plan With Employees

Based on the hazard assessment, plans need to be written for the identified potential emergency situations. All employees must be told what actions they are to take in the various emergency situations that may occur in the workplace. For emergency evacuation, the use of floor plans or workplace maps that clearly show the emergency escape routes and safe or refuge areas need to be included in the plan. These also need to be posted throughout your facility.

This plan needs to be reviewed with employees initially when the plan is developed, whenever the employees' responsibilities under the plan change, and whenever the plan is changed. A copy should be kept where employees can refer to it at convenient times. As a best practice, the employer should provide each employee with a copy of the plan during initial training. Copies of the plan should be given to all new employees when hired. An emergency response plan and template to use is given in the *On-Site Emergency Response Planning Guide,* available from the National Safety Council.

Chain of Command

A chain of command should be established to minimize confusion, so that employees will have no doubt about who has authority for making decisions. Select responsible individuals to coordinate the work of the emergency response team. In larger organizations, there may be a facility coordinator in charge of all facility operations, department coordinators who supervise particular department operations, and others. Because of the importance of these functions, adequate backup must be arranged so that trained personnel is always available. The duties of the *Emergency Response Team Coordinator* should include the following:
- assessing the situation and determining whether an emergency exists that requires activating the emergency procedures
- directing all efforts in the area including evacuating personnel and minimizing property loss

- ensuring that outside emergency services, such as medical aid and local fire departments are called in when necessary
- directing the shutdown of plant operations when necessary (Figure 18-4).

Assembly Areas and Command Centers

During a major emergency involving a fire or explosion, it may be necessary to evacuate offices in addition to manufacturing areas. A prearranged assembly area communicated to employees in advance is essential. This is known as the *primary assembly area*. In the event the primary assembly area is unavailable it is necessary to have an alternate assembly area to which employees can report and be accounted. Designated assembly areas for employees need to be in areas that outside emergency response service agencies will not be utilizing. The assembly site becomes a focal point for incoming and outgoing communications. The person designated as being in charge will make the assembly site the command center. The command center serves as an area to meet with outside emergency agencies and account for the evacuees.

Communications Warning Alarm System

Lines of communications between outside emergency response agencies and the on-site employees are the most critical part of an emergency response plan and one of the first to fail.

Emergency communications need to include primary and backup equipment such as portable phones, alarm systems, amateur radio systems, public address systems, or portable radio units for notifying employees of the emergency and for contacting local outside emergency agencies.

The warning alarm system is a method of communication needed to alert employees of the evacuation or to take other action as required in the plan. Alarms should be audible or seen by all people in the plant. These systems need to have an auxiliary power supply in the event electricity is affected. The alarm should be distinctive and recognizable as a signal to evacuate the work area or perform actions designated under the emergency action plan. A backup portable air-horn system may have to be considered in the event the main alarm is in inoperable. Another consideration is the alarm system may not be heard in all areas of a facility due to the noise levels of the activities conducted there or by employees with physical impairments. As a best practice, repeat the alarm several times followed by public address announcements. Additionally, facilities should install visual alarm systems for the hearing impaired. Another best practice is to have preappointed search teams of two to enter areas where alarms may not be heard and alert the employees there of the emergency response action to take.

The employer needs to explain to each employee the means for reporting emergencies, such as manual pull box alarms, public address systems, or calling a certain in-house telephone number. Emergency phone numbers should be posted on or near telephones or other conspicuous locations. It may be necessary to notify other key personnel such as the facilities manager or physician during off-duty hours. An updated written list of these numbers needs to be available to key personnel at all times (Figure 18-5).

Accounting for Personnel

In the event of an evacuation, management will need to know when all personnel have been accounted for. This can be difficult during shift changes or if contractors are on site. A responsible person should be appointed to account for personnel and to inform the results to the person who is designated as the person in charge of the site during an emergency. The information of those believed missing will be forwarded by the person in charge to the proper outside agencies.

Special Response Teams

In emergencies, *special response teams* are the first line of defense for a facility. Before assigning personnel to these teams, the employer must assure that employees are physically capable of performing the duties that may be assigned to them. Once assigned to these duties these employees need to be trained and have the opportunity to practice their skills. Depending on the size and particular operations, there may be one or more of these special teams as noted in these examples:
- use of various types of fire extinguishers
- first aid, including cardiopulmonary resuscitation (CPR)
- emergency shutdown procedures
- confined space, overhead structure or under water rescue
- chemical spill control procedures
- use of self-contained breathing apparatus (SCBA) and other respirators

---

CHAIN OF COMMAND

Director of Crisis Management

The highest ranking on-site "Facilities and Office Services" employee during an emergency becomes the *"Director of Crisis Management"*. The *Director of Crisis Management* is the highest ranking emergency decision-maker on-site. This person is the play-maker who would activate the proper response action for employees to follow. This person also interacts with outside response agencies. Emergency Coordinators will update the Director of Crisis Management on the status on evacuation of personnel and other needed information.

Emergency Coordinator (ECO)

Each department manager or director is to designate a responsible person as their departments *"Emergency Coordinator"* or (ECO). The ECO coordinates the emergency duties of a particular department. The ECO appoints as needed *Assistant Emergency Coordinators, Department Searchers* and *Stairwell Monitors* for the department.

Assistant Emergency Coordinator (AEC)

Assistant Emergency Coordinator appointed by the ECO will assist or take over for the ECO of a department as needed during an emergency.

Stairwell Monitors

ECO assigns employees as stairwell monitors who are closest to fire doors by stairways. They are to check fire doors for the presence of heat and smoke (protecting stairs) to determine if they are safe to utilize.

Searchers

Searchers are employees who are assigned in pairs by their department ECO. Searchers check through department areas of offices, cubicles, storage areas, and rest room areas to notify employees who may have not heard the emergency announcement.

First Aid Responders

First Aid Responders are trained and certified in first aid/CPR procedures. Teams are established by floors. They are activated by the receptionist using the public address system in the event of a medical emergency. They provide assistance until the arrival of professional medical services.

Figure 18-4. Sample of an Emergency Chain of Command.

- incipient and advanced-stage firefighting.

The type and extent of the emergency will depend on the plant operations. The response will vary according to the type of process, the material handled, the number of employees, and the availability of outside resources. Special response teams need to be trained in the types of emergency actions to be performed. For example they need to be informed about special hazards, such as storage and use of flammable materials, toxic chemicals, radioactive sources, and water-reactive substances, to which they may be exposed during a fire and other emergencies. It is important to determine when not to intervene. For example, a firefighting team must be able to determine if the fire is too large for them to handle or whether search-and-emergency rescue procedures are too dangerous for them to perform. If members of the special response team are in danger of receiving fatal or incapacitating injuries, they need to wait for the response of an appropriate professional agency.

TRAINING

Training is important for the effectiveness of an emergency plan. Before implementing an emergency action plan, a sufficient number of key people must be trained to assist in administering the key elements of the plan. All other employees need to be trained on how to respond to each type of emergency. This will enable employees to know what actions are required. Employees also need to understand that they must use common sense when responding in any emergency situation. In addition to the specialized training for key people other employees should be trained in the following:

- evacuation plans and routes
- when elevator use needs to be avoided
- alarm identifications
- reporting emergency conditions
- shutdown procedures
- responses for identified emergencies and the use of common sense when necessary
- smoke inhalation dangers

EMERGENCY REPORTING GUIDELINES

All emergencies, fire, bomb threat, medical or any other shall be reported immediately to the operator at extension ____, call for appropriate outside assistance.

- When reporting an emergency by phone, start out by giving your name, department, extension number, and your exact location. Report the emergency event as clearly and as accurately as possible. Remember, in any situation, it is important to remain calm.

- The Receptionist/Dispatcher may need additional information. Stay on the line, do not hang up. The Receptionist/Dispatcher will either tell you to hang up or will hang up on you.

- If the emergency is fire-related, describe the equipment involved, smoke, fire and size. After your call to the Receptionist/Dispatcher, only attempt to extinguish a fire when your personal safety is not in jeopardy.

SITE SPECIFIC EVACUATION ALARM INFORMATION

THE ALARM SYSTEM FOR THIS LOCATION IS: _____

IT IS LOCATED: _____

Figure 18-5. Effective communications is the most important single element of emergency responding and the first critical element to be lost during an emergency. All employees need to be instructed on how to report emergencies.

- crawling under smoke to escape to safety
- how to check closed doors for the presence of heat and smoke.

When should training occur? Training programs need to be held as follows:

- initially when the plan is developed, then each year
- for all new employees
- when new equipment, materials, or processes are introduced
- when procedures have been updated or revised
- when a drill or actual emergency event determines an improvement needs to be communicated with employees.

The emergency response procedures should be written in concise terms and made available to all personnel. A drill should be held for all personnel at random intervals and at least once annually. An evaluation of performance should be made immediately by management and employees. When possible, a drill should include groups supplying outside services such as fire and police departments. In buildings or industrial parks with several places of employment, the emergency plans should be coordinated with other companies and employees in the building or industrial park. Finally, the emergency plan should be reviewed periodically and updated to maintain adequate response personnel and program efficiency (Figure 18-6).

DRILLS

Drills are used by an emergency response planner for training employees and as a tool to measure planning effectiveness. A well-designed drill allows the participants to actually walk through and practice responses they have been trained to perform (Figure 18-7). Drill documentation is extremely important and should be maintained as part of the safety records for a facility (Figure 18-8). Overall actions of employees need to be evaluated. Recording details of the drill, such as time that alarms are sounded as well as responses by employees, will aid the emergency response planner to determine the effectiveness of training (Figure 18-9).

Drills need to be carefully controlled in facilities where critical procedures are performed. Key operations personnel need to be involved in the planning process of the drill. They need to be informed when the drill will occur and conclude. Before a drill takes place, other employees may also need to be informed a drill will be taking place. These are the employees who will need to maintain critical operations during an emergency. These employees must be identified before any drill begins. What are *critical operations*? These will vary for each facility. One employee whose job is categorized as a critical operation is that of the receptionist—to a facility maintaining communications is critical.

It is a *best practice* for the emergency response planner to inform all potential off-site emergency

---

EMERGENCY RESPONSE EXAM

1. Who makes the determination to evacuate at your place of employment? _____

2. What do you hear if a General Evacuation of the building is sounded? _____

3. During an emergency, who is the highest ranking emergency person at your location? Give the title: _____

4. That person's back-up or assistant is ? _____

5. To report a bomb threat you dial what number? _____

6. Fire breaks out in your work area and it is out of control, after reporting the fire what should you do? ____ A) Evacuate immediately ____ B) Evacuate after the evacuation alarm is sounded?

7. Remembering that smoke rises, your work area starts to fill up with smoke. You go to your primary exit and find it is impassable. The smoke is now waist high. Check the correct response. ____ A. Hold your breath & run to the alternate exit. ____ B. Get under the smoke & crawl to your alternate exit.

8. How are you informed that a tornado emergency exists and you are to go to your safe areas? By _____.

9. You find yourself next to a stairway exit door on an upper floor when the "Evacuation Alarm" sounds. Check the correct response. ____ A. I will check the door for heat and smoke to be sure it is safe to use. ____ B. I will exit immediately through door without checking.

10. "Safe Areas" are part of what emergency plan? _____
    "Safe Areas" are located where in your building? _____

11. My department assembles where outside during an evacuation? _____

12. To report a fire you dial what number? Internally _____ Externally _____

13. To report a medical emergency you dial? Internally _____ Externally _____
    Answers to Questions 1, 2, 3, 4, 5, 8, 10, 11, 12 and 13 are site specific to your workplace.
    Question 6 (A)..... Question 7 (B).......Question 9 ....(A)

Figure 18-6. The Emergency Response Exam is used at the National Safety Council. The form assists employees understanding their responsibilities.

response agencies in advance when a drill will be held and the nature of the drill. In facilities where radios are utilized this is very important. Radio exchanges between employees of a facility who are responding to a drill may be heard by those outside the facility and misinterpreted by those who are listening. The drill may be reported by these listeners as an actual emergency to outside emergency agencies or news agencies (Figure 18-10).

PERSONAL PROTECTIVE EQUIPMENT

Effective personal protection is essential for anyone who may be exposed to potentially hazardous substances. In emergency situations employees can be exposed to a variety of hazardous circumstances, including these:

1. chemical splashes or contact with toxic materials
2. falling objects and flying particles
3. unknown atmospheres that may contain toxic gases, vapors, mists, or inadequate oxygen to sustain life
4. fire and electrical hazards.

It is extremely important that employees be adequately protected in these situations. Some of the safety equipment that can be used includes the following:

1. safety glasses, goggles, or face shields for eye protection
2. hard hats and safety shoes for head and foot protection

EMPLOYEE EMERGENCY RESPONSE
SECTION IV

On-Site Emergency Response Procedures

DRILLS AND TESTING SUGGESTIONS

Drills are the ultimate test for determining emergency preparedness. It is essential that you plan your drills with intended goals. The following are drill suggestions and responses to observe. Use the "Exercise Documentation Form" to record your responses.

EVACUATION DRILLS: Sound evacuation alarm. Observe employee response. Did Searchers operate in pairs? Did employees know their assembly area? Was a roll call taken? Record start and end times of drills.

TORNADO DRILL: Test the communication process of informing your personnel. (Alarm not used in tornado emergencies.) Did searchers operate in pairs? Did everyone get to "safe area"? Were people assigned the task of shutting off utilities? Did they have their tools present?

BOMB THREAT: *Called in Threat*—Let person know this is a drill from the beginning and identify yourself. Proceed with the drill. Observe if person taking calls recorded exact message and used the check-off procedure to identify caller.

Suspicious Package—Let employees know this is a drill. Observe actions taken when they find the "bomb." Ask what their next action would be (evacuation, contacting local police, etc).

MEDICAL EMERGENCY: This exercise involves the use of employees demonstrating their First Aid and CPR skills under emergency-like conditions. CPR skills can be practiced on a mannequin. Note if participants "called" for outside help.

HAZARDOUS MATERIAL EMERGENCIES: Time and practice shutting down air system and isolating building. Time how long it would take it to go to your designated off-site meeting area. Make employees aware of the possibility of hazardous material releases from industry or transportation lines (highways, railroads, etc).

ARMED ROBBERY: In a meeting setting— test employees' ability to recall descriptions of person(s) witnessed. Question them on steps to take in sounding alarm securing the scene, contact notifications, etc.

Figure 18-7. A well-designed drill allows the participants to actually walk through and practice responses they have been trained to perform.

3. proper respirators for breathing protection
4. whole-body coverings, gloves, hoods, and boots for body protection from chemicals
5. body protection for abnormal environmental conditions such as extreme temperatures.

The equipment selected must meet the criteria contained in the OSHA standards. The choice of protective equipment is not a simple matter. Therefore, consult with health and safety professionals before making any purchases. Manufacturers and distributors of health and safety products may be able to answer questions if they have enough information about the potential hazards involved.

RESPIRATORY PROTECTION

Consult with professionals when providing adequate respiratory protection. Respiratory protection is necessary for toxic atmospheres of dusts, mists, gases, or vapors and for oxygen-deficient atmospheres. These are three basic categories of respirators:

1. *Air-purifying devices* remove contaminants from the air through filters and sorbents (filters, gas masks, and chemical cartridges), but cannot be used in oxygen-deficient atmospheres (less then 19.5% oxygen).
2. *Atmosphere-supplying respirators* provide breathable air from an air source outside the contaminated work area (e.g., airline, SCBA).
3. *Combination air-purifying and atmosphere supplying* is a mechanical combination of an air-line respirator and an auxiliary air-purifying attachment that provides protection if the air-line fails.

The OSHA standard minimum compliance requirements must be met before assigning or using

EMPLOYEE EMERGENCY RESPONSE
SECTION IV

On-Site Emergency
Response Procedures

EMERGENCY RESPONSE EXERCISE DOCUMENTATION FORM

Initial Notifications/Comments: Describe initial employee notification process and effectiveness, also evaluate communication process during drill.
_____
_____

Recommendation: _____
_____

Assessment: Observances of employee transition from normal operations to emergency operations.
_____
_____

Recommendation: _____
_____

Command & Coordination: Effectiveness of ECO and/or Employees charged with decision-making responsibility during drill.
_____
_____

Recommendation: _____
_____

Protective Action: Familiarization with Emergency Response Procedures, of Employees, Contractors, etc.
_____
_____

Recommendation: _____
_____

Parallel Action: Involvement or use of concerned parties that can aid crisis manager during drill.
_____
_____

Recommendation: _____
_____

Figure 18-8. Drill documentation is extremely important and should be maintained as part of the safety records for a facility. Overall actions of employees need to be evaluated.

respiratory equipment. These requirements are as follows:
- written respiratory protection program with worksite-specific procedures
- proper respiratory protection selection procedures
- medical evaluations of employees required to use respirators
- employee training of respiratory hazards in routine and emergency situations
- employee fit testing for tight-fitting respirators
- routine use and emergency use procedures
- respiratory cleaning disinfection, storage, inspection, repairing, discarding, and other maintenance procedures
- atmosphere-supplying respirator procedures to ensure adequate air quality, quantity, and flow of breathing air
- employee training of proper use and limitations of respirators
- program evaluating procedures
- air sampling.

Self-contained breathing apparatus (SCBA) offers the best protection to employees involved in controlling emergency situations. SCBA should have a minimum rating oxygen supply of 30 minutes. Conditions that require use of a SCBA include the following:
- leaking cylinders or containers
- smoke from chemical fires

CHAPTER 18: EMERGENCY RESPONSE PROGRAMS

EMPLOYEE EMERGENCY RESPONSE
SECTION IV

On-Site Emergency Response Procedures

EMERGENCY RESPONSE EXERCISE DOCUMENTATION FORM

All locations shall test one or several of their Emergency Response Procedures once a year. Actual emergencies can count as drills. Any questions — contact the Emergency Planning Committee.

Location _____ State: _____ ext. _____
Address: _____ City: _____ County: _____
Manager: _____ Phone: _____

Drill Coordinator(s): _____

Date of Drill: _____ Start Time: _____ End Time: _____
# of Employees Assigned to Station: _____ # Participating: _____ (Approx.)

Type of Drill(s):
    Evacuation ( )   Tornado ( )   Bomb ( )   Medical Emergency ( )
    Hazardous Material Emergency ( )   Earthquake ( )   Robbery ( )   Other ( )

Brief Description of Drill: _____

Areas of Operation Was Drill Held	Initial Response of Personnel	Time Responded By	# of Employees Participating
_____	_____	_____	_____

Post Drill Critique: Date _____   Time _____   Location _____

Attendees: _____

Initial Comments and/or Suggestions: _____

IMPORTANT DOCUMENTATION IS REQUIRED
cc: File Originals On-Site
Attach copies of check-off sheets from roll calls, etc.

Figure 18-9. Recording details of the drill, such as time that alarms are sounded and responses by employees, will aid the emergency response planner to determine the effectiveness of training.

- chemical spills that indicate high potential for exposure to toxic substances
- atmospheres with unknown contaminants
- atmospheres' unknown contaminant concentrations
- confined spaces or oxygen-deficient atmospheres.

CONFINED SPACE ENTRY

Detailed plans need to be in place prior to entry into any confined spaces. All on-site confined spaces need to be identified. Personnel are never to enter a confined space unless the components of the confined space entry program have been followed and appropriate permits have been completed. These include atmospheric testing of oxygen, upper and lower explosive level limits, and toxics.

Confined spaces can contain a variety of hazards, including toxic gases, explosive atmospheres, and oxygen deficiency. Other hazards that also must be considered are electrical hazards and hazards created by mixers or impellers and exhaust fumes from combustible engines. Hazards need to be controlled, deactivated, *locked-out,* or removed. Communications between all workers within and outside the confined space need to be maintained. Life lines are to be unobstructed and untangled. In the event of an emergency situation, those rescue procedures written and practiced in advance need to be activated.

POWER PLANT DRILL EXAMPLE

Drill Objectives
1st Time the initial and backup fire responders response time.
2nd Observe first aid response in realistic station conditions.
3rd Test the stations evacuation procedure and accounting of personnel.

1105 Hundred Hours - Coal Mill Area Activities: The on-shift supervisor (OSS) is contacted that a fire has broken out in the coal mill area. The initial fire response alarm and announcement is made requesting initial incipient fire fighters to respond to #4/5 coal mill area. In less than two (2) minutes, a number of mechanics and operators were on-site at the scene of the "fire". It was noted that a number of these initial responders had brought fire extinguishers with them into the area. These responders were then informed by the Drill planning committee that, as part of the drill, the Relief Supervisor, was injured, needed immediate first aid attention and would also have to be evacuated from the plant. The operators and mechanics quickly located a first aid kit and stretcher. One individual referred to his "First Aid Pocket Guide Book" for burn treatment, and administered first aid. By 1125 hundred hours, the Relief Supervisor had been treated and successfully transported to an outside assembly area.

TESTED	RESULTS
Initial Fire Fighting	2 minute response
First Aid Skills	Excellent
Evacuation	Immediate

1110 Hundred Hours - Mechanical Maintenance Department Activities: Alarm and announcement are sounded twice, requesting back-up fire fighters to assemble in machine shop to await further instructions. Immediately 20 mechanics and electricians assembled. At 1113 hundred hours the employees were informed that this was a drill, and that as part of the drill, an injured employee was at the west end of the shop and needed first aid treatment. Personnel responded from several other departments who assisted the injured employee.

TESTED	RESULTS
Back-up Fire Fighters Response	1–3 minute response
First Aid Skills	Excellent

1118 Hundred Hours: The evacuation alarm is sounded throughout station. Guard force is notified, and internal gates and turnstiles are opened. All personnel proceeded to outside assembly area where operator took roll call, and radioed results from each department back to the control room. 1126 Hundred Hours: All department personnel are accounted. 1128 Hundred Hours: Drill concluded. Normal activity resumed.

TESTED	RESULTS
Back-up Fire Fighters Response	1–3 minute response
First Aid Skills	Excellent
Evacuation	Immediate evacuation of 137 personnel
Roll Call Taking	4 minutes

Improvement Recommendations
- Have an assigned area marked for each department in the assembly area.
- Place a First Aid Pocket Guide and several pairs of gloves in each First Aid Kit.
- Searchers need to work in pairs.
- Emphasize meaning of various alarms to employees.

Figure 18-10. A team of power station employees planned a drill at their location with built-in safeguards so as not to disrupt critical operations. As personnel became involved with the exercise, they were informed it was a drill. Note response times and areas of improvement when you hold a drill.

MEDICAL ASSISTANCE

In a medical emergency, time is a crucial factor in minimizing injuries. Most small businesses do not have a formal medical program, but they are required to have the following medical and first aid services reasonably available:

1. In the absence of an infirmary, clinic, or hospital in close proximity (6 to 10 minutes) to the workplace that can be used for treatment of all injured employees, the employer must ensure that a person or persons are adequately trained to render first aid.

2. The employer must ensure the ready availability of medical personnel for advice and consultation on matters of employee health. This does not mean that health care must be provided, but

rather that, if health problems develop in the workplace, medical help will be available to resolve them.
3. Survey the medical facilities near the place of business and make arrangements to handle routine and emergency cases. A written emergency medical procedure needs to be included as part of an emergency response plan.
4. If the business is located far from medical facilities, at least one and preferably more employees on each shift must be adequately trained to render first aid. The National Safety Council, American Heart Association, local safety councils, fire departments, and others may be contacted to provide this training.
5. First-aid supplies need to be provided for emergency use. This equipment should be ordered through consultation with a physician familar with the particular workplace hazards.
6. Area ambulance services need to be surveyed to determine response time and emergency handling capabilities and hospital locations. Advance contact with ambulance services also needs to familiarize them with a location and access routes.

SECURITY

During an emergency, it is often necessary to secure the area to prevent unauthorized access and to control the event. An *off-limits* area must be established. Access by emergency vehicles should not be impeded. Mechanical gates should be secured in an upright position to facilitate their movement. Security needs to assist emergency personnel to the area where they are needed. All entries and exits need to be documented by time. If there is an electronic entry system it can supply information on the employees who are on-site. (Also note that if there is an electronic entry system, a power failure coupled with the emergency can freeze the entry system. It may be necessary for security to know how to manually activate entry operations.)

NEWS MEDIA

If the emergency is newsworthy, representatives from the press also will be present. Establish an area for the press so they can be briefed on a regular basis as facts are established. This keeps the press together and allows for the accurate release of information. Also see chapter 19, Community Involvement Programs for more information about communicating risk information to the media.

BUSINESS RECORDS

Certain records also may need to be protected, such as essential accounting files, legal documents, and lists of employees' relatives to be notified in case of emergency. These records can be stored in duplicate outside the plant or in protected secure locations within the plant. Computer records and other transactions need to be backed up and stored off-site.

RESPONSE AGREEMENTS

Any type of emergency mutual aid agreement needs to be in writing. These agreements need to be in place with suppliers and the providers of specialized services that a facility may be depending on during a disaster or an emergency.

REFERENCES

Emergency Management Institute National Training Center. *Introduction to Emergency Management.* SM 230,-Unit 2, pp 15, 18, 19, 29, 30 31, 32 & 33.

U.S. Department of Labor-Occupational Safety and Health Administration 1991-OSHA 3088 (Revised). *How to Prepare for Workplace Emergencies.*

National Safety Council. *Accident Facts®, 1997.* Itasca IL: National Safety Council, 1998.

Vulpitta R. *On-Site Emergency Response Planning Guide* (and diskette). Itasca: NSC Press, 1999.

chapter 19

Community Involvement Programs

by L. Darryl Armstrong, PhD, APR, CCMC

345 Introduction
345 Objectives
346 Thomas Jefferson—Father of Public Participation Principles
346 Reasons for a Communications Program
348 Public Participation
   Who is the public? ■ Trend toward stakeholders ■ Advantages and disadvantages
349 Strategies for an Effective Communications Program
   Building rapport with stakeholders ■ Encourage employees to value departmental meetings ■ Use graphics and factoids in your newsletters ■ E-mail flashes for breaking news ■ Annual reports—keep them simple and readable ■ Brown bag lunches are effective with employees ■ Benefits of employee training sessions
352 Establishing Your Community Presence
353 Participating in Environmental Organizations
353 Value of Business and Civic Organizations
353 Case for Regulatory Advocacy
353 Why Community Outreach Can Work
   High-profile projects—cost versus benefits
357 Using Public Information Correctly
   Related effective communication strategies and techniques ■ Media relations strategy is critical ■ Using news releases, photographs, and feature stories ■ When to use paid advertising ■ Learning to speak in "sound-bites" ■ Why news releases can work for you ■ Media briefings and conferences: Use sparingly ■ Written communications strategies and tools
360 Four Critical Skills for Effective Communication
   Presentations ■ Avoid using technical jargon ■ How to communicate risk to the public ■ Covello and Sandman's effective ways to present risk ■ How to handle public outrage
365 Cultural Sensitivity and Communications Require an Exchange
365 "Active Listening" Is Critical to Your Success
366 Handling an Upset Audience
367 Summary
367 References

INTRODUCTION

Corporate and government environmental, safety, and health (ESH) managers are being required by federal and state mandates and regulators to involve members of the public in their decision-making processes. Therefore, an understanding of the components and strategies of an effective environmental, safety, and health communications program that incorporates a public involvement process and the critical skills required to be successful are an essential part of an environmental, safety, and health manager's toolbox. When public involvement and participation is successful, the public has provided meaningful input into the process and may develop ownership in the decisions that affect them. Therefore, they also have the reasons to see a project or proposal fulfilled. This ownership allows the process and decision making to go forward efficiently and effectively. It lessens the chances of conflict, litigation, costly delays, and public and political outrage.

OBJECTIVES

This chapter will acquaint environmental, safety, and health managers with the federal mandates and regulations that require you to understand, develop and integrate the concepts and skills necessary for an effective environmental, safety, and health communications, public participation and involvement program. The concepts, activities, and skills required are outlined and explained. Research supports the concept that if this information is integrated into the environmental, safety, and health disciplines and then practiced, it will help managers and technical staffs communicate more effectively with the public. This is especially true in settings such as public meetings, advisory committee meetings, or media briefings. Such programs will assist the managers in achieving successful, laudable environmental, safety, and health communications, and community involvement and participation.

THOMAS JEFFERSON—FATHER OF PUBLIC PARTICIPATION PRINCIPLES

The origin of public participation, involvement, and public decision-making is based in the Jeffersonian philosophy of self-governance. Laws and mandates now require such processes be a part of the environmental, safety, and health decision-making processes.

It has only been during the 20th century, however, that those principles, which were touted as public involvement, became codified and given legal status through federal legislation. Laws such as the Administrative Procedures Act, which provided the public with public notices, hearings, and the right-to-comment provisions, opened the door to public and community involvement. The Freedom of Information Act provided avenues for citizens to get public documents and other "information" from their governing bodies. The Open Meetings Act required that decision making meetings conducted by appointed and elected officials, with few exceptions most notably personnel actions, be open and accessible to the public. State governments seized on this act and created their own "Sunshine Laws" that opened state meetings to the taxpayers and citizens (English, 1996).

Despite all the legislation and the supposed good intentions of these laws, the citizen was commenting and criticizing after decision makers had reached conclusions rather than actively and fully participating in the decision making process.

As the country progressed through the turbulence of the1960s, the public saw these procedures become standard ways of doing business within agencies. Although the public was consulted, albeit it usually after the fact, rarely did the consultation alter the way business was being conducted by the agencies.

As Dawn S. Ford, a public participation consultant says, "Agencies tend to comply with the letter of the law but often not the spirit of the law. If, after a public involvement process, the public has not convinced the agency to change something, it is quite possible the spirit of the law has been disregarded."

However, even after the fact, public consultation was a significant change for agencies, who previously had solicited advice and counsel only from politically appointed commissions. Surely then, the trend toward early and inclusive public and community involvement has been reinforced through such an approach.

REASONS FOR A COMMUNICATIONS PROGRAM

The compelling reasons to implement an environmental, safety, and health communications and public participation program are found in the increased regulatory requirements for disclosing information on environmental performance. These federally mandated requirements combined with voluntary state and international standards drive the need to engage the public in an ongoing proactive dialogue.

During the past few decades, numerous regulations have been enacted that provide communities access to information and involve them in the environmental, safety, and health decision-making processes regarding permitting and remediation of certain contaminated sites. Other regulations require the reporting or disclosure of information about wastes and emissions, chemical storage, and other environmentally significant activities at facilities. Among the mandates and regulations that directly affect environmental, safety, and health managers are the following:

- The Emergency Planning and Community Right-to-Know Act (EPCRA)
- The Comprehensive Environmental Response, Compensation and Liability Act (CERCLA/Superfund)
- The Resource, Conservation and Recovery Act (RCRA)
- The Toxic Substances Control Act (TSCA)
- The Occupational Safety and Health Act (OSHA)
- The Toxic Release Inventory (TRI)

The Emergency Planning and Community Right-to-Know Act (EPCRA) requires reporting of environmental information, such as emission levels and storage quantities of certain hazardous chemicals. The law is housed in Title III of the Superfund Amendments and Reauthorization Act (SARA May 30, 1999) of 1986. The best known of the data available to the public under EPCRA is the Toxic Release Inventory (TRI), which lists the amounts of regulated chemicals released to the land, water and air or injected into the ground by manufacturing facilities.

Reporting under the EPCRA does not require public involvement in the decision-making process. There are no public comment periods or public hearings required. However, the fact the public has this information available to them means that reporting facilities may be asked by community residents and environmental activists to explain why you release certain chemicals to the environment. These inquiries can lead to significant and substantial public dialogue.

Under EPCRA (40 *CFR* Part 355), if a facility has hazardous substances on site in quantities at or above threshold levels, management is required to report such information to the State Emergency Response Center and designate a facility representative to participate on a Local Emergency Planning Committee (LEPC). This provides management an excellent opportunity to learn about community concerns and views regarding the facility. Management's diligence in working with the LEPC demonstrates to other community representatives the organization's commitment to addressing potentially hazardous situations and to answering questions about their environmental, safety, and health program.

In mid-1999, certain facilities were required to disclose their "worst case" environmental disaster scenario and spill history under Section 112(r) of the Clean Air Act. This provision required substantial dialogue be conducted between elected officials and other leaders and interested parties in the community where such facilities exist either at the time of the release of this information or later. The regulation requires that facilities that store "threshold quantities" of certain chemicals prepare "risk management plans," which provides information on the extent of potential off-site consequences in the event of an accidental release.

This information, which must be made available to the public, may raise concerns, issues and questions that environmental, safety, and health and other management personnel are required to accommodate.

Managers should be aware that requirements for environmental communications and public involvement are especially prominent in the Comprehensive Environmental Response, Compensation and Liability Act (CERCLA/Superfund), and the Resource, Conservation and Recovery Act (RCRA).

The Comprehensive Environmental Response, Compensation and Liability Act regulations allow for public comment on the remedial alternatives being considered for the cleanup of Superfund sites. Guidance under CERCLA also requires a number of other communication activities, such as public notices, hearings, and the establishment of an information repository, to keep community residents and other interested parties informed about activities at CERCLA sites.

The Resource, Conservation and Recovery Act provides for expanded public participation. RCRA regulates the generation, transportation, treatment, storage and disposal of hazardous wastes. The regulations require community meetings before permit applications are submitted to environmental agencies. It also requires the posting of signs on facility property and placement of advertisements in local newspapers providing information on upcoming permitting activities. It requires the development and issuance of fact sheets and other communications and public involvement activities designed to inform and solicit input from area residents about the permitting activities.

The Toxic Substances Control Act (TSCA) regulations (40 *CFR* Part 720) and the Occupational Safety and Health Act (OSHA) hazardous communications regulations (29 *CFR* 1910.1200) require manufacturers and importers of new chemical substances to determine if they are hazardous.

Under OSHA, employees must be made aware of any hazardous substances in the workplace to which they might be exposed. Pursuant to TSCA, any chemical substance that is hazardous must be properly labeled and have a Material Safety Data Sheet (MSDS).

Such regulations can and do frequently have many effects on the management and the communications processes. The increase in questions and concerns voiced by the public after the Toxic Release Inventory (TRI) data were released in the late 1980s is a classic example. The TRI data have provided information that has raised the level of awareness of citizens about emissions and chemical use within their communities and has provided a cornucopia of information to activist groups. Frequently this information is used to interrogate facility managers about their operations and environmental, safety, and health performance. Industry leaders, the U.S. Environmental Protection Agency (EPA), state regulators and activist groups all credit the TRI with providing an early warning system for businesses to begin to address environmental concerns expressed by the public.

The public availability of information under regulations such as CERCLA and RCRA has prompted environmental activists and community residents to demand information about other facilities, sites and projects. This is often the case despite the fact these facilities do not fall under disclosure requirements. For example, when people get information about a local Superfund site through the CERCLA and RCRA processes, they are led to believe they can get similar information about other facilities and sites.

Because of the growing interest, awareness and in some cases concern, foresighted organizations are developing environmental, safety, and health communications programs that incorporate tenets that address the public's issues and concerns even when they are not required to do so. In many cases, they are taking this proactive step to ensure they have a good relationship with their communities and they may be able to forestall or even prevent serious opposition or even litigation.

PUBLIC PARTICIPATION

Public participation is a process housed within an effective environmental, safety, and health communications program by which the public helps you identify and incorporate public views, issues, concerns, and opinions into your decision-making process. An effective process provides information and opportunities for the public to be involved in a meaningful way so they can review, analyze, formulate and evaluate alternatives. It provides a forum that allows managers to listen to the public and then provides an avenue for the management to provide appropriate feedback.

Who Is the Public?

The *public* is defined as any party that is affected or interested in the decisions that potentially affects them or their community. Members of the public may be your next door neighbor, representatives of governmental units, regulators, politicians, members of your own family, and certainly your employees.

Trend Toward Stakeholders

The members of the public that participate in these processes have a *stake* of some kind in the decision making process. That stake may result from personal concerns about health and safety, concerns about how the decision impacts the economics of themselves or their community, issues about environmental, safety, and health impacts, or other factors that they perceive affect or impact them in some way.

Therefore, there has been a movement toward using the term *stakeholder involvement* and the use of the term *stakeholder.* Although there has been a building movement toward public participation or stakeholder involvement, and this author believes this is a movement in the appropriate direction, there also has been a growing concern that it is not fully adequate. Thus the adjustments toward stakeholder involvement as a key component to public participation has begun.

The concern for more effective public participation has been generated by the fact that many who profess to do public participation at best do only a token effort to ensure they can say they "involved the public." Some of the federal agencies we have worked with even go as far as to simply add the term "public involvement" to their checklist so they can affirm they involved the public.

Such shortsightedness frequently comes back to haunt the delinquent agency.

There are intrinsic problems associated with the concept of public participation. That is, public participation fails to recognize that people have important but different concerns and cannot be treated as a singular mass of people with similar views. Further, conceptually the model of public participation assumes there is something appropriate about a pyramidal, indirect consent model of decision making rather than an egalitarian, direct consent model (English, 1996).

Since the early 1980s, there has been a growing pressure to alter these assumptions, to recognize that the "public" has diverse interests and "stakes" that are often value driven. This pressure has created an environment that gives stakeholders the opportunity to interact in an equal status in the process, more as "collaborators" than just participants. This collaborative approach, an approach that incorporates the concept of co-laboring with management to reach decisions, rather than just co-operating, "being nicey-nice and going along," is gaining significant momentum in the government and corporate world (Armstrong, 1994).

Perhaps, one of the major reasons collaborative stakeholder involvement is becoming more of a norm comes from a pragmatic assessment by managers of the consequences. That is, if the public doesn't understand and accept the project concepts on the front-end of the process, then frequently on the back-end of the process, management has to regroup and conduct time-consuming explanations, and often encounter political and community dissent, project delays and in some cases extremely costly litigation.

With the movement toward stakeholder involvement, this process stresses and recognizes that all people, officials as well as citizens, have different "stakes" and sometimes conflicting "stakes" in the ultimate outcome. These views are driven by different values in the outcomes of the decision making process.

Stakeholder involvement clearly suggests that people are affected by the ultimate outcome, therefore, it helps to reorient the decision making process from the people who traditionally had the power, by virtue of their position or influence, to involvement and inclusion of others who must live with the ultimate decisions (English, 1996).

Advantages and Disadvantages

There are advantages and disadvantages of looking upon the public as stakeholders However, even within the concept of *stakeholder involvement* problems arise. Most frequently, and as readers of this article we suspect you are one, management suggests that such processes consume valuable time, are costly, create tension, heighten opportunities for conflict and are counter-democratic because they increase the influence of special interest groups.

Such criticism makes stakeholder involvement especially vulnerable to the objections raised by management that special interest groups are not representative of the public at large. This is compounded by the often-challenging problem of determining who is a stakeholder in any given process.

So, who are appropriate stakeholders in any given process? How do you overcome their histories of fear, anger, and distrust that make some stakeholders dislike others and dislike you? Who really speaks for institutions, if public agencies are supposed to speak for the public at large? How can adequate representation be made for the stakeholders who don't speak out? Should all stakeholders be given equal footing, or should the views of some outweigh those of others?

Frankly, these issues are very important and do require thought and strategy development. However, they are simply procedural challenges by nature. Although some may seem to be insurmountable, public participation professionals working with managers and technical staffs can with a focused approach, the right attitude, carefully designed processes, strategies, and techniques minimize and usually negate, or at the very least appropriately handle these issues.

It is critical in today's business environment to develop and implement an effective environmental, safety, and health communications program that employs different strategies to establish rapport and trust with the public and engage them in a proactive dialogue, if environmental, safety, and health managers are to be successful.

STRATEGIES FOR AN EFFECTIVE COMMUNICATIONS PROGRAM

Note: *Employees are your single most important audience.* An effective environmental, safety, and health communications program includes effective employee communications, community outreach, comprehensive public information, and an ongoing, proactive, and interactive public participation strategies.

Building Rapport With Stakeholders

The overarching goal of any environmental, safety, and health communications program should be to build rapport and trust with all stakeholders. Employees are an organization's singular most effective and believable avenue for disseminating information within a community. After employees, third-party, independent people such as local businessmen, school superintendents, ministers, local college faculty, and other professionals, such as those in health care professions are rated high on believability. They are followed in believability by local community

DECISION-MAKING OPTIONS CHART
Decision-making options that organizations can use in the public participation process and the time and level of involvement required

HIGH Level of organizational involvement

{**Delegation of consensus** - certain constraints & boundaries or criteria; however, everyone agrees to live with the decision and support it. They don't have to agree with all aspects of the decision.}

{**Negotiating to agreement** - consensus/compromise; all have a fair hearing; express what they think/feel; agree to disagree; willing to commit to outcome; majority opinion rules.}

{**Participatory approach** - gather input from **group(s)**, then decide or provide feedback.}

{**Consultative approach** - gather input from **individuals**, then decide.}

{**"DAD"** - decide, announce and defend and involve no one.}

LOW ————> Time commitment of organization ————> **HIGH**

Figure. The more you involve the public the greater your commitment to time and involvement. The seemingly easiest approach to decision-making is to decide, announce and then defend your decision. Commonly known as the "DAD" approach, many corporations and government agencies still try and approach public decision-making in this way. Unfortunately the lack of front-end investment in public participation can result in serious ramifications on the back-end of the process. Costly litigation and loss of political and public support are common using the DAD approach.

The consultative approach gathers input from individuals, or "key stakeholders" and then a decision is made. Somewhat more defensible, this approach assumes the key stakeholders you consult are the correct ones and they can assist in the defense of your decisions. This approach is often used by politicians.

The participatory approach has input gathered from a group and then a decision is made. Again, assuming you are talking to the appropriate group, such an approach is even more defensible than the DAD or consultative approach. This approach is often used in civic clubs.

The negotiating to yes approach assumes you are willing to work for a consensus or at least a compromise. Everyone has a fair hearing, irrespective of how long it takes, people are free to express their thinking and feeling, they agree to disagree when needed, are willing to commit to the outcome, and the majority rules.

The highest level of participation is when we agree to delegate consensus. We are willing to stay at the table with the stakeholders until an acceptable decision has been reached. All parties agree to live with the decision and support it. They don't have to necessarily agree with all aspects of the decision, but agree to not actively work against it once the decision is made.

activists and then the local environmental, safety, and health management. Least believable are your corporate executives, consultants, and officials from state and federal government.

Prudent environmental, safety, and health managers recognize the value of communicating all sides of environmental, safety, and health issues-the good and the bad. Employees should never read about a organization's environmental, safety, and health problems or accomplishments in the newspaper or hear it first on television. It is much better for an employee to know as soon as possible rather than later of an environmental, safety, and health situation that affects them or the public. If the employee is informed, many members of the public will be effectively informed. However, when employees are kept in the dark and only brought into service during a crisis, the overall dividends to an organization are greatly reduced.

Employee communications means engaging organization employees in the overall environmental, safety, and health communications program. This is done by involving, educating and informing your employees about the organization's environmental, safety, and health efforts. Employees must know the progress in managing and reducing the organization's environmental impacts and the status of the environmental, safety, and health program.

Employees frequently speak informally to their neighbors and friends about organization operations, issues and management and must be routinely informed and updated so they can speak authoritatively. What they say is often taken as the organization's position and is considered by their audiences as being factual. In other words, friends, neighbors and citizens perceive employees as the organization's experts.

One of the first steps in developing an effective environmental, safety, and health communications program is to have a strategy to ensure employees are aware of what the organization is doing to minimize its environmental impacts and be a good environmental citizen. This helps mitigate community fears and concerns.

Communication managers should implement environmental awareness training for selected employees and contractors. Other viable techniques include departmental meetings, employee newsletters, annual reports, brown bag lunches, and family days at the site to report on the organization's environmental issues.

Encourage Employees to Value Departmental Meetings

One of the most effective ways within corporate settings to inform employees is top-down through departmental or project staff meetings. Usually, the environmental, safety, and health manager, or a designated staff member, will personally report the environmental, safety, and health accomplishments, issues, concerns, and current situational analysis. However, in large facilities it may be more cost effective to prepare and use videotapes.

The videotapes should be 10 minutes or less in length and can be presented by an environmental, safety, and health staff member who can answer specific questions.

Rockwell International used such videotapes when they managed the Rocky Flats facility in Colorado. They used this strategy as a way to directly communicate to their employees and the employees' families. Many of these videos also made it to the school systems through employees where students gained a better understanding of the operations of the site (Rymer, 1999).

Use Graphics and Factoids in Your Newsletters

The in-house newsletter, if it is well written, informative, and short enough to be interesting yet long enough to cover the subject, is an effective, efficient and powerful tool for communicating environmental, safety, and health information to your employees. Newsletters should be creative in design and layout and written for the lay person to understand. It is not necessary to have a "slickly" produced newsletter. Readable, interesting, and understandable content combined with creativity in design and layout are much more important.

Regular articles placed in the newsletter to help employees speak more knowledgeably about the facility benefits everyone. Articles, charts, graphics, factoids, and short stories improve the flow of positive and accurate information to the community through the employee's networks. They also serve to improve morale since the employees feel better informed about their organization. Newsletters can serve to showcase the leadership your facility is giving to environmental, safety, and health efforts.

Employees should be kept current on pollution prevention programs, site cleanup efforts, recycling programs, and chemical use and reduction measures and waste minimization efforts. You should always inform employees about your environmental achievements, as well as the public or regulatory issues you are encountering.

E-mail Flashes for Breaking News

Gail Rymer reports that at Rockwell International when e-mail was first introduced, she used "e-mail flashes" to get important information out to all employees in a timely manner.

"We would title the e-mail as a "FLASH" in bold type and then the body of the message was succinct, and simply stated what the employee had to know to be informed. More details would be issued later in traditional media formats such as newsletters, news releases, and letters from management."

Annual Reports—Keep Them Simple and Readable

Developing and publishing an annual environmental, safety, and health report that is informatively written for your employees can help you place into perspective the accomplishments and the challenges of your program.

Lockheed Martin's Skunk Works facility in Palmdale, California published an environmental report and mailed it to each employee at home. Employees had the opportunity to read the report when convenient and share the information with family and friends. Some of the children of the employees used information in the report for school projects (Belsten & Weiland, 1999).

Brown Bag Lunches are Effective with Employees

Periodic brown bag lunches are also an effective technique to keep employees informed. If you are considering hosting a community outreach event and need to recruit volunteers for planting trees, cleaning up a vacant lot, building trails, collecting household hazardous waste or similar activities; hosting a lunch-time event offers an excellent opportunity to educate and recruit the employees. Such a venue also allows you the opportunity to identify and gain commitment from employees to assist in such efforts.

Benefits of Employee Training Sessions

These are forums for employees to better understand the organization's environmental philosophy, achievements, issues and future focus. New employees will be pleased to understand the environmental conscience of the organization. Veteran employees appreciate the updates. Such sessions should be held on a regular basis and always provide opportunity for questions. They should be interesting and informative. Consult with your communications managers on how best to structure such sessions.

ESTABLISHING YOUR COMMUNITY PRESENCE

Establishing an organization's presence in a community is critical for the success of any organization. Environmental, safety, and health managers are wise to get involved with organizations and institutions, which influence public dialogue and policy regarding environmental quality issues and standards of business performance.

Environmental, safety, and health managers working with their public affairs or community relations staffs can build effective and important long-term relationships with community leaders, elected officials and interested citizens. This is accomplished by joining and participating on a regular basis with the chambers of commerce, service organizations, environmental groups, school clubs and related activities. It is important to know and work with key constituents and understand their issues. Such involvement can help you gain a better understanding of the current issues and concerns of the community and address them before they become contentious issues.

Managers who are willing to serve as officers of such organizations can obtain even more visibility and will be in the stream of the community to hear insider and candid views and opinions from decision-makers and other constituents. This can lead to the creation of an informal dialogue that can ultimately benefit the organization. It assists you in building levels of rapport and trust within your community.

Environmental, safety, and health managers should consider how they can most effectively participate in such activities as "Earth Day" and the related activities mentioned previously. These types of activities build relationships within the community that are difficult to establish otherwise. Environmental, safety, and health managers may find themselves working along side environmental or community activists, whom they would normally only know within the context of a contentious environmental issue. Establishing a dialogue on such an issue from the common ground of a shared and positive experience can ease current and future relationships considerably.

"Simply, whenever and wherever there is serious and responsible dialogue about the community's environmental issues, your organization's environmental, safety, and health management should be present," says Dave Weiland, Lockheed Martin manager of Environmental, Safety, and Health Communications.

PARTICIPATING IN ENVIRONMENTAL ORGANIZATIONS

You can establish a community presence quickly and efficiently by regularly attending and actively participating in meetings of environmental groups. Because these groups work on a variety of issues ranging from land use to wildlife habitat protection and water conservation, your work with them can help them further enhance their goals while laying a foundation to develop rapport and trust with them over a broad spectrum of issues. People begin to see you as a real person that has real concerns and when the time comes to talk with them about your issues, they will be more open and willing to listen.

One noteworthy example of rapport and trust building resulted when Dr. Martin Rivers, formerly a director of environmental programs for the Tennessee Valley Authority, attended and actively listened to the issues and concerns of the environmental activists in the Kentucky District, a 27 county area serviced by his organization. Over a period of several months, Dr. Rivers and this author built enough rapport with the activists to convene an annual environmental constituency meeting to discuss their concerns and the utility's issues as well. A number of issues were resolved without threatened litigation or airing of negative issues in the media.

The Lockheed Martin Astronautics Group in Denver, Colorado recently worked closely with several civic and environmental groups to create a wetlands area and an adjacent facility that serves as a trail head for several major trails in Colorado. These projects helped Lockheed Martin mitigate and temper the tide of public opinion and sentiment that had been negative earlier in the decade during a period of alleged drinking water contamination (Belsten & Weiland, 1999).

VALUE OF BUSINESS AND CIVIC ORGANIZATIONS

It is helpful for environmental, safety, and health managers to also become actively involved with civic and business organizations that are dealing with environmental issues. These may include local chambers of commerce, city environmental quality boards, or regional economic development groups.

In most cases the environmental, safety, and health manager's participation should initially be one of careful listening to the issues and concerns. Only after there is an appreciation of these issues and the politics behind them should the environmental, safety, and health manager become actively involved in providing counsel and assistance.

This author worked closely with the Hopkinsville-Christian County Chamber of Commerce for two years and then became president of the organization. During the term of the presidency, we were able to convene the regional businesses into an environmental forum where we discussed how to deal with such programs as pollution prevention and waste and chemical minimization. The program began an open dialogue with the companies and subsequently the public.

Such sustained interaction can lead to positioning your organization and the organization's environmental, safety, and health commitments in such a manner as to be a leader for such programs. It is critical to be accessible to other professionals and lay people to provide them consultation and assistance when needed.

CASE FOR REGULATORY ADVOCACY

One strategy that can result in direct, measurable savings to your organization is to become actively involved in relevant regulatory boards and commissions in your state. Your direct involvement and monitoring of air and water commissions and other pertinent regulators where rules and decisions are made that directly impact your environmental, safety, and health program can result in better and more informed decisions that can lead to fairer, more cost-effective regulations.

One organization's work in this area resulted in a response to Title V of the Clean Air Act by the regulators that ultimately saved millions in capital retrofit costs, as well as annual recurring expenses. This was accomplished by having rapport established with the regulators that permitted the organization to present alternative methods of compliance that were negotiable and ultimately acceptable.

WHY COMMUNITY OUTREACH CAN WORK

To enhance two-way communications and build rapport and trust within a community, companies must be sensitive to designing a community outreach program that is inclusive of these objectives. It is not necessary to launch an extensive effort; rather it is better to do a few projects that are well coordinated and accomplishable. These can include such activities as tours and open houses, neighborhood coffees and visits, exhibits, school presentations, a 24-hour contact line, a community advisory panel, or a speaker's bureau.

It is critical that you work with all levels of management to ensure the programs you implement are appropriate for the organization. Analyze and focus carefully on the business your are engaged in and what level of representation you want in your community.

High-Profile Projects—Cost Versus Benefits

High-profile projects such as Martin Marietta's Environmental Fair in Oak Ridge, Tennessee, or Bechtel Jacobs Small Business Fair also in Oak Ridge, can be expensive and time-consuming. "However, they may be appropriate when balanced against your organization's need to correct or improve relationships within your community. The expenditure of the resources can go a long way in building rapport and trust," says public participation consultant Kay Armstrong.

Tours and Open Houses

Tours and open houses offer citizens the opportunity to visit your operations and see firsthand what you do and how you do it. They are effective in fostering a better understanding of your products and services and can mitigate the public's concerns about your environmental affairs. Tours and open houses can help to dispel citizen's fears and provide an avenue for you to hear the citizens' issues and concerns.

Every tour should be carefully rehearsed and your facility should be immaculate before inviting the public. Remember that all necessary safety equipment should be provided such as hard hats and safety glasses. You should include presentations that are short, entertaining, informative and educational. Fact sheets and other "take aways" should be provided. Small gifts or tokens with your name and telephone number make appropriate take aways as well. Light refreshments such as cookies and drinks are appropriate for such events.

Tour guides should use hand held public address systems to ensure that everyone can hear. There should always be plenty of time for questions. Schedule your tours so stakeholders will be able to attend. That is, don't schedule them on the night of the big football game in the community! Citizens may be invited by invitation only, or your invitation can be broadcast to an entire community through newspapers and radio and television.

If you open your facility to the community, it is prudent to have people who wish to attend to call in advance so that you may be prepared with an appropriate number of hosts and hostesses, refreshments, hard hats and safety glasses.

Even if citizens are unable to attend your tour and open house, the fact you gave them the opportunity, the fact that you opened your doors to them, sends an important message about your willingness to inform and educate citizens. Open houses should be held at least annually to be most effective with the goal of increasing the number of people attending at each one. The more people know about your operations first hand, the better the chances of building effective rapport and trust.

When M4 Environmental, Inc., a former start-up company in the southeastern United States, built a unique and new facility that would handle radioactive waste, they used an open house to acquaint community leaders with their operations. It also reassured citizens of how safe the operations were. More than 250 people attended the open house and informative and educational media coverage also resulted (Davis, 1999).

The first ever open house at the Department of Energy's Oak Ridge National Laboratory was billed and marketed as a "Community Day" and more than 1200 people attended the day long event. This was the first time in the 50 years history of the lab that the public had been invited behind the security fences to better understand what the laboratory was all about. It has become an annual event.

Neighborhood Visits and Community Coffees

One of the most effective strategies to establish rapport and build trust within a community is to use community coffees or neighborhood visits. These are low-cost ways to become known in your area. This is best accomplished by getting to know key constituents and community leaders and then having them assist you in hosting the event.

Working with the community leader, you can draw up a list of the people you would like to attend an informal coffee klatch. Be sure to include key elected officials, especially those that represent the area in which you are meeting. Serve pastries and coffee, juices, or soft drinks and simply get to know the attendees and "actively listen" to their concerns, answer their questions, and accommodate their requests where possible. It is appropriate to launch such meetings with a short briefing about your facility and any issues you wish to discuss. It is also appropriate to propose a set of "operating principles" for the meeting. An ongoing dialogue can be established in advance of a need. When your organization is in need of having access to the citizens, the platform is well established to do so.

Such informal meetings can be the focus of a community outreach strategy and no amount of written material or public information can substitute for these types of face-to-face interchanges. This process gives citizens a personal contact in your organization with whom they can discuss their issues with on an ongoing basis. It helps to humanize your organization and its management.

Dawn Ford, the former director of the Citizens Action Office at the Tennessee Valley Authority, pioneered such meetings for the utility in the 1980s. Working with residents around the utility's nuclear plants, her efforts enabled the organization to open and maintain dialogue that eased the fears of their neighbors about nuclear power.

"It is essential that the organization is responsive and willing to act when appropriate upon those issues and concerns that are raised in such meetings," said Ms. Ford.

This author would add that it is equally critical to let citizens know when you can't act upon their comments or suggestions and why. Our experience clearly demonstrates that stakeholders want the communications loop closed and respect management that is willing to do so.

You can supplement actual visits and coffees through periodic and regular telephone calls to citizens that have attended these sessions in the past. These calls need not be lengthy in duration, especially if you don't have an issue to discuss. Nevertheless, such a call reinforces to the stakeholder your interest in them and their issues and provides them with a real person within your organization they can relate to and with, and the opportunity to raise questions before they become issues.

This elevates the probability that if there is a problem in the future at the organization that this stakeholder will contact the representative they have come to know. Such contacts make it easier to start a dialogue with the stakeholders in the event of a planned expansion, a permitting activity, or in the event of an emergency. Routine coffees followed up with periodic telephone calls minimize the accusation: "I never hear from them unless there's a problem."

Speakers' Bureau: Only as Good as the Organizer and the Commitment

A speakers' bureau is only as effective as you choose to make it. It is an excellent strategy to get trained speakers into the community to convey your messages. Bureaus are composed of public affairs, environmental, safety, and health managers, and other selected managers who can speak knowledgeably on the environmental, safety, and health issues of your organization. Speakers visit schools, civic clubs, professional organizations, and nonprofit groups to provide information, answer questions, convey a positive image for your organization, and the pay off is increased development of rapport and trust with your constituents.

Effective speakers' bureau may have a coordinator who outlines, writes, develops visuals and otherwise assists in the development of environmental, safety, and health messages, or these services may be part of the public affairs staff. Effective speakers' bureaus are highly marketed using brochures, letters from the executive management offering speakers to selected groups, and personal phone calls to solicit speaking engagements.

However, all speakers who will participate in the "chicken and peas" circuit should be extensively trained to give presentations to lay audiences and those presentations should be rehearsed frequently. Under no circumstances should an untrained and unpracticed speaker be sent into the community to represent your organization.

Low-Cost Exhibits Sustain Your Presence

Portable exhibits are reasonably priced these days costing less than $2,500 for three portable panels with appropriate art and graphics. Exhibits explaining your organization's environmental, safety, and health program that can be displayed at civic activities, conferences and trade shows are an excellent investment and, if they are carefully designed, can be used extensively. They also can be displayed in public buildings such as libraries, city halls, senior citizen centers, schools, and malls. They make an excellent information addition to open houses and tours, and for special presentations to council meetings, and even neighborhood coffees. They can be used for public meetings and community events such as fairs. When displays are used at events a representative who can answer questions should always accompany them, or at a minimum, fact sheets should be available as a take away.

Exhibits combined with an aggressive conference schedule, where your organization sends a trained representative to make a presentation or present a technical paper, is an excellent venue for marketing environmental, safety, and health accomplishments to peers and the industry.

M4 Environmental Inc. used such an approach to become a nationally known organization in less than 24 months. They used a simple message on the display with creative graphics backed up with creatively designed fact sheets and brochures (Rymer, 1999).

TVA's "Land Between The Lakes" during the 1970s used such displays to inform and educate the public throughout the Tennessee Valley and the eastern United States about their environmental education efforts (Ford, 1999).

Work With Community Schools
Schools need your help and the return on the investment is worth the effort. Gail Rymer, the current director of Environmental Communications for Lockheed Martin Corporation is a firm believer in using school programs to convey educational and informative corporate messages. Ms. Rymer's organization actively participates in a national program called "Space Day" each spring.

Kay Armstrong coordinates the program for her. She explains, "Each year we select a school who can benefit from our assistance. We provide them with curriculum, videotapes, presenters, volunteers, hands-on activities, materials and other assistance to enhance their learning about space and related activities during this event. The response has been overwhelming from teachers, students and parents."

When companies can affiliate with such events such as "Earth Day" or "Space Day" and provide such assistance, they too can benefit from the direct exposure in the school systems. You might also want to consider tutoring opportunities, assisting with science fairs or science projects, donating equipment, or related activities. Find out what the school needs are and see if you can fulfill the needs. The payback is good for you and most appropriate for this segment of your community.

Toll-Free Telephone Numbers or 24-Hour Lines, Fax and E-Mail Services
You can provide a major convenience to the public by providing a telephone number, a fax line, and/or an e-mail address where they can lodge their complaints and issues or ask questions. If it is appropriate, the use of a toll-free number is recommended.

A number of facilities have established such services. Invariably the companies that have been successful in this effort have broadcast the services far and wide using the media, direct mail, and their speakers' bureaus. The purposes of the number are clearly explained and one manager, or his designate, is responsible for handling and closing the communications loop on all inquiries. It would be wise to establish a policy that all inquiries will be responded to within a reasonable period of time. One organization established that turn around time at 24 hours, another at 72 hours, and they are committed to meeting those requirements.

The Tennessee Valley Authority during the 1980s used a toll-free hotline that received on the average 10,000 calls monthly. Six full-time employees staffed the toll-free line. Regular public opinion and trend reports were prepared from these calls and forwarded to the Board of Directors for action. Members of the board staffed the lines each month (Ford, 1999).

Although you may get more inquiries about job opportunities than environmental, safety, and health concerns, the commitment to open such an avenue of communication is often very well perceived by the public. It demonstrates clearly the organization's values of being responsive to the public's inquiries and once again helps establish rapport and trust with the community.

Community Advisory Boards: New Paradigm
This strategy is becoming more and more common. Pioneered by the chemical and oil industry in the early 1980s, Community Advisory Boards (CABs) are groups of interested citizens, elected officials, representatives of environmental interests, religious leaders, educators, local business people and other stakeholders who meet on a regular basis and engage in a dialogue about environmental, safety, and health issues concerning your organization.

They provide a formalized mechanism to exchange information and get comments (Hastings, 1999).

Although properly designing and managing CABs can be challenging and labor intensive, they are one of the most effective strategies for establishing an ongoing public interaction that can ultimately benefit your organization. Community advisory boards can provide feedback and suggestions from other community members and help you to improve the design of your community outreach efforts.

These advisory boards can range in size from 5 to 30 people. We suggest you start small and then, if needed, increase the size of your board. Community advisory boards meet to discuss specific issues and the meetings provide forums for fairly detailed and sometimes more technical discussions of remediation plans and priorities, emergency response plans, and developing issues.

Community advisory board members have an ongoing opportunity to learn about your facility's operations, concerns, and issues. They take tours, engage in discussion about your issues and concerns, and reflect their own to you and your representatives. They can assist in hosting neighborhood coffees and in hosting site tours.

Setting up a CAB requires careful consideration of who should be members, who will be reasonable, who will be representative of the community, and who will actively participate.

CABs can be set up for a limited duration or be an ongoing standing board with a well-defined term limit and charter. It is also important that such advisory boards have an independent facilitator, set agendas, and have an agreed-to set of well-defined operating principles.

Once an organization sets up a CAB, that organization must be prepared to accept at least some of the advisors input and suggestions and act on all their comments to ensure the communications loop is closed. This does not mean the organization has to implement each and every suggestion, rather it means that when a organization finds suggestions impractical, impossible, or not in the best interest of the organization that an explanation is provided and the matter is closed. To do less than this is not honoring the investment of the CABs time and energy (Armstrong, 1994).

Therefore, it is important to set well-defined parameters so the CAB understands what to expect from the organization. It is not wise for a representative of the organization to universally agree to implement all suggestions from the CAB. Craft a well-defined charter that spells out exactly what the CAB's parameters are. The charter should also define carefully what the CAB can expect from the organization. This will ensure no misunderstandings will occur.

The purpose of a CAB is to establish on-going rapport and trust and provide opportunities for personal face-to-face communications. This type of communication is critical and too often underestimated. There is simply no substitute for personal communications.

Web Pages and the Internet

The newest avenue to get information to the public is through web pages and the Internet. We recommend you work with creative web page designers to outline carefully what information will be placed on such a page.

Communications managers should be consulted to ensure that the material is readable and is what the community needs. Be sure to list your page on as many search engines as possible, use keywords for searches the public may conduct, ensure you have appropriate security precautions in place to prevent hacking, and update the page regularly. Place nothing on the site that you don't want shared throughout cyberspace.

Finally, provide an e-mail link where the public can contact you directly. Check, read, and respond to e-mail on a regular schedule and let the public know in advance how long they can expect to wait for a reply. Provide telephone, fax and U.S. postal information on the site as well.

USING PUBLIC INFORMATION CORRECTLY

The public is informed using the concepts of public information. Public information supports the concepts of public participation by providing clear, objective, easily understood, and timely information. This enables the public to effectively participate in the public participation processes.

Information products include fact sheets, brochures, newsletters, exhibits and displays. Each product should clearly identify how the public can comment to management on the information. Avenues of communication that should be displayed prominently on each include telephone and fax numbers, e-mail addresses, and postal addresses. Where appropriate, a person should be identified as a point of contact so that the public can communicate with an individual.

Public information should set the stage for presentation of objective information. Public information differs from public relations in that public relations uses prepared information to "market or sell" a particular strategy or an idea to the public.

Community relations is the process by which a corporation or organization sets about to create a relationship within a community to demonstrate its commitment to the public. Often community relations programs are philanthropic in nature, that is money is given to the community to support projects and programs.

Related Effective Communication Strategies and Techniques

Additional techniques useful to the environmental, safety, and health communications program include having a well-defined media strategy: newspapers, radio, television, news releases and media conferences and use of written communications: fact sheets, newsletters, Q&A sheets, and brochures.

Please note that environmental, safety, and health managers should work closely with their public affairs staffs, or their public participation consultants, when dealing with the media. Often times relationships have already been established with the media and these staffs can guide you on how best to interact with them.

Media Relations Strategy Is Critical

The media is the most effective way to reach members of the general public. They provide a valuable "third party" viewpoint using communication mediums that appeal to those members of the public, who prefer not to review written material produced by your organization. Finally, whether you like it or not, environmental news—good or bad—is just that, news!

When you have a well-defined media strategy you can work more positively in promoting your messages and be proactive rather than reactive with the media. Although you will not be able to keep unfavorable stories out of print or off the air, by having a proactive and ongoing media relations strategy, you will be better positioned to present your organization's point of view. This will at least bring some balance to the story.

Media involvement is not for everyone, especially in times of crisis. Choose spokespeople carefully and any management that will deal directly with the media, at the very least, should be trained in communications awareness and media training.

Using News Releases, Photographs, and Feature Stories

News releases, creative photographs, and feature stories are the meat and potatoes of the newspaper business. You can develop and distribute stories about your facility's operations, your waste minimization activities, and other positive environmental, safety, and health messages and stories. Guest columns, editorials, and published letters to the editor are also effective. In certain situations, paid advertising is also appropriate.

As the former Director of Public Affairs for Oak Ridge National Laboratory, my staff and I developed and distributed 52 science features with photographs and more than 150 news releases each year. More than 75% of these items were used by publications on a regular basis. Companies should be aware that the media will use properly targeted and well-written news releases and features.

These stories also prompted reporters to contact us for additional information and other positive stories were generated as a result of the contact. We accomplished this aggressive task using journalism interns from a nearby university and limited professional staff.

Target your media efforts carefully. For example, if you have a local weekly or small daily newspaper you have greater opportunity to get more exposure than in a larger metropolitan area. Research and use trade specific publications to get specific messages published.

Take time to go to the newspaper and meet the reporters and especially the editors and publisher. Meet with the editorial board, if you have one. Go prepared with the key messages you want them to know and be flexible in these meetings. Offer to write guest columns, or to be a source to comment on technical aspects of environmental, safety, and health issues.

Take time to write letters to the editor and don't write them just when you have a problem or want to correct some information. When you see a well-written environmental, safety, and health story, write a letter to the editor commending the reporter on their work. The newspaper will be more open to running your clarifying and correcting letters if you have already established a working relationship with them.

One organization used letters to the editor as a principal strategy to educate the community about a particular issue. The letters were tightly written and focused on no more than three key messages. They were published in a weekly newspaper each week for 12 weeks. The result was a lessening of the public criticism of the organization for previously being unresponsive to citizens' concerns.

When to Use Paid Advertising

There are times when the most appropriate way to get your unedited message to the public is to use paid advertising. Paid advertising can be run in the media when you need to get your message out in a prompt and unedited fashion.

One chemical company, faced with serious opposition to its operations, bought weekly advertisements in their local newspaper for a year. Each advertisement provided easily understood information about their environmental, safety, and health efforts and regular progress reports on the plant's environmental issues and programs. They began to create community expectations about the next steps for the organization and explained how those steps would be undertaken and met. The community began to appreciate the complexity of the situation yet also began to understand the commitment that the organization had toward fixing the problem. The advertisement included all the conceivable ways the public could contact the organization (telephone, fax, mail, e-mail) and, when contacted, the organization responded quickly.

However, another organization decided instead of buying paid advertising they would tour the editor of the paper at their facility. Part of the agreement with the reporter was that they would provide a technical person that could speak in laymen's language. The reporter requested permission to take any pictures she wanted to take and asked that the company spokesperson follow up on any questions and get her answers within 48 hours. The company fulfilled the tour request, the reporter got her pictures and all her questions answered, and the newspaper ran an extensive well-written, factual, illustrated story on the facility.

Learning to Speak in "Sound-Bites"

Radio and television are viewed or listened to at some time during the day by 88% of the U.S. public, according to one national poll. Radio stations are required by the Federal Communications Commission (FCC) to broadcast periodic public service announcements or PSAs. Such announcements are one way you can get attention to your environmental, safety, and health efforts, such as upcoming public meetings, open houses, community tours, or exhibits you are displaying.

Before designing and employing a broadcast media strategy, ensure that all management involved has had media training and understands how the electronic media works. Managers working with the electronic media should be able to speak in laymen's language about technical subjects, be able to think quickly on their feet, and be able to maintain their composure under pressure. Critical to the success of working with the electronic media is understanding how to condense key messages into *sound bites*. Managers will learn how to do this in an effective media training course.

Radio and television both have paid advertising as venues as well. We recommend you work closely with media strategists to design this advertising and buy air time. In addition to PSAs and paid advertising, radio and television stations have talk shows that you can use to discuss local issues and programs at your facilities. One organization used the call-in shows monthly to update their community on the operations at the plant, discuss significant achievements, and answer questions from members of the public that call in. These situations are best handled by well-trained and experienced organization spokespeople because there are inherent risks in dealing with the broadcast media. Consider each situation on a case-by-case basis.

With the advent of cable television companies, community cable television stations are eager for information. You can provide them with interviews, news releases, or videotapes. They are an excellent resource to use in your community.

Why News Releases Can Work for You

The most common way to get your environmental, safety, and health information into the media is to issue a news release. The news release must be newsworthy, well written, and two pages or less. Announcements can be about national or local awards your plant receives, installation of new pollution prevention equipment, or an update on a remediation project.

News releases provide the initial foundation of the coverage of your story. Proactive news releases can bring attention to your organization about positive work you are doing. On those occasions when you need to react and respond to an issue, news releases provide your side of the story. News releases should be objective and follow the style of the *Associated Press Stylebook*.

Media Briefings and Conferences: Use Sparingly

Media and press briefings and conferences are usually conducted only during times of unusual events or emergencies. Briefings are generally held to update the media on an evolving situation, while a press conference is usually held to make a major announcement. These sessions are interactive, meaning they are open to questions and you provide the answers. Conferences and briefings allow the release of information to the media at the same time. Make sure that anyone participating in these sessions has extensive media training and substantial practice.

Written Communications Strategies and Tools

Written communications is one-way communication and provide one-way forums. This is not a substitute for face-to-face relationships with your stakeholders. Written materials are an important supplement to other forms of communications and often are used as take aways. Such materials should provide sufficient information to ensure a general understanding of a situation.

This material can be referenced by stakeholders and shared with other members of the community. Written materials should be written to allow for easy reading. Use graphics such as charts, graphs, and pictures to enhance the understanding of the reader, but these do not necessarily have to be "slickly" and expensively produced.

Fact Sheets

Facts sheets are designed to explain regulatory or technical issues, provide overviews or updates of activities, or provide background information on a specific program or issue. Used to help the public understand technical issues, they should be confined when possible to one page front and back, written in laymen's language, acronyms should be explained, and they should be simple and graphically appealing. They are distributed whenever needed and can be used as take aways and mail outs.

Newsletters

Newsletters are distributed periodically and are used to reach a wider audience than fact sheets. They can be used to report the work of CABs, provide overviews of planned activities, explain in more detail plans and strategies, and highlight accomplishments. They don't have to be slick to be informative. They also can be used with stakeholders who don't choose to attend meetings to keep them informed.

Question and Answer Sheets

Question and answer sheets can be used like fact sheets and can be put into brochures such as "The 10 Most Frequently Asked Questions About Project X." They can be distributed internally or externally as needed and are helpful to have as a take away from public meetings. Question and answer sheets ensure that clear, concise answers are developed for sensitive questions and are used to explain technical subjects in understandable language. Developing Q&A sheets help spokespeople identify key issues and determine how best to discuss them with the stakeholders. The Q&A sheets should be direct, truthful, concise when possible, and provide a complete understanding of the issues.

Brochures

Brochures are usually used to provide an overview of a organization's performance, in-depth information about a specific program, or focus on a particular topic. They are not necessarily specific to a community and often serve as background materials.

FOUR CRITICAL SKILLS FOR EFFECTIVE COMMUNICATION

Managers and technicians who will be involved in the public participation or stakeholder involvement programs should have four distinct skills in their repertoire.

- First, you must be skilled in making presentations to the public. Frequently, we see highly competent managers, who are quite capable of making presentations to their peers, fail when making presentations to the public.

    Why? Most often they have not adjusted their language and their demeanor to the level the public can understand and relate to; they have used poor or no metaphors or stories to place their information in perspective; their graphics are too technical or non-readable; and they haven't practiced the presentation adequately.

- Second, managers must be skilled in communicating risk to the public. It is one thing to communicate risk to other technical people, but when communicating such issues to the public, careful planning and practice of the skills is critical. Simply stated, the public doesn't know what ten to the minus two means and frankly, they don't care. Risk to the public is really an issue of safety; more about that later.
- Third, managers that participate in public participation programs ultimately will have to handle a hostile and upset audience. Interpersonal skills, the ability to center one's self, and the ability to direct the audience participation into a meaningful dialogue is critical to conduct a productive interaction.
- Finally, managers must be capable of interacting and communicating with people with diverse cultural backgrounds. This is especially true when management comes into rural areas from metropolitan communities and is equally the case when working with inner city issues, or tribal governments.

These four skills tend to be synergistic and build on each other so that when one is used, the manager's effectiveness with the other skills is improved. However, beyond mastering these skills, is the challenge of knowing which ones to use and when. The audience the manager interacts with actually determines which of the skills, if not all four will be used.

For example, if the public needs and wants information, then obviously presentation skills are called for, as is the need to understand risk communications. If on the other hand, the audience is outraged and expresses anger, or tries to convince the management of another alternative, the manager will have to be skilled at handling emotions while understanding risk communications.

Presentations

Not all presentations are equal. Those that will be made to the public require careful thought and planning and practice. For presentations to be effective, you must be simple. The rule to remember is *KISS: Keep It Super Simple.* Begin designing your presentation by clarifying to yourself what the presentation is about and what you want to convey to your audience.

The presentation must be presented in a logical sequence and have a progressive order of information. Help the audience understand how the information progresses and how the proposed decisions or alternatives were reached. Presenters should anticipate what the audience wants to hear and build the answers to meet those expectations in their presentation. We recommend that every presenter use outside counsel to develop the 20 most difficult questions that could be asked. The presenter then should develop succinct yet informative, easy to understand answers (Armstrong, 1996).

The audience that you expect to attend your presentation should be carefully analyzed and the appropriate level of detail should be factored into the presentation. It does little good to have an overly technical presentation for a group of neighbors. Ask yourself, and your colleagues or consultants, and then answer such questions as: How familiar is the audience with this topic? What are their interests and needs? What beliefs and attitudes do they hold? What stereotypes might exist? Are they going to be very interested and enthusiastic about your presentation, don't care, uninterested, or even hostile?

Write out your key messages. Remember the KISS approach. Write the presentation as if you were outlining a story. Develop your key points in a logical manner. Assemble your supporting information.

The manager should seek to incorporate stories, examples, and anecdotes that display a human side to the material and helps the audience identify with the presenter and the information. Create a common people experience, make yourself and your organization human in this process, after all we are all human and the public wants to see and relate to the human side of a presenter. Carefully craft the opening of your presentation because the first 30 seconds are the most critical. If you can't get their attention and interest within that 30 seconds you are in serious trouble. In closing your remarks, if it is appropriate, recommend action, and be sure to reinforce your basic points and key messages.

Presenters should seek effective uses of analogies, for example, one presenter announced his presentation would be equivalent to the length of the Gettysburg address, "Long enough to cover the subject, but short enough to be interesting." Relate your messages to something that people can understand. However, don't use analogies that belittle or patronize your audience.

The speaker should care about his subject. There is nothing worse for an audience to endure than a presenter who doesn't want to be there, has no interest in what they are speaking about, but is willing to make them endure their monologue.

Those speakers who can actually convey to the audience they want to be there and care about their listeners' reactions, opinions, and issues make effective presentations. This is most often conveyed not through what is said, but how it is said, and a person's use of body language, word choice, communications style, language matching, and position in the room.

Finally, there is a certain fluency and eloquence that must be part of any presentation for it to be effective. This fluency and eloquence comes from repeated sincere practice on the part of the presenter. In other words, the presenter must respect his audience enough to take the time to practice his presentation and anticipate and practice the answers to the toughest questions that could be asked. A failure to understand the need for such involvement on the part of the presenter is a recipe for disaster.

If you plan to use graphics, carefully choose and select them. Don't ever put facts and figures on an overhead and then say, "I know you can't read this but..." Use a minimum of 24 point type, bold and dramatic graphics and clear photographs and drawings.

You will get to Carnegie Hall, the old adage says, only if you practice, practice, and practice. And practice is the only way you will become a proficient presenter. Practicing aloud and with colleagues is essential. If you practice alone, use a tape recorder to listen to your voice.

We also recommend you use people outside your organization to critique and ask you the tough questions. Many people subscribe to the opinion that justified criticism is what you give while unjustified criticism is what you receive. We suggest otherwise. Don't take offense at the criticism you may get, use it to improve the presentation. As professionals, who deal with managers on presentations, we firmly believe everyone should do a minimum of two "dry-runs" before giving a public presentation.

Finally, when making your presentation stand erect, yet relaxed. Relax your knees and shoulders, and breathe. In fact, I recommend you place in large type above each page of your outline the words, "Breathe, slow down, breathe, relax, and breathe."

Maintain eye contact, move your focus around the room and maintain your energy and enthusiasm. Gestures should be natural and used according to what you are thinking, feeling and saying.

Avoid Using Technical Jargon

Most managers and technical presenters degrade even their most effective presentations by using technical jargon and acronyms that the public at large simply don't understand. When managers use the term "CERCLA" and don't explain what it is and what it does, the audience can hear that term as "Circle A."

To make effective presentations remember to:
- Eliminate as many technical terms as possible.
- Sparingly use initials, abbreviations, and acronyms and be sure to immediately explain them, if you use them.
- Eliminate use of such unfamiliar annotations, such as "The risk is ten to the minus two power."
- Don't assume that your audience has prior knowledge about the basic concepts of science because most audiences don't.
- Don't assume the public understands the importance of the scientific method, and therefore you think they understand the importance of certain practices and procedures (Creighton, 1998).

How to Communicate Risk to the Public

The ability to effectively communicate risk is a critical skill the manager must develop to interact successfully with the public. There is a science involved in communicating risk. It assumes that there is a logical manner in which the public perceives risk; however, it is usually quite different than how the technical risk communicator perceives the same risk.

The public perceives risk as being the probability that something bad will happen to them, or people they care about, combined with the things about that situation that upset them. To the public there are some very specific factors, or things, that create confusion, ill will, bad feelings and even outrage.

The public tends to believe that risk is greater when:
1. they don't understand or accept the projected benefits being presented about a project

2. they believe the risk can be avoided by using another solution or alternative
3. some of their neighbors, or other people, are singled out for negative impacts
4. the risk is imposed, rather than accepted voluntarily
5. the person being exposed to risk can't control the degree of that risk
6. children are involved and impacted
7. the emotion of "dread" is associated with the risk
8. the risk is caused by direct human actions
9. the risk has received a great deal of media attention
10. the people affected aren't familiar with the risk that is being discussed
11. they don't understand the nature of the risk being discussed
12. they don't trust the institutions taking the actions that will cause the risk (Covello & Sandman, 1987).

When several of these factors exist, the public often becomes highly agitated, anxious, confused, upset, outraged, and even morally indignant. As managers involved in the public participation process, you must become sensitive to the fact that when the public becomes morally indignant, your project can be in serious danger. At this stage the public doesn't care what it will cost you to "fix the problem," they just expect it to be fixed at any cost.

Simply stated, the public doesn't assess risk, as professional risk assessors do, rather they concern themselves with how safe an action is. They simply want to know if the action that is proposed is "safe," as their value system defines it. Therefore, the public acceptability of a proposed action is a direct consequence of their values and not the science, the facts, or the decision process that led to the action. If in the communications process the presenter has helped the public "feel" that the proposed action is "safe," most likely it will be accepted as so.

Successful presenters have long understood that to communicate effectively with the public, technical competency is about 20% of the mix of talent you need, while the remaining 80% is a function of the ability of the presenter to be dedicated, honest, empathetic and caring.

Kay Armstrong summarizes the situation well when she says, "Make a person feel what it is you are saying and you will remember it long after you have stopped talking!"

Covello and Sandman's Effective Ways to Present Risk

Covello and Sandman (1987) suggest several effective ways to compare risks that will assist presenters in their presentations. The most acceptable ways to present risk include these:

- comparison of the same risk at two different times: "Air emissions of nitric oxides have been cut by 20% since July of 1997."
- comparison of risks against a standard: "The release of contaminants into the water is well below the accepted standards of the state."
- comparisons with different estimates of the same risk: "Our estimate is 15%, while the state's estimate is 25%, and the Wildlife Society's is 30%.

Equally acceptable, yet according to Covello and Sandman, less desirable ways to express risk include these:

- comparisons of the risk of doing something versus not doing it
- comparing alternative solutions to the same problem
- comparing the same risk as experienced in other geographic locations.

The least desirable comparison approaches include these:

- comparing average risk with peak risk at a particular time or location, such as "The risk at the property line is 50% less than at the building itself."
- comparing risks from one source of a particular adverse risk with the risk from all sources of that same adverse risk, such as "Your chance of getting colon cancer from exposure to this chemical is approximately two-hundredths of one percent of total cancer risk."

How to Handle Public Outrage

There is a significant difference between reducing risk hazards and effectively handling and reducing public outrage. Just ask any manager who has faced an outraged audience! We know that we can effectively reduce hazards by reducing exposures, doing a more thorough and accurate job of assessing the risk situation, and by monitoring the risk.

There are equally efficient and effective ways to reduce public outrage. Some of the following actions, if employed properly, will help minimize the public's outrage.

To effectively reduce outrage, presenters must listen effectively, increase the level of the public's control of the situation, show genuine and believable concern about the risk situation, increase the public's familiarity with the risk situation, and continue to develop personal credibility with the public.

Presenters should demonstrate they also have a concern about the risk they are presenting. This demonstration is in the form of language and behavior, in other words, presenters must "walk their talk" when it comes to the issue of risk. This is best demonstrated through careful explanations of what the risk is and how it will be effectively handled.

Discussing why the project is needed rather than immediately jumping into the proposed actions will also reduce outrage. The public must understand why the actions are proposed and the basis for the proposal. This helps to satisfy the public's concerns about whether the project does indeed justify the risks associated with it. Presenters should frequently check with their audience, as part of their feedback process, to ensure the audience understands the reasons why the project is being undertaken.

It is always appropriate to discuss several alternatives; presenting only one alternative will usually create a situation where the public feels and thinks that the action is being imposed upon them. If they feel or think this initially, outrage builds and grows. All alternatives that have been considered should be presented and thoroughly explained, even if they are no longer viable. When the audience is educated on the thought processes and the alternatives that were considered, they feel/think more favorably toward the presenter and what the presenter is proposing—that is, they feel the presenter is being direct and candid with them, which he is.

One successful strategy to minimize outrage is to ensure the public participates in the decision making process. To accomplish this, involve them early in the process to get their input and their thinking and feeling and then reflect back to them on how you have integrated their views into the alternatives. This approach contributes to the public sense of inclusion and helps them feel they have some degree of control over the situation.

Assist the public by communicating thoroughly and effectively in familiarizing the public's understanding of the risk involved. For example, show the public how to take and read measures of risk. The more familiar they are with such processes, the less threatening the situation is to them, and you build credibility in the process.

"What they don't know won't hurt them!" Maybe, but it will eventually hurt you and your decision-making process. Give the public full and complete information, even when it hurts. Assuming you have been trustworthy in the past, people will accept what you say in the present. However, if the public ever feels misled by you, it will take you years to rebuild credibility.

Your presentation should couch information where possible in a context that makes emotional sense to them. Don't just feed them statistics, which are often meaningless and devoid of emotion, rather translate the figures into impacts that the public can relate to and understand. Frequently what appears to be arguments over risk are actually politically based discussions and disagreements. Simply stated, you must be astute enough to understand when it is a political issue and not an issue about science.

Ensure that you are clear about how decisions were made. Those decisions that are value based should be communicated as such, those that are technical should be equally well communicated. When value choices must be made, the public must be considered, or else the consequences in the legal court and that of public opinion can be dire.

Be clear with yourself that you as a presenter communicate with the public less often than the media, influential stakeholders in the community, friends and special interest groups. More than likely all these people will communicate with the public more often than you. Plan for this fact in advance and offer to be more communicative, if the public so desires.

Public outrage can be minimized if the public is educated in advance of a project being proposed. When an organization becomes a proponent for a project or action, it is no longer considered a source of credible information. Therefore, it becomes important in the planning process to get information into the public arena before they make up their minds.

Finally, remember the media is a resource to be used and not feared. All news media will make your project or proposed action "newsworthy" and frequently will present the controversial sides of the proposal. However, prepared presenters will understand and have practiced how to handle the media's questions and use the media as an outlet to get information broadcast. Remember that reporters don't write headlines, as often stories reflect well your message and the headline offsets the message. Also, more people watch television than read a newspaper. We recommend you get some good media training and learn to think and speak in "sound bites," if you are going to be involved with the media in any way.

CULTURAL SENSITIVITY AND COMMUNICATIONS REQUIRE AN EXCHANGE

To truly communicate with another person, there must be an exchange of common words, expressions and symbols. Critical to this success is the operative word-exchange. When the audience understands the meaning of your words, communication has taken place. However, each of us have different backgrounds, experiences, training and cultures. These differences affect the way in which we verbally and nonverbally communicate. The most effective communications with diverse cultures is achieved when the speaker understands both his audience and its cultural orientation (Creighton, 1997).

To effectively communicate with diverse cultural audiences, first accept that you must understand who the audience is and the reasons they are attending your meeting. Understand as much as possible about their interest levels, how much they know about your subject, how aware they are of the project, their education level, and their expected needs, desires and outcomes. Secure an understanding of their background, the history they have with your project, the issues they have about the proposal, and if possible, their attitude toward you as a speaker.

Practice diligently, as it is important to make a presentation with an appropriate tone and attitude. Ensure that you are comfortable enough to be at ease, convey to the audience that you are there to participate, don't lay blame and don't hide behind policy.

Because you're working within a diverse environment be careful about how you use objects and symbols, inappropriate gestures and body language. Avoid using terms such as "you people," and never, ever say "you don't understand."

Candidly and thoroughly in laymen's English, answer questions and refrain from ever saying such things as, "you don't want to know the answer to that question," or "what you are really asking is."

Your body language should always convey you are glad to be attending the meeting. Don't look at a clock or watch, but honor your time commitments; start and end the meetings on the times you agreed to, unless you negotiate to go beyond the ending time because of audience interest.

Try to not fold your arms across your chest, as this appears defensive. Stand near the group, even approach the group and walk among them, if feasible. Make eye contact with several people in your audience. Speak to the understanding level of the audience. Don't talk over their heads and don't be condescending by speaking beneath their level of understanding. And never, ever read a speech!

Avoid stereotypes in making assumptions about who your audience is. If appropriate, provide information such as documents, brochures and fact sheets in the language of the audience you are speaking with and never assume that one person speaks for the entire population of the group you are visiting.

Finally, avoid phrases such as "to tell you the truth," or "to be honest," or "trust me on this."

You will maintain the trust and confidence of any audience you speak to if you publicly commit to and follow up your presentations with concrete actions when appropriate and where warranted.

"ACTIVE LISTENING" IS CRITICAL TO YOUR SUCCESS

When people are not listened to and appropriately responded to, it makes no difference how effective the presenter is, communication simply doesn't exist. A presenter can quickly determine when people feel like they are being shut out of the process by the various behaviors they display. For example, when people feel shunned and unheard, they are often compelled to repeat whatever they felt was not acknowledged. And then, if still not acknowledged appropriately, the public will become more emotionally involved in the situation, their remarks often become sarcastic, their tone and volume changes, and they can become downright belligerent. Often they become more accusatory, their body language shows rigid and fixed postures, they become less open to alternatives, and they start seeing you and those associated with you as the enemy.

There are numerous impediments to listening that occur in meetings. These impediments shut down effective two-way communications. The unintentional language or the behaviors of the presenter often create these barriers. Among them are such language and behaviors as the following:
1. ordering and demanding language
2. warning and threatening language
3. admonishing and moralizing
4. persuading, arguing, or lecturing
5. advising, giving answers, or proposing solutions
6. criticizing, disagreeing, and contradicting
7. praising and agreeing
8. reassuring and sympathizing
9. criticizing, judging, and evaluating
10. interpreting and diagnosing
11. probing and questioning
12. sarcasm, kidding, and humor
13. diverting and avoiding.

Each of these behaviors when used inappropriately will close down effective communications (Covello & Sandman, 1987).

By simply acknowledging what his audience is saying and feeling, the presenter can effectively move into "active listening." It is not appropriate for the presenter to agree or disagree with his audience, since either behavior will most likely alienate someone. It is appropriate, however, to create an active listening environment and present problems objectively while moving from positions to the listeners interests. To do this, the presenter summarizes feelings and ideas rather than judging them.

It is also most important to match your language to that of the audience. If the audience says to you they "feel" a certain way, it is inappropriate of you to reflect back how you "think;" rather, it is appropriate for you to tell them how you "feel" about what you are saying and summarize it for the group. By matching the language pattern of the participants, you are honoring the communications process and communicating with integrity (Armstrong, 1998). Therefore, summarize feelings when voice tone and words become intense; when people repeat the same viewpoint; and when people say they are not being understood. Presenters or meeting facilitators should express and summarize feelings and concerns when: the group has gotten off the topic that was agreed to in the set-up of the meeting; when people are not able to complete their comments due to interruptions from other members of the group; when comments are excessively long, or go beyond an agreed-to time limit; when name-calling is involved; when it is necessary to remind the group of time limitations; and when it is appropriate to suggest a new technique be introduced into the process such as brainstorming.

Equally, however, presenters must be careful to avoid judging or admonishing the group for their feelings; using their position power to influence the situation; or proposing solutions to problems out of context or without giving explanations that the audience can integrate and understand.

The appropriate way to express concerns to the audience is to communicate the problem or feelings and not a solution or judgement. The presenter, or meeting facilitator, must "own" their own feelings. This is demonstrated by telling the audience how you feel, combined with a description of the behaviors you are feeling, and the proposal you will make to rectify the situation.

An example would be, "I am feeling a high level of frustration because I would like for everyone who wishes to, to have the opportunity to participate, but I am concerned you aren't getting an adequate opportunity to do so. I would like to propose that you raise your hand so I can call on you." Therefore, the equation is "I feel (you are accepting ownership) + a feeling word + a behavioral description + (if appropriate) a proposal to rectify the situation."

HANDLING AN UPSET AUDIENCE

As you gain experience as a presenter or meeting facilitator, the opportunity will arise for you to handle an audience that gets upset. Audiences, when they get anxious or emotionally charged, will begin to ask what is often perceived as difficult questions. When analyzed carefully these difficult questions are really emotionally charged statements, not requests for information.

People want their feelings to be acknowledged and prefer in such situations that presenters not try and give them more facts, figures, or arguments for why their feelings are wrong. They simply want those feelings recognized.

Therefore, when presenters answer questions, the technique of active listening, previously discussed, is critical to a productive interaction. Active listening provides a technique for the presenter to show empathy. An active listener listens closely to the stakeholder and if a statement is really a statement and not a question, many times it is sufficient to say, "I hear what you have said and I thank you for your comment." Often times by simply acknowledging the statement, closure can be brought to the exchange. On the other hand if it is a question, by listening carefully, then repeating the question, and getting clarification, the presenter has had the opportunity to demonstrate empathy and gain a short respite to gather their thoughts before answering. The presenter can show concern but doesn't have to agree a problem exists. It is sufficient for the presenter or facilitator to say, "I hear what you say and we too are concerned about the issue as well." As a meeting leader, always act on behalf of the interests of the group as a whole, and explain all your interactions, interventions, and proposals in terms of your concerns for the audiences' ability to participate fairly and comfortably.

SUMMARY

Public participation and involvement is increasingly becoming a way of doing business in the corporate and government world. It is essential to understand that this field of endeavor is always a dynamic and changing environment that will only be mastered when the critical skills are integrated, practiced and used appropriately by managers and technical staffs. As the 21st century dawns, we can expect more and more demands for involvement in the decision making processes. We should be prepared to embrace these demands and effectively use these processes for the betterment of our own decision making in business and government.

REFERENCES

Armstrong LD. A candid interview: P2 activities that work and those that don't based on our 25 years' experience. *Perspectives Newsletter,* 1994, Armstrong & Assoc, 455, Hillside Trail, Eddyville, Ky 42038, Tel. 502.388.0347, Fax 502.388.0348, e-mail: drdarryl@aol.com.

Armstrong LD. Managing a crisis: Observations on the DOE Hanford tank incident. *Newsletter,* June 1997.

Armstrong LD, Armstrong BK. Public Meeting Facilitation Skills: How to Communicate With Integrity Workshop, June 1996.

Armstrong LD. Communicating With Integrity Workshop developed for the International Public Participation Association (IAP2), 1998.

Armstrong BK. Communications Awareness Workshop developed for Martin Marietta Energy Systems, Inc. Environmental Restoration Program, 1990.

Belsten L, Weiland D. Lockheed Martin web page and draft ESH best practices document, 1999.

Covello VT, Sandman PM, Slovic P. Risk communication, risk statistics, and risk comparisons: a manual for plant managers. In Covello et al (eds). *Effective Risk Communication: The Role and Responsibility of Government and Nongovernment Organizations.* New York: Plenum Press, 1987.

Creighton J. Communicating with the Public: A Participant's Workbook Workshop developed by Creighton & Creighton, Inc., with assistance from the Waste Policy Institute Applied Sciences Laboratory, Inc. and presented to the U.S. Department of Energy, Office of Intergovernmental Accountability (EM-22), 1998.

English M. Stakeholder and environmental policymaking. *Center View Newsletter,* a publication of the Center for Applied and Professional Ethics, University of Tennessee, June 1996.

Hastings D. Manager of Communications Services, ABS Group, Knoxville, Tenn., Presentation to the Tennessee Valley Chapter of IAP2, May 1999.

Rymer G, Perkins S, Armstrong LD. Using Computer-Assisted Process Facilitation Techniques in Government Sponsored Public Meetings and Working Sessions: A Paper Addressing The East Fork Poplar Creek Working Group Experience, January 1994.

chapter 20

Program Assessment and Evaluation

by Gary R. Krieger, MD, MPH, DABT
Marci Z. Balge, RN, MSN, COHN-S

369 **Introduction**
    Components of Medical Assessment Management
    Programs ■ Language of assessment
370 **Pre-Evaluation Activities**
371 **Site Activities**
371 **Evaluate Findings**
371 **Summary**
371 **Appendix—Occupational Health Program Assessment**
    Overall areas ■ Background information and general facility information ■ Orientation tour
372 **Injury, Illness, and Incident Reporting**
    Understanding management systems ■ Facilty reporting procedures
372 **Worksite Hazards and Surveillance Information**
372 **Medical Examinations**
373 **Education and Training and Related Activities Directed Toward Risk Reduction**
373 **Occupational Health Administration Services**
375 **Urgent Care**
    First aid ■ Emergency response plan ■ Medical facility ■ Communications
376 **Fitness for Work**
377 **Hearing Conservation**
    Detailed worksheet for medical examinations and environmental exposure surveillance record keeping U.S. regulations

INTRODUCTION

Program evaluation is a critical part of any occupational health service. While there are a variety of ways to perform an evaluation of an occupational medical department, many organizations have developed protocols and processes that can be considered as *Medical Assessment Management Programs* (*MAMP*).

Components of Medical Assessment Management Programs

These programs provide a consistent and logical approach for benchmarking and evaluating medical department performance. Typically, the MAMP framework is keyed to six general objectives:

1. A systematic, objective method of verifying that standards—such as regulations, or company policies—are being met.
2. An evaluation of procedures and practices that lead to verification of a host facility's compliance with legal requirements, corporate policies, and/or accepted industry-wide practices.
3. A process to verify the existence and use of adequate internal and external control for staff and outside suppliers/providers.
4. A benchmarking process to help measure a medical department's performance against internal and external standards and practices.
5. An evaluation process that uses personnel interviews, facility visits, and records reviews to develop an accurate performance picture.
6. A final exit meeting and report that presents and discusses findings uncovered during the assess-

MAMP Process		
Pre-Evaluation Activities	**Key Activities at Site**	**Post-Evaluation Activities**
■ Benchmarking	Step 1: Understand Management Systems ■ Interviewing	■ Prepare Draft Report ■ Draft Report Reviewed
■ Pre-Evaluation Questionnaire ■ Pre-Project Conference	Step 2: ■ Gather Data	■ Issue Final Report
■ Schedule Facility Visit ■ Develop Detailed Timetable	Step 3: ■ Evaluate Findings ■ Summarize Findings	
■ Confirm Scope, Schedule and Budget	Step 4: ■ Report Preliminary Findings ■ Exit Meeting	

Figure. Components of the Medical Assessment Management Programs Process.

ment. These "findings" or exceptions are documented against internal or external performance/regulatory requirements.

Language of Assessment

Within the assessment world, there are many synonyms that are often interchangeably used. These are as follows:
- audit
- review
- compliance review
- survey
- surveillance
- appraisal
- assessment
- evaluation
- inspection.

Since many individuals feel that the term *audit* has a negative or punitive connotation, many organizations typically use the other listed terminology. However, there are two terms that are particularly important for any occupational health department review.
- assessment
- verification.

Assessment is the process of providing expert judgment/opinion on medical management and control measures.

Verification determines and documents performance by evaluating the application of, and adherence to, company policies and procedures. In addition, it certifies the validity of company data and reports and evaluates the effectiveness of management systems.

Verification can demonstrate that existing programs are in compliance and that adequate management systems are in place. It can also assist in identifying gaps in organizational policies and standards.

A crucial task in program development is to identify the appropriate mix of assessment and verification so that the host facility's evaluation goals and objectives are fulfilled.

The overall MAMP process is shown in the Figure and incorporates the basic building blocks necessary to fully evaluate any occupational health department.

PRE-EVALUATION ACTIVITIES

A *preassessment (evaluation) questionnaire* for the facility presents some of the basic information that should be obtained and reviewed prior to the formal audit. After the pre-evaluation questionnaire has been reviewed, the assessment team should construct a proposed timetable for the audit and confirm the overall content and scope with the host facility. It is critical that assessment scope, schedule, and cost be documented before the on-site assessment begins, so that reasonable expectations can be established.

Often, audits run into trouble because of differences in expectations regarding scope and schedule. Thus the project timetable should be jointly developed, discussed, and agreed upon. This schedule should indicate the planned number of days, sequence of activities, and number of staff required to perform the agreed upon scope of the evaluation.

SITE ACTIVITIES

The general purpose of these assessments is to perform a broad review of the activities and services provided by the facility's occupational health department. Data are obtained based on the following sources:

1. individual interviews with health and safety staff
2. interviews with site management
3. tour of facility
4. observation of workers
5. clinic audits and interviews with local medical providers.

In order to preserve confidentiality and obtain candid opinions from staff, specific source attribution is not provided in the final report. A framework for data collection is shown in the Appendix to this chapter.

A key feature of any audit that must always be considered when results are reviewed is the *temporality* of the report. By temporality, we mean that the audit assesses performance of an organization at a moment in time. While all organizations are changing and evolving, the audit process freeze-frames an organization during the assessment process.

Therefore we develop a series of short declarative present tense statements known as *findings*. Specific findings and observations are time-specific and are based on direct observations during the audit. The review, analysis, and corrective measures (if any) are typically the responsibility of the audited department and are not part of the audit report.

EVALUATE FINDINGS

After all of the on-site information has been gathered, there is a multistep process that the audit team follows for evaluating results. This detailed effort is necessary to provide adequate quality assessment and quality control (QA/QC) for both the information-gathering process and reporting results. Since the host facility plans to rely upon the results of the assessment report, it is critical that the assessment team carefully document all findings with sufficient detail and precision, e.g. location, date, etc.

Unless otherwise requested by the host facility, the assessors typically will inform the appropriate staff of any significant findings and deficiencies as soon as they are identified. Such communication is an integral part of a smooth, effective evaluation, and should be done both by the individual team member who investigated the particular aspect of the evaluation and by the team as a whole during informal discussions with the department's staff. This is a *preliminary findings meeting* and is not a substitute for the final report to senior management.

Preliminary findings should be documented both verbally and during daily preliminary close-out meetings. There should be a final close-out meeting where findings are presented and checked for accuracy. A *final report* is usually issued based upon these verified findings.

SUMMARY

The evaluation function is a critical part of assessing the performance of an Occupational Health Department. The assessment should be carefully defined according to a mutually agreed-upon scope, schedule, and cost. A variety of standardized background and compliance questionnaires are available that can be further refined for a given site or situation.

APPENDIX—OCCUPATIONAL HEALTH PROGRAM ASSESSMENT

Overall Areas

- injury, illness, and accident reporting
- worksite hazards and surveillance information
- promotion and prevention oriented clinical assessment and intervention services
- education and training and related activities directed toward health maintenance and risk reduction
- occupational health administration services
- urgent care
- fitness for work
- hearing conservation.

Background Information and General Facility Information

- applicable federal, state, and local regulations

- applicable company and facility policies, procedures, and standards
- operating manuals, standing orders for nurses
- organizational charts, job descriptions, performance standards, training records
- off-site locations which are supervised or are under control of facility management.

Orientation Tour
Meet with facility management to:
- obtain a brief overview of facility operations, key responsibilities, physical layout, etc.
- tour the facility to gain a general understanding of the areas of employee health.

INJURY, ILLNESS, AND INCIDENT REPORTING
Understanding Management Systems
Develop an understanding as to how the facility manages its injury, illness and accident reporting program. Considerations may include:
- injury, illness, and incident reporting and record keeping procedures
- investigative and follow-up procedures
- training of personnel.

Facility Reporting Procedures
Evaluate the effectiveness of the facility's injury, illness and incident reporting system by performing the following:

Review the facility's dispensary log, incident/injury reports/payroll reports, or, if applicable, worker's compensation reports. Prepare a listing of employees who have experienced injuries or illnesses requiring treatment in excess of normal first aid and/or who can reasonably be expected to have experienced an incident resulting in lost time. Determine whether those incidents/injuries that meet recordability criteria have been recorded and reported appropriately.

WORKSITE HAZARDS AND SURVEILLANCE INFORMATION
Document how the facility assures that worksite hazards, which have been detected, are evaluated and monitored. This encompasses the use of a qualified medical professional to monitor worksite hazards, compliance with the use of personal protective equipment, and medical surveillance examinations. Considerations may include:

- Are industrial hygiene studies done appropriately for hazard recognition?
- Are occupational exposure surveillance exams done?
- Is the worksite surveyed regularly by medical personnel?
- Are studies done to assess the nature, scope, and significance of hazards or patterns of worker illness and/or injuries which may reflect unidentified hazards (e.g., epidemiology or ergonomic studies)?
- Are management and employees informed of the need for these studies, the results and recommendations?
- Are new materials and processes that are introduced at the worksite evaluated by health professions for health impact?

MEDICAL EXAMINATIONS
Develop a listing of the types of medical examinations available through the local medical provider. Conclude, based on discussions with on-site staff and medical personnel, that all appropriate medical examinations are available given the nature of the facility's operations.

These include general health assessment, such as periodic examinations. Occupational-related examinations such as:
- preplacement
- occupational-special function evaluations
- vehicle operators
- hazmat teams
- chamber work
- confined space
- food handler
- other.

Occupational-environmental surveillance exposure surveillance, such as:
- noise
- specific chemical hazards (e.g. asbestos, benzene)
- multiple chemicals workers.

Using employee relations department records, select a representative number of employees included in occupational health program based on risk assessment, job classification, specific exposure or task requiring medical evaluation. Determine that exposed employees selected take part in periodic medical evaluations appropriate to their exposures and job requirements.

Determine that the frequency and content of medical evaluations are consistent with governmental or company requirements. Confirm that employees have been notified of the results of their examinations.

EDUCATION AND TRAINING AND RELATED ACTIVITIES DIRECTED TOWARD RISK REDUCTION

Document understanding as to how the facility manages its education and awareness programs. These activities should reflect the needs of the employee population as determined by the findings of the worksite evaluations, epidemiological studies, demographic and health-risk analyses. Promotional activities could be made through local and community courses. What health educational and training opportunities are promoted at the facility? Consider, who, what, why, when, and how, with the following:

- health education programs
- health counseling
- training, including CPR, first aid, and general health issues
- supervision training, including troubled employee, absence management, avoiding on-the-job injuries
- health publications distribution
- personal health risk assessment
- fitness programs
- troubled employee
- crisis intervention
- assessment/referrals
- consultation and advice
- internal training and awareness programs.

OCCUPATIONAL HEALTH ADMINISTRATION SERVICES

Review the facility's written procedures, protocols, records, plans and operations. Confirm through inquiry and/or review that health services provided locally are appropriate, effective and efficient in meeting the needs of management, population served, and governmental regulations. Considerations may include who, why, how, where, and when.

What is the mechanism used to assure that the following health service providers are qualified to perform their duties?

- physicians
- nurses
- EAP counselors
- medical technicians
- x-ray
- laboratory
- audio
- pulmonary function
- CPR
- first aid.

How does the site evaluate providers' continuing education/training related to their professional field?

- personal training records (e.g., CPR)
- training courses at facility
- professional affiliations.

Determine through interview and procedure review how the quality of health programs and services is evaluated for appropriateness, effectiveness, and efficiency, both in-house and those services provided by outside contractors.

- laboratory services
- employee assistance program (EAP)
- clinics
- individual health provider
- other.

Are there internal medical assessment procedures in place?

Are policy and procedure manuals available that cover acute and routine care?

- illness/injury absence guidelines
- short-term disability (STD)
- long-term disability (LTD)
- medical service data collection
- medical emergency plans
- emergency alert list—facility
- emergency alert list—community.

Is there a procedure for the protection of health care workers to reduce the incidence of injury and disease? Consider the following:

- handwashing
- cleaning, disinfecting, and sterilizing of patient-care equipment
- microbiologic sampling
- infective waste
- housekeeping
- waste containers properly labeled methods for handling sharp glass/needles human body fluid clean up
- lifting techniques
- fire hazards
- prophylactic immunization (i.e., hepatitis B vaccine).

Document how the facility manages reproductive hazards. Considerations include:
- listing of reproductive hazards in the workplace
- informing employees of such hazards in the workplace
- engineering controls, work practices and personal protective equipment.

Document how the facility manages substance abuse and testing. Considerations include:
- signed consent obtained prior to collection
- review chain of custody procedure
- intent—drug free workplace
- who manages communications
- how is quality assured for laboratory services
- is a counseling program available to employees
- how are services monitored and program quality assured
- what is the degree and extent of rehabilitation (inpatient vs. outpatient care)

Are policies and procedures in place for any other matters related to administrative issues? Considerations include:
- orientation and training program for new department personnel
- procedure for transportation of ill and injured employees
- procedure for reporting of communicable diseases
- procedure for the administration of allergy and other injections
- procedures for immunization and travel
- procedure for deaths occurring at worksite
- policy on confidentiality for medical department personnel
- policy on medical record keeping
- nursing protocol procedures
- protocol for prescribing and/or dispensing of drugs
  - nonprescription
  - prescription
  - controlled drugs
  - pharmacy regulations
  - record of narcotics
  - labeling regulations
  - tracking expiration dates
  - unit dose regulations.

Evaluate the facility's procedures for response to health concerns that are addressed to the health and safety department, regarding chemicals on operations or products. Confirm through inquiry or review that the appropriate action is taken for:
- telephone complaints of chemicals in the air from neighbors
- requests for toxic information of products
- allegations of adverse health effects from products
- facility employees' request for hazardous effect of chemicals that they are working with.

For off-site medical facilities utilized, is there a management system for inspection, calibration, and maintenance of medical equipment and drugs. Consideration includes:
- defibrillator
- ECG
- compressed liquid gas cylinders (oxygen, carbon dioxide, etc.)
- showers
- eyewash stations
- pulmonary function spirometer
- audiometer
- hydroculator
- x-ray machines
- physical therapy equipment
- emergency medication trays
- ventilation assurance
- wheelchairs/stretchers
- sphygmomanometers
- slit lamp
- emergency phones and radios
- controlled medications
- prescription medications
- nonprescription medications
- medical laboratory equipment.

Is there a management system and is it operational for the following:
- standard for order of contents of medical records
- security system for records
- legal access procedure
- confidentiality protocol
- record retention time
- employee access rights procedure
- release of information procedure
- standardized format for entering information into a medical record
- procedure for records of temporary, terminated, and retired employees
- format for reporting of occupational injury/illness
- format for reporting nonemployee illness/injuries.

URGENT CARE

Obtain and document your understanding as to how the facility assures access to health care services for employees who experience injury or illness while at work or who otherwise are urgently in need of assistance.

First Aid

- Is there immediate care for medical emergencies that occur at work?
- Are there persons adequately trained to render first aid?
- Are first aid supplies available, which have been approved by a consulting physician?
- Has the site management developed and maintained emergency medical plans, and posted emergency instructions?
- Are records for training and recertification maintained?
- Are personal protective equipment such as pocket mask and disposable gloves readily available?
- Do the contents of the first aid kit agree with the content list?
- Are first aid kits available? (Not locked up, out of sight.)
- Is definitive treatment of occupational injuries or illnesses within the scope of the professional skill and facilities available (or augmented by resources external to site medical facility as required to meet the patient's needs)?

Emergency Response Plan

- Is there a multiple/mass/disaster plan?
- Is it a practical plan?
- Is it easy to understand?
- Will it deal with any type of emergency?
- Has the plan been updated periodically?
- Are there emergency drills?
- Are responses to drills reviewed to determine areas where improvement is needed?
- Are all shifts included in drills?
- Has plan management worked with community leaders to develop an appropriate response plan?
- Has the plan been distributed to all key personnel in the company, the switchboard operator and local police, fire, hospital and ambulance officials?
- Are emergency personnel listed with phone numbers and the list updated and distributed?
- Are local emergency response personnel brought into the site periodically for familiarization?
- Are primary and backup two-way communication systems developed and in place?
- Is there a written spill control plan, i.e., containment, neutralization, disposal?

Are local hospitals, physicians, and other medical/paramedical staff provided with information regarding:

- List of hazardous chemical and material safety data sheets Acute symptoms? Delayed symptoms? Bioassay tests?
- Special treatment required?

Also see chapter 18, Emergency Response Programs.

Medical Facility

Are facility and equipment appropriate to the extent that services provided in-house for the clinical prevention/promotion sufficient to allow emergency on-site treatment and/or stabilization of ill or injured employees prior to transfer? Is the facility located to provide convenient access for employee/patient and emergency personnel?

Are policy and procedure manuals available that cover acute and routine care to the extent of services which are provided?

Are there guidelines related to referral, off-hour sources of care, liaison with employees' private physicians provided?

Is emergency equipment present and provided in adequate quantities? Considerations include:

- oxygen and administration equipment to treat several casualties
- showers
- eye irrigation and treatment
- emergency drugs
- wheelchairs and stretchers
- sphygmomanometers and stethoscope
- emergency phones and radios
- ECG and defibrillator
- surgical dressing and other supplies
- biosafety
- soap and water
- labeled waste containers/receptacles
- labeled sharp container
- label and bags for human body fluids, mask, bag, gown, gloves.

Is emergency equipment checked, tested, and calibrated periodically for operational readiness?

- Are personnel trained and experienced in its use?
- Are potential problem areas and processes pointed out and discussed?

- Are primary and backup two-way communication systems developed and in place?
- Is a published protocol with the details of the basic steps needed to handle emergencies in the workplace readily available? These emergencies include accidental release of toxic gases, chemical spills, fires, explosions, and personal injury.
- Is procedure periodically updated? The emergency response plan should be comprehensive enough to deal with all types of emergencies.
- Is it comprehensive? (Does it deal with all types of emergencies?)

Also see chapter 18, Emergency Response Programs.

Communications

Through interviews, determine if employees know how to report emergencies such as use of manual pull-box alarms, public-address systems, telephones. Emergency phone numbers should be posted on or near telephones, on employees' notice boards or other conspicuous locations. The warning plan should be in writing and management must be sure each employee knows what it means and what action is to be taken. An updated written list of key people such as plant manager and physician should be posted. When a medical facility is on plant site, a complete alert call-in list should be available.

Are there records of training of medical department personnel in specialized proceduces (i.e., cardiopulmonary resuscitation) available?

Are personal protective equipment (and personnel) available to medical personnel for incident response, such as:
- safety glasses
- goggles or face shield
- hard hats
- safety shoes
- proper respirator
- gloves
- hood
- boots
- rubber coats, etc.

Are books, lists, or other references on hazardous materials available to medical personnel? Also see chapter 18, Emergency Response Programs.

FITNESS FOR WORK

Document how the facility assures that provision of services are available to evaluate and assist employees in returning to and remaining effective in their jobs following an episode of illness or injury, or other event that has altered their performance capacity. Considerations include the following:

- Policies and procedures related to rehabilitative and follow-up care should be in place, current, and in evident use. These policies and procedures should reflect the extent and complexity of in-house services, and demonstrate a recognition of management's personnel policies and the program's role related to: employee performance and physical/mental capacity to meet the requirements of their job; the need for sound medical judgement regarding rehabilitative and recuperative progress; the need for consultation with management regarding employee needs and capabilities in the work setting; and liaison with the employee's private physician.
- When services are carried out on contract or otherwise off the premises of the program, the relationship of those services to the program, should be specified in writing. This encompasses:
  - dressing changes
  - blood pressure checks
  - work-related rehabilitation assistance
  - counseling related to above
  - disability evaluations
  - work restrictions.

Develop an understanding of the medical facility's management controls through completion of this questionnaire and discussions with medical department personnel.

Who at the facility is responsible for development, implementation, and administration of programs for compliance with applicable government and company requirements for each of the following safety and health issues:

- administration control programs (i.e., medical data collection, record retention)
- injury and illness record keeping
- personal protective equipment for Medical Department personnel
- emergency response plan
- hazard communications
- industrial hygiene
- education and training (i.e. CPR, First Aid)
- calibrating, testing, and maintenance of medical equipment
- hazard communication
- hearing testing
- medical surveillance evaluations and tracking
- drug abuse program.

Determine how the following are handled at the site (Medical Emergency Response):
- first aid
- eye baths and showers.

How does the site handle first aid situations at the worksite?
- transportation to site medical facility with additional first aid/medical treatment
- stabilization and transportation to physician or medical facility
- notification of family, department, private physician, hospital, etc.
- notification of family, department, private physician, hospital, etc.
- definitive intervention or referral for treatment
- evaluate and appropriately assist worker in returning to and/or remaining effective in his/her job
- policies related to rehabilitative and follow-up care
- medical emergency communication system (telephones with posted emergency numbers)
- off-hours sources of care
- medical internal assessment
- detailed worksheet for occupational injury/illness record keeping—U.S. regulations
  - objective—To verify compliance of record keeping of Workers' Compensation and OSHA regulations
  - population—Select a representative sample of reportable occupational injuries, illnesses, and lost workday cases

Documents to examine:
- payroll report of occupational injury/illness with lost work days. (NOTE. *Use this as part of your worksheet.*)
- site OSHA log.
- site workers' compensation report
- site official medical report. This may be the individual medical record or OSHA supplementary 101 record.
- Conduct sampling by using the payroll report of occupational injury/illness with lost work days and the OSHA 200 log
- Does the number of lost work days in each document agree? Note any deficiencies
- Now use the site official medical record and the OSHA 200 log
- Does information in OSHA 200 log, columns A, B, C, D, E, and F agree with medical record?

(Note: *Use decoding sheet to find nature, part of body and cause in column F. Note any deficiencies.*)
- Are OSHA reportable cases reported correctly?
- Are work restrictions noted in medical record reported on OSHA 200 log?
- Were reportable OSHA cases entered on OSHA 200 log within six workdays?
- Are OSHA reportable cases reported correctly as injury or illness?
- Were cases with a change in status shown on medical record updated on OSHA 200 log?
- Using the same medical record, determine if the state's workers' compensation regulations are being followed.

HEARING CONSERVATION

The following are selected requirements for hearing conservation. Refer to *CFR* 1910.95 for additional requirements.

Employers establish and maintain an audiometric testing program for all employees whose exposure is equal to or exceeding the action level. (29 *CFR* 1910.95 (g)(1))

Within six months of employee's first exposure at or above action level—a valid baseline audiogram is established. (29 *CFR* 1910.95(g)(5))

Annual audiograms. At least annually obtaining the baseline audiogram, the employer obtains a new audiogram annually for each employee at or above the action level. (29 *CFR* 1910.95 (g)(6))

If Standard Threshold Shift (STS) is indicated, employee is informed in writing within 21 days of the determination. (29 *CFR* 1910.95(g)(8))

Employees not using hearing protection shall:
- be fitted with hearing protection
- trained in use and care (29 *CFR* 1910.95(g)(8)).

Employees using hearing protection shall:
- be refitted
- retrained
- Provided with hearing protection offering greater attenuation if necessary (29 *CFR* 1910.95 (g)(8)).

If subsequent audiometric testing indicates that STS is not persistent, employers:
- inform employees of new audiometric interpretation
- may discontinue required use of hearing protection (29 *CFR* 1910.95(g)(8)).

Audiometric test requirements are pure tone, air conduction, hearing threshold examinations, with test frequencies, including as a minimum 500, 1000, 2000, 3000, 4000, and 6000 HZ, taken separately for each year. (29 *CFR* 1910.95(h))

Functional operation of audiometer checked before each day of use. (29 *CFR* 1910.95(h)(5))

Audiometer calibration checked acoustically at least annually. (29 *CFR* 1910.95(h)(5))

Exhaustive calibration performed at least every two years. (29 *CFR* 1910.95(h)(5))

Employer institutes a training program for all employees exposed to an 8-hour TWA of 85 dBA (ensure participation). (29 *CFR* 1910.95(k)(1))

Training is repeated annually. (29 *CFR* 1910.95(k)(2))

Employer ensures that each employee is informed of (29 *CFR* 1910.95(k)(3)):
- effects of noise on hearing
- purpose of hearing protection, advantages, disadvantages, various types of protection, fit, use and care.

Audiometric test records are retained for at least the duration of employee's employment. (29 *CFR* 1910.95(m)(3))

Detailed Worksheet for Medical Examinations and Environmental Exposure Surveillance Record Keeping U.S. Regulations

Objective—To verify compliance of record keeping with corporate standards, site standards, and OSHA regulations.

Population—Select a random or representative sample of employees from department records, job classifications and/or exposure areas.

Documents to examine:
- individual medical records
- special function evaluations and environmental exposure surveillance protocols
- site recall system (i.e., Health Appraisal Recall System (HARS)).

Conduct sampling by using the individual medical record and site recall system.

Does medical record show that employee took part in appropriate examination? Were employees advised of medical recommendations? Are records documented as to those employees who decline examinations? Are follow-up and referrals documented? Is there a baseline examination? Now, using the same medical record, review to see if appropriate examinations were done as indicated on the special function protocol. Note any exceptions.

Part 4

Human Resources Issues

chapter 21

Preplacement Testing

by Thomas Herington, MD
revised by Gary R. Krieger, MD, MPH, DABT
Marci Z. Balge, RN, MSN, COHN-S

INTRODUCTION

The medical screening of workers occurs in a variety of contexts and includes the following types of examinations:

- preplacement evaluations
- periodic evaluations (annual examinations)
- Department of Transportation (DOT) examinations
- Federal Aviation Administration (FAA) examinations
- return-to-work or fitness-for-duty evaluations
- medical surveillance examinations
- impairment (disability) evaluations
- executive physicals.

This chapter will discuss the role of the medical evaluation, its benefits, and limitations. It explores medical, legal, ethical, and administrative concerns as well as the mechanics of the examination itself.

THE ADA AND PREPLACEMENT EXAMINATIONS

Passed in 1990, the Americans with Disabilities Act (ADA) prohibits companies from giving job applicants any medical examination, until an offer of employment has been made. Once an offer of employment has been made, a job candidate can be given a preplacement evaluation. The preplacement examination can also be administered to any current employee being transferred to a new position.

The goal of a preplacement examination is to determine an employee's medical fitness for a given

---

- 381 Introduction
- 381 The ADA and Preplacement Examinations
- 382 Purpose of Preplacement Evaluations
- 382 Job Descriptions
  Contents of the job description ■ Preparing the job description
- 383 Occupational History
- 384 Physical Examination
  Testing techniques ■ Laboratory studies
- 386 Drug Screening
- 387 The Examiner
- 388 Administrative Issues
- 388 Legal and Ethical Issues
  The Americans with Disabilities Act ■ Confidentiality
- 389 Future
- 389 Summary
- 390 References

job activity. However, medical fitness is only one of many factors used to determine employability. The ADA underscores use of a medical examination for the purpose of placement not employability.

PURPOSE OF PREPLACEMENT EVALUATIONS

It is useful to be clear about the purpose of a preplacement evaluation so that the company can establish realistic expectations regarding the quality and the appropriate use of information obtained from such an examination. Some employers have perceived the preplacement evaluation as an opportunity to screen for the good employee, to select the super worker—or at least the one who will stay healthy and be at low risk for job-related injuries. But no one can predict the future with certainty, and the purpose of a preplacement evaluation is not to do so. The preplacement evaluation is an attempt to define a level of risk in a given applicant for a given job. Rothstein (1984) uses the term *selection screening* to define this process of selecting and maintaining a work force by applying medical criteria.

The goal of preplacement evaluation is not to discover hidden disease so that appropriate treatment can be initiated. The examination may uncover previously known or unknown problems (which must, of course, be managed appropriately), but the primary purpose of the examination is not to screen for medical problems in order to initiate therapy. This fact distinguishes the preplacement evaluation from the traditional physical examination and highlights the singular relationship between the examiner and the applicant in this context. Most often the examiner is not the patient's usual family physician or medical practitioner. He or she may, in fact, never see the patient again after the preplacement evaluation.

The evaluation emphasizes preventing the onset of a medical problem by identifying risk in an individual. In this way, the company can eliminate the cause of a potential problem through proper job placement, or job modification. The examiner must decide whether a worker can, from a medical standpoint, safely perform a job assignment without significant hazard to self or to others. The examiner must document worker fitness and ascertain degree of risk associated with a given job assignment so that management can make placement decisions. Hence, a preplacement evaluation has a dual purpose:

1. To document the current and past medical status of an applicant, providing a baseline against which to measure the effect of future hazardous exposure(s).
2. To determine current medical capability with or without restrictions for a proposed job.

JOB DESCRIPTIONS

As noted, occupational health practitioners are asked to judge an applicant's fitness for a given job and the potential risk of placing the employee in that job. The accuracy of this judgment depends, in part, on the examiner's understanding of the physical and emotional demands of the position and of attendant environmental risks in the workplace. Therefore, a preplacement evaluation begins with an accurate job description or job function analysis (JFA). A JFA is a more intensive review of a position in that it can include specific ergonomic analyses and measurements.

Do not assume that examiner and employer have the same understanding of a given job from its title. An office worker can sit all day or stand all day; can use a keyboard seven hours per shift or one hour per shift; can have a sedentary job or be required to lift and carry heavy file boxes regularly. Further, the examiner should not rely on the applicant, who has not yet worked in the new position, for an accurate chronicle of job tasks.

Ideally, the employer should give the examiner a detailed job analysis that identifies specific physical requirements. This does not mean a list of job responsibilities from a policies and procedures manual. It is not as helpful to know that one must type all correspondence, schedule appointments, and key financial data into the computer as it is to know that one must sit four hours without a break, must stand at a reception window six hours per shift, must lift 60 lb frequently, or must face temperature extremes.

Contents of the Job Description

In general, a job description for this purpose should include the information listed in Table 21-A. Directly measuring these demands is helpful but may not be practical because of cost and time considerations. For example, continuous ECG monitoring to determine stress and metabolic requirements, videotaping with subsequent ergonomic analysis, or environmental monitoring of heat, humidity, or noise levels can refine the

Table 21-A. Typical Job Description.

1. Job title and list of job tasks
2. Schedule of hours, rotating or split shifts, and overtime
3. Physical demands of job duties
   - Lifting
   - Carrying
   - Pushing
   - Pulling
   - Climbing
   - Bending
   - Crawling
   - Reaching
   - Handling
   - Fingering
   - Seeing
   - Standing
   - Sitting
   - Walking
   - Balance
4. Working conditions
   - Indoors
   - Noise
   - Altitude
   - Stress
   - Substance exposure
   - Dust
   - Chemicals
   - Fumes
   - Gases
   - Repetitive activity
   - Temperature extremes
   - Driving
   - Use of machinery
   - Protective equipment
   - Mask
   - Safety glasses
   - Respirator
   - Special clothing
5. Special equipment used
   - Tools
   - Hand-held
   - Weight
   - Vibration
   - Electric
   - Magnetic
   - Ergonomic
6. Infrequent activities
7. Faculties needed
   - Binocular vision
   - Color vision
   - Hearing acuity
   - Tactile skills

value of a job analysis to the examiner, but must be weighed against other factors such as degree of risk and cost versus benefit. At the least, the employer might indicate whether these activities are performed rarely (0%–25% of the time), moderately (25%–75%), or frequently (75%–100%). In general, the more risk inherent in a job, the more detail there should be in a job description.

A final and important part of a thorough job analysis involves determining which listed job tasks are considered to be essential. Subjective issues of function, productivity, and safety are central to the concept of essential requirements. Essential tasks are usually defined by degree of importance, by frequency of performance, and by the consequences of error. In the future, essential job tasks likely will be defined by quantitative and statistical rating scales, rather than by the commonsense, subjective means currently used. Nonetheless, the employer must make an honest attempt to determine which tasks, for any given job assignment, are considered essential.

Preparing the Job Description

Safety personnel, industrial hygienists, and occupational therapists are well equipped to help write job descriptions. However, the appropriate department manager is likely to have the most intimate understanding of the requirements of a particular job.

These people should complete job analyses for every position that requires a preplacement evaluation. A detailed and accurate job analysis will greatly help the examiner ascertain job fit.

OCCUPATIONAL HISTORY

Without question, the most important part of a preplacement evaluation is the occupational history. By answering a questionnaire, the applicant should provide this history before taking the physical examination. The medical examiner can identify any pertinent positive responses and detail them directly with the subject.

The questionnaire should ask for a standard medical history, including present complaints, medical history, family medical history, social history, and organ system review. The occupational history also can include the special information outlined in Table 21-B.

Special questionnaires can be used to evaluate positive responses on the general occupational/medical history or to help answer specific questions such as an employee's ability to lift or to wear a respirator. Table 21-C presents a sample back questionnaire. You can modify the information in these tables according to your specific needs. The revised OSHA Respiratory Protection Standard now requires a specific questionnaire for workers who will wear respirators. This questionnaire can be found in 29 CFR 1910.134.

Table 21-B. Special Information for an Occupational History.

Has a job offer been made to you? ❏ Yes ❏ No
1. Current job
   - Job title
   - Job description
   - Name/address/phone of employer
   - How long employed
   - Other jobs held in this company/dates
   - Known exposure to
     - Noise
     - Temperature extremes
     - Vibration
     - Chemical exposure
     - Dust
     - Gases, fumes, vapors
     - Radiation
   - Describe effects of above exposure(s)
   - Describe any protective equipment used
2. Past work history
   - List jobs, going back from current job, including
     - Years (from _____ to _____) in each job
     - Description of job tasks
     - Exposures
     - Adverse effects
     - Protective equipment used
     - Previous medical examination or treatment relating to job, including name/address of physician/clinic
3. Other jobs
   - List vacation jobs, part-time jobs, second jobs, including
     - Dates
     - Description of job tasks
     - Exposures
     - Adverse effects
     - Protective equipment used
     - Previous medical examination or treatment relating to job, including name/address of physician/clinic
   - List military service
     - Job title and description of tasks
     - Dates
     - Exposures
     - Adverse effects
     - Protective equipment used
     - Previous medical examination or treatment relating to military service, including name/address of physician/clinic
   - Work taken home or done at home (e.g., computer work)
     - Describe job tasks, dates, time spent, exposures, adverse effects
4. Other hazardous exposure(s)
   - Household member(s) who may bring home hazardous materials (chemicals, lead, heavy metals, asbestos, etc.) on clothing, in hair, etc.
   - Residence near facility (chemical plant, refining, shipyard, smelter, etc.)
   - Hobbies (you or household member) with exposure to hazardous material (ceramics, solvents, glue, lead, etc.)
   - Smoking history
   - Use of drugs, alcohol, caffeine
   - Use of pesticides at home
5. Have you ever changed your occupation for medical reasons?
6. Have you ever had any ill effects from work not listed above?
7. Have you lost time from work for more than a few days due to illness during the past three years? If yes, give approximate number of days and state reasons.
8. Has your work ever been limited because of your health? Explain.
9. Do you have any condition that might require a special assignment?
10. Describe any exposure to loud noise at work.
11. Have you ever worked around the following?
    ❏ Chemicals ❏ Sprays ❏ Gases
    ❏ Solvents ❏ Pesticides ❏ Asbestos
    ❏ Cytotoxins ❏ Other
12. Have you worked around metals such as:
    ❏ Lead ❏ Arsenic ❏ Zinc
    ❏ Cadmium ❏ Beryllium ❏ Magnesium
    ❏ Other
13. Have you ever been employed in the following:
    ❏ Dusty trades ❏ Mining ❏ Foundries
    ❏ Refractories ❏ Sandblasting
    ❏ Asbestos ❏ Manufacturing

You can also refine the general medical history to provide additional information as warranted for specific jobs. For example, health care workers can provide an immunization record as outlined in Table 21-D.

PHYSICAL EXAMINATION

The physical examination is the responsibility of the health care practitioner. Details regarding the examination itself appear in physical diagnosis texts. The examination is a notable opportunity for the physician or other occupational practitioner to act as detective. A three-minute examination that reveals clear lungs, a regular heart rate, and no abdominal mass or hernia can tell you that a subject is alive and little else. However, the occupational history of past jobs or exposures as well as anticipated hazards of the new position should trigger a focused inspection, during which the medical practitioner should record pertinent negatives. For example, welders should be checked for cataracts and corneal scarring. Agricultural workers should be screened for pulmonary disease (farmer's lung) or effects of pesticide exposure. While the examination can, in fact, be brief, it must be performed with solicitude and thoughtful attention to purpose.

Table 21-C. Back Questionnaire.

Please use the back of this form to describe any item more fully.

1. Have you ever had any back/neck discomfort? Describe and give date or frequency
2. What was the site of discomfort?
   - ❏ Neck
   - ❏ Mid-back/Ribs
   - ❏ Arm(s)
   - ❏ Upper back/Shoulder
   - ❏ Low back
   - ❏ Legs
   - ❏ Other
3. How did you develop this discomfort?
4. Did you see a health care professional?
   - ❏ Physician (Give name, specialty)
   - ❏ Chiropractor
   - ❏ Acupuncturist
   - ❏ Holistic practitioner
   - ❏ Other (Describe)
5. Have you ever had any of the following tests? Provide date and location.
   - ❏ X rays
   - ❏ MRI scan
   - ❏ CT (CAT) scan
   - ❏ Bone scan
   - ❏ Myelogram
   - ❏ EMG (Nerve conduction test)
6. How was the discomfort treated?
   - ❏ Medication
   - ❏ Ice/heat
   - ❏ Bed rest
   - ❏ Brace/support
   - ❏ Proper body mechanics
   - ❏ Traction
   - ❏ Physical therapy
   - ❏ Surgery
   - ❏ Injection
   - ❏ Exercise
   - ❏ Biofeedback
   - ❏ Massage
   - ❏ Other
7. Does the discomfort return? How often? What sets it off?
8. How do you make it better?
9. How long does it take to get the discomfort under control?
10. Did you need time off work because of the injury? When and how long?
11. Do you have any limitation of muscle strength or joint range of motion due to any old injury or illness? Describe.
12. Do you ever suffer from any loss of balance or coordination? Describe.
13. Do you have symptoms now? What and where?
14. Have you ever attended body mechanics classes? If yes, when and how long?
15. Have you ever had body mechanics training/practice with an instructor giving you feedback?
16. Have you ever had any other bone, muscle, or joint problems? Describe.
17. Describe your current exercise program. How often per week? How long per session? How long continuously involved in exercise (months/years)?

Testing Techniques

At times, the examiner may need to use special examination techniques or studies to determine level of risk or medical status. Certain high-risk jobs can call for adjunctive functional capacity testing. For example, a job with extreme musculoskeletal demands, such as frequent heavy lifting, can account for a high rate of worker injuries. Screening for this type of job can necessitate more detailed objective testing than can be obtained in a routine physical examination.

A wide variety of equipment and testing techniques are available. These should be selected on the basis of individual need and with the cooperation, guidance, and participation of the medical officer. Factors to weigh include reliability of the test, reproducibility of results, skill and experience of the person administering the test, cost, and extent to which the test actually assesses the demands of the job (Figures 21-1 and 21-2). Tests can identify the rare person who, at the time of the examination, cannot lift the required weight or perform the required bending or reaching. However, such testing cannot guarantee that those who pass these tests will avoid injury. Remember that no test can predict the future. An applicant who walks or climbs normally today can still sprain an ankle tomorrow. In short, any lift test or functional capacity test provides, at best, a snapshot of limited data. Maintain realistic expectations when applying these results to placement decisions. The ADA also requires that accommodation be considered when testing a disabled person.

Data from such testing can be useful to the examiner who must make a refined recommendation regarding job modification. Further discussion of functional capacity testing appears in chapter 16, Ergonomics Programs.

Table 21-D. Immunization Record.

Please provide dates:

Tetanus _____   Rubella _____

Hepatitis B vaccination _____

Date of last TB skin test _____
   Positive _____   Negative _____

Have you taken INH?   ____ Yes ____ No

Have you ever received BCG?   ____ Yes ____ No

Other vaccinations received and dates _____

Other situations also call for special evaluation techniques. For example, OSHA regulations may require medical clearance that determines a worker's ability to wear a respirator. Such an examination will emphasize the cardiopulmonary system. While not routinely required, chest x-ray films and/or pulmonary function tests can be indicated for some, if not most, candidates (Hodous, 1986). Other examples of specialized examinations include those for workers exposed to toxic materials such as asbestos and lead. For these workers, radiographs or blood work are required or recommended as outlined in the federal OSHA regulations.

Laboratory Studies

The National Institute for Occupational Safety and Health (NIOSH) and the Occupational Safety and Health Administration (OSHA) recommend screening tests for workers exposed to many of 386 chemicals listed in the 1974 Standards Completion Project.

Consider routinely obtaining some combination of a baseline complete blood count (CBC), chemistry panel (SMA), and urinalysis, or leave such testing to the discretion of the examiner based on the candidate's age, general state of health, and details elicited in the medical or occupational history. Again, special circumstances dictate special requirements. For example, medical personnel and hospital workers, as opposed to ironworkers, may be required to undergo rubella testing or tuberculin testing before receiving a medical clearance.

Routine screening back x-ray films are to be condemned. Research supports the view that such x-ray films are neither cost-effective nor predictive of future injury. They have absolutely no role as a routine part of a general screening medical examination. They are useful only where warranted by specific clinical indications (e.g., positive history, current symptoms, positive findings on physical examination).

DRUG SCREENING

While a complete discussion of preplacement drug screening is beyond the scope of this chapter (it is covered in chapter 23, Substance Abuse), a few salient points should be mentioned. These examinations, may be required by regulation in settings where a worker under the influence can do harm to self or to others, particularly in settings where pub-

Figure 21-1. Many types of equipment and testing techniques are available, such as this system, which tests progressive lift from floor to overhead.

lic safety is involved, as with bus drivers, airline pilots, and other transportation workers.

Before the company initiates a drug-screening program, representatives from management (personnel/safety departments, corporate lawyers, managers), from the work force (union leaders, lawyers, worker representatives), and from the medical department (clinic or private physician, medical office manager) should meet and agree on terms and methods of testing and follow-up. Policies and procedures should be in place so that everyone understands the conditions for obtaining specimens, what constitutes a positive result, under what conditions a test might be repeated or analyzed by an alternative technique, what happens if an applicant refuses the test, who is responsible for informing the applicant of test results, who gets the test result, where the information is kept, and so on.

A specific medical review officer (MRO) should be designated. This individual should have completed an MRO course specifically designed for

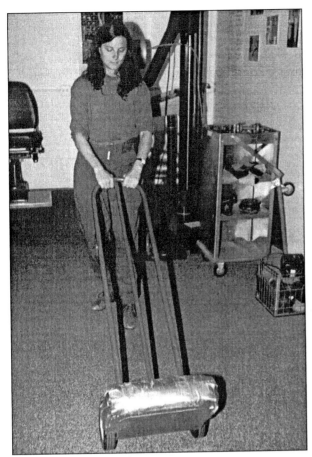

Figure 21-2. The type of test given should be selected on the basis of need and with the assistance of the medical officer. Here, a tape-wrapped cement bag is used to assess push-pull.

drug testing issues in the workplace. The American College of Occupational and Environmental Medicine (ACOEM) offers these courses on a regular basis.

Once a program is established, the facility obtaining the specimens must strictly adhere to a forensic chain of custody and must use a reputable, reliable, quality laboratory for speedy analysis of the specimens.

The ethical and legal questions surrounding drug testing are numerous and complicated. Detailed discussions regarding the legalities of drug screening, procedures for specimen collection, and the role of the medical review officer are available from numerous sources.

Remember that the context for drug screening ought to be health and safety, not crime and punishment. Therefore, any company that implements a drug-screening policy should have an established employee assistance program (EAP) referral mechanism available for those identified as substance abusers. Company referrals to the EAP are usually reserved for existing employees. Applicants who have not yet been hired should be referred to an appropriate outside resource as warranted. In either case, it is valid to identify a problem so that a solution can be proposed.

THE EXAMINER

Despite attempts to obtain objective information through an occupational history, a physical examination, and adjunctive lab or functional capacity testing, the medical decision regarding placement often reduces to a subjective judgment. The issues can be complicated, and there may be no simple correct answer.

Such decisions can be frustrating for the examiner, who may or may not have training in occupational medicine and who is asked to make a complicated decision in the context of a brief, low-bid examination. Additionally, companies may apply the pressure of a time deadline, since the much-needed employee is to start work tomorrow (or yesterday). There is often little interest in more detailed and expensive testing or in obtaining medical records. Further, the examiner may face an applicant whose opening remark pleads the need for work in order to support a family. Yet the medical decision cannot be taken lightly, since serious legal and ethical implications accompany any assertion that a candidate may or may not be placed in a job.

Companies without an internal occupational health department are advised, therefore, to use thought and care in selecting a medical examiner and to encourage second opinions as indicated. Consider the examiner's experience or training in occupational medicine; the extent to which his or her goals, expectations, and ethics match those of the company; and the examiner's willingness to communicate an honest medical opinion. Taking the cheapest bid for preplacement examinations may well get you what you pay for.

In many respects, the examiner represents the company to the applicant. The practitioner-patient interaction during an examination is quality time that can be used for health promotion. This is a time to discuss smoking cessation, weight loss, exercise, and workplace safety. Such instruction also sends the prospective employee a message regarding the company's concern for health and safety. The preplacement examination can be the first company benefit received by the new employee.

ADMINISTRATIVE ISSUES

The best intentions can be subverted by faulty logistics. Identify any issues concerning custody of medical records and lines of communication, and resolve them with the medical officer or the employer. Strict confidentiality must be maintained and will be compromised if completed examination forms are mailed to the personnel office, opened by unauthorized persons or left on a desk where passersby can inadvertently learn that an applicant had a biopsy for cancer or had venereal disease.

Ideally, the examining physician should keep the medical records and send the company a simple advice form. Any records returned to the company must be kept in a locked file, with limited access, and separate from the personnel department or personnel file. The physician or hiring officer need only report a conclusion with some variation of the following information:

- medical clearance for placement without restriction
- medical clearance for placement with the following restrictions: (list)
  - not medically fit for placement in the particular position
  - medical hold pending further evaluation.

LEGAL AND ETHICAL ISSUES

The legal issues surrounding preplacement testing touch on some of the ethical concerns of employment selection procedures. Title VII of the Civil Rights Act of 1964 prohibits employment discrimination based on race, color, religion, sex, or national origin. It applies to state and federal government as well as to private companies with 15 or more employees. The Americans with Disabilities Act of 1990, Rehabilitation Act of 1973, and the Vietnam Era Veterans Readjustment Assistance Act of 1974 seek to protect all disabled persons from discriminatory hiring practices (Hogan, 1987). The spirit of these acts is to provide the guarantee of equal opportunity to the disabled. For more information, see chapter 27, Workers with Disabilities and the ADA.

The 1978 Uniform Guidelines for Employment Selection Procedures assert that an applicant need not be required to perform all the tasks of a job but should be assessed on the ability to perform only job tasks defined as essential. Thus, as noted earlier, it is important to determine essential tasks in the job analysis. The thrust is to discourage employment discrimination in the form of screening tests that have a prejudicial impact on a particular class or group of persons. A variety of state and local laws also seek to address the issue of discrimination in employment.

The Americans with Disabilities Act

The Americans with Disabilities Act (ADA) of 1990 is sweeping federal legislation that addresses civil rights in the areas of employment, public services, public accommodations, transportation, telecommunications, and state and local government services. Title I details issues of employment and includes directives regarding employee medical examinations and inquiries. In most ways, the ADA does not significantly change what has been the good practice of occupational medicine.

In general, employers are prohibited from asking about the presence, nature, or severity of an applicant's disability. An employer can only ask about the applicant's ability to perform job-related functions. Preplacement examinations referred to in the ADA as *Post Offer Pre Employment Examinations* (a good argument for eschewing acronyms) can be conducted only after the applicant has been offered employment. This offer can be conditional on the results of the examination, but all entering employees in similar positions must be subject to medical examination. Individual applicants may not be sent selectively for medical examination with the notion of withdrawing an offer of employment if medical clearance is not obtained. It might therefore be prudent for the medical questionnaire to include an inquiry regarding whether or not the examinee has been offered a job. A negative response should caution the examiner about proceeding with the medical examination.

Employers can also conduct medical examinations on current employees. These examinations can be required, that is, involuntary, if they are job-related and consonant with business necessity. Of course, employees can be offered voluntary examinations as an aspect of health promotion. In either case, medical examinations can include any appropriate medical question.

Only job-related findings, however, can be considered in worker placement.

The ADA includes in the definition of discrimination the use of qualification standards, employment tests, or other selection criteria that tend to screen out persons with disabilities, unless such tests can be shown to be job-related for the position in question and consistent with business necessity. This statement emphasizes the need for the examiner to remain ethical and objective. The appropriate use of adjunctive testing, such as functional capacity testing, as well as the medical examination itself, is to make recommendations regarding placement or job accommodation. Such testing may not be used to discriminately disqualify the disabled.

The ADA protects qualified persons with a disability who with or without accommodation, are able to perform the essential functions of the job from discrimination. An employer is required to consider reasonable accommodation to qualified applicants or employees with disabilities. This reasonable accommodation can include modified work schedules; job restructuring; adjustment of equipment; modification of job tasks or work assignments; provision of assistants, readers, or interpreters; or actual reconstruction of the physical environment. An employer can be excepted from such accommodation if it creates undue hardship in terms of cost and difficulty. However, the law is strict and clear in its intent to provide employment opportunity for all persons with disabilities.

Confidentiality

As noted earlier, preplacement testing can contain sensitive information. For example, this can concern drug or alcohol use, AIDS antibody status, and personal medical or psychiatric problems. For this reason, companies should not have full medical records. A medical consultant is useful only if employees trust him or her. Lack of confidentiality quickly undermines excellence. Furthermore, in this context, confidentiality is more than a nice idea since there is legal liability around inappropriate disclosure of such information.

The ADA underscores the confidentiality of medical records. They must be maintained on separate forms and in separate medical files, with three exceptions:

1. Managers can be informed of a person's work restrictions or required job accommodations.
2. Safety or first aid personnel can be notified of a condition or disability that might require emergency care.
3. Government officials investigating compliance with the ADA can be provided relevant information as requested.

Conversely, results of the medical examination or lab testing should never be withheld from the applicant. Such information is paid for but not owned by the employer. The candidate must be aware of any medical problems or concerns uncovered during the evaluation.

FUTURE

Since the purpose of preplacement testing is to identify the high-risk person, there is great interest in research that aims to identify people who have a genetic susceptibility or predisposition to the toxic effects of chemical or physical agents, because of different abilities to metabolize and detoxify chemical exposures. *Ecogenetics* refers to the importance of genetic variation in the response to environmental agents in general. Occupational ecogenetics focuses on genetic or biochemical indicators used to predict a worker's susceptibility to chemicals or other industrial pollutants (Mulvihill, 1986). For example, persons with thalassemia or ALA dehydratase deficiency may have an increased risk of toxicity with exposure to lead (Calabrese, 1986). An entire area of molecular epidemiology has developed during the 1990s.

Unfortunately, life is never that simple. Genetic factors are likely to be just one of many factors (age, life-style, presence of other diseases, etc.) that affect susceptibility to environmental agents. Research into genetic or biochemical indicators of high risk continues. More than 50 genetic diseases are now identified as having risk consequences in the workplace. The legal and ethical issues will need to be addressed as the science of genetic testing expands (Nelkin & Tancredi, 1989; Bishop & Waldholz, 1990).

SUMMARY

Preplacement testing presents both an opportunity and a challenge. If the evaluations are approached with realistic expectations, are thoughtfully performed, and are targeted to the job, they can serve both employer and employee as an important step toward workplace health and safety.

REFERENCES

Calabrese EJ. Ecogenetics: Historical foundation and current status. *J Occ Med* 28(10):1096–1102, 1986.

Hans M. Pre-employment physicals and the ADA. *Safety & Health* 145:61–62, 1992.

Himmelstein JS. *Worker Fitness and Risk Evaluations.* State of the Art Reviews: Occupational Medicine Series. Philadelphia: Hanley & Belfus, 3(2): April–June 1988.

Hodous TK. Screening prospective workers for the ability to use respirators. *J Occ Med* 28(10):1074–1080, 1986.

Hogan JC. Developing job-related preplacement medical examinations. *J Occ Med* 23(7):469–476, 1987.

Mulvihill JJ. Occupational ecogenetics: Gene-environment interactions in the workplace. *J Occ Med* 28(10):1093–1095, 1986.

Nelkin D, Tancredi L. *Dangerous Diagnostics: The Social Power of Biological Information.* New York: Basic Books, 1989.

Ratcliffe JM. The prevalence of screening in industry: Report from the National Institute for Occupational Safety and Health National Occupational Hazard Survey. *J Occ Med* 28(10):906–912, 1986.

Rom WN. *Environmental and Occupational Medicine.* Boston: Little, Brown, 1983.

Rothstein MA. *Medical Screening of Workers.* Washington DC: Bureau of National Affairs, 1998.

chapter 22

Stress Management

by Cynthia D. Scott, MPH, PhD
Dennis T. Jaffe, PhD
revised by Marci Z. Balge, RN, MSN, COHN-S

391 **Introduction**
392 **Stress in the Workplace**
   Stress-prone work ■ Workers' perception of stress
395 **Effects of Stress**
   Stress claims, losses, and disability ■ Burnout
   ■ Personal responses
398 **Programs for Responding to Stress**
   Personal approaches to stress management
   ■ Organizational approaches to stress management
   ■ Implementing a comprehensive stress-management program
405 **Goals for the Future**
406 **Summary**
406 **References**

INTRODUCTION

We live in an era of unparalleled change. In the workplace, people are facing increasing demands and rapidly evolving means of communication and connectivity; upheaval caused by corporate reengineering and mergers and acquisitions activity; and pressure for continual innovation and adaptation. As a result of the pressures, changes, and demands of modern life, stress is increasing (Byers, 1987; Pelletier, 1977). Everyone experiences this problem, yet it seems so pervasive and global that many people believe they are unable to do much to overcome its negative effects.

Stress is the combination of adverse emotional and physical reactions people have to stressors (pressure, demands, and changes) in their environment. Their reactions can impair personal health and organizational effectiveness.

Stress has been linked directly to almost every common disease, from heart disease to flu (Cox, 1980). The National Institute for Occupational Safety and Health (NIOSH) predicts that our changing economy, movement toward increased technological advances, and the changing demographics of the work force will increase the rate of stress-related disorders.

Stress is not just a personal health problem. It is costly to organizations, and affects their ability to meet today's demands for productivity and innovation. The problem cuts both ways: Organizational factors affect personal health and stress, and personal stress affects organizational health and well-being.

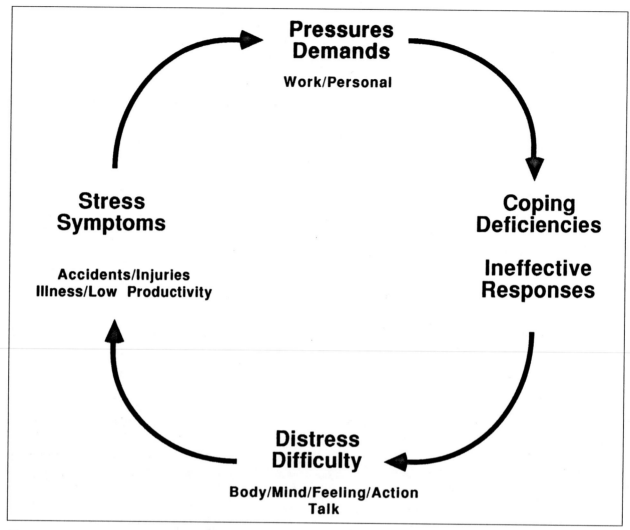

Figure 22-1. Model of interaction between individual and organizational stress. (Reprinted with permission from Jones JW. Corporate stress management. *The Risk Report,* International Risk Management Institute, 1985.)

Recognizing the costs of stress to individuals and organizations, many organizations have initiated stress-management programs. Their goals are to increase people's skills at managing life stress and to provide special help to people with problems.

STRESS IN THE WORKPLACE

Employees are exposed to many stressors: difficult working conditions, frustrating demands of customers, changing expectations in the workplace. Workers often lack the coping skills to manage the physical, mental, and emotional stress they feel in response to these pressures. In other words, without coping skills, they can easily become distressed. Thus, there is an interaction between individual and organizational stress. Organizational demands and pressures can push an employee with coping deficiencies and ineffective responses into distress/difficulty and stress symptoms, which can result in accidents, injuries, illness, low productivity, and workers' compensation lawsuits. These symptoms of distress then become stressors themselves, perpetuating the cycle of distress and increasing the effects on individual and organizational productivity. The result is more pressure, and the cycle continues.

Figure 22-1 illustrates the model of the interaction between individual and organizational stress. In the first phase, personal characteristics and job pressures come together to produce stress symptoms, including job dissatisfaction, headaches, overeating, smoking, ulcers, sleeplessness, messy surroundings, low productivity, inefficiency, absenteeism, and high turnover. In the second phase, the employees suffer injury, illness, emotional, and/or physical disability. In the last phase, the affected

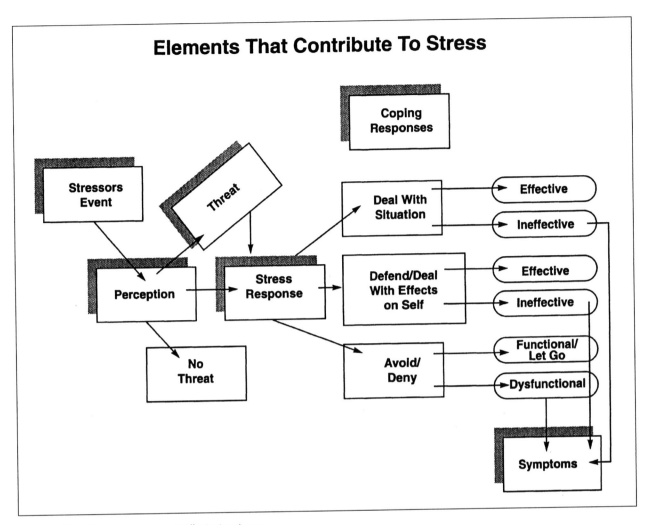

Figure 22-2. Elements that contribute to stress.

employees make stress claims in response to their symptoms.

This model implies that to break the cycle requires skills for individual and organizational stress management. To interrupt the distress cycle, the organization must identify and modify stressors in employees and in the organization, and it must increase personal and organizational ability to cope. Later, this chapter describes both types of stress-management programs.

Stress-Prone Work

Many aspects of the work environment are connected with stress. Certain types of jobs seem to be especially stress-prone. In its ranking of 130 occupations, NIOSH determined that the greatest levels of stress occur in 12 occupations: laborer, secretary, inspector, clinical lab technician, office manager, foreman, manager/administrator, waitress/waiter, machine operator, farm worker, miner, and painter.

Generally, stress rises when a person is faced with multiple demands but perceives few pathways to take control over them (Figure 22-2) (Shostak, 1980). More specifically, a number of job factors associated with stress include:

- work that does not allow the worker to participate in decisions about the work process
- jobs that place an employee between two groups, such as supervisor and shop floor employees, or management and customers/clients
- jobs that demand more or less skill than the person has
- being evaluated, or lack of clarity about expectations and standards
- changes in work demands, such as market shifts or restructuring
- lack of a clear career development path or opportunity for growth or advancement
- conflict with co-workers or supervisor.

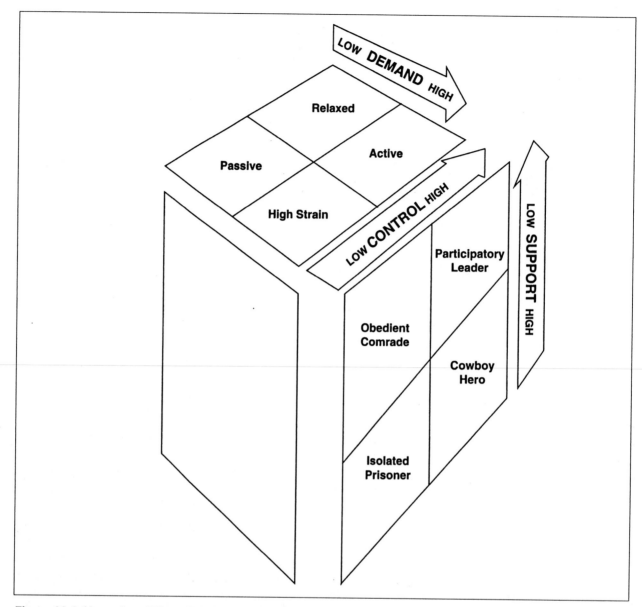

Figure 22-3. Karasek and Theorell (1990) model of the causes of work stress and its organizational implications: job demands, control, and social support. (Reprinted with permission from Karasek R, Theorell T. *Healthy Work: Stress, Productivity, and the Reconstruction of Working Life.* New York: Basic Books, 1990.)

All these factors have been related in research studies both to subjective (perceived) stress and to physical stress symptoms. These difficulties are common to all organizations, and few can be resolved quickly (MacLean, 1980).

Robert Karasek and Tores Theorell (1990) offer a comprehensive model of the causes of work stress and its organizational implications. In their analysis, based on several studies of their own, Karasek and Theorell examine relationships between health and certain qualities of the worker's job. They look specifically at three variables in relation to work stress:

- *job demands*—the extent of the pressure for results at work
- *control*—how much discretion the worker has over using skills, making decisions, and allocating work
- *social support*—positive personal relationships with supervisor and co-workers.

Their model (presented in Figure 22-3) looks at how these three factors interact on the job. They are especially concerned about the links between high-demand work and high and low control. High-demand/low-control environments cause increased health and stress problems, especially heart disease. This is the opposite of the high-demand/high-control, or active environments, where the rate of these health symptoms is much lower.

They suggest that the very design of work and workplaces, more than dysfunctional personal coping styles, creates ill health. Their extensive research at the Karolinska Institute of Sweden suggests that only when a workplace is redesigned to create more opportunities for all workers to exercise control and experience social support, workers' health will improve. Today's workplaces are increasingly high-demand environments. There is fast change, restructuring, downsizing, and reduced resources. Because Karasek and Theorell gathered much of the data a decade ago, it may underestimate the degree of current work demands and therefore the strain in current jobs.

As organizations continue to restructure and increase their emphasis on self-managing teams, empowerment, and risk taking, employees are faced with increased demands. This is especially true since many employees, especially at lower levels, have spent their entire lives in organizations that diminish their control and responsibility. Today, as organizations increasingly undertake work redesign, these efforts include more employee participation, choice, and involvement in the workplace. Those efforts, while undertaken to make companies more effective, may also have the unexpected positive consequences of making employees more healthy and able to resist stress.

Karasek and Theorell make a novel addition to their theory in the form of positing a feedback loop about learning and personal growth. Under threat and pressure, people get more rigid, less flexible, and find their perceptions and judgment blocked by intense emotions. In what they call the high-strain environment, people are inhibited from learning by the increasing demands, and therefore cannot master their environments. This leads to low morale, illness, lost productivity, and rigidity of response; these are the precursors to work burnout.

Contrarily, in an active work environment, people have more opportunity to experiment, learn new skills and apply them, leading to greater mastery over stress, and therefore greater satisfaction and health. Thus, the two environments, active and high-strain, get more extreme and differentiated over time, unless resources break the cycle and increase the learning opportunities for people in the high-strain environment. A workplace where people have control over demands is a learning environment, while the more traditional high-strain workplace tends to inhibit employee learning. This aspect of their model suggests again that the participatory workplace is more adaptive in a demanding environment.

In the second part of their book, Karasek and Theorell suggest that job redesign to provide people with more control and support is a key element in creating healthy workplaces. Research also suggests that greater self-determination and less division of labor lead to more effective work. They cite the extensive sociotechnical redesign efforts and research in Sweden and the United States and suggest that allowing greater discretion and control, and opportunities for continual learning, will lead to the following:

1. individual health
2. employee satisfaction
3. organizational effectiveness.

Workers' Perception of Stress

Workers perceive stress to be a major problem at work. Control Data Corporation reports that in its annual in-house survey of employee health, *stress/anxiety/tension* is the most-cited problem, checked by nearly 33% of employees. Control Data has used the same survey with more than 30 other companies, with the same pattern of responses: Stress always tops the list of health problems.

Work stress is a particular problem for blue-collar workers. June Fisher, MD, (reported by Bylinsky, 1986) found that bus drivers' blood pressure increased sharply when they were placed in high-pressure situations such as facing unrealistic schedules and factors beyond their control.

EFFECTS OF STRESS

Stress exacts a heavy toll on personal health. The multitude of effects include these problems (Cox, 1980, p. 92):

1. *subjective effects*—anxiety, aggression, apathy, boredom, depression, fatigue, frustration, guilt and shame, irritability and bad temper, moodiness, low self-esteem, threat and tension, nervousness, loneliness, inability to make decisions and concentrate, frequent forgetfulness, hypersensitivity to criticism, and mental blocks
2. *health effects*—asthma, amenorrhea, chest and back pains, coronary heart disease, diarrhea, faintness and dizziness, dyspepsia, headaches and migraine, neuroses, nightmares, insomnia, psychoses, psychosomatic disorders, diabetes

mellitus, skin rash, ulcers, loss of sexual interest, and weakness
3. *behavioral effects*—accident proneness, drug taking, emotional outbursts, excessive eating or loss of appetite, excessive drinking and smoking, excitability, impulsive behavior, impaired speech, nervous laughter, restlessness, and trembling
4. *organizational effects*—absenteeism, poor industrial relations, poor productivity, high accident rate, high turnover, poor organizational climate, low morale, antagonism at work, and job dissatisfaction.

Stress is a major concern of businesses because its effects extend to every area of an organization. First, work pressures can be a major contributor to a person's quota of stress and to related physical disease and personal distress. The environment, the structure, and the style of personal relationships in a business can create undue stress for employees. Finally, business must pay an increasing share of the costs of stress, both directly in the form of health care costs for employees and indirectly in the form of lowered productivity. For example, one study estimated the direct costs of excessive executive stress to exceed $19 billion a year (Everly & Feldman, 1985). For these reasons, worksites and health care providers increasingly initiate stress-management programs.

Although it is one of the major adult health problems today, job stress is one of the least understood areas of organizational cost. Stress affects both employee health and profits. Poor management of individual and organizational stress has been implicated as a contributor to workers' compensation claims. Poor employee health affects the economic health of the organization directly through absenteeism, turnover, accidents, and health care expenditures, and indirectly through diminished morale, productivity, and innovation.

Stress Claims, Losses, and Disability

Work stress leads to organizational costs in terms of workers' compensation, disability, and accident claims. For example, symptoms of stress that may predispose employees to accidents include fatigue, poor judgment, impaired physical coordination, inattentiveness, distorted perception, indecision, and intoxication from alcohol or drug abuse (Steffy et al, 1986). Student nurses who experienced several personal stressful events were apt to suffer from a rash of accidents and job-related errors in the following weeks (Sheehan, 1981). The St. Paul Companies, a major insurer, found that industrial incidents, productivity problems, health claims, and malpractice claims are clearly related to personal and work stress.

The annual cost of work-related incidents in the United States alone is well above $30 billion. The percentages of workers' compensation claims related to stress have steadily risen. However, stress is just one contributing cause in a chain that leads to illness, so it is hard to estimate how much of the cost of these diseases should be charged off to stress. The task of measuring the costs and effects of stress is further complicated because it is impossible to measure the effects of different sources of stress on a person—how much, for example, years of job stress or family stress contribute to a heart attack.

A growing legal trend is for employers to be found liable for work stress. Ivancevich et al (1985) did extensive work on the legal implications of organizational stress management. They suggest that any company is liable unless it learns to monitor, diagnose, and treat a stressful situation before it goes to court. They also suggest that managers must learn to identify the potential stress trouble spots in their organizations, try to relieve these when necessary, and document their efforts.

In an attempt to find nonlegislative alternatives to increased workers' compensation costs, the Michigan Bureau of Workers' Disability Compensation compared organizations with excellent workers' compensation loss experience to firms with very poor experience. They discovered some very striking differences. Within a given industry, the best firms had a claim frequency that was up to 10 times lower than that of the worst firms. The size, type, and location difference accounted for only 25% of the total variation. The remaining factors are under the organization's control, having to do with the nature and culture of the workplace.

Burnout

When work stress is extreme and continuous, the result can be burnout. *Burnout* is an end point of work stress characterized by a sharp drop in motivation, satisfaction, commitment, and effectiveness. Burnout was first noticed in the helping professions, in which workers are subject to continual and often conflicting people-related demands, and now is documented in a variety of workplaces where the work is primarily with other people (Maslach, 1982). Burnout is a common problem, affecting 45% of the 4,300 employees in one study (Golembiewski, 1985).

According to Maslach and Jackson (1981), burnout has three components:
1. *depersonalization*—seeing clients, customers, and colleagues as less than human
2. *emotional exhaustion*—having nothing more to give
3. *reduced sense of personal accomplishment*—feeling no sense of competence or satisfaction from work.

Like stress, burnout impairs individual health and productivity. The literature cites a long list of emotional and physical symptoms of distress associated with the burned-out employee. Some of the signals of burnout are regular and robust deterioration of working conditions—for example, job tension increases, participation in work-related decisions decreases, and satisfaction with work decreases. Employees experience a dramatic increase in such physical symptoms as stomach pains, headaches, persistent cough and colds, fatigue, poor appetite, nervousness, and feeling worn out. From the organizational standpoint, employees' performance appraisals drop, and productivity plummets.

Types of Burnout
Based on a study of over 4,000 employees in businesses, federal agencies, and public organizations, Golembiewski (1986) identified two different types of burnout: chronic and acute. Chronic burnout is the most prevalent. It builds over time and lasts for indefinite periods. Acute burnout is triggered by sudden major strain. Its onset is rapid, and it improves over time with social support.

Roots of Burnout
According to Golembiewski (1986), chronic burnout is rooted in the immediate work group and in the behavior of its direct supervisor. Therefore, it can no longer be thought of as an individual phenomenon. Rather, burnout is an organizational concern. Several signs of work groups that are prone to burnout are low cohesiveness, low supervisory support, high pressure to produce, low standards of performance, and unclear roles and goals.

Intervention
The finding that burnout is more directly related to the work group environment than to the person's coping skills or personality supports interventions that focus on burned-out work groups, rather than burned-out individuals. Getting stress and burnout under control in the work settings can play a key role in reducing the risks for workers' compensation claims.

The first step in reversing burnout in the workplace is to monitor the level of burnout and to act when the percentage of burned out employees in any work group rises. It is most effective to begin by working with low-level burnout cases by enlarging jobs and restructuring how the work is done. These interventions can have a dramatic effect. Increasing participation and choice in the work often enhances the relationship with the supervisor.

Intervene next with those in middle phases. Introduce flex time, job rotation, and expanded roles. To intervene in highly burned-out work groups often requires changing supervisory relationships as well as implementing the previously mentioned strategies. It is best to avoid one-track remedies.

Personal Responses
While much stress results from external conditions over which people have no control (excessive noise, working conditions, and competing demands), a large portion of the stress that people experience is due in part to the way that they respond to stressful demands. Some people respond unusually well to pressure; they are labeled "stress resistant" or optimal performers. Other people experience the dysfunctional responses discussed so far in this chapter.

The most studied of the dysfunctional responses to stress is the Type A personality style, first formulated by Rosenman and Friedman (1974) and the subject of countless studies. This style of response to stress contains elements of hurrying, doing many activities at once, needing to maintain control over everything, feeling frustration and resentment, and having difficulty collaborating with others. Type A behavior was connected with heart disease in scores of studies. As the concept of Type A behavior was refined, further research showed that after people with this behavior style experience a heart attack, they can learn new styles of behavior that reduce the risk of further health crises.

Other aspects of personal style are linked to ineffective stress management and health problems. Emotional unresponsiveness—the inhibition of anger and difficulty in expressing feelings—is linked to various physical ailments. Other studies linked a sense that life is meaningful and has purpose to the ability to resist stress (Antonovsky, 1979).

Positive Responses to Stress

For certain people at certain times, work stress is a source of challenge and inspiration. Indeed, studies of peak performers and thriving work environments suggest that there is a level of stress that is productive and stimulating at work. That level exists when employees feel challenged, are involved and committed to the organization's goals, and feel they have a reasonable degree of control over work decisions (Jaffe & Scott, 1989).

This two-sided nature of stress has created a controversy (Moss, 1981). Many top managers, who have a stress-resistant style and tend to be healthier than their employees, believe that their experience shows stress is good for performance. Yet people in the middle and lower levels of an organization experience greater distress and more frequent health problems, due in part to difficulty managing stress and a less-developed set of coping skills.

What is it about some people that makes them relatively stress-resistant? Regular exercise has been identified as an important resistance resource. Personality can make a difference by influencing how a person reacts to the stressful situations with which he or she is faced. Thus, there seem to be qualities that enable a person to resist stress, and these are the same qualities that are associated with increased performance at work.

Personal control. One possible source of stress resistance is related to the personality trait of *hardiness,* an orientation characterized by commitment (vs. alienation), control (vs. powerlessness), and challenge (vs. threat) as a resource against stressful life events (Kobasa, 1979). In a comparison of high-stress executives with similar jobs in the same company, those who had the fewest illnesses and health claims were more likely than their less healthy colleagues to have this trait of hardiness (Kobasa et al, 1982).

The common denominator of Type A research, research on stress-resistance resources (Kobasa, 1979; Antonovsky, 1979), and observations of work stress seems to be that the more perceived and actual control a person has over work situations, the more effective the person can be at managing stress. As a result, stress-management programs now stress the possibility of enhancing personal power—the ability to get what one needs accomplished—as a central capacity for effectiveness in managing stress (Jaffe & Scott, 1987).

Social support buffer. The other key factor that enables a person to manage stress is the help and support of other people. Social support (House, 1981) is a stress-inoculating variable that protects persons from the negative effects of high stress. The more a person feels that he or she can turn to other people, seek help, and get the information and resources needed to complete a task, the greater that person's ability to manage stressful situations.

PROGRAMS FOR RESPONDING TO STRESS

There are many ways that people and organizations can influence personal and organizational ability to manage stress. As data concerning the negative effects of stress began to emerge, health and business groups responded by formulating programs that might prevent future illness, organizational difficulties, and costs related to stress. As research identified skills and pathways to manage stress, several types of programs were formulated. In the past decade, the first stress-management programs, which taught relaxation methods, gave way to much more complex interventions. While programs proliferate, research on the effects of programs lags far behind. However, based on program experience and informal feedback, a consensus is emerging about goals, format, and style of stress-management programs.

Personal Approaches to Stress Management

Golembiewski (1986) reports on a company that surveyed all its employees for their level of burnout, then advertised a program on managing stress and burnout. To his surprise, almost none of the employees who were assessed as highly burned out attended the program. When asked why, they responded that such programs were a waste of time, there was nothing they could do about burnout, and they couldn't spare the time. Golembiewski suggests that these people are so fatalistic that only personal attention and counseling could influence them. In contrast, mildly burned-out employees felt that training programs were helpful, and did attend them when offered.

The basic theme of stress-management programs is that a person is not a victim of stress. While there are many sources of stress that a person cannot change, a person can act in many ways to reduce or manage the negative effects of stress. This is the theme underlying all health promotion activities: A person can make a difference in his or her health and well-being. The interventions then teach the self-management skills needed to achieve that goal.

Types of programs

Jaffe et al (1986) conducted a survey of 50 companies nationwide to determine what organizations were doing to implement stress-management programs. There were four major delivery formats, with some programs using multiple formats. There were four main types of programs:

1. education/awareness-building lecture or self-instructional format
2. assessment focused
3. skill building
4. therapeutic/counseling.

Of these programs:

- 80% used a lecture format for presentations
- 60% had skill-building components
- 60% had personal coaching or counseling
- 20% had a self-instructional format.

Preventive stress-management programs usually conform to one or more of these types. They are most frequently designed to identify people with particular difficulties and teach them skills to help them manage the negative effects of stress, learn to be more effective at responding to stressful situations, and overcome the physical effects of chronic life stress, before it becomes a health problem. Another kind of program focuses on *high-risk persons*, who have particular stress-related symptoms, teaching them self-management of stress as a way to overcome or reduce pain and other physical ailments.

Educational/awareness building. The goal of educational or awareness-building programs is to make people aware of the links between stress, illness, and personal behavior and of ways they can make a difference. Such programs can be lectures or presentations. They often use written materials—for example, booklets, paycheck stuffers, etc.,—that make employees aware of the nature of stress, the danger signs of stress problems, and what they might do to prevent those problems.

Educational and awareness programs are usually the first step in a company-wide program. They are often used to initiate and recruit for more intensive programs. They help people decide that they can make a difference, as well as suggest some of the pathways and resources that are available. Of the companies surveyed by Jaffe et al (1986), 80% of their programs included an education/awareness-building component.

Assessment focused. Many companies and health care settings offer personal stress assessments. Of the companies surveyed, 15% of their programs had a formalized assessment process. This type of program helps the organization target its efforts to the most in need.

People who learn about stress often want to know whether they have a particular problem with stress. Since everybody perceives himself or herself as under stress, and most people assume that they "handle" it well, various assessment tools are important to identify people with special difficulties, and to pinpoint specific skills deficits and problem areas to focus on.

Companies use a number of common risk appraisal and stress scales. For example, under the direction of John W. Jones, the St. Paul Companies, a large insurer, developed the Human Factors Risk Management Program to help companies install effective stress-management programs, control losses, and increase profits. Included in that program is a Human Factors Stress Inventory (HFI), consisting of three areas:

1. *general job stress*—emotional and temperamental, physical and psychological symptoms
2. *organizational stress*—turnover, accidents, and counterproductivity
3. *personal stress*—stressful life changes.

Each employee's total scores on the scales are compared to national norms. The results of the inventories are fed back according to company by administrative job levels and departments. The inventory delineates the departments with the highest stress. Based on the results, the company can plan possible intervention programs such as stress management, employee assistance programs (EAP), conflict management, and organizational development.

The Human Factors Stress Inventory (HFI) is effective in showing that company insurance claims and losses are indeed related to stress. High-risk departments, as assessed by high HFI scores, typically exhibit higher rates of accidents, injuries, illness, negligence, and counterproductivity (theft, vandalism, drug abuse). Companies use information from the survey to correct the organizational factors that lead to these stress-related losses. Company representatives also attend the St. Paul's International Human Factors Institute to learn how to implement a variety of programs designed to control stress-related losses.

Another approach to assessment is taken in StressMap® (Orioli et al, 1987). The instrument was designed as a self-care tool for a person to learn about positive and negative aspects of his or her own style of coping with stress. This instrument was designed by integrating previously validated scales that were used to measure different aspects of stress. Factors such as life change, work and family stress, personal power, time management, self-esteem, emotional expression, anger, flexibility, and traditional stress symptom checklists are combined into 21 scales that assess specific aspects relating to stress management.

The tool is self-scoring and confidential, allowing the employee to gain information about his or her specific performance in each area. Then the person reads the interpretation for each scale and goes through an action planning process to create a program for personal change. The tool was norm-tested in seven companies and health settings, and each of the skill areas is highly correlated with stress symptoms.

Skill building. In classes or workshops, workers learn specific skills that help them manage stress and the negative effects of excess stress on their health. These programs often include additional self-study material such as tapes and workbooks for home practice. Another model is the one- or two-day retreat, which is usually offered to upper management. Of the programs reported in the Jaffe et al (1986) survey, 65% focused on skill-building efforts in one of the following areas: changing one's physiological responses, relaxation training, changing response to stress, effective coping, and changing interpersonal styles, conflict resolution, or building a support network.

Skill-building programs focused on three types of skills:

1. *Relaxation skills*—teaching people to reduce the negative physical effects of repeated stress. A person can practice physical self-care, using techniques like relaxation and meditation, to maintain personal health. Various methods to achieve relaxation are such a universal part of stress-management programs that they are often perceived to be the whole program. Herbert Benson (1975), who conducted some of the original research on relaxation techniques and stress relief, suggests that relaxation methods produce a state that is opposite to the stress response and that helps the body release built-up chronic stress. Techniques taught include meditation, progressive muscle relaxation, self-hypnosis, guided imagery, and autogenic training. A study of a meditation program in the New York Telephone Company (Carrington et al, 1980) found that these techniques are all easily learned and, when practiced regularly, have clear psychological and physiological effects on work stress. These skills take care of the body after it has experienced stress, and are very useful to people who work in high-stress environments.

2. *Coping skills*—Coping skills include all the ways that people respond to stressful situations. People develop certain styles of managing pressure situations. Certain styles, such as Type A behavior, and types of response, such as avoidance, inflexibility, or lack of personal priorities, have been connected with difficulty managing stress. People can learn and practice effective behavioral responses to stressful events. The skills include cognitive skills (changing the way one perceives and defines stressful events), and the personal beliefs, internal conversations, expectations, and evaluations one has about the pressures one faces. Other program emphases include employing and evaluating one's response to stressful situations and creating more effective ways to respond. Another area of focus is time management (setting personal priorities and planning and anticipating stressful events), and life planning (discovering one's personal priorities and acting upon them). These lead to changes in behavior that meet personal needs, enhance well-being, and prevent chronic stress from building. A final area of focus is managing the emotional aspects of stress, especially anger. People learn to label feelings and respond to them effectively.

3. *Interpersonal skills*—A majority of the stress-producing situations involve relations with other people. Managing stress, then, involves learning the skills to work harmoniously with other people. This begins with teaching how to build personal support—getting others to help, sharing one's feelings, and seeking out others—which is one of the key methods of inoculating oneself against the negative effects of stress. Stress-management classes teach assertiveness, active listening, effective communication, conflict resolution, and team-building. Since these skills are often covered in other management training efforts, and are often initiated by different divisions of the company (e.g., training rather than health promotion), companies sometimes do not explicitly consider them when defining their stress management efforts.

Therapeutic/counseling. Programs with a therapeutic/counseling focus provide special help for people who react inappropriately to stress, to prevent more serious problems. The people who are at highest risk or are experiencing the gravest difficulty can be coached, through employee assistance programs (EAPs) or counseling, to modify their self-defeating responses. Personal counseling, usually for several sessions and often featuring technologies such as biofeedback, helps people with special stress problems to make changes. In the Jaffe et al survey, 70% of the programs had some kind of crisis intervention counseling for high-risk populations.

After the pioneering efforts at Equitable Life Insurance Company in New York (Manuso, 1978), many companies now include stress management counseling and consultation as part of their executive counseling or EAPs. Employee assistance programs, which were originally designed to handle drug and alcohol problems that impair performance, now are expanded in scope to include stress, financial, career, and family problems.

Current practices. Stress-management programs seem to be implemented because employees perceive a need and request them. In most of the 50-plus companies surveyed for this survey (Jaffe et al, 1986), stress-management programs were specifically requested. When the programs were offered, many employees attended them, and those who conducted the programs received positive feedback when evaluations were done. The programs were attended by employees at all levels. All ages attended and the programs were popular with employees' families when families were included.

The programs offered have many similarities. Many companies buy packaged programs that include instructor's guides, workshop curricula, audiotapes and videotapes, and self-instructional workbooks. Often, the companies use their internal trainers or health promotion specialists to adapt these materials to their own needs. Companies often hire outside instructors to teach classes or workshops. Hospitals, health maintenance organizations (HMOs), health clubs, community centers, and educational institutions all offer stress-management classes, workshops, and seminars.

Organizational Approaches to Stress Management

Teaching correct relaxation and breathing is not enough to manage work stress. Personally focused health practices—relaxation, diet, and exercise—are important but often not enough. These strategies buffer the person, but do not directly change the relationship between the person and the organizational environment. It is often necessary to change environmental factors to reduce the stressors and enhance employees' ability to adapt to the environment.

If the environment presents chronic sources of stress and the workplace is inadvertently promoting distress in employees, then teaching individual skills will be of only limited effectiveness. The company needs to devise a systemic approach to stress management that focuses on several levels and types of intervention. These comprehensive interventions ultimately focus on stress-producing factors within the organization.

Responses to Workplace Stressors

Companies can identify aspects of the work setting that contribute to stress—whether environmental, structural, or interpersonal—and create changes designed to reduce stress. Although many organizational interventions that look at such factors are not designated or thought of as stress-management programs, their goal is to reduce stress and raise productivity. For example, many companies now have extensive quality-of-work-life programs, which increase employee participation and thereby reduce stress. Quality circles are one aspect of these programs. Employees meet in small groups with a peer facilitator to consider how the work can be done more effectively. While such processes are designed to make the organization more effective, they also increase workers' self-esteem and their feeling of control and participation in their work (Crocker et al, 1984).

One of the most pressing sources of stress is conflict between personal or family demands and those of the workplace. For example, working parents, especially single parents, have to balance their children's needs with job requirements. Often both the family and work suffer. In response, companies have begun to offer alternative work arrangements to help employees fit their personal and work demands more harmoniously. These new arrangements include flextime, flexible work hours, job-sharing, and permanent part-time jobs, as well as various leave and sabbatical policies. Many companies include child-care benefits or provide company-sponsored child-care services.

Types of Programs

Organizations have developed various types of programs to respond to workplace stressors. These include behavioral safety programs, management and culture programs, and disability prevention and management programs.

Behavioral safety programs. A behavioral safety program focuses on physical safety—accident prevention, lifting, handling of hazardous materials, etc. These programs are part of a comprehensive stress-management strategy, because these aspects of work often "stress" employees.

Management and culture programs. Other programs teach managers how to create a work environment that is less stressful for employees. Such a program could include some or all of the following components:

- *organizational analysis*—a survey of employee perceptions identifies key elements of organizational stress. The results show management which parts of their organization need attention, and targets recommendations for new policies, programs, and management activities.
- *interpersonal communication*—promotes skill-building sessions to share problem solving and decision making. Focuses on ways to communicate when under pressure.
- *managing organizational change*—teaches key skills for accepting and leading change. Includes understanding resistance to change, negotiating for commitment, building teams, and developing commitment to new ways of doing things.

Prevention/management programs. Organizations should have an active hazard-prevention program in place. Reducing hazards and preventing accidents means preventing occupationally caused disabilities. But if employees are injured, knowing the company has a proactive approach for placing and managing workers with disabilities can also help reduce their stress. Knowing that management is committed to avoiding hazards and resultant accidents can help alleviate stress among employees. The Americans with Disabilities Act of 1990 ensures that all U.S. disabled employees (and potential employees) will be given needed access and support to do the jobs for which they were hired. This can mean that the building structure itself may have to be modified to allow access; that the employee's workstation may have to be modified as needed; that the job functions must be modified, etc., to support all employees. Other company actions can include:

- *back care programs*—Since back injuries are a common type of disability, an important component of disability prevention is a program to prevent or reduce back injuries
- *wellness/fitness programs*—Another way to minimize disability is to help employees follow healthy practices. The company can teach workers about smoking cessation and weight management. It can also give employees an opportunity to participate in classes in aerobics, movement, yoga, and other exercise.
- *data tracking*—The company should collect data to track disability risks across departments over time.
- *screening of employee health risks*—The company can make available tests measuring various aspects of individual health, including weight, blood pressure, and cholesterol level (see chapter 9, Occupational Health Nursing Programs for more information).

Implementing a Comprehensive Stress-Management Program

A comprehensive stress-management program often develops over time. One intervention can build on another, and the program soon includes several interventions that define the desired change, modify workplace conditions and/or employee behavior, and assess changes at each stage of intervention. The steps in a comprehensive stress-management program are awareness building, assessment, program review, report to management, action, and prevention.

The responsibility for developing and implementing these programs can fall to the occupational health nurse, medical director, EAP director, or the health-promotion department (if available). Occasionally, this responsibility falls into training and development or to an employee union or employee association. All departments should cooperate on these efforts.

Building Management Awareness
Making management aware of the problem is best handled by conducting a briefing for senior management. Cover the relationship between employee stress and potential risk (and financial cost) for the organization. Focus on how stress-management programs can reduce risk and contain costs.

Assessment
Conduct an organizational survey of employees' emotional, attitudinal, and physical reactions to all types of personal, job, and organizational stressors. Analyze the type and level of organizational stressors and the level of coping skills among employees. By comparing these scores to national norms, determine which units are at high risk.

Program Review
Review current departmental efforts to address organizational wellness, stress management, health promotion, and related concerns. Review internal resources to provide these services inside the company.

Report
Deliver the results of the organizational survey to management, highlighting the sources and levels of stress by department, job category, and work shift. The information in this report can be used by a management-staff task force in the following several ways:
- to identify organizational strengths and weaknesses
- to focus efforts on the highest-risk groups
- to evaluate current efforts in providing organizational health and stress-management programs.

Management often uses the assessment report to give the entire employee population feedback on the findings. This often improves morale, increases participation, and opens important channels of communication that can be used to introduce further improvement efforts.

Action
Once management is aware of the needs and resources within the organization, a comprehensive approach can be prepared (Figure 22-4).
1. *awareness building:* This is designed to increase employee awareness of the sources and levels of stress in their lives.
2. *assessment:* Collect baseline data on medical claims, accidents, illness, turnover, absenteeism, theft, and productivity by department. After stress-management programs are introduced, collect these data and compare them with the baseline data.
3. *program review:* Look at existing in-house stress-management programs and previous efforts.
4. *report:* Create a management brief, highlighting areas of greatest risk, and target multiple paths of action.
5. *action:*
    a. Introduce programs focused on individual and organizational components of stress.
    b. More assessment can target more specific stress-management tools. These allow persons to focus on their level of skill or stressors, so they can take additional action (e.g., classes, life change, etc.) to reduce the impact of the stressors.
    c. Skill-building enables employees to reduce stress associated with lack of skills. Tools include the following classes: time management; stress management (i.e., relaxation techniques, exercise, nutrition/weight loss); assertiveness; communication/problem solving; smoking cessation; AIDS prevention; hypertension reduction.
    d. Organizational consultation is an approach whereby you can consult with high-risk departments to reduce the problems identified in the assessment phase as organizational concerns. Interventions can include help with redesigning work, building teams, managing conflict, communications skills, managing change, mediating and negotiating skills, how to effectively set and meet goals.

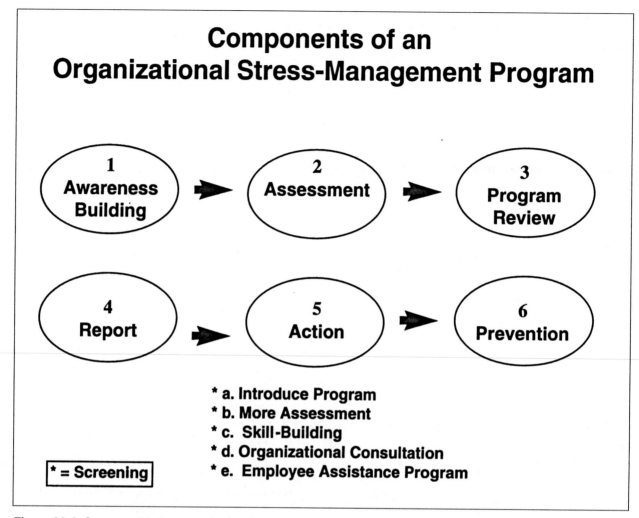

Figure 22-4. Components of an organizational stress-management program.

e. EAPs are programs to assess, refer, and offer ongoing therapeutic services to employees and their families. Problems can include work-related issues or personal problems (e.g., financial, substance abuse, child-rearing, taking care of an elderly relative, medical, etc.).

6. *prevention:* To maintain a highly functioning work force, the company can conduct pre-placement screening. Such efforts can place special emphasis on matching the employee's abilities with the specific physical and mental requirements of the job. The purpose of matching abilities with job requirements is to ensure that employees will "fit" with the job environment in which they will function. Health-promotion screening can identify medical conditions (such as diabetes or high blood pressure) at levels that can benefit from changes in life-style.

Impact of Stress-Management Programs

Worksite stress-management programs are feasible, and a variety of techniques can be effective in helping workers reduce physiological arousal levels and psychological manifestations of stress. Improvement appears to last over time, but the durability of physiological changes after training is questionable. Although not enough studies have been conducted to determine the relative merits of select techniques and to compute cost-benefit ratios, stress-management programs can improve worker well-being and partially offset the costs of occupational stress—losses in productivity and medical claims for stress-related disorders.

Rising workers' compensation claims for stress-related disability and the knowledge that behavioral factors play a significant role in seven of the 10 leading causes of death will likely prompt a significant growth in the use of worksite stress-management programs (Murphy, 1984a, p 11).

Most of the evaluation studies are based on programs that use classes to teach workers relaxation skills to reduce their chronic stress levels. Murphy (1984b), studying 52 highway maintenance workers using a combination of biofeedback training and audiocassette instruction, found that relaxation training significantly reduced muscle tension and perceived stress, and increased quality of sleep. Manuso (1978), studying 30 insurance company employees with headache and anxiety problems, found that physiological, psychological, and work factors improved with individual biofeedback training. Studies have yet to explore more fully the effects of stress interventions on job stress factors.

A study of 154 New York Telephone employees (Carrington et al, 1980) found that even a self-taught, home-study framework could dramatically reduce psychological and somatic symptoms six months after training, even in people who were not practicing the techniques regularly. Another unpublished study, by Control Data Corporation, surveyed 78 employees who elected to take a self-study course titled "How to Relax" Using self-report data, 78% reported using at least one stress management technique regularly, 77% reported paying more attention to the effects of stress, 49% reported using a relaxation technique regularly, and 82% rated the course as good or excellent. The study reported similar outcome data with trainer-led courses, the only difference being that more people complete the course when it is trainer-led. Stress management is Control Data's most popular health promotion course.

In evaluating the Institute for Labor and Mental Health's occupational stress support groups as an intervention to help blue-collar and union workers respond to stress, Lerner (1984) found that the program significantly improves physiological, psychological, and work productivity variables.

Research into the financial effects of stress programs have been performed. Two corporate programs attempted to calculate cost-benefit ratios for stress-management programs. Manuso (1978) calculated the pretreatment and post treatment costs of stress symptoms, clinic visits, and absenteeism. His Equitable Life Program provided personal counseling and biofeedback training to teach relaxation skills to people who had serious stress problems or were in high-risk categories. The cost-benefit ratio was $1 to $5.52. (For each dollar spent on treatment, the company saved $5.52.)

In an unpublished study at New York Telephone, former medical director Loring Wood reports that the meditation training program cited previously reached half of the 25% of employees who were assessed, using a symptom checklist, as having stress-related difficulty. Wood found a 10% decrease in absence among the 1,000 people who participated, saving the company $2,677 per person just for decreased absence. The cost of the intervention was $300 per person.

GOALS FOR THE FUTURE

Stress-management programs are fast becoming an integral part of training in human and personal skills. These programs are offered in companies and in health care and health promotion settings. The consensus is growing that people can benefit from training in basic personal management skills—skills that help a person manage stressful situations and practice self-care. As the awareness of these skills becomes more widespread, people presumably will begin to practice them, and the skills may become part of health-promotion programs in the work setting. Just as more people have learned to practice positive health habits in other areas, more and more people, through education, reading, and participation in programs, may begin to practice stress-positive coping skills.

Several short-term and long-term changes would maximize the potential of stress management. For the short term, we envision the continuing growth and development of stress management courses, seminars, programs, and other learning activities. Companies are likely to explore different formats besides the traditional classroom/seminar: interactive software programs, peer-led sessions, videodiscs, etc. Also, the design of learning materials for stress-management programs will become more sophisticated and more targeted toward skill development. Assessment of specific learning outcomes, and research on effects of specific program formats and lessons, will bring information about what works and what the long-term effects are.

However, the ultimate goal is not for the company to institute a stress management program or for the employee to learn stress-coping skills. The ultimate goal is to create living, working, and community environments that allow people to live and work together in ways that optimize their health and well-being. This means more than teaching employees to cope with stress. It involves the design of work environments and a culture of work relationships that support the most effective ways of persons to manage stress and work under pressure. Ultimately, stress management involves working in a place where the employee feels cared for, connected to decisions, a sense of personal efficacy, and has opportunities to pursue personal goals and health as well as the organizational goals. Organizational redesign is the long-term activity that will allow people to manage stress.

SUMMARY

The stress level of an organization is affected by the norms, the working culture, career pathing, evaluation procedures, workers' rights, benefits packages, and the working environment. These factors deeply influence an employee's sense of self-worth and capacity to resist or succumb to stress. Recently, people in health promotion and stress management have begun to suggest that an audit of the company culture and organizational interventions are perhaps the most effective long-term strategy for managing personal stress (Tager & Blanchard, 1985; Jaffe & Scott, 1989).

Management must become aware of the factors that reduce stress and increase health and productivity. Companies that lack this awareness will tend to continue a cycle of stress-related losses and claims. Many differences between low-loss and high-loss employers can be summed up in one word: management. Companies that manage safety carefully, involve employees in the workings of the organization, and emphasize disability prevention and management have significantly lower losses than their less-proactive counterparts. It is the management of people and the proactive orientation of the company that separate the excellent organizations from the also-rans.

REFERENCES

Antonovsky A. *Health, Stress and Coping.* San Francisco: Jossey-Bass, 1979.

Benson H. *The Relaxation Response.* New York: Morrow, 1975.

Blacklock E. Workplace stress: A hospital team approach. *Prof Nurse* 13(11):744–747, 1998.

Byers S. Organizational stress: Implications for health promotion managers. *Am J Health Promotion* 2(3):21–27, 1987.

Bylinsky G. The new assault on heart attacks. *Fortune,* March 31, 1986, pp 80–89.

Carrington P, Collings G, Benson H. The use of meditation-relaxation techniques for the management of stress in a working population. *J Occ Med* 22:221, 1980.

Cooper CL, Cartwright S. An intervention strategy for workplace stress. *J Psychosom Res* 43(1):7–16, 1997.

Cox T. *Stress.* University Park, MD: University Park Press, 1980.

Crocker O, Charney S, Chiu J. *Quality Circles.* New York: Mentor, 1984.

Everly G, Feldman R. The development of occupational stress-management programs. In *Occupational Health Promotion.* New York: Wiley, 1985.

Friedman M, Ulmer D. *Treating Type A Behavior and Your Heart.* New York: Fawcett, 1984.

Golembiewski R. Dealing with burnout: Emerging realities and challenges. *Corporate Commentary* 1(5), 1985.

Golembiewski R. *Stress in Organizations.* New York: Praeger, 1986.

House J. *Work Stress and Social Support.* Reading MA: Addison-Wesley, 1981.

Ivancevich JM, Matteson MT, Richards EP. Who's liable for stress on the job? *Harvard Bus Rev,* March–April 1985.

Jaffe D, Scott C. *Self-Renewal: Achieving Health and High Performance in a High Stress Environment.* New York: Simon & Schuster, 1989.

Jaffe D, Scott C. *Take This Job and Love It.* New York: Simon & Schuster, 1987.

Jaffe D, Scott C, Orioli E. Stress management: Programs and prospects. *Am J Health Promotion* 1(1):29–37, 1986.

Jones JW. Corporate stress management. *The Risk Report.* Dallas: International Risk Management Institute, 1985.

Lusk SL. Health effects of stress management in the worksite. *AAOHN J* 45(3):149–152, 1997.

Karasek R, Theorell T. *Healthy Work: Stress, Productivity, and the Reconstruction of Working Life.* New York: Basic Books, 1990.

Kobasa S. Stressful life events, personality and health. *J Personality Soc Psych* 37, 1979.

Kobasa S, Maddi S, Kahn S. Hardiness and health: A prospective study. *J Personality Soc Psych* 42(1):168–177, 1982.

Lazarus R. *Psychological Stress and the Coping Process.* New York: McGraw-Hill, 1966.

Lerner M. *Occupational Stress Groups and the Psychodynamics of Work.* Oakland CA: Institute for Labor and Mental Health, 1984.

MacLean A. *Work Stress.* Reading MA: Addison-Wesley, 1980.

Majumdar B, Ladak S. Management of family and workplace stress experienced by women of colour from various cultural backgrounds. *Can J Public Health* 89(1):48–52, 1998.

Manuso J. Corporate mental health programs and policies. In Ng L (ed). *Strategies for Public Health.* New York: Van Nostrand-Reinhold, 1978.

Maslach C. *Burnout: The Costs of Caring.* Englewood Cliffs NJ: Prentice-Hall, 1982.

Maslach C, Jackson SE. The measurement of experienced burnout. *J Occ Behavior* 2, 1981.

McLeroy K, Green L, Mullen K, et al. Assessing the effects of health promotion in worksites: A review of stress program evaluations. *Health Ed Q* 11(4):379–401, 1984.

Moss L. *Management Stress.* Reading MA: Addison-Wesley, 1981.

Murphy L. Occupational stress management: A review and appraisal. *J Occ Psych* 57:1–15, 1984a.

Murphy L. Stress management in highway maintenance workers. *J Occ Med* 26:6, 1984b.

Newsweek. Stress on the job. April 25, 1988.

Orioli E, Jaffe D, Scott C. *StressMap: Personal Diary Edition.* New York: Newmarket Press, 1987.

Pelletier KR. *Mind as Healer, Mind as Slayer.* New York: Delta, 1977.

Rosenman R, Friedman M. *Type A Behavior and Your Heart.* New York: Knopf, 1974.

Scott CD, Hawk J. *Heal Thyself: The Health of Health Care Professionals.* New York: Brunner-Mazel, 1986.

Sheehan D. Psychiatry in medicine. *Science News,* August 22, 1981.

Shostak A. *Blue-Collar Stress.* Reading MA: Addison-Wesley, 1980.

Steffy BD, Jones JW, Murphy LR, et al. A demonstration of the impact of stress abatement programs on reducing employees' accidents and their costs. *Am J Health Promotion* 1(2):1986.

Tager M, Blanchard M. *Working Well.* New York: Simon & Schuster, 1985.

Yandrick RM. Stress and the workplace. Conflict management can prevent behavioral health problems. *Behav Health Tomorrow* 8(3):23,26–27, 1999.

chapter 23

Substance Abuse

by Richard B. Seymour, MA
David E. Smith, MD
revised by Marci Z. Balge, RN, MSN, COHN-S
Gary R. Krieger, MD, MPH, DABT

409 **Introduction**
409 **Psychoactive Drugs**
410 **Nonmedical Drug Use**
Dynamics of drug abuse ■ Addiction and addictive disease
412 **Drugs in the Workplace**
Primary reasons for drug abuse in the workplace ■ Work impairment ■ Roles of occupational health and safety professionals ■ Observing substance abusers ■ Urine testing ■ Drug-testing laboratories
425 **Referral and Treatment**
Role of the employee assistance program (EAP) ■ Confidentiality ■ Safety
427 **Criteria for Referral into Treatment**
User-program relationship ■ Diversion programs for impaired professionals
428 **Reentry and Rehabilitation**
429 **Summary**
429 **References**

INTRODUCTION

In the past 25 years, the United States has evolved from a country with a drug-using subculture into a culture that uses drugs. At the same time, drug abuse has intruded into the workplace and is profoundly affecting work safety. According to such sources as the National Household Survey on Drug Abuse, the Urban Institute, the Justice Department, and the House Select Committee on Narcotics Abuse and Control, drug-related absenteeism and medical expenses cost businesses about 3% of their payroll, and employers report that as many as one worker in five has a problem with alcohol or drug abuse. The consequences include an increase in absenteeism, health care claims, and lost worker productivity.

To be able to combat drug abuse in the workplace, as well as any drug use that interferes with work performance and safety, a program requires two basic tools:

1. a clear understanding of the nature and effects of licit and illicit drugs
2. a knowledge of the reliable methods available for identifying individuals in the workplace who are abusing licit and/or illicit drugs.

This chapter provides both of these tools.

PSYCHOACTIVE DRUGS

The terms psychoactive drug, central nervous system drug, and CNS drug are used for drugs of abuse because these drugs have their principal action in the brain. In acting primarily on the central nervous system, these drugs differ from the medicines that are used primarily in treating disease symptoms, fighting disease, or correcting bodily malfunctions. Even though medicines can have side effects within the central nervous system, their primary effects are in other parts of the body.

Not all psychoactive drugs are drugs of abuse. Some, such as the antipsychotic and antianxiety medications, do not produce effects that would lead to their abuse. On the other hand, all generally abused drugs are psychoactive. Psychoactive drugs fall into four general types or categories: narcotic analgesics, sedative-hypnotics, stimulants, and psychedelics. Within each of these categories are found both licit drugs (legitimately manufactured pharmaceuticals that have medical indications for their use in treatment) and illicit drugs (those that are illegally harvested and/or made by underground chemists for the express purpose of supplying the illegal drug trade). Psychoactive drugs in each category are subject to abuse. Once a drug's category is identified, this identification will provide a great deal of information on what that drug can do and how to identify its use.

Psychoactive drugs are either composed of naturally occurring substances or are chemically constructed. Drugs in the latter group, known as synthetic, have been developed for either legitimate pharmaceutical research for their treatment properties or by underground chemists for their value as drugs of abuse.

NONMEDICAL DRUG USE

In defining nonmedical drug use, a variety of terms indicate a person's type and degree of involvement with drugs. While these terms are often used interchangeably, they actually have distinct meanings and can help to clarify the different stages and dynamics involved. The classifications of nonmedical drug use by the National Commission on Marijuana and Drug Abuse (1973) combines description with motivation:

- *experimental use*—a short-term, nonpatterned trial of one or more drugs, motivated primarily by curiosity or a desire to experience an altered mood state.
- *recreational use*—use that occurs in a social setting among friends or acquaintances who desire to share an experience that they define as both acceptable and pleasurable. Generally, recreational use is both voluntary and patterned and tends not to escalate to more frequent or intense use patterns.
- *circumstantial use*—use that is generally motivated by the user's perceived need or desire to achieve a new and anticipated effect in order to cope with a specific personal or work problem, situation, or condition. This classification includes students who use stimulants during preparation for exams, long-distance truckers who rely on similar substances to provide extended endurance and alertness, military personnel who use drugs to cope with stress in combat situations, athletes who attempt to improve their performance, and homemakers who seek to relieve tension, anxiety, boredom, or other stresses through the use of sedatives or stimulants.
- *intensified use*—drug use that occurs at least daily and is motivated by a person's perceived need to achieve relief from a persistent problem or stressful situation, or the desire to maintain a certain self-prescribed level of performance. This classification includes youth who have turned to drugs as sources of excitement or meaning in otherwise unsatisfying existences.
- *compulsive use*—a patterned behavior of high frequency and high level of intensity, characterized by a high degree of psychological dependence and perhaps physical dependence as well. The distinguishing feature of this behavior is that drug use dominates the person's existence, and preoccupation with drug-taking precludes other social functioning.

Dynamics of Drug Abuse

A person's pattern of drug use can change with time, and drug use beginning as experimental can eventually develop into compulsive drug abuse. A person's pattern of drug use results from an interaction between environmental factors, such as access, current fads of use, and peer pressure; the person's attitudes and beliefs regarding the use of drugs; the pharmacology of the drug; the psychological predisposition of the individual; and genetic factors. The psychological predisposition and genetic factors can be combined in the term psychobiological predisposition. People differ in their responses to drugs, and drugs that most people use without harm can produce addiction in others. We do not fully know why some people are more vulnerable to drugs than others. However, the notion of an *addictive personality* is generally considered inadequate and misleading.

The difference between *noncompulsive* and *compulsive* use is one of control over the amount and circumstances of use. When a person continues to use as long as the drug is available and the person is capable of using it, that person has lost control over use and needs treatment. An example is the person who intends to drink two beers but regularly consumes to a point of intoxication.

Alcoholics may claim control when they are able to stop drinking for periods of time, but within a drinking episode they may be unable to control the amount consumed except by passing out or running out of alcohol. Many alcoholics do stop drinking for periods of time to prove that they are not out of control: "I can quit drinking anytime I want to." It is easier for the alcoholic to stop completely for a time than to control the amount consumed. For the alcoholic, one drink is too many, and a hundred are not enough.

Addiction and Addictive Disease

Addiction is often equated with physical dependence. Drug abusers wrongly assume that if they are not physically dependent, they are not addicted. This notion leads drug addicts to seek detoxification as the sole treatment for their addiction. However, addiction is more than physical dependence. Physical dependence may or may not be present. Chronic, daily, compulsive abuse of sedative-hypnotics, including barbiturates and benzodiazepines, or opiates will lead to physical dependence. When the drug is stopped, a well-defined withdrawal syndrome will occur. Compulsive, chronic, daily abuse of amphetamine or cocaine produces tolerance and physical dependence, and a withdrawal syndrome that is primarily psychological in nature, including clinical depression and drug craving. However, all psychoactive drugs can produce addiction. The emphasis in treatment is on addiction as a primary problem that exists independently of physical dependence or the circumstances initiating drug abuse.

Addictive disease is a pathological state with characteristic signs and symptoms, as well as a predictable outcome if not treated. As we have mentioned, it is characterized by a compulsive desire for the drug, loss of control when exposed to the drug, and continued use in spite of adverse consequences.

Terms related to addiction can have a unique meaning when related to drug treatment. When used by members of Alcoholics Anonymous (AA), other 12-Step Programs, and an increasing number of programs for the treatment of chemical dependency, the term recovery refers to a person's state of being: acceptance of the inability to control the use of alcohol or any other psychoactive drug, acceptance of alcoholism and addiction as a disease, and the maintenance of sobriety through active avoidance of alcohol and other psychoactive drugs by using peer support and adherence to the 12-Step Programs of AA, Narcotics Anonymous, Cocaine Anonymous, and other recovery-oriented groups. Recovery as a dynamic, ongoing process is a pragmatically useful and valuable concept in the treatment of alcoholism and other drug dependencies.

Treatment can be narrowly defined to include only services rendered by a health professional or can be broadened to include a social model recovery or AA. Medical treatment of drug dependence usually emphasizes crisis intervention, detoxification using medications, psychotherapy, and treatment of physical complications of drug abuse. Treatment can also include maintenance on medications, such as methadone maintenance used for treatment of heroin dependence, prescription of disulfiram (Antabuse) for alcoholism, or treatment of heroin dependence with a narcotic antagonist, such as naltrexone.

Rehabilitation is the process of returning a person to his or her original level of functioning. Although usually referring to return to employment, rehabilitation can also include return to health or to economic or psychosocial functioning. Rehabilitation can be a goal of medical treatment, family therapy, physical therapy, and education, as well as financial, legal, and vocational assistance. In drug dependence disorders, recovery is a necessary step for rehabilitation, but recovery is not the same as rehabilitation. Someone can be in recovery but incapable of returning to work due to skill deficiency or educational obsolescence.

One concept that does not enter into the process of treatment, recovery, and rehabilitation of alcoholics and other drug-dependent persons is that of a cure. To "cure" an addict would mean that he or she could go back to controlled or social use of the drug of choice. By definition, any attempt to do so would result in relapse into active addiction, so such a return is not an option. Rather, addiction is considered to be a progressive, incurable disease that can be brought into remission through a process of recovery and abstention from all psychoactive substances.

DRUGS IN THE WORKPLACE

The use of drugs in the workplace can profoundly detract from work performance and safety. We all have heard the dire consequences of the train engineer who smoked a joint or the forklift operator who downed a pint during lunch break. A significant number of industrial accidents were the direct result of using alcohol or other drugs on the job.

We may be less aware of the workplace effects of alcohol and other drugs used off the job. Many drugs have long-term effects. Some last days, or even weeks or months, as is the case with marijuana and phencyclidine (PCP). Others have aftereffects that can be as dangerous as the drugs themselves.

Drug use by employees in industry—even of illicit drugs—is not confined to unskilled or blue-collar workers. Rather, it occurs among all levels of personnel.

Primary Reasons for Drug Abuse in the Workplace

Motivation for drug use in the workplace is varied and can change over time even for the same drug and person. The motivation for most drug use in the workplace is either to facilitate performance, relieve boredom, in response to addiction, or as a form of self-medication.

performance facilitation—Some drug use is an attempt by the employee to work harder or to be more productive. Pieceworkers may take amphetamines or opiates to enable them to work longer and faster without intolerable discomfort. Executives may use cocaine to allow them to work past their usual fatigue limit. Writers may use alcohol or cocaine to loosen up in an attempt to enhance their creative output.

relief of boredom—Many jobs are inherently routine and offer only rare opportunity for challenge. Many jobs are structured to have a person available in case equipment malfunctions or an atypical condition occurs. The increasing automation of routine tasks means that an increasing number of jobs will have this watchdog function. Workers in this type of situation are extremely vulnerable to drug abuse, and the employee's impairment may not be apparent until there is an emergency that requires alertness and good judgment from the employee to avert disaster.

drug addiction—As an employee's drug use progresses from episodic, weekend, or off-time use to chronic, maintenance use, the person who previously never used at work can begin to bring drugs to the workplace. For employees who value their job and work performance, use at work may not occur until late in the addiction process—long after disruption of home life and other areas of personal functioning.

self-medication for side effects of other drugs—With high-dose recreational drug use of alcohol or cocaine and other stimulants, hangovers, depression, and nervousness can persist throughout the next day. Users can attempt to offset these adverse effects by taking tranquilizers or other medications.

Work Impairment

No matter what the reason for drug use in the workplace, the inevitable result is the impairment of the worker's ability to perform, loss of productivity, and even danger to himself or herself and his or her co-workers. Drug-related work impairment is a reduction in work quality, capacity, or creativity, occurring as a result of an employee's alcoholism or other drug abuse.

In addition to impairment from drug intoxication or hangovers, drug-related work impairment can have several sources:

- chronic use of alcohol, marijuana, or cocaine that results in brain dysfunction persisting beyond the period of intoxication
- drug use away from work that results in family discord or personal problems that preoccupy the employee's mental activities while at work
- incidents while an employee is intoxicated, resulting in long hospitalizations, loss of productivity, or extended use of workers' compensation benefits
- chronic abusers' preoccupation with anticipation of obtaining drugs and their next episode of drug use, resulting in a reduction in motivation for work and all activities unrelated to drug use.

How much impairment is tolerable depends on the work task and the potential for situations requiring peak performance. A commercial airline pilot's impairment due to a hangover, for example, may not be apparent during routine flight conditions, but can be disastrous if the aircraft has an equipment malfunction.

Abuse of illicit drugs can result in additional problems for the employer. Some examples include theft or misuse of company resources to purchase drugs, vulnerability of the employee to blackmail related to the drug use, and association of the employee with criminals while buying drugs. In addition, certain long-acting drugs, such as marijuana, can produce impairment and increase error rate in complex tasks (e.g., a flight simulation), even when the person does not feel that he or she is high or under the influence of the drug.

Roles of Occupational Health and Safety Professionals

Although present standards of practice may allow the industrial physician to ignore surreptitious alcoholism and other drug dependence, the standard of practice is changing. People injured by the drug-impaired employee are suing industrial physicians for failure to diagnose the drug-impaired condition. The reluctance of physicians to make a primary diagnosis of alcoholism or other drug abuse has been reduced by two additional factors. First, respected public figures have acknowledged their personal drug abuse or alcoholism. Second, insurers are providing greater coverage for medical treatment of alcoholism and other drug abuse as primary illnesses.

Occupational health practitioners should routinely inquire about the use of alcohol and other drugs when they take a medical history. Early signs are behavioral. Patients who are concerned about their abuse may deny use or minimize the amount used, but an emotional response to routine questions about their drug use increases the probability that an employee has a problem with substance abuse. An indirect question, such as "Is anyone in your family concerned about your alcohol (or drug) use?" is a good screening question. Many drug users will acknowledge a family member's concern, followed with the reasons why the family member is wrong.

Unless the employee's abuse is out of control, he or she will not come to a medical examination intoxicated, even on a day off work. Most alcoholic employees will make a point of not drinking for a day before the examination, hoping that the company physician will be unaware of their addiction. Abstinence can result in their being in a mild state of alcohol withdrawal with tremor, anxiety, increased pulse rate, increased blood pressure, and perspiration. Besides history-taking and behavioral observation during the examination, information from work supervisors or family members and any history of unusual work injuries can suggest a diagnosis of alcohol or other drug dependence.

Arrests for driving under the influence are public record in many states and can be used to screen for abuse in employees. Anyone may unwisely drive once while intoxicated, but repeating the behavior after adverse consequences indicates that use is out of control.

Organ pathology and physical findings occur in advanced alcoholism. Hypertension (high blood pressure) can result from daily heavy drinking, and treatment of the hypertension is ineffective unless the alcoholism also is treated. The most sensitive laboratory finding for alcoholism is an elevation of a liver enzyme, gamma-glutamic transferase (GGT). Unless explained by liver disease unrelated to abuse or prescribed medication such as phenobarbital that activates GGT, an elevated GGT in an otherwise healthy person is presumptive evidence of alcohol abuse. Another presumptive indicator of chronic alcohol abuse is an increase in the size of red blood cells that occurs due to change in the metabolism of folic acid. The average size of red blood cells is usually measured in a complete blood count, a routine test in physical examinations. Elevations of uric acid and triglyceride are common in alcoholism but may have other causes.

Observing Substance Abusers

Aside from annual and other physical examinations by a physician, there are other approaches to identifying possible abusers. Supervisors and safety personnel can take a primary and direct role in these prescreening approaches. Prescreening methods of identifying possible substance abusers in the work force involve observing physiological and behavioral signs and social effects associated with drug use. These observations are known as presumptive screening. Tests that are simple, inexpensive, and frequently applied can indicate employees who should undergo confirmatory diagnostic screening tests, such as urine testing.

Physiological Signs

The first section of this chapter discussed the nature of the four drug groups and listed some signs of intoxication, overdose, and withdrawal. The outward manifestations of these symptoms can be observed as physiological signs of abuse. Other physical effects of substance abuse can include:

- skin lesions
- infections
- trouble with nose, sinuses, or lungs
- difficulty with balance and coordination.

Some forms of drug abuse produce marked physical effects and visible lesions. Intravenous injection of drugs such as narcotics, cocaine, and amphetamines causes puncture marks and produces scarred veins called tracks. These appear as long, thin lines on the arms or legs and are usually purplish, reddish, or maroon. Generally, the darker and longer the track marks appear, the longer the history of drug use.

Drug injections often cause skin infections, resulting in circular reddish areas that are swollen and tender. Infections can also travel to other areas of the body, such as the lungs or heart valves, where they can cause serious problems that require hospitalization. One common illness in people who inject drugs is hepatitis. The presence of serum hepatitis, as opposed to infectious hepatitis, is always a reason to suspect drug abuse. Further, intravenous (IV) drug users are highly vulnerable to HIV infection, both through sharing unsterile needles and through unsafe sex while under the influence of drugs.

Cocaine users often develop nose and throat trouble due to irritation of these passages by inhaling or snorting cocaine. This can take the form of bloody nasal discharge, a runny nose, infections of the nose or sinuses, or frequent coughing. Runny nose can also result from histaminic discharges in heroin use, and coughing can be a sign of chronic tobacco use. Cocaine users will sometimes lose all sense of smell, and this symptom is usually a tip-off for this type of abuse.

Another symptom suggestive of substance use, abuse, or addiction is drowsiness and trouble with balance and coordination. This commonly occurs in people who are taking sedatives, including alcohol, or opioids, including heroin, and can be very dangerous in the workplace. When other symptoms are not readily identifiable, two signs that often indicate active abuse are pupil size and nystagmus.

eye pupil diameter—The diameter of pupils is a useful, quick, presumptive indicator of recent opiate use. The diameter of normal pupils is 0.12–0.24 in. (3–6 mm). However, following opiate use, the pupils are constricted to 0.04–0.08 in. (1-2 mm) and are commonly called "pinned." Continued opiate use does not make a person tolerant to pupillary constriction, so pupillary constriction is a useful sign even in daily users.

Pupil size is normally altered by the amount of ambient light, so observe the eyes under low-light conditions. Also, pupil size varies substantially from person to person. After observing a person's pupil diameter under a low-light condition, if later you observe a very constricted pupil under similar lighting, there is strong presumptive evidence of opiate use.

Although opiate use is the most common reason for pinned pupils, medications used in the treatment of glaucoma (increased pressure inside the eye) will also produce extreme pupillary constriction that persists under low-light conditions.

However, in an opiate overdose producing unconsciousness, pupils can be dilated even though the overdose was due to opiates. Pupillary dilation during an opiate overdose occurs due to the lack of oxygen in the brain, which causes the pupils to dilate, overriding the pupillary constriction induced by the opiates.

nystagmus—Nystagmus is the persistent 0.04–0.08 in. (1–2 mm) back-and-forth eye movements occurring at eye positions of extreme lateral gaze. This useful presumptive sign of drug intoxication is often used by highway officers in roadside checks. The movements are induced by asking the subject to hold his or her head in a fixed position while tracking a finger, pen, or small flashlight. The object to be tracked is moved across the subject's visual field at a distance of 5–8 in. (12–20 cm) from the face.

It is easier to observe nystagmus if a light source diagonal to the subject's eyes reflects a small point of light from the white portion of the subject's eye. The eye movement has a quick component to the side of the gaze and a slow return movement toward the nose.

Most people will have one to three cycles of nystagmus after the eyes are moved to the extreme lateral position, but the movement stops. In persons who are intoxicated with alcohol or sedative-hypnotics, the back-and-forth eye movements persist—a condition known as sustained horizontal nystagmus. A similar disturbance in eye movements occurs in extreme vertical gaze (vertical nystagmus), which can include a slight rotation of the eyes (rotary nystagmus). Phencyclidine (PCP) intoxication produces severe disturbances in eye movements, and back-and-forth eye movements can occur even when the intoxicated person is looking straight ahead (central nystagmus). Sustained nystagmus is out of the person's voluntary control and is objective, reproducible, and strong presumptive evidence for alcohol, sedative-hypnotic, or PCP intoxication.

Behavioral Signs

Reports of the employee's behavior from spouse, friends, work supervisors, or others are a type of behavioral assessment. Behavioral and psychological effects of alcohol and other drug abuse can include the following:

- late to work or absent
- decrease in performance and productivity
- borrowing money
- change in mood or affect
- confusion.

Drug use sometimes causes no effects other than those produced shortly after the drug is taken; at other times, one or more effects can be seen long afterward. In the first case, even a highly trained observer will not notice any effects unless the drug is detected by a urine or blood test. In the second case, the symptoms of drug use will sometimes resemble those of fatigue, a cold, or other normal events. Thus, though there are definite symptoms of use, abuse, and addiction, they may not always be present or easily identified. One early symptom of addiction can be overachievement, which serves to mask the actual problem.

The first adverse effect of substance use, abuse, or addiction is often a behavioral or psychological change. One of the most common indications occurs when an employee who has been punctual begins arriving at work late, or even misses days, often with no notice. When confronted with this behavior, the employee usually conceals the real cause, providing a justification that can be more or less convincing. In extreme cases, the reasons given for absence can be extremely imaginative. When stimulant psychosis is a factor, the excuses can be increasingly bizarre and include such elements as kidnapping or brushes with the criminal underworld.

Another indication occurs when an employee's productivity declines and he or she performs less efficiently, quickly, and accurately than in the past. Again, when confronted by a supervisor about this behavior, the employee denies the underlying problem.

A third sign of trouble is borrowing money from other employees, often with no good reason, and failing to pay it back promptly. This behavior can often take bizarre forms, such as asking for $20 to get home on the train. If the employee has many friends, and especially if there are long-standing relationships, the friends may not seriously question such requests, even if the requests are unusual. After a substance-abusing employee has been uncovered and friends begin to compare notes, it is common to discover that each of them has loaned considerable amounts of money and not been repaid. This pattern of behavior is especially common with abuse of cocaine, opioids, alcohol, and amphetamines.

The employee's mood or affect can change as a result of drug abuse. The change can take the form of mood swings from low to high and back to low, or the development of persistent mood states such as depression, anxiety, anger, or paranoia. All drugs of abuse profoundly affect the user's mood, and these effects wax and wane according to the person's tolerance, the dose used, the frequency of use, and the duration of the drug's action. Most drug abusers or addicts go through frequent cycles of intoxication and withdrawal, often several times a day, which can produce frequent unexplained mood changes. Such changes are more likely to be seen in abusers and addicts than in casual users, and are more often seen around paydays or after weekends.

Certain drugs of abuse are associated with specific mood states or with mental confusion. Cocaine and other stimulants produce paranoia and anger. Persistent use of depressants, such as barbiturates, or high doses of benzodiazepines produce depression. Hallucinogens can produce lapses in concentration and memory and a range of serious psychiatric disorders.

Behavior in Recovery

In dealing with recovering drug abusers, a sensitive method of behavioral observation for detecting impending or actual relapse to drug use is peer group participation. When recovering drug abusers relate to one another in group settings, those recovering are alert to patterns of thinking that indicate waning commitment to a recovery life-style and may forecast relapse to drug use. Confrontation by peers is often effective in eliciting disclosure of drug use. Peer groups used as a monitoring tool have a sensitivity in detecting relapse that generally exceeds random urine testing (Buxton, 1990).

Social Effects
The social effects of abuse and addiction can include the following:
- family problems
- legal problems
- new "friends."

Drug abusers often develop family problems. These usually result from the financial, mental, and behavioral problems associated with drug abuse. Drug-related family problems can be severe and lead to frequent arguments, separations, and divorce. Domestic violence is often caused by impaired judgment, emotional instability, or negative affects such as anger that are magnified or caused by drug abuse. Conversely, a substance abuse problem in a family member can cause problems for the employee, who then can seek help through such self-help programs as Al-Anon or through the workplace's Employee Assistance Program (EAP). Legal problems are another indicator of drug abuse. Examples are repeated traffic violations and arrests for unspecified reasons, fighting, or disorderly conduct. These social and legal problems can usually be traced to the effects on mood, affect, judgment, and behavior of the specific drugs abused.

Another social danger sign is the employee's development of new "friends" who appear to be of questionable moral integrity. Such relationships are commonly known as *bad associates*. These people can be other drug users or even drug dealers who sometimes come to visit the employee unexpectedly during normal working hours (Woody, 1989; Smith & Wesson, 1984).

Urine Testing
Often there are no overt signs of drug abuse that can be identified with any degree of assurance without further testing. Many employees who use or abuse drugs do so off the job and manage to conceal the problem for a long time. What the observation, adequately documented, can provide is a reason for taking the next diagnostic step in identifying an abusing or addicted employee. That step is typically urine testing.

Medical Review Officer
Given the importance of urine testing and its results, the company may wish to establish a special medical review officer (MRO) position for dealing with positive urinalysis results. Figure 23-1 provides a schematic view of the procedures the MRO would follow in reviewing and acting on positive urinalysis results. Once the final step has been taken in this schematic, it is up to management to initiate discipline, ranging up to termination, or to refer the employee to the EAP for evaluation and referral into treatment.

The MRO's role is described in detail in the *Medical Review Officer Manual: A Guide to Evaluating Urine Drug Analysis* (DHHS, 1988) and is based on the *Mandatory Guidelines for Federal Workplace Drug Testing Programs*, published in the *Federal Register* on April 11, 1988. These two sets of guidelines form the basis for the following model description of an MRO provided by the AC Transit District of Oakland, Calif., which provides public transportation services to California's Alameda and Contra Costa counties.

> I. Qualifications
> The MRO shall be a licensed physician with clinical experience in the field of industrial medicine, drug and alcohol treatment programs, or drug testing programs, and shall have knowledge of substance abuse disorders, medical use or prescription drugs, and the pharmacology and toxicology of illicit drugs.
>
> II. Scope of Work
> A. Receipt and Transmittal of Urine Test Results
> 1. Review of Test Results
> The MRO agrees to review and interpret positive urine test results as received from the authorized antidrug program laboratory, i.e., laboratory results. These tests will be reviewed only for marijuana, cocaine, opiates, amphetamines, and phencyclidine. This review will be conducted in compliance with the provisions of "Procedures of Transportation Workplace Drug Testing Programs," 49 CFR Part 40, Section 40.33.

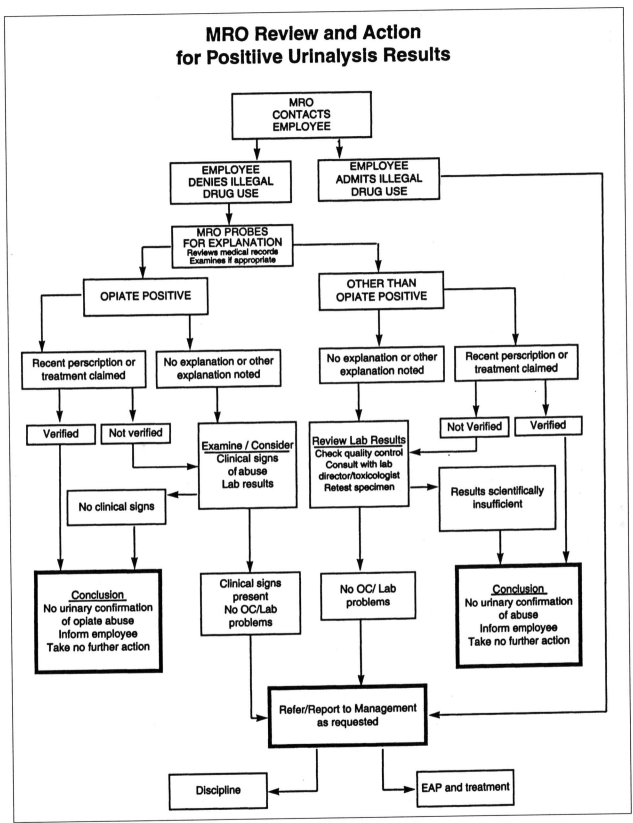

Figure 23-1. MRO review and action for positive urinalysis results. (Reprinted from U.S. Department of Health and Human Services [DHHS]. *Medical Review Office Manual: A Guide to Evaluating Urine Drug Analysis.* Washington DC: DHHS, September 1988.)

Test results will not be reported positive to the Agency Representative until after the Contractor has examined alternate medical explanations. The methods used to verify the validity of the test result shall be at the discretion of the professional judgment of the MRO.

In reviewing the laboratory results, the MRO may conduct a medical interview with the individual, review the individual's medical history, or review any other relevant biomedical factors. The MRO shall review all medical records made available by the tested individual when a confirmed positive test could have resulted from legally prescribed medication. Prior to making a final decision to verify a positive test result, the MRO shall give the individual tested an opportunity to discuss the test result with him or her. If the employee refuses or fails to discuss such issues or provide pertinent medical history for review in such cases, the MRO shall confirm the positive test result. The MRO shall not delay determination of test results beyond the time frames noted herein. It is the employee's responsibility to assist the MRO with the inquiry in a timely manner and failing that, to provide the Agency with such information after positive test result decision of the MRO.

If any question arises as to the accuracy or validity of a positive test result, the MRO should, in collaboration with the laboratory director and consultants, review the laboratory records to determine whether the required procedures were followed. The MRO will then make a determination as to whether the result is scientifically sufficient to take further action. If records from collection sites or laboratories raise doubts about the handling of a sample, the MRO may determine the urinary evidence insufficient and no further actions relative to individual employees would occur. In such situations, the MRO shall note indications of possible errors in laboratory analysis or chain of custody procedures and bring these to the attention of the Agency Representative.

Each DHHS-certified laboratory shall have a Laboratory Scientific Director or a consultant forensic toxicologist available to consult with the MRO on interpretations of laboratory reports.

The MRO can conduct interviews, in person if requested or appropriate, review pertinent medical records, and shall review any information provided by an employee attempting to show legitimate use of a drug. The MRO may perform limited physical examinations, seeking, for example, needle tracks and determining whether clinical signs of drug abuse are present.

Additionally, the MRO, based on review of inspection reports, quality control data, multiple samples, and other pertinent results, may determine that the result is scientifically insufficient for further action and declare the test specimen negative. The MRO can request reanalysis of the original sample before making this decision, according to the procedures specified herein.

If the MRO determines that there is a legitimate medical explanation for the positive test result, the MRO shall report the test result to the Agency Representative as negative.

2. Doctor-Patient Relationship
Physicians shall maintain confidential doctor-patient relationships. Relationships to third parties, i.e.,

insurance companies or employers, shall be governed under legal, procedural, and ethical controls. Information to be shared with such third parties is usually required to be specified in advance and disclosed only with the consent of the patient/employee.

Agency (AC Transit) Plans provide for written notice to employees in advance of specimen collection, describing Agency policy and the actions that will commence if laboratory findings are positive. When, and under what conditions, such positive results and any related information will be shared with management and other sources is also specified in writing.

When an employee with a positive laboratory result is referred, the MRO must define his/her role within the limits of the Agency Plan. The MRO must especially address conditions under which medical and related information will be disclosed as follows:

a. *The MRO's major function is to determine if the laboratory evidence indicating the use of illegal drugs is justified.*

b. *If there is not reasonable medical reason, such as legitimate prescription, or other reason, such as a breakdown in chain of custody or laboratory error, to account for the positive results, these results will be disclosed to management and others as required by the Agency Plan. Any medical information provided that is not specifically related to use of illegal drugs will be treated as confidential and not disclosed.*

c. *If it is determined with reasonable certainty that there is a legitimate medical or other reason to account for the positive laboratory findings, no information identifying the specific employee will be disclosed. Any medical information provided will be treated as confidential.*

d. *Although the MRO may assist in employee rehabilitation efforts related to substance abuse efforts, any such assistance must be in deference to the requirements of the agency's substance abuse policies and procedures.*

3. *Positive Result Standard*
 The MRO agrees to review and make a positive determination for the following group of drugs, which will be known as a "basic panel," only at the minimum levels of concentration.

 All tests will be reported by the laboratory to the MRO. No test results below these minimum concentration levels shall be reported as positive. The MRO shall report on the basic panel only. The MRO shall not report on the presence of any other illegal drug or any legal drug.

4. *The Agency reserves the right to change the minimum levels of concentration determining a positive result. The Agency will notify Contractor in advance of any change including the substances, new cut-off levels and proposed effective date of the change.*

Urine-Testing Instruments and Their Uses

The step of requesting a urine test should not be taken lightly, nor should it be used as the only grounds for disciplinary action or dismissal. Testing of urine and other body fluids is primarily a diagnostic tool. Even when used with utmost care, it can yield misleading results, so it should not be taken as proof positive of drug abuse. Such testing should provide a stage in understanding and possible referral of an employee into further evaluation. That can be accomplished through an EAP.

Urine and breath tests can be very helpful in precisely determining whether drug use or abuse is occurring and, if so, what drugs are involved. Breath tests have been used for many years. They test for alcohol levels in the blood and are usually reliable, though less accurate than a blood test. Several portable test devices in common use can be carried in the hand. Urine testing is much more complicated, as it usually involves an initial or screening test, followed by a confirmatory test, and different test methods are used for different drugs by different laboratories.

All positive tests must be confirmed by a second procedure. Most laboratories use a screening test that attempts to identify a variety of drugs of abuse if they are present. This is a general test designed to detect the presence of a class of drugs.

Because its aims are so general, this test cannot be trusted to confirm the presence of a specific drug. A specimen that is positive on this general screening test must be evaluated by a second procedure that uses a more specialized test directed at a particular drug.

The most commonly used screening technique is an immunoassay technique using a specific enzyme-antibody reaction, such as *Enzyme Multiplied Immunoassay Technique (EMIT), Fluorescences Polarization Immunoassay (FPIA),* or *Radioimmunoassay (RIA)*. The confirmatory test most commonly accepted uses chromatic technology, which is complex and cannot be "beaten" by enzyme inactivity. This is a highly specific, accurate, and expensive test and is generally considered the primary measure of urine-testing technology.

Chang (1987) points out that there are many applications for urine drug tests. These include drug rehabilitation programs, correctional facilities, routine screening of military personnel, preplacement screening, investigation of workplace accidents, and employee monitoring when public safety is involved. Each situation is different, so the requirements for test sensitivity and specificity vary. The variety of approaches available are designed to meet a variety of needs. Most methods for detecting drugs of abuse in urine can be categorized as either chromatographic or immunoassay methods.

With all forms of chromatography, the urine is first extracted with a reagent, and the extract is subjected to a procedure that causes the components to separate. With thin-layer chromatography (TLC), a spot of urine extract is put on a glass plate that has been coated with a thin layer of silica gel (Figure 23-2). The bottom edge of the coated plate is put into a solvent that moves up the plate by capillary action, carrying with it any drugs present in the extract. The drugs are identified by spraying the plate with a solution that reacts with the drugs, producing colored spots. By matching the color and position on the plate with known drug solutions, the technician can identify drugs present in the urine. The TLC is inexpensive, but interpretation of the results requires a skilled technician, and the method is subject to false-positives.

With gas-liquid chromatography (GLC), the urine extract is heated until it becomes a vapor, and the components are separated in a specially prepared column by movement provided by a flow of helium or nitrogen gas (Figure 23-3). The vapor then passes through a detector that may be photometric or mass spectrometric.

Gas chromatography with a mass spectrometry detector (GC/MS) is the most sensitive and specific test procedure commonly used for drug identification and is used primarily for confirmatory testing (Figure 23-4). The sample is first separated into components by gas chromatography, and the mass spectrometer is used to identify the substances emerging from the gas chromatograph. The mass spectrometer subjects the compounds to an electron beam that breaks them into fragments and accelerates them through a magnetic field. Because a molecule of a drug always breaks into the same fragments, known as its mass spectrum, the mass spectrum is like a fingerprint that is compared to known compounds. The mass spectrum is unique for each drug, so the detection by GC/MS is highly specific. The equipment for GC/MS is costly and requires a high degree of technical expertise to conduct the analysis and interpret the results.

In immunoassays, antigens compete for sites on antibodies. The antigens most frequently used for detecting drugs in urine are enzymes, as in EMIT, and enzymes tagged with radioactive materials, as in RIA. Immunoassays are sensitive and can detect small amounts of a drug. The RIA requires use of expensive equipment, radioactive materials, and trained, licensed technicians.

The sensitivity of various procedures in detecting a variety of drugs is given in Table 23-A. The values given are approximations, as the detection limit varies among laboratories. Detection technol-

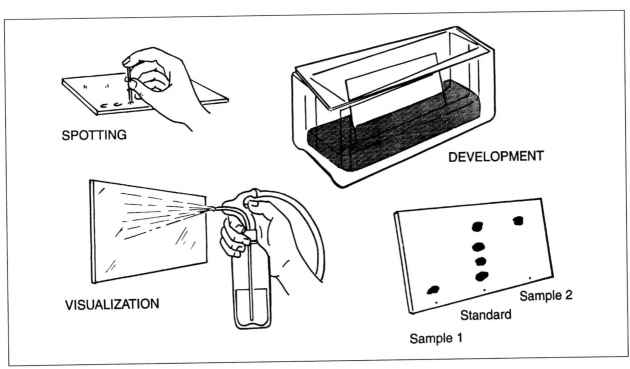

Figure 23-2. Thin-Layer Chromatography. How TLC Works (from left to right): After the urine sample is processed to extract, purify, and concentrate the drugs to be detected, these concentrates are spotted across the lower portion of a plate coated with silica gel. The plate is then placed into a tank containing an organic solvent, which migrates upward through capillary action, carrying along and separating the various drugs of interest. The drugs are visualized by the use of spraying reagents that react with the spots containing drugs and give characteristic colors. The drug is identified by visually analyzing the color and position of spots on the plate, relative to known drug standards.

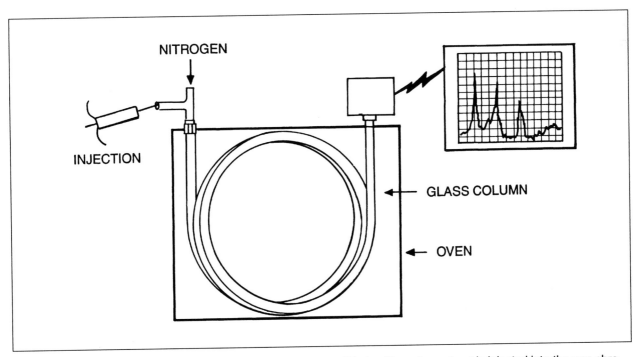

Figure 23-3. Gas Chromatography. How Gas Chromatography Works: The urine extract is injected into the gas chromatograph (GC), where it is vaporized by the high temperature of the oven. The various drugs are swept through a special glass column at different rates of migration. Drug identity is determined by the specific time taken for the drug to emerge from the column, a measurement know as "retention time."

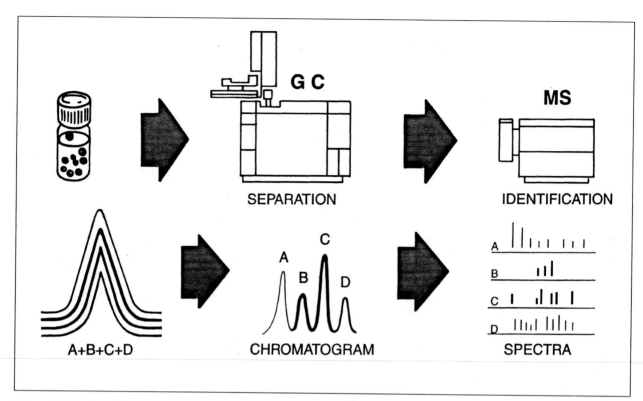

Figure 23-4. Gas Chromatography/Mass Spectrometry. How GC/MS Works (from left to right): The sample is separated into its components by the gas chromatograph, and then the components are ionized and identified by the characteristic spectra produced by the mass spectrometer. The mass spectrometer also allows the analyst to determine the molecular weight of the unknown drug and to confirm the identity of that drug by comparing its molecular weight and unique fragmentation spectrum to that of an analytical standard.

Table 23-A. Sensitivity of Commonly Used Urine Analysis Methods.*

Drug Group	Chromatography			Immunoassay	
	Thin-Layer Chromatography	Gas-Liquid Chromatography	Gas Chromatography/ Mass Spectrometry	EMIT†	Radio-immunoassay*
Radioimmunoassay‡					
Amphetamine	0.5 mg	0.7 mg	10 ng	0.7 mg	1.0 mg
Barbiturates	0.5 mg	0.5 mg	0.5 mg	0.5 mg	0.1 mg
Benzodiazepines			0.5 mg	0.5 mg	
Cannabinoids			1 ng	100 ng	100 ng
Cocaine	2.0 mg	0.75 mg	5 ng	0.75 mg	5 mg
Methadone	1.0 mg	0.5 mg	5 ng	0.5 mg	
Heroin/Morphine	0.5 mg		0.5 mg	0.5 mg	25 ng
Phencyclidine	0.5 mg	150 ng	5 ng	150 ng	100 ng

* All values are either micrograms (mg) or nanograms (ng) per milliliter. The values listed are not precise, as many variables alter the sensitivity of the test in a particular laboratory.
† EMIT is the trademark of Syva Company, Palo Alto, Calif.
‡ Radioimmunoassay levels listed are the lower limit of detection.

ogy is changing rapidly. Periodically ask the laboratory about its detection limits.

Besides sensitivity, another important factor in urinalysis is the length of time after a drug has been used during which it can still be identified in a urine sample. Table 23-B provides currently accepted information on the detection period of the most commonly abused drugs. (Note: There is some clinical evidence that the detection period for chronic use of smokable cocaine [freebase, crack, etc.] may be longer than is stated in this table.)

All urinalysis results are open to interpretation. Table 21-C indicates the most common interpretations placed on urinalysis results.

Table 23-B. Drug Detection Periods.

Drug	Category	Detection Period*
Amphetamines	(Stimulants)	
Amphetamine		2–4 days
Methamphetamine		2–4 days
Barbiturates	(Sedative Hypnotics)	
Amobarbital		2–4 days
Butalbital		2–4 days
Pentobarbital		2–4 days
Phenobarbital		Up to 30 days
Secobarbital		2–4 days
Benzodiazepines	(Sedative Hypnotics)	
Diazepam (Valium®)		Up to 30 days
Chlordiazepoxide (Librium®)		Up to 30 days
Cocaine	(Stimulants)	
Benzoylecgonine		12–72 hours
Cannabinoids (Marijuana)	(Euphoriants)	
Casual Use		2–7 days
Chronic Use		Up to 30 days
Ethanol	(Sedative Hypnotics)	Very short**
Methadone	(Narcotic Analgesics)	2–4 days
Methaqualone (Quaalude®)	(Sedative Hypnotics)	2–4 days
Opiates	(Narcotic Analgesics)	
Codeine		2–4 days
Hydromorphone (Dilaudid®)		2–4 days
Morphine (for Heroin)		2–4 days
Phencyclidine (PCP)	(Hallucinogens)	
Casual Use		2–7 days
Chronic Use		Up to 30 days

* Detection periods vary; rates of metabolism and excretion are different for each drug user. Detection periods should be viewed as estimates. Cases can always be found to contradict these approximations.
† Detection period depends on amount consumed. Alcohol is excreted at the rate of approximately 1 oz/hr (0.03 L/hr).

Special Testing of Designer Drugs

One result of the move toward increased testing for drugs in the workplace has been the increasing popularity of drugs that have either an extremely short drug-detection period or are active at such minute dosages that their use doesn't show up in standard urine tests. The designer analogs of the synthetic opioid fentanyl fit into both categories. The following description of special testing for fentanyl analogs is provided by Chemical Toxicology Institute of Foster City, Calif.

The potent narcotic analgesics fentanyl and sufentanil have relatively short elimination half-lives (approximately 2–6 hr) and are abused in rather low doses (usually 25–250 mg). For these reasons, for testing to detect usage, body fluids must be analyzed within a short period of time after the person last took the drugs.

Anticoagulated whole blood (sodium fluoride or EDTA) or serum (1.2–1.5 oz [4–5 mL] of either one) should preferably be drawn within six hours of the last suspected usage and certainly no later than 24 hours after use. Urine (minimum, 3 oz [10 mL]) should be collected within 48 hours of last usage and certainly no more than 72 hours afterward. Blood and urine fentanyl and sufentanil concentrations are usually in the range of 0.5–5 ng/mL in specimens collected within these time limits from known abusers; the detectability of the gas chromatography/mass spectrometry method used is 0.5 ng/mL.

Legal Challenges

A variety of court challenges to drug-testing programs are in process. The issues include right to privacy, unreasonable search and seizure, due-process violations, negligence, violation of collective bargaining agreements, and wrongful terminations. The courts have accepted drug testing in the workplace, if it is related to workplace safety. The challenges relate to inappropriate policy and procedure, such as violation of confidentiality or inaccurate laboratory procedure.

Urine-Testing Procedures

The U.S. Department of Health and Human Services (DHHS) has issued mandatory Guidelines for Federal Workplace Drug Testing Programs. These

Table 23-C. Interpretation of Drug Urinalysis Results.

Results Positive For:	Indicates:
1. Morphine	■ Heroin, morphine or codeine use. ■ Poppy seeds often contain morphine and can cause positive test results. ■ Some prescription drugs contain morphine or codeine.
2. Codeine	■ Codeine use alone.
3. Morphine and Codeine	■ Possible heroin use alone, since street heroin often contains codeine as an impurity in the preparation. ■ Possible codeine use, since morphine is produced by the body when codeine is ingested. ■ Any combination of morphine, codeine, or heroin use could lead to this result.
4. Meperidine (Demerol)	■ Meperidine use only.
5. Methadone	■ In patients with normal liver function, this result indicates that methadone was added directly to the urine sample and not ingested. ■ Pregnancy may alter metabolism so that a person taking methadone may show this result. (Note: Methadone is generally available only through narcotic treatment programs.)
6. Methadone Metabolite	■ Methadone use.
7. Methadone and Metabolite	■ Methadone use.
8. Propoxyphene (Darvon)	■ Propoxyphene use. ■ Propoxyphene appears in a number of prescription medications including Darvon and Darvocet–N.
9. Pentazocine (Talwin)	■ Pentazocine use. (Note: This prescription analgesic is popular in some regions as a substitute for heroin.)
10. Hydromorphone (Dilaudid®*)	■ Hydromorphone use only.
11. Quinine	■ Quinine use. ■ Also an ingredient in tonic water. ■ Quinine is present in some medications, including antimalarial drugs. (Note: Still used in some parts of the United States as an adulterant in heroin.)
12. Amphetamine	■ Amphetamine use. (Note: Very rarely used as a prescription drug.)
13. Methamphetamine	■ Methamphetamine use. (Note: Much more common as a street drug than amphetamine. Very rarely used as prescription medication.)
14. Amphetamine and Methamphetamine	■ Methamphetamine use. ■ Amphetamine and methamphetamine use.
15. Cocaine and/or Cocaine Metabolite	■ Cocaine use. (Note: Local anesthetics such as procaine and lidocaine will not result in a cocaine positive.)
16. Cocaine Metabolite	■ Cocaine use.
17. Phenylpropanolamine	■ Phenylpropanolamine use. (Note: Widely available over the counter in diet aids and decongestants. Common ingredient in street stimulants.)
18. Procaine (Novacaine)	■ Procaine use. (Note: Sometimes used as an adulterant in street cocaine and heroin.)
19. Barbiturates Amobarbital Pentobarbital Secobarbital Butabarbital Phenobarbital Butalbital	■ Barbiturate use. (Note: Many prescription drugs contain barbiturates.)
20. Benzodiazepines	■ Benzodiazepine use. (Note: By prescribed use only.)
21. PCP	■ PCP use. (Note: No prescription drugs contain PCP.)
22. Methaqualone (Quaalude®)	■ Methaqualone use only.
23. Alcohol	■ Alcohol use only.
24. Marijuana	■ Use of marijuana or other cannabis products. (Note: No known prescription or nonprescriptive drugs will give positive marijuana results.)

federal standards are considered a model for all urine-testing programs. DHHS specifies that an immunoassay be used for the initial screen, followed by GC/MS confirmation.

It is important to understand that urine must be tested twice to be confirmed and that the second test should be of a different type than the first. It is also important to regularly check the reliability of the laboratory, even when it uses two procedures. Some companies routinely divide the initial sample into two parts, sending one portion to the lab and freezing the second portion. If the first test turns out positive, the second sample is then tested. This provides yet another safeguard against a false-positive test result.

Urine must be collected under controlled conditions that ensure the proper identity of the sample, establish a valid chain of custody, and prevent adulteration, dilution, or substitution of the sample. The standards of the National Institute on Drug Abuse (NIDA) require samples to be collected unobserved, except in cases where there is reasonable suspicion that persons may attempt to adulterate their samples.

The urine sample must be properly labeled and securely transported to the lab. This is extremely important when someone's job is at stake. The methods for labeling and transporting urine specimens often resemble those developed for custody of evidence in criminal proceedings. The laboratory must demonstrate a "chain of custody." Records must prove that the specimen was securely transported from the clinic to the laboratory, the results were accurately measured and recorded, and the findings were accurately reported back to the clinic.

Forensic Standards for Collecting and Handling Samples

The forensic standards for collecting and handling samples may be more strict than the general guidelines, but they are more apt to stand up to a legal challenge. First, the transfer of urine, blood, or saliva from the subject to the container must be witnessed. If a physician collects the blood sample, the physician becomes the first link in the chain of custody. Most physicians do not understand the requirements of legal chain of custody procedures and must be instructed. Otherwise, physicians will handle the blood sample with usual clinical laboratory standards, which will not stand up to challenge by a knowledgeable lawyer. Because the physician may be an intern or a resident who will leave the geographical area within a year or so, and therefore not be available as a witness, the collection of the blood sample should be observed by someone who could give testimony regarding the collection of the sample. The observer must verify the accuracy of the container label (including the subject's name and other identifying information, date, time of collection, and type of collection receptacle) and must maintain the chain of custody of the sample until it reaches the laboratory.

Many drug users are skillful and creative in devising methods of deceiving an observer of urine samples. For example, men can tape a small tube to their penis and deliver a sample urine from an attached balloon. Users may also attempt to foil test results by adding an adulterant to their urine sample when they know drugs are present. The sophistication of some users is remarkable. Laboratory manuals listing reasons for false-negatives have value on the black market among some groups of drug users, and common means of foiling detection of drugs in urine are published in drug paraphernalia magazines.

In case it becomes necessary to replicate the urinalysis, the sample should be divided into two, with one-half of the specimen frozen and stored. Urine, blood, and saliva are subject to biological degradation over time, which can alter the concentration of the drug contained in the sample. In general, freezing the sample will retard biological degradation—however, stability of the sample is not assured.

The user of laboratory results has a need and a right to know how samples are handled within the laboratory as well as the analytical techniques used. In addition, the laboratory should make available information on the threshold sensitivity for qualitative results and should provide reasons for false-positives and false-negatives.

REFERRAL AND TREATMENT

There is a growing trend in industry backed by the American Society of Addiction Medicine (ASAM) to use the positive drug screen as a method for early identification of workers with a drug problem, and diverting them to treatment. This diversion to treatment is best accomplished by referral to the company's EAP, which can be in-house or an outside agency.

Role of the Employee Assistance Program (EAP)

The employee is referred initially to the company's Employee Assistance Program (EAP), and the EAP evaluator may refer the employee into counseling or some other form of treatment for chemical dependency. Many companies develop a set of EAP procedures, usually in conjunction with the EAP.

The EAP may be an internal entity, administered in-house, or an external organization that has been contracted to set up and administer a program for the organization. In choosing an EAP provider, an employer should be sure that the services offered fit the problems of the organization. One way to do this is to hold meetings at which the EAP provider explains the services it provides and answers any questions from the company. Once the company has adopted a program, the EAP provider should provide a general drug education and training program for supervisors, coupled with drug education and awareness for employees.

If the employer sets up an in-house program rather than contracting for an external program, the program can have as narrow or wide a scope as the employer chooses. It can have the sole objective of maintaining a drug-free workplace, or it can include in-house counseling of employees with alcohol and drug problems.

Whether the EAP is internal or external, the employer should select a counselor, usually a psychologist or other degreed and licensed professional, who will be available to employees 24 hours a day. The counselor and management need to agree on procedures:

- time during or outside of work hours
- location
- office facilities that will be available to employees
- general procedure for referrals from supervisors
- general procedures for employees who ask for help voluntarily
- confidentiality (Bureau of Business Practice, 1987).

Essentially, it is the role of the EAP to evaluate the employee, whether referred by the company directly or self-referred, and when deemed necessary, to refer the employee into treatment.

Confidentiality

Confidentiality is a critical issue in the EAP process. It is important to all levels, including management and safety officers. The following discussion is presented by Judith R. Vicary, PhD (1989).

The issue of confidentiality and release of information is always a great concern of EAP staff and is a critical aspect of any EAP, particularly one that intends to foster self-referrals. It becomes an even more important and sensitive concern when use of illegal drugs is involved. The EAP professional will be held accountable only for those files that become a part of an employee's treatment or referral record within the EAP. Some records maintained by the EAP and employer, e.g., medical records and job performance data, may be afforded legally enforceable confidentiality protections arising from a variety of sources, particularly state laws and union contracts and employment agreements. EAP records may also be afforded some protections arising out of professional ethics and the well-considered comprehensive and personnel policies of the employer. However, confidentiality is a complex matter, almost always involving the balance of competing rational interests in both disclosing and not disclosing certain personally identifiable information in particular circumstances. Unless that information and those circumstances are the subject of a thorough and detailed written policy founded on a commitment by the employer, a general affirmation of confidentiality will take on meaning only as real-life situations present themselves.

The records of alcohol and drug abuse clients in an EAP may be protected under detailed federal statutes and implementing regulations, if the corporation involved is federally conducted, regulated, or assisted, either directly or indirectly. These regulations prohibit disclosure of EAP alcohol or drug abuse patient records, with the following exceptions:

1. disclosure if the employee consents in writing in accordance with detailed provisions of the regulation.
2. disclosure without patient consent if the disclosure is to medical personnel to meet any individual's bona fide medical emergency or to qualified personnel for research, audit, or program evaluation. These qualified personnel may not disclose patient identifying information.
3. disclosure pursuant to a court order after the court has made a finding that "good cause" exists for purposes expressly set out in the regulations.

Federal confidentiality protections do not apply to the following:

1. reporting under state law of incidents of suspected child abuse and neglect
2. communications within an EAP or between a program and an entity having direct administrative control over an EAP

3. communications between an EAP and a "qualified service organization"
4. disclosures to law enforcement officers concerning a patient's commission of or threat to commit a crime at the EAP site or against personnel of the EAP.

Though the federal confidentiality regulations have exceptions to the prohibition on disclosure of alcohol and drug abuse patient information, they do not create a requirement to make any disclosure, that is, permitted disclosures are not compelled by the federal confidentiality provisions. There is a criminal penalty for making prohibited disclosures. Each alcohol or drug abuse client in a program covered by the federal regulations must be told about these confidentiality provisions and furnished a summary in writing.

While these regulations do not apply to all EAPs and the companies that they serve, their detailed provisions afford some insight into the range of consideration necessary to the development of a meaningful confidentiality policy. Company policy and procedures should make responsibilities regarding confidentiality clear to both EAP staff and employees, taking into consideration any applicable law or labor agreement.

Safety

Employers are responsible for providing a safe work environment—a position supported in a number of recent cases related to drug use. If any employee is clearly unfit for duty, the company is liable for the safety of other employees as well as for the public good in the event of an incident. The safety officers and EAP staff must be sure that supervisory training stresses the need to intervene with an employee whose on-the-job behavior suggests impairment or that the employee is a threat to self and/or others.

CRITERIA FOR REFERRAL INTO TREATMENT

Although referral of employees into treatment is specifically the task of the EAP, some sense of criteria can be helpful to management and to health and safety officers.

User-Program Relationship

The attitude of the user toward treatment and personal recovery is often the first criterion when seeking potential treatment referrals. Does the treatment program accept clients with employee substance abuse problems? Does the program staff discuss these freely? The program staff should be well versed on drugs and knowledgeable about their effects. Be wary of a physician or program that is all too ready to prescribe yet another drug for the employee's problem. An example of this is the physician who thinks he or she is solving the problem of a Valium dependency by shifting a patient to another benzodiazepine, such as Librium or Ativan. All sedative-hypnotic drugs produce cross-dependency, and such a strategy is like prescribing vodka for an alcoholic whose drug of choice is gin.

Do the program director and staff seem well versed in the available treatment options that provide maximum continuity of care? Do they know, for example, which recovery programs will be uniquely helpful for the particular patient? These can vary a great deal. For example, someone with a dual diagnosis—say, an alcoholic who is taking lithium for a bipolar depression should be involved with a recovery group that doesn't object to the patient's prescribed use of the lithium necessary for treating the bipolar disorder.

While there are many legitimate and effective treatment programs in the field, others can border on the sadistic in their approach or provide expensive amenities in a country club atmosphere without any real lasting treatment. The referring agent needs to know what procedures have worked with particular drug problems and what sorts of treatment programs offer them. For example, someone with a low-dose codeine dependency shouldn't go directly into methadone maintenance, a treatment better suited to hard-core opioid addicts. A person who has a history of chronic pain or who controls a seizure problem with phenobarbital shouldn't be placed in a drug-free therapeutic community (Seymour & Smith, 1987).

Diversion Programs for Impaired Professionals

In recent years, an increasing number of professionals have developed specific and highly evolved diversion and treatment programs for their alcohol-impaired and drug-impaired members. These are often implemented in the philosophy that a highly trained, highly motivated professional who is suffering from chemical dependence can be intervened upon, be treated, and reenter professional work with a very good chance for recovery, while loss of such a person through criminal action and/or dislicensure would be a disaster to him or her and the training and development of a replacement much more costly than the treatment program regimen.

A pioneer in this approach has been the medical profession, where state-mandated impaired-physician programs are providing a model for other health professions and for professions beyond the health field. In California, which developed one of the first state diversion programs for physicians, legislation enacted in 1979 established three primary supports for physician diversion:

1. It allowed physicians in a diversion program to keep their licenses intact, even though they may have violated the law.
2. It established confidentiality for physicians in the program.
3. It directed that appointments to the Diversion Evaluation Committee be made strictly on the basis of expertise.

On the basis of data and attitude, the committee decides pragmatically whether a physician is treatable. The decision is based on whether or not the physician is involved in his or her own recovery. The committee looks for an attitude that "I have a problem with addiction. I can overcome this problem with medical help, but I always have to work at it, keep at top priority abstinence from drugs or alcohol, and do what people tell me to do to maintain sobriety." That attitude ensures at least a chance of success.

Once a candidate is accepted, the committee combines treatment elements and a monitoring schedule based on the candidate's needs and circumstances into an itemized treatment contract. The following are examples of items that can be included in such a contract:

- Follow directions of the assigned treating physician.
- Abstain from alcohol or drug use.
- Consent to random testing of body fluid at specified time intervals.
- Participate in peer support group meetings and/or 12-Step meetings with a specified frequency and verified attendance.
- Participate in individual psychotherapy.
- Take Antabuse or participate in a naltrexone program if appropriate.
- Practice in a supervised work setting; get the approval of the committee before changing the work setting.
- Voluntarily relinquish DEA number (permit from the Drug Enforcement Administration for prescribing controlled medications).
- Live in a specified treatment facility.
- Participate in a certain number of hours (up to 25) of continuing medical education related to appropriate prescribing practices. (These hours are in addition to those required for renewal of license in California.)
- Cooperate with periodic exams by a psychiatrist, who will report findings to the Diversion Evaluation Committee (Shapiro, 1980; Smith & Seymour, 1983, 1985).

REENTRY AND REHABILITATION

A growing movement within the helping professions acknowledges that addictive disease is an occupational hazard in the health professions and many other professions as well. This movement is concerned about addictive disease and the disease process. Addiction is recognized as a treatable disease, and correct treatment requires total abstinence from all psychoactive substances. It is critical to allow the recovering person to return successfully to work armed with healthy, growth-oriented new responses. Early intervention based on care and concern for the health of an addicted person and treatment is the only appropriate response to this occupational and social problem.

Treatment for chemical dependency and addiction is multifaceted, often beginning with observation and testing to detect the employee in need of help, and continuing through intervention and referral to an EAP counselor for evaluation. The employee can then enter specific treatment for his or her drug problem. As we have pointed out, this treatment can take many forms: inpatient, outpatient, long-term residential, etc. In any event, it will probably mean an extended absence from the workplace.

After the employee has entered into supported recovery and abstinence from all psychoactive substances and has maintained that recovery for a prescribed length of time, the next aspect of treatment is reentry into the workplace. In the most successful treatment regimens, a carefully monitored reentry at the worksite is considered part of the ongoing spectrum of treatment, part of the continuity of care.

The need to monitor reentry at the workplace is especially critical for health professionals, who will once more have access to psychoactive medications, but it is important for all employees reentering the workplace. Recovering persons who have undergone a monitored reentry have had a two-times-higher compliance rate than those who have not.

Often, monitored reentry is initiated through a contract among the patient, the treatment program medical director, a monitoring and reentry counselor, and the monitoring and reentry coordinator. Experience shows that the more detailed and precise this document is, the better. Millicent Buxton, a pioneer in the field of monitoring and reentry, has developed a basic contract format for monitoring and reentry of health professionals (Buxton et al, 1985). That contract format could be adapted to the reentry of any employee or professional person. In such a contract, the reentering employee can agree to some or all of the following terms:

- to abstain from alcohol and all other psychoactive substances
- to meet regularly with a monitoring and reentry program or committee in order to review progress and compliance
- to attend and document attendance at a stipulated number of meetings based on a combination of 12-Step, Alcoholics Anonymous, Narcotics Anonymous, and Cocaine Anonymous meetings, aftercare meetings, and professional support group meetings. Planned absences from any of these meetings would need to be preapproved and made up as soon as possible.
- to get a 12-Step sponsor
- to submit to random analysis of body fluids regularly and as needed
- to take Antabuse or participate in a naltrexone program for the duration of the contract, and to keep his or her primary care physician apprised of the treatment. Besides providing an additional hedge against the cues and temptations presented by the availability of drugs at the workplace, naltrexone use acts as a protection for the opioid-using employee if questions of possible relapse and use arise in the workplace.
- to obtain a worksite monitor, who will agree to maintain regular contact with the monitoring and reentry program, review the employee's work, and promptly report any suspected use or unusual behavior.
- to maintain no controlled substances in the office and to remove any there, if the reentry is of a physician or other health professional. The monitor should make a detailed list of what substances are allowed to remain. Further, all medical activities, including patient charts, prescriptions, and dispensing, must conform to accepted medical practice and record-keeping requirements.

The contract should include provisions for what is to be done in the event of a relapse. Measures might entail notification of the program and the person's primary care physician, return to direct treatment, and a variety of consequences involving reevaluation, augmentation of the treatment plan, corrective action, possible loss of privileges, and other responses.

The agreement also should provide for periodic reevaluation and flexibility as the patient moves further into recovery and compliance with the contract's particulars, 100% compliance with the contract, and eventual completion of terms and fully accomplished reentry into the workplace.

SUMMARY

Workers' use of psychoactive drugs is costly for their employers as well as for their loved ones. Safety and health officers and others in the workplace who are responsible for the safety and well-being of employees must be able to identify potential abusers, who may be a danger to themselves and/or co-workers. Observation of employees and their behavior is the first step in detecting a problem. This may be followed up with a variety of formal testing methods.

It is possible to solve a drug abuse problem in a humane and pragmatic way. Identifying, treating, and rehabilitating a chemically dependent employee is more cost-efficient than simply firing him or her and makes the most sense in terms of legal, labor relations, and other employer/employee concerns. The process often has a happy ending for all concerned, vastly improving the life of the employee and providing the employer with a sober and recovering effective worker.

REFERENCES

Bureau of Business Practice. *Drugs in the Workplace: Solutions for Business and Industry.* Englewood Cliffs NJ: Prentice-Hall, 1987.

Buxton M. Monitoring, reentry, and relapse prevention for chemically dependent health care professionals. *J Psychoactive Drugs* 22(4): 447–450, 1990.

Buxton M, Jessup M, Landry MJ. Treatment of the chemically dependent health professional. In Milkman HB, Shaffer HJ (eds). *The Addictions: Multidisciplinary Perspectives and Treatments.* Lexington MA: Lexington, 1985.

Chang JY. Drug testing and interpretation of results. *PharmChem Newsletter* 16(1), May 1987.

Drury DL, Masci V, Jacobson JW, et al. Urine drug screening: Can counterfeit urine samples pass inspection? *J Occup Environ Med* 41(8):622–624, 1999.

Fraser AD, Worth D. Experience with a urine opiate screening and confirmation cutoff of 2000 ng/mL. *J Anal Toxiol* 23(6): 549–551, 1999.

Hamid R, Deren S, Beardsley M, et al. Agreement between urinalysis and self-reported drug use. *Subst Use Misuse* 34(1):155–159, 1999.

Hoffman BH. Analysis of race effects on drug-test results. *J Occup Environ Med* 41(7):612–614, 1999.

Jaffe JH. Drug addiction and drug abuse. In Goodman LS, Gilman A (eds). *The Pharmacological Basis of Therapeutics.* New York: Macmillan, 1980.

Joseph RE Jr, Hold KM, Wilkins DG, et al. Drug testing with alternative matrices. II. Mechanisms of cocaine and codeine deposition in hair. *J Anal Toxicol* 23(6):396–408, 1999.

National Commission on Marijuana and Drug Abuse. *Drug Use in America: Problem in Perspective.* Washington DC: U.S. Government Printing Office, 1973.

Schilling RF, Bidassie B, El-Bassel N. Detecting cocaine and opiates in urine: Comparing three commmercial assays. *J Anal Toxicol* 23(6):549–551, 1999.

Seymour RB. *MDMA.* San Francisco: Partisan, 1986.

Seymour RB, Smith DE. *The Physician's Guide to Psychoactive Drugs.* New York: Haworth, 1987.

Seymour RB, Smith DE. Drugfree. *A Unique, Positive Approach to Staying Off Alcohol and Other Drugs.* New York: Facts on File, 1987.

Shapiro R. California's diversion program for physicians. *California Society for the Treatment of Alcoholism and Other Drug Dependencies Newsletter* 7(4):1–3, 1980.

Smith DE, Gay GR. *It's So Good Don't Even Try It Once.* Englewood Cliffs NJ: Prentice-Hall, 1972.

Smith DE, Milkman HB, Sundeerwirth SG. Addictive disease: Concept and controversy. In Milkman HB, Shaffer HJ (eds). *The Addictions: Multidisciplinary Perspectives and Treatments.* Lexington MA: Lexington, 1985.

Smith DE, Seymour RB. Commentary. *Addictions Alert* 3(8), October 1989.

Smith DE, Seymour RB. Impaired health professionals. *Inter J Addictions* 20(5), 1985.

Smith DE, Seymour RB. Misprescribing: A question of perspective. In Morgan JP, Kagan DV (eds). *Society and Medication: Conflicting Signals for Prescribers and Patients.* Lexington MA: Lexington, 1983.

Smith DE, Wesson DR. Substance abuse in industry: Identification, intervention, treatment and prevention. In Smith DE, Wesson DR, Zerkin EL, et al (eds). *Substance Abuse in the Workplace.* San Francisco: Haight-Ashbury, 1984.

Smolle KH, Hofmann G, Kaufmann P, et al. Q.E.D. Alcohol test: A simple and quick method to detect ethanol in saliva of patients in emergency departments. Comparison with the conventional determination in blood. *Intensive Care Med* 25(5):492–495, 1999.

U.S. Department of Health and Human Services (DHHS). *Medical Review Officer Manual, A Guide to Evaluating Urine Drug Analysis.* Washington DC: DHHS, September 1988.

Vicary JR. Overview of drug abuse and employee assistance programs at the workplace. *In Drug Abuse Curriculum for Employee Assistance Program Professionals.* Rockville MD: National Institute on Drug Abuse (NIDA), 1989.

Weeks S, Flatt S, Singleton C. Drugs test cop-out? *Nurs Stand* 13(40):22–23, 1999.

Woody GE. Pharmacology of abused drugs. In *Drug Abuse Curriculum for Employee Assistance Program Professionals.* Rockville MD: NIDA, 1989.

Yesavage JA, Leirer BO, Denari M. Carry-over effects of marijuana intoxication in aircraft pilots' performance: A preliminary report. *Am J Psychiatry* 142:1325–1329, 1985.

chapter 24

Scheduling Shiftwork

by Richard M. Coleman, PhD
Joseph LaDou, MD
revised by Gary R. Krieger, MD, MPH, DABT
Marci Z. Balge, RN, MSN, COHN-S

431 **Introduction**
432 **Criteria for an Optimal Schedule**
Business needs ■ Employee preferences ■ Health requirements
435 **Creating a Safe Work Schedule**
Factors to consider ■ Schedule options
440 **Shiftwork Education**
440 **Shiftworkers and Health**
440 **Summary**
441 **References**

INTRODUCTION

Shiftwork is defined as work done primarily in other-than-normal daytime work hours. According to an article in the *Wall Street Journal* (1988), the effects of reduced alertness on productivity and safety are costing U.S. companies that use shiftwork an estimated $70 billion a year. Some catastrophic accidents that resulted in part from shiftworker errors were the wreck of a large oil tanker and the accidents at Three Mile Island (Pennsylvania), Bhopal (India), and Chernobyl (Ukraine). Concern about such large-scale disasters, coupled with the increasing expense and sophistication of modern operating technologies, has intensified the need to develop optimal work scheduling.

The starting point for understanding schedules is defining the work load of a business. Depending on the workload, shiftwork can be classified into four categories:

1. *continuous, balanced coverage*—24 hours a day, 365 days a year, with a constant workload (e.g. nuclear power plants, refineries).
2. *continuous but unbalanced coverage*—24 hours a day, 365 days a year (e.g., service industries, police forces, hospitals, maintenance departments). With a nonuniform workload more coverage is required on the daylight-hour shifts (Luna, 1997a, 1997b).

3. *shift coverage required by economic demand*—not necessarily 24 hours a day, seven days a week (e.g., automobile plants, manufacturing plants). Because shifts can be cut back depending upon business climate, they may not be needed at certain hours or days of the week.
4. *irregular shiftwork*—where occasional shiftwork is required and scheduling can be unpredictable (e.g., train crews, truck drivers, some service industries) (Prunier-Poulmaire et al, 1998).

CRITERIA FOR AN OPTIMAL SCHEDULE

Thousands of shiftwork schedules are in use. Most of them came about through negotiations, inheritance as part of a company's tradition, or duplication of a nearby facility's method. Thus, the same shiftwork schedules may function poorly at different facilities.

Companies rarely implement schedule recommendations based on health concerns alone. In contrast, an optimal schedule should be planned to meet three criteria. These include business needs, employee desires, and the facility's requirements for safety and health (Figure 24-1).

Business Needs

From a business viewpoint, the essential requirement of work scheduling is to match the workload, providing low-cost, reliable staff coverage (Coleman, 1989). In practice this can be quite difficult. Workloads can be unpredictable, seasonal, or vary from day to day. Because few schedules are designed specifically for a facility, most incorporate routine overstaffing and understaffing. As a result, safety can suffer. Overstaffing can lead to complacency and lack of alertness, while understaffing is often associated with increased risk-taking.

Staffing levels are critical. At one rolling mill, shiftworkers were required to work 2,500 hours per year—640 hours more than the average American day worker puts in (equal to 16 extra 40-hour workweeks). As far back as World War II, this level of work hours was found to be associated with decreased safety and productivity (Colquhon & Rutenfranz, 1980).

Employee Preferences

For shift employees, the number one prerequisite for a good schedule is adequate time off. Workers who do not get enough time off can suffer from

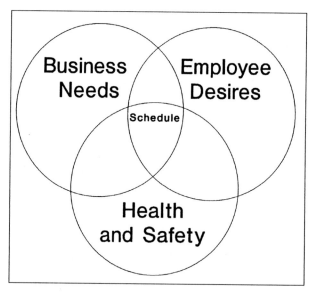

Figure 24-1. An optimal schedule should be planned to meet these three criteria: business needs, employee desires, and facility requirements for safety and health.

psychosocial problems, and these problems can be as difficult to cope with as physiological problems. The conditions and situations reported cover a broad spectrum, from divorce to spouse-battering, family violence, social isolation, and sexual dysfunction. On certain schedules, shiftworkers have their social contacts primarily with co-workers on the same crew. They can become isolated from family, friends, and community. Many workers resent their inability to attend their children's activities, sports events, religious services, and other activities that are convenient for daytime employees. Few schedules allow workers to trade shifts in order to obtain time off whenever needed. Companies with schedules that allow trading often require that the worker get permission before trading shifts; and then complete the trade by working two shifts in a row or returning after taking off only one shift.

Sleep deprivation and associated poor mood also can disrupt a worker's social and family life. A common schedule that provides for occasional long weekend breaks after a stretch of night shifts often results in poor-quality time off.

A schedule that builds in quality, quantity, and flexible time off can solve most of these problems. In fact, in many companies we have worked with, we have instituted new schedules with time off, which seems so attractive that day workers bid into shiftwork (for example, 6–10 weeks of vacation per year).

Health Requirements

A good shiftwork schedule promotes alertness and safety on the job, as well as good long-term health. Because this chapter is intended for occupational health and safety professionals, emphasis will be on this dimension of scheduling. The designer of the work schedule should be familiar with three key health variables:
1. daily sleep requirement
2. circadian rhythms
3. coping strategies.

Many shiftwork research scientists and schedule design consultants pay attention to only one or two of these critical variables; the results were less productive than expected.

Daily Sleep Requirement

It is possible to accurately measure the amount of sleep a person needs each day. Assuming the human body needs enough sleep to be alert the next day, researchers at the Sleep Disorder Center at Stanford University Medical School developed a measurement called the multiple sleep latency test (MSLT). The MSLT determines a person's need for sleep based on how fast the person falls asleep when given the opportunity to nap (Carskadon, 1989). For example, a day worker who slept for six hours at night would receive a series of nap "test" opportunities at 10:00 a.m., noon, 2:00 p.m., 4:00 p.m., and 6:00 p.m. Each nap opportunity took place in a dark, quiet room lasting up to 20 minutes. If the subject did not fall asleep during any of the nap opportunities, he or she would receive a score of 20, which indicates optimal alertness with six hours of daily sleep. If, in contrast, the subject fell asleep in three to four minutes on average, this low score would indicate serious sleep deprivation; six hours of daily sleep would not be enough for that person to maintain daytime alertness. Patients with narcolepsy, a genetically induced sleep-wake disorder, typically score within this low range. Thus, there are normative and pathological values for the test that reveal whether total sleep time is sufficient for daytime alertness.

The average U.S. day worker obtains seven to eight hours of sleep each night and scores between 10 and 15 on the test, indicating chronic mild sleepiness. When nighttime sleeping increases to nine hours, daytime sleepiness virtually disappears.

Therefore, sleep researchers suggest that close to nine hours of sleep each night is optimal for the average U.S. worker to achieve full daytime alertness. Of course, most of us can rarely afford the luxury of going to bed when we feel sleepy and waking up when we are refreshed. Most U.S. day workers wake up and go to sleep according to external schedules—alarm clocks, television shows, and work schedules—rather than the way their body feels.

The problem is even greater for shiftworkers. The average U.S. shiftworker obtains 6.7 hours of sleep on the day shift, 7.1 hours on evening shift, 6.6 hours on night shift, and 8.3 hours on extended-time-off breaks (Coleman & Associates, Inc., U.S.A. National Shiftworker Database). Sleep researchers at Stanford documented that effects of mild sleep deprivation are cumulative (Carskadon & Dement, 1987). Although one night of reduced sleep may not significantly impair alertness, the impact grows on each succeeding night. Thus, the sleep patterns of shiftworkers cause sleep deprivation to build up. Work schedules should include frequent days off to prevent this buildup of chronic sleep deprivation (Knauth, 1997; Gillberg, 1998).

Circadian Rhythms

Extensive research reveals that, like plants and animals, human beings have a natural biological clock, a pacemaker that controls more than 100 physiological variables, including cycles of sleep and alertness. A day worker can fall asleep at 11:00 p.m. because his or her biological clock is set for sleep at that hour; it is "in sync" with the clock on the wall. The reason shiftworkers cannot fall asleep at 8:00 a.m. after their first night shift is not that the shift has ended, but that their internal alarm clock is trying to wake them up. It is easy to synchronize one's internal clock with a regular day work schedule, but nearly impossible to synchronize it with a poorly designed shiftwork schedule.

As this example suggests, one of the most important questions facing occupational health and safety professionals is the degree to which shiftwork schedules will disturb the daily (circadian) internal rhythms of the body, that is, the human biological clock. Studies performed in special time-free environments (apartments, bunkers, underground caves) to understand how the natural biological clock functions reveal three major findings relevant to work scheduling (Moore-Ede, 1993):

1. There is a natural tendency for the human body to stay up later and wake up later each successive day (natural drift).

2. The biological clock responds to time cues from the outside world, and can be prevented from drifting if we synchronize ourselves with stable routines and schedules.
3. Our biological clock is adjustable, but only within strict limits—only by two to three hours per day.

These circadian principles suggest that shiftworkers trying to adjust to schedules should try to stay on the same shift for as long as practical, allowing time to synchronize the body's rhythms. In addition, clockwise shift rotation (days to evenings to nights to days) is easier for the body and is associated with decreased risk of heart disease (Scott, 1990; Scott & LaDou, 1993; Boggild, 1999). A carefully designed rotating shift schedule with these features built in can promote adequate sleep and alertness. On the other hand, the tendency for shiftworkers to invert their sleep-wake pattern on days off limits the circadian adaptation.

In one of our earlier controlled studies (Coleman, 1986), 81% of shiftworkers in a chemical and mining facility who were phase-advanced every seven days to an earlier shift (day to night to swing to day) indicated it took half a week or longer after each change for their sleep habits to adapt. (This group included 26% who were unable to adapt at all.) The researchers concluded that a rotating shift that fails to take circadian rhythms into account will result in sleep disturbances, which, of course, can lead to lowered job performance. The entire area of which rotational schedule is the "most physiologic" is quite complex and has been reviewed by several scientists (Knauth, 1997, Askenazi, 1997).

Coping Strategies

Even if management has designed a schedule that provides for business needs and adequate time off, the shiftworker's coping strategy still needs to be considered. There are five major strategies: *zombie, tough-it-out, circadian, anchor-sleep,* and *fixed-shift* strategies.

	M	T	W	T	F	S	S		M	T	W	T	F	S	S
(1)	D	D	D	D	D	—	—	(2)	D	D	D	D	D	—	—
(3)	E	E	E	E	E	—	—	(4)	E	E	E	E	E	—	—
(5)	N	N	N	N	N	N	N	(6)	N	N	N	—	—	—	—

Figure 24-2. An example of a six-week sample from a circadian schedule.

- *zombie*—When the existing schedule is impossible for the body to adjust to or the shiftworker has no strategy at all, the result is the *zombie strategy*. For example, one refinery had a schedule that required a 16-hour shift, followed by eight hours off, followed by another eight-hour shift. A worker who started at 11:00 p.m. would be off from 3:00 p.m. to 11:00 p.m., then on from 11:00 p.m. to 7:00 a.m. Other shiftworkers on better schedules might work a second or even third job, get little sleep, and/or sleep in a noisy, well-lighted room.

  The zombie shiftworkers typically claim to be adjusting well to shiftwork. They are proud of the fact they can fall asleep whenever they want, not realizing this is a sign of pathological sleepiness, like narcolepsy. An alert shiftworker should have his or her body clock setting an alert and a sleepy segment in each 24-hour period.

- *tough-it-out strategy*—In the tough-it-out coping strategy, shiftworkers try rotating so rapidly through the night shift that the biological clock has no chance of adjusting to the shift change. In fact, the shiftworker's intention is not to adjust to shift rotation, but to tough it out.

  For this type of strategy to be effective, the shiftworker should be well rested before starting the night shift. Only a few consecutive night shifts should be scheduled. Shifts can be eight or 12 hours long. At one South American mine, the shiftworkers wanted a 12-hour shift schedule (to maintain more days off), but they lived two hours from the mine. To prevent the buildup of sleep deprivation, we implemented a new schedule that gave three days off per week and required working only one consecutive night shift.

  The biological clock will not invert with only one night shift. However, a well-rested shiftworker can tough out a single night's shift without much difficulty. Tough-it-out schedules and coping strategies are generally preferred by younger shiftworkers.

- *circadian strategy*—The concept behind the circadian coping strategy is that shiftworkers try to adjust their biological clock to each shift sequence. Thus, schedules in this category are always eight-hour shifts with a medium-to-slow clockwise rotation. For example, a six-week sample from a circadian schedule may look like the one in Figure 24-2.

The first two weeks of days (D) and two weeks of evenings (E) are easy for the body clock to adjust to. The shiftworker starts night (N) shifts only once every six weeks. Typically, the body clock will take about three nights to start synchronizing to these new hours. The shiftworker on a circadian schedule tries to invert his or her sleep schedule and social life during this 10-night stretch. When nights end, the shiftworker has 32 calendar days without any night shifts. The stretch of 32 days followed by 10 nights enables the circadian clock sufficient time to adjust. This adaptation is generally not favored by shiftworkers, except those who are extremely health conscious or older.

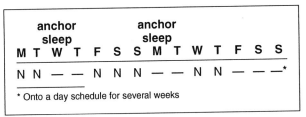

Figure 24-3. An example of an anchor-sleep coping strategy schedule.

- *anchor-sleep strategy*—A fourth coping strategy is the anchor-sleep strategy, which compromises between the tough-it-out and circadian approaches. Although shiftworkers work only two or three consecutive night shifts, they "anchor" their body clock in a night-owl position on days off. This allows them to easily readjust to nights when they go back to work.

  Figure 24-3 provides an example. On Wednesday and Thursday of Week 1 and on Monday and Tuesday of Week 2, the shiftworker stays up late to 3:00 a.m. and sleeps in until 11:00 a.m. Workers can develop specific sleeping schedules to match the exact work schedule. This staffing is preferred by "night owls" and disliked by those workers who need to get up early in the morning for family-social reasons.

- *fixed-shift strategy*—The fixed-shift coping strategy is good for day workers and evening workers. Unfortunately, workers on a fixed night shift can never adapt. When working, they try to adjust their body clock for alertness at night and sleep during the daytime. On days off, they try the opposite. Anchor sleep is not effective on permanent night shift. The result is typically that one-third of the shiftworkers are "zombies."

Fixed shifts are more appropriate with unbalanced staffing when proportionally only a small night crew is needed. Older (at least 50 years old) shiftworkers—not necessarily those with more tenure—should be given an opportunity to work a straight day shift or straight evening shift. This is because the ability to cope with sleep deprivation and frequent phase shifts appears to decrease with age (Moore-Ede, 1982).

CREATING A SAFE WORK SCHEDULE

In general, more recordable safety events occur on the day shift, when more work is performed (e.g., maintenance, special projects, construction, engineering). On the evening and night shifts when daytime engineers and managers are at home, more emphasis is placed on maintaining the status quo, as opposed to making major operational changes (Figures 24-4 and 24-5). In addition, the work load, especially discretionary work load, is typically lower on these shifts. However, safety incidents per total manpower effort can be higher on the back shifts.

Factors to Consider

To determine whether your company's shift schedule has an impact upon safety, you need to look at safety statistics by shift sequence, not simply by shift. To see how this works, consider the data in Figures 24-5 through 24-9, which are from a continuous-processing chemical facility with 400 shiftworkers. The schedule requires a weekly counterclockwise rotation. If alertness plays an important role in safety, the first few night shifts are likely to be the worst, since this is the most difficult schedule transition. Also, the first few day shifts would be problematic, since the evening-to-day transition is against the natural biological clock. The results of Figures 24-7 through 24-9 indicate that safety is worst on the first few shifts when difficult transitions are made. A single change to a clockwise rotation should reduce the "transition-related" events. Follow-up data being collected at this plant support this trend.

On 12-hour schedules, safety records have varied depending upon the nature of work being performed, the exact schedule implemented and the quality of the safety program in effect (Gillberg, 1997). With proper education and employee involvement, two surface Amax coal mines in Wyoming implemented tough-it-out 12-hour shift schedules, which have worked over 1 million hours without a lost-time

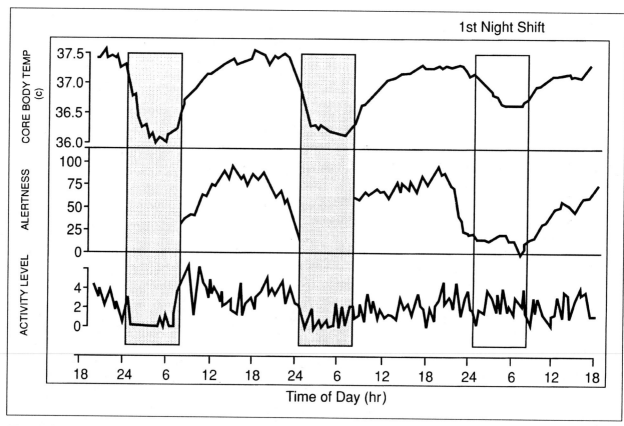

Figure 24-4. If we trace the biological rhythms of a shiftworker, we can learn something about how the inner clock works. The rhythms are the output of the internal clock, like the hands of the clock on the wall. This shows the rhythms of alertness, body temperature, and activity for a shiftworker who has two days off before his or her first night shift. Each physiological function has its own unique pattern that cycles over a 24-hour period. During the daytime (wake period), body temperature is high, but during the night (sleep period), it drops by 2–3°F (1.1–1.7°C). Alertness cycles closely follow the body temperature curve. We feel most alert when our body temperature is highest, and sleepiest when it is lowest. As shown in the cycles on the night shift, although the shiftworker is active between midnight and 8:00 a.m., the body clock sets a low temperature pattern, which makes the employee sleepy on the job (Czeisler, 1999).

accident (*Coal Mining Newsletter,* 1988). Scheduling regular safety meetings and strategic rest breaks enabled the mine to win the Mine Safety and Health Association's Sentinels of Safety Award in 1987, the first full year on the new schedule.

Schedule Options

The designer of a facility's shift schedule can either copy a previous schedule or design one that fits the work load and goals of the particular organization. Thousands of schedules are available (Figure 24-10). These are the major variables to consider in designing a schedule (Tucker, 1999; Knauth, 1997):

- number of coverage hours per week
- balanced staffing (levels that stay the same during all hours of the workweek) versus unbalanced staffing (levels that vary across the hours of the workweek)
- number of crews, typically from 0 to 21
- length of shift
- shift sequence: rotating (workers switch from days to evenings to nights), fixed (workers remain on days, evenings, or nights), oscillating (one shift is fixed and others rotate)
- pay week (typically 40 or 42 hours)
- rotation of shifts: rapid (occurs after a few days), medium (from week to week), or slow (after two or more weeks)—counterclockwise rotation is not a viable option
- day-off pattern: 3-2, 4-2, 4-3, 5-2, 6-3, 7-4, 10-4, other
- day-off pattern: balanced versus unbalanced
- starting time for day shift: 5, 6, 7, 8, 9, 10, 11, 12
- cross-training built in versus rigid job requirements
- team concept vs. crew concept vs. solo concept
- coping strategy: zombie, tough-it-out, anchor sleep, fixed, circadian.

Hourly Shift Schedule

	M	T	W	T	F	S	S
WEEK 1	-	-	E	E	E	E	E
WEEK 2	E	E	-	-	D*	D	D
WEEK 3	D*	D*	D*	D*	-	N	N
WEEK 4	N	N	N	N	N	-	-

E to D Break = 56 Hours
D to N Break = 32 Hours
N to E Break = 128 Hours

N = Night (1st shift), 10:00pm to 6:00am
D = Days (2nd shift), 6:00am to 2:00pm
E = Eves (3rd Shift), 2:00pm to 10:00pm
D* = OT relief day, Each person is assigned one of these days off be seniority.

For pay purposes, the work week begins Sunday night at 10:00pm

Figure 24-5. Sample hourly shift schedule.

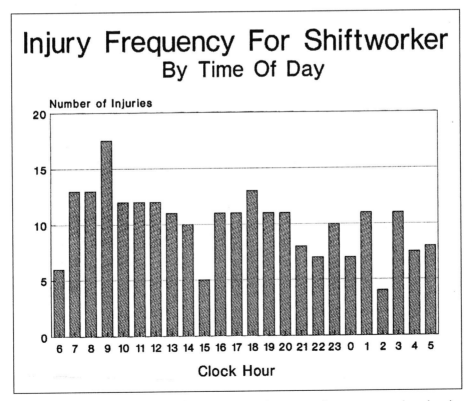

Figure 24-6. Safety statistics by shift schedule from a continuous-processing chemical facility with 400 shiftworkers—injury frequency for shiftworker by time of day.

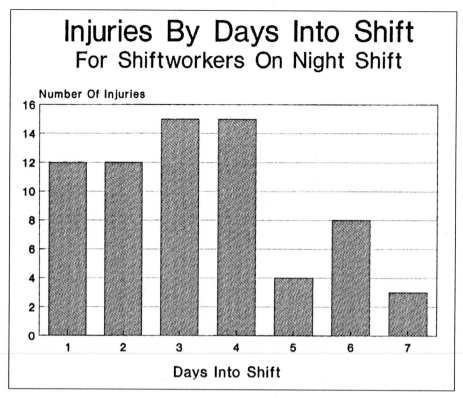

Figure 24-7. Safety statistics by shift schedule from a continuous-processing chemical facility with 400 shiftworkers—injuries by days into shift for shiftworkers on night shift.

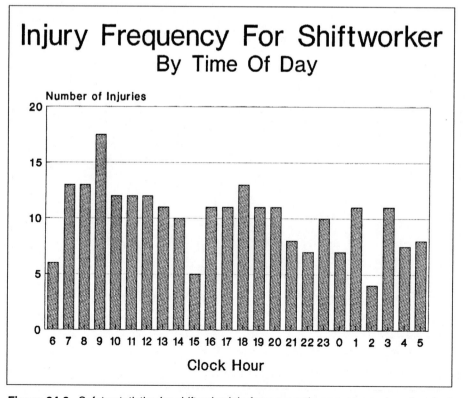

Figure 24-8. Safety statistics by shift schedule from a continuous-processing chemical facility with 400 shiftworkers—injuries by days into shift for shiftworkers on day shift.

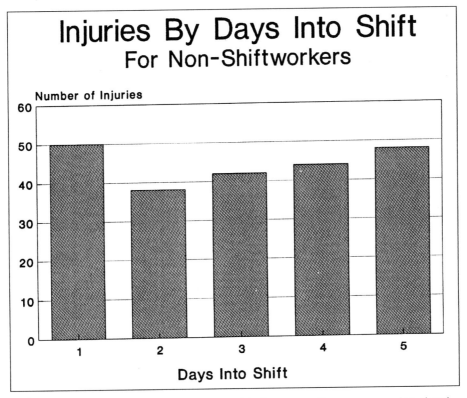

Figure 24-9. Safety statistics by shift schedule from a continuous-processing chemical facility with 400 shiftworkers—injuries by days into shift for non-shiftworkers.

SAMPLE SCHEDULE

Crew/Week	M	T	W	T	F	S	S
1	D	D	D	D	-	-	-
2	t	t	t	t	-	-	-
3	-	-	-	-	D	D	D
4	-	N	N	-	-	N	N
5	N	-	-	N	N	-	-

D = 7:00 am–7:00 pm
t = 7:00 am–3:00 pm, training, sick and vacation relief, work overload
N = 7:00 pm–7:00am
- = day off

Figure 24-10. A sample shift schedule.

From the health and safety viewpoint, advanced schedules can string together 24 consecutive day sequences, and reduce the number of night shifts by 28%.

SHIFTWORK EDUCATION

Occupational health and safety professionals should help employees adapt to shiftwork. They can educate workers about the following topics:

- *caffeine*—when used in large quantities may have a paradoxical effect and actually make the shiftworker sleepier. Shiftworkers should use caffeine sparingly and strategically.
- *alcohol*—results in frequent awakening during sleep.
- *sleeping pills*—cannot reset the biological clock, and can be addictive if not used properly.
- *exercise*—does not play a major role in resetting the body clock.
- *diet*—There is currently no convincing evidence that any special diet will help people adjust to shiftwork.
- *sanctioned napping*—during shiftwork makes sense and is used in some industries.
- *lighting*—can impart biological rhythm and adjustment to shiftwork, but must be precisely timed.

Unfortunately, many shiftworkers do not fully accept the fact that their company is an around-the-clock operation. This can lead to resentment and unhappiness, both at work and at home. Family members should understand the shift schedule when planning events and social activities. Putting together the worker's family and social life, work, and sleep times can be difficult. However, shiftwork also has certain advantages, such as time off during the week and long weekend breaks. Learning more about biological clocks and sleep cycles can help make shiftwork easier to cope with for the entire family.

SHIFTWORKERS AND HEALTH

Since all humans have the same physiological clock, our adjustment capabilities to shiftwork should roughly be the same. For example, hundreds of years ago, travelers from Europe to the United States did not get "boat lag." The ships traveled so slowly that everyone's biological clock had a chance to catch up with this slow rotation. Now all air travelers from Europe to the United States get jet lag (even if they claim the contrary), because the human biological clock cannot adjust to these rapid changes.

Thus, the only persons who are poor risks for shiftwork are those with pathologies. Relevant health problems include the following:

- a history of gastrointestinal disorders—Shiftwork disrupts meal schedules and makes it difficult to obtain hot, nutritious meals.
- diabetes and thyrotoxicosis—These diseases have been found to be rhythmic in nature, so people with these conditions need to balance food intake and drug administration at proper intervals. This can be difficult to manage during shift hours.
- poorly controlled seizure disorders—The incidence of seizures apparently rises steadily from 3:00 a.m. to 6:00 or 7:00 a.m., which is the peak seizure period. The second highest seizure period occurs between 10:00 p.m. and midnight.
- sleep disturbances unrelated to shiftwork—narcolepsy, untreated sleep apnea, or chronic insomnia.
- major psychiatric disorders, such as manic-depression or seasonal-affective disorders—Shiftwork schedules can exacerbate these conditions by disrupting the circadian clock.

SUMMARY

Occupational health and safety personnel will need to give more attention to shiftworkers and their schedules. Companies, for the sake of the health and performance of their employees, should insist on periodic evaluation and examination of these workers. All potential shiftworkers should receive medical exams before being assigned to shiftwork. Once assigned to a shift, they should be examined on a regular basis. Those older than 50 years of age should be examined more frequently.

However, the work schedule is the key to employee health. A poorly designed schedule will make adaptation a problem for everyone, and greater health problems will be likely. A good schedule with good time off, which allows the body to cope, can result in living well with shiftwork.

REFERENCES

Ashkenazi IE, Reinberg AE, Motohashi Y. Interindividual differences in the flexibility of human temporal organization: pertinence to jet lag and shiftwork. *Chronobiol Int* 14(2):99–113, 1997.

Bioryatn B et al. Bright light treatment used for adaptation to night work and re-adaptation back to day life. A field study at an oil platform in the North Sea. *J Sleep Res* 8(2):105–112, 1999.

Boggild H et al. Shift work, risk factors and cardiovascular disease. *Scand J. Work Environ Health* 25(2): 85–99, 1999.

Carskadon MA. Measuring daytime sleepiness. In Kryger MH, Roth T, Dement WC (eds). *Principles and Practice of Sleep Medicine.* Philadelphia: Saunders, 1989.

Carskadon MA, Dement WC. Daytime sleepiness: Quantification of a behavioral state. *Neuroscience and Biobehavioral Reviews* 11:(3): 307–317, 1987.

Coal Mining Newsletter (National Safety Council). Better shiftwork schedules can lower production costs. National Safety Council-*Coal Mining Newsletter,* November–December, 1988.

Coleman RM. Shiftwork scheduling for the 1990s. *AMA Personnel,* January 1989.

Coleman RM. *Wide Awake at 3:00 a.m., by Choice or by Chance.* Stanford CA: Stanford University Press, 1986.

Colquhon WP, Rutenfranz J (eds). *Studies of Shiftwork.* London: Taylor & Francis Ltd, 1980.

Czeisler CA, Duffy JF, Shanahan TL, et al. Stability, precision, and near-24-hour period of the human circadian pacemaker. *Science* 2177–2181, 1999.

Gillberg M. Subjective alertness and sleep quality in connection with permanent 12-hour day and night shifts. *Scand J Work Environ Health* 24(suppl 3):76–80, 1998.

Knauth P. Changing schedules: Shiftwork. *Chronobiol Int* 14(2):159–171, 1997.

Luna TD. Air traffic controller shiftwork: What are the implications for aviation safety? A review. *Aviat Space Environ Med* 68(1):69–70, 1997.

Luna TD, French J. Mitcha JL. A study of USAF air traffic controller shiftwork: Sleep, fatigue, activity, and mood analyses. *Aviat Space Environ Med* 68(1):18–23, 1997.

Moore-Ede M. *The Clocks That Time Us.* Cambridge MA: Harvard University Press, 1982.

Prunier-Poulmaire S, Gadbois C, Volkoff S. Combined effects of shift systems and work requirements on customs officers. *Scand J Work Environ Health* 24(suppl 3): 134–140, 1998.

Scott A (ed). *Shiftwork.* State of the Art Reviews: Occupational Medicine Series. Philadelphia: Hanley & Belfus 5(3), July–September, 1990.

Scott A, LaDou J. Health and safety of shiftworkers. In Zenz C (ed). *Occupational Medicine: Principles and Practical Applications,* 3rd ed. St. Louis: Mosby-Yearbook, 1993.

The Wall Street Journal. Problem proves costly on the job as productivity and safety decline. July 7, 1988.

Tucker P et al. Bright light treatment used for adaptation to night work and re-adaptation back to day life. A Field study at an oil platform in the North Sea. *J Sleep Res* 8(2):105–112, 1999.

chapter 25

Employee Education

by Barbara J. Burgel, RN, COHN
revised by Gary R. Krieger, MD, MPH, DABT
Marci Z. Balge, RN, MSN, COHN-S

443 **Introduction**
443 **Legal, Professional, and Ethical Frameworks for Employee Education**
OSHA standards ■ Healthy People 2000 ■ Ethical considerations
445 **Issues in Employee Education**
446 **Training Guidelines**
Determining whether training is needed ■ Identifying training needs ■ Identifying goals and objectives ■ Developing learning activities ■ Conducting the training ■ Tracking training ■ Evaluating program effectiveness ■ Improving the program
449 **Summary**
449 **References**

INTRODUCTION

The major goals of employee education are to promote understanding of workplace hazards and to promote a safe workplace. Employee education encourages workers to engage in safe practices and stimulates a level of understanding for employees to recognize and report potential hazards to employers. The key variable is the scope and depth of training programs, with an emphasis on planned behavior change and risk communication.

LEGAL, PROFESSIONAL, AND ETHICAL FRAMEWORKS FOR EMPLOYEE EDUCATION

The Occupational Safety and Health Act of 1970 (OSHAct) assures "so far as possible every working man and woman in the nation safe and healthful working conditions and to preserve our human resources." The OSHAct clearly identifies that it is the legal responsibility of the employer to provide a safe and healthful workplace. It states that employers and employees have "separate but dependent responsibilities and rights with respect to achieving safe and healthful working conditions." The responsibilities of the employer include, among others, complying with mandatory standards, informing employees of OSHA, minimizing and reducing hazards, ensuring that employees have and use safe tools and equipment, and posting signs when needed to warn employees of potential hazards. Employees are required to read the OSHA poster at the worksite, comply with all applicable OSHA standards, follow all employer safety and health rules and regulations, including use of prescribed protective equipment, report hazardous conditions, and report any job-related injury or illness to the employer (DOL, 1985).

These responsibilities need to be communicated to employees. Ways of doing so can involve providing printed information to meet basic legal requirements or instituting a targeted educational program designed to meet the educational background, literacy, and cultural needs of the work force.

OSHA Standards

More than 100 OSHA standards require the employer to train employees in health and safety. This training requirement can be explicitly stated, as in the Cotton Dust Standard, Section 1910.1043(i)(1) and (2). To meet that standard, training must include the nature of operations that could lead to cotton dust exposure, protective measures, the proper use of respirators, the role of medical surveillance, and the requirement that this training program be offered initially and then annually. Other OSHA standards are less specific in defining the scope of the training, limiting jobs to those who are "qualified" or "competent." For example, the Standard for Overhead and Gantry Cranes, Section 1910.179(b)(8), states that "only designated personnel shall be permitted to operate a crane covered by this section."

Many standards do not contain training requirements, although OSHA recognizes that employees need appropriate training to do the job safely (DOL, 1987). For this purpose, OSHA established voluntary training guidelines to help employers determine when training is needed. These training guidelines, which offer employers a model for planning their own programs, include the following steps (DOL, 1987):

1. Determine whether training is needed.
2. Identify training needs.
3. Identify goals and objectives.
4. Develop learning activities.
5. Conduct the training.
6. Evaluate the effectiveness of the program.
7. Improve the program.

Healthy People 2000

In 1991, the U.S. Department of Health and Human Services published *Healthy People 2000—National Health Promotion and Disease Prevention Objectives*. These objectives for the nation include 15 objectives specific to occupational safety and health, with many other objectives outlined for additional priority areas, such as physical activity and fitness, nutrition, tobacco use, and HIV infection.

In general, education is a method to meet all of the objectives. Objectives 10.12 and 10.13, however, specifically address the implementation of worker health and safety programs, with back injury prevention noted as a program priority.

> *10.12 Increase to at least 70% the proportion of worksites with 50 or more employees that have implemented programs on worker health and safety.*

> *10.13 Increase to at least 50% the proportion of worksites with 50 or more employees that offer back injury prevention and rehabilitation programs.*

These objectives recognize that the worksite is an ideal location for teaching people about positive health practices, with the potential for influencing the health practices of dependents and retirees. Back injury prevention and rehabilitation programs, in addition to smoking control activities, are some of the more prevalent health promotion programs offered in the workplace (U.S. DHHS, 1991). These objectives provide a professional framework by which to prioritize occupational health and safety education efforts. Not only are employees a target audience, but management and primary health care providers must be included as target audiences.

Ethical Considerations

Employers must recognize that employees have a right to a complete and accurate disclosure of risks. The major ethical principle that applies to employee education is the principle of autonomy. Autonomy is self-directing freedom and moral independence. Education gives the individual the knowledge and the freedom to act (Rest & Patterson, 1986). Privacy rights and confidentiality of personal health information are additional ethical

Factors Influencing Workers' Health by Locus of Control

	Employer Control	
Worker Control	**High**	**Low**
High	**[Cell 1]** Work practices Use of personal protective and safety equipment Workplace hygiene and housekeeping Maintenance and upkeep of machines and equipment	**[Cell 2]** Lifestyle, personal health habits
Low	**[Cell 3]** Work environment and process: substances used, design of machinery, hazard controls, job design	**[Cell 4]** Biological and genetic characteristics Physical and mental impairment Cultural characteristics Climate

Figure. KL Green (1988) developed an analysis of the factors influencing workers' health. These factors were distributed into four cells according to how much control workers and employers could exercise. (Reprinted with permission from Green KL. Issues of control and responsibility in workers' health, *Health Ed Q* 15 (4):475, Winter 1988.)

considerations when designing educational programs. Employees need to clearly understand whether educational offerings are voluntary or mandatory, and whether their participation or lack of involvement has any consequences (Ashford & Caldart, 1985). What may be viewed as voluntary in a democracy differs sharply from the meaning of voluntary in a bureaucracy or a corporate structure (Conrad, 1987).

Another ethical consideration from the employer's perspective is the principle of beneficence, or actions taken to improve the welfare of others. Examples of management's beneficence are funding in-depth training and making engineering controls of hazards a priority.

ISSUES IN EMPLOYEE EDUCATION

Over the past decade increasing health care benefit costs have led businesses to place increasing emphasis on the prevention and control of personal risk factors at the worksite. This trend has engendered considerable debate over personal responsibility versus organizational responsibility for health. Many occupational health professionals maintain that health promotion programs have a limited role in the worksite, because these are voluntary behaviors, outside of the responsibility and control of employers. Furthermore, employers shift resources and attention away from hazardous worksites to health promotion programs, thereby blurring the focus on occupational health (Conrad, 1987; Green, 1988). In contrast, individuals argue for an integrative approach recognizing that health promotion, in concert with hazard control programs, provides a balanced approach to significant environmental, social, and individual health variables (Jordan-Marsh et al, 1987; Vojtecky, 1988).

One way to set priorities is with Green's (1988) analysis of the factors influencing workers' health. Green sorted those factors into four cells according to how much control workers and employers could exercise (Figure). Both the employer and the worker have high control, and therefore shared responsibility, over the factors in cell 1, which include work practices, use of personal protective and safety equipment, workplace hygiene and housekeeping, and maintenance and upkeep of machines and equipment. In cell 2, which includes factors related to life-style and personal health habits, the employer has a low degree of control and the worker has a high degree of control. In cell 3, the employer has a high degree of control, and the worker has a low degree of control over the work environment and work processes, such as the design of machinery, job design, and hazard controls. In cell 4, variables over which the employer and the worker have a low degree of control include cultural characteristics and physical and mental impairment. According to this model, the priority areas for employers should be the two cells where employers have a high degree of control and hence responsibility (cells 1 and 3). Likewise, employees have responsibility for the areas identified in cells 1 and 2.

Risk perception also has gained increased attention over the past decade. Not only is the standard for hazard communication broadened in scope, but there have been many environmental cases—notably those involving communities adjacent to industrial facilities—that have prompted a reexamination of how best to communicate information to groups at risk. A group of NIOSH researchers has designed a set of guidelines for the development of printed materials aimed at communicating health risk (Table 25-A). Additionally, these researchers have formulated a tool for critiquing printed materials to ensure that the materials accurately and sensitively define risk (Table 25-B). Our understanding of learning and of risk perception indicates that it is very important to involve the target group in preparing these written health risk messages (Knowles, 1978; Slovic, 1986; Slovic, 1987). Also see chapter 19, Community Involvement Programs, for more information on the risk communication process.

TRAINING GUIDELINES

Earlier, this chapter listed OSHA's voluntary training guidelines, which provide employers with a basic approach to program planning. Let's take a closer look at some of those guidelines.

Determining Whether Training Is Needed

There are several ways to determine the need for training. It may be mandated by law. For example,

Table 25-A. Listing of Health-Risk Message Guidelines by Category.*

Orientation and Perspective

Guideline 1	Acknowledging workers' right to know workplace hazards and risks and needs to disclose such information
Guideline 2	Integrating health-risk messages with other informational/educational activities aimed at promoting workplace health and safety

Background Preparation

Guideline 3	Establishing target work group's level of knowledge about workplace hazards
Guideline 4	Ascertaining target group's perceptions of hazards in everyday operations and appreciation of protective measures to reduce risks
Guideline 5	Fitting messages to the demographic makeup of the target audience

Message Content/Structure

Guideline 6	Designating credible sources of information
Guideline 7	Identifying high-risk jobs, operations, or work conditions with particularity
Guideline 8	Balancing fear arousal in hazard messages with actions that can control risk
Guideline 9	Making delayed, insidious health threats more imaginable
Guideline 10	Enhancing the meaning of quantitative measures of health risk
Guideline 11	Structuring messages to stress main points and using expressions familiar to the target group

Delivery

Guideline 12	Using multiple messages, varying in form and presented intermittently, rather than mass single-dose communications
Guideline 13	Directing messages through as many channels of the existing social/communications networks as feasible

Evaluation

Guideline 14	Providing means for evaluating the effect of the message in meeting its intended goals

* Reprinted with permission from Cohen A, Colligan MJ, Berger P. Psychology in health risk messages for workers. *J Occ Med* 27(8):545, 1985.

the Respiratory Protection Standard, Section 1910.134(b)(3), states that the user must be instructed and trained in the proper use of respirators and their limitations. Need can be stimulated by an increased rate of accident reports from a specific department or from a specific work process.

Table 25-B. Checklist in Gauging the Merits of Health-Risk Informational Materials for Increasing Worker Understanding of Workplace Hazards and Responsiveness to Preventive Actions.*

1. Do the information materials as developed show good-faith efforts by the employer to inform workers of known suspected health hazards in their work environment?
2. Where uncertainties exist about possible health risks, is such information still disclosed to workers who may be affected? Are efforts made to update worker information about such risks as it becomes available?
3. Are the critical points, terms, or expressions found in health-risk informational messages consistent with those used in job safety training and on labels or other communications that workers are apt to receive in regard to the recognition and control of workplace hazards?
4. Is the informational material in language and content appropriate for the intended worker group in terms of their educational/reading level, age, and length of job service?
5. Have workers or worker representatives been involved or consulted with regard to developing the informational materials?
6. Are the sources to whom data about workplace hazards and risks are attributed in informational messages considered credible to the affected worker groups?
7. Does the informational material distinguish between high- and low-risk job conditions? Are high-risk job operations, locations, and exposure factors appropriately detailed to be appreciated by new workers?
8. Do the messages informing workers at risk to known or suspect workplace hazards contain elements that address the risk as well as preventive or control actions?
9. Do the informational materials attempt to portray workplace hazards and health risks in various ways such as to make them more imaginable and meaningful to the target groups of concern?
10. Are the health-risk messages to workers organized in ways that highlight main points and display a writing style that invites interest and easy reading?
11. Are initial health-risk messages to workers followed up by subsequent communications that serve to reinforce or amplify critical points in hazard recognition and control?
12. Are a variety of paths used for delivering the informational material to the target worker groups to ensure its being received and read?
13. Do the activities for composing and delivering health-risk messages to workers also include a plan for evaluating their impact in terms of worker responsiveness to the information and the attainment of certain goals?
14. Do follow-up communications take account of shortcomings in worker response to the original material?

* Reprinted with permission from Cohen A, Colligan MJ, Berger P. Psychology in health risk messages for workers. *J Occ Med* 27(8):550, 1985.

The occupational health and safety professional can see an increasing trend in lost workdays from a specific type of injury or work process. A review of the most hazardous job classifications can help this person target a cyclical retraining program before noting any increases in the accident rate. Training needs can also be stimulated by external events. For example, a recent community disaster can illustrate the need for review and training about the company disaster plan. Gleason and Golden (1988) identify additional organizational and community data sources to use in determining worksite needs.

Once the problem is initially identified, the next step is to assess the needs more thoroughly to determine the full scope of the problem and whether training is the appropriate response. According to OSHA, training is most helpful when workers lack knowledge of a work process, are unfamiliar with equipment, or are executing a task incorrectly.

OSHA does recognize that training does not always solve the problem and emphasizes the role of engineering controls and the need for a comprehensive health and safety program (DOL, 1987). For example, if an interview and/or questionnaire survey of the target audience makes it clear that employees are discarding needles appropriately and that they are sticking themselves because the disposal units are too full, training may not be indicated. Rather, administrative policy must change to service the needle disposal units more often. Likewise, if the investigation into the problem revealed that needlesticks were occurring because the needle disposal units were located above eye level, the solution would not be one of training, but rather of engineering control—lowering the disposal units to eye level. Training would be indicated if a review of needlestick injuries revealed that sticks were occurring when staff were recapping needles. In that case, the execution of the task appears to be incorrect and points to a gap between what the staff should be doing and what is actually occurring. This gap can be filled by education.

Identifying Training Needs

After or during the process of determining whether training is needed, the occupational health and safety professional should identify training needs. This step includes following guidelines 3, 4, and 5 in Table 25-A. It focuses on assessing interests, needs, values, beliefs, and knowledge of the target population. OSHA recommends that this step incorporate a review of OSHA standards and a thorough job analysis.

A valuable way to collect data is to observe employees doing their jobs. Identify behaviors that can be changed for improved health and safety; this is the essence of education. Examine individual behaviors and organizational variables, and clearly distinguish who is responsible for change (Figure).

Identifying Goals and Objectives

Based on the knowledge or behavior deficits identified, the health and safety professional writes goals and objectives. OSHA standards can outline what needs to be in a specific training. However, it is still very important to determine whether additional objectives or content areas need to be included for the particular target area. For example, the company may already offer a respirator protection program, yet several employees are not wearing the respirator at appropriate times. An objective that responds to this problem might focus on perceptions of risk.

Objectives should be clear and measurable. To that end, they should focus on behaviors and include a time frame, a specified direction of change, and a way to measure the change. Here is an example of goals and objectives for a respiratory protection program:

Goal: To educate those who use respirators to wear them correctly and appropriately.

Objectives: At the end of the session, employees will:
1. State three work practices that would require use of the respirator.
2. Discuss the barriers to wearing a respirator.
3. State the frequency of changing the filters.
4. Demonstrate qualitative fit testing of their respirator.

A useful way to identify goals and objectives in the work setting is to convene planning groups. This approach allows for broad involvement in the program, often including union representation. Involvement, in turn, improves the success of the training.

Developing Learning Activities

The design of the teaching sessions is crucial to the success of the training. An important criterion is to choose teaching methods that are acceptable to the target group (Dignan & Carr, 1987). Also critical is to examine the literacy of the target population (Bruening, 1989) and explore ways in which education has successfully been provided to this group in the past.

Match the teaching strategies to the goals of the training session. If increased knowledge is the goal, a lecture format can be most effective. If the training seeks to change attitudes and beliefs, small-group guided discussion can be successful. If skills need to be refined, the educational sessions should simulate the work environment and occur in small groups. Additional criteria include size of the group, time allowed, resources available in the setting, feasibility of learning activities and convenience of use, as well as the values, mores, and folkways of the target population (Dignan & Carr, 1987). Guidelines 6–13 in Table 25-A are helpful in choosing possible approaches. Dignan and Carr (1987) systematically analyze various teaching methods and activities.

Conducting the Training

Perhaps the most important component is the introduction of a training session. The instructor should see himself or herself as facilitating learning, that is, as assisting others in accomplishing their learning goals (Knowles, 1978; Ricks, 1977). Employees must be convinced of the relevance of the training. Therefore, the trainer should share the full scope of the problem, including recent environmental data, aggregate biological monitoring data, and engineering controls that have been done or are in progress. Outlining how employee behavior contributes to a safe and healthy workplace is also valuable.

Management and union behavior can influence how employees benefit from training. Participation levels will depend on whether time spent in the educational sessions is paid or unpaid. The visible presence of management usually shows support for the program but may inhibit sharing of employee concerns. Union presence usually shows support for the program and will help the training succeed.

Adults learn by immediately applying the material covered, so hands-on practice and opportunities for role play are recommended teaching strategies. Having the group pool its expertise to discuss cases also is supported by adult teaching-learning theory. Another way to facilitate learning is to have participants formulate their own learning objectives (Knowles, 1978; Ricks, 1977). Using feedback in the worksite—posting aggregate monitoring data, for example—reinforces the adoption of safe behavior (Vojtecky, 1988). Guidelines 12 and 13 in Table 25-A offer additional suggestions for conducting the training.

Tracking Training

A significant challenge in managing employee training programs is keeping track of who should participate, how often, and whether the training course is required or recommended. Often the courses required are driven by regulatory standards, which are open for interpretation. Not all state-driven health and safety requirements are consistent, i.e., California, Texas, etc.

Given the variations in regulatory requirements, it is wise to develop and maintain a tracking system for employee training. The tracking system should link to personnel databases—many software packages are available for this purpose. For smaller organizations a simple spreadsheet will accommodate this need.

Evaluating Program Effectiveness

Evaluation methods focus on the results of the planning process, and the results of the learning process. Evaluation helps in reassessing needs, interests, and values (Knowles, 1978). Short-term and long-term evaluation methods, established before the training session, will determine whether the objectives and, ultimately, the goal have been met after the training session. Ideally, to clearly document the effectiveness of the training provided to the target audience, evaluation methods should include a similar comparison group of workers who did not receive the training (American Public Health Association, 1987). Observation in the worksite will show whether employees have integrated new behaviors into the work setting. Cost data also can be used to determine the efficiency of the intervention.

Improving the Program

Evaluation data indicate ways to improve the program and when to offer retraining to reinforce new behaviors. If the problem persists, the health and safety professional should repeat the needs assessment phase to determine whether training is in fact the appropriate solution. Chapter 20, Program Assessment and Evaluation, presents a method for assessing an overall medical occupational program.

SUMMARY

Employee education, in conjunction with engineering controls, helps to ensure a safe and healthy workplace. OSHA standards, which specifically require training, provide a basic content outline for training programs. The U.S. DHHS further identifies national objectives, which assist in setting educational priorities. The ethical principle of autonomy, in concert with the expansion of the hazard communication standard, recognizes that workers have a right to complete and accurate disclosure of risks. The OSHA voluntary training guidelines suggest an approach to program planning. The role of the employer in correcting hazardous worksite conditions is of primary importance. Educational programs, with a focus on risk communication and planned behavior change, can successfully promote a safe and healthy workplace.

REFERENCES

American Public Health Association. Criteria for the development of health promotion and education programs. *APHA Journal* 77(1):89–92, 1987.

Ashford NA, Caldart CC. The "right to know": Toxics information transfer in the workplace. *Ann Rev Pub Health* 6:383–401, 1985.

Biener L, Glanz K, McLerran D, et al. Impact of the Working Well Trial on the worksite smoking and nutrition environment. *Health Educ Behav* 26(4):478–494, 1999.

Bruening JC. Workplace illiteracy: The threat to worker safety. *Occ Haz* 10:118–122, 1989.

Cohen A, Colligan MJ, Berger P. Psychology in health risk messages for workers. *J Occ Med* 27(8):543–551, 1985.

Conrad P. Wellness in the work place: Potentials and pitfalls of work-site health promotion. *Milbank Quarterly* 65(2):255–275, 1987.

DHHS (U.S. Department of Health and Human Services). *Healthy People 2000—National Health Promotion and Disease Prevention Objectives*. DHHS Pub. No. (PHS) 91-50212. Washington DC: 1991.

DOL (U.S. Department of Labor), Occupational Safety and Health Administration (OSHA). *All About OSHA*, OSHA Report 2056. Washington DC: OSHA, 1985.

DOL, OSHA. *Training Requirements in OSHA Standards and Training Guidelines*, OSHA Report 2254. Washington DC: OSHA, 1987.

Dignan MB, Carr PA. *Program Planning for Health Education and Health Promotion.* Philadelphia: Lea & Febiger, 1987.

Fisher TF. Preventing upper extremity cumulative trauma disorders. An approach to employee wellness. *AAOHN J* 46(6):296–301, 1998.

Gleason SE, Golden J. Data sources: Selecting worksite health and safety programs. *AAOHN J* 36(10):408–416, 1988.

Green KL. Issues of control and responsibility in workers' health. *Health Ed Q* 15(4):473–486, 1988.

Heany CA, Goldenhar LM. Worksite health programs Working together to advance employee health. *Health Educ Q* 23(2):133–136, 1996.

Jordan-Marsh M, Vojtecky MA, Marsh DD. Workplace health promotion/protection: Correlates of integrative activities. *J Occ Med* 29(4):353–356, 1987.

Knowles M. *The Adult Learner: A Neglected Species.* Houston: Gulf, 1978.

Melhorn JM. The impact of workplace screening on the occurrence of cumulative trauma disorders and workers' compensation claims. *J Occup Environ Med* 41(2):84–92, 1999.

Peterson M, Dunnagan T. Analysis of a worksite health promotion program's impact on job satisfaction. *J Occup Environ Med* 40(11): 973–979, 1998.

Reardon J. The history and impact of worksite wellness. *Nurs Econ* 16(3):117–121, 1998.

Rest KM, Patterson WB. Ethics and moral reasoning in occupational health. *Seminars in Occupational Medicine* 1(1):49–57, 1986.

Ricks DM. Making the most of the first 20 minutes of your training. *Training* 8:13–14, 1977.

Slovic P. Informing and educating the public about risk. *Risk Analysis* 6(4):403–415, 1986.

Slovic P. Perception of risk. *Science* 236:280–285, 1987.

Sorensen G, Hunt MK, Cohen N, et al. Worksite and family education for dietary change: The Treatwell 5-a-Day program. *Health Educ Res* 13(4):577–591, 1998.

Stokols D, Pelletier KR, Fielding JE. The ecology of work and health: Research and policy directions for the promotion of employee health. *Health Educ Q* 23(2):137–158, 1996.

Vojtecky MA. Education for job safety and health. *Health Ed Q* 15(3):289–298, 1988.

Wilson CK. Team behaviors: Working effectively in teams. *Semin Nurse Manag* 6(4):188–194, 1998.

Wilson MG, Jorgensen C, Cole G. The health effects of worksite HIV/AIDS interventions: A review of the research literature. *Am J Health Promot* 11(2):150–157, 1996.

chapter 26

Gender Issues in the Workplace

by Dana Headapohl, MD
Linda Hawes Clever, MD, FACP
revised by Gary R. Krieger, MD, MPH, DABT

451 **Overview**
   Gender distribution in the labor force ■ Employment segregation ■ Pay inequities
453 **Gender-Specific Vulnerabilities**
   Physical strength and work capacity ■ Heat exposure ■ Low back injuries ■ Coronary heart disease
457 **Gender Gap in Health Research**
457 **Disability**
459 **Chemical Exposure and Reproductive Risk**
   General mechanism of action ■ Male reproductive hazards ■ Female reproductive hazards
462 **Assessing and Communicating Risk**
464 **Reproductive and Developmental Hazard Policy**
465 **Protection or Discrimination?**
467 **Harassment**
467 **Family-Related Issues**
467 **Summary**
468 **References**

OVERVIEW

Gender issues in the workplace are increasingly recognized and should be considered in all occupational health and safety programs. Historically, companies have considered gender issues to include little more than fetal protection policy. But all workers must be protected from reproductive toxicants. There are many important physical and physiological differences between men and women—in the way toxic substances are absorbed, distributed, metabolized, and excreted; tolerance for heat and humidity; risk for heart disease; and ability to perform heavy labor. Gender as well as individual differences must be taken into account for job placement and ergonomic accommodations. Gender-related company policies should address discrimination, harassment, and family issues.

Gender Distribution in the Labor Force

While the percentage of men in the paid labor force has remained fairly stable during this century, the contribution of women has been steadily increasing. Women account for 45.5% of the labor force and 52% of the total population. While the female share of the population has been relatively stable, the overall female participation in the labor force has doubled since 1970. Over 60% of the growth in the labor force between 1979 and 1990 was due to the influx of women. In the 18 to 64-year-old age group, 69%

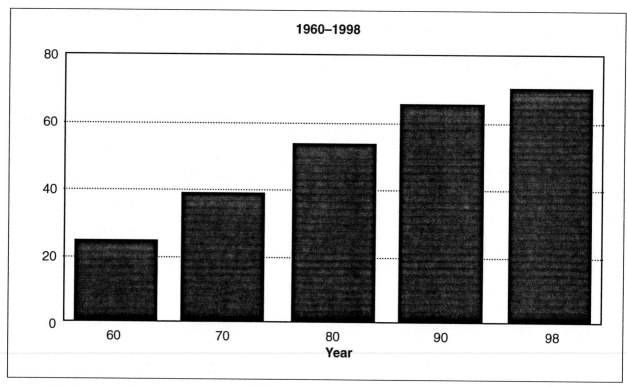

Figure 26-1. Labor force participation rates for married women with children, 1960–1998. Source: Statistical Abstract for the United States 1998 and Bureau of Labor Statistics.

of women participate in the civilian labor force compared to 88% of men (DOL, October 1990). There are some projections that women will make up almost half of the workforce by the year 2020. This shift is a result of the dramatic rise in participation of married women with children in the overall labor force—from 28% in 1960 to more than 70% in 1999 (Figure 26-1).

Employment Segregation

Although women represented almost 50% of all workers in 1999, there is not an even distribution throughout different occupations (Table 26-A). Women tend to be concentrated in low-paying, unskilled jobs that anticipate high turnover rates. For example, women account for 94% of employees in private household positions and almost 80% of those in administrative support positions. Women account for fewer than 10% of workers in transportation and other physically demanding jobs such as materials moving. Although women's manual dexterity is used in the sewing and garment industry, less than 10% of higher-paying precision production, craft, and repair workers are women. Women have made significant progress into some higher paying jobs, but they are still underrepresented in many areas—only 14% of professionals are women. Although more than 50% of working women are in administrative jobs, most are in support positions. Women account for 1%-2% of the senior executive positions (Collins & Thornberry, 1989; DOL, 1989).

Pay Inequities

There is a considerable earnings gap between men and women, although it is slowly narrowing. The difference is least for hourly pay workers, who represent 60% of all workers. Women in this category earn 77.5% of the male hourly rate. Women who

Table 26-A. Proportion of Female Workers in Selected Occupations, 1975, 1985, and 1989.*

Occupation	Women as Percent of Total Employed		
	1975	1985	1991
Architect	4.3	11.3	17.3
Child care worker	98.4	96.1	96.0
Computer programmer	25.6	34.3	33.6
Lawyer, judge	7.1	18.2	18.9
Mail carrier	8.7	17.2	27.7
Physician	3.0	17.2	20.2
Registered nurse	97.0	95.1	94.8

* This information was abstracted from the Bureau of Labor Statistics, 1989, 1991; and "Facts on Working Women," U.S. Department of Labor, Women's Bureau, No. 90-3, October 1990.

earn weekly wages make 74% of what their male counterparts are paid. Those women on annual salaries earn 69.9% of male annual wage earners. In 1991, women working full time earned an average of 71% of what men earned, according to the National Committee on Pay Equity. Differences in skill and experience account for less than 50% of the differential for college graduates and only about 25% of the differential for nonhigh school and high school graduates. Women tend to be paid less than men for comparable work. Even with the same job title, women are given different kinds of responsibilities and authority.

GENDER-SPECIFIC VULNERABILITIES

Depending on the job, workers can be exposed to biological, chemical, physical, and/or ergonomic hazards. These exposures can cause acute or chronic health problems that are seldom recognized as work-related. Too little is known about the interrelationship of work and health, especially with regard to genetic effects in both men and women. Research in this area is highly complex, often demanding personal and family histories extending back several decades or even generations. Acute problems, such as massive exposure to a toxic substance resulting in liver damage, are easy to identify as work-related. But diseases that develop slowly over many years can go unnoticed for long periods and are difficult to ascribe to the work environment. The effects of chronic low-dose exposures are often not linked to work. Moreover, there can be more than one contributing exposure. For example, industrial solvents, hepatitis, and alcohol abuse—all can cause liver disease. Which, then, is the cause in a person who is occupationally exposed to solvents, is also a heavy drinker, and has intimate relations with a carrier of hepatitis B? In addition, the worker's personal physician may be of minimal help when determining whether a health problem is work-related, since many physicians may be unaware of their patient's exposures to specific occupational health hazards. Consequently, they cannot relate a patient's symptoms to his or her work.

Moreover, there are gender and individual differences in how toxic chemicals are absorbed, distributed, metabolized, and excreted, and how physical stressors such as heavy lifting, heat, and humidity are tolerated. While animal studies show toxicokinetic gender differences, relatively little is known about human gender differences. In the few studies that have been done, neither gender has a distinct advantage. Women retain benzene and trichloroethylene longer and are more susceptible to lead-induced blood chemistry changes, but men are more susceptible to developing aflatoxin-caused liver cancer and cigarette-smoking–induced lung cancer (Calabrese, 1986). Existing exposure standards seldom take gender vulnerabilities into account.

Physical Strength and Work Capacity

The average woman is smaller and has 66% of the muscle strength of the average man (Table 26-B) (Chaffin & Anderson, 1984). Strength differences vary with muscle groups and tasks. For example, women on average have only slightly less lower extremity strength than men, but considerably less upper extremity lifting strength. In addition to muscle differences, there are skeletal differences; the female lumbar spine is able to withstand 15%–20% less compressive force (Messite & Welch, 1987).

The isometric strength of 25 different muscle groups is optimal at 20 years of age in women and 30 years of age in men. With increasing age, there is a gradual decrease in muscle strength, which progresses more rapidly after age 40 years. Oxygen consumption with exercise, a key indicator of work capacity, also decreases in both men and women with aging. Findings from subjects of both sexes and various age groups tested on bicycle ergometers accompanied by heart recordings indicate that aerobic capacity and heart rate at exercise are optimal at around age 20 years and decline gradually thereafter. (Mean oxygen uptake was 3.6 L/min for the men and 2.7 L/min for the women.) Size and physical fitness accounts for some, but not all, of the divergence, since there is a 10%–15% deficit in women after correction for the weight difference (Table 26-B, Figure 26-2).

Blood volume for an average man is normally 75 mL/kg; for an adult woman, 64 mL/kg. Women at the same age and body weight also have 10% less total lung volume and 10% less hemoglobin. The average hematocrit, the relative amount of plasma and red blood cells, is 42 for women and 47 for men. Arterial oxygen content is 16.7 mL/100 mL for women versus 19.2 mL/100 mL for men. Thus, for submaximal work (oxygen uptake of 1.5 L/min), women need 9.0 L of cardiac output to transport 1.0 L of oxygen, while men require only 8.0 L.

Table 26-B. Predicted Standard Strength for Various Cases of Stature, Body Weight, Gender, and Age.*

Torso Strength

Age	Height	Male Body Weight		Female Body Weight	
		100 lb (45 kg)	200 lb (90 kg)	100 lb (45 kg)	200 lb (90 kg)
20 yr	5'0" (150 cm)	75.3 (33.8 kg)	123.8 (55.7 kg)	50.3 (22.6 kg)	78.8 (35.5 kg)
	6'0" (180 cm)	88.4 (39.8 kg)	155.2 (68.9 kg) (max)	63.5 (28.6 kg)	105.3 (47.4 kg)
50 yr	5'0" (150 cm)	56.4 (25.4 kg)	91.0 (41 kg)	31.4 (14.1 kg) (min)	41.1 (18.5 kg)
	6'0" (180 cm)	69.8 (31.4 kg)	117.5 (52.8 kg)	44.6 (20 kg)	67.5 (30.4 kg)

Arm Strength

Age	Height	Male Body Weight		Female Body Weight	
		100 lb (45 kg)	200 lb (90 kg)	100 lb (45 kg)	200 lb (90 kg)
20 yr	5'0" (150 cm)	72.5 (32.6 kg)	88.3 (39.7 kg)	41.2 (18.5 kg)	55.9 (25.1 kg)
	6'0" (180 cm)	67.8 (30.5 kg)	96.7 (43.5 kg) (max)	44.4 (20 kg)	64.3 (28.9 kg)
50 yr	5'0" (150 cm)	64.6 (29 kg)	72.4 (32.6 kg)	32.3 (14.5 kg) (min)	40.0 (18 kg)
	6'0" (180 cm)	58.8 (26.5 kg)	80.8 (36.4 kg)	36.5 (16.4 kg)	48.4 (21.8 kg)

Leg Strength

Age	Height	Male Body Weight		Female Body Weight	
		100 lb (45 kg)	200 lb (90 kg)	100 lb (45 kg)	200 lb (90 kg)
20 yr	5'0" (150 cm)	117.5 (52.8 kg)	226.9 (102.1 kg)	82.4 (37.1 kg)	131.8 (59.3 kg)
	6'0" (180 cm)	187.4 (84.2 kg)	246.7 (110 kg) (max)	92.2 (41.5 kg)	151.5 (68.2 kg)
50 yr	5'0" (150 cm)	151.7 (68.3 kg)	175.4 (78.9 kg)	56.6 (25.5 kg) (min)	80.2 (36 kg)
	6'0" (180 cm)	156.5 (70.4 kg)	184.8 (83.2 kg)	61.3 (27.6 kg)	89.7 (40.4 kg)

* Adapted with permission from *Journal of Occupational Medicine*, Vol. 20, No. 6, June 1978.

There are significant physiological differences not only between the genders, but also among the population in height, weight, strength, training, and physical fitness. It is important to carefully match individual workers and jobs. Women should not be excluded from physically intensive jobs solely on the basis of their gender.

Physical capacity can be determined by tests of aerobic capacity (oxygen uptake) and work capacity, in which the person exercises on a bicycle ergometer to maximum capacity, while heart rate, blood pressure, and other important physiological variables are measured. A functional capacity evaluation (FCE) is a more cost-effective and job-specific method of assessing an employee's ability.

To properly match a worker to a job, the physical requirements for job tasks must be clearly described in a job analysis. In addition to careful placement of workers, workstations and tasks should be ergonomically designed and the National Institute for Occupational Safety and Health (NIOSH) Lifting Guidelines should be followed (NIOSH, 81-122). (Note: NIOSH recently revised the 1981 guidelines—they will be published later in 1993.)

Heat Exposure

In environmental heat studies, women have higher skin temperature and pulse rates and lower sweat rates than men under the same heat condition (Shapiro, 1980). NIOSH has established an allowable heat stress of level of 79°F (43.8°C) (WBGT) for unacclimated men and 76°F (42.2°C) (WBGT) for unacclimated women, since women do not tolerate heat as well as men. (Wet bulb globe temperature [WBGT] is a generally accepted index of heat stress, but does not necessarily correlate well with heat strain.) In a hot, moist environment, unacclimated women have less heat tolerance and higher pulse rates and core temperatures compared to men. In a hot, dry environment, there is no difference between the heat tolerance of men and women until relative humidity reaches 80%. At the higher humidity, core and skin temperatures remain equal, but the sweat rate of women does not increase to the extent that it does in men.

Gender Differences

Women are at a disadvantage in heat stress, as determined by oxygen uptake capacity and other

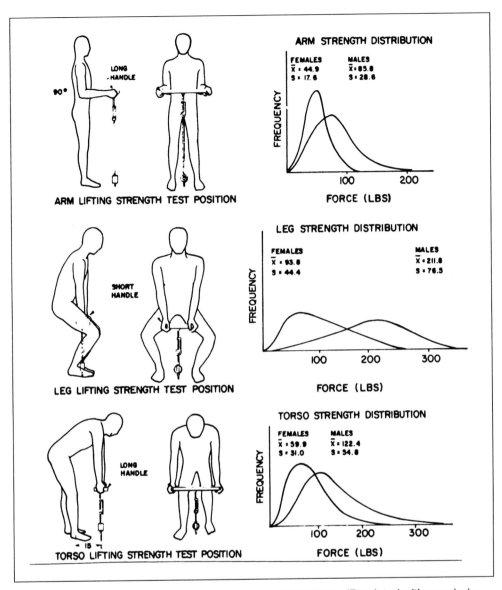

Figure 26-2. Standard strength test postures and distributions. (Reprinted with permission from *Journal of Occupational Medicine* 20(6), June 1978.)

measures. Although both men and women acclimatize to heat at equal rates, men maintain their relative heat-tolerance advantage. The more physically fit persons acclimatize most rapidly. A person's aerobic capacity also has an important bearing on heat tolerance, since a smaller person with limited aerobic capacity will use more reserve capacity than a larger, stronger person.

The circulation of blood is the body's first defense against heat. At higher temperatures, blood flow is shifted away from the interior of the body to the skin, thereby creating a loss of body heat by convection or radiation across the skin. The amount of cardiac output delivered to the skin during heat adaptation can rise to 20% as compared with 5% under normal conditions. With this change, blood pressure drops, blood flow to organs other than the skin decreases, and heart rate increases to increase cardiac output. Women have less skin surface area per lean body mass, so they tolerate heat less effectively.

The body's second, more efficient defense against heat is sweating, which cools the skin with evaporation. In healthy, acclimated men given adequate salt and fluid replacement, the sweat rate can exceed 1 L/hr for several hours. Women have fewer sweat glands than men, and trigger sweating at a higher temperature, helping to explain their disadvantage in tolerating heat.

Heat exhaustion and Heatstroke
These compensations of heat exposure are often not sufficient to protect some persons, and heat exhaustion or heatstroke can occur.

- *Heat exhaustion* results from dehydration and inadequate blood flow to the brain. It is characterized by nausea, dizziness, and, ultimately, fainting. Usually, it is not a serious medical problem, because fainting places the person in a position in which the blood flow is automatically returned to the brain. However, if it occurs in those people who are operating machines, equipment, or vehicles, there can be serious problems.
- *Heatstroke,* on the other hand, is a medical emergency. It is usually accompanied by a high body temperature (107.6°F [42°C] or higher) and severe disturbance of the central nervous system. The skin is dry because sweating stops. A person suffering from heatstroke can become irritable or aggressive, showing hysterical or disoriented behavior, or can become completely apathetic.

Treatment and Acclimation
If the person remains untreated, he or she will die. Consequently, heatstroke victims must be identified and treated rapidly. Rectal temperature is the only accurate indicator of heatstroke. If the rectal temperature is 105.8°F (41°C) or higher, he or she should be treated for heatstroke. Treatment consists of cooling the body continually for at least one hour, preferably until rectal temperature reaches 100.4°F (38°C). Cooling is best accomplished by spraying a combination of water and air over the naked body. Ice and ice water are not recommended as coolants, because vasoconstriction of the peripheral blood vessels can occur, blocking heat transfer and eventually resulting in hypothermic shock when cooling is stopped.

All employees required to work in hot, moist environments should be acclimatized. Acclimation consists of exposing the person to progressively longer periods (one to four hours daily) in the hot environment. It can be accomplished over a four- to seven-day period. Physically fit persons acclimate more rapidly than those less physically fit. Moreover, working or exercising in heat acclimates an employee faster than resting in heat. Finally, body salt and fluid content must be maintained during the acclimation, or the process will be ineffectual. Once a worker is acclimated, the effects will last only about two weeks. Thus, if an employee is on leave for two weeks or longer, he or she will have to be reacclimated.

Low Back Injuries
Low back injuries are extremely costly to industry. It is estimated that low back pain costs industry $16 billion annually. When cases that were not caused by work-related incidents, but involved work restriction or absence are included, the costs soar to above $50 billion (Andersson et al, 1991). Industry has attempted to control costs by screening out workers who might be likely to sustain a back injury. Studies have shown that workers who are less fit or strong have an increased risk of back injury. In some industries, women were not hired for this reason. However, under the 1990 Americans with Disabilities Act (ADA), an employer must show that an employee is being matched for the specific physical requirements essential for a specific job. Although most studies show low back pain to be as frequent in women as in men, because men have more often been placed in heavy lifting jobs, there are more compensable back injuries in men (Kelsey & Golden, 1988).

Coronary Heart Disease
Heart disease is the leading cause of death for both men and women. Risk factors such as smoking, family history, hypertension, diabetes, and lipid disorders do not account for all cases of heart disease; stress and occupational and environmental exposures are also implicated (Benowitz, 1990). The relationship between occupation and heart disease in men has been established (Haynes & Feinlieb, 1980), but there have been fewer studies looking at this relationship in women.

Mortality statistics do not suggest that women have increased their risk of coronary heart disease (CHD) over the past 10 years of increasing employment outside the home. Mortality rates for CHD have declined in the past decade. Trends in CHD-caused illness and disability are less clear, since morbidity rates do not provide information indicating whether women in the work force have a higher incidence of CHD over time than men. The 1980 Haynes & Feinlieb study sought to clarify this question. The study found that women working in clerical positions had twice the incidence of CHD as did other white-collar or blue-collar workers, and that women with children were at greater risk than single women without children. The incidence of CHD also rose with the number of children each woman had, so that working women with three or more children were more likely to develop CHD (11%) than working women with no children (6.5%) or housewives with three or more children (4.4%) (Figure 26-3). Women doing clerical work, married to blue-collar workers, and with three or more children, had a higher risk of CHD than any of the other groups, including Caucasian male executives. The investigators concluded that interpersonal relationships, coping styles, and the occupations of some of the employed women, coupled with family responsibilities, can be involved in the development of CHD (Haynes & Feinleib, 1980).

Results from the previous Framingham studies indicated that two of the most important psychological predictors of CHD among both working women and white-collar men are Type A behavior and the occupational stressor of suppressed hostility.

More recent studies do not show an increase in cardiac risk factors for working women. The Framingham Offspring Study found no significant differences in blood pressure, cholesterol, or blood lipids over an eight-year period between women who worked outside the home and homemakers. In fact, two other studies found working women to have more favorable risk factors (HDL, HDL/cholesterol, triglycerides, fasting and postchallenge glucose, and blood pressure). The differences were not explained by other potentially confounding factors such as age, socioeconomic status, obesity, exercise, smoking, alcohol consumption, and use of estrogen (Hazuda et al, 1986; Kritz-Silverstein et al, 1992).

Carbon disulfide, carbon dioxide, and possibly lead are associated with coronary artery disease; chlorinated hydrocarbons and arsenic can cause arrhythmias; and certain substances (antimony, arsenic, arsine, cobalt, and lead) can cause direct myocardial damage (Benowitz, 1990). A review of the chemicals listed in the Registry of Toxic Effects of Chemical Substances (RTECS) has identified 1,466 as potentially cardiotoxic (NIOSH 89-132).

Men of working age are more prone to develop heart disease than women, and they are more likely to be exposed to cardiotoxins. Workers of both genders should be protected from cardiotoxins.

GENDER GAP IN HEALTH RESEARCH

Despite the fact that women visit medical offices more often and have higher rates of hospitalization than men, women's health issues traditionally have received less attention. Clinical trials for drugs rarely include women. The Multiple Risk Factor Intervention Trials studying CHD factors included 15,000 men, but no women; the Physician's Health Study looking at the benefit of aspirin as preventive therapy for heart disease included 22,000 men but no women. However, research directed toward women's issues has been enhanced. In response to this problem, the Women's Health Equity Act was introduced in 1991 (HR 1161 and S 514). One of the bills, the National Institute of Health Reauthorization Act (HR 4, S 1) authorizes women's health research and codifies National Institutes of Health policy to include women in clinical trials. Workplace standards are based on studies of worker populations that are primarily male. NIOSH should follow the lead of NIH by including women workers in research on workplace standards.

DISABILITY

While men have slightly more work limitations due to chronic conditions (9.5%) compared to women (8.4%), women have more disability in terms of restricted-activity, bed-disability, and lost workdays. The increased disability occurs particularly in women employed in the transportation, materials handling, machine operation, assembly, and inspection jobs (Collins & Thornberry, 1989).

Hospital utilization rates vary by gender. Hospital discharge rates, days of care, and average hospital stays are shown in Table 26-C. Women had 33% more hospitalizations and more total hospital days than men, but shorter stays. When hospitalizations for pregnancy and related problems are taken away, the hospital utilization rate is less divergent, although women still show a greater use (DHHS, PHS, 1991). Smaller groups of workers show more variable patterns. A study of female coal miners showed that these highly selected women had fewer

Table 26-C. Gender and Hospitalization.*

	Male	Female
Discharges per 1,000 population	99.6	126.9
Days of care per 1,000 population	681.0	738.7
Average length of stay in days	6.8	5.8

Source: From DHHS, PHS. Health United States 1991 and Prevention Profile. Hyattsville MD: PHS, 1992.

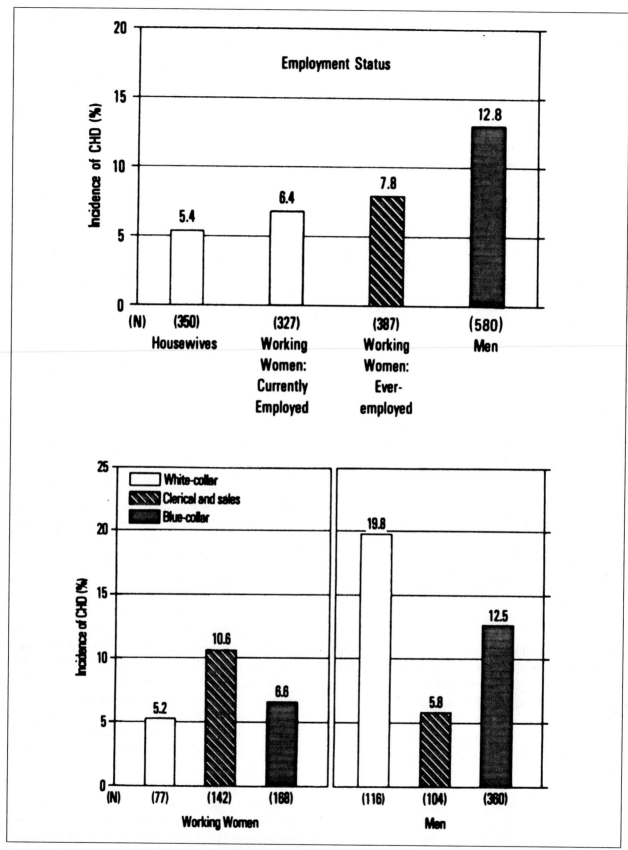

Figure 26-3. Upper section: Eight-year incidence of coronary heart disease by employment status among men and women aged 45–64 years. Lower section: Eight-year incidence of coronary heart disease by occupational status among working women and men aged 45–64 years. (Reprinted with permission from Haynes SG, Feinleib M. Women, work, and coronary heart disease: Prospective Findings from the Framingham Study. *Am J Pub Health* 70(2):133, 1980.)

injuries and fewer lost workdays than men (Watson & White, 1984). A study of Hawaiian working women showed a higher rate of hospital utilization than working men, but a lower rate than housewives. In fact, this and other studies have shown better health status for working women compared to their unpaid counterparts (Halfon et al, 1991).

A study of the sales and office personnel of a major life insurance company showed a significantly higher rate of work loss in women compared to men, even excluding maternity-related losses. The reasons and significance of why women tend to use the health care system more frequently than men should be investigated.

CHEMICAL EXPOSURE AND REPRODUCTIVE RISK

The use of chemicals has increased exponentially since World War II. There are more than 4 million identified chemicals and more than a thousand new chemicals added each year. Of these, more than 60,000 are in widespread use. Unfortunately, few chemicals have been adequately studied for their reproductive or gender-specific toxic effects.

Companies have not given enough attention even to known reproductive toxicants. A 1987 study of 198 chemical and electronics firms found that although 75% of the companies used at least one known or suspected reproductive hazard, only 40% were aware that the chemical required special care because of the reproductive toxicity. Fewer than 66% of these companies trained workers about reproductive risk, and only 25% had worker restrictions (Paul et al, 1989).

General Mechanism of Action

Human sexual function requires coordinated interactions between the neuroendocrine axis, sex organs, and accessory sex glands, and the central nervous system and psychological and behavioral factors. Disruption of reproductive function can result in any of the following:

- loss of libido
- impotence
- infertility
- spontaneous abortion
- congenital malformations
- growth deficits
- childhood cancer
- genetic diseases

Reproductive and developmental toxicants can include infectious agents, drugs, chemicals, physical agents, and maternal illness. Reproductive toxicants can act directly or indirectly on tissues (Table 26-D). Alkylating agents and metals can directly damage reproductive tissues; other substances like polycyclic aromatic hydrocarbons (PAHs) are toxic only after metabolic transformation. Toxicants can act indirectly by altering the production, secretion, or clearance of steroids; some toxicants act at multiple sites. There has been effort directed at chemicals that have or mimic the effects of estrogen. This area of research is very contentious and is the subject of many EPA discussions and publications.

Toxicants can produce reproductive or developmental effects. Reproductive effects are those that interfere with conception and may be reversible. Developmental effects include structural abnormalities, functional deficits, growth problems, or death; these effects are usually irreversible. Some exposures cause both effects. For example, ionizing radiation causes sterility and fetal death and is both a reproductive and developmental hazard.

Male Reproductive Hazards

In 1974, a group of male pesticide workers was found to have fertility problems caused by exposure to dibromochloropropane (DBCP) (Whorton & Foliart, 1988). Tragically, a study done more than a decade earlier associating DBCP with testicular atrophy in rats was ignored. Even today, companies are more likely to have reproductive protection programs only for women (Paul, 1989).

Chemicals can interfere with male reproduction by affecting the neuroendocrine system, the testes, post-testicular sites, or sexual function. Figure 26-4 shows the sites of action of some specific toxicants.

Some chemicals act by disrupting the hypothalamic-pituitary-testicular balances. Lead, for example, appears to reduce testosterone production by the testes, resulting in elevated levels of sex-hormone-binding globulin (SHBG) and luteinizing hormone (LH). Other toxicants such as DDT and PCB can act as hormonal agonists or antagonists (Schrader & Kesner, 1993). Heavy metals and chlorinated hydrocarbons have been found in seminal plasma, but it is not known whether there are any adverse developmental effects.

There are a number of chemicals that interfere with sperm production, maturation, or transport

Table 26-D. Reproductive Toxicants.*

Direct-Acting Reproductive Toxicants (Chemical Reactivity)		
Compound	Effect	Site
Alkylating agents	Altered menses Amenorrhea	Ovary
Lead	Abnormal menses Hypothalamus? Ovarian atrophy Decreased fertility	Pituitary? Ovary?
Mercury	Abnormal menses Hypothalamus?	Ovary
Cadmium	Follicular atresia Persistent diestrus Hypothalamus	Ovary Pituitary

Direct-Acting Reproductive Toxicants (Structural Similarity)		
Compound	Effect	Site
Oral contraceptives	Altered menses Hypothalamus	Pituitary
Azathioprine	Reduced follicle numbers	Ovary Oogenesis

Halogenated hydrocarbons
  Chlordecone
  DDT (dichlorodiphenyltrichloroethane)
  2,4-D (2, 4-dichlorophenoxyacetic acid)
  Lindane
  Toxaphene
  Hexachlor
	Sterility
	Hypothalamus?
	Altered menses
	Infertility
	Amenorrhea
	Hypermenorrhea
		Pituitary?

Indirect-Acting Reproductive Toxicants (Metabolic Activation)		
Compound	Effect	Site
Cytoxan	Amenorrhea Premature ovarian failure	Ovary
Polycyclic aromatic hydrocarbons	Impaired fertility?	Ovary Liver
Cigarette smoke	Altered menses Impaired fertility Reduced age at menopause	Ovary
DDT metabolites	Altered steroid metabolism	Liver

Indirect-Acting Reproductive Toxicants (Disrupted Homeostasis)		
Compound	Effect	Site
Halogenated hydrocarbons DDT	Abnormal menses Hypothalamus?	Pituitary?
PCBs, PBBs (polychlorinated and polybrominated biphenyls)	Abnormal menses Hypothalamus?	Pituitary?
Barbiturates	Increased steroid clearance	Liver

* Adapted with permission from Paul M (ed). *Occupational and Environmental Reproductive Hazards: A Guide for Clinicians.* Baltimore: Williams & Wilkins, 1993.

(Table 26-E). Because spermatogenesis is an ongoing process (unlike oogenesis) the male germ cells are more vulnerable to toxicants that interfere with cell division. Sperm count, motility, or morphology can be affected. Results of three studies of workers exposed to chlordecone (Kepone), a chlorinated hydrocarbon pesticide, showed decreased sperm count and increased abnormal and nonmotile sperm (Sever & Hessel, 1985). The substance DBCP has been found to interfere with spermatoagonia, resulting in oligospermia (sperm count below 20 million/mL) or azospermia.

Sperm damage can lead to infertility or postfertilization effects. In animal studies, dioxin, ethanol,

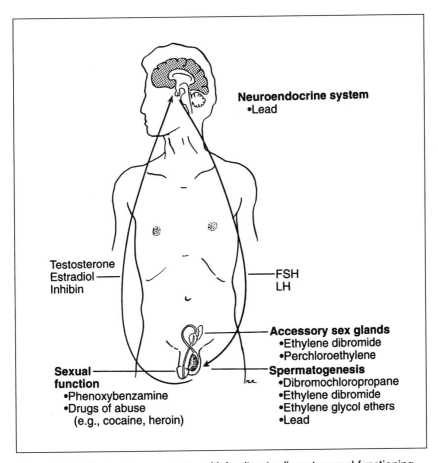

Figure 26-4. Toxicants can act at multiple sites to disrupt normal functioning of the male reproductive system. Primary targets include the neuroendocrine system, the testes, the accessory sex glands, and sexual function. Examples of toxicants that act at these various sites are shown. (Reprinted with permission from Schrader SM, Kesner JS. Male reproductive toxicology. In Paul M (ed). *Occupational and Environmental Reproductive Hazards: A Guide for Clinicians.* Baltimore: Williams & Wilkins, 1993.)

and anesthetic gases cause fetotoxic effects (Letz, 1990), but few studies link paternal exposures to postfertilization changes. One study evaluating brain tumors in 92 children in California found an association with parental employment in the aircraft industry and maternal exposure to chemicals (Peters et al, 1981). Another associates childhood leukemia to the father's occupational exposure to radiation (Aldhous, 1990). Wives of lead-exposed men show increased stillbirths, spontaneous abortion, and childlessness (Winder, 1985). Chemicals can also affect male reproduction by interfering with sexual response-erection, ejaculation, and orgasm; and loss of libido can result from endocrine imbalances.

Female Reproductive Hazards

Historically, companies have taken a fairly protectionist stance toward women. Multiple sites and processes of the female reproductive system can be affected by workplace exposures. Normal functions including oogenesis, folliculogenesis, follicular maturation, ovulation, luteinization, and luteus formation are orchestrated by hypothalamic-pituitary-ovarian interactions. PCBs, PBBs, DDT, and other insecticides can interfere with hormonal feedback loops, resulting in abnormal menses. Amenorrhea or fertility problems can also result from disruption of the neuroendocrine axis. Lead, cadmium, and manganese are associated with decreased fertility and menstrual problems in animals (Plowchalk et al, 1993). However, the effects on humans are less certain.

Oogenesis and the first phase of meiosis are completed before birth. The second meiotic division occurs after fertilization. Oocytes are less sensitive to substances that cause genetic damage than sperm; however, oocytes can be potentially damaged by

Table 26-E. Human Sperm Studies of Occupational Exposures.*

Sperm Indicator

	Count	Motility	Morphology	F Bodies
Detrimental effects				
Carbon disulfide	+	+	+	
Dibromochloropropane (DBCP)	+	+?	+?	+
Lead	+	+	+	
Inconclusive				
Boron	+?			
Cadmium	+?			
Carbaryl	–		+?	–
Ethylene dibromide	+?			
Kepone	+?	+?	+?	
Methylmercury	+?			
Toluenediamine and dinitrotoluene	+?		–	
No effects				
Anesthetic gases	–		–	
Epichlorohydrin	–		–	
Ethylene glycol monomethyl ether	–			
Formaldehyde	–	–	–	–
Glycerine production	–	+?	–	
p-Tertiary butyl benzoic acid	–			
Polybrominated biphenyls	–	–	–	
Wastewater treatment plant	–		–	

+ = Detrimental effect observed.
+? = Detrimental effect with unknown or marginal statistical significance or not clearly related to the exposure.
– = No detrimental effect observed.
* Adapted from Wyrobek. Reprinted with permission from LaDou J. (ed). *Occupational Medicine*. East Norwalk CT: Appleton & Lange, 1990.

high-dose exposure to lead, mercury, cadmium, benzopyrene, and polycyclic aromatic hydrocarbons.

Physiological changes of pregnancy

Pregnancy alters pulmonary, cardiovascular, gastrointestinal, hepatic, and renal functions (Table 26-F), resulting in increased pulmonary and gastrointestinal absorption, diluted concentration, increased distribution, and increased elimination of toxicants (Paul, 1993).

Exposure to lead, benzene, ionizing radiation, or other substances that inhibit red blood cell production can exacerbate the physiological anemia of pregnancy, and decrease the amount of oxygen transferred to the developing fetus.

Developmental toxicology

Depending on the developmental stage, the transplacental transfer of toxicants can be fetotoxic or teratogenic, resulting in malformations, anomalies, or functional deficits (Table 26-G). Developmental toxicants can cause chromosomal damage, genetic changes, or structural defects. As many as 50% of all pregnancies fail; most failed pregnancies go unrecognized. Up to 90% involve chromosomal abnormalities (Paul, 1993). Fetal damage can result from toxic exposure of either parent before conception. For example, exposure to high levels of anesthetic waste gas is associated with fetal loss in women and in wives of exposed men.

During the first trimester, the embryo is growing at a rapid rate. During this period of rapid organ development, the embryo is most vulnerable to teratogens. However, the central nervous system remains sensitive to teratogenic effects beyond birth, because of its relatively slow development. Known and suspected teratogens are listed in Table 26-H. Exposures during the second and third trimesters can cause intrauterine growth retardation. Childhood cancer can result from transplacental exposures, as in the case of DES, or from damage to parental germ cells. Ionizing radiation can produce a spectrum of reproductive and developmental damage.

Lactation

While few infants accompany parents to the workplace, they can receive exposures through breast milk. The percentage of employed mothers with very young children is increasing and this source of exposure for nursing infants must be considered. Solvents, lipophilic agents such as PCBs, heavy

Table 26-F. Hypothesized Occupational Health Impact of Physiological Changes in Pregnancy.

Some Known Physiological Changes	Agent or Condition	Example of Possible Impact	Suggested Job Accommodations
I General			
Fatigue or stress	Inflexible hours Shift work	May be aggravated	Scheduling flexibility Frequent rest breaks
Nausea	Ketones or acrylates Exhaust fumes	Sensitivity to chemicals with strong, unpleasant odors	Improve ventilation Respiratory protection
Metabolic rate	Carbon tetracholoride Protective gear	Hepatotoxicity (especially if metabolically activated) Discomfort or heat intolerance	Minimize exposure
II Cardiovascular			
Uteroplacental flow	Hemolytic agents (eg, arsine) Asphyxiants (eg, CO or agents metabolized to CO, eg, methylene chloride)	Maternal 02-carrying capacity Fetal oxygenationhypoxia	Minimize exposure
Myocardial irritability	Chlorinated hydrocarbons (eg, TCE)	Arrhythmias or MI	Minimize exposure
Autonomic control of vasomotor tone	Anesthetic agents Organic solvents	Arterial pressure Preeclampsia	Minimize exposure
Renal blood flow	Cadmium	Renal toxicity	Minimize exposure
III Respiratory			
Respiratory rate Tidal volume	All airborne chemicals	Absorbed dose per unit time	Minimize exposure
Hyperemic engorgement, capillary dilation	Formaldehyde Sulfur dioxide	Sensitivity to irritants, allergens	Minimize exposure
IV Musculoskeletal			
Lower back pain	Heavy lifting	Difficulty lifting	Mobility for postural changes
Lumbar lordosis, symphyseal and sacroiliac loosening	Ergonomically poor chairs and workstations	Aggravation of pain	Maximum lifting 20%–25% in last trimester
Shifted center of gravity	Well-designed chairs and workstations		

Reprinted with permission from LaDou J (ed). *Occupational Medicine*. East Norwalk CT: Appleton & Lange, 1990.

metals, and viruses have all been shown to be transmitted through human milk.

ASSESSING AND COMMUNICATING RISK

One of the most challenging tasks facing occupational health and safety professionals is assessing and explaining reproductive risks. Preplacement examinations rarely include consideration of reproductive effects, and these issues are usually discussed only when there are problems.

For women, evaluating reproductive effects can include a review of reproductive history (including menstrual cycles, spontaneous abortions, stillbirths, congenital malformations, genetic diseases, and childhood cancers in offspring), measurement of hormone levels, endometrial biopsies, and fertility evaluations. For men, reproductive evaluation can include a reproductive history (including postconception events such

Table 26-G. Reproductive Outcomes Associated with Female Exposures to Workplace Hazards.

Menstrual disorders
Altered fertility
Single-gene defects
Chromosomal defects
Spontaneous abortion
Congenital malformations
Intrauterine growth retardation
Late fetal deaths
Altered gestational time
Altered sex ratios
Perinatal deaths
Developmental disability
Behavioral disorders
Cancer
Childhood cancer
Cancer, female sex organs
Reduced libido
Premature menopause

Table 26-H. Human Teratogens.

KNOWN TERATOGENIC AGENTS
Ionizing Radiation
- Atomic weapons
- Radioiodine
- Therapeutic

Infections
- Cytomegalovirus (CMV)
- Herpes virus hominis ? I and II
- Parvovirus B-19 (Erythema infectiosum)
- Rubella virus
- Syphilis
- Toxoplasmosis
- Venezuelan equine encephalitis virus

Metabolic Imbalance
- Alcoholism
- Endemic cretinism
- Diabetes
- Folic acid deficiency
- Hyperthermia
- Phenylketonuria
- Rheumatic disease and congenital heart block
- Virilizing tumors

Drugs and Environmental Chemicals
- Aminopterin and methylaminopterin
- Androgenic hormones
- Busulfan
- Captropril; Enalapril and renal damage
- Chlorobiphenyls
- Cocaine
- Coumarin anticoagulants
- Cyclophosphamide
- Diethylstilbestrol
- Diphenlhydantoin and trimethadione
- Etretinate
- Lithium
- Methimazole and scalp defects
- Mercury, organic
- Penicillamine
- Tetracyclines
- Thalidomide
- Trimethadione
- 13-cis-retinoic acid (isotreninoin and Accutane)
- Valproic acid

POSSIBLE TERATOGENS
- Binge drinking
- Carbamazepine
- Cigarette smoking
- Disulfiram
- High vitamin A
- Lead
- Primidone
- Streptomycin
- Toluene abuse
- Varicella virus
- Zinc deficiency

UNLIKELY TERATOGENS
- Agent Orange
- Anesthetics
- Aspartame
- Aspirin (but aspirin in the second half of pregnancy may increase cerebral hemorrhage during delivery)
- Bendectin (antinauseants)
- Birth control pills
- Marijuana
- Lysergic acid diethylamide (LSD)
- Metronidazole
- Oral contraceptives
- Rubella vaccine
- Spermicides
- Video display screens

Reprinted with permission from Shepard TH, Fantel AG, Mirkes PE. Developmental toxicology: Prenatal period. In Paul M (ed) *Occupational and Environmental Reproductive Hazards*. Baltimore: Williams & Wilkins, 1993, pp 37–51.

as malformations, spontaneous abortion, childhood cancers, and genetic diseases), blood hormone levels, testicular histology, and semen analysis. Semen analysis is one of the most commonly used ways of measuring the effects of male reproductive hazards, but there are a number of difficulties including lack of standardized techniques and normal standards.

Differentiating occupationally caused reproductive disorders from those caused by nonoccupational factors is difficult; workers are exposed to a multitude of chemicals, physical stressors, and biological substances at work and away from work. Moreover, human sexual function is complex and there are many potentially adverse reproductive outcomes (Table 26-I and Figure 26-5). The background rate for reproductive and developmental problems is high, and large populations are required to show an increase.

REPRODUCTIVE AND DEVELOPMENTAL HAZARD POLICY

Since current programs to evaluate and control reproductive hazards are inadequate, NIOSH has identified several major needs in laboratory research, surveillance, and epidemiology and control strategies (DHHS, NIOSH 89-133). Company programs to control reproductive hazards should include the following elements:

1. elimination of reproductive hazards or substitution of less hazardous materials whenever possible. All reproductive hazards in the workplace should be identified. However, because of the lack of adequate reproductive toxicological information, exposure to all toxic substances should be minimized.

2. the use of engineering controls such as ventilation, enclosure of sources, and isolation of workers from exposure

Table 26-I. Observable Disorders of Reproduction Following Exposure to Various Types of Reproductive Hazards.*

Potential outcome	Reproductive Toxicants				Developmental Toxicants
	Nonmutagenic		Mutagenic		
	Male	Female	Male	Female	Female
Infertility	X	X	X	X	X
Spontaneous abortion					
normal karyotype	X	X	X	X	X
abnormal karyotype			X	X	
Prematurity					X
Birth defect			X	X	X
Delayed development					X
Low birth weight					X
Postmaturity					X
Semen abnormality	X		X		X**
Sexual dysfunction	X	X			X**
Dysfunction of other organ systems					X**
Childhood cancer			X	X	X**

* Reprinted from DHHS (NIOSH). *Proposed National Strategies for the Prevention of Work-Related Diseases and Injuries: Disorders of Reproduction.* Cincinnati: NIOSH Pub. No. 89-133, 1988.
** Abnormalities in children or adults resulting from *in utero* exposure

3. the use of administrative controls such as work scheduling and not allowing untrained workers to enter areas where reproductive toxicants are used
4. safe work practices, including employee education and careful supervision of workers. Employee training in reproductive risk is often not adequate for a number of reasons:
   a. There is not enough reproductive toxicological data for most chemicals.
   b. Material Safety Data Sheets (MSDSs) seldom contain reproductive warnings.
   c. An employer can mistakenly assume that no mention of reproductive toxicity on the MSDS implies that the substance has none. Absence of information typically means that no reproductive testing has been done.
   d. Many employers are not aware of the requirements of the OSHA Hazard Communication Standard and do not train employees on the specific hazards of each chemical.
5. the use of personal protective equipment. All workers should be properly fitted and trained in the use of personal protective devices.
6. careful disposal of waste materials.

PROTECTION OR DISCRIMINATION?

Our society attempts to protect women and fetuses from occupational hazards. But in the view of many, this protection is accompanied by discrimination. Recent court decisions have brought into question the assumptions underlying fetal protection. Title VII of the Civil Rights Act of 1964 made it illegal to discriminate on the basis of race, religion, color, sex, or national origin; and the 1978 Pregnancy Discrimination Act defined sex discrimination as including discrimination on the basis of pregnancy, childbirth, or related conditions. But many companies continued to exclude women of childbearing age from jobs with potential exposure to fetotoxicants such as lead. This practice is no longer legal. In 1991 the Supreme Court ruled in the International Union versus Johnson Controls case that employment opportunities are not to be limited on the basis of possible or actual pregnancy. Women are to be given a choice of working with these hazards. However, there is nothing to prevent an affected child from suing the company for health problems resulting from *in utero* exposures. This threat of litigation may encourage companies to reduce their use of reproductive toxicants (Annas, 1991).

Although there is evidence linking childhood leukemia to paternal occupational exposure to radiation (Aldhous, 1990), exclusion of fertile male workers from jobs with radiation exposures is unheard of. One author suggests that this is because companies view women as "primarily biologic (sic) actors not economically responsible for their families" and men "only as economic actors" (Becker, 1990). In fact, most women need to work. The majority of women in the work force are single, divorced, widowed, separated, or have husbands with annual incomes below $15,000 (DOL, 1990). Of families headed by single mothers, 44% are below the poverty level (DHHS, PHS, 1992).

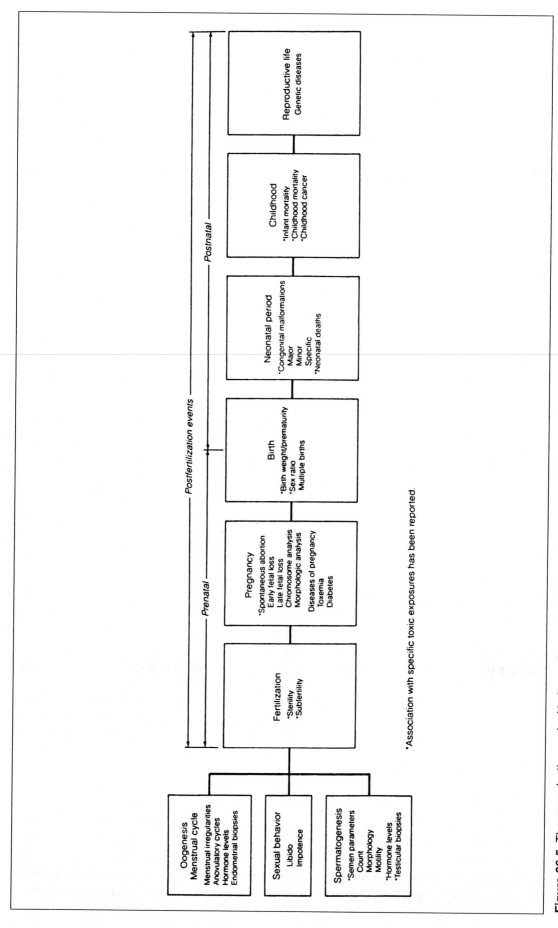

Figure 26-5. The reproductive cycle. Under each stage are listed the parameters that could be monitored in relation to toxic exposures. In the male, end points that have been investigated include reproductive history, blood hormone levels, testicular histology, and semen analyses, all of which reflect effects prior to conception. Theoretically, any postconception event (spontaneous abortion, congenital malformation, etc.) that is mediated by abnormal sperm could also be related to exposures occurring in the male prior to conception. The few existing studies make the available data difficult to interpret. Of the above end points, semen analysis is by far the most important in terms of medical surveillance. (Reprinted with permission from LaDou J (ed). *Occupational Medicine*. East Norwalk CT: Appleton & Lange, 1990.)

HARASSMENT

Sexual harassment is found throughout U.S. business and industry. The EEOC defines sexual harassment as conduct on the part of a superior or co-worker that requires submission for obtaining or retaining a job, creates an intimidating, hostile, or offensive work environment, or affects personal decisions or job performance of the harassed person. Employers have an affirmative duty to prevent and eliminate such sexual abuse (Collins & Blodget, 1981).

Sexual harassment can consist of off-color remarks, body contact when none is necessary, threats of job loss, poor personnel reports for not cooperating, or even attempted rape. Many workers hesitate to speak out against this form of harassment. It is not always clear what constitutes sexual harassment and workers may fear publicity, embarrassment, or job loss. While harassment is difficult to prove, workers must be encouraged to report instances of sexual harassment to their supervisors. Only by this means will it be stopped.

Occupational health and safety professionals should be alert to sexual harassment among employees, because:
- for employees it can result in decreased work performance and increased psychological problems.
- for companies it can result in decreased productivity and morale.

FAMILY-RELATED ISSUES

As the work force becomes more diverse, including many workers with "dual shifts" involving dependent care and household management, traditional assumptions underlying family benefits must be reexamined. Despite their increasing presence in the full-time labor force, working women are responsible for a disproportionate amount of home maintenance and child and elder care responsibilities. While biology can explain the advantage women have for the care of nursing infants, it does not explain the overwhelming delegation of other care-giving and home-maintenance responsibilities to women. Even with the woman earning more than the man, family responsibilities did not significantly shift (England & McCreary, 1987).

With an increasing percentage of mothers entering the work force childcare has become a major issue. A majority of mothers with children under the age of 1 year and 65% of mothers with children under 18 years of age work. Difficulties with childcare can increase tardiness and absenteeism and decrease productivity (DOL, 1989). A recent study identified the strongest predictive factor for absenteeism to be a mother with small children (Leigh, 1991). Elder care is a responsibility for an increasing number of workers, and finding affordable, quality dependent care is becoming more difficult. The Family Medical Leave Act (FMLA) of 1993 is an attempt to provide a solution to the issues of giving time off, job security, and economic cost-effectiveness. The FMLA is a mandatory federal leave act intended to protect employees who need to take time away from work to deal with specific family medical problems. The FMLA enables eligible employees to take a job-protected leave for up to 12 weeks. The FMLA applies to employers with 50 or more employees, all public agencies, and schools.

Another serious problem for all workers is health care. Over 12% of the population is without health insurance. Of the uninsured, 80% are employed or dependents of an employed worker. There are more uninsured persons working in the areas of retail and service trades and administration or sales. Approximately 66% of large firms and less than 50% of small firms (employing 40% of all workers) have paid sick leave.

SUMMARY

This chapter has touched upon the clear need for new policies and programs that will more completely address issues such as protection from reproductive and developmental hazards, physiological and physical gender differences, discrimination and harassment, the availability of health care, and other family-related issues. Changing demographics of the workplace are causing us to rethink many of our current practices. For example, workplace exposure standards, based upon data from male subjects, do not address the different protective needs of women. Gender-dependent differences no longer can be accommodated simply by excluding women from hazardous workplaces. Moreover, as paternal exposures are linked to problems in offspring, the definition of "safe" exposure levels for men is in question. Discrimination and harassment can profoundly affect morale and productivity, and are recognized as important issues to address for both genders. There is an increased recognition of the importance of addressing employees' family needs, such as dependent care and health care coverage. In the United States, federal statutes are being enacted to meet this need (e.g., the Family Medical Leave Act (FMLA) of 1993).

(Note: The material in this chapter is based on Women in the Workplace, a chapter written by Joseph LaDou, MD, and Mary Phares. This chapter appeared in the first edition of this book, *Introduction to Occupational Health and Safety*, 1986.)

REFERENCES

Aldhous P. Leukemia cases linked to fathers' radiation dose. *Nature* 343:679, 1990.

Andersson GB, Pope MH, Frymoyer JW, et al. Epidemiology and cost. In Pope MH, Andersson GB, Frymoyer JW, et al (eds). *Occupational Low Back Pain: Assessment, Treatment and Prevention*. St. Louis: Mosby-Yearbook 1991, pp 95–113.

Andrews MM. Transcultural perspectives in nursing administration. *J Nurs Adm* 28(11):30–38, 1998.

Annas GJ. Fetal protection and employment discrimination—the Johnson Controls Case. *N Engl J Med* 235(10):740–743, 1991.

Becker M. Can employers exclude women to protect children? *JAMA* 264(16):38–43, 1990.

Benowitz NL. Cardiovascular toxicology. In LA Dou (ed). *Occupational Medicine*. East Norwalk, CT: Appleton & Lange 1990, pp 233–246.

Calabrese EJ. Sex differences in susceptibility to toxic industrial chemicals. *Brit J Ind Med* 43: 577–579, 1986.

Chaffin DB, Anderson G. *Occupational Biomechanics*. New York: Wiley, 1984.

Collins ECG, Blodget TB. Sexual harrasment . . . Some see it . . . Some won't. *Harvard Bus Rev* 59(2):76–95, 1981.

Collins J, Thornberry O. Health characteristics of workers by occupation and sex: United States, 1983–85. Advance Data From *Vital and Health Statistics*. Washington DC: DHHS Pub. No. 168, April 25, 1989.

Congressional Caucus for Women's Issues. The Women's Health Equity Act of 1991. HR 1161, S 514 102nd Congress, February 1991.

Deyo RA (ed). *Back Pain in Workers*. State of the Art Reviews: Occupational Medicine Series. Philadelphia: Hanley & Belfus 3(1), January–March 1988.

DHHS (U.S. Department of Health and Human Services), NIOSH. *Proposed National Strategies for the Prevention of Work-Related Diseases and Injuries: Disorders of Reproduction*. Cincinnati: NIOSH Pub. No. 89–133, 1988.

DHHS, NIOSH. *Proposed National Strategies for the Prevention of Leading Work-Related Diseases and Injuries: Occupational Cardiovascular Diseases*. Cincinnati: NIOSH Pub. No. 89–132, 1986.

DHHS, NIOSH. *Technological Report: Work Practices Guide for Manual Lifting*. Cincinnati: NIOSH Pub. No. 81–122, 1981.

DHHS, PHS. *Health United States 1991 and Prevention Profile*. Hyattsville MD: PHS, 1992. (Available from U.S. Government Printing Office)

DOL (U.S. Department of Labor), Women's Bureau. *Facts on Working Women*. Washington DC: U.S. Government Printing Office Pub. No. 90–2, September 1990.

DOL, Women's Bureau. *Facts on Working Women: Women in Management*. Washington DC: U.S. Government Printing Office Pub. No. 89–4, 1989.

DOL, Women's Bureau. *Facts on Working Women*. Washington DC: U.S. Government Printing Office Pub. No. 90–3, October 1990.

DOL, Women's Bureau. *Working Women and Their Children*. Washington DC: U.S. Government Printing Office Pub. No. 89–3, August 1989.

England P, McCreary L. Integrating sociology and economics to study gender and work. In Stromberg AH, Larwood L, Gutek B (eds). *Women and Work*. Beverly Hills CA: Sage, 1987.

Halfon S, Kodama AM, Arbiet W. The health status of working women in Hawaii. *Hawaiian Med J*. 50(1):18–23, January 1991.

Haynes SG, Feinlieb M. Women, work, and coronary heart disease: Prospective findings from the Framingham Heart Study. *Am J Pub Health* 70(2):133, 1980.

Hazuda HP, Haffner SM, Stern MP et al. Employment status and women's protection against coronary heart disease. *Am J Epidemiol* 123: 623–640, 1986.

Hughes C, Kim JH, Hwang B. Assessing social integration in employment settings: Current knowledge and future directions. *Am J Ment Retard* 103(2):173–185, 1998.

Kelsey JL, Golden AL. *Occupational and Workplace Factors Associated with Low Back Pain.* State of the Art Reviews: Occupational Medicine Series. Philadelphia: Hanley & Belfus, 3(1):7–16, 1988.

Kritz-Silverstein D, Wingard DL, Barrett-Connor E. Employment status and heart disease risk factors in middle-aged women: The Rancho Bernadino study. *Am J Pub Health* 82:215–219, 1992.

Lee B. Steps to valuing diversity. *Nurs Stand* 13(29):17–19, 1999.

Lee B. Tackling racism. *Nurs Stand* 13(28):14–16, 1999.

Leigh JP. Employee and job attributes as predictors of absenteeism in a national sample of workers: The importance of health and dangerous working conditions. *Soc Sci Med* 33(2):127–187, 1991.

Letz G. Male reproductive toxicology. In LaDou J (ed). *Occupational Medicine.* East Norwalk CT: Appleton & Lange, 1990, pp 288–296.

Messite J, Welch L. An overview: Occupational health and women workers. In Stroberg A, Larwood L, Gutek B (eds). *Women and Work: An Annual Review.* Beverly Hills CA: Sage, 1987, pp 21–43.

Paul M, Daniels C, Rosofski R. Corporate response to reproductive hazards in the workplace: Results of the family, work, and health survey. *Am J Ind Med* 16:267–280, 1989.

Perney S. Embracing workplace diversity. *J Long Term Care Adm* 24(1):16–18, 1996.

Peters JM, Preston-Martin S, Yu MC. Brain tumors in children and occupational exposure of parents. *Science* 213:235–236, 1981.

Plowchalk D, Meadows MJ, Mattison DR. Female reproductive toxicology. In Paul M (ed). *Occupational and Environmental Reproductive Hazards: A Guide for Clinicians.* Baltimore: Williams & Wilkins, 1993, pp 18–24.

Schrader SM, Kesner JS. Male reproductive toxicology. In Paul M (ed). *Occupational and Environmental Reproductive Hazards: A Guide for Clinicians.* Baltimore: Williams & Wilkins, 1993.

Sever IE, Hessel NA. Toxic effects of occupational and environmental chemicals on the testes. In Thomas JA, et al (eds). *Endocrine Toxicology.* New York: Raven, 1985.

Shapiro Y, Pandolf KB, Goldman RF. Sex differences in acclimation to a hot-dry environment. *Ergonomics* 23(7):635–642, 1980.

Watson AP, White CL. Workplace injury experience of female coal miners in the United States. *Arch Environ Health* 39(4):284–293, 1984.

West JC. Sexual harassment: An ominous liability for healthcare professionals. *J Health Risk Manag* 19(3):14–25, 1999.

Whorton D, Foliart D. DBCP: Eleven years later. *Reproductive Toxicol* 2:155–161, 1988.

Winder C. Reproductive effects of occupational exposures to lead: A review. *Yale J Biol Med* 58:9–17, 1985.

chapter 27

Workers with Disabilities and the ADA

by Jean Spencer Felton, MD

471	Introduction
471	Legislation and Policy of The Rehabilitation Act of 1973
472	Current Legislation
472	Number of Disabled Persons
473	The Essence of the ADA
474	Title I Requirements Commonly used accommodations ■ Job accommodation network ■ Assistive services/sensory aids
476	Medical Examination Matching the worker to the job ■ Supportive procedures and follow-up
478	Supported Employment
479	The Language of Disability
479	Corporate Social Conscience
480	Other Legislation
481	Current Legislative Action
482	Summary
483	References

INTRODUCTION

Although people have experienced disabling injuries and illnesses since history began, employing disabled persons in commerce and industry has been a relatively recent occurrence. World War I was the first time that efforts were initiated to return so-called "handicapped" servicemen to gainful employment. As early as 1917 men were being taken out of civilian life, where they were wage-earners in normal civil capacities, and were required to provide a service to their country that could prove permanently disabling. Physical and occupational restoration was the Army's goal at war's onset.

World War II further emphasized the need for the fullest return to a productive life of those persons deprived by their military experience of certain customary living or working capabilities. With the functional restoration provided during postwar efforts in rehabilitation and the imperatives delivered through subsequent federal and state legislation, disabled persons have been placed in business and industrial positions and have proved to be as effective in their jobs as their fellow workers.

LEGISLATION AND POLICY OF THE REHABILITATION ACT OF 1973

The Rehabilitation Act of 1973 (Public Law 93-112, 93rd Congress, HR 8070, September 26, 1973) and its subsequent implementing regulations have been the basic policy underlying any effort to employ that segment of society with physical or behavioral disabilities. The purpose of the act was to provide a statutory basis for the Rehabilitation Services Administration and to authorize programs that would prepare disabled persons to engage in gainful employment. It also called for barrier-free construction of public facilities and the removal of architectural and transportation barriers that impede the mobility of disabled persons.

CURRENT LEGISLATION

In 1990, a more recent effort to strengthen employment opportunities for the disabled emerged. The Americans with Disabilities Act of 1990 (ADA) was passed by both houses of the U.S. Congress, and President Bush signed the bill on July 26, 1990. His Committee on Employment of People with Disabilities proceeded in a positive mode, despite strong lobbying activities by opposing groups.

The 1990 ADA called for sweeping legal protection against discriminatory actions in employment, public services, and public accommodations (Title I) for a then estimated 43 million disabled Americans. The act defines a disabled person as one with a physical impairment, a record of such impairment, or one who is perceived as having an impairment that substantially limits one or more major life activities. Specifically, the new act states that an employer cannot limit, segregate, or classify a disabled person in a way that hurts his or her job opportunities. Reasonable accommodation must be made unless, as in previous legislation, it imposes an undue hardship on business. Reasonable accommodation will be discussed later in this chapter.

The 1973 legislation was the specific defining of the "evaluation of rehabilitation potential." Although vocational rehabilitation counselors had tried to establish such an evaluation in previous decades, the procedure had not been delineated in sufficient detail. An evaluative study of this type was required to determine if a substantial barrier to employment was present, and to determine if an employee required vocational rehabilitation services to enable his or her entry into the labor market. This information also provided the prospective employer with all the necessary information needed for proper job placement of the applicant.

The evaluation comprised diagnostic studies of "pertinent medical, psychological, vocational, educational, cultural, social, and environmental factors which bear on the individual's handicap to employment," and, to the degree needed, "an inquiry into the client's work experience, vocational aptitudes and interests, personal and social adjustments, employment opportunities, and other pertinent data," that would prove helpful in selecting the scope of necessary rehabilitative services (Public Law, 1973).

In addition, to assess and develop a person's capacities to perform adequately in a work environment, it is necessary to appraise the person's work behavior patterns and his or her ability to acquire an occupational skill and to develop work attitudes, work habits, work tolerance, and social and behavior patterns suitable for successful job performance (Berkeley Planning Associates, 1982).

This kind of detailed study is not an esoteric exercise formulated to carry out some arcane administrative function, but is the basis for future referrals to prospective employers, who may never have faced the task of adding disabled workers to their work force. Further, many disabled people have not worked before and are unaccustomed to the ethos of business or corporate life. While most persons with full physical capacity move from home to school to job, many of those with disabilities never had the experience of summer, intermittent, or even sheltered or supported (postplacement) employment, let alone task-specific training for commonplace work. The entry into the labor force can produce extreme anxiety so that full preparation for this change in lifestyle must be made knowledgeably.

NUMBER OF DISABLED PERSONS

Before reviewing the ADA, one should be apprised of the number of persons with disabilities in the United States, a population to whom the legislation is applicable. A disabling condition can result from a variety of causes—a birth defect, an endocrine disorder, a malignancy (with or without metastases and treated or not treated), the residual impairment following an injury or a disease, the aging process, or a chronic disorder. The unintentional injury may be job-related or nonoccupational in origin.

Data are available for the period 1994–95, wherein it is seen that of a total population of 261,749,000, nearly 21% (53,907,000) showed some disability. Of this group, almost 10% had a severe disability (25,968,000) and another 11% (27,938,000) had a disability that was not severe.

Of the total number of whites, 21.4% presented with a disability, 21.6% of African-Americans were impaired, 23.9% of Native-Americans, Eskimos or Aleuts were disabled, while 12.1% of Asian-Americans or Pacific Islanders and 16.2% of persons of Hispanic origin showed some disability (U.S. Census Bureau website).

In the same period, 14.9% of the 95,000,000 persons aged 22 to 44 years old were limited. A wheelchair was used by 281,000; 7,359,000 had difficulty with one or more functional activities (difficulty or inability in seeing words and letters; in hearing normal conversation; in having speech understood; in lifting or carrying 10 pounds; or in walking three city blocks).

In the same group personal assistance was needed with one or more of the activities of daily living (ADL) by 663,000. The activities of daily living relate to an individual's ability to maintain his or her immediate environment, and include such entities as obtaining food, cooking, dressing, bathing, using/getting to toilet, or administering one's own medications. The older group, aged 45 to 54 years of age, demonstrated 24.5% as having disability, or 7,412,000 persons. These two groups are cited, for these are the age segments composing the workforce, in the main. Of the persons aged 55 to 64 years old (20,647,000), 36.3% had some disability.

The employment rate of persons age 21 to 64 years was reviewed, and it was seen that 82.1% of those with no disability had a job or business, 76.9% with nonsevere disability were employed, while only 26.1% of individuals with a severe disability were at work. In all instances, fewer women were employed than men, although among the severely disabled the rates approached each other with males employed at 27.8%, and females at 24.7%.

THE ESSENCE OF THE ADA

The essence of the ADA, Title I, lies in the requirement that any employer of 15 or more persons shall take affirmative action to employ and advance qualified disabled persons. This mandate applies to private employers, state and local governments, employment agencies, labor unions, and joint labor-management committees. The ADA mandates apply not only to the initial employment of disabled persons, but to all personnel actions such as promotions or demotions, transfers, layoffs, and rehirings, and the provision of benefits, such as sick leave, training programs, leaves of absence, or any other fringe benefit offered by the organization.

A job cannot be denied merely because an employer does not want to make the necessary accommodations. Preplacement tests cannot screen out disabled persons unless the tests are job-related and a business necessity. An interviewer cannot ask questions in interviews about disabilities—only about the person's ability to do the job. Further, one cannot use tests for employment that penalize those persons with sensory, manual, or speaking disabilities, unless these skills need to be measured for the job. Preplacement medical examinations cannot be required, but a conditional offer of employment can be made pending the results of the medical evaluation, if an employer does the same for all persons to whom a job has been offered. Also, an employer cannot discriminate against an employee in hiring or administering benefits because of his or her relationship with a disabled person, such as a spouse or child.

A person desirous of filing a complaint of employment discrimination can seek redress through the actions of the Equal Employment Opportunity Commission and subsequently through the courts, with the remedies being the same as under Title VII of the 1964 Civil Rights Act—injunctive relief and back pay. It must be pointed out that complaints registered under existing legislation, particularly as they relate to medical rejection for employment, have nearly always been ruled in favor of the plaintiff. An additional requirement calls for the posting of notices that explain the law, and such notices must be accessible, as needed, to persons with visual or other reading disabilities. A new equal employment opportunity (EEO) poster, containing ADA provisions and other federal employment nondiscrimination provisions, can be obtained by writing EEOC at 1801 L Street NW, Washington DC 20507, or calling 1-800/669-EEOC or 1-800/800-3302 (TDD) [Teletex Device for the Deaf].

Title II of the ADA covers all activities of state and local governments in that they give people with disabilities an equal opportunity to benefit from all available services, programs, or activities, such as public education, employment, transportation, recreation, health care, social services, courts, voting and town meetings (Guide to Disability Rights

Laws). Care must be taken in new construction and communities with hearing, vision, or speech disabilities, as examples.

Title III concerns public accommodations such as privately operated entities offering courses and examinations, privately operated transportation, and commercial facilities. Included are such areas as restaurants, retail stores, hotels, convention centers, physicians' offices, and sports centers, among other facilities utilized by persons with disabilities. There cannot be such actions as exclusions, segregation, or unequal treatment (discrimination) in these public areas.

While the disabled who sense or experience discriminating action can file claims, as noted above, the Department of Justice can sue a public entity for failing to comply with the ADA regulations. As recently as Jan. 29, 1999, the Department of Justice sued American Multi-Cinema, Inc. and AMC Entertainment, operators of one of the nation's largest chain of movie theaters, for not providing stadium-style seating to individuals using wheelchairs. To reach such seats, patrons had to climb stairs. Stadium-style seating is located on stepped 18-inch risers and provides comfortable, unobstructed viewing of the screen (Department of Justice document). Information concerning Title II and III regulations may be obtained on the ADA Home Page at http://www.usdoj.gov/crt/ada/adahom1.htm. The page offers information concerning the ADA Information Line, the Department of Justice's enforcement activities, and changes in ADA regulations and requirements.

TITLE I REQUIREMENTS

Title I of the 1990 Act became effective on July 26, 1992, and for the first two years, employers with 25 or more employees were covered. As of July 26, 1994, employers with 15 or more employees were covered by the act. Most significant in Title I regulations—the segment of the ADA most important in the employment of persons with disabilities—is reasonable accommodations, which indicates that, a recipient is required to make "reasonable accommodation to the known physical or mental limitations of an otherwise qualified disabled applicant or employee" unless the accommodation imposes an undue hardship on the operation of its program. Examples of accommodation might include these:

- modification of work schedules
- job restructuring
- reassignment to a vacant position
- appropriate adjustment or modification of examinations, training materials, or policies
- physical modifications or office relocation so that facilities become accessible to, and usable by, disabled persons; or the provision of readers or interpreters.

It was this segment that evoked the loudest response by employers who anticipated great costs in meeting the requirement. They could be exempted, as indicated, under "undue hardship," which would consider the factors of business necessity and financial cost and obligation.

To explore the problem, the U.S. Department of Labor supported a special study in which a national survey was conducted of private sector employers who contracted with the federal government to relay information about their accommodation practices for disabled employees. The project comprised a survey of 2,000 contracting entities; telephone interviews with 85 firms that had provided at least one accommodation; a survey of 145 disabled workers, 87 of whom had accommodations made for them; and intensive case studies of 10 firms whose responses indicated exemplary accommodation practices (Berkeley Planning Associates, 1982). The primary conclusion was that accommodation was extremely important to the employment of many disabled persons, but to the firms most actively involved, it was "no big deal." It took place as needed, rarely entailed much cost, and was not done out of a sense of charity, but with the objective of making the worker more productive on the job. Rather than the extensive remodeling or costly additions of new equipment, which was given big hype by the news media, the changes consisted of minor adjustments of the job and workplace, such as transferring a worker to a job or site where his or her disability did not give rise to a handicap, or transferring certain tasks to other workers, or moving furniture, raising a desk, and the like. The goal actually was to bring the disabled worker up to the company's standard of productivity for that particular job. As for costs, 51% of the accommodations cost absolutely nothing, while 41% cost less than $500 each. In short, meeting federal regulations by altering the worksite or job tasks was not a great problem.

Commonly Used Accommodations

The following commonly used accommodations should be considered when designing a safe workplace for employees with disabilities:

- Provide special parking spaces for employees with mobility impairments.
- Enlarge toilet facilities and install handrails for employees with mobility impairments.
- Provide levers instead of knobs for employees with limited use of their hands.
- Widen doorways to make sure they are not barriers for employees with mobility impairments.
- Provide special communication equipment, such as flashing strobe lights to smoke detectors, to make sure employees with hearing impairments are made aware of emergencies.
- Use strongly colored lights or markers to help guide employees with impaired vision around the workplace.
- Use the buddy system, appointing a specific employee or employees to alert workers with disabilities of emergencies and help them to exit the facility.

Job Accommodation Network

To facilitate the employment or retention of qualified workers with disabilities, the President's Committee on Employment of People with Disabilities (formerly the President's Committee on Employment of the Handicapped) created a service wherein information is available on solutions to accommodation problems. The *Job Accommodation Network (JAN)* is available via toll-free numbers, 1-800-526-7234 (V/TDD), 1-800-232-9675 (V/TDD), and from Canada 1-800-526-2262 or via FAX: (304) 293-5407 or e-mail: jan@jan.icdi.www.edu, so that problems involving accommodation can be reviewed with its human factors consultants. The JAN has assembled information from a broad variety of sources about practical ways of accommodating employees and job applicants with disabilities. The service is used successfully by employers, rehabilitation professionals, and people with disabilities themselves. Entered into its data bank are case examples of effective means of accommodation. Many of these applications were reported by employers in the private sector—a fact that should defuse concerns regarding government regulation.

If a person is being considered for employment and has a disability, the placement official of the company can call JAN and relate the functional requirements of the specific job, the functional limitations of the worker, the environmental factors, and other data germane to the problem. The computer database is searched and the consultant provides the information reflecting "close-to-matching" situations that were identified. There is no cost of the service—only a commitment that the person seeking assistance will complete a brief input data form so that the accommodation effected can become part of the data base and be available for future users. The JAN is valuable to employers in the use of available programs such as the Job Training Partnership Act, Projects with Industry, Supported Employment, Targeted Jobs Tax Credit, and Barrier Removal Incentives.

A few examples of accommodations and their low costs are as follows (JAN, 1989):

- providing a drafting table, page turner, and pressure-sensitive tape recorder to a sales agent paralyzed from a broken neck—$300
- renting a headset phone that allows an insurance agent with cerebral palsy to write while talking—$6.01 per month
- removing turnstiles in the cafeteria and installing lighter weight doors, as part of a general renovation, and having the cafeteria deliver lunch to a payroll technician disabled from polio—$40 per month.

Assistive Services/Sensory Aids

To compensate for physiological changes or losses that accompany aging, organ function decrement, injury, postillness residuals, or even birth defects, a large group of assistive devices is available commercially to compensate for the particular diminished capacity or sensory decrease. The following aids are readily obtainable (Russell et al, 1997):

- standard wheelchairs
- special-model wheelchairs (battery-powered, standup, elevating, etc.)
- special-feature wheelchairs (detachable arms, power drive, pneumatic tires, a narrowing device, computer-aided, etc.)
- writing aids
- telephoning aids
- speaking aids
- electronic aids (removable control devices, hearing aids)
- speech aids, electronic (transoral electronic larynx)

- vision aids (telescopic lenses, electronic magnifying reading devices)
- miscellaneous aids (wheelchair-accessible drinking fountains)
- screen-reading equipment (aloud and into braille).

MEDICAL EXAMINATION

Preexisting impairments of disabled people for which job accommodation can be made should not automatically be considered as evidence of an inability to perform the job. Following an offer of employment, medical examinations that match people's capabilities to jobs they can perform will contribute significantly to increased employment of qualified people with disabilities.

Under existing legislation, once an offer of employment is made, disabled persons can undergo medical examinations and based upon the results of such evaluations, provided they are administered to all entering applicants and employees in a nondiscriminatory manner (i.e., all are evaluated against the same medical standards) and the results are treated confidentially, a suitable job assignment can be made.

Matching the Worker to the Job

The physicians or nurses conducting such examinations must be provided with sufficiently detailed job descriptions, critical job task factors, and their medical standards, so that intelligent, clinically sound, and legally defensible decisions can be made concerning the matching of an applicant's capacities to the specific demands (physical, emotional, and environmental) of the particular position being sought (Figure 27-1). It must be constantly kept in mind that the objective of the examination is to determine whether the candidate is able to perform all job-related functions as outlined by the job description. Any activities or tasks that do not relate to the job should not affect placement, whether they are within or outside the capabilities of the person. Reasonable accommodation can be made, as indicated earlier, to allow for completion of the job's requirements.

At no time in the hiring process can the concept of future risk enter into employment considerations. If there are factors present such as a strong family history of heart disease, obesity in the examinee, borderline hypertension, etc., the applicant cannot be rejected because the evaluator prognosticates future illness or shortened work life. The examination is conducted with these goals in mind: (1) To place the applicant in a position matching his or her capacities where no harm will befall the worker or fellow workers, and no damage to company property, equipment, etc. will occur; (2) to initiate a preventive health program that will forestall worsening of the present disability or the onset of other disease processes; (3) to facilitate entry into the health care system if such entry has not already been made—particularly pertinent in the case of refugees or immigrants with disabilities; (4) to establish a medical status baseline so that the health effects of exposure, job-connected or accidental, to a toxic substance can be measured against precontact findings or levels; (5) to determine special needs in the area of reasonable accommodation, such as sensory aids, special work apparatus or additional devices, relocation of workstations, and special parking; (6) to identify coexisting diseases that may be both contagious to others and challenging to or destructive of the person's health; (7) to obtain knowledge of family health problems so that counsel can be provided, thus reducing periods of employee absence, stress, or lessened productivity; (8) to allow effective liaison with the employee's supervisor regarding potential emergency situations, such as insulin-dependent diabetes mellitus or a convulsive disorder; (9) to comply with state or local statutory requirements (when mandated) for certain job categories such as those in primary education or health care; and (10) to measure emotional status so that through referral to counseling sources, any job-related stress can be averted.

A placement examination, particularly when conducted at the worksite health facility, if one exists, can cement relationships between an apprehensive job applicant and a caring facility within the corporate structure, so that future visits, whether preventive, surveillant, therapeutic, or rehabilitative, will be welcoming occasions for the disabled worker.

Supportive Procedures and Follow-Up

As indicated earlier, entry into the world of work can be a traumatizing event for one who has never been employed or for a person who has been repeatedly rejected by society. The first contact with the health facility of the employer—within the

PHYSICAL DEMANDS FORM

Job Title Data Processing Entry   Occupational Code   4-44.110

Dictionary Title Data Processing Entry

Firm Name & Address

Industry_____ Industrial Code

Branch_____ Department

Company Officer_____ Analyst Wetzel   Date

Physical Activities				Working Conditions			
x 1	Walking	16	Throwing	x 51	Inside	66	Mechanical Hazards
2	Jumping	17	Pushing	52	Outside		
3	Running	18	Pulling	53	Hot	67	Moving Objects
4	Balancing	x 19	Handling	54	Cold	68	Cramped Quarters
5	Climbing	x 20	Fingering	55	Sudden Temp. Changes	69	High Places
6	Crawling	21	Feeling			70	Exposure to Burns
7	Standing	22	Talking	56	Humid	71	Electrical Hazards
8	Turning	23	Hearing	57	Dry	72	Explosives
9	Stooping	x 24	Seeing	60	Dirty	73	Radiant Energy
10	Crouching	25	Color Vision	61	Odors	74	Toxic Conditions
11	Kneeling	26	Depth Perception	62	Noisy	x 75	Working With Others
12	Sitting			63	Adequate Lighting		
x 13	Reaching	27	Working Speed	64	Adequate Ventilation	76	Working Around Others
x 14	Lifting						
x 15	Carrying	28		65	Vibration	77	Working Alone

Details Of Physical Activities:

Sits at computer most of the day, reads copy, and fingers keyboard to enter data. Periodically walks short distances, reaches for, lifts, and carries small stacks of billing materials. Reaches for, handles, pushes, and pulls when organizing and filing paperwork at desk level.

Figure 27-1. Physicians or nurses who examine a disabled employee must be given a detailed job description, critical job task factors, and their medical standards, so that intelligent clinically sound, and legally defensible decisions can be made concerning the matching of an applicant's capacities to the specific demands (physical, emotional, and environmental) of the position being sought.

facility or with an outside consultant—should be as warm and receptive as is feasible in a professional setting. A person with a disability has probably had innumerable contacts with health professionals and can readily detect those who want to assist and those who wish little to do with deformity, physical difference, or emotional volatility.

The physical examination should go beyond organ inventory, determination of functional capacities (Figure 27-2), performance of diagnostic laboratory procedures, and classification for placement. Details should be obtained about such factors as interrelational difficulties, acceptance of a wheelchair, problems in transportation, or experiences with prosthetics, for example. If there are visible physical variants such as amputation stumps, or healed traumatic or surgical incisional scars, they should be felt or hefted, or the applicant should show the method used to place an artificial limb or to transfer from a wheelchair to bed, and so on. Further, old burn scars, operative sites, or visible birth defects can be photographed and made a part of the permanent confidential record. Not only does this procedure manifest interest and solidify bonding even further, but it offers another form of baseline for comparison in the event of injury, aggravation, or superimposed disease.

Visits to a disabled employee's workstation will show continuing interest on the part of the company, and allow validation of job placement and the suggested mode of job accommodation. If the applicant was referred through a rehabilitation agency or facility, reports of the placement should be sent to the source, and contact should be maintained over

PHYSICAL CAPACITIES FORM

Disability: Leg amputation 5" below Knee

Artificial leg - good fitting

Name _____ Sex M Age 31 Height 5'9 1/2" Weight 155

	Physical Activities					Working Conditions			
x	1	Walking		16	Throwing		51	Inside		x 66	Mechanical Hazards
O	2	Jumping	x	17	Pushing		52	Outside			
O	3	Running	x	18	Pulling		53	Hot		x 67	Moving Objects
x	4	Balancing		19	Handling		54	Cold		O 68	Cramped Quarters
x	5	Climbing		20	Fingering		55	Sudden Temp. Changes		69	High Places
O	6	Crawling		21	Feeling					70	Exposure to Burns
x	7	Standing		22	Talking		x 56	Humid		71	Electrical Hazards
	8	Turning		23	Hearing		57	Dry		72	Explosives
	9	Stooping		24	Seeing		60	Dirty		73	Radiant Energy
O	10	Crouching		25	Color Vision		61	Odors		74	Toxic Conditions
	11	Kneeling		26	Depth Perception		62	Noisy		x 75	Working With Others
x	12	Sitting					63	Adequate Lighting			
x	13	Reaching		27	Working Speed		64	Adequate Ventilation		76	Working Around Others
x	14	Lifting									
x	15	Carrying		28			65	Vibration		77	Working Alone

Blank Space = Full Capacity   x = Partial Capacity   0 = No Capacity

May work ____ # of hours per day/ ____ days per week.

(IF TB, cardiac or disability requiring limited working hours.)

May lift or carry up to ___ pounds

Details Of Limitations For Specific Physical Activities:

Should not be required to walk, balance, climb, stand, kneel for prolonged periods of time. Should not lift heavy weights continuously. Should not carry long distances.

Figure 27-2. The examining physician should assess the physical abilities of all employees. Using this "Physical Capacities Form" the physician can answer these questions: (1) What physical activities can the employee perform? (2) Under which working conditions can the employee work safely? (3) How many hours per day can the employee work? (4) How many days per week can the employee work? (5) How many pounds can the employee lift or carry?

time. It must be emphasized that no one outside the company health department needs diagnostic specifics. The only information that is to be given to the human resources authority is the confirmation that the applicant is physically fit to undertake the job applied for, accompanied by recommendations for specific job accommodations or certain work restrictions (not to work at heights, in contact with certain toxic substances, etc.). With the person's permission, it is wise to apprise the immediate supervisor, as alluded to previously, of any condition that might present an emergent situation. On too many occasions panicked fellow workers have called in outside ambulances or rescue squads following a convulsive episode of an employee, when simple knowledgeable care would have allowed the person to rest and return to work without creating departmental disruption.

SUPPORTED EMPLOYMENT

A new concept in the employment of persons with severe disabilities is seen in the federal initiative of supported employment. This move represents paid work occurring in normal business environments for a target population not only of persons with severe disabilities, but also for those persons unable to gain and maintain employment. Differing from other forms of vocational rehabilitation, supported employment provides long-term permanent support throughout the person's duration of employment.

As defined by the President's Committee on Employment of Persons With Disabilities (PCEPD) supported employment facilitates competitive work in integrated work settings for those individuals with the most severe disabilities such as psychiatric disorders, mental retardation, learning disabilities, and traumatic brain injury, for whom competitive employment has not traditionally occurred. These persons need ongoing support services, and through supportive employment there are assistive entities such as job coaches, transportation, assistive technology, specialized job training, and individually tailored supervision (President's Committee).

Recent studies have demonstrated that through the provision of ongoing support services for people with severe disabilities there is a significant increase in their rates of employment retention. This type of supported work encourages people to work within their communities and results in a greater work and social interaction. This approach to placement may be effected through the support of a single individual or a small group of individuals. The benefits to the employer include the absence of any fee, the provision of screened applicants, the matching of the worker's abilities to the job requirements, on-site job training by professionals, additional training if needed, and follow-up services for the duration of employment (President's Committee). Additional information may be obtained from the Rehabilitation Services Administration, Switzer Building, 330 C Street SW, Washington DC 20202, 202/205-9297, or from the President's Committee, 1331 F Street NW, Washington DC, 20002, 202/376-6200 or 202/376-6205 TDD/TTY (teletypewriter). The outlook of supported employment is bright because the effort is bringing severely disabled persons into the nation's work force. Employers wishing to strengthen their commitment to social and community goals should contact their state vocational rehabilitation agencies.

THE LANGUAGE OF DISABILITY

An employer should maintain as steady a work group as possible, because turnover, recruitment, hiring, and training are costly. One practice that alienates persons with disabilities—as it does with ethnic groups—is the inappropriate use of terms that denigrate disabled persons. Long remembered are the deprecating nicknames given by classmates to children with crossed eyes, gait impairments, skin defects, or braces. Fortunately, today's youngsters have become more sensitive to differences and are helpful to friends with physical disabilities. Even orthodontic appliances are commonplace status symbols.

Certain terms or phrases should be omitted from conversation or reports because they are demeaning and can be emotionally upsetting to the person with a disability. The California Governor's committee for Employment of Disabled Persons has listed some preferred terms and expressions (*Language Guide on Disability*, 1997). The emphasis is on the person and not the disability. It is readily seen that such fringe derogations as *crippled, confined to a wheelchair*, and *afflicted* are absent. The terms listed are currently preferred usage:

blind	multihandicapped
deaf	nondisabled
developmentally disabled	paralyzed, paralysis persons with cerebral palsy
disabled	
hearing impaired	persons with disabilities
mentally/emotionally disabled	persons with paraplegia
mentally restored	visually impaired
mentally retarded	wheelchair-user
mobility impaired	

Such designations as *afflicted, crazy, deaf and dumb, defective, deformed, epileptic, fits, invalid, lame, maimed, poor unfortunate, stricken, suffers, victim* and *withered* should be eliminated. As witnessed in this writing, *handicapped*, which appeared in early legislation, has been replaced with *disabled* in later enactments.

CORPORATE SOCIAL CONSCIENCE

Anyone who views contemporary television programs, or attends conferences or concerts is well aware of business and industry's support of the arts. Sculpture and painting exhibits are frequently found in headquarters' plazas and corporate offices. Foundations support research in both the soft and hard sciences and participate in the public employment efforts of previous administrations. There are usually representatives of local manufacturing facilities or companies on the boards of community agencies. Corporate interests, despite certain media blitzes to the contrary, have swung in part to rectify the human condition, and policy development concerning the disabled matches other humanistic goals.

Many studies have demonstrated the worth and productivity of persons with disabilities, and it has been shown repeatedly that it is "good business" to add such applicants to worker rosters. Experience has also shown a desire on the part of fellow employees to help their disabled colleagues. The participation makes them feel good, particularly as they watch the newcomers grow occupationally, socially, and emotionally.

One of the last holdouts in the acceptance of disabled persons was the motion picture industry, which, for years, dared not to include the physically and emotionally different in their offerings. But the scene has changed. The film and video, television, and print markets now include story lines built around a variety of persons with alcoholism, mental retardation, psychiatric disorders, Down's syndrome, "idiot-savants," congenital anomalies, cerebral palsy, skin disorders, heart disease, hypochondriasis, cancer, various sensory disorders, substance abuse, or who are wheelchair users. If a box-office-conscious industry can take this giant step, smaller employers can enhance their proactivity. One can recall a host of popular films depicting various physical and mental disorders that create disability: Alcoholism in *Lost Weekend* (1945), *Leaving Las Vegas* (1995), *When a Man Loves a Woman* (1994); paraplegia in *Coming Home* (1978); deafness in *Johnny Belinda* (1948), *Children of a Lesser God* (1986); post-encephalitic catatonia in *Awakenings* (1990); Lou Gehrig disease (amyotrophic lateral sclerosis) in *The Pride of the Yankees* (1942); autism, "idiot savant" in *Rain Man* (1988); and mental retardation in *Sling Blade* (1996), *What's Eating Gilbert Grape* (1993). From television, *Ironside* (wheelchair user) played from 1967 to 1975. Real life examples of disabled public figures include such eminent figures as Franklin D. Roosevelt, Robert Dole, and Daniel K. Inouye.

Because of heated discussions among members of the laity, the health care professions, and industrial leaders concerning the acquired immunodeficiency syndrome (AIDS), attention had to be given to the employment of persons with AIDS. The issue was faced in 1998 when the Supreme Court held in a 5-4 decision in Bragdon v. Abbott that persons who are infected with HIV are protected against discrimination under the ADA even if they have no AIDS-related symptoms (Muhl, 1998). HIV infection was regarded as a "physiological disorder with an immediate, constant, and detrimental effect on the hemic and lymphatic systems." This newer view of *impairment* may be applied in the future by the courts in addressing other diseases and limitations in addition to AIDS (Muhl, 1998).

OTHER LEGISLATION

The ADA is not the sole federal act that affects persons with a disabling condition. The Family and Medical Leave Act (FMLA), Public Law 103-3, passed in 1993, was an effort by Congress to balance the demands of the worksite with the needs of families. The FMLA requires covered employers to give up to 12 weeks of unpaid, job-protected leave to "eligible" workers for certain family and medical reasons. Specific points of eligibility must be met, but beyond this base, there are areas of interaction between the FMLA and the ADA. A condition that qualifies as a serious health condition may or may not meet ADA's definition of disability. As examples, a temporary short-duration impairment may represent a serious health condition, but does not match ADA's view of disability; and a person with a condition such as quadriplegia does not constitute a serious health condition. In the same light an individual undergoing chemotherapy for a malignancy may ask for a modified work schedule as a reasonable accommodation under ADA, unless an undue hardship would result. The same employee could request FMLA time off as there is no undue-hardship exception (*What You Should Know*, 1996).

The Occupational Safety and Health Act (OSHA) Public Law 91-596, passed in 1970 requires that an employer provide a safe working environment. There may be interaction between the ADA and OSHA. When complying with OSHA standards, the employer may wish to consider the use of ADA's reasonable accommodation requirements. Under OSHA, toxic substances must be labeled along with having written materials available descriptive of the potential danger. An ADA reasonable accommodation could comprise the use of the universal symbol for poison and the provision of verbal warning regarding the potential hazard presented by the work substance (*What You Should Know*, 1996).

In relation to the federal and state workers' compensation laws and their second-injury fund, the company's return-to-work policies and procedures must be consistent not only with the workers' compensation legislation but also with ADA, FMLA, and OSHA regulations. The ADA makes no distinction between reasonable accommodations for persons sustaining work-related injuries and illnesses and those that are nonoccupational in origin (*What You Should Know*, 1996). There may be a claim made that a job-connected injury resulted in an ADA disability. If the affected employee requests an accommodation to continue working, the employer should (Ray, 1998):

- Take the request seriously.
- Act promptly.
- Ask for documentation.
- Determine the essential functions of the job.
- Examine reasonable accommodations.

It must be remembered that if one claims disability, it does not mean that he or she can no longer work (Ray, 1998). Most important in post-trauma disability is the soon-as-possible institution of physical and occupational therapy, conducted in keeping with the specific demands of the job the injured individual is returning to, or undertaking, if a different position (Kedjidian, 1996).

Further, the awarding of workers' compensation benefits or the assignment of a high workers' compensation disability rating does not imply automatically that the individual is protected by the ADA. Not every worker who sustains a job injury, even with some residual loss of function, will meet the ADA definition. The employer must review work-related injuries on a case-by-case basis to determine if there is protection under the ADA (*Workers' Compensation*, 1995).

CURRENT LEGISLATIVE ACTION

Much is accomplished on the behalf of persons with disabilities by the PCEPD, whose mission is to facilitate the communication, coordination, and promotion of public and private efforts to enhance the employment of such individuals. The committee provides information, training, and technical assistance to industry, labor, rehabilitation providers, advocacy organizations, and families and individuals with disabilities. The committee reports directly to the President and has done so since its inception in 1947 (President's Committee's Mission, 1999).

Each year, October is declared by the President as National Disability Employment Awareness Month, and the most recent was so proclaimed officially Sept. 29, 1998 (Clinton, 1998). Even though the economy is strong, as over 16 million new jobs were created in the past five years, the President related that "[W]e cannot consider ourselves truly successful until all Americans, including the 30 million working-age adults with disabilities, have access to the tools and opportunities they need to achieve economic independence" (Clinton, 1998). In proclaiming the month for 1998, the President called upon government officials, educators, labor leaders, employers, and the people of the United States to observe the month with appropriate programs and activities that would reaffirm the basic intent of the ADA.

On March 13, 1998—a date alluded to in the Proclamation—the President signed into law Executive Order 13078, with the objective of increasing

> the employment of adults with disabilities to a rate that is as close as possible to the employment rate of the general adult population and to support the goals articulated in the findings and purpose section of the [ADA of 1990]" and thereby ordered "the establishment of the Presidential Task Force on the Employment of Adults With Disabilities

(Executive Order 13078). The areas to be reviewed include reasonable accommodation; inadequate access to health care; lack of consumer-driven, long-term supports and services; transportation; accessible and integrated housing; telecommunications; assistive technology; community services; child care; education; vocational rehabilitation; training services; employment retention, promotion, and discrimination; on-the-job supports; and economic incentives to work. A final report is due July 26, 2002, ending nearly a four-year period of activity, but notably the 10th anniversary of the initial implementation of the ADA.

At the time of the first report (April 22, 1998) these data was shown that (Coehlo 1998):

- The employment rate of the general population was 82%.
- The employment rate of people with severe disabilities was 36%.
- The employment rate of African-Americans with severe disabilities was 18%.
- The employment rate of Hispanics with severe disabilities was 21%.

In the growth in diversity of our population and workforce, it is obvious that much needs to be done, and it is hoped that the Presidential Task Force will reach its goals.

Early 1998 saw the creation of the "Job Links" page on the web site of the PCEPD. The committee thus provides direct Internet links to the employment pages of employers that request to be on the site. The list includes large corporations (e.g., IBM Corporation and Procter and Gamble), private organizations, and government agencies. The list is lengthy and should bring job applicants and employers together with considerable rapidity. Job Links is available through the committee's home page at www.pcepd.gov or directly at www.pcepd.gov/joblinks.htm. Employers who wish listing need only to provide web site and mailing address information for posting to Mary Kaye Rubin at mrubin@pcepd.gov, 202/376-6200 (V), 202/376-6205 (TDD), or 202/376-6859 (Fax) (President's Committee, 1998).

Certain judicial opinions, while not of the same weight as positive action taken by the Congress, do establish precedents that can alter or clarify legislation that has not hitherto been questioned. Despite efforts by the Equal Employment Opportunity Commission (EEOC) of the Department of Justice to have mental conditions considered the same as physical disorders in long-term disability plans, it was not successful until early 1998. A New York bank finally agreed to eliminate distinctions between the two categories, adopting a long-term "parity" policy, which provides equal benefits (EEOC Settlement, 1998).

In a unanimous decision, the U.S. Supreme Court ruled on June 15, 1998, that the ADA applies to state prisons. In Pennsylvania Department of Correction v. Ronald Yeskey, a state prison inmate with a history of hypertension, was denied admission to a motivational boot camp program. He filed suit, alleging the exclusion was a violation of the ADA. The Court indicated that the ADA does not distinguish between the programs, services, or activities of a prison from those provided by other public entities. A bill introduced by Senators Helms (R-NC) and Thurmond (R-SC) was intended to exempt state and local agencies operating prisons from provisions covering public services, as outlined in the ADA (Supreme Court, 1998).

Tax incentives are available to businesses to aid in covering the cost of improving accessibility to their premises. The tax credit, established under Sec. 44 of the Internal Revenue Code, was created in 1990 to help small businesses cover ADA-related eligible access expenditures. IRS publications 535 and 334 provide information on tax incentives, available via 800/829-3676 (V), and 800/829-4059 (TDD) (Tax Incentives, 1999).

One final decision needs inclusion. Many persons use service animals that assist persons with disabilities in their daily activities. "Seeing eye dogs" are one example, used by some individuals who are blind. Other service animals alert persons with hearing impairments to sounds; some pull wheelchairs or carry and pick up things for persons with mobility impairments; and others assist persons with mobility impairments with their balance.

A service animal is not a pet, frequently will wear special harnesses or collars, and some are licensed or certified and have identification papers. Under the ADA, privately owned businesses serving the public are prohibited from discriminating against individuals with disabilities, and the ADA requires these businesses to allow people with disabilities to bring their service animals onto business premises in whatever areas customers are generally allowed. An owner, with a service animal at the worksite, may not be segregated from other customers (Commonly Asked Questions, 1996).

One more recent effort to aid persons with disabilities was undertaken by President Clinton in a proposal to use tax credits to help about 2 million disabled Americans, many of whom depend on relatives for long-term care. The issue is pending (Weinstein, 1999).

SUMMARY

Although the United States has yet to mature as a service society, some progress has been made in promoting human values. The disabled have faced society's lack of concern for decades. World War II brought workers with disabilities, along with women and the elderly, into our manufacturing facilities. War veterans found their place in the labor force despite having to experience destructive physical and emotional trauma. As vehicular accidents do not seem to diminish, more survivors have bodily deformities or loss of limb. While the obstetrical arts and science are improving, babies will still be born with lifelong impairments. And until all disease is genetically reengineered, there will be side effects from illness. This composite—the millions of people here or to come—will seek employment. In the past and currently, U.S. Congress has vigorously tried to eliminate the barriers to employment of persons not always previously welcomed by industry. As the century closes, there is hope that new opportunities will open for those persons with disabilities.

In closure, the final words in a request for attitudinal change are fitting:

> *If we could all act in a spirit of solidarity, recognizing the principles of human equality, if we could bring services to all in need, if we could contribute to a better quality of life, reduce the dependency and transfer power to them, then we would restore to disabled people their right to a life in dignity (Peat, 1997).*

REFERENCES

Berkeley Planning Associates. *Study of Accommodations Provided to Disabled Employees by Federal Contractors.* Contract No. J-9-E-1-0009." Prepared by U.S. Department of Labor, Employment Standards Administration, as submitted by Berkeley Planning Associates. Berkeley CA: The Associates, June, 1982.

Clinton WJ. National Disability Employment Awareness, 1998, by the President of the United States of America. A Proclamation. http://www50.pcepd. gov. Accessed Feb 8, 1999.

Coehlo T. Employment rates are lower for people with a disability and much lower for those with severe disability. April 22, 1998. http://www2.dol.gov. Accessed Jan 21, 1999.

Commonly Asked Questions About Service Animals in Places of Business. Washington DC: U.S. Department of Justice, Civil Rights Division, Disability Rights Section, July 1996.

DOJ Document. Justice Department sues major movie theater chain for failing to comply with ADA. January 29, 1999. http://www.usdoj.gov. Accessed Feb 4, 1999.

EEOC Settlement. ACOEM Report, March 1998; [n.v.]:6.

Executive Order 13078. *Fact Sheet.* http://www 2.dol.gov. Accessed Jan. 21, 1999.

Guide to Disability Rights Laws, Washington D.C.: U.S. Department of Justice, Civil Rights Division, Disability Rights Section, May 1997.

JAN, West Virginia University, Morgantown WV 26506: Job Accommodation Network 1989.

Kedjidjian CB. Bold steps to help injured workers and their employers. *Safety + Health* 154:28–30, 1996.

Language Guide on Disability - A Primer on How to Say What You Mean to Say. Sacramento CA: The Governor's Committee for Employment of Disabled Persons, October 1997.

Muhl CJ. The law at work - Americans with Disabilities Act. *Mo Labor Rev* 1998; 121:32.

Peat M. Attitudes and accesses: advancing the rights of people with disabilities. *Can Med Assoc J* 1997; 156:657–9. Cited from Relander E. Prejudice and dignity: an introduction to community based rehabilitation. New York: United Nations Development Program, 1993.

President's Committee on Employment of Persons with Disabilities. *Supported employment.* http://www50.pcepd.gov. Accessed February 8, 1989.

President's Committee's Mission. http://www50. pcepd.gov. Accessed Feb. 8, 1999.

Public Law 93-112, Title V, Sec. 7 (4 b & c), Sept. 26, 1973.

Ray N. Workers Comp and the ADA: Where laws intersect. *Safety + Health* 1998; 157:100–102.

Russell JN, Hendershot GE, Le Clere F, et al. Trends and differential use of assistive technology devices: United States, 1994. Hyattsville MD: Centers for Disease Control and Prevention, National Center for Health Statistics, Advance Data, No. 292, Nov. 13, 1997.

Supreme Court: ADA covers prisons. *Contemporary Rehab* 1998; 54:1.

Tax Incentives Packet on the Americans with Disabilities Act. http://www.usdoj.gov/crt/ada/taxpack. Accessed Feb. 4, 1999.

U.S. Census Bureau. Americans with disabilities: 1994–95. Tables 1, 4, 5. http://www.census.gov/ hhes/www/disable. Accessed June 8, 1998.

Weinstein MM. (New York Times). Tax credit for disabled debated; Clinton proposal seen as innovative, but with problems. *Press Democrat,* Jan. 17, 1999, p. E2.

What You Should Know About Workplace Laws. http://www50.pcepd.gov. Accessed Feb 8, 1999.

Workers' compensation: Developing company policies, 1995. http://www50.pcepd.gov. Accessed Feb. 8, 1999.

chapter 28

Outsourcing Occupational Health and Safety Services

by George Benjamin, MD
Susan McKenzie, MD
revised by Marci Z. Balge, RN, MSN, COHN-S

485 **Overview**
　　The small company ■ Health and safety coverage
　　■ Issues to consider when outsourcing
491 **Design of the Occupational Health Program**
　　On-site professional services ■ Types of providers
　　■ Protocol of services and procedures ■ Supervision
　　■ In-house medical facility ■ Records
494 **Summary**
494 **References**
495 **Suggested Readings**

OVERVIEW

Outsourcing has become a common term in business and industry, both large and small. Several factors effect outsourcing practices. These include the following:
- rapid technological change
- increased risk and job flexibility
- emphasis on core corporate values
- globalization.

Each of these factors can encourage companies to do more work outside the firm. Corporations no longer provide jobs "for life." The idea of companies caring for employees and their families unconditionally has dwindled in the global economy. In addition, corporate downsizing to improve profit margins most often affects "overhead" positions that are not perceived as profitable, i.e., health and safety services.

This chapter will assist the person responsible for starting or maintaining an occupational health program in either a small industrial organization or a large company searching for specific expertise. This information provides an approach to establishing a program that is suitable for the company's size, industrial activities, and budget, and that is adequate to assure the health and safety of its workers. This includes assessing the company's needs, designing and staffing an appropriate medical program, and securing assistance from qualified professionals when needed. This chapter suggests the elements of a comprehensive occupational health program, which must then be tailored to meet needs based on company size.

The Small Company

The services provided by a small company can be limited in scope, but they should not differ in substance or quality from those provided by a larger one. Similarly, the goals of an occupational health program should be alike in both large and small organizations. The Occupational Safety and Health Act of 1970 (OSHAct) mandates that employers have a responsibility to provide their employees with "a place of employment which is free from recognized hazards that are causing or likely to cause death or serious physical harm." Subsumed by this broader requirement are the more specific goals of providing for medical treatment and rehabilitation of occupational illness and injury, and for education and training of employees. In addition, a company should take seriously its obligation to protect the neighboring community from harmful effects of plant emissions and other sources of industrial contamination.

Establishing an occupational health program can be a complicated task, especially for a small company. In managing occupational health and safety, one must draw upon expertise in many fields, including occupational medicine, toxicology, industrial hygiene, safety, ergonomics (job design), epidemiology, health physics, and other related areas. The person in charge of the company's program must gain familiarity with the areas that are relevant to the company's needs and must know where to turn for assistance.

The occupational health and safety manager in a small company must also be familiar with the many regulations and recommendations governing exposure monitoring, medical surveillance, medical record keeping, illness and injury reporting, hazard communication, waste disposal, and workers' compensation case management. In addition, there are many nonregulatory components of an occupational health program, such as health promotion and substance abuse counseling that a company can offer to its employees and with which the manager may need to become familiar. Above all, the occupational health and safety manager must be able to identify potential occupational hazards and must possess the decision-making skills necessary for effective hazard management.

There is no exact numerical dividing line between "small" and "large" companies. The Small Business Administration (SBA) considers any commercial business or industrial enterprise to fall within its purview if it employs fewer than 500 workers. Such a company can be either an independent business organization or a local division of a larger industrial enterprise with corporate headquarters elsewhere.

The Occupational Safety and Health Administration (OSHA) designates as small those companies with fewer than 10 employees, based on its legislative mandate regarding routine plant inspections and record keeping. The National Institute for Occupational Safety and Health (NIOSH) broadens its definition to include companies with up to 50 employees. Local providers of occupational health care services can choose an operational definition of fewer than 100 or 200 employees as best suiting their marketing strategies.

Within any numerical cutoff are companies with diverse business and industrial activities. Some of these companies are engaged in "low-hazard" work (for example, banks and insurance agencies) and require limited occupational health services. Others are engaged in more hazardous work (for example, foundries and pesticide manufacturers) and may require formal occupational health programs.

The number of employees in a company is therefore only one determinant of its need for services. The company must also consider other equally important factors, such as the nature of its business activity, the potential for hazardous workplace exposures, the physical demands of its jobs, and its geographic location. No single generic program will meet the needs of all small companies; and no two companies are likely to have the same occupational health program, even within the same industry.

For the purposes of this chapter, *small* will be defined both by the nature of the work in which the company is engaged and by the size of the company's work force. Generally, the chapter will focus on the facility that does not require the services of a full-time physician but has enough employees or potential for hazardous exposures to warrant delivering formal occupational health services. Such a facility will most likely employ fewer than 500 employees, and usually fewer than 200.

Health and Safety Coverage

Most workers in the United States are employed by small companies. According to the SBA, the 5.7 million companies in the United States with fewer than 500 employees represent 99.8% of all U.S. business enterprises. So it is not surprising that these smaller workplaces report most of the U.S. work-related injuries. In fact, an examination of BLS statistics shows consistent annual peaks in the incidence of reported illness and injury rates for companies with 50–250 employees.

Despite these reported rates, many small companies lack adequate health and safety programs and are unprepared to deal with health and safety issues. Of the companies surveyed in a study by the National Federation of Independent Businesses, 23% of those engaged in hazardous work with fewer than 10 employees had a written safety plan, while 94% of similar companies with more than 100 employees reported that they did have such a plan. Similarly, companies with a large work force engaged in hazardous work were more likely to designate an employee for safety than were those with fewer employees.

Issues to Consider When Outsourcing

Depending on its resources, a small company can either have the in-house capability to develop a health and safety program or have to seek outside assistance. In either case, an internal assessment of the company's needs will enable the occupational health and safety manager to develop a more effective program and to deal more knowledgeably with consultants when the need arises.

This section outlines several self-assessment questions the company can use to evaluate existing information about its operations, its work force, and the potential for hazardous workplace exposures. If the company is already offering its employees some health and safety services, the answers to these questions can be used to improve the performance of the existing program.

Business Activity

In what business activity is the company engaged? Manufacturing? Agriculture? Mining? Some other industrial or service-oriented activity? Can the work be broadly characterized as "high-risk" or "low-risk" with regard to anticipated work exposures?

Corporate Affiliation

Is the company independent, or is it a branch of a large corporation with corporate headquarters elsewhere? If it is the latter, can corporate headquarters or a corporate medical department provide policy direction and technical expertise? Many large corporations consist of a number of geographically dispersed large and small firms, each of which requires a health and safety program appropriate to its own specific needs. Although the corporate medical department can provide direction, the smaller unit's health and safety program must comply with state and local laws and regulations.

Location of Worksite

Where is the worksite located? Is it near adequate medical services (especially emergency and acute care), or is it relatively isolated? A small, isolated unit engaged in hazardous work (for example, mining) may need to provide more comprehensive on-site medical and safety services than a similarly sized unit located in a large city. In how many buildings (or at how many sites) do manufacturing or other potentially hazardous activities take place? Are they clustered or geographically dispersed? The company must plan for provision of emergency medical services to each separate unit.

Employees

How many employees does the company have? How many are engaged in production, maintenance (often a forgotten group with potential for significant hazardous exposures), administration, office work, and other jobs? Does the company have demographic data on the work force? If so, are the workers mainly younger (less than 40 years of age) or older? Young workers may change jobs or terminate employment more frequently. Older workers engaged in hazardous work may require more frequent and more comprehensive medical services.

What is the rate of employee turnover? Do some members of the work force have needs that deserve special evaluation, such as women of childbearing age or persons with preexisting medical conditions? (See also chapters 21 through 27, which cover other special issues in the workplace.)

Expected Growth
What is the projected rate of growth for the company? If management expects growth, it will be wise to plan ahead and to lay the foundation for a program that will serve the company's future needs.

Organization and Management
We cannot overstate the importance of understanding management's expectations for an occupational health program. For the program to succeed, senior management (the president and other senior officers) must actively support the occupational health and safety program and accept responsibility for its mission.

Management Perceptions
What does senior management perceive to be the health and safety needs of the company? For what reasons has senior management chosen to develop a health and safety program?

Program Administration
Who will manage the day-to-day activities of the program? Will the company designate a current employee, or will it have the resources to hire a person to manage the program and advise the company?

To whom will this person report? In the larger company, the maxim "the higher the better, and the simpler the organization, the better" is true. In the smaller company, the occupational safety and health manager should report or at least have direct access to the president and other members of senior management. The reporting structure should also minimize conflicts of interest. For example, it makes little sense for a health and safety officer to report to a production manager who works on a tight budget, with deadlines, while overseeing the use of toxic agents.

If there are additional health and safety personnel, to whom will they report? For the program to succeed, medical, safety, and hygiene personnel must be able to solve problems together. The organizational setting in which they operate will determine the extent to which this is possible. Medical personnel will need adequate and timely access to industrial hygiene data job descriptions. Access to information can be facilitated by carefully delineated reporting relationships.

Information Flow
Are lines of communication to senior management and to supervisors and employees open? The occupational health and safety manager will need to keep senior management personally and routinely informed of the program's operations.

Supervisors, together with employees, will need to be included in policy-making decisions and informed of matters that affect them. If employees are represented by unions, these organizations will need to be involved in policy direction and program development. Health and safety personnel will need to keep informed of any contemplated changes in processes, products, or raw materials so that the company can address potential health effects before the changes are completed.

Time Constraints
Finally, does the manager have enough time to oversee an effective program? In a small company, the occupational health and safety manager may have other corporate responsibilities. The manager must have enough time to engage in prospective planning as well as in retrospective "fire fighting." Otherwise, the company's problems will outstrip the program's ability to solve them.

Inventory of Potential Work Exposures
Effective hazard management must include the identification of potentially hazardous exposures within the company. Experience has shown that the potential for harmful effects generally increases with the toxicity and severity of the exposure. However, an inventory of work exposures should include all potential hazards, even those to which exposure is expected to be slight.

Operations
The occupational health and safety manager must be familiar with the company's operations. Begin with a list of the company's products. What raw materials are used in producing them? What intermediate products are generated during production? If not already available, it may be helpful to construct a simple operations flowchart.

Potential Hazards
Develop a list of the chemical, biological, physical, and other agents to which employees are regularly exposed, indicating the number of employees exposed to each and the length of time exposed. The inventory of hazardous chemicals prepared to comply with the OSHA Hazard Communication Standard is a useful starting point. Quantities of

chemicals stored can serve as one indicator of potential for exposure.

Hazardous exposures can potentially occur from the following sources:

- *Chemical hazards*—raw solids, liquids, gases vapors, and dusts used in or generated by the manufacturing process, office operations, or maintenance of the facility
  - For each chemical used, it will be necessary to secure a Material Safety Data Sheet (MSDS) from the manufacturer or importer of the chemical. This sheet is required by law to provide information on the toxicity of the chemical. Many MSDSs will need to be supplemented by more comprehensive data from additional sources.
  - Each chemical on the list should be assessed for its potential to cause toxicity, both under normal working conditions and in the event of an unexpected exposure or spill.
- *Biological agents*—exposures to infectious agents, animals, and other organisms
- *Physical agents*—exposures to noise, ionizing and nonionizing radiation, vibration, heat and cold, and altitude
- *Ergonomic considerations*—exposures to hazards associated with physically demanding jobs (lifting, hauling, materials handling), repetitive motion (typing, assembly line work, checkout stand operations), and stationary postures that can cause muscle strain (see chapter 16, Ergonomics Programs)
- *Shiftwork*—an occupational exposure with its own set of associated health consequences Employees who work overtime can exceed limits for workplace exposures established by OSHA, which are based on an eight-hour workday (see chapter 24, Scheduling Shiftwork).
- *Sources of occupational stress*—exposures to the psychological demands of a job, which can result from machine paced work, high production requirements, piecework and incentive pay, conflicting administrative requirements, and job insecurity (DeHart et al, 1987). The number of claims for occupational stress has risen in some areas in the past few years. An effort should be made to identify and alleviate these potential sources of work-related illness (see chapter 22 Stress Management).
- *Safety hazards*—electrical and mechanical hazards that can predispose to injury (see chapters 7–20 in part III, Occupational Health & Safety Programs).

After listing potential hazards, it will be useful to characterize the regulatory environment in which the company works. Exposure to many of the agents listed here is regulated by federal agencies such as OSHA and the Environmental Protection Agency (EPA). States with OSHA-approved state plans may have more stringent standards than those that rely on federal OSHA. Some states may have passed special legislation or initiatives (such as SB 198 and Proposition 65 in California) that affect industrial programs. Cities and counties may have established their own local regulations, often using fire departments for enforcement.

Worksite Evaluation

Personally review the general state and appearance of the worksite(s). Are engineering controls functioning properly?

Are they tested periodically and reviewed for improvement? Are administrative controls used where appropriate to control employee exposures? Is personal protective equipment used only as a last resort? Is it maintained appropriately and used properly by workers?

Are chemicals stored safely and labeled in accordance with the Hazard Communication Standard (OSHA Report No. 3084, 1988). Is the worksite clean and free from hazards that predispose to accidents?

Are workers knowledgeable about the performance of their jobs? Jobs should be designed to minimize awkward, strenuous, and repeated movements. Employees should be given breaks and opportunities to change their work in order to vary mental and physical routines.

There is no substitute for firsthand knowledge of the workplace. The occupational health and safety manager and the industrial hygiene, safety, and medical personnel should visit the worksite regularly.

Review of Health and Safety Services

Many small companies offer some health and safety services, even if no formal program exists. An examination of health and safety data can provide useful information on where illness and injury occur in the company and on the kinds of occupational health services the company should make available.

Illness and Injury Statistics

The organization's past illness and injury statistics serve as a useful indicator of areas in which potentially hazardous work occurs. All companies covered by the OSHAct, except very small companies, must maintain records of and report occupational illness and injuries. In states with approved state OSHA plans, the very small companies also can be required to report.

Has the company been the subject of workers' compensation claims? If so, for what type of illness or injury? Have health or safety requests or comments been filed by workers, supervisors, or management? A meeting with the company's supervisors should help to enumerate potentially hazardous exposures.

Additional assistance in profiling hazardous exposures for an industry can also be obtained from OSHA, NIOSH (which has published numerous documents detailing hazards in various industries), workers' compensation and insurance carriers, and a company's corporate medical department.

Industrial Hygiene and Safety Evaluations

Will the company have an in-house industrial hygiene capability? Many small companies cannot afford an in-house industrial hygienist and will have to rely on outside sources to meet this need. OSHA's business consultation service can provide this service free of charge and will offer advice to identify and control specific hazards without issuing citations or proposing penalties (OSHA, 1985).

All potentially hazardous exposures must be identified and sampling measurements taken where appropriate (see chapter 11, Industrial Hygiene Programs). What do existing hygiene and safety reports indicate? Are exposures within legal and acceptable limits? Remember that not all exposures are airborne. Work practices—such as dipping hands in chemicals, eating in the workplace, and carrying hazardous material out of the workplace on clothing—can contribute to ill health and must be evaluated.

Medical Services

Most companies understand their obligation to provide acute medical care for workers who are injured on the job, so they have established a relationship with a local physician or clinic to provide that care. Beyond this, state workers' compensation laws generally require companies to provide complete medical treatment and rehabilitation for all work-related injury and illness, including that which accumulates over a long period of time.

The options available to the small company for securing medical services are described in the next section. If the company has already contracted for acute care, it will be useful to determine whether the providers can deliver a fuller menu of occupational health services and whether they are familiar with government record-keeping requirements. Excellent and detailed lists of the components of a comprehensive occupational health program can be found in part 3 of this book. Which of these services should a small company provide?

At a minimum, a company of any size should offer prompt diagnosis, treatment, and rehabilitation of work-related illness and injury, and should comply with all pertinent medical laws and regulations. For each regulated or potentially hazardous exposure, a compliance plan should be developed that allows the company to accomplish the following:

- Identify areas in which exposures occur.
- Collect monitoring data to indicate the level of exposure.
- Identify employees whose exposure levels require inclusion in a periodic medical evaluation program.
- Establish appropriate medical surveillance protocols.
- Comply with record keeping, reporting, and educational requirements of the law.
- Periodically reassess exposures and work practices and take correct action in the workplace when indicated.

Small companies should offer other minimum occupational health services. These services should facilitate proper placement of workers in potentially hazardous work through preplacement and return-to-work evaluations, medical reassessment on transfer to hazardous jobs, and medical clearance for the use of personal protective equipment; promote personal health by informing and counseling employees regarding all medical findings; provide educational programs on potential occupational hazards and the means by which to identify and prevent adverse health effects; and maintain adequate and confidential medical records. The company should consider providing other important services, including health promotion and referral for substance abuse counseling. Although it may be tempting to offer health promotion and physical fitness programs at the expense of traditional occupational health services, a company's most important health-promotion task is to assure a safe and healthful workplace.

Written Policies and Procedures

Does the company have written policies and procedures for health and safety? Can corporate headquarters provide guidelines for policies and protocols? Are all affected employees aware of the policies? If no such documents exist, creating them will be an important task.

Does the occupational health program have a clear statement of prioritized goals and objectives? Do these goals and objectives meet legal standards? Do they reflect the goals of senior management? Does the company have a plan to deal with natural and human-made disasters? Has it identified all significant medical and safety problems that could occur in the facility and planned for them? Do all employees understand their role in the event of such an emergency?

Education

Is the company currently providing periodic educational programs for its employees on health and safety issues and work practices? Is the company in compliance with federal, state, and local hazard communication requirements? A well-designed educational program can communicate the employer's interest in promoting occupational health and safety, and can motivate employees. Are MSDSs available for each chemical used, and are they accessible to all employees? Are toxicity information, regulations, and company policies accessible to employees in an understandable form?

Are health care personnel (even those who work only outside the facility) knowledgeable about the agents with which the employees work? Baseline and periodic medical examinations provide an excellent opportunity for employee education and for one-on-one discussion of potentially hazardous exposures and work practices.

DESIGN OF THE OCCUPATIONAL HEALTH PROGRAM

Once the small company's needs have been assessed, it is possible to design an occupational health program that appropriately meets those needs. Remember, this program should reflect the company's business and industrial activities and budget, as well as its size (See part 3 of this book).

On-Site Professional Services

Because of special toxic or radiation hazards, some operations with 200 or fewer workers employ a full-time physician. However, such medical coverage is usually not needed until the number of employees is well over a thousand. Indeed, some light industries and white-collar operations employ many thousands but have no physician on the premises. With this caveat in mind, a rule of thumb for determining the need for part-time physician services is that the company needs two hours a week for the first 100 employees and one hour a week for each subsequent 100 (Felton, 1990). Where the law permits, some physician activities-in particular, routine examinations can be assumed by a nurse practitioner or physician's assistant.

Within the setting of constantly changing work environments, corporate cost cutting strategies, and increased knowledge transfer among health and safety professionals, there are several options for staffing health and safety manger positions. The chapters in part 2 of this book describe the capabilities and role of occupational health nurses, physicians, safety and industrial hygiene professionals. The role of the occupational health nurse, in particular, has expanded such that not only do these professionals possess specific health care/emergency response expertise but hold a well grounded general knowledge of industrial hygiene and safety practice management systems. In addition, occupational health nursing consultants may be used to set up the program, contract with local medical providers, and periodically review and evaluate the program. The degree of hazard and the experience of the facility should guide the decision as to when a part-time or full-time nurse is indicated, but the safety and health budget frequently limits options.

Types of Providers

The data indicate that the vast majority of small businesses depend upon outside providers for primary support of their occupational health needs. A variety of such providers are available in large metropolitan areas, but the choice is limited in many parts of the country. Under the traditional agreement, a local practitioner provides the company with services such as treatment of emergencies, preplacement examinations, and disability evaluations. The provider/consultant is sometimes compensated by a retainer for any routine services given on the facility premises plus a fee for service for procedures at the clinic. Some authorities believe that a salary based on projected time is more desirable, but since treatment fees for worker's compensation cases can be paid through compensation insurance rather than directly from operating funds, fee-for-service arrangements continue to be attractive.

As fewer and fewer physicians remain in solo practice, the arrangements tend to be with a clinic or medical group rather than with an individual practitioner. Various types of clinics and consultants offer occupational health services. Many of these clinics are adjuncts to a hospital outpatient department or emergency room. Such clinics may not have full expertise to develop an occupational health program but simply offer a menu of services. With the development of occupational medicine and occupational health nursing as specialties more expertise has become available, but there are not yet enough specialists to meet the needs of U.S. industry. The American College of Occupational and Environmental Medicine, the American Association of Occupational Health Nurses, and Educational Resource Centers sponsored by NIOSH provide training for occupational health professionals.

A group of companies can support a central facility with satellite facilities on the participants' premises as indicated. Such enterprises are rare but can come about as the result of trade association agreement with a hospital-centered program. Large companies with expensive occupational health facilities have attempted to recover some operating costs by marketing services to other firms in their vicinity. Freestanding facilities have even been built for this purpose, but success has been limited, and investment in such ventures has not been feasible during the past decade. More often, large firms offer portions of their programs in the form of consultative services, even setting up separate departments or subsidiaries to manage these activities. Several corporations market safety or health promotion services as offshoots of their own internal programs, and insurance companies provide both consultation and specific loss-prevention services.

Protocol of Services and Procedures

Regardless of the nature of the business, the occupational health program must meet certain basic requirements. To this end, the company should have a structured program spelled out in a written protocol. The person who has responsibility for the program generally develops the protocol under the guidance of a qualified consultant. If not associated with local practitioners, the consultant can assist the company staff in locating suitable providers.

Some of the services to be covered are obvious. Emergencies must be attended. First-aid facilities will depend upon the availability of local medical facilities. OSHA regulations (29 *CFR* 1910.151) require first-aid capability on the premises if a medical facility is not in the immediate vicinity. A U.S. Court of Appeals has ruled that "immediate vicinity" means a response time of three minutes! (Santa Fe Trail Transport Company 20SHC 1274, 1974). The details of this capability are to be prescribed by a physician. A basic first-aid kit (American National Standards Institute, 1978) consists essentially of bandaging materials to control bleeding, to keep wounds clean, and to provide simple immobilization of injured extremities. Personnel trained in first aid and cardiopulmonary resuscitation (CPR) should be present whenever the facility is operating. However, oxygen or medications, even aspirin, should only be available through a trained health care professional.

All employers should be familiar with OSHA's Bloodborne Pathogen Standard (29 *CFR* 1910.1030). Employees assigned health care duties come under various provisions of this regulation, and certain protective equipment and training may be required. Procedures should be detailed in the protocol for the facility's occupational health program.

The protocol details arrangements for treating injured or acutely ill workers at a medical facility. The director of the medical facility should be involved in planning these arrangements. The protocol must clarify a number of points. A basic concern is how workers will be transported to the medical facility. In certain circumstances, transportation may be by company or private vehicle, in which case the employee should be accompanied by a responsible person. What are the available ambulance services and response times? Also, how will the facility know that the company has accepted responsibility for the treatment? Who will be the medical facility's contact at the company?

The medical facility staff should be willing to treat and return non-disabled cases expeditiously and to make available prompt reports concerning disposition and the need for follow-up. If the facility lacks certain specialized capabilities, the protocol should specify how to obtain these.

Supervision

All medical diagnostic and treatment aspects must be supervised by a physician. This person assumes responsibility for medical procedures, is available for consultation, and arranges for alternative coverage when he or she is not available, or if specialized services are needed. The physician issues written orders or guidelines for the nurse, physician's assistant, and first-aid providers.

If medical personnel are to undertake environmental monitoring tasks, such as noise dosimetry or ergonomic surveys, they should have appropriate training or operate under the supervision of a consulting industrial hygienist, occupational health nurse, or safety specialist. The responsible person must be careful not to operate outside the scope of their license, which varies from state to state.

In-House Medical Facility

When management has decided to have part-time or full-time services available at the facility, the occupational health consultant and the plant engineer should collaborate in designing the physical setup. The facility is most convenient in a central location, but other needs take precedence. Among these needs is accessibility to disabled employees and stretcher and ambulance crews. Traffic should be limited to those who have business in the facility. Good lighting and ventilation are important.

The medical literature provides sample layouts and lists of suggested equipment for an on-site medical facility. Textbook guidelines for size are at least 175 square feet for a facility of 200 employees, 300 square feet for 500 employees, plus one square foot for each employee over that number. Limited space can be greatly enhanced by a functional layout and attractive use of color.

As basic functions, the facility should provide for treatment, storage of supplies, and maintenance of records. Space must also be included for the staff, reception of employees, toilets, dressing areas, and perhaps special programs such as the fitting of safety glasses. It is difficult to reduce such a facility to less than 300 square feet. Specialized equipment is needed occasionally; audiometer booths, x-ray, and pulmonary function testing are now available in mobile units.

Records

As maintenance of records is an essential component of the occupational health program, the protocol should describe all procedures pertaining to the retention, storage, and accessibility of medical records. The medical record is an important tool of the physician and the nurse in ensuring adequate diagnosis and treatment. In fact, adequacy of medical records is a major parameter of quality assurance. Records also confirm that the employer has fulfilled its obligations to the worker and are therefore a resource in litigation. Records play a key role in surveillance, both for early detection of health problems of the individual worker and for early detection of hazards in the workplace. Unless a standard for a specific exposure states otherwise, medical records must be retained for at least 30 years (29 *CFR* 1910.20d) There are also various federal and state regulations regarding confidentiality and limited access to records.

If a medical facility is located on the premises, record storage, with adequate security, will be in that medical facility. Medical records must be physically secured with immediate access limited to medical personnel; i.e., the nurse, the doctor, and clerical support who have been indoctrinated regarding confidentiality. This means that if a medical facility is located on the premises, record storage will be within the physical confines of that medical facility in locked files. Only information directly related to the job, restrictions, conditions determined to be job related, surveillance data, and exposure information need to be kept in the personnel file. Title I of the 1990 Americans with Disabilities Act explicitly mandates this limitation (19 *CFR* 1630.14b). The protocol should prescribe procedures for release of information to persons such as workers or their representatives who may be entitled to specified information in the records.

There can be a problem if services are provided by an outside medical facility. Ideally, the medical records should be maintained on the premises of that provider, with medical clearance forms only kept in the personnel file. If medical records cannot be maintained by the provider, they should be in the custody of the company official responsible for the occupational health program; this person must maintain them separate from personnel files, according them the same confidentiality and security as would a health professional.

SUMMARY

Most workers in the United States are employed by small companies. Studies indicate that small companies are often sites of high injury and illness rates, and are usually the least adequately served by occupational health and safety programs. These parties must therefore take the initiative to assess their own needs and/or to reach out to consultants or providers (outsource) for assistance. Most workers see a well-designed health and safety program as an employee benefit. In addition to assuring compliance with various regulatory requirements, such a program gives the company an opportunity to show that it cares about the welfare of its employees.

REFERENCES

American Medical Association. *Guide to Developing Small Plant Occupational Health Programs.* Chicago: Environmental and Occupational Health Program of the AMA, 1983.

American National Standards Institute (ANSI). *Minimum Requirements for Industrial Unit-Type First Aid Kits,* ANSI Z308.1–1978. New York: ANSI, 1978.

American Occupational Medical Association, Medical Practice Committee. Scope of occupational health programs and occupational medical practice. *J Occ Med* 21:497–499, 1979b.

Block DL. Occupational health and safety programs in the workplace. In Levy BS, Wegman DH (eds). *Occupational Health: Recognizing and Preventing Work-Related Disease,* 2nd ed. Boston: Little, Brown, 1988.

DeHart RL, Robinson TR, Welter ES. *Guidelines for Establishing an Occupational Medical Program.* Arlington Heights IL: American Occupational Medical Association, 1987.

Deubner DC, Imbus HR. Models of occupational health practice. *Seminars Occ Med* 1:33–42, 1986.

Felton JS (ed). *Occupational Medical Management. A Guide to the Organization and Operation of In-Plant Occupational Health Services.* Boston: Little, Brown, 1990.

Kalina CM, Fitko J. Successful business process design. *AAOHN J* 45(2), February 1997.

Guidotti TL et al. *Occupational Health Services: A Practical Approach.* Chicago: American Medical Association, 1989.

Guidotti TL. The changing role of the occupational physician in the private sector: The Canadian experience. *Occ Med* (Lond.) 47(7):423–431, 1997.

Howe HE. Organization and operation of an occupational health program. *J Occ Med* 17(6):360–400, June 1975; 17(7):433–440, July 1975; 17(8):528–540, August 1975.

Konzen JL. Design and operation of an occupational health program. In. *The Industrial Environment: Its Evaluation and Control.* Washington DC: National Institute for Occupational Safety and Health, 1973, pp 693–701.

Mastroianni K, Machles D. What Are Consulting Services Worth? *AAOHN J* 45(1), January 1997.

National Federation of Independent Businesses (NFIB). *Small Business Decisions to Manage Workplace Hazards.* Washington DC: NFIB, November 1984.

Occupational Safety and Health Administration (OSHA). *Access to Medical and Exposure Records,* OSHA Report 3110. Washington DC: OSHA, 1988.

OSHA. *Chemical Hazard Communication,* OSHA Report 3084. Washington DC: OSHA, 1988.

OSHA. *Consultation Services for the Employer.* OSHA Pub. No. 3047 (revised). Washington DC: OSHA, 1985.

OSHA. Occupational Safety and Health Act of 1970. 84 Stat.1590. Section 5(a)(1). Dec. 29, 1970.

Putnam L, Langerman N. *OSHA Bloodborne Exposure Control Plan.* Chelsea MI: Lewis, 1992.

Richmond HW. Health care delivery in Cummins Engine Company. *Arch Environmental Health* 29:348–350,1974.

Skinner G. Outsourcing. *Health Estate* 52(6):34–35, 1998.

Warshaw LJ. Toward the year 2000-Challenge to the occupational physician. *J Soc Occup Med* 39:13, 1989.

Wood J. An argument for outsourcing. *Occ Health* (Lond.) 48(3):84, 1996.

SUGGESTED READINGS

American Industrial Hygiene Association. *Standards, Interpretations and Audit Criteria for Performance of Occupational Health Programs.* Akron OH: AIHA, 1975.

American Occupational Medical Association. Code of ethical conduct for physicians providing occupational medical services. *J Occ Med* 18: front cover, 1979.

Felton JS. The in-plant occupational health service. In Rom VIN (ed). *Environmental and Occupational Medicine,* 2nd ed. Boston: Little, Brown, 1992, pp 1405–1419.

Part 5

Future Issues

chapter 29

Infectious Diseases and Occupational Health in Developing Countries

by Barnett L. Cline, MD, MPH, PhD

499 **Introduction**

500 **Background**
Disease burden ■ Transmission of infectious diseases
Cultural context ■ Health-related behaviors and risks

505 **Health Services**
Governmental health services ■ Nongovernmental organizations ■ International organizations ■ Other local resources

507 **Environmental Approach to Public Health and Occupational Health**

510 **Appendix 1: Directories of NGOs Available Only in Written Form**

510 **Appendix 2: Directories of NGOs Accessible Via the Internet**

511 **Appendix 3: Selected National and Multinational Governmental Organizations Accessible via the Internet**

INTRODUCTION

The main objective of this chapter is to describe some of the key differences between the practice of occupational health in tropical and developing countries as compared with that in countries with *established market economies*. While the fundamental public health principles are identical, the disease exposures and types of risks can vary greatly, and the cultural, governmental, and environmental context is typically quite different from that which a U.S.-based employee would anticipate.

Finally, the chapter provides information on valuable resources, including data sources and organizations with expertise and ongoing programmatic activities in a wide range of developing countries. It also provides examples of global disease control initiatives, which may offer unique opportunities for worker protection in certain tropical regions.

While this chapter deals with infectious disease risks in developing countries, it does not cover "travel" medicine. Travel medicine is covered in chapter 15, Travel Health and Remote Work Programs.

BACKGROUND

The chapter begins by contrasting the *disease burden* in developed and developing countries, with an emphasis on the relative importance of infectious diseases in the latter. The transmission cycles of important infectious diseases are described to provide a conceptual framework for disease prevention and control, and cultural differences are explored to heighten awareness of this important consideration. Health-related behaviors are discussed for both expatriate and indigenous workers.

Disease Burden

For decades public health researchers have worked to develop and improve ways to objectively measure the *burden of disease* in a defined population. During recent years this effort has led to widespread acceptance of new approaches, and to the five-year Global Burden of Disease Study, published in May 1997 as a series of papers in *The Lancet*, a British medical journal. The studies were initiated by the World Bank and carried out in collaboration with the World Health Organization, and are considered the most current and definitive information available on disease patterns on a global scale. Comparisons between causes of death and disability in economically developing countries and in more developed countries such as the established market economies are found in this valuable study. The Global Burden of Disease Study is used by national health planners to help allocated limited health resources in a more effective and rational manner.

With respect to causes of death (*mortality*), in the developing world, about one death in three is caused by infectious and parasitic diseases, including respiratory infections. In contrast, in the established market economy countries this figure is about one in twenty. The risk of a death being caused by infectious agents, whether bacteria, viruses, parasites, or others, is clearly many-fold higher for populations living in the nations of the developing world. As will be discussed in the section under the heading Transmission of Infectious Diseases, this is true mainly because of the lower standards of sanitation and personal hygiene associated with poverty, malnutrition, and substandard living conditions in the developing world. In a typical developing country, respiratory infections, diarrhea caused by intestinal infections, and malaria are the most common causes of death, with a disproportional impact on the very young.

The newer methods created to measure disease burden also help estimate the economic impact of disease. This approach, known as *DALY (disability-adjusted life years)*, considers factors in addition to causes of death, for example, the degree of disability, the frequency and severity of a given disease category, and the years of expected life lost because of premature death. As such, DALY is a composite measure that is widely considered to reflect the net impact, not just death, caused by disease in a population. Using this approach, the impact of infectious and parasitic diseases is about 10 times greater in developing than in developed countries. For sub-Saharan Africa, for example, the burden caused by infectious diseases is about 15 times greater than that in the economically developed countries.

These differences are highlighted to focus attention on the fact that *the greatest single increase in exposure and disease risk while living and working an a developing country comes from infectious diseases, which are transmitted commonly in that setting.* Therefore special emphasis should be placed on appropriate measures to minimize these risks in the work force.

Transmission of Infectious Diseases

Infectious diseases can be named or classified in a number of ways such as symptoms caused (Yellow fever), agent responsible (malaria), organ system involved (hepatitis), geographical location (Ebola), and others. When dealing with public health and disease prevention, however, it is particularly useful to classify infectious diseases with respect to their transmission cycles, because this provides a rational framework for public health interventions. In short, if it is known how a disease is transmitted, the chain of transmission can be interrupted by specific and appropriate actions, as described in the following material.

While infectious diseases may be transmitted in remarkably diverse and complex manners, two broad transmission categories can be described:
1. diseases transmitted exclusively *person-to-person*
2. those which involve *complex* cycles.

Person-to-person cycles

Person-to-person transmission of infectious agents occurs mainly by one of three routes:
- respiratory
- intestinal (fecal-oral)
- sexual.

Examples of diseases transmitted by the respiratory route include influenza, tuberculosis, measles, diptheria, the common cold, and meningitis. Examples of diseases acquired via the fecal-oral route are poliomyelitis, typhoid fever, cholera, shigellosis, amebiasis, and pinworms. The number of known sexually transmitted diseases has increased rapidly during recent decades, and of course the Acquired Immunodeficiency Syndrome (AIDS) is the most dramatic example. Others include syphilis, gonorrhea, genital herpes, chancroid, and granuloma inguinale. It is now known that certain viruses transmitted sexually, such as human papilloma virus (of which more than 70 types are identified), can cause cervical, vaginal, and penile cancers.

From a public health perspective the person-to-person diseases share certain features.

1. They *travel* essentially wherever people go, and for this reason they are found in all parts of the globe, from the equator to the poles. Geographically they are usually not restricted, but are found in all parts of the world.
2. The most important modes of transmission are those mentioned previously: respiratory, fecal-oral, and sexual.
3. Many of these diseases have the potential to occur in outbreaks, and such outbreaks are relatively easy to detect by routine disease detection (surveillance) systems.
4. Immunization of a large proportion of a population group, whether by natural infection or vaccine, may prevent transmission because of the phenomenon of *herd immunity,* the low probability of one infection passing to another susceptible person in the population.

Complex Cycles

Diseases with *complex* life cycles include many of those that we think of as exotic or tropical, although they are by no means limited to the tropics. They also include diseases known as *zoonoses* that occur primarily in animals but which also may be transmitted to humans. Further, the *complex* category also includes diseases, which are transmitted to humans (and to animals) by the bite on an insect vector. It is useful to conceptualize several different complex cycles, as follows:

- *human to vector to human,* for which malaria, dengue fever and onchocerciasis (River blindness) are examples
- *animal to animal* (human infection is "accidental," with rabies and Lassa fever as examples)
- *animal to vector to animal* (human infection is "accidental," exemplified by a viral encephalitis such as *Japanese encephalitis* in which pigs serve as a reservoir and mosquitoes as vectors) (Figure)
- *intermediate hosts*, such as snails, supporting the multiplication of schistosome parasites *soil-transmitted parasites,* such as hookworm, whipworm, and ascaris (giant roundworm) for which transmission to humans can occur only after the parasite eggs have embryonated in warm, moist soil for a period of time determined by temperature.

Like the simpler person-to-person cycles, diseases that depend upon *complex cycles* share certain important characteristics that may be helpful in planning prevention and control measures. First, unlike their travel-prone cousins, they tend to be more restricted geographically. This is because the ecological requirements of one or more of the elements required for transmission are more focal in distribution. *Schistosomiasis (snail-fever)* transmission cannot occur in the absence of specific snail species. *Malaria transmission* depends upon many factors, including the ecological requirements of its *mosquito vectors (anopheline species)* to breed and propagate, and temperature requirements of malaria parasite (which can not multiply in the mosquitoes below critical temperatures). *Onchocerciasis,* a parasitic disease, is also known as *River blindness* because it occurs only near to rivers and streams that support the breeding of certain species of blackflies that have limited flight ranges. Thus, the distribution or potential distribution of many of these diseases is known, and can be mapped.

Second, humans acquire most of these diseases via the skin, usually by the bite of an insect, but also by the penetration of an aquatic larval stage of a parasite such as *schistosome cercariae,* or hookworm larvae in the soil. Risk of exposure can often be reduced dramatically by vector control, improved housing and screening, and a variety of personal protective measures, including insect repellants and clothing impregnated with insecticides.

Third, surveillance for, and detection of, diseases with complex transmission cycles tend to be more difficult than for diseases with person-to-person transmission. This is because the transmission

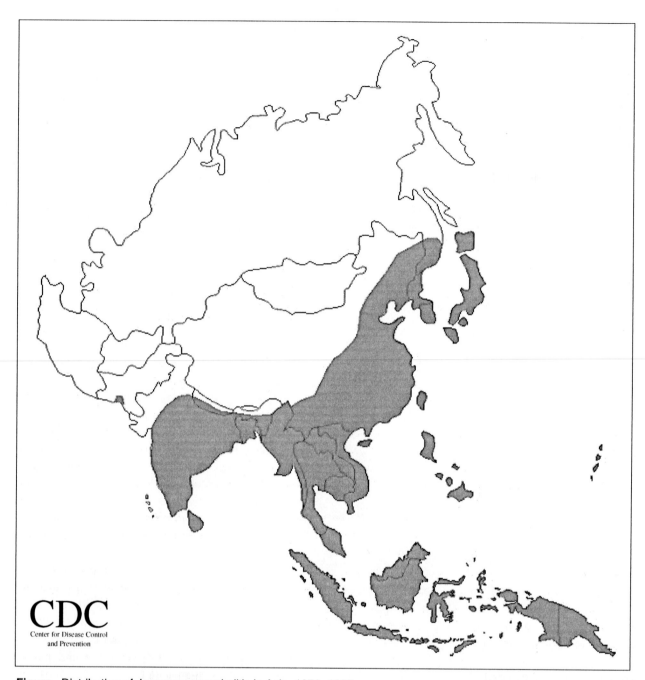

Figure. Distribution of Japanese encephalitis in Asia, 1970–1998.

typically occurs between wild or domestic animals and insect vectors, and often does not produce obvious disease in the infected animal source. For example, some of the vector-transmitted encephalitis viruses depend on bird/mosquito cycles, with neither showing signs of illness. Similarly, Lassa fever virus is transmitted among certain species of rodents in West Africa without causing illness, but infected rodents, which live a normal life span, excrete virus in their urine. This is the source of human infection through contamination of food or direct contact. On the other hand, an unexpected increase in monkey deaths in a tropical forest environment may signal the beginning of a Yellow fever or Ebola epidemic. Thus, early detection of outbreaks may require special approaches. Also, disease transmission often occurs on a seasonal basis, facilitating increased surveillance during critical periods of the year.

Fourth, the principle of *herd immunity* is often not operative with diseases transmitted in complex cycles. In other words, the probability of becoming infected is not influenced by the immune status of other humans in the community because transmission typically occurs independently of people, in the wild.

It is important to note that the *person-to-person category of diseases* contributes disproportionately to the burden of disease compared with the *complex category*. Respiratory and intestinal infections are the major killers of children in the developing world, with malaria as the main exception to the rule that the exotic diseases, while important, do not contribute to mortality rates like the common, poverty-associated infections. The coexistence of *both* categories of infectious diseases in tropical developing countries—where poverty *and* the ecological circumstances that can support transmission of complex cycles coexist—helps explain why the burden of disease, largely infectious, is so great relative to that of industrialized nations (Table 29-A).

Cultural Context

An individual engaged on a long-term basis in a different cultural environment can experience tensions and anxieties that go beyond physical separation from home or loved ones. We usually take our own cultural environment for granted; it is part of the world in which we function. Not until people find themselves in different or alien cultures do they realize that in their everyday interactions decisions are made based upon mutually understood clues.

For example, if we want more privacy in the home or workplace, others pick up on subtle clues that we want to be alone, at least for the moment. In other cultures, these "clues" may not be recognized; concepts of *privacy* may be lacking or drastically different. One of the important sources of anxiety in Peace Corps volunteers early in their assignments to a host country is their sacrifice of privacy. It may become difficult, if not impossible, to be alone within our own private space. The concept of personal space may simply not exist in the new culture; people in this culture may consider many people sleeping in one room to be the norm. The idea of a visitor being accustomed to something different, for example having their own bedroom, may not occur to them; personal space is not a universal entitlement.

Language can be another barrier when entering a new culture. Communicating our simplest needs can become a stressful chore. Repeated encounters

Table 29-A. Summary of Epidemiologic Features of Infectious Diseases by Transmission Cycle.

Feature	Person-to-Person	Complex
geographic distribution	global	more restricted
common modes of transmission	respiratory, fecal-oral & sexual	via skin (insect bite, larval penetration
ease of monitoring disease transmission	relatively simple	often difficult
effect of mass immunity	often reduces risk of transmission	minimal benefit
disease burden in developing countries	most important source	less important source

with bureaucratic barriers to accomplishing professional or personal tasks can lead to a sense of helplessness and frustration. Accumulated stress and anxiety typically manifest themselves by degrees of anger and irritability. The good news is, unpleasant as it may seem, this response is normal and, in the experience of many organizations that relocate U.S. citizens to countries with greatly differing cultures, it is part of a valuable learning process that ultimately enhances the ability to function between cultures. It often is a rewarding experience, offering opportunity for significant personal growth and self-knowledge.

Typically, after two to three months in the new environment, when the novelty of a new place has subsided and the prospect of a long stay seems more palpable, individuals commonly experience what is referred to as *culture shock*. It is usually transient, lasting perhaps weeks, and is manifested by the main symptom of anger. Family members such as spouses not buffered by the more familiar work environment can often suffer more severe symptoms. Knowing it is expected and normal makes it easier to pass through this stage of encounter with a new culture. Once passed, most individuals become happier in their private lives and more effective and satisfied in their workplace activities.

Health-Related Behaviors and Risks

Diseases do not occur in a cultural vacuum any more than they occur in an ecological vacuum. The risk of acquiring virtually any infectious disease is influenced to a greater or lesser degree by human behavior. While it is obvious that engaging in unprotected sex is a high-risk behavior, it is less obvious that ignoring the need to take antimalarial drugs on a regular basis, or swimming in a river where schistosomiasis may be transmitted, are examples of inappropriate health-related behavior.

We should guard against the common tendency to conclude that human behavior, driven by habits, cannot be modified with respect to disease risks. This in not the case; in fact, many examples exist to the contrary. The key to successful health education is to do the following:
1. Determine which health-related behaviors are beneficial and which are detrimental with respect to a given disease or group of diseases.
2. Beneficial behaviors already existing in the culture and its body of knowledge can be reinforced.
3. Those behaviors that are detrimental and need to be modified should be precisely identified, and clear, simple, and feasible messages constructed to modify the disease-enhancing behaviors.

Experience shows that modifying knowledge and behavior often can be accomplished with considerable success if it is undertaken in a culturally sensitive and culture-specific manner. Just as important—don't undertake this effort in a rigid top-down distribution of educational material. It is essential first to understand what the target population knows and believes before appropriate messages can be crafted. Collaborating with medical anthropology experts can help.

Expatriate Workers

In addition to the culture shock described previously, the infectious disease risks faced by expatriate workers are quite different from those faced by their host-country counterparts and the indigenous workforce. If this is a first assignment to a developing country, the expatriate worker may never have had prior exposure to common local diseases, such as malaria. These diseases are often severe and life-threatening. But for the native population, symptoms are often minimal or absent because of their protective immunity generated by previous exposure and infection.

In addition, irrational fear of exposure to diseases such as leprosy, AIDS, or even malaria can be a barrier to healthy adjustment. Through appropriate orientation, communication, and education on disease risks many fears and anxieties of newly arrived expatriates can be eliminated or minimized. The mental health status of the work force must not be ignored, especially during the critical first months of assignment in a developing country.

Machismo or other factors often can prevent newly arrived personnel from adhering to health requirements or to responding appropriately to health education messages. Taking antimalarial medications, reporting mild illness, or engaging in safe sexual practices may be considered unnecessary or signs of softness. However, the experience of missionaries, the military, and organizations like the Peace Corps clearly demonstrates the potential risks associated with prolonged residence in the developing country environment. For example, data from the Peace Corps shows that the top reported health-related events in Peace Corps volunteers are acute diarrhea, upper respiratory tract illness, dental problems, infectious skin disease, mental health problems, febrile illnesses, and unintentional injuries. The percent of volunteers affected annually was nearly 80% for acute diarrhea, more than 40% for upper respiratory illness, and about 20% or more for each of the other categories. In Africa, malaria rates in volunteers have increased significantly in recent years. About two thirds of all hospitalizations of volunteers were caused by infections. These experiences highlight the need for comprehensive and rigorous health programs and health-related education and communications for expatriate workers. When feasible, it may be helpful to reward compliance (or penalize noncompliance) to health requirements through administrative, financial, or other means.

A subset of expatriate workers may rotate between the worksite and their home country residence on a regular, often monthly, basis, introducing additional issues related to the potential that an infectious disease acquired abroad can be imported with the worker. This may pose a threat to the worker and, under some circumstances, to his or her family and community, requiring some system of monitoring for such unlikely events.

Indigenous Populations

In many overseas projects, such as building a pipeline or other infrastructure facilities, the bulk of the work force is comprised of local workers and others from the general region. Workers typically live in towns or villages near the project site and, as residents of these communities, are exposed to those infectious diseases, such as malaria, sexually transmitted diseases, and cholera, which may occur in the region. Creation of permanent project sites (such as an oilfield) and temporary sites (such a construction of an underground pipeline) may result in migration of workers, and their families or companions to these new sites. Project-related overland trucking may also introduce transient workers such as truck drivers who, in Africa, account for much of the long-distance spread of HIV virus, the cause of AIDS. During the workday intermingling of expatriate and indigenous workers occurs, introducing the potential for transmission of infectious diseases. Importantly, housing and food handling facilities for workers also create potential sites for disease transmission.

An additional consideration is the remote but alarming potential for outbreaks of deadly diseases such as Ebola fever, Lassa fever, or Yellow fever. The potential for this to occur in many parts of the developing world points to the need for appropriate disease-monitoring systems to detect such outbreaks rapidly, and to have contingency plans in place to effectively deal with them. Although such events are quite unlikely, inaction or inadequate response could have severe consequences.

HEALTH SERVICES

The health services in developing countries are considered, making a clear distinction between governmental and nongovernmental organizations; and the implications this has for program planning.

Governmental Health Services

Despite their critical importance to the health and economic well-being of a nation, the ministries of health in many developing countries lack the resources needed to provide or to oversee provision of basic health services to their populations. With few exceptions, ministries of health in developing countries find themselves overwhelmed with the curative and preventive needs of the citizens they serve. Budgets are small, typically representing a much lower percentage of the gross national product (GNP) than in economically developed nations. Whereas more than 10% of the GNP in the United States is spent in the health sector, in many developing countries it is 1% to 2%, which may translate into only a few dollars per capita for annual health expenditures. In addition to meager financial resources, ministries of health often must function with inadequate human resources. Salaries are low compared with those of the private sector; job security is jeopardized by political factors and frequent changes; assignments are often to isolated undesirable locations; and limited financial resources add to the frustration of governmental health officials.

Conversely, experience teaches that in developing countries we encounter a remarkable number of dedicated and able professionals, highly trained and motivated to improve the health status of their nation through appropriate public health measures and medical services. However, these officials typically face a constant barrage from international organizations and programs, which seek to initiate programs under their own terms. These may not be consistent with the developing nation's priorities. Most ministries of health operate under a set of policy objectives, and often are directed by endorsed health plans, such as five-year plans. The national priorities and approaches must be respected if we are to establish a meaningful partnerships with the governmental health authorities, whether at the national or health district level. For example, many countries place high priority on building the primary health care system on a sustainable, decentralized basis. This would be an important consideration in approaching a national ministry of health for project collaboration or endorsement.

Nongovernmental Organizations

Nongovernmental organizations (NGOs), also known as private volunteer organizations (PVOs) in the United States, are critically important players in the health sector of most developing countries. In a number of very low-income developing countries a de-facto recognition exists that the government does not have the resources to provide basic health services. In some countries virtually all of the health sector is nongovernmental, in that it is provided by organizations other than the national government. In the typical developing country in Africa, for example, there exists a blend of the two systems, but in each country the way in which this is achieved is quite different. In some the NGOs are given great freedom of operation, with relatively little official oversight or supervision, whereas in other countries the ministry of health is keen to exercise a high level of oversight and coordination of such a blended system. Further adding to the complexity in some countries is the fact that the health services of different provinces (states) in the *same* country may be largely supported through binational agreements with different industrialized nations, each with its own unique approach to the delivery and financing of health services.

Therefore, working effectively in a developing country is facilitated by awareness of the NGOs, which typically are committed to helping serve "underserved populations." For convenience of understanding their respective roles, NGOs can be grouped into two broad categories: international and indigenous/local. These are described in the following sections.

International NGOs

Examples of U.S.-based international NGOs (or PVOs) are CARE, Project HOPE, Africare, Save the Children Foundation, World Vision Relief and Development, Catholic Medical Mission, Interchurch Medical Assistance, Inc., International Eye Foundation, and many more. Some are very large, with thousands of workers in many countries; others are tiny, with perhaps a handful of individuals focusing on a particular country or mission within the country. The number of such organizations is impressive. For example, the *Mission Handbook 1998–2000: U.S. and Canadian Christian Ministries Overseas* is a 512-page directory of information on North American mission agencies, and *The Directory of Global Health: U.S.-Based Organizations Working in International Health* lists 450 U.S.-based organizations working in international health (see Appendix 1 of this chapter).

With respect to their primary missions, NGOs vary widely. Many have church affiliations. Some focus on child health and survival, others on operating remote general hospitals and clinics, and some limit their interest to specific diseases such as leprosy or prevention and treatment of blindness. Other NGOs define as their mission the provision of essential medical supplies and equipment at minimal or no cost. Industrialized countries besides the United States also support a broad array of NGOs, such as Doctors Without Borders supported by a number of European countries. Individuals involved with occupational health programs should become aware of the international NGOs active in their geographical areas. An example is the Mectizan Donation Program, supported mainly by a pharmaceutical corporation and based at the Carter Presidential Center, Atlanta, Georgia. Now active for more than 10 years, this program provides the drug which treats some 20 million persons yearly for the debilitating and potentially blinding parasitic disease onchocerciasis (or River blindness). If working in an onchocerciasis-endemic region, participation in the program could be beneficial to the work force.

The appendixes to this chapter offer information for ordering detailed directories. Those available only in written form are shown in Appendix 1. Appendix 2 lists the website addresses of directories of international NGOs and PVOs that are available on-line.

Indigenous (Local) NGOs

While usually smaller and more modestly funded than the large international NGOs supported by the wealthier industrialized nations, *indigenous or local NGOs* are expanding and flourishing in many developing countries. During recent decades these local NGOs have grown in number, support, and sophistication. Staffed primarily by individuals from the country itself, these NGOs complement the larger international organizations, addressing needs that can better be addressed by persons who intimately understand the local culture, language, and "politics." Their programs are as varied as those of the international organizations, including those that emphasize maternal and child health, improved housing, care for urban "street children," and countless other health and social needs.

Local NGOs in some countries have learned to acquire funding from bilateral donors such as the U.S. Agency for International Development, sometimes replacing the role of large international NGOs. Others derive support from local governments as well as from local individual donors and groups. While some degree of competition has evolved between the two bodies of nongovernmental organizations, this is generally a healthy type of competition, with each group learning from the other, and functioning in a mutually complementary manner.

General, nondogmatic advice is to determine whether indigenous NGOs can assist in health-related programs in a developing country before seeking international partners. In the long run, the people living in developing countries must sustain their own programs, without depending on outside support.

Systematic directories of indigenous (local) NGOs are harder to find, but usually this information can be obtained through ministries of health, finance, or planning. Some information can be found on the PVO *Executive Contact Lists* described in Appendix 2 of this chapter.

International Organizations

Perhaps the most important resources in this category are the United Nations agencies, which include the World Bank, the World Health Organization (Pan American Health Organization in the Americas), and the United Nations International Children's Emergency Fund (UNICEF). The World Bank invests heavily to assist development in poor countries; and many of these programs have a strong health component or health implications. An example is the planned Chad-Cameroon Oil Pipeline Project, a large-scale ($3.5-billion, 30-year) infrastructure project supported in large part by the World Bank in collaboration with the governments of Chad and Cameroon and a consortium of private oil companies. The partners in this undertaking, including the World Bank, are keen to insure that the project will not expand the HIV/AIDS epidemic in the region. This is a risk because of the some 2,000 construction workers and 500 truckers who will travel along the pipeline. The World Bank has played a lead role in assuring that a package of vigorous preventive measures are put into place in the project area, and is assisting the two countries in developing the capacity to implement the program and to monitor the health impact of the project.

The World Health Organization (WHO) is headquartered in Geneva with a number of regional headquarters is another valuable source of health-related information. It also has potential for collaboration in programs that relate directly to the health status of a work force. For example, there is a new program with WHO as a partner to control and eventually eliminate *lymphatic filariasis*. This is a mosquito-transmitted parasitic disease that infects about 120 million people (and threatens perhaps 1 billion) in the tropics; and may lead to dreaded *elephantiasis*, a disease whose symptoms include enormous and debilitating swelling of limbs or other body parts. This program, the WHO Collaborative Program to Eliminate Lymphatic Filariasis, has other major partners from the pharmaceutical industry and a range of other organizations, including many NGOs. (For more information about this program visit: http://www.filariasis.org.) A corporation undertaking a project in a region of the world where filariasis occurs would be well advised to be aware of this program as a source of valuable information, resources including drugs, and potential collaboration. As an example, a group of mining corporations working in the region of New Guinea are actively collaborating to undertake filariasis control efforts to protect their workers as well as the nonemployee indigenous population. Their gains include cost-effective disease prevention in the work force and considerable corporate good will derived from the local population, an important public relations gain.

The United Nations' International Children's Relief Fund (UNICEF) is a well-funded and highly active member of the United Nations family of agencies. It operates in virtually every country in the world. While the primary mission of UNICEF emphasizes infants, children, and young people, national programs often address health infrastructure needs. An important feature of UNICEF is that traditionally it has functioned on a rather decentralized basis, granting a large degree of freedom in national programming based on local needs and conditions. Visit the UNICEF office in any country where industry is working for current information and potential coordination of health-related efforts. The website addresses for selected national and multinational organizations active in health in developing countries are provided in Appendix 3 of this chapter.

Other Local Resources

National universities and schools of medicine where they exist are often overlooked as valuable resources in developing countries. Many developing countries have good medical schools, with faculty members who may be engaged in research into the prevention and control of locally important diseases, especially the major infectious diseases of the region. They may be the source of the most current information on the status of these diseases in the country; and also may have useful linkages to other important organizations or individuals. In some countries collaboration between schools of medicine and the ministry of health may be ongoing. Another resource is private corporations in the country or region that may have a vested interest in the control of a particular disease. An example noted previously is the collaboration of mining companies to combat filariasis in and around their areas of overseas involvement.

ENVIRONMENTAL APPROACH TO PUBLIC HEALTH AND OCCUPATIONAL HEALTH

Approaches to public and occupational health and safety are described with an emphasis on the *environmental approach* to disease prevention. Specific diseases are considered in the framework of their main modes of transmission, i.e., via the respiratory route, insect vectors of disease, sexual transmission, waterborne, foodborne, etc.

Surprisingly, there has not been a set of generally accepted guidelines to assist individuals responsible for planning and implementing public health and occupational health programs for large infrastructure projects in developing countries. The usual approach has been to focus on specific diseases, one by one, using the best available scientific knowledge about preventing that disease in the work force. For example, if working in an area where malaria is a known problem, a categorical *malaria control program* was implemented.

In recent years public and occupational health planners and economists, assisted by the World Bank and other organizations, have developed a more comprehensive and potentially more effective approach. This approach, the so-called *environmental approach,* emphasizes the broader environmental consequences of a large infrastructure project; and develops *interventions* (measures to prevent or mitigate negative consequences) within an environmental framework. This approach recognizes that large infrastructure projects, such as building dams, roads, or pipelines, have potentially positive and negative impacts on health. The challenge is to recognize, quantify, and maximize the benefits and minimize the negative impacts. Rather than start with the *disease,* this approach looks at the potential *impact of the project* on the disease, or *incident-related factors* of (1) housing, (2) water and sanitation, and (3) transportation. Job-specific occupational risks are considered separately. This approach has been described in a 1996 World Bank document, *Bridging Environmental Health Gaps* (BEHG).

Housing factors are closely linked to the risk of acquiring respiratory-transmitted infections, described earlier as person-to-person transmitted diseases. Perhaps surprisingly, the risk of acquiring many (complex transmission cycle) insect-transmitted diseases (including malaria and many others) also are largely a function of housing conditions—since most of them are acquired at night by the bite of a mosquito or other insect. Water and sanitation are the key factors in preventing diseases transmitted by the fecal-oral route, another set of diseases transmitted person-to-person, often via contaminated water or food (Table 29-B).

Following the recommendations of the BEHG document, environmental assessment of potential project impacts can be visualized in a matrix in which the three factors: (1) housing, (2) water and sanitation, and (3) transportation are examined with respect to the six following major environmental health areas:
- respiratory diseases
- water and foodborne diseases
- vector-related diseases
- sexually transmitted diseases
- incidents and injuries
- exposures to potentially hazardous materials.

Rather than attempt to provide examples for each of the cells of the matrix for a hypothetical project, a few examples are offered from a large infrastructure project in sub-Saharan Africa. With regard to housing for temporary and permanent workers, assess potentially adverse and beneficial impacts on the transmission of vector-related diseases. *Potential adverse impacts* include increased incidence of diseases such as malaria, schistosomiasis, and filariasis. This can be prevented (mitigated) by a series of measures including appropriate screening and use of insecticides in the housing, use of insecticide-impregnated clothing when appropriate, and routine use of antimalarial drugs for those at risk, and other measures (related to water and sanitation), which limit the breeding sites of responsible vectors. *Potential beneficial impacts* for vector-related diseases from the project include measures learned by local contractors regarding design measures for improved housing, and increased levels of knowledge regarding preventing vector-related diseases.

Another example is offered regarding transportation and sexually transmitted diseases. While the potential adverse impact of spread of HIV/AIDS between truckers and the work force exists, it can be mitigated by aggressive means including worker education (with special emphasis on truckers), active marketing of affordable condoms, community education, close coordination with national government HIV/AIDS program (particularly directed towards female commercial sex workers and other vulnerable women), active surveillance and treatment of sexually transmitted diseases, and surveillance to monitor trends in HIV/AIDS. It is believed that these measures will serve to minimize (mitigate) the adverse impact of the project, and will provide the important benefit of greatly enhanced knowledge

Table 29-B. Showing Known Cases and Outbreaks of Ebola Hemorrhagic Fever, in Chronological Order.

Year	Ebola species	Country	No. of human cases	Percentage of deaths among cases	Situation
1976	Ebola-Zaire	Zaire	318	88%	Occurred in Yambuku and surrounding area. Disease was spread by close personal contact and by use of contaminated needles and syringes in hospitals/clinics. This was the first recognition of the disease.
1976	Ebola-Sudan	Sudan	284	53%	Occurred in Nzara, Maridi and the surrounding area. Disease was spread mainly through close personal contact within hospitals. Many medical care personnel were infected.
1976	Ebola-Sudan	England	1	0%	Laboratory infection by accidental stick of contaminated needle.
1979	Ebola-Sudan	Sudan	34	65%	Occurred in Nzara. Recurrent outbreak at the same site as the 1976 Sudan epiemic.
1989	Ebola-Reston	USA	0	0%	Ebola-Reston virus was introduced into quarantine facilities in Virginia, Texas, and Pennsylvania by monkeys imported from the Philippines. Four humans developed antibodies to Ebola-Reston virus but did not become ill.
1990	Ebola-Reston	USA	0	0%	Ebola was introduced once again into quarantine facilities in Virginia, and Texas by monkeys imported from the Philippines. Four humans developed antibodies but did not get sick.
1992	Ebola-Reston	Italy	0	0%	Ebola-Reston was introduced into quarantine facilities in Sienna by monkeys imported from the same export facility in the Philippines that was involved in the episodes in the United States. No humans were infected.
1994	Ebola-Zaire	Gabon	44	63%	Occurred in Minkebe, Makokou and gold-mining camps deep in the rain forest. Initially thought to be yellow fever; identified as Ebola hemorrhagic fever in 1995.
1994	Ebola-Ivory Coast	Ivory Coast	1	0%	Scientist became ill after conducting an autopsy on a wild chimpanzee in the Tai Forest. The patient was treated in Switzerland.
1995	Ebola-Zaire	Zaire	315	81%	Occurred in Kikwit, democratic Republic of Congo. Traced to index case-patient who worked in forest adjoining the city. Epidemic spread through families and hospitals. The outbreak impacted not only Kikwit but also surrounding communities.
1996	Ebola-Zaire	Gabon	37	57%	Occurred in Mayibout area. A chimpanzee found dead in the forest was eaten by people hunting for food. Nineteen people who were involved in the butchery of the animal became ill; other cases occurred in family members.
1996	Ebola-Zaire	Gabon	60	75%	Occurred in Boue area with transport of patients to Libreville. Index case-patient was a hunter who lived in a forest camp. Disease was spread by close contact with infected persons. A dead chimpanzee found in the forest at the time was determined to be infected.
1996	Ebola-Zaire	South Africa	2	50%	A medical professional traveled from Gabon to Johannesburg, South Africa, after having treated Ebola virus-infected patients and thus having been exposed to the virus there. He was hospitalized, and a nurse who took care of him became infected and died.
1996	Ebola-Reston	USA	0	0%	Ebola-Reston virus was introduced into a quarantine facility in Texas by monkeys imported from the Philippines. No human infections were identified.
1996	Ebola-Reston	Philippines	0	0%	Ebola-Reston virus was identified in a monkey export facility in the Philippines. No human infections were identified.

1999 Special Pathogens Branch
Division of Viral and Rickettsial Diseases, National Center for Infectious Diseases
Centers for Disease Control and Prevention
Public Health Service, U.S. Department of Health and Human Services

about AIDS, its transmission, and means of prevention in the work force and surrounding communities. Building the public health infrastructure to carry on the monitoring and treatment of sexually transmitted diseases would also provide a lasting benefit beyond the life of the construction project.

While it is beyond the scope of this chapter to explore the evolving concepts of an environmental health approach as compared with those of a traditional disease-specific focus, this discussion introduces some of the issues. However, this shift does not mean that all disease-specific approaches are discarded. Quite the contrary, diseases that require special attention in a project, such as HIV/AIDS and vector-transmitted diseases as mentioned in the previous example, will continue to be addressed in a more vertical, traditional manner. Since it has been estimated that nearly half of the disease burden in sub-Saharan Africa can be reduced by infrastructure investments, it is important to attempt to measure health impacts in a more comprehensive and environmentally sound manner.

APPENDIX 1: DIRECTORIES OF NGOs AVAILABLE ONLY IN WRITTEN FORM

The Directory of Global Health: U.S.-Based Organizations Working in International Health (1996–1997) is a comprehensive list of 450 U.S.-based organizations working in international health. It is compiled by the National Council of International Health (NCIH). The directory includes contact information, organizational mission and objectives, areas of work, and countries served. NCIH Members: $35; Nonmembers: $70. Contact NCIH, 1701 K. St., NW, Washington DC 20006; Tel. (202) 833-5900; e-mail: ncih@ncih.org

Mission Handbook 1998–2000: U.S. and Canadian Christian Ministries Overseas is a 512-page directory of information on North American mission agencies, including fax, e-mail, and website addresses. It is available for $49.95 from Mission Advanced Research and Communications (MARC), 800 W. Chestnut Avenue, Monrovia CA 91016, Tel. (800) 777-7752; e-mail: MARCpubs@wvi.org/

Voluntary Foreign Aid Programs-1998 is a free booklet giving summary information about more than 400 U.S. PVOs. To obtain a copy, contact the Office of Private and Voluntary Cooperation, U.S. Agency for International Development, Washington DC 20523-7600; Tel. (202) 712-0840; e-mail: VOstrich@usaid.gov/

APPENDIX 2: DIRECTORIES OF NGOs ACCESSIBLE VIA THE INTERNET*

The PVO Executive Contact Lists are online searchable databases for 800 PVOs around the world registered with USAID. The lists include U.S., international, third-world private volunteer organizations and nongovernmental organizations. Check out the list at: /http://intranet.dimen-intl.com/usaid/

Country Health Profiles are available at www.cihi.com/hthpub.htm/ These downloadable files include *Health Statistical Reports* with basic national-level demographic and health data for most developing countries, and *Country Health Profiles* with a more comprehensive analysis of health conditions and trends in selected countries (mostly sub-Saharan Africa). The information is maintained by The Center for International Health Information, 1601 N. Kent St., Suite 1014, Arlington VA 22209; Tel. (703) 524-5225; e-mail: info@cihi.com

InterAction, a major U.S. development PVO coalition of over 150 PVOs contains interesting information on humanitarian aid, situation updates, and highlights of their newsletter. Website: http://www.interaction.org/

The Internet NonProfit Center is home to donors and volunteers. This website contains a wealth of information about U.S. NGOs in all sectors: http://www.nonprofits.org/

Association of Voluntary Service Organizations (AVSO) is an international nongovernmental organization, forming a European platform of national and international nonprofit organizations active in the field of longer-term voluntary service. Website: http://village.agoranet.be/~avso/

The Union of International Associations was founded in Brussels in 1907 as an independent, nongovernmental, apolitical, nonprofit body with the aim of facilitating the evolution of the activities of the worldwide network of nonprofit organizations, especially nongovernmental or voluntary associations. Website: http://www.uia.org

The Nonprofit Website Directory of the Action Without Borders, a New York-based nonprofit organization founded in 1994 to help people take action on the issues that concern them, is an excellent site to do a search on international development of NGOs. Website: http://www.idealist.org/

NGO REPORTER is published quarterly by EXE-COM, The NGO/DPI Executive Committee. A full 50 new organizations were accredited to the Department of Public Information in March, 1998, bringing the total to 1,514! Website: http://www.ngo.org/

---

* Most of the contents of Appendices 1 and 2 were kindly provided by Franklin C. Baer, Baertracks, Website: www.bemorecreative.com, e-mail: fcbaer@bemorecreative.com

APPENDIX 3: SELECTED NATIONAL AND MULTINATIONAL GOVERNMENTAL ORGANIZATIONS ACCESSIBLE VIA THE INTERNET

United States Agency for International Development: http://www.info.usaid.gov/
United Nations: http://www.unsystem.org/
World Bank: http://www.worldbank.int/
World Health Organization: http://www.who.int/
United Nations International Children's Emergency Fund: http://www.unicef.org/

chapter 30

Information Management

by Becky L. Randolph, MS, CIH

513 **Introduction**
Technical components ■ Information vs. document management
514 **Benefits of Managing Information**
Clear, concise communication ■ Opportunity to change processes ■ Integration
515 **Buy vs. Build**
Buying "new" ■ Types of technology
516 **Summary**

INTRODUCTION

The objective of this chapter is to educate the reader on information management as a tool for occupational health and safety professionals. *Information management is the collection of specific data from a work process that is transformed from individual data points to useful information.* It is the collection of specific data to be managed for a grander delivery—to solve a problem, to determine compliance, to identify areas of concern, to reduce or eliminate risk. Examples of data that would be collected for the health and safety profession include incident information, industrial hygiene sample collection, medical information on individuals, environmental emission points, emergency-response equipment inventory, audit findings, chemical inventories, performance metrics, or Material Safety Data Sheet information, just to name a few. The benefit of an information management system is to turn data into information.

Technical Components

The two major technical components of an information management system is a *database* and a *program or application*. The two components can be from different vendors, or from the same vendor. The *program or application* is the technical component that the health and safety professional uses or interfaces with, the *front end*. This is the component that we want to be user-friendly. The *database* is the technical component or structure that houses the data. This is also referred to as the *back end*. The last component is the *work process*. Work process improvement is a potential result of an information management system. Work process and technology working together is where efficiencies can be gained.

Information vs. Document Management

Many times information management systems are confused with information searches or document management systems. Information searches also involve pieces of data points, but the tough decision is:

- Do you want that information in your database?
- Is the information necessary for your analysis?
- How important is the information?
- Is the information going to be used in your analysis or part of your reports?
- Is the information important to the work process?

Document management systems are systems that manage whole documents with links to specific data points. Where should the information reside, in a database or in a document management system? Consider this situation as an example—if your information system captures audit findings and the associated regulation that you have violated. If you want to read the actual citation, should that information be part of your database or part of your document management system or search information available on a government web site? These are the types of difficult questions that need to be resolved when designing, building, or buying information management systems.

BENEFITS OF MANAGING INFORMATION

Information is power. The information age is here. If your manager asks the question, how much did that incident cost, you need to know the total cost of the incident, not just the workers' compensation benefits, but also the costs associated with the lost production, the cleanup, the fines, and the indirect cost (employee replacement, supervisor investigation time, etc). Information management is the tool to provide the answers. Actually, your manager shouldn't have to ask the question. We, as health and safety professionals should be proactively merging health and safety data with business information. Technology is here and we need to incorporate it into our profession. We also need to expand beyond our historic "silo-like measures," like incident rates, and provide information integrating health and safety into business decision support.

Clear, Concise Communication

Information management is a method to communicate vital information in a clear and concise manner. The financial profession historically has been very successful at developing elaborate information-management systems as a tool to generate information that communicates performance and provides measurable metrics. The health and safety profession is now gaining interest in aligning with the business side by providing health and safety information. Therefore, health and safety information needs to be incorporated with business information with measurable performance metrics.

Opportunity to Change Processes

Another benefit of information management systems or applications is the opportunity to *change business processes*. Many build an application to fit a process. A process that was instituted before the concept of technology even existed. Managing information through technology provides the opportunity to review processes and determine the best way to incorporate technology into your business process. Less time should be spent on collection and input of the information and more time should be spent on analysis of the information. The results of a health and safety information management system should be improved processes using technology, not new technology with the same old forms or processes.

Integration

Integration is another benefit of information management systems. Integrated data is shared data. If a business has totally integrated all systems, your name is in the system one time and used by everyone through that one system. No system is duplicating your name. Integration eliminates duplication and islands of information or silos of information. Very few companies have systems that are totally integrated. The advantage of integration is the efficiency of having the information once. If it needs to be updated, only one source needs updating.

While integration is a benefit, it also has some shortcomings. One shortcoming is *ownership*. Having one group *own* a specific piece of information sounds great in concept. One group needs to own the data and be responsible for the information. And that group has specific interest. For example, human resources normally owns personal data and they have specific interests. A safety and health professional, for example, may have other interests. Job titles important to human resources deal with career development. Job titles important to an industrial hygienist developing homogenous exposure group data deal with exposure potentials. These are two very different uses for job title. And sharing information is not always easy. The more integration within a business enterprise system, the more inflexible some data elements become. Therefore, the ownership issue can cause a reduction in the flexibility of a system.

Many confuse *integration* with *interface*. An *interface* transfers the same information from one application to another. With *integration*, the information is shared, while with *interfacing* the information is duplicated. Most large enterprise systems today have interfaced three or four systems together to form a centralized business system. They have duplicated information, but the duplication is limited and controlled.

BUY VS. BUILD

In the 1970s, if a company had a health and safety information management system, they built it. Today, the trend is to buy a vendor product whenever possible. The cost associated with the maintenance and support of internally built applications and the technology supporting that applications can be expensive to maintain. Companies are moving back to core business products; and supporting a health and safety application internally is not their core business product. In addition, many companies have moved toward outsourcing their information technology departments. Therefore, internal builders are no longer available.

Buying "New"

With the new-buy approach, health and safety professionals have had to develop an evaluation process to compare different vendors. This requires companies to define the functional requirements as well as the technical requirements. Examples of *functional requirements* are as follows:
- calculating an incident rate
- generating an OSHA log
- generating a first report of injury
- collecting specific information about an industrial hygiene sample
- calculating a standard threshold shift from an audiogram and/or generating the specific compliance letter.

Examples of *technical requirements* are as follows:
- The systems should run in a client/server environment.
- The database must fit with the predetermined vendor.
- The application must be able to handle 50 users at one time, and/or the response time from user to server must met specific requirements.

Once the requirements are defined, they are weighted and evaluation criteria are determined. For example, we must met 90% of our technical requirements and 75% of our functional requirements. Then decisions must be made regarding how the unfulfilled requirements will be handled. This selection process is long and involved.

Types of Technology

There are three basic types of technology—*point solutions, integrated enterprise solutions,* and the *middle ground*.

Point Solutions

The simplest type of technology is a program or application with *limited data points fulfilling a single function*. These are called *point solutions* and are normally short-term solutions. With this tactical approach, one person inputs the information and uses the information to help him or her complete the job more effectively and efficiently. The program can be internally built or provided by a vendor. *Point solutions* are specific to a specific area and are normally positioned as plant or facility solutions. The hardware and software requirements are fairly basic. With point solutions, problems can occur when newer versions of local software become available. The migration of existing data to the newer environment is normally an additional burden. Again, the duplication of information can also cause inconsistencies. The implementation of the solution is normally minimal.

Integrated Enterprise Solution

On the other end of the spectrum is the *large, integrated enterprise solution*. The objective of an enterprise solution is to *share information across the various business systems*. The program covered many facets of the business from finance to human resources to inventory control. To date, enterprise providers have not included health and safety information as part of the total business enterprise. Some providers are moving in that direction, but currently health and safety information management has not been a major player in enterprise solutions. Many companies with enterprise solutions as a strategic direction are purchasing individual pieces; for example, a human resource application, a financial application, a health and safety application, and an environmental application and then *bolting* the pieces together with transparent interfaces to generate an enterprise solution.

Enterprise solutions are normally situated for large, global, corporate solutions that have a long-term, strategic approach to information management. The hardware and software requirements are large and complex. The advantages of an enterprise solution are lower maintenance and support with version control migration plans, reduced number of application vendors, and elimination of redundant data. The implementation of the solution is extensive. Implementations can include interfaces for data conversions from an existing system that is normally a consolidation of information, reengineering work processes to work with the technology, and interfaces to other systems.

Middle Ground

The third and most common type of technology lays somewhere between the point solution and the enterprise solution with some *controlled interfaces*. Most companies define the most common points of integration and build interfaces to connect the individual systems. The most common individual functional systems to-date are health and safety applications combining safety, medical, and industrial hygiene, environmental applications, and Material Safety Data Sheet applications.

SUMMARY

If your organization is moving toward using information management as a more strategic tool, here are some tips:

- Develop a tactical plan and a strategic plan. Implement existing technology and plan for new technology. Learn the difference between the cutting edge and the bleeding edge.
- Know your company's information technology strategic direction and work within that plan.
- Define your functional requirements. Make distinctions between what you need to have and what is nice too have. Include cost elements in your requirement definitions.
- An ideal work team includes both functional and technical professionals. Depending on the size of your project, dedicated resources work better than partially committed resources.
- Develop examples of the types of reports you want to generate even if you hand-draw the reports. Think of integrating health and safety information with existing business reports.
- Incorporate your information management system into your budget for the long term. It is not a one-time cost. As long as you use information management systems, you will need it as a line item in your budget. It is a tool that is ever-improving.
- Information management is a tool that can improve your personal and organizational performance.

Remember, if you can measure it, you can control it.

chapter 31

Sources of Help

518	**Associations and Organizations** Technical societies ■ Trade associations ■ Scientific and service organizations ■ Energy information service ■ Other organizations
524	**Governmental Organizations** Federal agencies ■ State agencies ■ Other agencies
527	**Databases** General ■ MEDLARS on-line network
529	**International Occupational Safety and Health Internet Sites**
531	**Basic Reference Books** General principles ■ Risk assessment ■ Sampling methods ■ Toxicology ■ Medical ■ Dermatological ■ Physical stresses ■ Ergonomics ■ Biological ■ Chemical ■ Pollution and hazardous waste ■ Control ■ Encyclopedias and handbooks
550	**NIOSH Publications** Criteria documents
551	**Data Sheets and Guides**
551	**Newsletters and Reports**
552	**Journals and Magazines**
554	**States with On-Site OSHA Consultation Agreements**
556	**Selecting and Using an Occupational Health & Safety Consultant** Define the problem ■ Sources ■ Selection ■ Compensation ■ The proposal ■ Working with the consultant

The occupational health and safety professional in routine activities frequently must make a decision as to the degree of health hazard arising from a process or operation. In emergency situations and in the absence of an industrial hygienist, immediate corrective action must be taken toward the recognition, evaluation, and control of occupational health hazards. If the unit is part of a multifacility corporation, consult with the home office. The supplier of materials that are suspected of being toxic can also be consulted for information on the potential hazards (this information also appears on the MSDS).

Other specialized health and safety information and assistance are available from a number of sources.

- Most employers are likely to have trained professional or scientific personnel who can provide some technical assistance or guidance, even though their primary interest is in a field other than industrial hygiene.
- Many insurance companies that carry workers' compensation insurance provide safety consulting and industrial hygiene services.
- Professional consultants and privately owned or endowed laboratories are available on a fee basis for concentrated studies of a specific problem or for a facility-wide or company-wide survey, which can be undertaken to identify and catalog individual environmental exposures and/or hazards.
- Many states have excellent industrial hygiene departments. Some state laboratories are extremely well equipped, and have available sophisticated sampling and analyzing equipment that an individual company might not justify purchasing, because of cost, need for specialized operator training, or infrequent or intermittent use.

ASSOCIATIONS AND ORGANIZATIONS

Many associations have an occupational safety, health, or industrial hygiene subcommittee that can be called upon for assistance.

Technical Societies

Scientific and technical societies that can help with health conservation or with a specific problem area are listed in this section. Some are prepared to provide consultation service to nonmembers as well as members; they all have a wealth of available technical information.

Air and Waste Management
1 Gateway Center, 3rd floor
Pittsburgh, PA 15222
http://www.awma.org
Industrialists, researchers, equipment manufacturers, governmental control personnel, educators, meteorologists, and others seeking economical answers to the problem of air pollution. Sponsors continuing education courses and maintains a library.

American Academy of Industrial Hygiene
6015 W. St Joseph, Ste 102
Lansing, MI 48917
Professional society of industrial hygienists.

American College of Occupational and Environmental Medicine
55 W Seegers Rd
Arlington Heights, IL 60005
http://www.acoem.org
Physicians who devote full time to some phase of occupational medicine. Promotes maintenance and improvement of the health of industrial workers.

American Academy of Ophthalmology
655 Beach St
San Francisco, CA 94109
http://www.eyenet.org
Ophthalmologists concerned with high quality eye care and the continuing education of members.

American Academy of Optometry
6110 Executive Blvd
Rockville, MD 20852
Professional society of optometrists, educators, and scientists interested in conserving human vision, clinical, and experimental research in visual problems.

American Academy of Otolaryngology—Head and Neck Surgery
1 Prince St
Alexandria, VA 22314-3357
http://www.entnet.org
Professional society of medical doctors specializing in otolaryngology, disease of the ear, nose, and throat, and head and neck surgery.

American Association of Occupational Health Nurses
50 Lenox Pointe
Atlanta, GA 30324
Registered professional nurses employed by business and industrial firms, nurse educators, nurse editors, nurse writers, and others interested in industrial nursing.

American Board of Industrial Hygiene
6015 W St Joseph, Ste 102
Lansing, MI 48917
This specialty board is authorized to certify properly qualified industrial hygienists. The overall objectives are to encourage the study, improve the practice, elevate the standards, and issue certificates to qualified applicants.

American Board of Medical Specialties
1007 Church Street, Ste 404
Evanston, IL 60201-5913

American Board of Preventive Medicine
9950 W Lawrence Ave, Ste 106
Schiller Park, IL 60176
http://www.abprevmed.com
The Board is authorized to certify properly qualified specialists. Overall purposes are to encourage the study, improve the practice, elevate the standards, and advance the cause of preventive medicine. Issues certificates of special knowledge to duly licensed physicians specializing in the fields of public health, aerospace medicine, occupational medicine, and general preventive medicine.

American Chemical Society
1155 16th St, NW
Washington DC 20036
http://www.acs.org
Scientific, educational, and professional society of chemists and chemical engineers.

American College of Occupational and Environmental Medicine (ACOEM)
1114 N. Arlington Heights Rd
Arlington Heights, IL 60004

American Conference of Governmental Industrial Hygienists
1300 Kemper Meadow Dr
Cincinnati, OH 45240

Professional society of persons responsible for full-time programs of industrial hygiene, who are employed by official governmental units. Primary function is to encourage the interchange of experience among governmental industrial hygienists, and to collect and make available information of value to them. Also promotes standards and techniques in industrial hygiene and coordinates governmental activities with community agencies.

American Industrial Hygiene Association
2700 Prosperity Ave, Ste 250
Fairfax, VA 22031
http://www.aiha.org

Organization of professionals trained in the recognition and control of health hazards and the prevention of illness related thereto. Industries are eligible for associate membership. Promotes the study and control of environmental factors affecting the health of industrial workers; provides information and communication services pertaining to industrial hygiene.

American Institute of Chemical Engineers
3 Park Ave
New York, NY 10016
http://www.aiche.org

Professional society of chemical engineers. Establishes standards for chemical engineering curricula.

American Institute of Chemists
501 Wythe St
Alexandria, VA 22314-1917
http://www.theaic.org

Chemists and chemical engineers. To elevate the professional and economic status of chemists and chemical engineers.

American Medical Association
515 N State St
Chicago, IL 60610
http://www.ama-assn.org

Professional association of persons holding either a medical degree or an unrestricted license to practice medicine. Its purpose is to promote the science and art of medicine and the betterment of public health. Consequently, it provides information on medical health topics; represents the medical profession before Congress and governmental agencies; cooperates in setting standards for medical schools and training programs; keeps members informed on significant medical and health legislation.

American College of Occupational and Environmental Medicine
55 W Seegers Rd
Arlington Heights, IL 60005
http://www.acoem.org

Professional society of medical directors and plant physicians specializing in industrial medicine and surgery, established to foster the study of the problems peculiar to the practice of industrial medicine and surgery, to encourage the development of methods of conserving and improving the health of workers, and to promote a more general understanding of the purpose and results of the medical care of these workers.

American Optometric Association
243 N Lindbergh Blvd
St. Louis, MO 63141
http://www.aoanet.org

Professional society of optometrists. Supports a library on ophthalmic and related sciences, with emphasis on optometry, its history, and socioeconomic aspects.

American Public Health Association
1015 15th St, NW
Washington DC 20005
http://www.apha.org

Professional association of physicians, nurses, educators, engineers, environmentalists, social workers, industrial hygienists, and others who have an interest in personal and environmental health. Its services include promulgation of standards, development of the etiology of communicable diseases, establishment of minimum educational qualifications for public health workers, accreditation of schools of public health, and research in public health.

American Society of Heating, Refrigerating, and Air Conditioning Engineers
1791 Tullie Circle, NE
Atlanta, GA 30329
http://www.ashrae.org

Professional society of heating, ventilating, refrigeration, and air conditioning engineers carries out a number of research programs in cooperation with universities and research laboratories.

American Society of Safety Engineers
1800 E Oakton St
Des Plaines, IL 60018
http://www.asse.org

Professional society of safety engineers, safety directors, and others concerned with accident prevention and safety programs. Compiles statistics.

American Society for Testing and Materials
100 Barr Harbor Dr
West Conshohocken, PA 19428
http://www.astm.org

Engineers, scientists, and skilled technicians holding membership as individuals or as representatives of business firms, government agencies, educational institutions, and laboratories. Establishes voluntary consensus standards for materials, products, systems, and services.

American Society for Training and Development
Box 1443
1640 King St
Alexandria, VA 22313
http://www.astd.org

Professional society of persons engaged in the training and development of business, industrial, and government personnel.

Board of Certified Safety Professionals
208 Burwash Ave
Savoy, IL 61874
http://www.bcsp.com

The principal purpose of the board is to establish minimum academic and experience attainments necessary to qualify as a certified safety professional and to determine the competence and issue certificates to qualified applicants.

Health Physics Society
1313 Dolly Madison Blvd
McLean, VA 22101-3926

Professional society of persons active in the field of health physics—the profession devoted to the protection of humans and the environment from radiation hazards. Aids research in the field, improves dissemination of information to individuals in the profession and in related fields, works to improve public understanding of the problems which exist in matters of radiation protection, and promotes the profession of health physics. Sponsors the American Board of Health Physics for the voluntary certification of health physicists.

Human Factors and Ergonomics Society
P O Box 1369
Santa Monica, CA 90406-1369

Professional society of psychologists, engineers, physiologists, and other related scientists who are concerned with the use of human factors in the development of systems and devices of all kinds.

Illuminating Engineering Society of North America
120 Wall St, 17th floor
New York, NY 10005-4001
http://www.iesna.org

Professional society whose members include engineers, architects, designers, educators, students, contractors, distributors, utility personnel, scientists, and manufacturers dealing with the art or science of illumination. Provides assistance with technical problems, reference help, speakers, and training aids. Maintains liaison with schools and colleges.

Institute of Electrical and Electronics Engineers
345 E 47th St
New York, NY 10017

Engineers and scientists in electrical engineering, electronics, and allied fields. Holds numerous meetings and special technical conferences. Conducts lecture courses at the local level on topics of current engineering and scientific interest. Assists student groups.

Institute of Environmental Sciences
940 E Northwest Hwy
Mount Prospect, IL 60056
http://www.intenvsci.org

Engineers, scientists, and management people engaged in the simulation of the natural environment and the environments induced by equipment operation and the testing of workers, materials, and equipment in the simulated environments.

Medical Review Officer Certification Council
9550 W. Lawrence Ave, Ste 106A
Schiller Park, IL 60176

Society of Automotive Engineers International
400 Commonwealth Dr
Warrendale, PA 15096-0001
http://www.sae.org
Professional society of engineers in field or self-propelled ground, flight, and space vehicles; engineering students are enrolled in special affiliation. To promote the arts, sciences, standards, and engineering practices related to the design, construction, and use of self-propelled mechanisms, prime movers, components thereof, and related equipment.

Society of Manufacturing Engineers
P O Box 930
One SME Drive
Dearborn, MI 48121-0930
http://www.sme.org
Professional society of manufacturing engineers and management executives concerned with manufacturing techniques. To advance the science of manufacturing through the continuing education of manufacturing engineers and management.

Society of Toxicology
1767 Business Center Dr
Reston, VA 20190-5332
http://www.toxicology.org
Persons who have conducted and published original investigations in some phase of toxicology and who have a continuing professional interest in this field.

System Safety Society
P.O. Box 70
Unionville, VA 22567-0070
The System Safety Society, Inc., is a nonprofit organization dedicated to the advancement of system safety principles, techniques, and methods in all technical endeavors where potential of injury, damage, or loss is present. Additional information on the society and membership requirements can be obtained by writing to the society.

Trade Associations

Another group of associations come under the broad heading of trade associations They are concerned with furthering the aims of their field of productive enterprise, including health preservation of employees and the public. These associations have trained personnel, cooperating committees, and publications that can be extremely helpful.

Alliance of American Insurers
1501 Woodfield Rd, Ste 400 W
Schaumburg, IL 60173-4980
An organization of leading mutual property and casualty insurance companies that promotes loss prevention principles amongst its members as well as disseminates safety information.

American Foundrymen's Society
505 State St
Des Plaines, IL 60016-8399
http://www.afsinc.org
Technical society of foundrymen, patternmakers, technologists, and educators. Maintains a Technical Information Center that provides literature-searching and document-retrieval service.

American Insurance Association
1130 Connecticut Ave, NW, Ste 1000
Washington DC 20036
Represents companies providing property and liability insurance. Seeks to promote the economic, legislative, and public standing of its participating companies through a broad spectrum of activities.

American Iron and Steel Institute
1101 17th St, NW
Washington DC 20036-4700
http://www.steel.org
Basic manufacturers and individuals in the steel industry. Conducts extensive research programs and workshops.

American Petroleum Institute
1220 L St, NW
Washington DC 20005
Producers, refiners, marketers, and transporters of petroleum and allied products. Seeks to maintain cooperation between government and industry, fosters foreign and domestic trade in American petroleum products, promotes the interests of the

petroleum industry, and provides extensive publication and information services.

AMT—Association for Manufacturing Technology
7901 Westpark Dr
McLean, VA 22102
http://www.mfgtech.org

Manufacturers of power-driven machines that are used to shape or form metal. Seeks to improve methods of producing and marketing machine tools; promotes research and development in the industry.

Compressed Gas Association
1725 Jefferson Davis Hwy, Ste 1004
Arlington, VA 22202-4100
http://www.cganet.org

Firms producing and distributing compressed, liquefied, and cryogenic gases; also manufacturers of related equipment. Submits recommendations to appropriate government agencies to improve safety standards and methods of handling, transporting, and storing gases; acts as advisor to regulatory authorities and other agencies concerned with safe handling of compressed gases; collaborates with national organizations to develop specifications and standards of safety.

Lead Industries Association
295 Madison Ave
New York, NY 10017
http://www.leadinfo.com

An association of mining companies, smelters, and refiners; and manufacturers of products of which lead is a component. Provides technical service and information to consumers; gathers statistical information.

Pulp Chemicals Association
P O Box 105113
Atlanta, GA 30348

Executives, managers, engineers, research scientists, superintendents, and technologists in the pulp, paper, and allied industries. Conducts conferences and develops testing procedures for laboratory analyses and process control.

Soap and Detergent Association
475 Park Ave
New York, NY 10016

Manufacturers of soap, synthetic detergents, fatty acids, and glycerine, and raw materials suppliers. Activities include cleanliness promotion, consumer information, environmental and human safety research, and government liaison.

Scientific and Service Organizations

These organizations have a significant interest in occupational health and safety.

American National Standards Institute
11 W 42nd St
New York, NY 10036
http://www.ansi.org

This federation of industrial, trade, technical, labor, and professional organizations; government agencies; and consumer groups coordinates development of standards in multiple subject areas, and oversees their publication.

International Atomic Energy Agency
Vienna International Center
Wagramstrasse 5
Postfach 100
A-1400 Vienna, Austria

Established in 1956 as an autonomous intergovernmental organization under the aegis of the United Nations for the purpose of accelerating and enlarging the contribution of atomic energy to peace, health, and prosperity. Required to ensure that assistance provided by it or under its suggestion, supervision, or control is not used to further any military purpose. Provides technical assistance and advice on developments in the use of nuclear power in electricity generating, and the use of radiation and radioisotopes in medicine, agriculture, and industry; on health and safety aspects of its use; and on the management of radioactive wastes.

International Radiation Protection Association
Postbus 662
NL-5600 AR Eindhoven, Netherlands

Organization of individuals and nationally affiliated societies. Provides a medium for contacts and cooperation among scientists engaged in radiation protection work; encourages establishment of radiation protection societies throughout the world; encourages establishment of standards and recommendations; and promotes research and education.

National Council on Radiation Protection and Measurements
7910 Woodmont Ave, Ste 800
Bethesda, MD 20814
http://www.ncrp.com
Congressionally chartered, nonprofit organization. Collects, analyzes, and disseminates information about radiation protection and measurements. Members are nationally recognized scientists who volunteer their services to the Council's program.

National Fire Protection Association
Batterymarch Park
P O Box 9101
Quincy, MA 02269-9101
http://www.nfpa.org
Membership drawn from fire service centers, business and industry, health care, educational and other institutions, insurance companies, government at all levels, architects and engineers, and others. Serves as a clearinghouse of information; compiles annual statistics on causes and occupancies of all fires.

National Safety Council
1121 Spring Lake Drive
Itasca, IL 60143-3201
http://www.nsc.org
Independent, nonprofit organization with the goal of reducing the number and severity of all kinds of accidents and industrial illnesses by collecting and distributing information about the causes of accidents and illnesses and ways to prevent them. Gathers and analyzes statistics, performs research in various areas of accident prevention and safety and health program effectiveness, sponsors special-interest conferences and committees, and provides research consultant services.

Underwriters Laboratories Inc.
333 Pfingsten Rd
Northbrook, IL 60062
http://www.ul.com
Independent, nonprofit organization for public safety testing. Operates laboratories for examination and testing of devices, systems, and materials. Product services include listing, classification, recognition, certification, and inspection. Fact-finding and research services are also conducted on a contract basis for manufacturers whose products meet UL safety requirements.

Energy Information Service
Emergency response system that can help.

CHEMTREC
Emergency information about hazardous chemicals involved in transportation accidents can now be obtained 24 hours a day. It is the Chemical Transportation Emergency Center (CHEMTREC), and it can be reached by a nationwide telephone number—800/424-9300. The Area Code 800 WATS line permits the caller to dial the station-to-station number without charge. CHEMTREC will provide the caller with response/action information for the product or products and tell what to do in case of spills, leaks, fires, and exposures. This informs the caller of the hazards, if any, and provides sufficient information to take immediate first steps in controlling the emergency. CHEMTREC is strictly an emergency operation provided for fire, police, and other emergency services. It is not a source of general chemical information of a nonemergency nature.

Other Organizations
Information concerning hazardous materials can also be acquired from such organizations as:

American Association of Poison Control Centers
c/o Dr. Ted Tong
Arizona Poison and Drug Information Center
Health and Sciences, Rm 3204K
1901 N Campbell
Tucson, AZ 85725
This association keeps up-to-date information on the ingredients and potential acute toxicity of substances that may cause accidental poisoning and of the proper management of such poisonings. It has established standards for poison information and control centers. Publishes a listing on Poison Centers that is updated annually.

National Response Center
2100 Second St, SW
Washington DC 20593

U.S. Coast Guard
2100 Second St, SW
Washington DC 20593

Chemical Manufacturers Association
1300 Wilson Blvd
Arlington, VA 22209

GOVERNMENTAL ORGANIZATIONS

There is an overwhelming amount of information available from the federal government that concerns all aspects of occupational safety and health, environmental problems, pollution, statistical data, and other industry problems.

Federal Agencies

Because of constant changes in government agency activities and frequent reorganizations within the government, it is recommended that the reader consult the latest issue of the United States Government Organization Manual, published annually by the U.S. Government Printing Office, Washington DC 20402. This paperbound book also can be found in most libraries.

Bureau of Mines
Department of the Interior
810 Seventh St, NW
Washington DC 20241

The Bureau of Mines is primarily a research and fact-finding agency. Its goal is to stimulate private industry to produce a substantial share of the nation's mineral needs in ways that best protect the public interest. Applied and basic research are conducted to develop the technology for the extraction, processing, use, and recycling of the nation's mineral resources at a reasonable cost without harm to the environment or the workers involved. Typical areas of research are mine health and safety, recycling of solid wastes, improvement of coal production technology, abatement of pollution and land damage caused by mineral extraction and processing operations, and development of ways to use domestic low-grade ores as alternative sources of critical minerals that must currently be imported.

Center for Devices and Radiological Health
Food and Drug Administration
9200 Corporate Blvd, # 100
Rockville, MD 20850

The center (1) carries out programs designed to reduce the exposure of man to hazardous ionizing and nonionizing radiation, (2) develops standards for safe limits of radiation exposure, (3) develops methodology for controlling radiation exposures, (4) conducts research on the health effects of radiation exposure, and (5) conducts an electronic product radiation control program to protect public health and safety, including the development and administration of performance standards to control the emission of radiation from electronic products and the undertaking by public and private organizations of research and investigation into the effects and control of such radiation emissions.

Consumer Product Safety Commission
East West Towers
4330 East West Hwy
Bethesda, MD 20814
http://www.cpsc.gov

The purpose of the Consumer Product Safety Commission is to protect the public against unreasonable risks of injury from consumer products, to assist consumers to evaluate the comparative safety of consumer products, to develop uniform safety standards for consumer products and minimize conflicting state and local regulations, and to promote research and investigation into the causes and prevention of product-related deaths, illnesses, and injuries.

Department of Labor
200 Constitution Ave, NW
Washington DC 20210
http://www.dol.gov

The purpose of the Department of Labor is to foster, promote, and develop the welfare of the wage earners of the United States, to improve their working conditions, and to advance their opportunities for profitable employment. In carrying out this mission, the department administers more than 130 federal labor laws guaranteeing workers' rights to safe and healthful working conditions, including OSHA, a minimum hourly wage and overtime pay scale, freedom from employment discrimination, unemployment insurance, and workers' compensation. The department also protects workers' pension rights; sponsors job training programs; helps workers find jobs; works to strengthen free collective bargaining; and keeps track of changes in employment, prices, and other national economic measurements. As the department seeks to assist all Americans who need and want to work, special efforts are made to meet the unique job market problems of older workers, youths, minority group members, women, the handicapped, and other groups.

Environmental Protection Agency
401 M St, SW
Washington DC 20460
http://www.epa.gov

The purpose of the Environmental Protection Agency (EPA) is to protect and enhance our environment today and for future generations to the fullest extent possible under the laws enacted by Congress. The EPA's mission is to control and abate pollution in the areas of water, air, solid waste, pesticides, noise, and radiation. EPA's mandate is to mount an integrated, coordinated attack on environmental pollution in cooperation with state and local governments.

Mine Safety and Health Administration
Department of Labor
4015 Wilson Blvd, Rm 601
Arlington, VA 22203
The Mine Safety and Health Administration (MSHA) conducts programs to control health hazards and reduce fatalities and injuries in the mineral industries through inspection, investigation, and enforcement; assessment of penalties for violations; technical support; and education, training, and safety motivation. Mandatory health and safety standards and regulations are developed or revised as warranted by new technology or by changing conditions. MSHA replaces the Mining Enforcement and Safety Administration.

National Center for Toxicological Research
3900 NCTR Rd
Jefferson, AR 72079
http://www.fda.gov/nctr/index.html
The National Center for Toxicological Research conducts research programs to study the biological effects of potentially toxic chemical substances found in the environment, emphasizing the determination of the health effects resulting from long-term low-level exposure to chemical toxicants and the basic biological processes for chemical toxicants in animal organisms, and the development of improved methodologies and test protocols for evaluating the safety of chemical toxicants and the data that will facilitate the extrapolation of toxicological data from laboratory animals to humans.

National Institute of Standards and Technology
Rte I-270 and Quince Orchard Rd
Gaithersburg, MD 20899
http://www.nist.gov
The National Institute of Standards and Technology (NIST) provides the basis for the nation's measurement standards. These standards are the means through which people and nations buy and sell goods, develop products, judge the quality of their environment, and provide guidelines for the protection of health and safety. The bureau's overall goal is to strengthen and advance the nation's science and technology and facilitate their effective application for public benefit. NIST is involved in over 1,500 projects aimed at dealing with such national concerns as energy conservation and research, fire protection and prevention, and consumer product safety.

National Technical Information Service
Department of Commerce
5285 Port Royal Rd
Springfield, VA 22161
http://www.ntis.gov
NTIS was established to simplify and improve public access to Department of Commerce publications and to data files and scientific and technical reports sponsored by federal agencies. NTIS is the central point in the United States for the public sale of government-funded research and development reports and other analyses prepared by federal agencies, their contractors, or grantees.

Nuclear Regulatory Commission
Washington DC 20555-0001
http://www.nrc.gov
The Nuclear Regulatory Commission (NRC) licenses and regulates the uses of nuclear energy to protect the public health and safety and the environment. It does this by licensing persons and companies to build and operate nuclear reactors and to own and use nuclear materials. The NRC makes rules and sets standards for these types of licenses. The NRC also carefully inspects the activities of the persons and companies licensed to make sure that they do not violate the safety rules of the Commission.

Occupational Safety and Health Administration
U.S. Department of Labor
200 Constitution Ave, NW
Washington DC 20210
http://www.osha.gov
The Assistant Secretary of Labor for Occupational Safety and Health has responsibility for occupational safety and health activities.

The Occupational Safety and Health Administration, established pursuant to the Occupational Safety and Health Act of 1970 (84 Stat. 1590), develops and promulgates occupational safety and health standards; develops and issues regulations;

conducts investigations and inspections to determine the status of compliance with safety and health standards and regulations; and issues citations and proposes penalties for noncompliance with safety and health standards and regulations.

Occupational Safety and Health Review Commission
1120 Twentieth St, NW
Washington DC 20036
http://www.oshrc.gov

The Occupational Safety and Health Review Commission (OSHRC) is concerned with providing safe and healthful working conditions for both the employer and the employee. It adjudicates cases forwarded to it by the Department of Labor when disagreements arise over the results of safety and health inspections performed by the department.

Office of Energy Research
Department of Energy
1000 Independence Ave, SW
Washington DC 20585

The Office of Energy Research advises the Secretary of Energy on the physical and energy research and development programs of the department. It is responsible for conducting basic research in (1) basic energy (physical) sciences; (2) high energy and nuclear physics; (3) biological and environmental research; and (4) magnetic fusion energy.

Office of Hazardous Materials Safety
Department of Transportation
400 Seventh St, NW
Washington DC 20590

The Office of Hazardous Materials Transportation develops and issues regulations for the safe transportation of hazardous materials by all modes, excluding bulk transportation by water. The regulations cover shipping and carrier operations, packaging and container specifications, and hazardous materials definitions. The office is also responsible for the enforcement of regulations other than those applicable to a single mode of transportation. It reviews and analyzes reports made by the industry and by field staff bearing upon compliance with the regulations, and conducts training and education programs. The office is the national focal point for coordination and control of the departments multimodal hazardous materials regulatory programs, ensuring uniformity of approach and action by all modal administrations.

U.S. Fire Administration
Federal Emergency Management Agency
500 C Street, SW
Washington DC 20472

The primary mission of the U.S. Fire Administration is to reduce the loss of life and property through better fire prevention and control with a program coordinated to support and reinforce the fire prevention and control activities of state and local governments.

State Agencies

As mentioned previously, many states have excellent health and safety departments; these are a fertile source of help.

Some state laboratories are extremely well equipped and have numerous devices for sampling and analysis. Not only are state services helpful in solving day-to-day problems, but they can also assist in unusual or complex problems. Contact your state agencies for more information on these services.

Other Agencies
Centers for Disease Control
Department of Health and Human Resources
Public Health Service
1600 Clifton Rd, NE
Atlanta, GA 30333
http://www.cdc.gov

Health Resources and Service Administration
Dept of Health & Human Services
5600 Fishers Ln
Rockville, MD 20857
http://www.dhhs.gov/hrsa

Library of Congress
Science and Technology Division
Washington DC 20540
http://www.loc.gov

National Aeronautical and Space Administration
600 Independence Ave, SW
Washington DC 20546

National Cancer Institute
National Institutes of Health
9000 Rockville Pike
Bethesda, MD 20892
http://www.nci.nih.gov

National Center for Health Statistics
Public Health Service
6525 Belcrest Rd
Hyattsville, MD 20782
http://www.cdc.gov/nehswww.test.htm

National Council on Radiation Protection and Measurement (NCRP)
7910 Woodmont Ave, #800
Bethesda, MD 20814
http://www.ncrp.com

National Heart, Lung, and Blood Institute
National Institutes of Health
7550 Wisconsin Ave
Bethesda, MD 20892

National Institute of Environmental Health Sciences (NIEHS)
P O Box 12233
Research Triangle Park, NC 27709
http://www.niehs.nih.gov

Substance Abuse and Mental Health Services Administration
Department of Health and Human Services
5600 Fishers Ln
Rockville, MD 20857
http://www.samhsa.gov

National Institute for Occupational Safety and Health (NIOSH)
1600 Clifton Rd, NE
Atlanta, GA 30333
—also located at
4676 Columbia Pkwy
Cincinnati, OH 45226,
and
944 Chestnut Ridge Rd
Morgantown, WV 26505
http://www.cdc.gov/niosh

National Library of Medicine
8600 Rockville Pike
Bethesda, MD 20894
http://www.nlm.nih.gov
Note: See details under MEDLARS On-Line Network in the next section.

U.S. Government Printing Office
Washington DC 20402

DATABASES
General

Books in Print. Information source on U.S. book publishing; from R.R. Bowker Co., 121 Chanlon Rd, New Providence, NJ 07974.

International Dissertation Abstracts. Doctoral dissertations from accredited universities (predominantly in the U.S.); from University Microfilm International, 300 North Zeeb Rd, Ann Arbor, MI 48106.

GPO Monthly Catalog. The machine equivalent of the printed Monthly Catalog of U.S. Government Publications, from U.S. Government Printing Office, 732 Capitol St, NW, Washington DC 20401.

National Newspaper Index. An Index to The Christian Science Monitor, The New York Times, and The Wall Street Journal; from Information Access Corporation, 362 Lakeside Dr, Foster City, CA 94404.

National Technical Information Service. Citations to government-sponsored research, mostly federal, but also contains some state and local reports; from National Technical Information Service, 5285 Port Royal Rd, Springfield, VA 22161.

TSCA Initial Inventory. A nonbibliographic dictionary listing of chemical substances in commercial use in the United States as of June 1, 1979. It is not a list of toxic chemicals. From Dialog Inc., 3460 Hillview Ave, Palo Alto, CA 94304.

Ulrich's International Periodicals and Irregular Serials. Information on periodicals and serials; from R.R. Bowker Co., 121 Chanlon Rd, New Providence, NJ 07974.

MEDLARS On-Line Network

Health and safety professionals will find the databases in the MEDLARS on-line network most useful. MEDLARS is an on-line network of approximately 20 bibliographic databases covering worldwide literature in the health sciences. References may be retrieved by searching one or a combination of the 14,000 designated Medical Subject Headings used by NLM in indexing and cataloging materials. Requestors may obtain a complete record for each reference retrieved—including subject headings and abstracts—or a less detailed format including only the elements necessary to locate the item: author, title, and publication source.

Records: MEDLARS contains more than 20 databases covering in excess of six million documents published after 1965. The system is updated continuously.

Individuals can obtain direct access to MEDLARS via the Internet at http://www.nlm.nih.gov/.

AVLINE. AVLINE (Audio Visuals on-Line) is one of the databases maintained by the National Library of Medicine.

Subject:	Educational aids in the health sciences.
Sources:	From audiovisuals and computer software acquired from producers from all over the United States.
Content:	The database contains citations to over 31,000 audiovisual teaching packages. Subject coverage includes: dentistry, nursing, allied health, and other disciplines. Descriptive review information regarding review data is included in some cases. This information includes: rating, audience levels, instructional design, specialties, and abstracts. Procurement information on titles is provided. Most items have been produced within the last ten years.
Producer:	National Library of Medicine MEDLARS Management Section 8600 Rockville Pike Bethesda, MD 20894 301/496-6193

CANCERLIT. CANCERLIT (Cancer Literature) contains more than 785,000 citations, with abstracts, to the worldwide literature on oncological epidemiology, pathology, treatment, and research. Before 1976, the database corresponded to *Cancer Therapy Abstracts* from 1967, and *Carcinogenesis Abstracts* from 1963.

Subject:	Biomedicine.
Sources:	NIH's National Cancer Institute (NCI), more than 300,000 U.S. and foreign journals, books, reports, and meeting abstracts.
Content:	The database contains English abstract references dealing with various aspects of cancer. The database abstracts over 3,500 U.S. and foreign journals, as well as books, reports, and meeting abstracts.
Date of information:	1963 to present.
Producer:	U.S. National Institutes of Health National Cancer Institute International Cancer Research Data Bank Program 9000 Rockville Pike, Bldg 82 Bethesda, MD 20894 301/496-7403

CANCERNET (International Database on Oncology)

Type:	Reference (Bibliographic).
Subject:	Biomedicine.
Producer:	CANCERNET/Centre National de la Recherche Scientific (CNRS).
On-line service:	University of Tsukuba.
Content:	Contains citations, with abstracts, to literature on oncology. Covers clinical and experimental carcinology, epidemiology, public health, and fundamental sciences (e.g., immunology, virology). *Corresponds to Cancer/Oncology, Bulletin Signaletique 251.*
Language:	English and French.
Coverage:	International.
Time span:	1968 to 1984; abstracts from some articles, from 1981 to 1984.
Updating:	Not updated.

MEDLINE. MEDLINE (Medical Literature Analysis and Retrieval System On-Line) contains references to citations from 3,000 biomedical journals. It is designed to help health professionals find out easily and quickly what has been published recently on any specific biomedical subject.

Subject:	References to biomedical literature.
Sources:	Indexed articles from over 3,600 international journals. Forty percent of the records added since 1975 include author abstracts taken from published articles. Each year over 250,000 records are added.
Contents:	Approximately 600,000 references to biomedical journal articles published in the present year and two years before. MEDLINE corresponds to three printed indexes: Index Medicus, Index to Dental Literature, and International Nursing Index. Indexes using NLM's (National Library of Medicine)

controlled vocabulary MeSH (Medical Subject Headings). Can be used to update a search periodically. Information from 1966 to present totals more than 3.8 million references.

Producer: Office of Inquiries and Publications Management
National Library of Medicine
8600 Rockville Pike
Bethesda, MD 20894

RTECS. RTECS (Registry of Toxic Effects of Chemical Substances) is an on-line, interacting version of the National Institute of Occupational Safety and Health (NIOSH) publication, *Registry of Toxic Effects of Chemical Substances*. It contains basic acute and chronic toxicity data for more than 106,000 potentially toxic chemical identifiers, exposure standards, and status under various federal regulations and programs. The file can be searched by chemical identifiers, type of effect, or other criteria.

Subject: Toxicology.
Sources: Based on the National Institute of Occupational Safety and Health (NIOSH) publication.
Content: The direct search on this database allows the user to display, for a given list of compounds, the CAS Registry Number, and the details of each published toxicity measurement for each compound, including literature references. The user may ask for all entries relating to specific end effects on specific classes of animals for specific means of application, having dosage within a given range. The result of a query such as this is a list of Registry Numbers which can then be used to display RTECS data or to obtain information from other modules of the CIS.
Producer: U.S. National Institutes of Health
National Institute for Occupational Safety and Health
1600 Clifton Rd
Atlanta, GA 30333
404/329-3771

TOXNET (Toxicology Data Network) is composed of approximately 4,000 comprehensive, peer-reviewed chemical records. Compounds selected for the network include highly regulated chemicals, high-volume production/exposure chemicals, and drugs and pesticides exhibiting high toxicity potential.

Subject: Toxicology.
Sources: Approximately 4,000 chemical records.
Content: This database contains toxicological, pharmacological, environmental, occupational, manufacturing, and use information as well as chemical and physical properties.
Producer: National Library of Medicine
Toxicology Information Program
8600 Rockville Pike
Bethesda, MD 20894
301/496-6193
Available on: National Library of Medicine.

INTERESTING INTERNATIONAL OCCUPATIONAL SAFETY AND HEALTH INTERNET SITES

Government Agencies

Australia

http://www.wt.com.au/safetyline

Worksafe Western Australia. This occupational safety and health site contains over 8,000 pages of information. One of the features of Safetyline is the interactive self-test packages, whereby the user can become a "Graduate in Occupational Safety and Health Laws." It offers a minimal amount of material in other languages. It offers an on-line inquiry service and its publications are available for downloading including its quarterly magazine.

Worksafe Australia, National Occupational Health & Safety Commission

http://www.worksafe.gov.au/worksafe/home.htm

Provides information on standards, chemicals, small businesses, and managing workplace hazards.

Canada

http://www.ccohs.ca/

Canadian Centre for Occupational Health and Safety (CCOHS). These web pages provide useful information about the organization and its services. It has lists of other related health and safety sites. Most of its material is in both English and French. CCOHS offers an on-line inquiry service as well as a substantial amount of prepackaged information.

Finland
http://www.occuphealth.fi/
Finland Institute of Occupational Health. Informative site, which links to many other worldwide sites. It charges for its publications and offers information in both in English and Finnish. Does not offer an on-line inquiry service; however, it has a unique "Computer Aided Advice in Occupational Health—Selection of personal protector" (select the most appropriate personal protective equipment based on a description of your current work environment). Includes extensive listing of safety and health databases maintained by EU members' countries. This is also a cross reference from the European Agency for Safety and Health at Work site.

United Kingdom
http://www.open.gov.uk/hse/hsehome.htm
Health and Safety Executive. Informative site that offers an on-line inquiry service and publications that can be downloaded. Has an excellent HSE statistics site and links to worldwide sources. Listing of electronic products includes a multimeida training course on welding fumes and consultative expert system software on dust explosions.

International Organizations
European Committee for Standardization
httu://www.cenorm.be
Covers all aspects of EU standards making. Does not offer a search engine and offers information in English only.

International Labour Office (ILO) Health and Safety Centre
http://www.ilo.org/public/english/90travai/cis
This site lists CIS products and services, including addresses of CIS collaborating centres worldwide. Extensive links to other occupational safety and health sites. In particular, links are provided to the CIS national centres, the Asian Pacific Information network, and to the Global Information Network on Chemical Safety (GINC). Offers information in English, French, and Spanish. Has added "Datasheets on Occupations" (so far only firefighters, roofers [non-metal], and divers [indigenous fishermen]) to its site. Note, in particular, profiles on the Global Programme on Occupational Safety and Health being established by ILO) and a front page link to the XVth World Congress on Occupational Safety and Health.

International Agency for Research on Cancer (IARC)
http://www.iarc.fr
Not exclusively an occupational safety and health site. It charges for its printed publications and does not offer an online inquiry service. Free access to their 3 databases—P53 Mutation Database (over 9000 somatic cell mutations and tumors), Epidemiology Database, and Monograph Database. Some information in French.

Institute for Systems, Informatics and Safety (Joint Research Center—European Commission)
http://ulisse.etoit.eudra.org/Ecdin.html
Provides access to their Environmental Chemicals Data Information Network (ECDIN) and the European Community Pharmaceutical Information Network.

European Agency for Safety and Health at Work
http://www.eu-osha.es
Is exclusively an occupational safety and health site and provides information and links to other European Member States web sites. Does not offer a search engine and does not offer an inquiry service. English only. Provides links to legislation in force in the EU. Many areas still under development.

International Atomic Energy Agency
http://www.iaea.org
Links to worldwide sources and databases. However, it is not an exclusive occupational safety and health site but does offer a search engine.

International Organization for Standards (ISO)
http://www.iso.ch/
Is not exclusively an occupational safety and health site; however, it useful for checking latest standards information. Gives details about members, how standards are made, and the catalogue of ISO standards. Password required to use the search engine and offers information in both French and English.

United Nations Environment Programme
http://www.unep.org
Is not exclusively an occupational safety and health site; however it provides access to other UN environment sites and links to a wide range of other sites around the world. Offers a search engine and an extensive calendar of environmental-related events.

World Health Organization (WHO)
http://www.who.int/

This web site provides information on the organization's program, reports, statistical information, publications, and health advice; however it is not exclusively an occupational safety and health site. Offers a search engine and an on-line inquiry service. Has an excellent "Cool Links Just for Kids" site (click on Links) and a link to the "Intergovernmental Forum on Chemical Safety." Some information in French.

British Occupational Hygiene Society (BOHS)
http://www.bohs.org

BOHS is a multidisciplinary learned society, which has members in over 30 countries. Charges for its publications and databases. No inquiry service or search engine. English only and events calendar limited to the United Kingdom. Also has very good links to other occupational safety and health sites.

International Occupational Hygiene Association
http://www.bohs.org/ioha/index.html

This organization conducts a wide range of activities to promote and develop occupational hygiene worldwide. From its creation in 1987 the IOHA has grown to more than 20 member organizations, representing over 20,000 occupational hygienists worldwide. Promotes the exchange of information and offers many of its publications in an HTML format for downloading. English only.

OTHER

Christie Communications (Business)
http://www.mrg.ab.ca/christie/safelist.htm

Presents a very wide range of occupational safety and health services, information, training and products. Offers an extensive number of link sites as it has more than 2700 entries on their "Safety Related Internet Resources" site.

SilverPlatter Information Ltd. (Business)
http://www.silverplatter.com/oshinfo.htm

The "Occupational Health and Safety World" contains editorials, diary of events, etc. Does not offer a search engine; however, it does offer an alphabetical subject index and a country index. Has occupational safety and health news from other countries.

OSHWEB (Tampere University of Technology, Finland)
http://www-iea.me.tut.fi/cgi-bin/oshweb.pl

An outstanding site. Has an extensive collection of links to material safety data sheets, conferences, government agencies, international organizations, professional organizations (including trade unions), radiation safety, universities, government safety and health agencies world wide, and other resource lists.

BASIC REFERENCE BOOKS

The basic reference books listed in this section supplement the specific references appended at the end of the other chapters in this manual. The reference material cited in this bibliography was selected to provide health and safety professionals with sources of information that are likely to prove most useful in coping with problems of worker health protection and hazard assessment. This compilation is not to be viewed as a comprehensive coverage of the abundant literature on this subject, nor is any endorsement implied. The reference books are listed according to the following outline:

A. General principles
B. Risk assessment
C. Sampling methods
D. Toxicology
E. Medical
F. Dermatitis
G. Physical stresses
H. Ergonomics
I. Biological
J. Chemical
K. Pollution and hazardous waste
L. Control
M. Encyclopedias and handbooks

The intent of this section is to provide occupational health and safety professionals with brief descriptions of the basic reference books and publications in the field of occupational health and safety. A handy compendium of occupational health and safety references can also be obtained from the Manager of the NSC Press at the National Safety Council—e-mail to schonfej@nsc.org (phone: 800/621-7615, ext 2374). Out-of-print books can usually be obtained from a library or by interlibrary exchange.

A. General Principles

Accident Prevention Manual for Business & Industry, 11th ed. vol 1: *Administration & Programs;* vol 2: *Engineering & Technology* (©1996); vol 3: *Environmental Management* (2nd ed.—©2000), and vol 4: *Security Management* (©1997). Itasca: National Safety Council.

The *Administration & Programs* volume contains information on hazard control, organizing a program, governmental regulations and workers' compensation, computers and information management systems, training, motivation, maintaining interest, publicizing safety, environmental concepts, emergency planning, personal protective equipment, industrial sanitation, handicapped workers, and product safety.

The *Engineering & Technology* volume is more nuts and bolts oriented, covering heavy and light construction (including tools and machines), materials handling of all types, safeguarding of both woodworking and metalworking machinery, automated processes, welding and cutting, electrical hazards, flammable liquids hazards, and fire protection.

Adams EE. *Total Quality Safety Management.* Des Plaines, IL: ASSE, 1995.

American Conference of Governmental Industrial Hygienists. Industrial Ventilation: A Manual of Recommended Practice. Cincinnati: ACGIH, 1998.

Annino R, Driver R. *Scientific and Engineering Applications with Personal Computers.* New York: Wiley, 1986.

How to use the personal computer in the laboratory; both hardware and software are covered, as are a number of examples for the Apple, IBM-PC, and CPM. Chapters are devoted to programming, storage and program management, computer graphics, numerical analysis and modeling, and analysis of experimental data.

Baetjer AM. *Women in Industry: Their Health and Efficiency.* (Reprint of 1961 edition). Salem, MA: Ayer, 1977.

Boissenvain AL, Henderson RE, Claybaugh DJ. *Corporate Health and Safety: Managing Environmental Issues in the Workplace.* Southampton PA: Ergonomics Inc, 1996.

Box GER, Hunter WG, Hunter JS. *Statistics for Experimenters.* New York: Wiley, 1978.

As an introduction to design, data analysis, and model building, this text focuses on applications in physical, engineering, biological, and social sciences. It shows step-by-step how to set up experimental programs of high statistical and engineering efficiency. Useful to experimenters in all fields.

Burgess WA. *Recognition of Health Hazards in Industry,* 2nd ed. New York: Wiley, 1995.

Discusses how industrial operations can affect worker health. An example of a plant survey is given, along with the information needed for evaluating common unit operations. Materials and equipment are described, the physical form and origin of air contaminants are identified, as well as physical stresses encountered in the process or industry.

Campbell RL, Langford RE. *Substance Abuse in the Workplace.* Boca Raton, FL: CRC Press, 1995.

Cheremisinoff NP, Graffia M. *Safety Management Practices for Hazardous Materials.* New York: Dekker, 1996.

Clayton GD, Clayton FE (eds). *Patty's Industrial Hygiene and Toxicology,* 4th ed; vol 1: *General Principles.* New York: Wiley, 1991.

This authoritative handbook and reference provides a complete guide to methods of evaluation, record keeping, and control, as well as historical background. Written more for the industrial hygienist.

Cohen B (ed). *Human Aspects in Office Automation.* New York: Elsevier, 1984.

This book illustrates the relationship of employee health and job satisfaction to the creation and maintenance of proper working conditions.

Cralley LJ, Cralley LV (eds). *Patty's Industrial Hygiene and Toxicology;* vol 3: *Theory and Rationale of Industrial Hygiene Practice,* 2nd ed. (in two parts: A and B). New York: Wiley 1985.

Part 3A covers health promotion in the workplace, health surveillance programs in industry, occupational exposure limits, pharmacokinetics, and the effects of unusual work schedules.

Part 3B discusses the biological responses of the body to chemical and environmental hazards likely found in industrial workplaces. Text tells how the safety of chemicals, biological agents, ionizing radiation, and noise levels is evaluated.

Dinardi SR. *The Occupational Environment: Its Evaluation and Control.* Fairfax, VA: AIHA, 1997.

Dobin D. *Microcomputer Applications in Occupational Health and Safety.* Chelsea, MI: Lewis, 1987.
Derived from a symposium sponsored by the ACGIH in March of 1986, scope of this work includes information systems and communications, integration of data, use of electronic spreadsheets, materials inventory and MSDS information, database management, toxic substances research, expenditure prediction, and health and safety computer-aided design.

Eller PM, Cassinelli ME. *NIOSH Manual of Analytical Methods,* 4th ed. Cincinnati: NIOSH, 1994.

Ferry TS. *Modern Accident Investigation and Analysis,* 2nd ed. New York: Wiley, 1988.

Firenze RJ. *The Process of Hazard Control.* Dubuque, IA: Kendall/Hunt, 1978.
This guide has been specifically prepared to be used by both instructor and student in courses and seminars in order to guide discussion, stimulate interest, and direct the study of occupational safety, occupational health, and industrial hazard control. It explores areas of engineering, management, occupational health, hazard analysis, and fire protection as they relate to effective hazard reduction. The term hazard control is used throughout to familiarize management more thoroughly with the full dimension of hazards occurring from failures in techniques, equipment, systems, and operations that are responsible for dollar and humanpower losses, repair or replacement of tools, equipment, litigation expenses, and the like.

Geller ES. *The Psychology of Safety: How to Improve Behaviors and Attitudes on the Job.* Radnor PA; Chilton, 1996.

Giustina D. *Safety and Environmental Management.* New York: Van Nostrand Reinhold, 1996.

Gunn A, Vesilind PA. *Environmental Ethics for Engineers.* Chelsea, MI: Lewis, 1986.
Environmental ethics apply to interactions of engineers with clients or employers, and to interactions between engineers and nature; that is, the effect of their work on the natural environment. Book discusses the issue of environmental ethics in-depth.

Johnson WG. MORT *Safety Assurance Systems.* New York: Dekker, 1980.
MORT (management oversight and risk tree) is a proven guide to the generic casual factors in an accident, a formal disciplined logic that integrates a wide variety of safety concepts systematically.

Kohn JP, Friend MA, Winterbergere CA. *Fundamentals of Occupational Safety and Health.* Rockville, MD: Government Institutes, 1996.

Krause TR. *Employee-Driven Systems for Safe Behavior: Integrating Behavioral and Statistical Methodologies.* New York: Van Nostrand Reinhold, 1995.

Levin L. *An Investigative Approach to Industrial Hygiene: Sleuth at Work.* New York: Van Nostrand, 1996.

Lowry GG, Lowry RC. *Handbook of Hazard Communication and OSHA Requirements.* Chelsea, MI: Lewis, 1986.
Explains how to meet OSHA's right-to-know requirements. Chapters include discussion of physical and health hazards characteristics, written hazard communication program, employee training, label design, consequences, and legal penalties.

Moran M. *Construction Safety Handbook: A Practical Guide to OSHA Compliance and Injury Prevention.* Rockville, MD: Government Institutes, 1996.

National Safety Council. *On-Site Emergency Response Planning Guide.* Itasca, IL: NSC Press, 1999.
Strategy and systems behind emergency planning. Develop your own emergency plan using the enclosed templates on the computer disk supplied with the book.

National Safety Council. *Out in Front: Effective Supervision in the Workplace.* Chicago: National Safety Council, 1990.

For all supervisors, this book covers the multi-ethnic workforce of the 90s in terms of effective communication and safe and healthful conditions.

National Safety Council. *Protecting Workers Lives,* 2nd ed. Chicago: National Safety Council, 1992.

Provides an overview of workplace occupational health and safety. This book is especially helpful to members of occupational safety and health committees.

National Safety Council. *Safety Culture and Effective Safety Management.* Itasca, IL: NSC Press, 2000.

Focuses on what an effective safety culture is, how it affects the overall safety management programs, and how to develop an effective safety culture.

National Safety Council. *Safety Through Design.* Itasca, IL: NSC Press, 1999.

Learn about building safety into the job right through the production phase.

National Safety Council. *Supervisors' Safety Manual,* 9th ed. Itasca, IL: National Safety Council, 1997.

Written especially for first-line supervisors. This book covers occupational health and safety as it applies to supervisory personnel.

O'Donnell MP, Ainsworth TH (eds). *Health Promotion in the Workplace.* Albany, NY: Delmar, 1984.

Book covers the past, current, and future status of the evolving health promotion movement. A major portion is devoted to in-depth analyses of program design, management, evaluation, facility design, and an assessment of general health.

Otway HJ. *Regulating Industrial Risks: Public, Experts & Media.* Stoneham, MA: Butterworth-Heinemann, 1986.

Regulations covering industrial plant licensing, operation, and work conditions define to a great extent the risks to which society is exposed. This book deals systematically with regulatory processes and the factors that influence them.

The book looks at the influence of national styles, organizational forces, international aspects, and problems of implementation, and examines the roles played by public policy groups, expert testimony and analysis, and the communications media. This book will interest all concerned with risks to the environment and to public health safety.

Parmeggiani L (ed). *Encyclopedia of Occupational Health and Safety,* 4th rev. ed. Washington DC: International Labour Office, 1998.

Reference work containing 900 articles prepared by 700 specialists in more than 70 countries and 10 international organizations. Covers all aspects of occupational safety and health. Emphasizes the safety precautions to be taken against the main hazards encountered in each branch of industry. Examples are quoted from international standards rather than national legislation. Articles are arranged alphabetically and each article includes bibliographic references. The second volume contains nine appendixes, a list of authors, and a comprehensive analytic index.

Perkins JL, Rose VE. *Case Studies in Industrial Hygiene.* New York: Wiley, 1987.

Plog B (ed). *Fundamentals of Industrial Hygiene,* 4th ed. Itasca, IL: National Safety Council, 1996.

Putnam LD, Langerman N. *OSHA Bloodborne Pathogens Exposure Control Plan.* Chelsea, MI: Lewis, 1992.

Revoir WH, Bien C. *Respiratory Protection Handbook.* Boca Raton, FL: Lewis, 1997.

Rothstein MA. *Occupational Safety and Health Law,* 4th ed. St. Paul, MN, West, 1998.

B. Risk Assessment

Andelman JB, Underhill DW (eds). *Health Effects from Hazardous Waste Sites.* Chelsea, MI: Lewis, 1987.

This information and data for evaluating health effects from hazardous waste sites are the result of efforts of specialists representing research centers, hospitals, and government agencies. Consultant, as well as corporate viewpoints are presented. The work evolved from the Fourth Annual Symposium on Environmental Epidemiology.

Conway RA (ed). *Environmental Risk Analysis for Chemicals.* Melbourne, FL: Van Nostrand Reinhold, 1982.

This book provides information for preparing an overall plan for basic testing. Step-by-step procedures are provided for determining the environmental risk of any chemical. The best methods for determining environmentally safe entry quantities of chemicals are presented.

Hallenbeck WH, Cunningham KM. *Quantitative Risk Assessment for Environmental and Occupational Health.* Chelsea, MI: Lewis, 1986.

Of value wherever chemicals are manufactured or used, this book treats a complicated subject in a straightforward and understandable way. The reader needs only a basic knowledge of toxicology, epidemiology, and statistics to understand and perform the calculations presented. Sophisticated computer programs are not required.

C. Sampling Methods

Hesketh HE. *Fine Particles in Gaseous Media.* Chelsea, MI: Lewis, 1986.

Both theory and practice on the behavior and control of particles, their collection and measurement, are presented. The understanding of particle behavior becomes increasingly important with the development of hazardous emissions regulations. A useful book for environmental, mechanical, chemical, and civil engineers as well as regulatory officials in source testing, research, and designing and using control equipment.

Hinds WC. *Aerosol Technology.* New York: Wiley, 1982.

This book gives a complete exposition of the basic principles of aerosol science, including particle motion, forces on particles, the interaction of particles with the suspending gas, other particles and electromagnetic radiation, and the application of these principles to aerosol measurement.

Intersociety Committee. *Methods of Air Sampling and Analysis,* 3rd ed. Chelsea, MI: Lewis, 1988.

Represents the methods adopted by The Committee for a Manual of Methods of Air Sampling and Analysis, according to its established procedures.

Jungreis E. *Spot Test Analysis.* New York: Wiley, 1985.

This book provides procedures for conducting quick tests for a wide range of investigations, including biochemical components, forensic substances, geochemical composition, air and water pollutants, soil chemistry, and food adulteration and composition.

Katz M. *Measurement of Air Pollutants: Guide to Selection of Methods.* Geneva, Switzerland: World Health Organization, 1969.

Leithe W. *The Analysis of Air Pollutants.* Philadelphia: Coronet, 1970.

Linch AL. *Evaluation of Ambient Air Quality by Personnel Monitoring,* 2nd ed. Boca Raton, FL: CRC, 1981.

Personnel monitoring is a term designating the determination of the inhaled dose of an airborne toxic material or of an air-mediated hazardous physical force by the continuous collection of samples in the breathing or auditory zone, or other appropriate exposed body area, over a finite period of exposure time. A personnel monitor is a self-powered device worn by the monitored individual to collect a representative sample for laboratory analysis, or to provide accumulated dose or instantaneous warning of immediately hazardous conditions by visible or auditory means while being worn.

D. Toxicology

AMA Handbook of Poisonous and Injurious Plants. Chicago: American Medical Association, 1985.

Arena JM. *Poisoning: Toxicology, Symptoms, Treatments,* 5th ed. Springfield, IL: Thomas, 1986.

Topics include general considerations of poisoning; insecticides, rodenticides, and herbicides; industrial hazards; occupational hazards; drugs; soaps and detergents; poisonous plants, insects, and fish; and miscellaneous compounds and topics, including radioactive isotope poisoning, rocket fuels, and welding hazards. Appendix of normal laboratory values used in the diagnosis and treatment of poisoning.

Caldwell J (ed). *Amphetamines and Related Stimulants,* vol 1. Boca Raton, FL: CRC, 1981.

The aim of this volume is to explore historical, chemical, biological, clinical, and sociological aspects of the amphetamines and related stimulants with reference both to legitimate medical use and to their abuse.

There is at the present time an enormous literature on the amphetamines and related stimulants, particularly in the area of neuropsychopharmacology, but it is extremely difficult to distill from this the information of relevance to the problem of the abuse of these compounds. The aim of this volume is to draw together those aspects of the chemical, biological, clinical, and sociological studies that have the maximum impact on the abuse problem.

Carson BL, Ellis HV III, McCann JL. *Toxicology and Biological Monitoring of Metals in Humans Including Feasibility and Need.* Chelsea, MI: Lewis, 1986.

Persons in many disciplines besides toxicology need information about toxic effects of substances related to exposure and how to monitor exposure. This book contains toxicological, exposure, and monitoring information about metals in a brief, uniform format.

Clayton GD, Clayton FE (eds). *Patty's Industrial Hygiene and Toxicology;* vols 2A, 2B, 2C, 2D: *Toxicology.* New York: Wiley, 1993, 1994.

Volume 2 of *Patty's Industrial Hygiene and Toxicology* is divided into three parts. Part A contains a historical summary of toxicology, followed by a thorough description of the handling of 40 metals. Part B encompasses the hundreds of newly released data appearing since the previous edition. Topics discussed include occupational carcinogenesis, the halogens, and the nonmetals. Part C gives a thorough treatment of glycols, alcohols, and ketones.

Cohen, GM. *Target Organ Toxicity.* Boca Raton, FL: CRC, 1986.

The volume provides essential information on the general principles of target organ toxicity. The general principles are then illustrated using specific examples of toxicity in different target organs and systems. Modification of DNA and repair in tumor induction, and specificity in tumor initiation are also explained. This book is of primary interest to toxicologists, pharmacologists, biochemists, and environmental toxicologists.

Dominguez GS (ed). *Guidebook: Toxic Substances Control Act.* Boca Raton, FL: CRC, 1977.

The book is not only a complete guide to the law, but it is also a source on how to prepare for and respond to the many newly instituted federal requirements. The book provides practical and clear advice on anticipated compliance approaches, as well as suggestions for early organizational preparation and planning.

Dunnom D. *Health Effects of Synthetic Silica Particulates.* Philadelphia: American Society for Testing and Materials, 1981.

Focused on health problems caused by synthetic silica dust, this book gives the picture of current understanding of the physiological effects of silica particulates.

Dutka BJ, Bitton G. *Toxicology Testing Using Microorganisms.* Boca Raton, FL: CRC, 1986.

A compendium of new and traditional technology for microbiological toxicity testing procedures. Procedures, apparatus, degree of reliability, advantages, and pitfalls of each technology are outlined, with references to original literature.

Flamm WG, Lorentzen RJ (eds). *Mechanisms and Toxicity of Chemical Carcinogens and Mutagens.* Princeton, NJ: Princeton Scientific, 1985.

This volume provides both a status report and an indication of future directions of research on the genetic mechanisms of carcinogenesis. The introduction reviews two etiological factors involved in cancer causation and shows, where possible, how various theories on the mechanism of carcinogenesis relate to what is known about the etiology of cancer. The experimental evidence supporting the somatic mutation theory of cancer induction is discussed.

Friberg L, Elinder C-G, et al. *Cadmium and Health: A Toxicological and Epidemiological Appraisal.* Boca Raton, FL: CRC, 1985.

A comprehensive review and critical evaluation of studies conducted in many nations, including important research in Japan. The most significant studies are discussed and analyzed at length.

Gleason MN, Gosselin RE, et al. *Clinical Toxicology of Commercial Products: Acute Poisoning,* 5th ed. Baltimore: Williams & Wilkins, 1984.

This book assists physicians in dealing quickly and effectively with acute chemical poisonings in the home and on the farm, arising through misuse of commercial products. It provides a list of trade-name products together with their ingredients when these have been revealed, addresses and telephone numbers of companies for use when ingredients are not listed, sample formulas of many types of products with an estimate of the toxicity of each formula, toxicological information including an estimate of the toxicity of individual ingredients, recommendations for treatment, names and addresses of manufacturers, and a system of standard nomenclature for the clarification of poisonings. Medical libraries, pharmacies, industrial medical departments, public health nursing centers, and any other agency frequently called upon for emergency help should also find it helpful as a quick source of information on first aid, treatment procedures, and other questions.

Goldsmith JR. *Environmental Epidemiology: Epidemiological Investigation of Community Environmental Health Problems.* Boca Raton, FL: CRC, 1986.

The experiences of practicing epidemiologists in solving worldwide community environmental health problems are discussed. Emphasis is placed on problems facing the community, methods of analysis, and means and results of action. Actual case histories of various complexity provide exercises in solving community health problems using applicable elementary concepts of statistics. Selected tables offer quantitation of problems and authoritative references direct interested students to further reading.

Grant WM. *Toxicology of the Eye,* 3rd ed. Springfield, IL: Thomas, 1986.

Grover PL (ed). *Chemical Carcinogens and DNA.* (2 vols). Boca Raton, FL: CRC, 1979.

Deals with the chemical modification of the DNA molecule and the ways these modifications can be detected.

Halpern S. *Drug Abuse and Your Company.* Ann Arbor, MI: University Microfilms International, Books-on-Demand, 1972.

Hardin JW, Arena JM. *Human Poisoning from Native and Cultivated Plants.* Durham, NC: Duke University, 1973.

Most of the existing literature on poisonous plants deals with those that are poisonous to livestock. A real need exists for a source of information on just those plants poisonous to humans—particularly children. Physicians, health officers, nurses, scout leaders, camp counselors, teachers, parents, and many others should not only know the dangerous plants of their area but have a ready reference in case of emergencies. This book has been written with these people in mind and has grown out of a number of years' experience with poisonous plants accumulated by both of the authors in the field, laboratory, and clinic.

Kopfler F, Crawn G (eds). *Environmental Epidemiology.* Chelsea, MI: Lewis, 1986.

This book is valuable to a broad spectrum of individuals active in the environmental and health sciences as well as those involved in the measurement and effects of numerous kinds of drinking water contamination and indoor and ambient air pollution. Environmental researchers involved in human exposure to toxic substances, regulators, and administrators also will find this book of value.

Krewski D, et al. *Toxicological Risk Assessment: Biological and Statistical Criteria,* vols 1 & 2. Boca Raton, FL: CRC, 1985.

This book discussed measurements of certain environmental risks and their application to the relatively low levels to which humans are generally exposed.

Matheson Chemicals Company. *Effects of Exposure to Toxic Gases First Aid and Medical Treatment.* Lyndhurst, NJ: Matheson, 1983.

This edition includes a chapter on industrial hygiene in an effort to stress the necessity of incorporating good safety and health control in the design of work areas where hazardous gases will be used. Workers' attention to the need for good work practices, special emergency measures, and equipment that should be available to minimize the risk and consequences of accidents is also discussed.

Milman HA, Weisburger EK (eds). *Handbook of Carcinogen Testing.* Park Ridge, NJ: Noyes Data, 1985.

This book considers short-term and long-term testing for carcinogens, and all other facets of the operation. The volume affords a total view of a bioassay from initial phases to application.

Monson RR. *Occupational Epidemiology.* Boca Raton, FL: CRC, 1980.

The detection and control of long-term effects, and a review of general epidemiologic methods and their application in the context of occupational settings are included. The book discusses how data from occupational populations can be used to prevent illness, and principles to be followed in setting up and running comprehensive programs in occupational epidemiology are included.

Neely WB, Blau GE. *Environmental Exposure from Chemicals.* Boca Raton, FL: CRC, 1985.

This two-volume set includes estimation of physical properties, pollution sorption in environmental systems, air/soil exchange coefficients, air/water exchange coefficients, biogradation, hydrolysis, photochemical transformations, and equilibrium models for initial integration of physical and chemical properties.

O'Donoghue JL (ed). *Neurotoxicity of Industrial and Commercial Chemicals.* Boca Raton, FL: CRC, 1985.

A collection of up-to-date information on the neurotoxicity of chemicals used in industry or having commercial value. Chemicals reported to cause a variety of effects on the nervous system are thoroughly reviewed. Exposure data, clinical manifestations, pathology, experimental neurology, metabolism, and structure activity correlates are integrated and presented by the anatomical and functional areas of the nervous systems affected, and by chemical classes with neurotoxic effects. Much of the information is presented in tabular format.

Ottoboni MA. *The Dose Makes the Poison.* Berkeley, CA: Vincente, 1984.

This book explains, in nontechnical language, what makes chemicals harmful or harmless. Dr. Ottoboni states that all living organisms have to deal with exposure to many noxious substances; but only when we overwhelm our bodily natural defense mechanisms by taking in too much too often, do we get into trouble.

Plunkett ER. *Handbook of Industrial Toxicology.* New York: Chemical, 1987.

Randolph TG. *Human Ecology and Susceptibility to the Chemical Environment.* Springfield, IL: Thomas, 1981.

Most illnesses were originally believed to have arisen within the body. Only recently has this age-old concept been challenged. The importance of the outside environment as a cause of sickness was first demonstrated in respect to infectious diseases approximately 80 years ago and to allergic diseases approximately 50 years ago. Although the general principles of infectious disease are now fully accepted and applied, the medical profession has been slow in learning and applying the necessary techniques to demonstrate cause-and-effect relationships between the nonmicrobial environment and ill health.

Stich HF (ed). *Carcinogens and Mutagens in the Environment;* vol 4: *The Workplace: Monitoring and Prevention of Occupational Hazards.* Boca Raton, FL: CRC, 1985.

This volume includes chapters on cancer as an occupational hazard; tissues in cancer prevention; improvement for worker protection; government, employers, and labor relations through information and participation; risk assessment; exposed populations; and hazards and safety of the modern office environment.

Sunshine I. *Methodology for Analytical Toxicology,* 3 vols. Boca Raton, FL: CRC, 1975, 1982, 1985.

E. Medical

Alderman MH, Hanley MJ (eds). *Clinical Medicine for the Occupational Physician.* New York: Dekker, 1982.

This book covers the interdisciplinary topics related to personal health and medical care. It presents the workplace as site for health promotion, stressing the management of total personal health by occupational physicians and nurses. The book provides focused examinations of the special characteristics of women, elderly and disabled workers, and employees with chronic illnesses, and features in-depth coverage of toxicology and both chemical and physical hazards, emphasizing occupational cancer and problems in dermatology.

Alderson M. *Occupational Cancer.* Stoneham, MA: Butterworth-Heinemann, 1986.

The increased risks of cancer resulting from exposure to particular agents in the workplace or from working in a particular industry have now been extensively studied. The book reviews the current information on the occurrence and causes of cancer, concentrating on material drawn from epidemiological studies.

The greater portion of the book discusses each chemical and physical agent associated with an increased risk of cancer and is an essential reference for those involved with the health-care of workers. Chapters include: epidemiological methods; agents that cause occupationally-induced cancers; occupations associated with increased risk of cancer; aetiology of cancer; and towards control of occupational cancer.

Cataldo MF, Coates TJ. *Health and Industry.* New York: Wiley, 1986.

A behavioral medicine perspective, this book covers the principles of behavioral medicine in the industrial setting, plus the day-to-day problems affecting workers including obesity, stress, hypertension, cardiovascular disease, smoking, dental care, cancer, and accidents.

Craun GF. *Waterborne Diseases in the United States.* Boca Raton, FL: CRC, 1986.

Water-related illness is examined, emphasizing transmission of infectious diseases through contaminated drinking water supplies and deficiencies in water supply systems that allow waterborne outbreaks. Also included are important etiologic agents, surveillance activities, regulations, preventive measures, procedures for investigating disease outbreaks, and laboratory methods for identifying pathogens and diseases that result from ingesting, bathing or wading, and inhalation. This volume is a good reference for public health investigators, environmental engineers, and sanitarians.

Felton JS. *Occupational Medical Management.* Boston: Little Brown, 1990.

Feneis H, Kaiser HE. *Pocket Atlas of Human Anatomy.* New York: Thieme, 1985.

Convenient reference as an aid in explaining diagnosis to patients. Provides a systematic look at anatomy through the use of international nomenclature and illustrations.

Guides to the Evaluation of Permanent Impairment, 3rd rev. ed. Chicago: American Medical Association, 1990.

The AMA has provided authoritative material to assist physicians and others in determining levels of impairment of patients, clients, or applicants who are seeking benefits from the various agencies and programs serving the disabled.

Himmelstein JS, Pransky GS (eds). *Occupational Medicine: Worker Fitness and Risk Evaluations. State of the Art Reviews* 3 (2). Philadelphia: Hanley & Belfus, 1988.

Hunter D. *The Diseases of Occupations,* 7th ed. London, UK: English Universities Press, 1987.

Includes historical outline of occupational diseases and their treatment. Discusses hazardous materials, such as metals and noxious gases, describes the processes that led to their being recognized as hazards, symptoms, preventive methods, treatment, and some case histories.

Lampe KF, McCann MA. *AMA Handbook of Poisonous & Injurious Plants.* Chicago: American Medical Association, 1985.

This botanical reference work outlines diagnosis and treatment of medical conditions caused by plant toxicity as well as contamination from pesticides, herbicides, and fertilizers. Useful for physicians, hikers, campers, and others. Full-color photos of plants with common and scientific names aid in identification.

Lefevre MJ. *First Aid Manual for Chemical Accidents: For Use with Nonpharmaceutical Chemicals.* New York: Van Nostrand Reinhold, 1989.

First aid for chemical accidents requires specific, quick, correct action to obviate and minimize potentially serious, harmful effects. This first aid manual is dedicated totally to emergency care at the workplace in which industrial chemicals are used.

A color-indexing scheme makes proper treatment information easily accessible. Each compound of the nearly 500 industrial products is referred to a color-coded section: toxicology; symptoms of overexposure; and first aid procedures in cases of inhalation, ingestion, skin contact, and eye contact.

This quick-reference manual for emergencies can be equally valuable for the prevention of accidents in the training of anyone who works with industrial solvents, insecticides, commercial chemicals, fertilizers, or chemical intermediates.

Nichols PJ. *Rehabilitation Medicine,* 2nd ed. Stoneham, MA: Butterworth-Heinemann, 1980. *Safety Guide for Health Care Institutions,* 5th ed. Chicago: National Safety Council/American Hospital Association, 1994.

This guide is intended to recognize and identify hazards in health care facilities, provide information for their elimination, and stimulate each hospital's personnel to improve its safety program. To be most effective, this book should be used with other safety books from the NFPA, National Safety Council, and the AIHA.

Schilling RSF (ed). *Occupational Health Practice,* 2nd ed. London, UK: Butterworths, 1981.

Sunderman EW, Sunderman FW, Jr (eds). *Laboratory Diagnosis of Diseases Caused by Toxic Agents.* St. Louis: Warren H. Green, 1970.

Zenz C (ed). *Occupational Medicine: Principles and Practical Applications,* 3rd ed. Chicago: Year Book Medical Publishers, 1994.

F. Dermatological

Adams RM. *Occupational Skin Disease,* 2nd ed. Philadelphia: Saunders, 1990.

This book guides dermatologists, allergists, and industrial physicians in the successful management of occupational dermatoses. The information on prevention and treatment will enable workers, industrial hygienists, and plant superintendents to plan and implement methods that should lower the incidence of such disease.

Steigleder. *Pocket Atlas of Dermatology.* New York, NY: Thieme, 1984.

A concise survey of dermatologic and venereologic disease patterns emphasizing differential diagnosis. Clinicians will find illustrations useful in identifying disorders and anomalies.

G. Physical Stresses

Attix FH, Roesch WC (eds). *Radiation Dosimetry,* 2nd ed. San Diego, CA: Academic, 1967–1972.

This second edition fills the need for a comprehensive treatise, written primarily as a reference work for radiation workers. Many useful tables, curves, illustrations, formulas, and references to the literature have been included. Every effort was made to present the material as clearly as possible, for those just entering the field.

Brodsky A. *CRC Handbook of Radiation Measurement and Protection.* Boca Raton, FL: CRC, 1979–1982.

This handbook is a comprehensive guide to methods and data used for measuring or estimating radiation doses and provides information on radiation protection for industrial, research, and medical installations. This book will assist in solving problems of radiation dose estimation, biological risk evaluation, and facility design.

Cember H. *Introduction to Health Physics,* 2nd ed. New York: Pergamon, 1983.

Contents deal with a review of physical principles, atomic and nuclear structure, radioactivity, interaction of radiation with matter, radiation dosimetry, biological effects of radiation, radiation protection guides, health physics instrumentation, external and internal protection, criticality, and evaluation of protective measures.

Harris CM (ed). *Handbook of Noise Control,* 2nd ed. New York: McGraw-Hill, 1979.

In general, the material presented relates to properties of sound, effects of noise on humans, vibration control, instrumentation and noise measurement, techniques of noise control, noise control in buildings, sources of noise and examples of noise control of machinery and electrical equipment, noise control in transportation, community noise, and the legal aspects of noise problems.

This manual is a handy reference source because of the large number of references given. It should prove helpful to those concerned with almost any kind of noise problem, legal or technical.

Heating and Cooling for Man in Industry, 2nd ed. Akron, OH: American Industrial Hygiene Assoc., 1975.

Written for the working industrial hygienist and heating and ventilation engineer, this manual contains information to obviate the need for extensive research when attempting to solve a problem. It has been written with one objective: describing the means of controlling the working environment to conduct a variety of operations with fluctuating outdoor conditions. Methods of varying temperature, air motion, and humidities within the work space are described. A worker's space is the primary area of interest.

Kryter KD. *The Effects of Noise on Man,* 2nd ed. San Diego, CA: Academic, 1985.

Treats auditory system responses, subjective responses, and nonauditory system responses to noise; includes an extensive set of references.

Miller DG. *Radioactivity and Radiation Detection.* New York: Gordon & Breach, 1972.

Miller KL, Weinder WA. *CRC Handbook of Management of Radiation Protection Programs.* Boca Raton, FL: CRC, 1986.

This guidebook organizes the profusion of rules and regulations surrounding radiation protection into an easy-to-use, single-volume reference. Employee and public protection, accident prevention, and emergency preparedness are included in this comprehensive coverage. Whenever possible, information is presented in convenient checklists, tables, or outlines that enable readers to locate information quickly.

This book is ideal as a starting point for organizations that are establishing a program, and as a self-evaluation tool for existing programs.

Morgan KZ, Turner JE (eds). *Principles of Radiation Protection: A Textbook of Health Physics.* Melbourne, FL: Krieger, 1967 (Rep. 1973).

Contents include history of health physics; passage of heavy charged particles, gamma-rays, and x rays through matter; radiation quantities and units; physical basis of dosimetry; detection and measurement of ionization; dose from electrons and beta-rays; dose from external sources; internal exposure; and radiation biology and biophysics.

Okress EC (ed). *Microwave Power Engineering.* San Diego, CA: Academic, 1968.

This book introduces the electronics technology of microwave power and its applications. This technology emphasizes microwave (and eventually quantum) electronics for direct power utilization and transmission purposes rather than exclusively for information and communications applications. Essentially, microwave power can be divided into microwave heating, microwave processing, microwave dynamics, and microwave power transmission involving generation and power amplification, direct power utilization, and closed waveguide or radiation beam propagation for remote utilization and rectification.

Pearce B (ed). *Health Hazards of VDTs.* Ann Arbor, MI: University Microfilms, Books on Demand, 1984.

The author takes an objective look at the highly emotional debate of the health, safety, and quality of working life of those using the new VDT technology. It examines the evidence for the alleged direct health hazards, such as facial rashes, radiation emissions, and cataracts, and indirect hazards, such as vision and lighting problems and emotional stress. The authors also consider some of the ways by which the working conditions of VDT users might be improved.

Polk C. *Handbook of Biological Effects of Electromagnetic Fields.* Boca Raton, FL: CRC, 1986.

This book presents the current knowledge about the effects of electromagnetic fields on living matter. The three-part format covers dielectric permittivity and electrical conductivity of biological materials; effect of direct current and low frequency fields; and effects of radio frequency (including microwave) fields. The parts are designed to be consulted independently or in sequence, depending upon the needs of the reader. Useful appendices on measurement units and safety standards are also included.

Radiation Protection Procedures. Vienna, Austria: International Atomic Energy Agency, 1973. (Available from UNIPUB, Inc., P O Box 433, New York, NY 10016. Order No. STI/PUB-257.)

Text reviews the fundamentals of nuclear physics and interactions of ionizing radiations with matter and living cells; the basic concepts governing the formulation of units for the measurement of radiations; methods used for measurements of radiations; selection, calibration, and maintenance of instruments used for monitoring; shielding; protective clothing; decontamination measures; radioactive waste management; the transport of radioactive materials; and emergency procedures for radiation accidents.

Discusses various administrative and technical measures that could form the basis for establishing a successful radiation protection program.

Sataloff J, Michael PL. *Hearing Conservation.* Springfield, IL: Thomas, 1973.

Shapiro J. *Radiation Protection: A Guide for Scientists and Physicians,* 3rd ed. Cambridge, MA: Harvard University Press, 1990.
Provides the radiation user with information on protection and compliance with governmental and institutional regulations regarding the use of radionuclides and radiation sources. Designed to obviate the need for reviews of atomic and radiation physics; the mathematics is limited to elementary arithmetical and algebraic operations.

Sliney D, Wolbarsht ML. *Safety with Lasers and Other Optical Sources.* New York: Plenum, 1980.
This book thoroughly reviews current knowledge of biological hazards from optical radiation and lasers, presents current exposure limits, and provides a wealth of information required for the control of health hazards.

Stewart DC. *Data for Radioactive Waste Management and Nuclear Applications.* Melbourne, FL: Krieger, 1991 (Rep. of 1985 ed.)
This comprehensive reference provides essential information for solutions to radioactive waste disposal problems. It provides the data needed for safely disposing of high-level liquid wastes, intermediate-level wastes, and packaged radioactive wastes.
Tables and charts provide data on waste container sizes, nuclear migration through rock, leachability, and exposure rates. There is information on nuclear reactor operational data concerning shielding materials, radiation damages, and decontamination.

Sullivan JB Jr, Krieger GR (eds). *Hazardous Materials Toxicology: Clinical Principles of Environmental Health,* 2nd ed. Baltimore: Williams & Wilkins, 1999.

H. Ergonomics

Alexander DC, Pulat BM. *Industrial Ergonomics: A Practitioner's Guide.* Atlanta: Industrial Engineering & Management Press, 1985.
A very practical, job-oriented approach.

Astrand PO, Rodahl, K. *Textbook of Work Physiology,* 3rd ed. New York: McGraw-Hill, 1986.

Barnes R. *Motion and Time Study,* 7th ed. New York: Wiley, 1980.

Chaffin DB, Andersson JB. *Occupational Biomechanics,* 2nd ed. New York: Wiley, 1991.
This book provides an understanding of musculoskeletal mechanics to assist in modifying potentially debilitating conditions in the workplace. The book contains more than 200 information-packed diagrams demonstrating the mechanics of the musculoskeletal system as well as solutions to mechanical workload stress.

Chapanis A. *Research Techniques in Human Engineering.* Ann Arbor, MI: University Microfilms International, Books-on-Demand, 1959.

Eastman Kodak Company. *Ergonomic Design for People at Work,* vol 1. Belmont, CA: Lifetime Learning, 1989.
This book offers a practical discussion of workplace, equipment, environmental design and of the transfer of information in the workplace. It summarizes current data, experience, and thoughts assembled from the published literature, internal research, and observation by the Human Factors Section of the Eastman Kodak Company. The guidelines and examples are drawn from case studies. These principles have been successfully applied in the workplace to reduce the potential for occupational injury, increase the number of people who can perform a job, and improve performance on the job, thereby increasing productivity and quality.

Eastman Kodak Company. *Ergonomic Design for People at Work,* vol 2. New York: Van Nostrand-Reinhold, 1989.
Effective ergonomic design must take every aspect of human physiology into account. This volume shows how to design a work environment that is in conformance with basic human physiology. It goes well beyond factors such as height and reach and into the effects of muscular contractions, biological rhythms, and heart rate on worker capabilities. Human mechanics and job demand; the effects of work patterns on worker efficiency; and manual materials handling are discussed.
The book also provides methods to measure maximum human capacities, interpret heart rate, and analyze timed activities. Practical explanations of basic psychological factors help in

taking steps to reduce unnecessary stress in the workplace.

Fitts PM, Posner MJ. *Human Performance.* Westport, CT: Greenwood, 1979.

Geldard F. *The Human Senses,* 2nd ed. New York: Wiley, 1973.

Grandjean E. *Ergonomics in Computerized Offices.* London, UK: Taylor & Francis, 1987.

Kroemer K, Grandjean E. *Fitting the Task to the Human: A Textbook of Occupational Ergonomics,* 5th ed. London, UK: Taylor & Francis, 1997.

This book contains chapters on man-machine systems and the questionnaire for controlling working conditions; a number of topical factors such as seating at work, heart rate as an indication of physical stress, monotony, daytime lighting, environmental climate in offices, and some recent advances in the assessment of heat stress are also considered.

Helander MG (ed). *Handbook of Human-Computer Interaction.* New York: Elsevier, 1989.

Konz S. *Work Design: Industrial Ergonomics,* 3rd ed. Columbus, OH: Grid, 1990.

Parson HM. *Man-Machine System Experiments.* Baltimore: Johns Hopkins Press, 1972.

Pulat BM, Alexander DC (eds). *Industrial Ergonomics: Case Studies.* Norcross, GA: Institute of Industrial Engineers, 1991.

Salvendy G (ed). *Handbook of Human Factors and Ergonomics.* New York: Wiley, 1997.

Sanders MS, McCormick EJ. *Human Factors in Engineering and Design,* 7th ed. New York: McGraw-Hill, 1993.

Singleton WT, Fox JF, Whitfield D (eds). *Measurement of Man at Work: Papers.* Philadelphia: Taylor & Francis, 1971.

Steindler A. *Kinesiology: Of the Human Body under Normal and Pathological Conditions.* Springfield, IL: Thomas, 1977.

Thompson CW. *Kranz Manual of Structural Kinesiology,* 10th ed. St. Louis: Mosby, 1988.

Tichauer ER. *The Biomechanical Basis of Ergonomics Anatomy Applied to the Design.* New York: Wiley, 1978.

Now professionals in manufacturing and service industries concerned with the health, welfare, and performance of people at work can apply functional anatomy to increase productivity while reducing on-the-job hazards.

Tichauer ER. Human Factors Engineering, in 1984 McGraw-Hill *Yearbook of Science and Technology.* New York: McGraw-Hill, 1983.

Woodson WE, Tillman P. *Human Factors Design Handbook,* 2nd ed. New York: McGraw-Hill, 1991.

Woodson WE, Conover DW. *Human Engineering Guide for Equipment Designers,* 2nd rev. ed. Berkeley, CA: University of California Press, 1964.

The greatest expansion in this new revision has occurred in the first parts of the Guide. The first chapter, Design Philosophy, is entirely new, having replaced the former introductory section. Chapter 2 is a considerably expanded version of the original material; however, an attempt has been made to retain the original direct format, which seems to have been appreciated by most designers.

The chapter on Body Measurement has been revised appreciably and made more practical from the designer's point of view. This change is a reflection of the application experience of the writers in working very closely with aerospace and weapon system designers since the beginning of the Jet Age. Revisions in the remaining parts of the book are less extensive, but reflect many of the changes brought about by more recent research, especially in the area of man-in-space and in industrial applications.

I. Biological

Hubbert WT. *Diseases Transmitted from Animals to Man,* 6th ed. Springfield, IL: Thomas, 1975.

Laskin AL, Lechevalier H (eds). *CRC Handbook of Microbiology,* 2nd ed: vol 1, Bacteria; vol 2, Fungi, Algae, Protozoa, and Viruses. Boca Raton, FL: CRC, 1978.

J. Chemical

AT & T Bell Laboratories. *Chemical Hygiene Plan.* Itasca, IL: National Safety Council, 1991.

Bretherick L. *Handbook of Reactive Chemical Hazards,* 3rd ed. Stoneham, MA: Butterworth, 1985.

The majority of the book is devoted to specific information on the stability of the listed compounds, or the reactivity of mixtures of two or more under various circumstances. Each description of an incident or violent reaction gives references to the original literature. Each chemical is classified on the basis of similarity in structure or reactivity and each class is described in a separate section. Many cross references throughout both sections emphasize similarity in compounds or incidents not obviously related, and an introductory chapter identifies the underlying principles which govern the complex subject of reactive chemical hazards. In many cases, quantitative thermodynamic data on energies of decomposition or reaction are included. An appendix lists the fire-related properties of the higher risk materials and there are comprehensive indexes and a glossary of specialized terms.

Chissick SS, Derricott R (eds). *Asbestos: Properties, Applications, and Hazards.* New York: Wiley, 1983.

This volume is a comprehensive compendium of published information on asbestos from world sources. It is a useful sourcebook on the literature of the subject for both experts and nonexperts.

Cold Cleaning with Halogenated Solvents (STP 403A). Philadelphia: American Society for Testing and Materials, 1981.

This book has important information on flammability, health hazards (such as inhalation, skin contact, and ingestion), and container handling and storage. The manual would be useful to industries, such as electrical, electronic, electromechanical, automotive and equipment maintenance, metal and metal fabrication, aerospace, and nuclear.

Collings AJ, Luxon SG. *Safe Use of Solvents.* Orlando, FL: Academic, 1982.

Solvents have varying uses in different areas of modern industry, ranging from the extraction of petroleum products, foods, and natural products, to their use as coupling agents in adhesives and surface coatings, and as carriers for perfumes, essences, and insecticides. Thus, government, industry, and trade unions have an active interest in how solvents are manufactured, stored, transported, used, and disposed of in a safe manner.

Compressed Gas Association. *Handbook of Compressed Gases,* 3rd ed. New York: Van Nostrand-Reinhold, 1990.

Discusses 49 widely used compressed gases in terms of their properties, methods of manufacture, commercial uses, and physiological effects. Includes data relative to the materials of construction required for all types of compressed gas installations, equipment, and containers. Also includes a chapter on safe handling of compressed gases, as well as information on hazardous materials regulations.

Cote R, Linville J (eds). *Fire Protection Handbook,* 18th ed. Quincy, MA: National Fire Protection Association, 1997.

Much new material has been added, recognizing the many advances made in fire protection technology since the previous edition was published. New fire problems, and the solutions to them, that in the last decade were only then beginning to make themselves known, are now deserving of extensive attention. (Highrise buildings, for example, received only passing mention in the last edition with no direct reference to their potential as a source of hazard to life.)

Dean AE, Tower K. *Fire Protection Guide on Hazardous Materials.* Quincy, MA: National Fire Protection Association, 1991.

Fawcett HH (ed). *Hazardous and Toxic Materials: Safe Handling and Disposal,* 2nd ed. New York: Wiley, 1988.

The book presents a balanced view of the latest scientific information about hazardous and toxic materials, their containment, and their availability to humans, animals, and plants. It takes a close look at the laboratory where most of these materials originate and then proceeds into fire and explosion hazards and their detection. It details the personal protection that is necessary in the waste site investigation and clean-up environments. Assuming a positive problem-solving approach, the book emphasizes the importance of alternative disposal methods and shows how to control and prevent future environmental disasters.

Forsberg K, Keith LH. *Chemical Protective Clothing Performance Index Book*. New York: Wiley, 1989.

Green AE (ed). *High Risk Safety Technology*. New York: Wiley, 1982.

The techniques for evaluating the safety of technological systems are described in a logical and quantitative form. Risk evaluation and criteria are treated in such a way as to make this a sourcebook for developing ideas in a scientific and analytical manner in those areas of technology where high risks of any nature are involved.

James D. *Fire Prevention Handbook*. Stoneham, MA: Butterworth, 1986.

This practical handbook explains clearly and without technical jargon the causes of fire and the processes by which it spreads. It sets down the basic rules of fire prevention and shows how good housekeeping, training, and general motivation can contribute to the prevention of fire.

Lefevre MJ. *First Aid Manual for Chemical Accidents, for Use with Pharmaceutical Chemicals*, 2nd ed. New York: Van Nostrand-Reinhold, 1989.

This manual deals entirely with first aid at workplaces that use industrial chemicals.

Levadic B (ed). *Definitions for Asbestos and Other Health-Related Silicates*. Philadelphia: American Society for Testing and Materials, 1984.

This book addresses the need for clarification of terminology used to refer to groups of minerals; namely, asbestos, silica, and talc. The volume includes a comprehensive glossary for health-related silicates, defining 46 terms. A practical handbook for the scientific community, regulatory agencies, medical and legal professions, industry management, and the media.

O'Connor CJ, Lirtzman SI. *Handbook of Chemical Industry Labeling*. Park Ridge, NJ: Noyes Data, 1984.

The need for informative labeling in workplace, transportation, distribution, and disposal operations has been recognized by all levels of government agencies. Society-at-large has demanded increased information on chemical products, and organized labor has actively pursued an improved hazardous label communication program.

Rappe C, Choudhary G, Keith LH. *Chlorinated Dioxins and Dibenzofurans in Perspective*. Chelsea, MI: Lewis, 1986.

This book provides the latest human exposure data and the most advanced analytical techniques developed in the continuing effort against contamination by chlorinated dioxins and dibenzofurans.

Stricoff RS, Walters DB. *Laboratory Health and Safety Handbook: A Guide for the Preparation of a Chemical Hygiene Plan*. New York: Wiley, 1990.

Zabetakis MG. *Safety with Cryogenic Fluids*. Ann Arbor, MI: University Microfilms International, Books-on-Demand, 1967.

This monograph was prepared in an effort to present in concise form the principles of safety that are applicable to the field of cryogenics. Thus, while it includes safety rules, design data, first aid, and hazard control procedures, emphasis has been placed on basic principles. An appreciation of these principles permits an individual to conduct a safe operation under a wider variety of conditions than is possible if he or she is familiar only with a list of safety rules.

K. Pollution and Hazardous Waste

Bhatt HG, Sykes RM, Sweeney TR. *Management of Toxic and Hazardous Wastes*. Chelsea, MI: Lewis, 1985.

The demand for cleaning of hazardous waste disposal sites has grown since the passage of Superfund. This book presents the important aspects of hazardous waste management. Attention is focused on waste treatment and recycling, risk assessment, public participation, and land disposal. A section on legal considerations provides pointers on precautions to be taken to minimize legal liabilities.

Calvert S, Englund HM. *Handbook of Air Pollution Technology*. New York: Wiley, 1984.

The text provides an up-to-date guide for defining, analyzing, and controlling a wide variety of air pollution problems. Experts in the field discuss methods for the control of gases, gaseous pollutant characteristics, the effects of air pollutants on the atmosphere and materials. It contains valuable material for the professional concerned with reducing and controlling the often-invisible pollutants existing in our atmosphere.

Canter LW. *River Water Quality Monitoring.* Chelsea, MI: Lewis, 1985.

This practical guide gives the information required to plan and conduct river water quality studies. These studies are necessary to establish baseline conditions, set water quality criteria and standards, monitor temporal changes, and determine the impacts of specific projects and developments.

Canter LW, Know RC. *Ground Water Pollution Control.* Chelsea, MI: Lewis, 1986.

This comprehensive work thoroughly covers technologies for ground pollution in part one and deals in depth with aquifer restoration decision-making in part two. Part three gives detailed abstracts of 225 selected references and a range of case studies.

Canter LW, Knox RC, Fairchild DM. *Ground Water Quality Protection.* Chelsea, MI: Lewis, 1986.

Considered by the EPA to be one of the major environmental issues of the 1980s, groundwater supplies a large majority of the water we use. The book deals with this critical problem and the action to be taken to prevent despoliation of the aquifers where this water is now found, because once contaminated, an aquifer is difficult to decontaminate.

Dawson GW, Mercer B. *Hazardous Waste Management.* New York: Wiley, 1986.

This reference work deals with hazardous waste materials. The book gives cost-effective practical approaches as well as technological and policy considerations of facilities design, waste site reclamation treatment processes, incineration alternatives, land fill, and salt dome disposal, as well as survey techniques.

Godish T. *Air Quality,* 2nd ed. Chelsea, MI: Lewis, 1990.

This book on air quality provides comprehensive coverage of the subject with special treatment of atmospheric effects, effects on humans, plants, and materials, as well as pollution regulation and motor vehicle emission control.

Lioy PJ, Daisey JM (eds). *Toxic Air Pollution.* Chelsea, MI: Lewis, 1986.

The book deals with the characteristics and dynamics of noncriteria pollutants. It stems from the work of the authors on the Airborne Toxic Element and Organics Substances (ATEOS) project, which is the first successful attempt to develop an understanding of the subject. The text reports the results and major conclusions from a two-year study in three different urban environments.

Martin EJ, Johnson JH. *Hazardous Waste Management Engineering.* New York: Van Nostrand-Reinhold, 1987.

The book includes discussions on the following topics: hazardous waste and chemical substances; exposure and risk assessment; chemical, physical, and biological treatment of hazardous waste; incineration of hazardous waste; storage of hazardous waste; land disposal of hazardous waste; hazardous waste leachate management; and hazardous waste facility siting.

Robinson WD. *The Solid Waste Handbook, A Practical Guide.* New York: Wiley, 1986.

The handbook covers all aspects of solid waste management, including public policy and implementation of hazardous waste administration guidelines. Changes in waste management practices mandated by RCRA through 1984 are covered. It also examines the state-of-the-art practices for land disposal, resource recovery, and decision making in both the public and private sectors.

Wadden RA, Scheff PA. *Indoor Air Pollution.* New York: Wiley, 1982.

This book provides guidance to evaluate and control indoor air pollution caused by reduced ventilation and energy saving measures. The emphasis is on the environment of domestic and public buildings, although the material is equally applicable to many indoor spaces. The text also includes state-of-the-art information on indoor pollution hazards, methods for measuring, and predictive models.

L. Control

Air Pollution Reference Library. Cincinnati: ACGIH, latest edition.

This is a compilation of references on air pollution and includes lists of books, handbooks, journals, periodicals, and other references.

Alden JL, Kane JM. *Design of Industrial Ventilation Systems,* 5th ed. New York: Industrial Press, 1982.

This edition covers the interrelated areas of general exhaust ventilation and air makeup supply. The manual contains the information required to design, purchase, and operate an exhaust system that will comply with government regulations for the protection of employees and that will minimize atmospheric pollution. Like the earlier editions, the book also deals with local exhaust systems.

American Industrial Hygiene Association (AIHA). *Heating and Cooling for Man in Industry,* 2nd ed. Akron, OH: AIHA, 1975.

A wealth of information is presented here on methods of evaluation and control of heat and cold stress conditions.

Constance JD. *Controlling In-Plant Airborne Contaminants.* New York: Dekker, 1983.

This book covers the control of each of the various types of in-plant airborne contaminants. Emphasizing the interaction of theory with practical considerations, the book details proven control methods, allowing for the evaluation, selection, design, and implementation of the most economic and effective option.

Cralley LJ, Cralley LV (eds). *Industrial Hygiene Aspects of Plant Operations* (in three volumes). New York: Macmillan, 1982, 1984, 1986.

Volume 1: Process Flows details all of the basic operations involved in the process flow of the material from inception through production in the manufacture of a cross section of representative industrial products.

Volume 2: Unit Operation and Products Fabrication fills an especially important and urgent need with its flowsheet style of presentation. Contributors discuss unit operations as distinct entities along an industry-wide concept and then cover the operations and procedures for assembling parts and materials into final products.

Volume 3: Engineering Considerations in Equipment Selection, Layout, and Building Design provides a comprehensive, up-to-date, and useful reference on recognizing, measuring, and controlling job-related health hazards.

Deisler PF Jr. *Reducing the Carcinogenic Risks in Industry.* New York: Dekker, 1984.

This work examines various private and government policy approaches to risk reduction. It discusses government regulatory and industry self-regulatory methods in the United States and Europe, spotlighting specific examples and accomplishments.

Hemeon W (ed). *Plant and Process Ventilation,* 2nd ed. Ann Arbor, MI: University Microfilms International, Books-on-Demand, 1963.

This book was written to assist those individuals charged with the responsibility of designing exhaust ventilation systems. The first half is concerned with methods for analyzing a factory ventilation problem and explains the dynamics of the air-polluting process to determine in what manner the air is to be channeled through the space in question.

McDermott H. *Handbook of Ventilation for Contaminant Control,* 2nd ed. Stoneham, MA: Butterworth, 1985.

The volume covers the following topics: indoor air pollution, OSHA ventilation standards, hazard assessment, how local exhaust systems work, hood selection and design, air cleaner selection, ventilation system design, fans, ventilation for high toxicity or high-nuisance contaminants, saving ventilation dollars, testing, and solving ventilation system problems.

Patterson JW (ed). *Metals Speciation, Separation and Recovery.* Chelsea, MI: Lewis, 1987.

By and for engineers and scientists, this book covers all aspects of metals chemistry, separation chemistry, and metals separation processes. State-of-the-art papers give new and recent developments and future research needs.

Pleiffer JB (ed). *Sulfur Removal and Recovery from Industrial Processes.* (Advances in Chemistry Series No. 139.) Washington DC: American Chemical Society, 1975.

Sixteen chapters form a consolidated reference source of sulfur removal and recovery methods concentrating on recovery techniques from sources other than power plant stacks. Emissions from smelter gas streams and Claus units are discussed, and seven scrubbing processes are described. Companion volume is No. 140 New Uses of Sulfur.

Pipitone DA (ed). *Safe Storage of Laboratory Chemicals.* New York: Wiley, 1984.

This volume provides full information on fundamental principles of chemical storage. Readers will learn the variety and type of chemical storage hazard, degrees of hazards, and practical guidelines for analyzing and correcting deficiencies. Problems related to hazardous, flammable, unstable, and incompatible chemicals are discussed. Information on facts about surveys and inspections, storage systems, computerized warehousing functions, and choosing a computer for its information retrieval functions is also included.

Plog B. *Fundamentals of Industrial Hygiene,* 4th ed. Itasca, IL: National Safety Council, 1996.

Rajhans G, Blackwell DS. *Practical Guide to Respirator Usage in Industry.* Stoneham, MA: Butterworth, 1985.

This practical reference will greatly simplify the task of anyone responsible for worker protection in industry. The table of contents includes overview of respiratory hazards and evaluation, respirator types and limitations, criteria for selection and fitting, administration and training, maintenance and care, medical supervision, criteria for respiratory protection program industrial applications, and research needs.

One chapter of particular interest makes use of case studies to point out common mistakes in judgment or choice that can lead to severe health problems or even death. Each study is followed by an explanation of how the specific situation could have been avoided or remedied.

Salvato JA. *Environmental Engineering and Sanitation,* 3rd ed. New York: Wiley, 1982.

This guide covers virtually every problem encountered in sanitary and environmental engineering and administration. The third edition includes updated and expanded coverage of alternate on-site sewage disposal; water reclamation and reuse; protection of groundwater quality; control and management of hazardous waste; resource recovery; and energy conservation. There are also sections on food sanitation and integrated pest management.

Stoner DL, et al. *Engineering a Safe Hospital Environment.* New York: Wiley, 1982.

The book focuses on safety: electrical, general and building, mechanical and laboratory, and radiation; control: infection and environmental; fire protection; and safe use and operation of medical equipment. The book provides summaries of standards, codes, and regulations as well as safety checklists, which enable readers to assimilate the information easily and install routine practices essential to an adequate safety program.

M. Encyclopedias and Handbooks

Calvert S, Englund HM (eds). *Handbook of Air Pollution Technology.* New York: Wiley, 1984.

Essential technical principles, design ideas and methods, practical examples, and other valuable information are included. This book is of interest to all concerned with reducing and controlling the often invisible pollutants that exist in our atmosphere.

Considine DM (ed). *Encyclopedia of Chemistry,* 4th ed. New York: Van Nostrand-Reinhold, 1984.

The revised and enlarged edition incorporates the advancements in chemistry achieved during the past decade and fully updates the traditional aspects of the field. Nearly 90% of the text is new—approximately 1,300 alphabetically arranged entries, each prepared by an expert in the field, give you valuable facts on processing and use of natural and synthetic chemical solids, liquids, and gases.

Biomass energy, chromatography, exosmosis, freeze-concentrating, graphite structures, hypobaric systems, immunochemistry, macromolecules, recombinant DNA, and many other vital topics are covered.

Encyclopedia of Occupational Health and Safety. 4 vols. Geneva, Switzerland: International Labour Office, 1998.

The encyclopedia is a reference tool providing information about causes of accidents, resultant illnesses, descriptions of occupational terms, and recommendations for preventive measures. This information is presented in a practical manner within the context of economic and social conditions, covering specific hazards encountered in every working environment and their prevention.

Handbook of Hazardous Waste Regulation; vol 2: *How To Protect Employees During Environmental Incident Response—Official EPA Health and Safety Guidance.* Madison, CT: Bureau of Law and Business, 1985.

The guidance manual provides needed information for establishing and maintaining health and safety protection for response personnel.

Hawley G. *The Condensed Chemical Dictionary,* 11th ed. New York: Van Nostrand Reinhold, 1987. (Now called Hawley's Condensed Chemical Dictionary. Revised by: N. Irving Sax and Richard J. Lewis, Jr.)

This edition defines many new terms covering thousands of chemicals and chemical phenomena. It provides information on an abundance of topics, including pollution and waste control, chemical manufacturing equipment, and energy sources and their potential.

Three distinct types of information are provided: (1) Technical descriptions of chemicals, raw materials, and processes, (2) Expanded definitions of chemical entities, phenomena, and terminology, and (3) Descriptions of identifications of a broad range of trademarked products used in the chemical industries. Aspects of hazardous materials, such as toxicity and flammability, explosion risks, radioactive hazards, oxidizing and corrosive properties, and tissue irritants are described.

Hazardous Waste Compliance Checklists for Supervisors. Madison, CT: Bureau of Law and Business, 1985.

This book contains checklists on a number of different areas associated with hazardous waste. There are over 35 checklists provided, including: hazardous spills checklist, ethylene oxide and toluene checklist, benzene checklist, and decontamination checklist.

King R, Hudson R. *Construction Hazard and Safety Handbook.* Stoneham, MA: Butterworth, 1985.

From the earliest times, construction has been one of the most dangerous activities undertaken by man. Despite efforts to make construction work safer, real improvements have lagged behind those of other industries. This book was written in an attempt to identify and analyze the hazards of construction, not only in buildings, but also industrial plant construction, offshore work, and civil engineering projects. It covers hazards of health, accidental injury, fire and explosion, and much more. The authors also offer general guidance in improving safety, training, and hazard monitoring.

Lederer WH. *Regulatory Chemicals of Health and Environmental Concern.* New York: Van Nostrand Reinhold, 1985.

This sourcebook contains detailed guidance to important regulations, standards, and other related information regarding chemicals of health and environmental concern. The book covers, in alphabetical order, approximately 2,000 chemicals that are listed in federal regulations, voluntary standards, and consent decrees.

Lewis RJ Sr. *Rapid Guide to Hazardous Chemicals in the Workplace.* New York: Van Nostrand-Reinhold, 1986.

This guide is designed to afford easy access to information on the adverse properties of commonly encountered industrial materials. The information provided allows for a quick assessment of the relative hazards of the material and the types and nature of the hazards likely to be encountered. Reference codes included with each entry refer to sources of additional information.

Lowry GG, Lowry RC. *Handbook of Hazard Communication and OSHA Requirements.* Chelsea, MI: Lewis, 1985.

This book provides guidance, explanation, and critical evaluation for the thousands of companies required to comply with the OSHA Hazard Communication Standard. The volume will help companies of all sizes to comply with the standard, which was created to protect employee health and provide safe working conditions for individuals who may come in contact with hazardous substances.

McGraw-Hill *Encyclopedia of Science and Technology,* 6th ed, in 20 vols. New York: McGraw-Hill, 1987.

The Merck Index: An Encyclopedia of Chemicals and Drugs, 11th ed. Rahway, NJ: Merck, 1989.

An essential work that includes information on toxicology.

Miller DE (ed). *Occupational Safety, Health and Fire Index,* vol 1. Ann Arbor, MI: University Microfilms International, Books-on-Demand, 1976.

This is a comprehensive and unified reference listing of the many safety, health, and fire codes, standards, guides, and publications in the field.

Sax NI. *Dangerous Properties of Industrial Materials,* 7th ed. New York: Van Nostrand-Reinhold, 1988.

This handbook includes single-figure toxic hazard ratings for all the chemicals discussed. The edition includes a 50,000-synonym index, and NIOSH and CAS numbers for nearly all entries.

Sittig M. *Handbook of Toxic and Hazardous Chemicals and Carcinogens,* 2nd ed. Park Ridge, NJ: Noyes Data, 1985.

This handbook presents concise chemical, health, and safety information on nearly 800 toxic and hazardous chemicals so that responsible decisions can be made by those who may have contact with or interest in these chemicals due to their own or third-party exposure.

Stedman's Medical Dictionary, 25th ed. Baltimore, MD: Williams & Wilkins, 1990.

Steere N. *Handbook of Laboratory Safety,* 2nd ed. Boca Raton, FL: CRC, 1971.

This book provides useful and accurate information for preventing or controlling accidents, injuries, fires, and losses in laboratories. Some of the topics included in the handbook are responsibility for laboratory safety, protective equipment, legal liability for accidents, working alone, fire hazards, and ventilation and exhaust systems.

Weast RC (ed). *Handbook of Chemistry and Physics.* Boca Raton, FL: CRC, latest edition.

Weast RC. *Handbook of Chemistry and Physical Data on Organic Compounds,* 2nd ed. Boca Raton, FL: CRC, 1988.

This handbook presents chemical and physical data on 24,000 organic compounds encountered in academic, medical, and industrial research, manufacturing and processing, and environmental control. Compounds are listed in alphabetical sequence by IUPAC name and accompanied by other information including CAS number, boiling and melting points, and synonyms.

NIOSH PUBLICATIONS

The National Institute for Occupational Safety and Health (NIOSH) has published many useful publications in the field of industrial hygiene. Consult the current Publications Catalog, listing all NIOSH publications in print and their prices, for further information. The NIOSH publications can be obtained by requesting single copies from:

National Institute for Occupational Safety and Health
Division of Technical Services
Publications Dissemination
4676 Columbia Pkwy
Cincinnati, OH 45226

Many of these publications are also available from:
Superintendent of Documents
U.S. Government Printing Office
Washington DC 20402

Some NIOSH publications can also be obtained from:
National Technical Information Service (NTIS)
Springfield, VA 22161

Criteria Documents

NIOSH is responsible for providing relevant data from which valid criteria for effective standards can be derived. The recommended standards for occupational exposure, which are the result of this work, are based on the health effects of exposure.

The single most comprehensive source of information on a particular material will probably be found in the NIOSH "Criteria Document" for that substance. The Table of Contents for a *Criteria Document* is as follows:

I. Recommendations for an Occupational Exposure Standard
    Section 1—Environmental (workplace air)
    Section 2—Medical
    Section 3—Labeling and posting
    Section 4—Personal protective equipment and clothing
    Section 5—Informing employees of hazards
    Section 6—Work practices
    Section 7—Sanitation practices
    Section 8—Monitoring and record keeping
II. Introduction
III. Biological Effects of Exposure
    Extent of exposure
    Historical reports
    Effects on humans
    Epidemiological studies
    Animal toxicity
    Correlation of exposure and effect
    Carcinogenicity, mutagenicity, teratogenicity, and effects on reproduction
    Summary tables of exposure and effect
IV. Environmental Data
    Environmental concentrations

Sampling and analysis
Engineering controls
V. Work Practices

DATA SHEETS AND GUIDES

AIHA Ergonomics Guide Series. Fairfax, VA: American Industrial Hygiene Association.
Contents: No.1—Guide to Manual Lifting; No.2—Ergonomics Guide for the Assessment of Human Static Strength; No.3—Guide to Assessment of Metabolic and Cardiac Costs of Physical Work; No.4—Guide to Assessment of Physical Work Capacity; No.5—Guide to Carpal Tunnel Syndrome; No.6—Ergonomics Principles Basic to Hand Tool Design; No.7—Ergonomics of VDT Workplaces. Each provides basic information on new concepts.

AIHA Hygienic Guide Series: Fairfax, VA: American Industrial Hygiene Association.
Separate data sheets on specific substances giving hygienic standards, properties, industrial hygiene practice, specific procedures, and references.

ASTM Standards with Related Material. Philadelphia: American Society for Testing and Materials.

CIS Information Sheets. Geneva 20, Switzerland: International Occupational Safety and Health Centre (CIS), International Labour Office.
Pamphlets on general or specific occupational safety and health topics; published irregularly.

Technical Bulletins: Chemical-Toxicological Series. Pittsburgh: Industrial Health Foundation. No. 1. Industrial Air Sampling and Analysis (1947). No. 2. Hygienic Guides (1963). No. 3. Threshold Limit Values: recommended revisions and additions for 1966 (1965). No. 4. Emergency Exposure Limits and Hygienic Guides (1965). No. 6. Range Finding Toxicity Data for 43 compounds (1967). No. 7. Threshold Limit Values; a report of progress in 1967 and announcement of 1968 workshop on TLVs (1967). No. 8. Suggested Principles and Procedures for Developing Data for Threshold Limit Values for air (1969).

Technical Bulletins: Medical Series. Pittsburgh: Industrial Health Foundation.
Reports on various topics in occupational health and medicine. Representative titles include: Emphysema in Industry, Asbestos Bioeffects Research for Industry, the Pneumoconioses, Acute Radiation Syndrome, and Proteolytic Enzymes in Cotton Mill Dust—A Possible Cause of Byssinosis.

Threshold Limit Values and Biological Exposure Indices. Issued Annually. Cincinnati: American Conference of Governmental Industrial Hygienists.
Threshold limits and exposure indices based on information from industrial experience and experimental human and animal studies; they are intended for use in the practice of industrial hygiene.

NEWSLETTERS AND REPORTS

ACOEM Report. American College of Occupational and Environmental Medicine, 55 W Seegers Rd, Arlington Hts, IL 60005.

Environmental Health Letter. Business Publishers, 951 Pershing Dr, Silver Springs, MD 20910-4464.

Environmental Health and Safety News. Department of Environmental Health, School of Public Health and Community Medicine, University of Washington, F-461 Health Sciences Bldg, Seattle, WA 98195.

Industrial Hygiene News Bimonthly. Rimbach Publishing, 8650 Babcock Blvd, Pittsburgh, PA 15237.

Newsletters. Bimonthly. National Safety Council, 1121 Spring Lake Dr, Itasca, IL 60143-3201.

Occupational Health and Safety Letter. Semimonthly. Business Publishers, 951 Pershing Dr, Silver Springs, MD 20910-4464.

Occupational Safety and Health Reporter. Bureau of National Affairs, 1231 25th St, NW, Washington DC 20037.

OSHA Up-to-Date. Monthly. National Safety Council, 1121 Spring Lake Dr, Itasca, IL 60143-3201.

World Health Organization Technical Report Series. WHO, Distribution and Sales, CH-1211 Geneva 27, Switzerland.

JOURNALS AND MAGAZINES

Many articles on industrial hygiene can be found in the following journals and magazines.

Across the Board (formerly *Conference Board Report*). Conference Board, 845 Third Ave, New York, NY 10022.

Air Pollution Control (Novi). Eagle Publications, 42400 Nine Mile Rd, Ste B, Novi, MI 48375. Bimonthly.

Air Pollution Control (Washington). The Bureau of National Affairs, 1231 25th St, NW, Washington DC 20037. Biweekly.

American Industrial Hygiene Association Journal. American Industrial Hygiene Association, 345 White Pond Dr, Box 8390, Akron, OH 44320-1155. Monthly.

American Journal of Epidemiology. Johns Hopkins University, School of Hygiene and Public Health, 2007 E Monument St, Baltimore, MD 21205.

American Journal of Medicine. Cahners, Medical Health Care Group, York Medical Journal, 249 W 17th St, New York, NY 10011-5301.

American Journal of Public Health. American Public Health Association, 1015 Fifteenth St, Washington DC 20005. Monthly.

American Medical News. American Medical Association, 515 N State St, Chicago, IL 60610.

American Review of Respiratory Disease. American Lung Association, 1740 Broadway, New York, NY 10019.

Analytical Chemistry. American Chemical Society, 1155 16th St, NW, Washington DC 20036.

Annals of Occupational Hygiene. Pergamon, Journals Division, 660 White Plaines Rd, Tarrytown, NY 10591-5153.

Applied Occupational and Environmental Hygiene (formerly *Applied Industrial Hygiene*), 6500 Glenway Ave, D-7, Cincinnati, OH 45211. Monthly.

Archives of Environmental Health. Heldref Publications, 1319 18th St, NW, Washington DC 20036-1802. Bimonthly.

Archives of Internal Medicine. American Medical Association, 515 N State St, Chicago, IL 60610.

Archives of Pathology and Laboratory Medicine. American Medical Association, 515 N State St, Chicago, IL 60610.

ASHRAE Journal. American Society of Heating, Refrigerating and Air-Conditioning Engineers, 1791 Tullie Circle, NE, Atlanta, GA 30329.

British Journal of Industrial Medicine. British Medical Association, Tavistock Sq, London WC1 H9JP, UK. Monthly.

British Medical Journal. Tavistock Sq, London WC1 H9JP, UK.

Chemical and Engineering News. American Chemical Society, 1155 16th St, NW, Washington DC 20036.

Environment. Scientists' Institute for Public Information, Heldref Publications, 1319 18th St NW, Washington DC 20036-1802.

Environmental Health and Safety News. University of Washington, Department of Environmental Health, School of Public Health and Community Medicine, F-461 Health Sciences Bldg, Seattle, WA 98195. Monthly.

Environmental Health Perspectives. National Institute of Environmental Health Sciences. P O Box 12233, Research Triangle Park, NC 27709.

Environmental Research. Academic Press, Journal Division, 1250 Sixth Ave, San Diego, CA 92101.

Environmental Science and Technology. American Chemical Society, 1155 16th St, NW, Washington DC 20036. Monthly.

Ergonomics. Taylor & Francis, 242 Cherry St, Philadelphia, PA 19106.

Excerpta Medica. Section 35: Occupational Health and Industrial Medicine. Elsevier, P O Box 882, Madison Sq Sta, New York, NY 10159. Monthly.

Health Physics. Williams & Wilkins, 428 E Preston St, Baltimore, MD 21202. Sponsored by the Health Physics Society. Monthly.

Industrial Hygiene Digest. Industrial Health Foundation, 34 Penn Circle W, Pittsburgh, PA 15206. Monthly.

Industrial Hygiene News. Rimbach Publishing, 8650 Babcock Blvd, Pittsburgh, PA 15237.

International Archives of Occupational and Environmental Health. Springer-Verlag, 175 Fifth Ave, New York, NY 10010. Quarterly.

Job Safety and Health. The Bureau of National Affairs, 1231 25th St, NW, Washington DC 20037. Biweekly.

Journal of the Acoustical Society of America. American Institute of Physics, 335 E 45th St, New York, NY 10017.

Journal of the American Medical Association (JAMA). American Medical Association, 515 N State St, Chicago, IL 60610.

Journal of Aviation, Space and Environmental Medicine (formerly: *Aerospace Medicine*). Aerospace Medical Association, 320 S Henry St, Alexandria, VA 22314-3579. Monthly.

Journal of Toxicology: Clinical Toxicology. Marcel Dekker, 270 Madison Ave, New York, NY 10016. Bimonthly.

Journal of Occupational Medicine. American Occupational Medicine Association, Williams & Wilkins, 428 E Preston St, Baltimore, MD 21202. Monthly.

Journal of Toxicology & Environmental Health. Hemisphere, 1900 Frost Rd, Ste 101, Bristol, PA 19007-1598.

Lancet. Williams & Wilkins, 428 E Preston St, Baltimore, MD 21202.

New England Journal of Medicine. New England Journal of Medicine, 1440 Main St, Waltham, MA 02254.

Noise Control Engineering Journal. Institute of Noise Control Engineering, Department of Mechanical Engineering, Auburn University, Auburn, AL 36849.

Noise Regulation Report (formerly *Noise Control Report*). Business Publishers, 951 Pershing Dr, Silver Springs, MD 20910-4464.

Noise/News. Noise Control Foundation, 2469 Arlington Branch, Poughkeepsie, NY 12603.

Occupational Hazards. Penton, 1100 Superior Ave, Cleveland, OH 44114.

Occupational Health. Available from U.S. National Technical Information Service, 5825 Port Royal Rd, Springfield, VA 22161.

Occupational Health & Safety. Stevens Publishing Corp, 225 N New Rd, Waco, TX 76710.

Occupational Health and Safety Letter. Business Publishers, 951 Pershing Dr, Silver Springs, MD 20910-4464.

Occupational Medicine (formerly *Journal of the Society of Occupational Medicine*). Butterworth-Heinemann, Linacre House, Jordan Hill, Oxford OX2 8DP, UK.

Occupational Safety & Health Reporter. Bureau of National Affairs, 1231 25th St, NW, Washington, DC 20037. Weekly.

Occupational Safety & Health Subscription Service. Standards, Interpretations, Regulations, and Procedures. U.S. Occupational Safety and Health Administration, Washington DC, 5 vols. (OSH 01 through OSH 05). Available from Government Printing Office. Catalog No. L35.6/3-1 through -5.

OSHA Up-to-Date. National Safety Council, 1121 Spring Lake Dr, Itasca, IL 60143-3201. Monthly.

Plant Engineering. Division of Reed Publishing, 1350 E Touhy Ave, Box 5080, Des Plaines, IL 60017-5080.

Professional Safety (formerly *American Society of Safety Engineers Journal*). ASSE, 1800 E Oakton St, Des Plaines, IL 60018.

Public Health Reports. U.S. Public Health Services, Department of Health and Human Services, Parklawn Bldg, Rm 13C-26, 5600 Fishers Ln, Rockville, MD 20857.

Safety + Health. National Safety Council, 1121 Spring Lake Dr, Itasca, IL 60143-3201.

Scandinavian Journal of Work, Environment & Health. Finnish Institute of Occupational Health, Topeliuksenkatu 41aA, SF-00250 Helsinki, Finland.

Skin and Allergy News. International Medical News Group, 12230 Wilkins Ave, Rockville, MD 20852.

Sound and Vibration. Acoustical Publications, PO Box 40416, Bay Village, OH 44140.

State of the Art Reviews: Occupational Medicine. Hanley & Belfus, 210 S 13th St, Philadelphia, PA 19107. Quarterly.

Toxicology and Applied Pharmacology. Academic Press, Journal Division, 1250 Sixth Ave, San Diego, CA 92101.

STATES WITH ON-SITE OSHA CONSULTATION AGREEMENTS

Rules were promulgated under section 7(c)(1) of the Occupational Safety and Health Act of 1970 (84 Stat. 1590), which authorizes OSHA to reimburse state agencies for use of their personnel to provide on-site consultation services to employers. The service will be made available at no cost to employers to assist them in providing their employees a place of employment that is safe and healthful. Consultants will identify specific hazards in the workplace and provide advice on their elimination. On-site consultation will be conducted independently of any OSHA enforcement activity, and the discovery of hazards will not mandate citation or penalties.

Here is a listing by state to request on-site consultation:

OSHA On-site Consultation Project Directory, 1992.

State and Telephone:

Arizona
602/255-5795

Arkansas
501/682-4522

California
415/982-8515

Colorado
970/ 491-6151

Connecticut
860/566-4550

Delaware
302/761-8219

District of Columbia
202/576-6339

Florida (IV)
904/488-3044

Georgia
404/8942643

Guam
671/475-0136

Hawaii
808/586-9100

Idaho
208/385-3283

Illinois
312/814-2337

Indiana
317/232-2688

Iowa
515/965-7162

Kansas
913/296-7476

Kentucky
502/564-6895

Louisiana
504/342-9601

Maine
207/624-6460

Maryland
410/880-4970

Massachusetts
617/727-3982

Michigan (Health)
517/332-8250 (health)

Minnesota
612/297-2393

Mississippi
601/987-3981

Missouri
573/751-3403

Montana
406/444-6418

Nebraska
402/471-4717

Nevada
702/486-5016

New Hampshire
603/271-2024

New Jersey
609/292-2424

New Mexico
505/827-4230

New York
518/457-2481

North Carolina
919/662-4644

North Dakota
701/328-5188

Ohio
614/644-2246

Oklahoma
405/528-1500

Oregon
503/378-3272

Pennsylvania
412/357-2561

Puerto Rico (II)
787/754-2188

Rhode Island
401/277-2438

South Carolina
803/734-9614

South Dakota
605/688-4101

Tennessee
615/741-7036

Texas
512/440-3809

Utah
801/530-7606

Vermont
802/828-2765

Virginia
804/786-6359

Virgin Islands
809/772-1315

Washington
360/902-5638

West Virginia
304/558-7890

Wisconsin
   608/266-8579 (health)
   414/521-5063 (safety)

Wyoming
   307/777-7786

SELECTING AND USING AN OCCUPATIONAL HEALTH & SAFETY CONSULTANT

At times, an occupational health and safety problem will defy your best efforts to find a solution: those are the times to seek the advice of an expert. Someone who has studied occupational safety, industrial hygiene, and/or occupational medicine for many years can often find a more economical solution than would occur to a newcomer or inexperienced person.

This section covers what a consultant does, thus enabling you to make better use of consulting services.

Many companies selling personal protective equipment, personal protective wear, and sampling and measuring instruments are staffed by knowledgeable professionals. Often, they can advise you on a specific health and safety issue at no cost.

An occupational health and safety consultant, however, is an independent professional, a group of professionals, a private consultant, or a state government agency that performs services for clients on a fee basis. Consulting services can be as broad or narrow as needed. Using a consultant who is certified usually guarantees that the project, whether long-term or occasional, will be addressed with maximum skill and economy.

A consultant can offer the following:

- creativity born of diverse experience with a wide range of problem-solving applications
- professional independence—An independent consultant is in a position to divorce judgment of methods and materials from all secondary interests and keep a purely objective viewpoint.
- successful solutions at a reasonable cost
- organizational stability—The consultant reduces the need for the organization, administration, and eventual disbandment of large internal staffs formed to meet peak load projects or specialized problems.
- complete accreditation—Consultants are usually qualified by education, experience, and certification to give complete consulting services.
- flexibility and mobility—The consultant is available whenever special counsel is required.

Before a decision is made about hiring a consultant, question whether the consultant can:

- Do the job faster?
- Do it better?
- Do it at less overall cost?

Typical situations when a consultant can be used are as follows:

- when you do not know exactly what to do
- when you do not know how to do it
- when you want third-party advice
- when you need an unbiased viewpoint
- when you need access to special facilities—Special facilities or equipment may be needed for one project and cannot be justified as a purchase, or acquired in the time available.
- when you need to pursue alternate solutions—Staff may not be available internally to analyze many possible solutions to a problem.
- when you need to convince management to take a particular course of action. An in-company solution to a problem may exist, but management may refuse to believe the recommendations of its own people.

In the final analysis, the decision to hire an industrial hygiene consultant is determined by the economics of the situation.

Consultants can be used to keep management aware of both current and proposed federal and state regulations in the area of occupational health and safety. Occupational safety consultants can assess the organization and administration of the safety and health program. They can inform management when medical examinations of employees may be recommended or required by regulation. They should be able to recommend appropriate physicians or clinics specializing in occupational medicine. Industrial hygiene consultants can play a valuable role in providing the examining physician with information on the occupational exposures of employees.

Consultants can design employee training programs. A consultant can serve as an expert witness if your company is involved in a lawsuit and data must be obtained, interpreted, and presented by a disinterested third party.

To maximize use of consulting services, the purpose of the investigation or study must be defined. In any problem-solving situation, it is important to find out what the problem really is—before you try to solve it. This is the essence of what a good consultant does. Company personnel may not always be able to describe the problem clearly.

The consultant is obligated to pinpoint the problem, regardless of what the client thinks it is. This can be difficult, to say the least, but it must be done. Nothing is so disconcerting as solving the wrong problem or a nonexistent problem.

Define the Problem
First, try to formulate and define the problem; then list the specific qualifications and experience a consultant should have to solve this problem.

Currently any person can legally offer services as an occupational safety and health consultant; consequently, it is important to avoid hiring those who are unsuitable because of lack of training, experience, or competence. Individuals or firms billing themselves as occupational health and safety consultants can be broadly classified according to (1) whether they recommend a particular procedure, product, service, or control process, or (2) whether they are independent consultants.

Product-oriented individuals or firms vary in their backgrounds from nontechnical product sales personnel to experienced occupational health and safety professionals. In this case, *consulting consists* mainly of recommending appropriate equipment and facilities. This type of consultation can include assistance in soliciting proposals for the design and installation of control equipment, such as ventilation control systems or respirators.

The advantage of using this group directly is that you avoid consultant costs and pay only for the product or service. The disadvantage in dealing with a product-oriented consultant is that these consultants may not consider all options available. Thousands of dollars could be spent in purchasing a particular control system, only to discover later that the desired results cannot be obtained or that another solution could have been obtained for less money.

If there are any doubts as to the proper method for solving a problem, then an independent consultant (one free from ties to a particular service or line of products) should be called in. It is this type of industrial hygiene consultant that will be discussed in the remainder of this section.

Sources
There are several sources one can go to for information and names of consultants. One source of information is professional associations and public-service organizations related to occupational safety and health, such as the National Safety Council, American Industrial Hygiene Association (AIHA), and the American Society of Safety Engineers (ASSE). These three associations have local chapters, sections, or offices in major cities that can be a source of consultation information and assistance. In fact, the National Safety Council offers consulting services—phone (800) 621-7615 for more information.

Many insurance companies now have loss prevention programs that employ occupational health and safety professionals. Ask your present insurer; you may want to compare the services they offer with those of other insurance companies. Finally, there may be a university or college in your area that has an environmental health program (see the listing earlier in this section).

Selection
Selecting a consultant should be guided by one primary consideration—the qualification of the consulting staff for the project to be undertaken. The size of a consulting firm is seldom a reliable single determinant, nor is the length of time in practice a major factor for consideration.

A good line of action to follow is to consider the qualifications of a number of individuals or firms that appear to be capable of meeting the requirements of the project to be undertaken.

Select a limited number of individuals or firms that appear to be best qualified for the particular project. Write each of them individual letters—briefly describe the project and inquire as to their interest in it. Invite the companies expressing interest to come in for separate personal interviews. At the interview, go over the qualifications and record of each firm. Have the firms submit up-to-date data and available staff information, a brief description of work on hand that might possibly conflict with your project, and the qualifications of specific key personnel who will be assigned to your company's project.

A series of questions for the consultant is given here. They should not be given equal weight since some are minor in importance. (The list is organized roughly in descending order of importance.)

EXPERIENCE
1. How many years have you been professionally active in the occupational health and safety field?

2. Please supply a list of recent clients that you have served, preferably in this geographical area and on problems similar to those in which I am interested.
3. What teaching have you done or training have you had in your professional specialty? What groups were involved—university, industry, trade associations, civic groups, engineers, symposia?

Consultation Status

1. Are you now an independent consultant? How many years? Full time or part time?
2. If part time:
   a. Who is your chief employer or in what other business ventures are you involved?
   b. Is your employer aware and does he/she approve of your part-time activity as a consultant?

Education

1. What schools did you attend and what courses did you take related to occupational health and safety?
2. What degrees did you receive and when?
3. What special conferences, seminars, symposia, or short courses have you attended (especially recently) to stay abreast of technical information and governmental regulations?
4. What other sources of information do you use to keep up-to-date with your field?

Professional Affiliations

1. What professional associations do you belong to? (Representative ones are the American Industrial Hygiene Association, National Safety Council, American Conference of Governmental Industrial Hygienists, and the American Society of Safety Engineers.) What is your present grade of membership and length of time in that grade for each association?
2. Are you certified by any of the following?
   a. American Board of Industrial Hygiene (specify area of certification)
   b. Board of Certified Safety Professionals
3. Are you a registered professional engineer? In what states and disciplines?
4. Of what professional engineer associations are you or your firm a member?
5. Of what trade associations, chambers of commerce, or similar business groups are you or your firm a member?

Special Capabilities

1. In what areas of occupational health and safety do you specialize?
2. What equipment do you have for conducting occupational health and safety evaluations? Other types of inspections?
3. What laboratories do you use for the analysis of your exposure measurement samples?
4. Can you serve as an expert witness, either for your client or as a friend of the court?
5. What experience have you had as an expert witness?

Business Practices

1. Please indicate your fee structure. Do you work by hourly charges, estimates for the total job, retainer charges, or any combination of these?
2. In your charges, how do you treat such expenses as travel, subsistence, shipping, report reproduction, and computer time?
3. Can you supply a list of typical laboratory analytical fees?

Compensation

Compensation for consultant services may be calculated and established by a variety of methods.

- Fixed lump sum
- Cost plus a fixed amount
- Salary cost times a factor, plus incurred or out-of-pocket expenses
- Per diem

As with medical, legal, and other professional services, consulting services should never be secured purely on the basis of a price comparison. Competitive proposals for professional consulting services are undesirable because there is no direct cost basis for comparison of services that involve judgment and creative thinking. These factors cannot be evaluated precisely in advance of performance.

The Proposal

Once you have selected a consultant, you can arrange to obtain services in several ways. A verbal commitment is sometimes all that is necessary. However, you may wish to request a written proposal that spells out the steps to be taken in the solution of your problem.

Aside from background qualifications of the consultant, the proposal should answer the questions:

1. How much is the service going to cost? Smaller jobs are often bid on an hourly basis, with a minimum of one-half day's work, plus direct expenses commonly specified. Larger jobs are usually bid at a fixed amount.

2. What is the consultant going to do? The answer to this question may range from a simple agreement to study the problem to a comprehensive step-by-step plan to solve it.
3. What will be the result? The answer to this question is all too often not clearly understood; the result is usually a report that specifies the consultant's recommendation. If you do not want to pay for the preparation of a written report because a verbal one will do, specify this in advance.

Since recommendations often call for construction or other operations to be carried out by others whose work is not subject to the consultant's control, results can usually not be guaranteed by the consultant. Rather, an estimate of the exposure control to be attained is all that can be expected.

If the consultant is to provide drawings from which the contractor will work, the proposal must specify sketches or finished drawings. If special materials are required, the consultant should agree to specify alternative selections, if possible. If you want a guaranteed result, experimental work will usually be necessary and will have to be paid for.

The consultant can monitor construction to determine compliance with specifications. The consultant can also measure after installation to confirm predictions and supply oral briefings as needed.

Even if your consultant is to serve as an expert witness for you, the consultant is not automatically on your side. Rather, the consultant is more like a friend of the court, devoted to bringing out the facts, with careful separation of fact from expert opinion.

Working with the Consultant

Once an occupational health and safety consultant is hired, the problem should be defined as exactly as possible. There should be no guesswork unless there is no other choice. All the data bearing on the problem should be provided at the beginning of the study or investigation—it does no good to leave out embarrassing or unpleasant facts. (The consultant will find them anyway, and the extra time spent will cost you money.) The consultant should be introduced to key members in your organization, and appropriate personnel should be available when needed.

Before hiring a consultant, consider the following items. Be sure to have any necessary equipment available and a place for the consultant to work. Do not let the consultant wander around alone trying to find information or things. Do not hold back information from the consultant.

Monitor the consultant's progress regularly. Check to see that corporate personnel are cooperating fully, that there is no personality conflict, and that the consultant is getting the help desired.

Review particularly the beginning phases, including the formulation of the problem and the analysis of the problem. Review at the end of each phase.

The three phases of consulting work. Most consulting work can be divided into three phases—problem definition, problem analysis, and solution.

The problem definition phase lays the groundwork for subsequent stages. The need for thoroughness and accuracy at this stage is obvious.

The problem analysis phase deals with the facts. The consultant defines the problem and identifies opportunities for improvement, determines the causes of the problem, determines the objectives to be met by the solution, and develops alternative solutions.

The solution phase involves selecting the most effective, workable, timely, and practical solution. The details of the solution must be worked out carefully. The solution should be the equivalent of a blueprint describing what needs to be done, how it is to be done, by whom, and in what sequence the actions are to take place.

Problems and pitfalls. The successful consulting job is always a team effort between the company and the consultant. Too often a consultant is literally challenged to do a successful job as though someone were saying, "We couldn't do it and neither can you."

Some typical problems and pitfalls that you may run into are described here. Some of these items are negative versions of items on the checklist. Some of the failures listed are the company's fault, some are the consultant's, and some involve both.

These pitfalls include the following failures:
- to clearly define the problem
- to set specific objectives
- to establish financial arrangements in the beginning
- to establish realistic time requirements
- to select the best qualified consultant
- to make all pertinent information available to the consultant
- to obtain a clear and complete proposal from the consultant

- to review the oral or written proposal for clarity, completeness, creative approaches, and qualification
- to make a cost/benefit analysis
- to review progress periodically
- because of exceeding the scope of the assignment.

Achieving the greatest value from a consultant is largely a matter of providing full information, defining responsibilities clearly, and establishing workable lines of communication. The consultant should have full information regarding the project from the outset. The where, when, how, and why of the task are the very tools needed to solve the problem.

Appendix—
Contributors

Mary Amann, COHN-S, is currently a faculty member at the University of Illinois College of Nursing. She is also a NIOSH Educational Resource Center continuing education instructor as well as a private consultant in the areas of occupational health management and information systems. She is the chair of the American Board of Occupational Health Nurses and section editor for the *American Association of Occupational Health Nurses Journal.* She was a manager of Occupational Health Services for AT&T from 1985–1999. Mary Amann can be contacted at maryamann@worldnet.att.net.

L. Darryl Armstrong, PhD, APR, CCMC, is a corporate and government facilitator, practitioner, and consultant in public participation and stakeholder involvement, organizational development, strategic planning, conflict resolution, and crisis and issues management. He is a former board member of the International Association of Public Participation (IAP2); past president of the Tennessee Valley Chapter of IAP2; is certified by the Institute of Crisis Management, and is accredited by the Public Relations Society of America (PRSA). He and his wife Kay operate Armstrong and Associates, the Commonwealth of Kentucky's only full-service public participation and public relations consulting firm. He has developed several workshops that he conducts throughout the United States. He can be contacted via e-mail at DRDarryl@aol.com.

Marci Z. Balge, RN, MSN, COHN-S, is an occupational/environmental health specialist and strategic planner for occupational and public health programs and services. She is a board-certified Occupational Health Nursing Specialist. Ms. Balge's experience over the past 20 years includes planning, implementation, and evaluation of all phases of occupational and environmental health programs; public health planning, assessment and evaluation; and food and housing sanitation audits, assessments, and training programs. She is currently a consultant at NewFields, Inc., in Denver, Colorado. As a professional consultant, Ms. Balge has been involved in public and project health planning for a variety of projects in multiple international settings. Most recently she has participated in project and public health planning for a major pipeline through Chad and Cameroon, using newly revised World Bank guidelines. This has included focused programs addressing project health and safety organization, personnel issues with medical impacts, sanitary engineering, water/sewage/waste programs, preventive food services, laundry services, housing sanitation, pest control, on-site medical services, community health resources, health education, medical surveillance program management, medical emergency response plans, and specific disease prevention strategies. She has also participated in preliminary disease status assessments for several other countries in Africa, Latin America, and Southeast Asia. Ms. Balge has designed, implemented, managed, and evaluated medical surveillance programs for a variety of businesses and industry. These work environments include petrochemical plants, hazardous waste operations, EPA superfund sites, wastewater treatment facilities, health care facilities, DOE facilities, and construction operations. Areas of focus include in-depth analyses of potential employee exposure, establishment of systematic biological sampling programs, evaluation of employee work practices and schedules, and establishment of employee tracking systems. Ms. Balge is active in several professional organizations She has published a number of articles and book chapters and has presented papers at several scientific conferences. She can be contacted via e-mail at mbalge@newfields.com.

Charles E. Becker, MD, director, Center for Occupational and Environmental Health, San Francisco General Hospital Medical Center, San Francisco. Dr. Becker also serves San Francisco General as chief of toxicology and is professor of medicine and pharmacy at the University of California School of Medicine. He was an advisor on acute toxics to the U.S. Environmental Protection Agency Science Advisory Board, a special consultant to the National Academy of Science Institute of Medicine, and serves on several national, state, and local committees and panels devoted to environmental and occupational health. Mailing address: Charles E. Becker, MD, COEH, San Francisco General Hospital Medical Center, Bldg 30, 5th Floor, San Francisco, CA 94110.

George S. Benjamin, MD, medical director, Managed Health Care, Liberty Mutual Insurance Group, Boston, Mass. Dr. Benjamin's career encompasses 12 years' surgical experience in the U.S. Army Medical Corps, including a special assignment to the National Aeronautical & Space Administration, as well as private, corporate, and association medical and surgical practice, as an instructor for Tufts University and Harvard Medical School; and as professor at the University of Illinois' schools of Medicine and Public Health. Prior to joining Liberty Mutual, Dr. Benjamin was director of Occupational, Environmental, and Consumer Health Programs for the National Safety Council; he now serves as Council liaison to the American College of Occupational Medicine. Mailing address: George S. Benjamin, MD, Managed Health Care, Liberty Mutual Insurance Group, 175 Berkeley St, Boston, MA 02117-0140.

Kimberly Doherty Bradley is a senior consultant at Kazarians & Associates. Ms Bradley is an expert in environmental compliance and has specialized in clean water and underground storage tank-related regulations. She has served as project manager and account manager for numerous private and public sector clients. These projects have included facility audits, risk management, pollution prevention and waste minimization, site assessment and investigation, remedial action evaluation and implementation, storm water permitting, industrial compliance documentation and training, underground storage tank removal and replacement, and construction surveillance. Ms. Bradley has accumulated extensive experience in compliance with local, state, and federal regulations. Ms. Bradley provides strategic planning assistance and support to private sector facilities and industry trade associations an a variety of compliance issues, including those related to the CAA, CWA, RCRA, and CERCLA. She has extensive experience in dealing with various regulatory agencies and attorneys and in representing clients in regulatory negotiations. She can be contacted at Kazarians & Assoc., 425 E. Colorado St., Ste 545, Glendale, CA 91205.

Stephen R. Burastero, MD, MPH, is director of the Foreign Travel Clinic, Lawrence Livermore National Laboratory and assistant clinical professor at University of California San Francisco, School of Medicine. Dr. Burastero has done public health research in the Malaria Branch at the Centers for Disease Control and has practiced medicine in Africa and the Middle East. He was involved in preparing members of the United Nations peacekeeping force for travel to Iraq; and is currently involved in planning health emergency medical response Department of Energy programs in the former Soviet Republics. He is conducting research in ensuring business traveler compliance with health precautions and travel medicine for Russia. Dr Burastero is board-certified in Internal Medicine and Occupational Medicine, and holds membership in the International Society of Travel Medicine. Dr. Burastero can be contacted at Health Services Department, Lawrence Livermore National Laboratory, P O 808, L-723, Livermore, CA 94551 or via e-mail at burastero1@llnl.gov.

Barbara J. Burgel, RN, COHN, MS, associate clinical professor, Department of Mental Health, Community and Administrative Nursing, School of Nursing, University of California, San Francisco. Ms. Burgel also practices as an occupational health nurse practitioner at the University of California Occupational Medicine Clinic. She is certified by the American Board for Occupational Health Nurses, of which she is a director, and is active on several committees for the American Association of Occupational Health Nurses. Ms. Burgel's former affiliations include the University of Michigan Medical Center, Ann Arbor; St. Mary's Medical Center, San Francisco; and San Francisco General Hospital. In 1990, she was named "Teacher of the Year" by Master's Students at the UCSF. Mailing address: Barbara J. Burgel, RN, MS, University of California School of Nursing, DMHCAN, N511F, Box 0608, San Francisco, CA 94143-0608.

Regina M. Cambridge, RN, COHN-S, has been an occupational health nurse for 18 years. She was certified in Occupational Health Nursing in 1989. She now works for National Renewable Energy Laboratory, 1617 Cole Blvd., Golden, CO 80401.

Barnett L. Cline, MD, MPH, PhD, is professor emeritus of Tropical Medicine, Tulane University School of Public Health and Tropical Medicine. Dr. Cline began his career as a U.S. Peace Corps Physician in Bolivia and Costa Rica. Since that time he has served as director, San Juan Laboratories, Bureau of Laboratories, CDC, U.S. Public Health Service; assistant director, Helminthic Diseases Branch, Division of Parasitic Diseases, CDC; professor and chairman, Tulane University School of Public Health and Tropical Medicine; and director of the Center for International Community-Based Studies, Tulane School of Public Health and Tropical Medicine, New Orleans, Lousiana. Dr. Cline currently serves as a member of the Program Review Group, WHO Collaborative Global Programme to Eliminate Lymphatic Filariasis. The recipient of the Distinguished Service Award, U.S. Dpeartment of Health and Human Services in 1997, Dr. Cline has also published extensively in professional journals, including *Advances in Parasitology,* the *American Journal of Tropical Medicine and Hygiene,* and *Acta Tropica*. He can be contacted via e-mail at blchome@moment.net.

Thomas J. Coates, PhD, professor of medicine, University of California, San Francisco, with joint appointments in the departments of Psychiatry, Epidemiology, and Biostatistics; Director of the Behavioral Medicine Unit in the Division of General Internal Medicine; and Director of the Center for AIDS Prevention Studies. Dr. Coates previously was on the staff at Johns Hopkins University School of Medicine and on the faculty of the Stanford Heart Disease Prevention Program. His interests and experience have focused on the study of disease-related behavior, with an emphasis on interventions to modify high-risk behaviors in adolescents and adults. He serves on several committees of the National Research Council and National Academy of Sciences, and chairs the Standing Committee for Social and Behavioral Research for the World Health Organization Global Programme on AIDS. Mailing address: Thomas J. Coates, PhD, Box 0320, Division of General Internal Medicine, University of California, San Francisco, San Francisco, CA 94143.

Bertram Cohen, JD, is presiding judge, Workers' Compensation Appeals Board, Stockton, California. Dr. Cohen's career in workers' compensation extends across education, publishing, and private practice as an attorney-at-law. A member of the California State Bar and a Certified Specialist in workers' compensation law, Cohen lectures at the Berkeley and Davis campuses of the University of California. He also serves as lecturer, moderator, and a panelist on case law development, judge training, industrial claims, and medical-legal programs for many other educational organizations and seminars. Mailing address: Bertram Cohen, Presiding Judge, Workers' Compensation Appeals Board, 31 E. Channel St., Room 344, Stockton, CA 95202.

Richard M. Coleman, PhD, president of Coleman & Associates, a shiftwork operations consulting firm located in Ross, Calif. A psychologist, he is an assistant professor at Stanford University Medical School. He has helped more than 100 companies improve their work schedules, advised the U.S. Olympic Team on improved alertness and performance, and has been a frequent guest on a number of television talk shows. His most recent book is *Wide Awake at Three A.M.: By Choice or by Chance,* W. H. Freeman (1986). Mailing address: Richard M. Coleman, MD, P O Box 128, Ross, CA 94957.

Nita Drolet, RN, MPH, CIC, is Health Care Educator, Certified Infection Control Practitioner. She has developed infection control policies and education plans for hospitals, extended care, private industry, fire, police, EMS, and county governments. Ms. Drolet can be contacted via e-mail at nmd@worldnet.att.net.

Edwin B. Fisher, Jr., PhD, professor of psychology and director of the Center for Health Behavior Research, Department of Medicine, Washington University, St. Louis, Mo. His work has included the psychological aspects of health and health promotion with such subjects as smoking, diabetes, asthma, and cardiovascular diseases. In addition, Dr. Fisher's interest in interpersonal support for risk reduction and chronic disease management has led to studies of smoking cessation, cardiovascular risk reduction, and asthma management in worksite and community settings. Mailing address: Edwin B. Fisher, Jr., PhD, Center for Health Behavior Research, Washington University, 33 S Euclid, St. Louis, MO 63108.

Jean Spencer Felton, MD, a graduate of Stanford University, is clinical professor of medicine (Occupational Medicine), University of California, Irvine, and clinical professor emeritus of preventive medicine, University of Southern California. Dr. Felton has served as Medical Director of the Oak Ridge National Laboratory, Oak Ridge, Tennessee, and has held academic appointments at the University of Oklahoma, and University of California, Los Angeles. He has written extensively for the occupational medicine literature and is, in addition to being certified by the American Board of Preventive Medicine, a fellow of the Faculty of Occupational Medicine, Royal College of Physicians, London. He can be reached via e-mail at jfelton@mcn.org.

Douglas P. Fowler, PhD, CIH, is owner and principal consultant of Fowler Associates, Redwood City, California. Dr. Fowler provides occupational and environmental health services that help private organizations and public agencies to recognize, evaluate, and control occupational and environmental health hazards. His consultancy includes: asbestos evaluation and control; exposure assessment; toxic emissions; pesticides; electronic industry health and safety; indoor air quality; and occupational epidemiology. He also is a lecturer and course director on industrial hygiene for several courses on industrial hygiene for Programs in Environmental Hazard Management at the University of California. He has served on many committees for the American Industrial Hygiene Association and other professional organizations. His recent article on exposures to asbestos arising from bandsawing gasket material appears in *Applied Occupational Environmental Hygiene.* He can be contacted at Fowler Associates, Bayport Marina Plaza, 643 Bair Island Rd, Ste 305, Redwood City, CA 94063-2757.

Dana Headapohl, MD, president, Environmental Associates, Missoula, Mont. In addition to consulting on occupational and environmental health, Dr. Headapohl is clinical professor of the School of Pharmacy, University of Montana, and clinical instructor at the University of California's Division of Occupational and Environmental Medicine, San

Francisco. Her specialties include human biology, environmental engineering, and occupational medicine, in which she is board certified. She also serves as medical director of TEAMWORK Occupational Health Service at St. Patrick Hospital, Missoula. She consults on occupational safety and health programs, hazardous waste, environmental concerns, health risk assessment, indoor air pollution, ergonomics, and training. Mailing address: Dana Headapohl, MD, Occupational Health Services, St. Patrick Hospital, 500 W Broadway, Missoula, MT 59802.

Joan M. Heins, project director, Center for Health Behavior Research, Department of Medicine, Washington University, St. Louis, Mo. Heins' interests lie in organizational psychology applied to health care, with emphasis on alternative and innovative approaches to disease prevention and health promotion. In addition to research in worksite-based risk reduction, she has studied organizational factors associated with adoption of integrated patient care programs in health care settings. Ms. Heins serves on the Board of Directors of the American Diabetes Association. Mailing address: Joan M. Heins, Center for Health Behavior Research, Washington University, 33 S Euclid, St. Louis, MO 63108.

Thomas Herington, MD, medical director, Occupational Injury Clinic, California Pacific Medical Center, San Francisco. Dr. Herington also is vice chair of the Department of Occupational Medicine at California Pacific and is an assistant clinical professor of medicine at the University of California School of Medicine, San Francisco. He is board certified in internal medicine and occupational medicine. Mailing address: Thomas Herington, MD, Occupational Injury Clinic, California Pacific Medical Center, 2351 Clay St, Ste 134, P O Box 7999, San Francisco, CA 94120.

John S. Hughes, MD, is physician and president of Occupational Medicine Physicians of Colorado, PC. Dr. Hughes is board-certified by the American Board of Preventive Medicine in Occupational Medicine. He is an assistant clinical professor of Medicine at the University of Colorado School of Medicine and serves as a primary care preceptor for medical students as well as a resident preceptor. Dr Hughes is a particular expert in the area of workers' compensation, participating as a consultant to the Colorado Division of Workers' Compensation and as a member of several task force groups that developed rules of procedure pertaining to evaluation and treatment of occupational conditions. Dr. Hughes received operational medicine training through the U.S. Navy in diving, submarine, and radiation medicine, continuing into a tour of duty as medical officer for the Mobile Diving and Salvage Unit and SEAL Teams in Little Creek, VA. This culminated in Dr. Hughes' participation as contributory author of the *SEAL Team Physical Fitness Manual* in 1997; and he is currently a captain/medical corps in the U.S. Naval Reserve, assigned to Fleet Hospital 23 in Minneapolis, MN. He can be reached at Occupational Medicine Physicians, 20 W. Dry Creek Circle, Ste 111, Littleton, CO 80120.

Dennis T. Jaffe, PhD, consultant, HeartWork, Inc, San Francisco. Dr. Jaffe uses his training and background in organizational design and sociology to assist executives and professionals in organizational innovation leadership. He is a professor of Organizational Inquiry at the Saybrook Institute, San Francisco, and the coauthor with Cynthia D. Scott of numerous articles and books on management. Mailing address: Dennis T. Jaffe, PhD, HeartWork, Inc, 461 Second St, Ste 232, San Francisco, CA 94107.

Ira L. Janowitz, PT, CPE, is on the staff of the University of California San Francisco/Berkeley Ergonomics Program in Richmond, where he has consulting and teaching responsibilities. He has degrees in Industrial Engineering, Administration, and Physical Therapy. Mr. Janowitz has been involved in occupational health and safety programs since 1977, and is board-certified as a Professional Ergonomist. He has published several articles and book chapters on the prevention and management of work-related musculoskeletal problems, and has assisted many employers and insurance companies in addressing these issues. He can be contacted via e-mail at janowitz@earthlink.net; his home page address is http://www.me.Berkeley.edu:80/ergo/ira.html.

Sara Joswiak, MPH, is an industrial hygienist whose areas of expertise include industrial hygiene monitoring, training, and OSHA compliance audits. She has worked as a staff representative at the National Safety Council and as a corporate industrial hygienist for a large hazardous waste recycling corporation. She has performed ergonomic evaluations, indoor air quality investigations, noise dosimetry studies, and noise maps, in addition to monitoring a wide variety of air contaminants for industrial hygiene studies. She has developed and presented training programs on a variety of occupational health and safety topics to managers, facility personnel, and safety employees. She has experience in loss control, and was health and safety coordinator at a chemical plant where she also wrote technical material, including Material Safety Data Sheets and product labels. She can be contacted via e-mail at thejoswiaks@juno.com.

Mardy Kazarians, PhD, is a principal of Kazarians & Associates. He is a recognized expert in risk and safety with considerable experience in developing safety programs and procedures, in training personnel on safety practices, conducting and training process safety management programs, hazard and operability studies, safety audits, and risk assessments. Mr. Kazarians has published numerous articles on the subjects of risk management, risk assessment, fire analysis, reliability, and risk management, and is a lecturer of industry courses. He has contributed to or has been the principal investigator for risk and safety studies for refineries, chemical manufacturing plants, water treatment facilities, refrigeration systems, aerospace manufacturing plants, semiconductor manufacturing plants, nuclear fuel processing plants, nuclear power plants, and transportation systems. Mr. Kazarians, in addition to consulting, has developed and taught courses in risk and safety at the University of Southern California. He can be contacted via e-mail at mardy.kazarians@rmpcorp.com.

Keith Keller is the office safety coordinator for the Dames & Moore/URS Corporation staff at the West Valley Demonstration Project, a U.S. Department of Energy-operated facility. He has been involved in occupational health and safety programs in various capacities throughout his 20-plus years as an environmental consultant, including safety coordinator positions at various U.S. Environmental Protection Agency Superfund sites. He can be contacted via e-mail at kellerk@wv.doe.gov.

Sharon D. Kemerer, RN, MSN, COHN-S, is executive director, American Board for Occupational Health Nurses, Inc. Ms. Kemerer is responsible for all aspects of program management for the national certification board for occupational health nurses, and acts as its primary spokesman. Prior to this position, she was a consultant and program manager for occupational health with major consulting firms, providing occupational health project services to large national clients. She has experience in the design and management of hospital-based occupational health delivery systems and in nursing education. She can be contacted via e-mail at skemerer@abohn.org.

Gary R. Krieger, MD, MPH, DABT, is a board-certified medical toxicologist with over 15 years' experience regarding hazardous materials toxicology in both a workplace and environmental setting. He is currently a principal at NewFields, Inc., in Denver, Colorado, and also an associate professor of Toxicology (adjunct), University of Colorado, Department of Toxicology, School of Pharmacy. As a professional consultant, Dr. Krieger has been involved in hundreds of projects involving the risk assessment and toxicology of potentially hazardous materials in both workplace and general environmental (community) settings. He has extensive experience with Superfund, RCRA, and OSHA; and has successfully completed numerous projects in every EPA region in the United States. In addition to his U.S. experience, Dr. Krieger has been an advisor and has worked on projects with the U.S. Peace Corps, U.S. Agency for International Development (U.S. AID), Harvard International Institute for International Development (HIID), World Bank, and numerous international governments in Europe and Asia-Pacific and Africa. He has been an invited speaker at many U.S. and international scientific forums and meetings. Dr. Krieger is recognized as an expert in toxicology, occupational medicine, and internal medicine at the federal and state court levels. He is board-certified in Internal Medicine, Occupational Medicine, and Toxicology. He is a member of numerous professional societies and serves on the editorial boards of several scientific journals. Dr. Krieger is the co-editor of the widely used toxicology book, *Hazardous Materials Toxicology: Clinical Principles of Environmental Health* (1st and 2nd editions). In addition, he is the co-editor of four books for the National Safety Council covering environmental affairs, occupational medicine and industrial health, and safety practice. Dr. Krieger has written numerous book chapters, review articles, and original scientific papers in the areas of toxicology, occupational medicine, and public health. He serves on the editorial board of the *Journal of Safety Research*. His recent public health work has centered on the potential for disease amplification and magnification associated with large infrastructure projects in sub-Saharan Africa. Dr. Krieger can be contacted via e-mail at gkrieger@newfields.com.

Richard Lack, PE, CSP, CPP, CHCM, RSP (UK), is consultant, Safety & Protection Services Consulting, Cheyenne, Wyoming. Prior to starting his consulting business in 1977, Lack pursued a 36-year career in safety management in various industries and organizations, including the San Francisco International Airport Commission, the Great Plains Coal Gasification project in North Dakota, Reynolds Metals Company, and Kaiser Aluminum & Chemical Corporation in Jamaica, West Indies. He is a registered Professional Engineer in Safety Engineering and has certifications as a Safety Professional, Protection Professional, Hazard Control Manager, Safety Manager and is a Registered Safety Professional (UK). He has served on the board of directors of the American Society of Safety Engineers, and was recently elected as fellow of the society. He has served on the governing board of the National Safety Management Society and currently is a member the executive committee of the International Air Transport Section of the National Safety Council. Lack is a member of many safety- and security-related organizations, is a regular speaker at professional conferences, and author of numerous papers. He has edited the *Essentials of Safety and Health Management* (CRC Press, 1996) and the *APM: Security Management* (National Safety Council, 1997). He can be reached via e-mail at rwlack@lonetree.com.

Joseph LaDou, MD, director of the International Center for Occupational Medicine, University of California School of Medicine, San Francisco. He directs the continuing education programs in occupational medicine at UCSF where more than 1,000 physicians have received short-course instruction in the specialty of occupational medicine. Dr. LaDou was chief of UCSF's Division of Occupational and Environmental Medicine until assuming his current post. His research interests are in the areas of shiftwork, microelectronics, and international issues in occupational and environmental medicine. He is an officer in the Society for Occupational and Environmental Health and the International Commission on Occupational Health. He has trained more than 100 occupational physicians from newly industrialized countries. Dr. LaDou is editor of *Occupational Medicine,* Appleton & Lange (1990), and has written numerous chapters in textbooks and scientific articles. He is chief editor of this revision of *Occupational Health & Safety.* Mailing address: Joseph LaDou, MD, International Center for Occupational Medicine, University of California, 350 Parnassus Ave, Ste 609, San Francisco, CA 94117-0924.

Michael Larsen has more than 25 years of experience managing environmental and technology-based business. He has started and operated several successful companies. He is a skilled communicator with special expertise in strategic planning, risk management, and productivity. He is a public speaker and management consultant with his own firm Larsen Environmental Associates. For more information see the website: www.larsenenviron.com.

Connie S. Lawson, RN, BSN, MS, C, COHN-S/CM, CHES, CUSA, is director, Disease Prevention and Safety, Health, and Environment Promotion, Bell Atlantic Network Services, Inc. Ms. Lawson has 20 years of diverse experience in safety and health care, with extensive professional experience in strategic planning, design, education, and communications. She has implemented innovative health and safety awareness and prevention programs for widely dispersed populations. She is the author of several articles on health and safety notably about the *whole person model,* which is an interdisciplinary approach to safety and health. She is certified by the National Safety Council as a Certified Utilities Safety Administrator (CUSA), by the National Commission for Health Education Credentialing, Inc., as a Certified Health Education Specialist (CHES), by the board of directors of the American Board for Occupational Health Nurses as a Certified Occupational Health Nurse Specialist and as a Case Manager COHN-S/CM, by the

American Nurses Credentialing Center in Nursing Continuing Education and Staff Development-C, and by the Joyce Institute to teach ergonomic skills. She is also certified in "ADULT" Learning, Delivery, and Development and holds a certificate in professional human resource management. She is currently the chairperson for the Occupational Health Nurses Advisory Committee to the National Safety Council. She can be contacted via e-mail at connie.s.lawson@bellatlantic.com.

Susan R. McKenzie, MD, consulting physician on occupational and environmental medicine. Dr. McKenzie is certified by the American Board of Preventive Medicine in occupational medicine. She has provided medical services to companies and health care facilities, including development of occupational health care programs, design and review of medical surveillance systems, provision of on-site physician care and education, and evaluation of work-related illness. She also served as district medical director on the Industrial Medical Council, California Department of Industrial Relations, which regulates medical-legal evaluations within the California Workers' Compensation System. Mailing address: Susan R. McKenzie, MD, P O Box 620215, Woodside, CA 94062.

Leela I. Murthy, MSc, PhD, has been with the U.S. National Institute for Occupational Safety and Health (NIOSH) since 1975 and has experience in dealing with occupational safety and health problems, specifically medical surveillance, biological monitoring, surveillance activities, criteria document writing, etc., for chemicals being used in the workplace. Dr. Murthy also has been an Equal Employment Opportunity (EEO) Counselor at NIOSH for over 15 years. In addition, Dr. Murthy is the only U.S. researcher involved in WHO's International Chemical Safety Card (ICSC) preparation. She was an sssistant professor of Chemistry for eight years in various colleges, including head of department at Mount Carmel College, Bangalore, India. She received her PhD in Bio-Organic Chemistry from the University of Cincinnati, Ohio, and was an assistant professor in the departments of Chemistry, Physiology, and Research Pediatrics for eight years. She conducted research on kidney transplant rejection at Shriner's Hospital, Cincinnati. She received individual grants from NIH to conduct research at Children's Hospital Research Foundation on phenylketonuria, a disease that causes mental retardation among children. Dr. Murthy can be reached via e-mail at lim1@cdc.gov.

Kenneth R. Pelletier, PhD, clinical associate professor of medicine and director of the Stanford Corporate Health Program, Stanford Center for Research in Disease Prevention, Department of Medicine, Stanford University School of Medicine; and senior clinical associate, Johnson & Johnson Health Management, New Brunswick, N.J. The Stanford Corporate Health Program is a collaboration between Stanford and 22 major corporations. Dr. Pelletier's specialty is developing and evaluating corporate health promotion and disease prevention programs. He is an advisor to the U.S. Department of Health & Human Services, the Canadian Ministry of Health, and the World Health Organization, as well as to many private firms. He is a widely published author and makes frequent television appearances on health and business. Mailing address: Kenneth R. Pelletier, PhD, Stanford University School of Medicine, 1000 Welch Rd, Stanford, CA 94304-1885.

Dan Petersen, ED, PE, CSP, is a consultant in safety management and organizational behavior. The list of Dr. Petersen's industrial and governmental clients is a roster of major corporations and high level agencies. Prior to setting up his own practice, he headed the graduate program in Safety Management at the University of Arizona and was associate professor at Colorado State University. In industry, he served in training and safety administration for several major corporations. He has published 12 books, 47 articles, and 13 videotapes on safety management. Dan Petersen can be contacted at 8031 E. Birwood Rd., Tucson, AZ 85750; phone & fax: 520/885-8992.

Becky L. Randolph, MS, CIH, is business development manager for Clarity Solutions, Inc. in Houston, Texas. Ms. Randolph has been practicing Industrial Hygiene for the past 20 years, with emphasis on information management for the last 10 years. She is currently helping large corporations implement software solutions for occupational health and safety. Ms. Randolph can be reached via e-mail at randolphbl@aol.com.

Cynthia D. Scott, MPH, PhD, consultant and founding principal of HeartWork, Inc, a San Francisco-based organizational development firm. A licensed clinical psychologist, Dr. Scott's training and experience in anthropology, health planning, and administration make her a recognized leader in helping businesses and associations manage change and bridge the gap between individual and organizational performance. She has taught at the University of California-San Francisco's School of Medicine; John F. Kennedy University, Orinda, Calif.; and the University of San Francisco. She is the author or coauthor, with Dennis T. Jaffe, of numerous books and articles on worker empowerment and personal and organizational change. Dr. Scott is a consultant and speaker for businesses and makes frequent television and radio appearances. Mailing address: Cynthia D. Scott, PhD; Heartwork, Inc, 461 Second St, Ste 232, San Francisco, CA 94107.

Richard B. Seymour, MA, president and chief executive officer, Westwind Associates, Sausalito, Calif. Seymour oversees the development, production, and publishing of professional educational and training materials for physicians, counselors, and other health professionals who work in addiction medicine, recovery, and allied fields. He also is an assistant professor/instructor at: Sonoma State University, Rohnert Park, Calif.; John F. Kennedy University, Orinda, Calif., and Downtown Community College Center, San Francisco. He is a consultant or advisor on drug education and abuse prevention to many academic institutions and government organizations, both domestic and foreign, as well as editor and a journalist for publications that report on addiction and drugs. Mailing address: Richard B. Seymour, 90 Harrison Ave, Sausalito, CA 94965.

Joyce A. Simonowitz, RN, MSN, is a nursing consultant with the Cal/OSHA Medical Unit in Van Nuys, California. She authored the Cal/OSHA publication, *Guidelines for Security and Safety of Health Care and Community Service Workers,* and was part of the team that developed the federal OSHA publication, *Preventing Violence in Health Care and Social Service Workers.* Ms. Simonowitz can be reached at Cal/OSHA, 5160 Van Nuys Blvd., Suite 410, Van Nuys, CA 91401.

Ara Tahmassian, PhD, Radiation Safety Program manager/officer, University of California, San Francisco. Dr. Tahmassian is responsible for radiation safety for a Broad Scope Radioactive Materials Licensee Program which includes three medical centers and major program on biomedical research. He specializes and consults on radiological health and safety and nuclear medicine, including the development of radiation safety procedure manuals. Mailing address: Ara Tahmassian, PhD, Radiation Safety Program Manager, University of California, 50 Medical Center Way, San Francisco, CA 94143-0942

David A. Thompson, PhD, CPE, is professor emeritus of Industrial Engineering and Engineering Management, Stanford University, Palo Alto, CA; and is a member, Clinical Faculty, Department of Occupational Medicine, University of California Medical School, San Francisco, California. Dr. Thompson holds degrees in industrial engineering and mechanical engineering, including a doctorate from Stanford University. He is president of Portola Associates, a small engineering consulting firm in Incline Village, Nevada. Research interests include human-computer interaction, especially computer graphics and keyboard design. Consulting clients include major silicon valley companies, the U.S. State Department, NATO, and the Government of Mexico. He can be contacted via e-mail at overjoy@compuserve.com.

R.T. Vulpitta, CEM, CUSA, is the manager of the National Safety Council Utilities Division. Rick has over 25 years of utility safety and security experience. In his career, Rick has developed a number of safety procedures that have been published and produced as safety videos. Mr. Vulpitta has spoken at several national safety conferences concerning on-site emergency planning. He has been recognized by the American Red Cross for serving as the northwest Indiana disaster service chairman for two years, and chairman for CPR/First Aid Training Committee for over five years. Rick is a board member of the Certified Utility Safety Administrator program and serves as the National Safety Council's Emergency Response manager. He can be reached via e-mail at vulpittr@nsc.org.

Index

A

Accident investigation report form, 163–164
Accidents, in workplace. *See* Safety programs
Action level (AL), 180
Activities of daily living (ADL), 473
Addams, Jane, 40
Addiction, substance abuse, 411–412
Addictive disease, 411
Administrative controls
    and health hazards, 103
Administrative Procedures Act, 346
Aflatoxin, 273
Africare, 506
Aging, in the workplace, 111–112
Agricola, Georgius, 29
AIDS. *See* Human Immunodeficiency Virus
Air pollution, 272
Air-purifying respirators, 180
Airborne contaminants, 96–97, 168, 187
    respiratory protection, 174
    sampling and analysis, 96, 168
ALARA concept, 231–232
Alarm system, 335
Alcohol hazards, 274–275
Alcoholics Anonymous (AA), 411, 416
Altitude sickness, 272–273
Amebiases, 501
American Association of Industrial Nurses, 41
American Association of Industrial Physicians and
    Surgeons, 41
American Association for Labor Legislation (AALL), 34
American Association of Occupational Health Nurses
    (AAOHN), 42, 61–62, 64, 69–71, 492
  code of ethics, 70
  Governmental Affairs, 70
  Member Services and Public Affairs, 71
  Professional Affairs, 70
  standards of practice, 72
American Board of Industrial Hygiene (ABIH), 89
American Board of Medical Specialties (ABMS), 52
American Board for Occupational Health Nurses, Inc.
    (ABOHN), 62, 64
American Board of Preventive Medicine, 52–53
American College of Occupational and Environmental
    Medicine (ACOEM), 53, 59, 387
  code of ethical conduct, 54

American Conference of Governmental Industrial Hygienists (ACGIH), 88, 126, 168
American Industrial Hygiene Association (AIHA), 42, 87–88
American National Standards Institute (ANSI), 137, 173, 179
American Occupational Health Conference, 42
American Occupational Medicine Association, 292
American Society of Addiction Medicine (ASAM), 425
American Society of Heating, Refrigerating and Air Conditioning Engineers (ASHRE), 187
American Society of Safety Engineers, 38, 77, 78–79, 81
Americans with Disabilities Act (ADA), 17–18, 55, 471–483
   confidentiality of records, 389
   and employee selection, 292
   essence of, 473–474
   gender issues in workplace, 456
   and occupational medicine
      direct threat, 55
      reasonable accommodation, 56
   and occupational nursing, 63
   physical examination, 128
   and preplacement testing, 381, 385, 388–389
   President's Committee on Employment of People with Disabilities, 472
   small businesses, 22
   and stress management, 402
   and substance abuse testing, 142
   Title I, 473, 493
      requirements, 474
   Title II, 473–474
   Title III, 474
   web page, 474
Ames test, 119
Andrews, John B., 34, 35, 38
Animal model, 118
Anthrax poisoning, 14, 260
AOE/COE concept, 240, 245
Area sampling, for airborne contaminants, 168
Arsenic poisoning, 121
Asbestos exposure, 41, 187
   sampling and analysis of airborne contaminants, 96–97
Asbestosis, 14
Assault, 316
Assessment, 370
Associated General Contractors of America (AGC), 10
   and OSHA partnership, 10
Association of Occupational Health Professionals in Healthcare, 71
Athlete's foot, 272
Atmosphere-supplying respirator, 180
Audiometric testing program, 181–182
   annual audiogram, 181
   baseline audiogram, 181

Audit, hazard, 333–334
Audit program
   design, in workplace, 116
   ergonomic, 303, 305
   management of results, 116
   and safety management, 156–157
      alternatives, 156–157
      packaged, 156
      self-built, 156
Autonomy principle, 444

B

Babylonians, occupational health and safety development, 27
Back injuries or pain
   administrative controls, 291–292
   back school, 288–291
   and gender differences, 456
   lifting, 283–285
   materials handling, 281
   risk factors, 281–282
   sitting, 285–286, 287–288
   spinal degeneration, 282–285
   and workplace design, 286–287
Back school, 288–291
   criteria, 289
   effectiveness, 289
   engineering controls, 289–290
   NIOSH lifting guidelines, 290–291
Baseline examination
   of employees, 128
Behavior-based safety, 165–166
Behavioral safety programs, 402
Benefits
   death/survivor, 242
   social security, 242
   of workers' compensation, 240–241, 245
Benzodiazepine medications, 262
Best Available Control Technology (BACT), 210
Best Available Retrofit Technology (BART), 211
Bioassay program, 229–231
Biomedical research, 222
Black lung disease, 34
Bloodborne pathogens, exposure to, 143–144
   Centers for Disease Control and Prevention (CDC), 143
   hepatitis B, 143
   HIV, 143
   immunizations, 144
   OSHA standard, 143
   personal protective equipment, 144
   post-exposure follow-up, 144
   record keeping, 144
   safety requirements, 143–144
   universal precautions, 143
   viruses, 143, 144

Index

Board of Certification in Professional Ergonomics, 277
Board of Certified Safety Professionals, 77, 81
    bachelor's or master's degree listing, 80
    requirements, 80
Board of Hazard Control Management, 81
Bridging Environmental Health Gaps (BEHG), 508
Brucellosis, 14
Burden of disease, 500
Bureau of Labor Statistics (BLS), 10
    fatal occupational injuries, 10
    Survey of Occupational Injuries, 316
    terms for reporting, 10–11
Bureau of Mines, 38
Burnout, 396–397
    acute, 397
    chronic, 397
    depersonalization, 397
    emotional exhaustion, 397
    intervention, 397
    reduced sense of personal accomplishment, 397
    symptoms of, 397
Byssinosis, 14

C

Cabot, Richard C., 36
Cadmium, 187
Cal/OSHA Injury and Illness Prevention Program, 78
Canal caps, 182
Canute (King), 29
Carbon monoxide, 209
CARE, 506
Case management
    nursing programs, 136
Case Management Society of America, 71
Catholic Medical Mission, 506
Census of Fatal Occupational Injuries (CFOI), 11
Centers for Disease Control and Prevention (CDC), 143
    and travel health, 250, 275
    and tuberculosis, 144
Certification
    of industrial hygienists, 89
    of nurses, 64, 69
    of physicians, 52–53
Certified Safety Professional (CSP), 77, 82
Chadwick, Edwin, 32
Chair design, 287–288
Chancroid, 501
Chemical hygiene officer (CHO), 186
Chemical hygiene plan (CHP), 186
Chemical use
    inventory, 140
    labeling, 140
    and occupational health nurse placement, 63
    and skin disorders, 173
    primary irritants, 173
    sensitizers, 173
Childcare, 131
    and stress management, 402
    and working mothers, 467
Children, in work force, 37
Chloroquine, 267
Cholera, 504
    epidemiology studies, 124
    transmission, 501
    and travel health, 259–260
Ciprofloxacin, 265
Circadian rhythms, 433–434
Civil Rights Act
    Title VII, 142, 388, 465, 473
Clean Air Act, 209–211, 347
    acid rain, 211
    Best Available Control Technology (BACT), 210
    Best Available Retrofit Technology (BART), 211
    criteria pollutants, 209
    Lowest Achievable Emissions Rate (LAER), 211
    Maximum Achievable Control Technology (MACT), 211
    Montreal Protocol, 211
    National Ambient Air Quality Standard (NAAQS), 209–210
    New Source Performance Standards (NSPS), 210
    operating permit program, 211
    OSHA standard, 209
    pollution problems, 211
    Prevention of Significant Retention (PSD), 210
    reasonably available control technology (RACT), 210
    State Implementation Plan (SIP), 209–210
    tailpipe emissions, 211
Clean Water Act (CWA), 211–213
    history of, 212
    National Pollutant Discharge Elimination System (NPDES), 212
    New Source Performance Standards, 213
    objectives of, 212
    point source discharges, 212
    Toxic Hot Spots program, 212
Clothing, protective, 174
Clotrimazole solution (Lotrimin 1%), 272
Coal Mine Health and Safety Act (CMHSAct) (1969), 34
Coal minor's pneumonokoniosis, 33
Cocaine Anonymous, 411
Coccidiomycosis, 14
Code of ethics
    industrial hygiene, 88
    nursing, 69
        autonomy, 69
        beneficence, 69
        justice, 69
        nonmaleficence, 69
    safety, 82–83

Code of King Rothari, 28–29
Codeine, 265
Collective bargaining agreement
  and drug and alcohol testing, 141
Commercial drivers license (CDL), 142
Communication
  program, 346–361
    annual reports, 352
    brown bag lunches, 352
    e-mail, 351–352
    employee training, 352
    in-house newsletters, 351
    public participation, 348–349
    reasons for, 346–348
    skills, 360–361
    strategies, 349–352
  and small businesses, 488
  strategies for program design, 113
  warning alarm system, 335
Community Advisory Boards (CABs), 356–357
Community involvement programs, 345–367
  civic organizations, 353
  communication strategies, 358
  communications program, 346–348
    Comprehensive Environmental Response, Compensation and Liability Act (CERCLA), 346, 347
    Emergency Planning and Community Right-to-Know Act (EPCRA), 346–347
    OSHAct, 346, 347
    Resource, Conservation and Recovery Act (RCRA), 346, 347
    Toxic Release Inventory (TRI), 346, 347–348
    Toxic Substances Control Act (TSCA), 346, 347
  community presence, 352
  cultural sensitivity, 365
  effectiveness of program, 349–352
    annual reports, 352
    brown bag lunches, 352
    departmental meetings, 351
    e-mail, 351–352
    employee training, 352
    in-house newsletters, 351
    stakeholder rapport, 349–351
  environmental participation, 353
  high-profile projects, 354–357
  media relations, 358–360
  origin of, 346
    Administrative Procedures Act, 346
    Freedom of Information Act, 346
    Open Meetings Act, 346
  public participation, 348–349
    advantages and disadvantages, 349
    stakeholder involvement, 348–349
  regulatory advocacy, 353

  skills for communication, 360–365
  written communications, 360
Competency
  of nurses, categories, 64
Compliance
  and occupational health nursing, 63
Comprehensive Environmental Response Compensation and Liability Act (CERCLA), 217
  Hazardous Substance Superfund, 217
  Superfund Amendments and Reauthorization Act (SARA), 217
Computer technology, in workplace, 112
  and ergonomics, 292–295
    keyboard design, 294
    visual fatigue, 294–295
    work space layout, 293
Confidentiality, 48
  and audit program, 116
  and educational programs, 444–445
  and preplacement testing, 389
  and substance abuse, 426–427
Confined space entry, 341
Construction industry
  death statistics, 12
  incidence rates of disease or injury, 16
Consumer Product Safety Commission (CPSC), 13
Contact dermatitis, eczema, 14
Continuing education programs, 115
Coronary heart disease (CHD), 456–457
  Framingham Offspring Study, 457
  and gender differences, 456–457
Corporate medical director, 51–52
Coumarin, 273
Council on Environmental Quality, 208
Counseling services, 135–136
Critical incident, 147–148
Critical incident stress, 148
Critical incident stress debriefing, 148–149
  purpose of, 149
Crowley, John, 30
Culture shock, 503
Cumulative trauma disorders (CTDs), 155, 296, 300, 301

D

Data collection
  and quality management, 115
  for stress management, 138
Death, occupational, 11–12. *See also* Occupational death
Debriefing, 148–149
  education, 149
  event description, 149
  follow-up, 149
  introduction, 148–149
  reaction, 149

Dengue fever, 260, 501
Department of Transportation
   commercial drivers license (CDL), 142
   guidelines, 142–143
   regulations, 142
   and substance abuse testing, 142
Design, of occupational health and safety program, 109–116
   accountability, 115–116
   assessment of program needs, 113
   audit program, 116
   communication strategies, 113
   education programs, 115
   goal of, 112, 113–114
   metrics accuracy, 114
   proactive program, 112
   productivity management, 114
   program considerations, 112–115
   quality monitoring, 115, 116
   records retention, 114–115
   risk assessment, 113
   role of professional, 110
   role in workplace, 110
   small businesses, 491–494
   trends in workplace demographics, 110–112
Diarrhea, traveler's 262–265
Didier, Pierre, 32
Diet and nutrition
   and employee health promotion, 130
Diodorus, 26
Diphenoxylate hydrochloride, 265
Disability. *See also* Workers with disabilities
   definition, 57
   permanent partial, 242
   permanent total, 241, 245
   temporary partial, 242
   temporary total, 241–242
   workers with, 471–483
Disability-adjusted life years (DALY), 500
Disability evaluations, 57
Disaster, 326–327
Discrimination
   and workers' compensation, 244
Disease, occupational. *See* Occupational disease
Disease-prevention programs, 132
   benefits of, 133
Diseases, infectious
   in developing countries, 499–511. *See also* Infectious diseases
District of Columbia Workers' Compensation Act, 237
Disulfiram (Antabuse), 275
Division of Labor Standards, 315
Doctors Without Borders, 506
Dose-response relationship, 117–118, 119
Doxycycline, 268
Drills, emergency, 337–338

Drug abuse. *See* Substance abuse
Drug and alcohol testing program, 140–142. *See also* Substance abuse
   and collective bargaining agreements, 141
   policy for, 141
   purpose of, 140
   regulations, 142
      Americans with Disabilities Act, 142
      Department of Transportation, 142
      Drug Free Workplace Act, 142
      Title VII, 142
   supervisory training, 141–142
   testing types, 141
      annual or semiannual, 141
      for-cause/reasonable, 141
      preemployment, 141
      random, 141
Drug Free Workplace Act, 142
Drug screening, 386–387
Dust diseases of lung, 33–34
   black lung disease, 34
   progressive massive fibrosis (PMF), 34

E

E-mail, 351–352, 356, 357
Earmuffs, 182
Earplugs, 182
Eastman, Crystal, 36
Ebers, Papyrus, 26
Ebola, 502, 505
   transmission, 500
Education and training, 187, 443–450
   communications programs, 352
   of emergency action plan, 336–337, 338
   goals of, 443
   hazard communication, 140, 186
   hearing conservation, 182
   industrial hygienist, 89
   issues in, 445–446
      and risk perception, 446
   legal, professional, and ethical frameworks, 443–445
      ethical considerations, 444–445
      Healthy People 2000, 444
      OSHA standards, 444
      OSHAct, 443
   NIOSH, 20–21
   nurses, 64
   physicians, 52
   respirator use, 176
   safety procedures, 160
   safety professionals, 79–81, 83–84
   on shiftwork, 440
   and small businesses, 491
   stress management, 399
   substance abuse detection program, 141

training guidelines, 446–449
  conduct of, 448–449
  evaluation of effectiveness, 449
  goals and objectives, 448
  learning activities development, 448
  need determination, 446–447
  need identification, 447–448
  tracking of, 449
workplace violence, 322
Ecogenetics, 389
Educational and Research Centers (ERC), 21, 64
  programs listings, 65–68
Educational Resource Centers, 492
Edwin Smith Papyrus, 26
Egyptians, occupational health and safety development, 26–27
18th century, occupational health and safety development, 29–30
Electrocardiogram, 118
Ellenbog, Ulrich, 29
Emergency, 326
Emergency Broadcast System (EBS), 329
Emergency and disaster management, 326
  goals of, 327
  mitigation, 326
  preparedness, 326
  recovery, 326
  response, 326
  for small businesses, 492
Emergency drills, 337–338
Emergency planning, 325–326
  lines of defense, 327–332
Emergency Planning and Community Right-to-Know Act (EPCRA), 217–218, 346–347
  Local Emergency Planning Committees (LEPCs), 218
  local emergency response plan, 218
  State Emergency Response Commission (SERC), 218
Emergency response programs, 325–343
  business records, 343
  confined space entry, 341
  disasters, 326–327
  emergencies, 326–327
  emergency drills, 337–338
  goals, 325, 327
  governmental responsibilities, 327–332
    Federal Emergency Management Agency (FEMA), 327, 328–329
    federal government, 328
    local government, 331–332
    state government, 329–331
  medical assistance, 342–343
  mitigation, 326
  news media, 343
  personal protective equipment, 338–339
  preparedness, 326
  respiratory protection, 339–341
  response, 326
  response agreements, 343
  response plans, 332–336
    accounting of personnel, 335
    assembly areas, 335
    chain of command, 334–335
    communication warning alarm system, 335
    effectiveness, 332–333
    elements of, 333
    employee review, 334
    hazard audit, 333–334
    material safety data sheets, 334
    purpose, 332
    risk evaluations, 334
    special response teams, 335–336
  security, 343
  training of staff, 336–337
Emergency response team coordinator, 334–335
Employee Assistance Programs (EAPs), 131–132, 133, 135–136, 148, 149, 320, 387, 416
  counseling services, 135–136
  goal of, 135
  and stress management, 399, 401
  and substance abuse treatment, 425–426
Employee safety and security programs, 315–324
  during emergencies, 343
  workplace violence, 316
    assaults, 316
    causes of, 316–317
    cost, 316
    homicides, 316
    OSHA guidelines, 316
    predictors, 318
    preventive measures, 318–322
    types, 317–318
Employment segregation, 452
Emporiatrics, 249–251
Enclosure
  controlling occupational health hazards, 102
Enclosure-type hearing protectors, 182
Engineering controls
  in industrial hygiene, 102–103
Environmental assessment (EA), 209
Environmental Impact Statement, 208
Environmental Protection Agency (EPA), 207, 348
  main function of, 207
Environmental regulations, 207–219
  Clean Air Act, 209–211
    Best Available Control Technology (BACT), 210
    Best Available Retrofit Technology (BART), 211
    criteria pollutants, 209
    Lowest Achievable Emissions Rate (LAER), 211
    Maximum Achievable Control Technology (MACT), 211

National Ambient Air Quality Standard
(NAAQS), 209–210
New Source Performance Standards (NSPS), 210
pollution problems, 211
Prevention of Significant Retention (PSD), 210
State Implementation Plan (SIP), 209, 210
Clean Water Act (CWA), 211–213
goals of, 212
National Pollutant Discharge Elimination System
(NPDES), 212, 213
Oil Pollution Act, 212
point source, 212
primary elements of, 212
Comprehensive Environmental Response Compensation and Liability Act (CERCLA), 217
benzene, toluene and xylene (BTX), 217
Hazard Substance Superfund, 217
Superfund Amendments and Reauthorization Act
(SARA), 217
Emergency Planning and Right-to-Know Act
(EPCRA), 217–218
local emergency response plan, 218
Pollution Prevention Act, 218
Environmental Protection Agency (EPA), 207
National Environmental Policy Act (NEPA),
208–209
compliance, 208
environmental assessment (EA), 209
Environmental Impact Statement, 209
Finding of No Significant Impact (FONSI), 209
Pollution Prevention Act (PPA), 218–219
Resource Conservation and Recovery Act (RCRA),
214–215
DOT regulations, 215
hazardous waste, 214–215
waste, 214–215
Waste Analysis Plan (WAP), 215
Toxic Substances Control Act, 213–214
and pesticides, 214
premanufacture notice (PMN), 213
underground storage tanks (USTs), 215–217
Epidemiology, 120, 124–126
biases, 125
descriptive, 124
experimental, 124
hypotheses formulation, 124
life table analysis, 126
prevalence studies, 124
prospective studies, 124
retrospective studies. 124–125
standard mortality ratio (SMR), 125
Equal Employment Opportunity Commission, 473, 482
Equipment design
information displays, 296, 297
operating controls, 296
Ergonomics, definition of, 277

Ergonomics programs, 277–314
Board of Certification in Professional Ergonomics,
277
definition of, 277
design, 277, 278–281
of equipment, 296–297
of hand tools, 297–300
repetitive work, 280–281
sitting versus standing work, 280
of workplaces, 278–280, 286–288
employee exercise programs, 305–307
appropriateness, 306–307
effect of strength, 307
and musculoskeletal stress, 306
environmental evaluation, 300–303
heat and humidity, 301
lighting, 300–301
noise levels, 300
physical hazards, 300
whole-body vibration, 303
equipment design, 296–297
operating machines, 297
user-machine communication and cooperation,
296
hand tool design, 297–300
motion trauma, 300, 301
and musculoskeletal stress, 298–299
working height and location, 299–300
light-duty and modified duty, 307–310
and alternative job selection, 308–310, 311
follow-up evaluation, 310
and injured workers, 305, 307, 310
materials handling, 281
and OSHA, 312
risk factors, 281–282
personal, 281–282
work-related, 281
spine biomechanics and biochemistry, 282–296
administrative controls, 291–292
back school, 288–291
and computers, 292–295
job inspection and monitoring, 295
musculoskeletal stress, 295–296, 298
sitting postures, 285–286
spinal degeneration, 282–285
workplace design, 286–288
task evaluations, 303–305
ergonomic audits, 303, 305
and human motion, 303
survey methods, 303
Esch Act, 35
Evaluations by physician
fitness-for-duty, 55–56
health hazard, 57
medical, 56–57
exit, 56–57

impairment and disability, 57
periodic, 56–57
preplacement, 56–57
Exercise programs, 305–307
and musculoskeletal stress, 306
and stress management, 398
Exit examination
of employees, 128

F

Fair Labor Standards Act, 37
Family Medical Leave Act (FMLA), 63, 467, 480
Farmer's lung, 14, 384
Federal Emergency Management Agency (FEMA), 327, 328–329
Emergency Broadcast System (EBS), 329
individual assistance, 329
National Flood Insurance Program, 329
Presidential Declaration of an Emergency or Major Disaster, 329
public assistance, 329
website, 329
Federal Employees Compensation Act (FECA), 237
Federal Employers' Liability Act, 237
Federal government, 327, 328–329
emergency management, 327, 328–329
Federal Emergency Management Agency, 328–329
Federal Insecticide, Fungicide, and Rodenticide Act (FIFRA), 214
Finding of No Significant Impact (FONSI), 209
Fit-testing, of respirators, 176, 179
Fitness for Duty examination
of employees, 129
Fitness programs, 132–135
behavior changes, 134
benefits, 133–134
cost effectiveness, 133–134
evaluations, 134
"hard-to-reach" employees, 134–135
health-promotion programs, 132–134
For-cause/reasonable suspicion testing, 141
Formaldehyde, 187
Freedom of Information Act, 346
Fulton, Frank Taylor, 40
Functional capacity evaluation (FCE), 454
Fungal infections, 272

G

Galen, 27
Gas and vapor sampling, 97–98
Gender issues in workplace, 451–469
assessing and communicating risk, 463–464, 465, 466
chemical exposure and reproductive risk, 459–463
female reproductive hazards, 461–462, 463
male reproductive hazards, 459–461
disability, 457, 459
distribution in labor force, 451–452
employee segregation, 452
family-related issues, 467
childcare, 467
elder care, 467
Family Medical Leave Act (FMLA), 467
insurance, 467
gender-specific vulnerabilities, 453–457
coronary heart disease, 456–457
heat exposure, 454–456
low back injuries, 456
physical strength, 453–454
work capacity, 453–454
and health research, 457
Multiple Risk Factor Intervention Trials, 457
Physician's Health Study, 457
Women's Health Equity Act, 457
pay inequities, 452–453
protection versus discrimination, 465
Civil Rights Act, 465
Pregnancy Discrimination Act, 465
reproductive and developmental hazard policy, 464–465
sexual harassment, 467
Genital herpes, 501
Global Burden of Disease Study, 500
Gloves, choices of, 174
Gonorrhea, 501
Government
and emergency management, 327–332
federal, 327, 328–329
Federal Emergency Management Agency (FEMA), 327, 328–329
local, 327, 331–332
state, 327, 329–331
Governmental health services, 505
Greeks, occupational health and safety development, 27–28
Grip strength, 298
Guidelines for Federal Workplace Drug Testing Programs, 423, 425

H

Hamilton, Alice, 34–35, 36, 38
Hammurabi, 27
Hand tool design, 297–300
factors causing musculoskeletal stress, 298–299
grip strength, 298
hand and arm vibration, 299
palmar flexion, 298

INDEX

repetitive finger action, 299
tool handles, 298
two-handed tools, 299
ulnar deviation, 298
wrist extension, 298
Hand-arm vibration syndrome (HAVS), 299
Handwashing, and skin disorders, 174
Hard, William, 36
Hargreaves, James, 30
Hazards
   audit, 333–334
   definition of, 120
   inhalation, 168
   monitoring program for radiation, 229–231
   physical, 300
   prevention and control, 320
   recognition and identification of, 129–130
   and safety programs, 153–166
Hazard communication (Hazcom) program, 7, 139–140, 183–186
   biological agents, 489
   chemical inventory, 140, 489
   citations, 140
   ergonomic considerations, 489
   federal standards, 139–140
   and industrial hygiene programs, 183–186
   labeling, 140
   MSDS, 140, 183
   occupational stress, 489
   OSHA standard, 139, 183
   physical agents, 489
   safety hazards, 489
   shiftwork, 489
   training, 140, 186
   written program, 183–186
Hazard communication standard, 6–7, 10, 23
   and industrial hygiene programs, 183
   violations, 7
Hazard monitoring program, 229–231
   environmental, 229
   instrumentation, 231
   personnel, 229–231
      external exposure, 229
      internal exposure, 229–231
Hazardous Substance Superfund, 217
Health club, on-site, 130
Health program, occupational, 20
Health promotion, 130–132
   benefits of, 133
   cost effectiveness, 133–134
   diet and nutrition, 130
   Employee Assistance Programs (EAPs), 131–132
   evaluations of programs, 134
   health-risk appraisal (HRA), 130
   on-site childcare, 131
   on-site health club, 130–131

physical exercise, 130
   programs, 132–134
Health risk appraisal (HRA), 130
Health and safety committees, 20
Healthy People 2000, 144, 444
Hearing Conservation Amendment, 180
Hearing conservation program (HCP), 137, 180–183
   action level (AL), 180
   audiometric evaluation, 137
      normal hearing, 137
      permanent threshold shift (PTS), 137
      standard threshold shift (STS), 137, 181
      temporary threshold shift (TTS), 137
   audiometric testing, 181–182
      annual audiogram, 181
      audiogram evaluations, 181
      baseline audiogram, 181
   baseline hearing testing, 137
   employee training, 182
   Hearing Conservation Amendment, 137, 182
   hearing protection, 182
   hearing protectors, 182–183
      canal caps, 182
      earmuffs, 182
      earplugs, 182
      enclosure-type hearing protectors, 182
   and industrial hygiene programs, 180–183
   noise measurements, 137
   permissible exposure limit (PEL), 180
   personal noise dosimetry, 180
   personal protective equipment, 180–183
   record keeping, 182
   time weighted average (TWA), 180
Hearing protection device (HPD), 137, 182
   canal caps, 182
   earmuffs, 182
   earplugs, 182
   enclosure-type hearing protectors, 182
Heat exhaustion, 272, 456
Heat exposure, 454–456
   gender differences, 454–455
   heat exhaustion or sun stroke, 14
Heat and humidity
   in workplace, 301
Heatstroke, 272, 456
   treatment, 456
Henderson, Charles R., 35
Hepatitis, 14, 258, 414
   transmission, 500
Hepatitis B, 143, 258, 259, 271
Hill Criteria, 119
Histoplasmosis, 14
HIV, 143–144, 269, 271, 414, 505, 508, 510
Hoffman, Frederick, 35
Holmes, Joseph E., 38
Homicide, 316

House Select Committee on Narcotics Abuse and Contol, 409
Human error
  and workplace safety, 154–155
    overload, 154
    workplace design, 155
Human Factors Stress Inventory (HFSI), 399–400
Human immunodeficiency virus (HIV), 143, 505, 508, 510
  and ADA, 480
  and substance abuse, 414
  transmission, 501
  and travel health, 269, 271
  and tuberculosis, 144
Human resources
  and industrial hygiene, 91
  and occupational health nurse placement, 63
  and workplace violence prevention, 147
Human resources issues. *See also* specific listings
  employee education, 443–450
  outsourcing, 485–495
  preplacement testing, 381–390
  program assessment and evaluation, 369–378
  stress management, 391–407
  substance abuse, 409–430
  workers with disabilities, 471–483
  workplace gender issues, 451–469

I

Illegal drug use, 275
Illinois Occupational Disease Commission, 35
Illness, occupational, 10, 14–15. *See also* Occupational illness
Illumination, 300–301, 302
Immigrant work force, 40
Immunizations, 144, 252–260
  cholera, 259–260
  dosing schedule, 255
  hepatitis A, 258
  hepatitis B vaccine, 144, 258, 259
  influenza, 259
  Japanese encephalitis, 258
  measles, 254, 256
  meningococcal vaccine, 258–259
  mumps, 257
  plague, 260
  pneumococcal vaccine, 259
  poliomyelitis, 257
  rabies, 258
  rubella, 257
  tetanus-diphtheria, 254
  and travel, 251, 252–260
  tuberculosis, 259
  typhoid fever, 258
  yellow fever, 254, 256–257

Impairment, 57
Incidence rates, 11
  of disease or injury, 15–16
  and safety programs, 155–156
Independent contractor, physician, 51–52
Indoor air quality (IAQ), 187–188
  problem sources, 188
Industrial ergonomics, 277. *See also* Ergonomics programs
Industrial Health Foundation, 34
Industrial hygiene, 42, 120
Industrial hygiene profession, 87–105. *See also* Industrial hygienist
  American Industrial Hygiene Association (AIHA), 87–88
  certification and licensure, 89–90
  code of ethics, 88
  consulting services, 91–92
  definition, 87
  education requirements, 89
  in-house services, 90–91
  industrial hygienist, 87–88
  practicing of, 92–101
    anticipating health hazards, 92
    controlling health hazards, 101–104
    evaluating health hazards, 94–98
    recognizing health hazards, 92–94
  professional organizations, 88–89
    American Conference of Governmental Industrial Hygienists (ACGIH), 88
    American Industrial Hygiene Association (AIHA), 88
Industrial hygiene programs, 167–205
  employee training, 187
  hazard communication, 183–186
    program (HCP), 183–186
  indoor air quality (IAQ), 187–188
  laboratory safety, 186
    chemical hygiene officer (CHO), 186
    chemical hygiene plan (CHP), 186
  objectives, 167
  personal protective equipment, 173–183
    hearing conservation, 180–183
    respiratory protection, 174–180
    skin protection, 173–174
  sampling, 168
    analytical method, 168, 169–172
    area, 168
    objectives, 168
    personal, 168
    for surface contamination, 189–205
    wipe, 168, 169
  substance-specific standards, 186–187
Industrial Hygiene Survey Checklist, 94, 95
Industrial hygienist. *See also* Industrial hygiene profession
  certification and licensure, 89–90

code of ethics, 88
definition, 87–88
education requirements, 89
occupational health and safety team, 90–92
consulting services, 91–92
in-house services, 90–91
and safety programs, 159
Industrial radiography, 222
Industrial revolution, occupational health and safety development, 30–31
Infectious diseases in developing countries, 499–511
behaviors and risks, 503–505
expatriate workers, 504
indigenous populations, 504–505
cultural context, 503
culture shock, 503
disease burden, 500
disability-adjusted life years (DALY), 500
Global Burden of Disease Study, 500
environmental approach to prevention, 507, 509
Bridging Environmental Health Gaps (BEHG), 508
governmental health services, 505
international NGOs, 506
international organizations, 507
UNICEF, 507
World Bank, 507
World Health Organization (WHO), 507
local NGOs, 506
nongovernmental organizations (NGOs), 505–506
transmission, 500–503
complex cycles, 501–503
person-to-person, 500–501
Influenza
transmission, 501
vaccine, 259
Informal networks, 135
Information management, 513–516
buying versus building, 515–516
controlled interfaces, 516
functional requirements, 515
integrated enterprise solution, 515–516
point solutions, 515
technical requirements, 515
communication, 514
data base, 513
document management systems, 514
integration, 514–515
material safety data sheet (MSDS), 513, 516
process changes, 514
program/application, 513
Injury, occupational, 10, 13. *See also* Occupational injury
definition of, 11
and travel, 274
Insect repellents, 269, 501
Institute for Labor and Mental Health, 405

Institution of Occupational Safety and Health (IOSH), 82
Insurance
and traveling abroad, 252
and workers' compensation, 238–239
cost incentives, 245–246
self-insurance, 238–239, 245
state funds, 238
Interchurch Medical Assistance, Inc., 506
Internal job-transfer evaluation, 129
International Association for Medical Assistance to Travelers, 252
International Commission on Radiological Protection (ICRP), 222
International Eye Foundation, 506
International Labor Office, 42
Iodochlorhydroxyquin (Entero-Vioforme), 265
Ionizing radiation, 223
Isolation
control of occupational health hazards, 102

J

Japanese encephalitis, 258, 501
Jet lag, 262–262
symptoms, 261
treatment, 261–262
Job Accommodation Network (JAN), 475
Job descriptions, 382–383
job function analysis (JFA), 382
contents of, 383
Job design, 278–281
repetitive work, 280–281
sitting versus standing, 280
work space layout, 278–280
avoidance of static holding, 279–280
elimination of "waist motion", 278–279
Job Links, 482
Job safety analysis (JSA), 162
Jock itch, 272
Journal of the Association of Occupational Health Professionals in Healthcare, 71
Justice Department, 409

K

Karasek, Robert, 394–395
Kober, George, 36

L

Laboratory safety, 186
chemical hygiene officer (CHO), 186
chemical hygiene plan (CHP), 186
components of, 186
OSHA standard, 186

Language diversity, in the workplace, 111
Latency
    and toxicity, 120
Lead exposure, 187, 209
    and toxicity, 120, 121
Lead poisoning
    early documentation, 33
Lethal Dose 50, 121–122
Lewis, John L., 35
Licensed practical nurses (LPNs), 64
Licensed vocational nurses (LVNs), 64
Licensure
    of industrial hygienist, 89–90
    of nurses, 64, 69
    of physicians, 52–53
    for radioactive material usage, 223
Lifting methods
    deep squat, 284, 285
    diagonal lift, 284
    power lift, 284
Light-duty work, 207–310
Lighting, in workplace, 300–301
Litigation, in the workplace, 111
Local Emergency Planning Committees (LEPC), 218, 330, 334, 347
Local government
    emergency management, 331–332
        functions of, 331–332
        local laws, 332
Loperamide, 265
Lost workday, 10
Lowest Achievable Emissions Rate (LAER), 211
Lucretius, 28

M

Malaria, 265–269, 504
    mosquito protection, 268–269
    prevention of, 266
    prophylactic medication, 266–268
    transmission, 500, 501
Management and culture programs, 402
Manufacturing industry
    incidence rates, 15
    occupational illnesses, 14–15
Material safety data sheet (MSDS), 6, 23, 94, 140, 183, 347, 465
    example of, 184–185
    and toxic materials, 334
Maximum Achievable Control Technology (MACT), 211
McCready, Benjamin W., 34
Measles, 254, 256
    transmission, 501
Meclizine, 261
Media relations, 358–360

Medical Assessment Management Programs (MAMP), 369–370
    framework for, 369
    process, 370
Medical assistance
    in emergencies, 342–343
Medical exams, and disabilities, 476–478
Medical and industrial hygiene services, 20
Medical kit, for traveling, 251–252, 253
Medical review officer (MRO), 386–387
    and urine testing, 416–419
Medical screening, 128
Medical surveillance, 17–18, 128
    OSHA standards, 129
Mefloquine (Lariam), 267
Melatonin, 262
Meningitis, 501
Meningococcal vaccine, 258–259
Mercury poisoning, 33, 41, 122
Mesothelioma, 118
Methanol poisoning, 118–119
Methods engineering technique, 303
Methylene chloride, 187
    sampling example, 169–172
Middle ages, occupational health and safety development, 28–29
Mine Safety and Health Administration (MSHA)
    and respirator selection, 175
Mining industry
    death statistics, 11–12
    illness statistics, 15
    incidence rates of disease or injury, 15
Minors, employment of
    and workers' compensation, 243
Mosquitoes and diseases, 254, 258, 260, 266, 268–269, 502, 507, 508
Motion sickness, 260–261
Moulder, Betty, 41
Multiple sleep latency test (MSLT), 433
Mumps vaccine, 257
Musculoskeletal stress, 295–296
    cumulative trauma disorder (CTD), 296
    and exercise programs, 306
    and hand tool design, 298–299
    symptoms, 295
Mycobacterium tuberculosis, 145
Mycotoxin, 273

N

Narcotics Anonymous, 411
National Ambient Air Quality Standard, 209
    criteria pollutants, 209
National Cash Register Company, 40
National Commission on Marijuana and Drug Abuse, 410

National Committeeon Pay Equity, 453
National Council on Compensation Insurance (NCCI), 16, 237
National Crime Victimization Survey (NCVS), 316
National Electronic Injury Surveillance System (NEISS), 13
National Environmental Policy Act (NEPA), 208–209
   compliance, 208
   Council on Environmental Quality, 208
   environmental assessment (EA), 209
   Environmental Impact Statement (EIS), 208
   Finding of No Significant Impact (FONSI), 209
   main purpose of, 208
National Federation of Independent Businesses, 487
National Governors' Association (NGA), 328
National Household Survey on Drug Abuse, 409
National Institute on Drug Abuse (NIDA), 425
National Institute of Health, 457
National Institute of Health Reauthorization Act, 457
National Institute for Occupational Safety and Health (NIOSH)
   creation of, 4
   education and training programs, 20–21
      Educational and Research Centers (ERC), 21, 23, 64, 65–68
      educational resource centers, 20
      models and partnerships, 21
      multimedia and Internet courses, 21
      occupational safety and health (OSH) curricula, 21
      Training Programs Grants (TPG), 21
   and health risk communication, 446
   heat stress level, 454
   indoor air quality (IAQ), 188
   informational resources, 6
   laboratory studies, 386
   lifting guidelines, 290–291, 454
   monitoring of occupational injury deaths, 12
   National Electronic Injury Surveillance System (NEISS), 13
   National Traumatic Occupational Fatalities (NTOF), 12
   reproductive and developmental hazard policy, 464–465
   research conduct, 4, 6
   respirator selection, 175
   small businesses, 10, 21–22, 490
   stress management, 391, 393
   Title IX, 9
   Title XII, 9–10
National Labor Relations Act, 138
National Pollutant Discharge Elimination System (NPDES), 212
National Occupational Exposure Survey (NOES), 18
National Safe Workplace Institute, 147
National Safety Council (NSC), 4, 11, 82, 160, 162, 332
National Safety Management Society (NSMS), 81
National Traumatic Occupational Fatalities (NTOF), 12
New Source Performance Standards (NSPS), 210
News media
   and emergency management, 343
Newsletters, 351
Nicander, 27
NIOSH. See National Institute for Occupational Safety and Health
Nitrogen oxides, 209
Noise levels, 300
Nonfatal recordable injuries and illnesses, 10
Nongovernmental organizations (NGOs), 505–506, 510–511
   international, 506
   local, 506
Nonmedical drug use, 410–412
   addiction, 411–412
   circumstantial use, 410
   compulsive use, 410
   drug abuse, 410–411
   experimental use, 410
   intensified use, 410
   recreational use, 410
Normal hearing, 137
Nuclear Regulatory Commission (NRC), 223, 328
Nurse practitioner, 64
Nursing, occupational health, 41–42, 61–76. See also Occupational health nurse
Nursing programs, 127–151. See also Occupational health nursing programs
Nystagmus, 414–415

O

Occupational deaths, 11–12
   Census of Fatal Occupational Injuries, 11
   National Traumatic Occupational Fatalities (NTOF), 12
   statistics, 11, 12
Occupational disease
   early observations, 32–33
      English reformers, 32
      lead poisoning, 33
      mercury poisoning, 33
   incidence rates, 15–16
   recognition of, 17
   and workers' compensation, 240
Occupational health
   in developing countries. See Infectious diseases
   growth of, 3–24
   nursing, 41–42, 61–76
   program evaluation, 369–378
Occupational health hazards, 92–104
   anticipation, 92
   control, 101–104
      engineering controls, 102–103

over human behavior, 103
personal protective equipment, 103–104
substitution, 102
evaluation, 94–101
airborne contaminants, 96–97
gas and vapor sampling, 97–98
Industrial Hygiene Survey Checklist, 94, 95
occupational epidemiology, 101
particulate material sampling, 98
physical agent, 99
process variables, 100
standard comparison, 100–101
surface evaluation, wipe sampling, 98
recognition, 92–94
data review, 93–94
walk-through survey, 92–93
Occupational health nurse, 61–76
American Association of Occupational Health Nurses, Inc. (AAOHN), 61–62, 64, 69–71
code of ethics, 70
American Board for Occupational Health Nurses, Inc. (ABOHN), 62, 64, 69
Certified Occupational Health Nurse Specialist (COHN-S), 69
Certified Occupational Health Nurses (COHN), 69
goals, 69
certification and licensure, 64, 69
ABOHN goals, 69
American Nurses Credentialing Center, 69
Case Management, 69
certification requirements, 69
Certified Occupational Hearing Conservationist (COHC), 69
NIOSH Approved Spirometry Certification, 69
code of ethics, 69, 70
collaborative practice, 74, 76
definition, 61, 62
emerging roles, 74
goal of, 62
nurse practitioner, 64
placement in organization, 62–63
within environmental affairs, 63
within human resources, 62–63
within production, 63
professional organizations, 69–71
AAOHN, 69–71
Association of Occupational Health Professionals in Healthcare (AOHP), 71
Case Management Society of America, 71
Governmental Affairs, 70
Member Services and Public Affairs, 71
Professional Affairs, 70
role of, 61–62, 72–74
case manager, 73–74
consultant, 74
direct care, 72

educator/advisor, 73
manager/coordinator/director, 72–73
and safety program, 159
scope of practice, 71–72
summary matrix, 74, 75
training and educational requirements, 63–64
competency categories, 64
licensed practical nurses (LPNs), 64
licensed vocational nurses (LVNs), 64
Occupational health nursing program, 127–151
bloodborne pathogens exposure, 143–144
immunizations, 144
OSHA standard, 143
personal protective equipment, 144
post-exposure follow-up, 144
precautions, 143
record keeping, 144
risk classification/task identification, 145
virus information, 143, 144
case management, 136
Department of Transportation guidelines, 142–143
commercial drivers license (CDL), 142
licensure, 143
physical examination, 142
random drug testing, 143
drug and alcohol testing program, 140–142
policy for, 141
purpose of, 140
regulations, 142
supervisory training, 141–142
testing types, 141
employee assistance programs (EAP), 135–136
counseling, services, 136
goal of, 135
fitness programs, 132–135
behavior changes, 134
benefits, 133–134
cost effectiveness, 133–134
evaluations, 134
"hard-to-reach" employees, 134–135
health-promotion programs, 132–134
hazard communication program, 139–140
chemical inventory, 140
federal standards, 139–140
labeling, 140
MSDS, 140
training, 140
health promotion/adult education, 130–132
diet and nutrition, 130
Employee Assistance Programs (EAPs), 131–132
health risk appraisal (HRA), 130
health screening, 130
on-site childcare, 131
on-site health club, 130–131
physical exercise, 130
programs, 132–134

hearing conservation program (HCP), 137
    audiometric evaluation, 137
    Hearing Conservation Amendment, 137
    noise measurements, 137
    testing of baseline hearing, 137
injury and illness prevention, 129–130
    hazard recognition and identification, 129–130
    safety diligence, 129
medical screening/surveillance, 128
    and OSHA standards, 129
    screening, definition, 128
    surveillance, definition, 128
physical examinations, 128–129
    Americans with Disabilities Act (ADA), 128
    baseline, 128
    exit, 128
    Fitness for Duty, 129
    internal job-transfer evaluation, 129
    periodic, 128
    preplacement, 128
    return-to-work evaluation, 129
stress management, 137–139
    stress prevention, 137–139
tuberculosis exposure, 144–146
    bacteria, 145–146
    best work practice, 146
    Health People 2000, 144
    history of, 145
    and HIV, 144
    mycobacterium tuberculosis, 145
    TB standard, 146
violence prevention, 146–149
    crisis management team, 147
    critical incident stress debriefing, 147–149
    high-risk characteristics, 147
    National Safe Workplace Institute, 147
    organizational measures, 147
    profile, 147
work/life strategies, 136
    programs, 136

Occupational health and safety
  history of, 25–43
  20th century developments, 40–42

Occupational Health and Safety Lead Body (OHSLB), 78

Occupational health and safety professionals, 18
  responsibilities of safety management, 159
  and substance abuse, 413

Occupational health and safety professions, 45–105
  industrial hygiene, 87–105
  occupational health nurse, 61–76
  occupational medicine, 47–60
  safety, 77–86

Occupational health and safety programs. *See also* specific program listings
  assessment, 371–378
    administration services, 373–374
    communications, 376
    education and training, 373
    hearing conservation, 377
    medical examinations, 372
    orientation tour, 372
    rehabilitative care, 376–377
    reporting procedures, 372
    urgent care, 375
    worksite hazards and surveillance information, 372
  community involvement programs, 345–367
  creation of, 20
  design, 109–116
    accountability, 115–116
    assessment of program needs, 113
    audit program, 116
    communication strategies, 113
    education programs, 115
    goal of, 112, 113–114
    metrics accuracy, 114
    proactive program, 112
    productivity management, 114
    program considerations, 112–115
    quality monitoring, 115, 116
    records retention, 114–115
    risk assessment, 113
    role of professional, 110
    role in workplace, 110
    trends in workplace demographics, 110–112
  disease recognition, 17
  emergency response programs, 325–343
  employee safety and security programs, 315–324
  ergonomics programs, 277–314
  importance, 17
  industrial hygiene programs, 167–205
  medical and industrial hygiene services, 20
  medical surveillance programs, 17–18
  nursing program, 127–151
  occupational medicine, 117–126
    association/causation, 117–120
    epidemiology, 124–126
    medical surveillance, 126
    risk assessment, 122–124
    toxicology principles, 120–122
  professionals, 18
  program assessment and evaluation, 369–378
  radiation safety programs, 221–233
  safety programs, 153–166, 315–324
  staff competency, 17
  starting of, 20
  team, 18–20
  travel health and remote work programs, 249–276
  workers' compensation management programs, 235–247

Occupational health and safety team, 18–20
  requirements, 18

Occupational history questionnaire, 383–384
Occupational illness, 10, 14–15
   categories, 14
   physical examinations, 40
   prevention programs, 129–130
   statistics, 14, 15, 490
Occupational injury, 10, 13
   data systems, 13
   incidence rates, 15–16
   National Electronic Injury Surveillance System (NEISS), 13
   prevention programs, 129–130
   statistics, 13, 490
   and worker's compensation, 50
Occupational medicine, 47–60. *See also* Physician
   confidentiality of information, 48
   historical trends, 49–51
   patient rights, 48
   physician responsibilities, 48–49
   program, 117–126
   role of, 48–49
   scope of practice, 55–59
      corporate medical direction, 58–59
      designated care, 55
      fitness-for-duty evaluation, 55–56
      general health promotion and preventive medicine, 57–58
      health hazard evaluation, 57
      medical evaluation of impairment and disability, 57
      medical role in drug and alcohol testing, 57
      preplacement, periodic, and exit medical evaluations, 56–57
Occupational medicine program, 117–126
   association/causation, 117–120
      animal models significance, 118–120
      dose-response relationship, 117–118
      Hill Criteria, 119
      sensitivity and specificity, 118
   epidemiology, 124–126
      biases, 125
      descriptive, 124
      experimental, 124
      investigative methods, 124–125
      life table analysis, 126
   medical surveillance, 126
      adverse effect measurement, 126
      sensitive indicators of exposure, 126
      susceptibility, 126
   risk assessment, 122–124
   toxic effect, 120–121
      environmental factors, 121
      exposure factors, 120
      latency, 120
      transgenerational, 120
   toxicity ratings, 121–122
      lethal dose 50, 121–122, 123
   toxicokinetics, 122
   toxicology principles, 120
      definition, 120
      and risk, 120
Occupational Safety and Health Act (OSHAct) (1970)
   black lung disease, 34
   and communications program, 346
   and disabilities, 480
   and employee education, 443
   history of, 25–43
   and industrial hygiene programs, 167
   NIOSH, 20–21
   passage of, 3, 4, 6, 18, 42
   safety and health standards, 6–7
      appeals, 6
      hazard communication standard, 6–7, 10
      inspections, 6
      material safety data sheet (MSDS), 6
      permissible exposure limits, 6
      record keeping, 6
   small businesses, 486
   Threshold Limit Values (TLV), 100
   violations, 7–10
   Williams-Steiger (1970), 4–10
   and workers' compensation, 236
   and workplace violence, 319
Occupational Safety and Health Administration (OSHA)
   creation of, 4
   and disabilities, 480
   employee training guidelines, 444, 446–449
   Hearing Conservation Amendment (HCA), 180
   and injury reporting, 244
   and medical surveillance, 129
   periodic medical evaluations, 56–57
   and Religious Freedom Restoration Act, 111
   safety and health standards, 6
   sampling for surface contamination, 189–205
   Security for Health Care and Social Service Workers, 319
   small businesses, 22, 486, 490
   and tuberculosis, 145
   website, 100
   and workplace litigation, 111
   and workplace violence, 316
   Workplace Violence Prevention Programs for Retail Establishments, 319
Occupational violence prevention. *See* Workplace violence
Open Meetings Act, 346
Operational errors, 153
OSHA standards. *See* Standards, OSHA
Outsourcing, 485–495
   business activity, 487
   communication lines, 488
   corporate affiliation, 487

design of occupational health program, 491–494
    in-house medical facility, 493
    on-site services, 491
    outside providers, 491–492
    protocol of services, 492–493
    record maintenance, 493–494
    supervision, 493
employee education, 491
employee statistics, 487
growth, 488
hazards, 488–489
health and safety coverage, 487
illness and injury statistics, 490
industrial hygiene and safety evaluations, 490
medical services, 490–491
organization and management, 488
policies and procedures, 491
small companies, 486–487
    NIOSH, 486
    OSHA, 486
    OSHAct, 486
    Small Business Administration (SBA), 486
time constraints, 488
work exposures, 488
worksite evaluation, 489
worksite location, 487
Ozone, 209

P

Packaged audit, 156
Palmar flexion of wrist, 298
Paracelsus, 29
Paregoric, 265
Particulate material sampling, 98
Patient advocate, 59
Patient rights, 48
Pepto-Bismol, 265
Perception survey, 157
Periodic examination
    of employees, 128
Permanent threshold shift (PTS), 137
Permissible exposure limits (PELs), 6, 97, 100, 168
Personal protective equipment (PPE)
    and bloodborne pathogen exposure, 144
    emergency management, 338–339
    and health hazards, 103–104
    hearing conservation, 180–183
        amendment (HCA), 180
        audiometric testing, 181–182
        employee training, 182
        hearing protection, 182
        hearing protectors, 182–183
        permissible exposure limit (PEL), 180
        record keeping, 182
    and industrial hygiene programs, 167, 173–183
    respiratory protection, 174–180
        air quality, 179
        employee training, 176
        fit-testing, 176, 179
        medical evaluations, 175
        OSHA requirements, 174
        program evaluation, 179
        record keeping, 180
        respirator cleaning, inspection, and storage, 179
        selection procedures, 175
        voluntary use, 180
        written procedures, 174–175
    skin protection, 173–174
        hand gloves, 174
        handwashing, 174
        protective clothing, 174
        skin disorders, 173
Personal air sampling, 168
Pharyngitis, 14
Physical agents
    evaluation, 99
Physical examinations
    of employees, 48
    and preplacement testing, 384
    types, 128–129
Physician. *See also* Occupational medicine
    certification and licensure, 52–53
    coordination of care, 59
    health and safety team member, 59
    interaction with medical community, 55
    patient advocate, 59
    placement in organization, 51–52
    professional organizations, 53, 55
        American College of Occupational and Environmental Medicine (ACOEM), 53
    role definition, 48–49
    and safety programs, 159
    scope of practice, 55–59
    training and educational requirements, 52
Pinworms, 501
Plague
    epidemiology studies, 124
    and travel health, 260
Plasmodium parasite, 266
Pliny the Elder, 28
Pneumococcal vaccine, 259
Pneumoconioses, 14
Pneumonitis, 14
Point source discharges, 212
Poisoning, 14
Poliomyelitis, 257, 501
    oral polio vaccine (OPV), 257
Pollution Prevention Act, 218
Pott, Dr. Percival, 49
Preassessment questionnaire, 370–371
Preemployment substance testing, 141

Pregnancy Discrimination Act, 465
Preplacement examination
  of employees, 128
Preplacement screening or testing, 291–292, 381–389
  administrative issues, 388
  Americans with Disabilities Act (ADA), 381–382
  drug screening, 386–387
    employee assistance program (EAP), 387
    medical review officer (MRO), 386
  ecogenetics, 389
  examiner, 387
  job descriptions, 382–383
    contents of, 382–383
    job function analysis (JFA), 382
    preparation of, 383
  legal and ethical issues, 388–389
    ADA, 388–389
    confidentiality, 389
  occupational history, 383-384
    OSHA Respiratory Protection Standard, 383
    questionnaire, 384
  physical examination, 384–386
    laboratory studies, 386
    testing techniques, 385–386
  purpose of evaluations, 382
President's Committee on Employment of Persons with Disabilities (PCEPD), 472, 475, 478–479
Prevalence epidemiological studies, 124
Prevention/management programs, for stress, 402
Prevention of Significant Retention (PSR), 210
Primary irritants
  and chemical skin disorders, 173
Proactive program, design, 112
Program administrator
  and respiratory protection, 175
Program assessment and evaluation, 369–378
  assessment, 370
  findings evaluation, 371
  Medical Assessment Management Programs (MAMP), 369
    framework of, 369
    process, 370
  occupational health program assessment, 371–378
  preassessment questionnaire, 370–371
  site activities, 371
  verification, 370
Progressive massive fibrosis (PMF), 34
Project HOPE, 506
Proquanil (Paludrine), 268
Prospective epidemiological studies, 124
Psychoactive drugs, 409–410
Public outrage, 363–365

Q
Quality control inspector, 295

Quality improvement
  and safety diligence, 129
Quality management, 115, 116
  requirements, 115

R
Rabies vaccine, 258, 501
Radiation hazards protection, 224–229
  administrative controls, 226–229
    procurement and disposal, 226
    reporting and investigation, 229
    training of staff, 229
    usage, 226
    waste disposal, 226
  external radiation hazards, 224–226
    distance, 225
    shielding, 225–226
    timing, 225
  internal contamination, 224
    absorption, 224
    ingestion, 224
    inhalation, 224
    punctures, 224
Radiation safety officer, 223
Radiation safety programs, 221–233
  ALARA concept, 231–232
  biological effects, 222
    genetic, 222
    International Commission on Radiological Protection (ICRP), 222
    somatic, 222
  facility and equipment design, 232–233
  hazard monitoring program, 229–231
    environmental, 229
    instrumentation, 231
    personnel, 229–231
  hazard protection, 224–229
    administrative controls, 226–229
    external radiation hazards, 224–226
    internal contamination hazards, 224
  natural sources, 222
    cosmic, 222
    internal sources, 222
    radon, 222
    terrestrial radiation, 222
  radiation and materials, 221–222
    biomedical research, 222
    industrial, 222
    industrial radiography, 222
    medical, 221
    utilities, 221
  regulatory controls, 222–224
    licensing, 223
    radiation safety officer (RSO), 223
    radiation safety programs, 223-224

U.S. Nuclear Regulatory Commission (NRC), 223
Radioactive materials, use of, 221–222
Radioisotope laboratory safety audit, 230
Radium poisoning, 41
Radon, 222
Raleigh, Sir Walter, 31
Ramazzini, Bernardino, 29–30, 266
Rameses II, 26
Random substance testing, 141
Reasonably available control technology (RACT), 210
Records retention, 114–115
    and emergency management, 343
    and small businesses, 493
Registered Safety Professional (RSP), 82
Registry of Toxic Effects of Chemical Substances (RTECS), 457
Rehabilitation, substance abuse, 411, 428–429
Rehabilitation Act of 1973, 388, 471–472
Religious Freedom Restoration Act, 111
Repetitive work, 280–281
Reproductive hazards, 459–465
Resource Conservation and Recovery Act (RCRA), 214–215, 346, 347
    DOT regulations, 215
    objective of, 214
    hazardous waste, 214
    Publicly Owned Treatment Works (POTW), 214
    waste, 214
    Waste Analysis Plan (WAP), 215
Respiratory protection, 174–180
    air quality, 179
    emergency management, 339–341
    employee training, 176
    fit-testing, 176, 179, 180
    maintenance, cleaning, inspection, and storage, 179
    medical evaluations, 175
        questionnaire example, 176–178
    OSHA requirements, 339–340
    program evaluation, 179
    record keeping, 180
    respirators, 180
        air-purifying, 180, 339
        atmosphere-supplying, 180, 339
    selection procedures, 175
        IDLH atmospheres, 175
        self-contained breathing apparatus (SCBA), 175, 179, 180, 340–341
    voluntary use, 180
    written procedures, 174–175
Retrospective epidemiological studies, 125
Return-to-work evaluation, 129
Rhinitis, 14
Risk, definition of, 120
Risk assessment
    and occupational medicine programs, 122–124
        toxicologist, 122
    and program design, 113
Risk evaluations, 334
Risk factors
    personal, 281–282
        low back pain, 281–282
    work-related, 281
        equipment vibration, 281
Risk perception, 446
River blindness, 501
Romans, occupational health and safety development, 28
Rubella vaccine, 257

S

Safety
    bloodborne pathogen exposure, 143–144
    definition of, 120
    and substance abuse, 427
    and workers' compensation, 245
    and workplace violence, 146–149
Safety committees, 161
Safety diligence, 129
    quality tools of, 129
Safety inspections, 161–162
Safety management, 153–155
    human errors, 154–155
    principles, 153
    system failures, 154
Safety Officer (5177), 85–86
Safety Organization (WSO), 82
Safety professional, 77–86
    American Society of Safety Engineers (ASSE), 78–79
    appendix, 85–86
        safety officer duties, 85–86
    Cal/OSHA Injury and Illness Prevention Program, 78
    certification and licensure, 81, 82
        American Society of Safety Engineers (ASSE), 81
        Board of Certified Safety Professionals (BSCP), 81
        Board of Hazard Control Management, 81
        Institution of Occupational Safety and Health (IOSH), 82
        National Safety Council (NSC), 82
        National Safety Management Society (NSMS), 81
        Safety Organization (WSO), 82
    certification maintenance, 82
    code of ethics, 82–83
    definitions, 77–78
    goal of, 78–82
    Occupational Health and Safety Lead Body (OHSLB), 78

placement in the organization, 79
role of, 78–79, 86
scope of practice, 83–84
    action elements, 84–85
    administrative elements, 84
    chief function of professional, 84
    hazards examples, 83
training and educational requirements, 79–81, 83–84
    Board of Certified Safety Professionals (BCSP), 80
UK Management of Health and Safety at Work, 78
Safety programs, 153–166. *See also* Employee safety and security programs
audit alternatives, 156–157
behavior-based, 165–166
    behavioral measurements, 165
    criteria, 166
    statistical validity, 165
company safety policy, 157–158
criteria for, 153
and employee safety, 160–164
    accident investigation, 162
    accident investigation report form, 163–164
    follow-up training, 160
    job safety analysis (JSA), 162
    National Safety Council publications, 160
    safety committees, 161
    safety inspections, 161–162
    training and motivation, 160–161
evaluation of effectiveness, 155–156
    accident/incidence rate, 155
    audit, 155
packaged audit controversy, 156
perception survey, 157
requirements, 155
rules and regulations, 158–159
    accountability of management, 159
    organization, 158–159
    responsibilities, 159
safety management principles, 153–155
    human error, 154–155
    system failures, 154
self-built audit, 156
Salmonella, 262
Sampling
    air, 168
        area, 168
        personal, 168
    surface, 168, 173
        wipe sampling, 168, 173
Save the Children Foundation, 506
Scheduling shiftwork, 431–441
    categories of, 431–432
    circadian rhythms, 433–434
    coping strategies, 434–435
    criteria for, 432–435
        business needs, 432
        employee preference, 432
        health requirements, 433–435
    daily sleep requirements, 433
    definition of, 431
    education of, 440
    safe work schedule, 435–440
        hourly shift schedule, 437
        injuries by days into shift, 438, 439
        injury frequency, 437, 438
        sample schedule, 439
        schedule options, 436, 440
    shiftworkers and health, 440
Schereschewsky, 41
Screening, 128
Security guards, 295
Security programs. *See* Employee safety and security programs
Self-contained breathing apparatus (SCBA), 175, 179, 180, 335, 340–341
Self-insurance
    and workers' compensation, 238–239, 245
Selikoff, Irving J., 51
Sensitivity
    in association/causation, 118
Sensitizer
    and chemical skin disorders, 173
Sexual harassment, 467
Sexually transmitted diseases
    HIV, 269, 271
    and travel health, 269–271
Shiftwork scheduling. *See* Scheduling shiftwork
Shigellosis, 501
Short-term exposure limits, 186
Siderosis, 14
Silicosis, 14, 33
    Gauley Bridge, 34
Sitting positions
    chair design, 287–288
    forward, 286
    reclining, 286
    upright, 285–286
    and workplace design, 280, 286–287
Sixteenth century, occupational health and safety development, 29
Skin disorders, 173–174
    biological agents, 173
    botanical agents, 173
    chemical agents, 173
        primary irritants, 173
        sensitizers, 173
    mechanical agents, 173
    physical agents, 173
Skin protection, 173–174
    clothing protection, 174
    glove use, 174
    handwashing, 174

Sleep requirements, 433
   multiple sleep latency test (MSLT), 433
Small Business Administration (SBA), 486
Small businesses, 10, 21–22, 23. *See also* Outsourcing
   Americans with Disabilities Act (ADA), 22
   OSHA regulations, 22
Spargo, John, 37
Speakers' Bureau, 355
Special response teams, 335–336
Specificity
   in association/causation, 118
Spinal degeneration, 282–285
Staffing, 432. *See also* Scheduling shiftwork
Stakeholders, and communication programs, 348–349
   advantages and disadvantages of, 349
Standard mortality rate (SMR), 125
Standard threshold shift (STS), 137, 182
Standards, OSHA
   bloodborne pathogens, 143, 492
      risk classification/task identification, 145
   Clean Air Act, 209
   emergency equipment, 339
   employee education and training, 444, 446, 448
   hazard communication, 139, 183, 488–489
   laboratory safety, 186
   machine operation, 297
   personal protective equipment (PPE), 173
   quality management, 115
   respiratory protection, 174, 180, 339–340, 383
   risk assessment, 113
   safety and health, 6–7
   substance-specific, 186–187
   tuberculosis, 146
Standing positions
   and workplace design, 280
State Emergency Response Commission (SERC), 218, 330
State government
   emergency management, 327, 329–331
      and federal government, 331
      governor responsibilities, 331
      OSHA requirements, 330
      SARA, 329–330
      state emergency management agencies, 330
      State Emergency Response Commission, 330
State Implementation Plans (SIPs)
   Clean Air Act, 210
Static positions, avoidance of, 279–280
Stewart, Ada Mayo, 41
Stress management, 137–139, 391–407. *See also* Stress management programs
   definition of, 391
   effects of, 395–398
      behavioral effects, 396
      burnout, 396–397
      health effects, 395–396
      organizational effects, 396
      personal responses, 397–398
      subjective effects, 395
      workers' compensation, disability, and accident claims, 396
   future goals, 405–406
   prevention, 137–139
      evaluation, 139
      intervention, 138–139
      problem identification, 138
   programs for management of, 398–405
      assessment focused, 399–400
      behavioral safety programs, 402
      educational/awareness building, 399
      Human Factors Stress Inventory (HFI), 399–400
      impact of, 404–405
      implementation, 402–405
      management and cultural programs, 402
      organizational approaches, 401–402
      prevention/management programs, 402
      skill building, 400–401
      StressMap®, 400
      therapeutic counseling, 401
   in workplace, 112, 392–395
      job factors, 393
      Karasek and Theorell model, 394
      model of interaction, 392
      stress-prone work, 393–395
      workers' perception, 395
Stress management programs, 398–405
   assessment focused, 399–400
      Human Factors Stress Inventory (HFSI), 399–400
      StressMap®, 400
   behavioral safety, 402
   educational/awareness building, 399
   impact of, 404–405
   implementation of, 402–404
      action, 403–404
      assessment, 403
      building management awareness, 403
      program review, 403
      reporting, 403
   management and culture, 402
      interpersonal communication, 402
      managing organizational change, 402
      organizational analysis, 402
   prevention/management, 402
      Americans with Disabilities Act, 402
      back care programs, 402
      data tracking, 402
      screening of employee health risks, 402
      wellness/fitness programs, 402
   responses to, 401–402
   skill building, 400–401
      coping, 400
      interpersonal, 401
      relaxation, 400

therapeutic/counseling, 401
    employee assistance programs (EAP), 401
StressMap®, 400
Substance abuse, 409–430. *See also* Drug and alcohol testing program
    criteria for referrals, 427–428
        diversion programs for impaired professionals, 427–428
        user-program relationship, 427
    drugs in workplace, 412–425
        behavioral signs, 415
        forensic standards for sample collecting, 425
        medical review officer (MRO), 416–419
        motivation for, 412
        observance of abusers, 413
        and occupational health practitioners, 413
        physiological signs, 414–415
        recovery behavior, 415–416
        social effects, 416
        urine testing, 416, 419–425
        work impairment, 412–413
    nonmedical drug use, 410–412
        addiction, 411–412
        circumstantial use, 410
        compulsive use, 410
        drug abuse dynamics, 410–411
        experimental use, 410
        intensified use, 410
        National Commission on Marijuana and Drug Abuse, 410
        recreational use, 410
    psychoactive drugs, 409–410
    reentry and rehabilitation, 428–429
    referral and treatment, 425–427
        American Society of Addiction Medicine (ASAM), 425
        confidentiality, 426–427
        employee assistance program (EAP), 425–426
        safety, 427
Sulfur dioxide, 209
Sullivan, Thomas F. P., 207
Sunburn, 271
Superfund Amendments and Reauthorization Act (SARA), 217, 347
Surface contamination, 168, 173
    and wipe sampling, 168, 173, 189–205
        goals of, 168
Surveillance, 128
Swimming and bathing hazards, 273–274
Swiss-Air Ambulance (REGA), 275
Syphilis, 501
System failures, in workplace, 154

T

Tailpipe emissions, 211
Task analysis, 303
Technology
    changes in workplace, 111
    computer, 112
Temporary threshold shift (TTS), 137
Tenosynovitis, 298
Tetanus
    early description of, 27–28
    immunizations, 254
Tetrodotoxin, 273
Thackrah, Dr. Charles Turner, 32–33, 49–50
Theorell, Tores, 394–395
Thompson, William Gilman, 36
Threshold Limit Values (TLV), 100, 168
Time-weighted-average (TWA)
    exposure levels, 97, 180
Tool handles, 298
Toxic hazards, and travel, 273
Toxic Hot Spots program, 212
Toxic Release Inventory (TRI), 346, 347–348
Toxic Substances Control Act (TSCA), 213–214, 346, 347
    and pesticides, 214
Toxicokinetics, 122
    definition of, 122
    problem of, 122
        mercury exposure, 122
Toxicologist, 120
    risk assessment, 122–123
Toxicology, 120
    definition of, 120
    principles of, 120
    problems of, 120
        environmental factors, 120
        exposure factors, 120
        latency, 120
        transgenerational, 120
    risk, 120
    toxic effect, 120–121
    toxicity ratings, 121–122
        lethal dose 50, 121–122, 123
Training and education. *See also* Education and training
    of emergency plan, 336–337
    employee education, 443–450
    hazard communication program, 186
    and respirators, 176
    and safety procedures, 160
    and workplace violence, 322
Training Programs Grants (TPG), 21
Transdermal scopolamine, 261
Travel health and remote work programs, 249–276
    bloodborne hepatitis, 269, 271
    Center for Disease Control International Traveler's Hotline, 275
    emporiatrics, 249–251

environmental hazards, 272–274
   air pollution, 272
   altitude sickness, 272–273, 274
   crime, 274
   heat exhaustion, 272
   heatstroke, 272
   injuries, 274
   swimming and bathing, 273–274
   toxic hazards, 273
immunizations, 252, 255
   anthrax, 260
   bacille Calmette-Guerin (BCG) vaccine, 259
   cholera, 259–260
   dengue fever, 260
   hepatitis A, 258
   hepatitis B, 258
   influenza, 259
   Japanese encephalitis, 258
   measles, 254, 256
   meningococcal vaccine, 258–259
   mumps, 257
   plague, 260
   pneumococcal vaccine, 259
   poliomyelitis, 257
   rabies, 258
   rubella, 257
   tetanus-diphtheria, 254
   tuberculosis, 259
   typhoid, 258
   yellow fever, 254
international air evacuation, 275
   Swiss-Air Ambulance (REGA), 275
jet lag, 261–262
   symptoms, 261
   treatment, 261–262
malaria, 265–269
   mosquito protection, 268–269
   prevention, 266
   prophylactic medications, 266–268
mental health and substance abuse, 274–275
   alcohol, 274–275
   illegal drug use, 275
motion sickness, 260–261
   medication for, 261
sexually transmitted diseases, 269, 271
   HIV-infected travelers, 271
skin conditions, 271–272
   fungal infections, 272
   sun exposure, 271–272
travelers' diarrhea, 262–265
   beverages, 263–264
   dietary precautions, 262
   prophylactic drugs, 264
   treatment for, 265
   water purification, 263–264, 265
travelers' diseases in tropics, 250
trip preparation, 251
   doctor consultation, 251
   health history, 251, 252
   insurance, 252
   medical kit, 251–252, 253–254
Traveler's diarrhea, 262–265
   beverages, 263–264
   dietary precautions, 262
   Montezuma's revenge, 262
   prophylactic drugs, 264–265
      Pepto-Bismol, 264
   treatment, 265
      antimotility agents, 265
      ciprofloxacin, 265
   water, 263–2640
Tropical travel, 250
Truncal ringworm, 272
Tuberculosis, 33, 144–146
   bacille Calmette-Guerin (BCG), 259
   best work practice, 146
   Centers for Disease Control, 144
   Healthy People 2000, 144
   history of, 145
   and HIV, 144
   mycobacterium tuberculosis, 145
   and OSHA, 145
   skin test, 259
   standard, 146
      purpose of, 146
   symptoms, 145
   transmission, 501
   and travel, 251
Typhoid fever
   transmission, 501
   vaccine, 258

U

UK Management of Health and Safety at Work, 78
Ulnar deviation of wrist, 298
Underground storage tanks, 215–217
   regulations and codes, 216–217
UNICEF, 507
Uniform Guidelines for Employment Selection
   Procedures, 388
Uninsured Employers' Fund, 243
Union of Trade Associations, 35
United Association of Casualty Inspectors, 38
United Mine Workers of America (UMWA), 35
Urban Institute, 409
Urine testing, 416–425
   forensic standards, 425
   instruments, 419–422
   legal challenges, 423
   and medical review officer (MRO), 416–419
   procedures, 423, 425

special testing, 423
U.S. Agency for International Development, 506
U.S. Department of Health and Human Services (DHHS), 423
U.S. Department of Labor, 474
U.S. Longshoremen's and Harbor Workers' Compensation Act (L&HWCA), 237–238

V

Ventilation
    control of occupational health hazards, 102
Verification, 370
Vibration
    hand-arm vibration syndrome (HAVS), 299
    whole-body, 303
Vietnam Era Veterans Readjustment Assistance Act, 388
Violence prevention, 146–149. *See also* Workplace violence
Virus, 143, 144
    hepatitis B, 143
    HIV, 143
Vitruvius, 28

W

Walk-through survey
    industrial hygiene, 92–93
Walsh-Healy Public Contracts Act, 39
Web sites
    ADA, 474
    American Association of Occupational Health Nurses (AAOHN), 71
    American Society of Heating, Refrigerating and Air Conditioning Engineers (ASHRE), 187
    Case Management Society of America (CMSA), 71
    CDC Travel, 275
    FEMA, 329
    International Society of Travel Medicine, 275
    Job Accommodation Network (JAN), 475
    Job Links, 482
    Journal of the Association of Occupational Health Professionals in Healthcare, 71
    National Institute for Occupational Safety and Health (NIOSH), 6
    Occupational Safety and Health Administration, 100
Whitney, Eli, 30
Williams-Steiger. *See* Occupational Safety and Health Act (OSHAct)
Wipe sampling, 98, 168, 173
    goals of, 168
    OSHA excerpt, 189–205
Women, in work force, 37, 451–469
    chemical exposure and reproductive risk, 459, 461–463
        developmental toxicology, 462
        lactation, 462–463
        and pregnancy, 462
    disability, 457, 459
    employment segregation, 452
    gender distribution, 451–452
    pay inequities, 452–453
    vulnerabilities, 453–457
        coronary heart disease, 456–457
        heat exposure, 454–456
        physical strength, 453–454
        work capacity, 453–454
Women's Health Equity Act, 457
Work/life strategies, 136
Work practice controls
    and health hazards, 103
Work stress. *See* Stress management
Workers' compensation, 16–17, 37–38
    claims statistics, 16
    costs, 4
    history of, 37–38, 50
    and low back injuries, 281
    management programs, 235–247
    National Council on Compensation Insurance (NCCI), 16
    and occupational health nurse placement, 63
    and small businesses, 16
    and stress management, 396, 404
Workers' compensation management programs, 235–247
    administration, 243–244
        discrimination charges, 244
        industrial/medical facilities, 243–244
        injury reporting, 244
    AOE/COE, 240
    benefits, 240–241
        cash or income, 241
        death, 241, 242
        medical, 241
        social security, 242
        subsequent injury funds, 242
        survivor, 242
        vocational rehabilitation, 241
    claims process, 244–245
    coverage laws, 239–240
        and employee status, 239
    compulsory or elective coverage, 243
        failure to insure, 243
        injury outside of jurisdiction, 243
        minors employment, 243
        Uninsured Employers' Fund, 243
    disability classifications, 241–242
        permanent partial, 242
        permanent total, 241
        temporary partial, 242
        temporary total, 241–242

federal laws and regulations, 237–238
   District of Columbia Workers' Compensation Act, 237
   Federal Employees Compensation Act (FECA), 237
   Federal Employers' Liability Act, 237
   U.S. Longshoremen's and Harbor Workers' Compensation Act (L&HWCA), 237–238
financial incentives, 245–247
   cash-flow plan, 246
   dividend plan, 246
   experience rating, 245–246
   of insureds, 245–246
   retrospective rating, 246
   of self-insureds, 245
history of, 235–236
National Council on Compensation Insurance, 237
occupational disease, 240
and OSHAct, 236
policy exclusions, 240
principles of, 236
self-insurance, 238–239
state funds, 238
   competitive, 238
   monopolistic, 238
system costs, 236–237
Workers with disabilities, 471–483. *See also* Americans with Disabilities Act
   Americans with Disabilities Act (ADA), 472, 480–481
      Civil Rights Act, 473
      Equal Employment Opportunity Commission, 473, 482
      essence of, 473–474
      Title I, 473, 474–476
      Title II, 473–474
      Title III, 474
      web site, 474
   corporate social conscience, 479–480
   disability language, 479
   Federal Medical Leave Act (FMLA), 480
   Job Links, 482
   medical examinations, 476–478
      physical capacities form, 478
      physical demands form, 477
      suitability of job, 476
      supportive procedures, 476–478
   number of disabled persons, 472–473
      activities of daily living (ADL), 473
   OSHA, 480
   Rehabilitation Act of 1973, 471–472
   supported employment, 478–479
      President's Committee on Employment of Persons with Disabilities (PCEPD), 478–479, 481
   tax incentives, 482

Workplace design
   layout of work space, 278
   and safety management, 155
Workplace Environmental Exposure Limits, 101
Workplace gender issues, 451–469. *See also* Gender issues in workplace
Workplace program design
   trends in demographics, 110–112
      aging employees, 110–111
      computer technology, 112
      cultural and religious needs, 111
      diversity of employees, 111
      language diversity, 111
      litigation, 111
      stress management, 112
      technological changes, 111
Workplace violence, 146–149, 316–324
   assaults, 316
   causes, 316–317
   cost, 316
   critical incident stress debriefing, 147–149
      purpose of, 149
   high-risk characteristics, 147
   homicide, 316
   National Safe Workplace Institute, 147
   and OSHA guidelines, 316
   predictors, 318
   prevention measures, 147, 318–322
      hazard prevention and control, 320
      post-incidence response, 320–322
      record keeping and evaluation, 322
      training and education, 322
      worksite and hazard analysis, 319–320, 324
   profile of perpetrator, 147
   types, 317–318
      Type I, 317
      Type II, 317–318
      Type III, 318
Worksite hazard analysis, 319–320, 324
World Health Organization, 42, 250, 260, 500, 507
   oral rehydration solution packets, 265
World Vision Relief and Development, 506
Wrist extension, 298

Y

Yellow fever, 502, 505
   immunization, 254, 256–257
   transmission, 500

Z

Zoonoses, 501